BIRKHÄUSER

Marek Kuczma

An Introduction to the Theory of Functional Equations and Inequalities

Cauchy's Equation and Jensen's Inequality

Second Edition

Edited by
Attila Gilányi

Birkhäuser
Basel · Boston · Berlin

Editor:

Attila Gilányi
Institute of Mathematics
University of Debrecen
P.O. Box 12
4010 Debrecen
Hungary
e-mail: gilanyi@math.klte.hue

2000 Mathematical Subject Classification: 39B05, 39B22, 39B32, 39B52, 39B62, 39B82, 26A51, 26B25

The first edition was published in 1985 by Uniwersytet Ślaski (Katowicach) (Silesian University of Katowice) and Pánstwowe Wydawnictwo Naukowe (Polish Scientific Publishers)
© Uniwersytet Ślaski and Pánstwowe Wydawnictwo Naukowe

Library of Congress Control Number: 2008939524

Bibliographic information published by Die Deutsche Bibliothek
Die Deutsche Bibliothek lists this publication in the Deutsche Nationalbibliografie;
detailed bibliographic data is available in the Internet at <http://dnb.ddb.de>.

ISBN 978-3-7643-8748-8 Birkhäuser Verlag AG, Basel – Boston – Berlin

© 2009 Birkhäuser Verlag AG
Basel · Boston · Berlin
P.O. Box 133, CH-4010 Basel, Switzerland
Part of Springer Science+Business Media
Printed on acid-free paper produced of chlorine-free pulp. TCF ∞

ISBN 978-3-7643-8748-8   ISBN 978-3-7643-8749-5 (eBook)

9 8 7 6 5 4 3 2 1   www.birkhauser.ch

Preface to the Second Edition

The first edition of Marek Kuczma's book *An Introduction to the Theory of Functional Equations and Inequalities* was published more than 20 years ago. Since then it has been considered as one of the most important monographs on functional equations, inequalities and related topics. As János Aczél wrote in Mathematical Reviews *"... this is a very useful book and a primary reference not only for those working in functional equations, but mainly for those in other fields of mathematics and its applications who look for a result on the Cauchy equation and/or the Jensen inequality."*

Based on the considerably high demand for the book, which has even increased after the first edition was sold out several years ago, we have decided to prepare its second edition. It corresponds to the first one and keeps its structure and organization almost everywhere. The few changes which were made are always marked by footnotes.

Several colleagues helped us in the preparation of the second edition. We cordially thank Roman Ger for his advice and help during the whole publication process, Karol Baron and Zoltán Boros for their conscientious proofreading, and Szabolcs Baják for typing and continuously correcting the manuscript. We are grateful to Eszter Gselmann, Fruzsina Mészáros, Gyöngyvér Péter and Pál Burai for typesetting several chapters, and we would like to thank the publisher, Birkhäuser, for undertaking and helping with the publication.

The new edition of Marek Kuczma's book is paying tribute to the memory of the highly respected teacher, the excellent mathematician and one of the most outstanding researchers of functional equations and inequalities.

Debrecen, October 2008

Attila Gilányi

Contents

Part II   Cauchy's Functional Equation and Jensen's Inequality

Part III   Related Topics

Introduction

The present book is based on the course given by the author at the Silesian University in the academic year 1974/75, entitled Additive Functions and Convex Functions. Writing it, we have used excellent notes taken by Professor K. Baron.

It may be objected whether an exposition devoted entirely to a single equation (Cauchy's Functional Equation) and a single inequality (Jensen's Inequality) deserves the name An introduction to the Theory of Functional Equations and Inequalities. However, the Cauchy equation plays such a prominent role in the theory of functional equations that the title seemed appropriate. Every adept of the theory of functional equations should be acquainted with the theory of the Cauchy equation. And a systematic exposition of the latter is still lacking in the mathematical literature, the results being scattered over particular papers and books. We hope that the present book will fill this gap.

The properties of convex functions (i.e., functions fulfilling the Jensen inequality) resemble so closely those of additive functions (i.e., functions satisfying the Cauchy equation) that it seemed quite appropriate to speak about the two classes of functions together.

Even in such a large book it was impossible to cover the whole material pertinent to the theory of the Cauchy equation and Jensen's inequality. The exercises at the end of each chapter and various bibliographical hints will help the reader to pursue further his studies of the subject if he feels interested in further developments of the theory. In the theory of convex functions we have concentrated ourselves rather on this part of the theory which does not require regularity assumptions about the functions considered. Continuous convex functions are only discussed very briefly in Chapter 7.

The emphasis in the book lies on the theory. There are essentially no examples or applications. We hope that the importance and usefulness of convex functions and additive functions is clear to everybody and requires no advertising. However, many examples of applications of the Cauchy equation may be found, in particular, in books Aczél [5] and Dhombres [68]. Concerning convex functions, numerous examples are scattered throughout almost the whole literature on mathematical analysis, but especially the reader is referred to special books on convex functions quoted in 5.3.

We have restricted ourselves to consider additive functions and convex functions defined in (the whole or subregions of) N-dimensional euclidean space $\mathbb{R}^N$. This gives the exposition greater uniformity. However, considerable parts of the theory presented can be extended to more general spaces (Banach spaces, topological linear spaces). Such an approach may be found in some other books (Dhombres [68], Roberts-Varberg [267]). Only occasionally we consider some functional equations on groups or related algebraic structures.

We assume that the reader has a basic knowledge of the calculus, theory of Lebesgue's measure and integral, algebra, topology and set theory. However, for the convenience of the reader, in the first part of the book we present such fragments of those theories which are often left out from the university courses devoted to them. Also, some parts which are usually included in the university courses of these subjects are also very shortly treated here in order to fix the notation and terminology.

In the notation we have tried to follow what is generally used in the mathematical literature[1]. The cardinality of a set A is denoted by $\operatorname{card} A$. The word *countable* or *denumerable* refers to sets whose cardinality is exactly $\aleph_0$. The topological closure and interior of A are denoted by $\operatorname{cl} A$ and $\operatorname{int} A$. Some special letters are used to denote particular sets of numbers. And so $\mathbb{N}$ denotes the set of positive integers, whereas $\mathbb{Z}$ denotes the set of all integers. $\mathbb{Q}$ stands for the set of all rational numbers, $\mathbb{R}$ for the set of all real numbers, and $\mathbb{C}$ for the set of all complex numbers. The letter N is reserved to denote the dimension of the underlying space. The end of every proof is marked by the sign $\square$. Other symbols are introduced in the text, and for the convenience of the reader they are gathered in an index at the end of the volume.

The book is divided in chapters, every chapter is divided into sections. When referring to an earlier formula, we use a three digit notation: (X.Y.Z) means formula Z in section Y in Chapter X. The same rule applies also to the numbering of theorems and lemmas. When quoting a section, we use a two digit notation: X.Y means section Y in Chapter X. The same rule applies also to exercises at the end of each chapter. The book is also divided in three parts, but this fact has no reflection in the numeration.

Many colleagues from Poland and abroad have helped us with bibliographical hints and otherwise. We do not endeavour to mention all their names, but nonetheless we would like to thank them sincerely at this place. But at least two names must be mentioned: Professor R. Ger, and above all, Professor K. Baron, whose help was especially substantial, and to whom our debt of gratitude is particularly great. We thank also the authorities of the Silesian University in Katowice, which agreed to publish this book. We hope that the mathematical community of the world will find it useful.

Katowice, July 1979

Marek Kuczma

---

[1]The notation in the second edition has been slightly changed. The following sentences are modified accordingly.

Part I

Preliminaries

Chapter 1

Set Theory

1.1 Axioms of Set Theory

The present book is based on the Zermelo-Fraenkel system of axioms of the Set Theory augmented by the axiom of choice. The axiom of choice plays a fundamental role in the entire book. The mere existence of discontinuous additive functions and discontinuous convex functions depends on that axiom[1]. Therefore the axiom of choice will equally be treated with the remaining axioms of the set theory and no special mention will be made whenever it is used.

The primitive notions of the set theory are: set, belongs to ($\in$), and being a relation type (τ) [$\alpha\tau\langle A, R\rangle$ means α is a relation type of $\langle A, R\rangle$; cf. Axiom 8]. The eight axioms read as follows.

Axiom 1.1.1. Axiom of Extension. *Two sets are equal if and only if they have the same elements:*

$$A = B \text{ if and only if } (x \in A) \leftrightarrow (x \in B).$$

Axiom 1.1.2. Axiom of Empty Set. *There exists a set $\varnothing$ which does not contain any element:*

$$\text{For every } x, \ x \notin \varnothing.$$

Axiom 1.1.3. Axiom of Unions. *For every collection[2] $\mathcal{A}$ of sets there exists a set $\bigcup \mathcal{A}$ which contains exactly those elements that belong to at least one set from $\mathcal{A}$:*

$$\left(x \in \bigcup \mathcal{A}\right) \Leftrightarrow (\text{there exists an } A \in \mathcal{A} \text{ such that } x \in A).$$

Axiom 1.1.4. Axiom of Powers. *For every set A there exists a collection $\mathcal{P}(A)$ of sets which consists exactly of all the subsets of A:*

$$\left(B \in \mathcal{P}(A)\right) \Leftrightarrow (B \subset A).$$

---

[1] R. M. Solovay has shown (Solovay [292]) that a model of mathematics (without axiom of choice) is possible in which all subsets of $\mathbb{R}$ (and consequently also all functions $f : \mathbb{R} \to \mathbb{R}$) are Lebesgue measurable.

[2] The word *collection* is, of course, a synonym of *set*.

Axiom 1.1.5. Axiom of Infinity. *There exists a collection $\mathcal{A}$ of sets which contains the empty set $\varnothing$ and for every $X \in \mathcal{A}$ there exists a $Y \in \mathcal{A}$ consisting of all the elements of X and X itself:*

$$\varnothing \in \mathcal{A} \text{ and for every } X \in \mathcal{A} \text{ there exists a } Y \in \mathcal{A} \text{ such that}$$

$$(x \in Y) \Leftrightarrow (x \in X) \text{ or } (x = X).$$

Axiom 1.1.6. Axiom of Choice. *The cartesian product of a non-empty family of non-empty sets is non-empty:*

$$\text{If } \mathcal{A} \neq \varnothing \text{ and for every } A \in \mathcal{A}, \; A \neq \varnothing, \text{ then } \underset{A \in \mathcal{A}}{\times} A \neq \varnothing.$$

Axiom 1.1.7. Axiom of Replacement. *Let $\psi(x, y)$ be a two-place propositional formula such that, for every x, there exists exactly one y such that $\psi(x, y)$ holds. Then for every set A there exists a set B which contains those and only those y for which $\psi(x, y)$, $x \in A$:*

If for every x there exists a z such that $\psi(x, y) \Leftrightarrow y = z$, then for every set A there exists a set B such that

$$(y \in B) \Leftrightarrow \text{ there exists an } x \in A \text{ such that } \psi(x, y).$$

Roughly speaking, if to every x there corresponds (according to ψ) a unique y, and if x runs over a set, then the corresponding y's run over a set.

Before stating the last axiom, we must introduce certain notions. Let A be a set and $R \subset A^2$ a relation in A. A couple $\langle A, R \rangle$ is called a relation system. Two relation systems $\langle A, R \rangle$ and $\langle B, S \rangle$ are said to be *isomorphic* iff there exists a one-to-one function f from A onto B (a bijection) such that for every $a, b \in A$ we have aRb if and only if $f(a)Sf(b)$.

Axiom 1.1.8. Axiom of Relation Types. *To every relation system $\langle A, R \rangle$ there corresponds an object α such that $\alpha\tau\langle A, R \rangle$ and if $\alpha\tau\langle A, R \rangle$ and $\beta\tau\langle B, S \rangle$, then $\alpha = \beta$ if and only if $\langle A, R \rangle$ and $\langle B, S \rangle$ are isomorphic.*

The Axiom of Replacement implies the following statement.

Axiom of Specification. *For every set A and for every propositional formula $\Phi(x)$ there exists the set $\{x \in A \mid \Phi(x)\}$ consisting of exactly those $x \in A$ for which $\Phi(x)$ holds true:*

$$x \in \{x \in A \mid \Phi(x)\} \Leftrightarrow x \in A \wedge \Phi(x).$$

It is enough to take $\psi(x, y) \Leftrightarrow \Phi(x) \wedge y = x$.

The Axiom of Empty Set can be replaced by the weaker

Axiom of Existence. *There exists a set.*

The empty set can be defined as (A being an existing set)

$$\varnothing = \{x \in A \mid x \neq x\}.$$

If we take into account the definition of the cartesian product of an arbitrary collection of sets, we can reformulate the Axiom of Choice as follows:

For every non-empty collection $\mathcal{A}$ of non-empty sets there exists a function $w :$ $\mathcal{A} \to \bigcup \mathcal{A}$ (the choice function) such that $w(A) \in A$ for every $A \in \mathcal{A}$.

The Axiom of Choice is usually used in this form.

The Axiom of Extension implies the uniqueness of sets whose existence is guaranteed by the remaining Axioms 2–7.

The Axiom of Relation Types can be omitted. The whole set theory can be built without a use of this axiom. The ordinal numbers (as well as cardinal numbers) must then be defined otherwise. (Cf., e.g., Halmos [130]).

From the Axioms 1.1.1–1.1.8 all the set theory can be built (cf. Kuratowski-Mostovski [198], Halmos [130], Rasiowa [262]). We assume that the reader is familiar with it. However, in the sequel we outline the theory of ordinal numbers, as the latter is often omitted in the university courses of the set theory.

1.2 Ordered sets

Let A be a set, and $\leqslant \subset A^2$ a relation which is reflexive, antisymmetric and transitive:

(*i*) $a \leqslant a$,

(*ii*) $(a \leqslant b) \wedge (b \leqslant a) \Rightarrow (a = b)$,

(*iii*) $(a \leqslant b) \wedge (b \leqslant c) \Rightarrow (a \leqslant c)$.

Such a relation $\leqslant$ is called an order[3] in A and the couple $(A, \leqslant)$ is called an *ordered set*. Clearly, every ordered set is a relation system in the sense of 1.1.

The strict relation $<$ is defined as

$$(a < b) \Leftrightarrow (a \leqslant b) \wedge (a \neq b).$$

Instead of $a \leqslant b$, $a < b$, we shall often write $b \geqslant a$, $b > a$.

If, besides (*i*), (*ii*) and (*iii*), also the trichotomy law holds:

(*iv*) For every $a, b \in A$, we have either $a < b$, or $b < a$ or $a = b$, then the set A is called *linearly ordered* or a *chain*.

Let $(A, \leqslant)$ be an ordered set. An element $a \in A$ is called *maximal* [*minimal*] iff there is no $b \in A$ strictly greater [smaller] than a. In other words, a is maximal iff

$$(b \in A) \wedge (a \leqslant b) \Rightarrow (b = a).$$

[a is minimal iff

$$(b \in A) \wedge (b \leqslant a) \Rightarrow (b = a)].$$

---

[3] In earlier texts *order* is often called a *partial order*, the word *order* being reserved for what is here called a *linear order*. This is due to the fact that, for arbitrary $a, b \in A$, we are often unable to decide whether $a \leqslant b$ or $b \leqslant a$. An illustrative example is the power set $\mathcal{P}(A)$ of a set A with the order relation defined as the inclusion:

$$a \leqslant b \Leftrightarrow a \subset b.$$

One ordered set may have several (or none) maximal [minimal] elements. If $a \in A$ is a maximal [minimal] element, then there may exist in A elements b which are not comparable with a, i.e., for which neither $a \leqslant b$, nor $b \leqslant a$ holds.

An element $a \in A$ is called the *greatest* [*smallest*] (or the *last* [*least*]) element iff $x \leqslant a$ [$a \leqslant x$] holds for every $x \in A$. The last [least] element, if it exists, is unique.

An element $a \in A$ is called the *upper bound* of a set $E \subset A$ iff $x \leqslant a$ holds for every $x \in E$. It is not required that $a \in E$, but it is possible. There may exist several (or none) upper bounds of a set $E \subset A$.

If $(B, \preccurlyeq)$ is another ordered set, then we say that $(A, \leqslant)$ and $(B, \preccurlyeq)$ are *similar* (and write $(A, \leqslant) \sim (B, \preccurlyeq)$) iff there exists a one-to-one order-preserving mapping f from A onto B. The relation of similarity is an isomorphism of relation systems $(A, \leqslant)$ and $(B, \preccurlyeq)$ as defined in 1.1:

$$(a \leqslant b) \Leftrightarrow \big(f(a) \preccurlyeq f(b) \big)$$

An ordered set, every non-empty subset of which has the smallest element, is called a *well-ordered* set, and the corresponding order is called a *well-order*. We have the following

Theorem 1.2.1. *Every well-ordered set is linearly ordered.*

Proof. This follows from the fact that for any $a, b \in A$, the pair $\{a, b\} \subset A$ has the smallest element. $\qquad\square$

Any finite linearly ordered set is well ordered and two such sets are similar if and only if they have the same number of elements. (The proof of these facts is left to the reader.) The set $(\mathbb{N}, \leqslant)$, where $\leqslant$ stands for the usual inequality between numbers, is well ordered.

1.3 Ordinal numbers

Let $(A, \leqslant)$ be a well-ordered set. Any set $P \subset A$ such that if $x \in P$ and $y \leqslant x$, then $y \in P$, is called an *initial segment* of A.

Theorem 1.3.1. *If P is an initial segment of a well-ordered set A[4], and $P \neq A$, then there exists in A an x such that*

$$P = P(x) = \{y \in A \mid y < x\}.$$

Proof. The set $A \setminus P \neq \varnothing$ has the smallest element x. We will show that $P = P(x)$.

Let $y \in P$. If we had $x \leqslant y$, then we would have $x \in P$, which contradicts the condition $x \in A \setminus P$. Thus $y < x$ and $y \in P(x)$. Consequently $P \subset P(x)$.

If $y \in P(x)$, then $y < x$, and since x is smallest in $A \setminus P$, we must have $y \in P$. Consequently $P(x) \subset P$.

Thus P and $P(x)$ have the same elements, and so they are equal: $P = P(x)$. $\quad\square$

---

[4] Instead of saying: A is the first component of the ordered set $(A, \leqslant)$, we often say simply: A is an ordered set.

The formulas, valid for arbitrary $a, b \in A$, are left to the reader as exercises:

1. $(a \leqslant b) \Leftrightarrow (P(a) \subset P(b))$,
2. $(a \neq b) \Leftrightarrow (P(a) \neq P(b))$,
3. For arbitrary well-ordered set $(A, \leqslant)$ put $P = \{P(x) \mid x \in A\}$. The ordered set $(P, \subset)$ is well ordered and is similar to $(A, \leqslant)$.

The relation types of well-ordered sets are called the *ordinal numbers*. If $(A, \leqslant)$ is a well-ordered set and if $\alpha \tau (A, \leqslant)$, then we write $\alpha = \overline{A}$. The $\overline{\mathbb{N}}$ is denoted by ω. If $(A, \leqslant)$ is a finite (well-ordered) set consisting of $n \in \mathbb{N}$ elements, then we assume $\overline{A} = n$. In particular, $\overline{\varnothing} = 0$.

If $(A, \leqslant)$ is a well-ordered set, then, by the Axiom of Infinity there exists a set B which contains all the elements of A and A itself. We order B by assuming additionally that $A \geqslant a$ for any $a \in A$. The ordinal number $\overline{B}$ is denoted by $\overline{A} + 1$. If an ordinal number cannot be written as $\alpha + 1$ with another ordinal number α, then it is called a *limit number*. An example of a limit number is ω.

If α and β are ordinal numbers, say $\alpha = \overline{A}$ and $\beta = \overline{B}$, then we say that $\alpha < \beta$ iff the set A is similar to an initial segment of B different from B.

Theorem 1.3.2. *For any ordinal number α, it is not true that $\alpha < \alpha$.*

Proof. For an indirect proof, suppose that $\alpha < \alpha$, i.e., A is similar to its initial segment different from A. Let f be a similarity function. Put

$$B = \{x \in A \mid f(x) < x\}.$$

By Theorem 1.3.1 there exist an $a \in A$ such that $A \sim P(a)$. Hence $f(a) < a$, i.e., $a \in B$. So $B \neq \varnothing$, and $B \subset A$, and hence there exists the smallest element, say b, in B. Then

$$f(b) < b, \tag{1.3.1}$$

and since f is order-preserving, $f\big(f(b)\big) < f(b)$. This means that $f(b) \in B$, which, by (1.3.1) contradicts the condition that b is smallest in B. $\qquad\square$

Theorem 1.3.3. *If $\alpha < \beta$ and $\beta < \gamma$, then $\alpha < \gamma$.*

Proof. Let A, B, C be well ordered sets such that $\overline{A} = \alpha$, $\overline{B} = \beta$ and $\overline{C} = \gamma$. Moreover, let $b \in B$ and $c \in C$ be such that $A \sim P(b)$ and $B \sim P(c)$, the similarity functions being f and g, respectively. Then it is easily checked that $g \circ f$ is a one-to-one, order-preserving mapping of A onto an initial segment of C, different from C. $\qquad\square$

Theorem 1.3.4. *If $\alpha < \beta$, then it is not true that $\beta < \alpha$.*

Proof. This follows from Theorems 1.3.3 and 1.3.2. $\qquad\square$

Theorems 1.3.2 and 1.3.3 imply that the inequality $\leqslant$ defined for ordinal numbers as follows: $\alpha \leqslant \beta$ iff either $\alpha < \beta$, or $\alpha = \beta$, is an order in the sense of 1.2.

1.4   Sets of ordinal numbers

We start with a lemma.

Lemma 1.4.1. *If $(A, \leqslant)$ and $(B, \leqslant)$ are similar well-ordered sets[5] and f is a similarity function, then f maps initial segments of A onto initial segments of B.*

Proof. Let f be a one-to-one order preserving mapping of A onto B. Then f^{-1} is a one-to-one order preserving mapping of B onto A. Let P be an initial segment of A. The thing to prove is that $f(P)$ is an initial segment of B.

Suppose that there exist $b_1 < b_2$, $b_2 \in f(P)$ and $b_1 \notin f(P)$. Put $a_i = f^{-1}(b_i)$, $i = 1, 2$. Then $a_1 < a_2$, $a_2 \in P$ and $a_1 \notin P$, which is impossible since P is an initial segment of A.   □

Corollary 1.4.1. *No two different initial segments of a well-ordered set are similar to each other.*

Proof. Let $(A, \leqslant)$ be a well-ordered set, and $P_1 \neq P_2$ initial segments of A, $P_1 \sim P_2$. Since $P_1 \neq P_2$, at least one of these segments, say P_1, must be different from A, and hence of the form $P(a)$ with an $a \in A$. If $a \in P_2$, then P_1 is an initial segment of $(P_2, \leqslant)$. Indeed, if $x \in P_1$ and $y < x$, then $y < a$ and hence $y \in P(a) = P_1$. And if $a \notin P_2$, then P_2 is an initial segment of $(P_1, \leqslant)$. Indeed, then $P_2 \neq A$ and hence of the form $P(b)$ with a $b \in A$. If $x \in P_2$ and $y < x$, then $y < b$ and $y \in P_2$.

Thus one of the sets P_1, P_2 is similar to its initial segment different from this set, which contradicts Theorem 1.3.2.   □

Theorem 1.4.1. *Of any two well-ordered sets, one is similar to an initial segment of the other.*

Proof. Let $(A, \leqslant)$ and $(B, \leqslant)$ be two well-ordered sets. Define the set Z,

$$Z = \{x \in A \mid \text{there exists a } y \in B \text{ such that } P(x) \sim P(y)\}.$$

By Corollary 1.4.1, such a y is unique. Thus we may define a function $f : Z \to B$ by putting $f(x) = y$ iff $P(x) \sim P(y)$. Again by Corollary 1.4.1 f is one-to-one.

The set Z is an initial segment of A. For if $x \in Z$, then $P(x) \sim P(y)$ for a $y \in B$. If $x' < x$, then $P(x') \subset P(x)$. Let g, mapping $P(x)$ onto $P(y)$, be a similarity function. Then g maps $P(x')$ onto an initial segment P of $P(y)$ and hence P is an initial segment of B. Since $y \notin P(y)$, also $y \notin P$, and thus $P \neq B$. So there exists a $y' \in B$ such that $P = P(y')$. Hence g establishes a similarity between $P(x')$ and $P(y')$. Thus $x' \in Z$ and $P(y') = P\big(f(x')\big)$ is an initial segment of $P(y) = P\big(f(x)\big)$, whence $f(x') < f(x)$. So as a by-product we have obtained the fact that f is order-preserving.

Similarly, $f(Z)$ is an initial segment of B. For if $y \in f(Z)$, then $P(y) \sim P(x)$ for an $x \in Z$. Let h be a similarity mapping. If $y' < y$, then h maps $P(y')$ onto an initial segment $P(x')$ of $P(x)$, whence $y' = f(x')$, $x' \in Z$ and $y' \in f(Z)$.

---

[5] It would be enough to postulate that one of these sets is well ordered. It follows then by the similarity that the other is well ordered, too. (Cf. Exercise 1.6)

We have already shown that $Z \sim f(Z)$. To complete the proof it is enough to show that either $Z = A$ or $f(Z) = B$. Suppose that $Z \neq A$ and $f(Z) \neq B$. Then there exist $a \in A$ and $b \in B$ such that $Z = P(a)$ and $f(Z) = P(b)$. Thus $P(a) \sim P(b)$ and $a \in Z$, which is incompatible with $Z = P(a)$. □

Theorem 1.4.1 implies the trichotomy law for ordinal numbers:
For any ordinal numbers α, β

$$\text{either } \alpha < \beta, \text{ or } \beta < \alpha, \text{ or } \alpha = \beta.$$

Let us note also the following

Theorem 1.4.2. *For every ordinal number α, there exists the set $\Gamma(\alpha)$ of all ordinal numbers $\beta < \alpha$.*

Proof. Let $(A, \leqslant)$ be a well-ordered set such that $\overline{A} = \alpha$. Let for $x \in A$, $\psi(x, y)$ be the propositional formula $y \tau P(x)$, i.e., $y = \overline{P(x)}$. By the Axiom of Replacement there exists a set B such that $y \in B \Leftrightarrow$ there exists an $x \in A$ such that $y \tau P(x)$. But this, in turn, is equivalent to the fact that $y < \alpha$. Thus B is the required set. □

Once we know that $\Gamma(\alpha)$ is a set, the formula $y = \overline{P(x)}$ for $x \in A$ defines a function from A onto $\Gamma(\alpha)$. This function clearly is one-to-one and order-preserving. It follows that $\Gamma(\alpha) \sim A$ and consequently $\Gamma(\alpha)$ is well ordered and

$$\overline{\Gamma(\alpha)} = \alpha. \qquad (1.4.1)$$

Actually, the fact that $\Gamma(\alpha)$ is well ordered is a particular case of the following statement.

Theorem 1.4.3. *Any set of ordinal numbers is well ordered by the inequality $\leqslant$.*

Proof. The thing to prove is that if $A \neq \varnothing$ is a set of ordinal numbers, then there exists the smallest element in A. Take an $\alpha \in A$. If α is smallest in A, there is nothing more to prove. If not, the set $A \cap \Gamma(\alpha) \subset \Gamma(\alpha) \neq \varnothing$. Since $\Gamma(\alpha)$ is well ordered, there exists the smallest element β in $A \cap \Gamma(\alpha)$. For an arbitrary $\gamma \in A$ we have either $\gamma \geqslant \alpha$, or $\gamma < \alpha$. In the first case, since $\beta \in \Gamma(\alpha)$, we have $\beta < \alpha \leqslant \gamma$, whence $\beta < \gamma$. In the other case $\gamma \in \Gamma(\alpha) \cap A$, so $\beta \leqslant \gamma$. Thus β is the smallest element in A. □

Next we have

Lemma 1.4.2. *For every ordinal number α we have $\alpha < \alpha + 1$.*

Proof. Let $\alpha = \overline{A}$. Then $\alpha + 1 = \overline{A^*}$, where A^* is the set consisting of all the elements of A and A itself ordered so that $A > a$ for all $a \in A$. Hence $P(A) = A$ in A^*. The function $f : A \to A^*$ defined as $f(a) = a$ for $a \in A$ establishes the similarity of A with $P(A)$ in A^*. Thus A is similar to an initial segment of A^* different from A^*, i.e., $\overline{A} < \overline{A}^*$. □

Theorem 1.4.4. *For every set of ordinal numbers there exists an ordinal number strictly greater than any number from the given set.*

Proof. Let A be a set of ordinal numbers. By the Axiom of Replacement there exists the collection of sets $\{\Gamma(\beta)\}_{\beta \in A}$, and by the Axiom of Unions there exists the set

$$B = \bigcup_{\beta \in A} \Gamma(\beta).$$

Then, for every $\beta \in A$, the set $\Gamma(\beta)$ clearly is an initial segment of B. Hence $\overline{B} \geqslant \overline{\Gamma(\beta)} = \beta$ (cf. (1.4.1)). In view of Lemma 1.4.2 the ordinal number $\overline{B} + 1$ has the desired property. $\qquad \square$

Corollary 1.4.2. *There does not exist the set of all ordinal numbers.*

Corollary 1.4.3. *For every set of ordinal numbers there exists the smallest ordinal number which does not belong to the given set.*

Proof. Let A be a set of ordinal numbers and let α be a number greater than any $\beta \in A$. If α is not the smallest number with this property, then the set $\Gamma(\alpha) \backslash A \subset \Gamma(\alpha)$ is non-empty, and consequently it has the smallest element γ. It is already seen that γ is the smallest ordinal number which does not belong to A. $\qquad \square$

1.5   Cardinality of ordinal numbers

If α is an ordinal number, then by definition any two well-ordered sets of type α are similar, i.e., there exists a one-to-one mapping from one set onto the other. Consequently these sets have the same cardinality. Consequently to any ordinal number α we may assign a cardinal number, the common cardinality of all well-ordered sets of type α. This cardinal number is called the *cardinality of* α and is denoted by $\overline{\alpha}$. In particular we have $\big($cf. (1.4.1)$\big)$

$$\overline{\alpha} = \operatorname{card} \Gamma(\alpha).$$

It is easy to check that for arbitrary ordinal numbers α, β we have

$$\alpha \leqslant \beta \Rightarrow \overline{\alpha} \leqslant \overline{\beta}, \tag{1.5.1}$$

Lemma 1.5.1. *There exists the set A which contains all ordinal numbers α with $\overline{\alpha} \leqslant \aleph_0$.*

Proof. If $\overline{\alpha} \leqslant \aleph_0$ for an ordinal number α, then there exists a one-to-one mapping $f : \Gamma(\alpha) \to \mathbb{N}$. If for $a, b \in f\big(\Gamma(\alpha)\big)$ we put

$$\big(a \preccurlyeq b\big) \Leftrightarrow \big(f^{-1}(a) \leqslant f^{-1}(b)\big),$$

the set $\big(f\big(\Gamma(\alpha)\big), \preccurlyeq\big)$ will become a well-ordered set similar to $\Gamma(\alpha)$ (f being the similarity function), and hence $\overline{\big(f\big(\Gamma(\alpha)\big), \preccurlyeq\big)} = \alpha$. Thus for any ordinal number α with $\overline{\alpha} \leqslant \aleph_0$ there exists a well-ordered set $(B, \preccurlyeq)$ such that $B \subset \mathbb{N}$, and $\overline{B} = \alpha$. The converse is also true. If $(B, \preccurlyeq)$ is a well-ordered set of a type α, and $B \subset \mathbb{N}$, then $\overline{\alpha} \leqslant \aleph_0$. Consequently we may describe ordinal numbers α such that $\overline{\alpha} \leqslant \aleph_0$ as order types of well-ordered subsets of $\mathbb{N}$. (However, the order in these sets may be different from the natural order in $\mathbb{N}$.)

Let $\mathcal{P}$ be the power set of $\mathbb{N}$ (Axiom 4) and, for $P \in \mathcal{P}$, let $\mathcal{R}_P$ be the power set of $P \times P$. Every element R of $\mathcal{R}_P$ is a relation in P. There exists the set (Axiom of Specification)

$$B = \{(P, R) \mid P \in \mathcal{P}, R \in \mathcal{R}_P\}$$

and the set

$$C = \{(P, R) \in B \mid R \text{ is a well order}\}.$$

By the Axiom of Replacement there exists the set of the types of all $(P, R) \in C$. This is the desired set. $\qquad\square$

The same argument shows that for any cardinal number $\mathfrak{m}$ there exists the set of all ordinal numbers α such that $\bar{\alpha} \leqslant \mathfrak{m}$. This set is denoted in the sequel by $M(\mathfrak{m})$.

Lemma 1.5.2. *For every cardinal number $\mathfrak{m}$, we have*

$$M(\mathfrak{m}) = \Gamma\overline{(M(\mathfrak{m}))}. \tag{1.5.2}$$

Proof. There exists an ordinal number β which is greater than any $\alpha \in M(\mathfrak{m})$. First we show that $M(\mathfrak{m})$ is an initial segment of $\Gamma(\beta)$.

Let $\gamma \in M(\mathfrak{m})$ and $\xi < \gamma$. Then by (1.5.1) $\bar{\xi} \leqslant \bar{\gamma} \leqslant \mathfrak{m}$, i.e., $\xi \in M(\mathfrak{m})$. So either $M(\mathfrak{m}) = \Gamma(\beta)$, or there exists an $\eta \in \Gamma(\beta)$ such that

$$M(\mathfrak{m}) = \{\xi \in \Gamma(\beta) \mid \xi < \eta\} = \Gamma(\eta).$$

At any case $M(\mathfrak{m}) = \Gamma(\alpha)$ for an ordinal number α. Since $\alpha = \overline{\Gamma(\alpha)} = \overline{M(\mathfrak{m})}$, we obtain hence (1.5.2). $\qquad\square$

Theorem 1.5.1. *For every cardinal number $\mathfrak{m}$, we have*

$$\operatorname{card} M(\mathfrak{m}) > \mathfrak{m}. \tag{1.5.3}$$

Proof. Put $\alpha = \overline{M(\mathfrak{m})}$ so that $M(\mathfrak{m}) = \Gamma(\alpha)$. Suppose that (1.5.3) does not hold, i.e., $\bar{\alpha} \leqslant \mathfrak{m}$. This means that $\alpha \in M(\mathfrak{m})$, i.e., $\alpha \in \Gamma(\alpha)$, which is impossible. $\qquad\square$

Remark 1.5.1. Here we have made use of the trichotomy law for the cardinals, which, however, cannot be proved at this stage. What we actually prove here is that the inequality $\bar{\alpha} \leqslant \mathfrak{m}$ is impossible. This is sufficient to prove the remaining theorems of the present chapter, and then the trichotomy law for the cardinals follows from Theorem 1.4.1 and 1.7.1, and hence also condition (1.5.3).

We define Ω to be the order type of the set $M(\aleph_0)$, and $\aleph_1$ to be the cardinality of $M(\aleph_0)$. Thus

$$\Omega = \overline{M(\aleph_0)}, \quad \aleph_1 = \bar{\Omega} = \operatorname{card} M(\aleph_0).$$

By Theorem 1.5.1 $\aleph_1 > \aleph_0$. Moreover

Theorem 1.5.2. *There is no cardinal number strictly between $\aleph_0$ and $\aleph_1$.*

In other words, $\aleph_1$ is the next cardinal number after $\aleph_0$.

Proof. The thing to show is that if for a cardinal number $\mathfrak{m}$ we have $\mathfrak{m} < \aleph_1$, then $\mathfrak{m} \leqslant \aleph_0$. Let $\mathfrak{m} < \aleph_1$ and let X be a set such that $\operatorname{card} X = \mathfrak{m}$. Since $\operatorname{card} X = \mathfrak{m} < \aleph_1 = \operatorname{card} M(\aleph_0)$, there exists a subset Y of $M(\aleph_0)$ with the same cardinality as X:

$$\text{there exists } Y \subset M(\aleph_0) : \operatorname{card} Y = \operatorname{card} X = \mathfrak{m}.$$

Let $\alpha = \overline{Y}$. Thus $Y \sim \Gamma(\alpha)$, and $\mathfrak{m} = \operatorname{card} \Gamma(\alpha)$. Of course, $\alpha < \Omega$, for otherwise we would have $\overline{\alpha} \geqslant \overline{\Omega}$, i.e., $\mathfrak{m} \geqslant \aleph_1$. Thus $\alpha \in \Gamma(\Omega) = M(\aleph_0)$ (cf. (1.5.2)), which means that $\mathfrak{m} = \overline{\alpha} \leqslant \aleph_0$. $\square$

On the basis of Axioms 1–8 it is impossible to compare $\aleph_1$ with the power of continuum $\mathfrak{c}$. The conjecture that $\aleph_1 = \mathfrak{c}$ is the contents of the celebrated

Continuum hypothesis: $\aleph_1 = \mathfrak{c}$.

P. J. Cohen [48] showed that the continuum hypothesis is independent of Axioms 1–8. In the present book, unless explicitly stated otherwise, we do not assume the continuum hypothesis.

Note that, as a result of (1.5.2), $\Gamma(\Omega) = M(\aleph_0)$. Consequently, for every ordinal number $\alpha < \Omega$ we have $\overline{\alpha} \leqslant \aleph_0$.

1.6 Transfinite induction

If α is an ordinal number, then any function defined on the set $\Gamma(\alpha)$ is called a *transfinite sequence of type* α. The values of this function are denoted by x_β (the value at the point $\beta \in \Gamma(\alpha)$) and the sequence itself by $\{x_\beta\}_{\beta<\alpha}$.

Theorem 1.6.1. *Let X be a set, $\{x_\beta\}_{\beta<\alpha}$ a transfinite sequence containing all the elements of X, Φ a propositional formula defined for $x \in X$, and α an ordinal number. If the following conditions are fulfilled*

  (i) $\Phi(x_0)$;

 (ii) If $\beta = \gamma + 1 < \alpha$ and $\Phi(x_\gamma)$, then $\Phi(x_\beta)$;

(iii) If $\beta < \alpha$ is a limit number and $\Phi(x_\gamma)$ for every $\gamma < \beta$, then $\Phi(x_\beta)$;

then $\Phi(x)$ for every $x \in X$[6].

Proof. Let $E = \{\beta \in \Gamma(\alpha) \mid \Phi(x_\beta)\}$, and let ξ be the smallest ordinal number in $\Gamma(\alpha)$ which does not belong to E. By (i) $\xi > 0$. If $\xi = \gamma + 1$, then $\gamma \in E$, and consequently $\Phi(x_\gamma)$ holds. Then by (ii) $\Phi(x_\xi)$. If ξ is a limit number, then $\Phi(x_\gamma)$ holds for every $\gamma < \beta$, whence by (iii) $\Phi(x_\xi)$. In both cases we arrive at $\xi \in E$. The contradiction obtained shows that there is not such a ξ in $\Gamma(\alpha)$, or, in other words, $E = \Gamma(\alpha)$. Thus $\Phi(x_\beta)$ for all $\beta < \alpha$, and since the sequence $\{x_\beta\}_{\beta<\alpha}$ contains all the elements of X, we have $\Phi(x)$ for all $x \in X$. $\square$

Theorem 1.6.1 extends the well-known induction principle for $\mathbb{N}$. The next theorem, known as definitions by transfinite induction, also is an extension of the corresponding theorem for natural numbers.

---

[6] Condition (ii) could be omitted, as it is contained in (iii), where β need not be a limit number. Similarly (i). Thus (i)–(iii) jointly can be replaced by
(iv) If $\beta < \alpha$, and $\Phi(x_\gamma)$ for every $\gamma < \beta$, then $\Phi(x_\beta)$.

Let $\mathcal{F}(\alpha, X)$, where α is an ordinal number and X is a set, be the collection of all transfinite sequences of types $\beta < \alpha$ and with values in X:

$$f \in \mathcal{F}(\alpha, X) \Leftrightarrow f : \Gamma(\beta) \to X \text{ for some } \beta < \alpha.$$

Theorem 1.6.2. *For every ordinal number α, for every set X, and for every function $h : \mathcal{F}(\alpha, X) \to X$, there exists exactly one transfinite sequence $f : \Gamma(\alpha) \to X$ such that*

$$f(\beta) = h\big(f \mid \Gamma(\beta)\big) \text{ for } \beta < \alpha.$$

The symbol $f \mid \Gamma(\beta)$ denotes the restriction of f to the set $\Gamma(\beta)$.

Proof. First we prove the uniqueness. Suppose that there exist two such functions f, g. Let Φ be the propositional formula for $\beta \in \Gamma(\alpha)$

$$\Phi(\beta) \Leftrightarrow [f(\beta) = g(\beta)].$$

For $\beta = 0$ we have $f(0) = h(f \mid \varnothing) = h(g \mid \varnothing) = g(0)$, so $\Phi(0)$ holds true. Let $\Phi(\gamma)$ for γ be less than a $\beta < \alpha$. Then $f(\beta) = h\big(f \mid \Gamma(\beta)\big) = h\big(g \mid \Gamma(\beta)\big) = g(\beta)$, i.e., $\Phi(\beta)$. By Theorem 1.6.1 $\Phi(\beta)$ for all $\beta \in \Gamma(\alpha)$, i.e., $f = g$.

To prove the existence, consider the set E of all those $\beta \leqslant \alpha$ for which there exists a transfinite sequence (by the first part of the proof necessarily unique) $f_\beta : \Gamma(\beta) \to X$ such that $f_\beta(\gamma) = h\big(f_\beta \mid \Gamma(\gamma)\big)$ for all $\gamma < \beta$. If $\delta \leqslant \gamma$, $\delta, \gamma \in E$, then the sequence $f_\gamma \mid \Gamma(\delta)$ fulfils for $\xi < \delta$ $f_\gamma \mid \Gamma(\delta)(\xi) = f_\gamma(\xi) = h\big(f_\gamma \mid \Gamma(\xi)\big) = h\big(f_\gamma \mid \Gamma(\delta) \mid \Gamma(\xi)\big)$ so that, by the uniqueness, $f_\gamma \mid \Gamma(\delta) = f_\delta$. Suppose that there exists a $\beta \leqslant \alpha$, $\beta \notin E$. We may assume that β is the smallest such number, i.e., $\gamma \in E$ for $\gamma < \beta$ (for otherwise we could choose the smallest such number in $\Gamma(\beta)$). Let β be a limit number. Then for every $\xi < \beta$ we have $\xi + 1 < \beta$. Put for $\xi < \beta$

$$f(\xi) = f_{\xi+1}(\xi).$$

Then $f : \Gamma(\beta) \to X$ is a transfinite sequence of type β and for $\xi < \beta$ we have

$$f(\xi) = f_{\xi+1}(\xi) = h\big(f_{\xi+1} \mid \Gamma(\xi)\big) = h\big(f \mid \Gamma(\xi)\big).$$

And if $\beta = \gamma + 1$, then we put for $\xi < \gamma$

$$f(\xi) = f_{\xi+1}(\xi)$$

and

$$f(\gamma) = h\big(f \mid \Gamma(\gamma)\big). \tag{1.6.1}$$

Then we have $f(\xi) = h\big(f \mid \Gamma(\xi)\big)$ for $\xi < \gamma$ as in the preceding case and $f(\gamma) = h\big(f \mid \Gamma(\gamma)\big)$ by (1.6.1). Thus $\beta \in E$, which shows that all $\beta \leqslant \alpha$ belong to E. $\qquad \square$

1.7   The Zermelo theorem

We start with a lemma.

Lemma 1.7.1. *Let X be a set. If there exists a transfinite sequence $f : \Gamma(\alpha) \to X$ (of a certain type α) such that $X = f(\Gamma(\alpha))$, then the set X can be well ordered.*

Proof. For $x \in X$ put

$$A_x = \{\beta \in \Gamma(\alpha) \mid f(\beta) = x\}$$

and let ξ_x be the smallest element in A_x. Define the order in X

$$(x \preceq y) \Leftrightarrow (\xi_x \leqslant \xi_y).$$

Relation $\preceq$ is a well-order in X.   □

Theorem 1.7.1. [Zermelo [326]] *Every set can be well ordered.*

Proof. It is enough to show that there exists a transfinite sequence whose range is X. Let $w : (\mathcal{P}(X) \setminus \varnothing) \to X$ be a choice function (cf. 1.1). Let $\operatorname{card} X = \mathfrak{m}$ and $\alpha = \overline{M}(\mathfrak{m})$. Define the transfinite sequence $f : \Gamma(\alpha) \to X$ by

$$f(\beta) = w\Big(X \setminus \bigcup_{\gamma < \beta} \{f(\gamma)\}\Big), \quad \beta < \alpha.$$

By Theorem 1.6.2 the sequence $\{x_\beta\}_{\beta < \alpha}$ is unambiguously defined. It remains to show that the range of f is X. Put

$$B = \bigcup_{\beta \in \Gamma(\alpha)} \{f(\beta)\}.$$

Clearly $B \subset X$. Moreover $\operatorname{card} B = \operatorname{card} \Gamma(\alpha) = \overline{\alpha}$. Hence $\mathfrak{m} = \operatorname{card} X \geqslant \operatorname{card} B = \overline{\alpha} = \operatorname{card} M(\mathfrak{m})$. This shows that f cannot be defined for all $\beta \in \Gamma(\alpha)$, i.e., for a certain $\beta < \alpha$ we must have $X \setminus \bigcup_{\gamma \in \Gamma(\beta)} f(\gamma) = \varnothing$. Thus f is of type β and its range is X.   □

In this proof of Zermelo theorem we have used the Axiom of Choice. This could not be avoided, the Axiom of Choice had to be used in the proof. Actually the Zermelo theorem is equivalent to the Axiom of Choice, as results from the following

Theorem 1.7.2. *If every set can be well ordered, then for every set there exists a choice function.*

Proof. Let X be an arbitrary set, and let $\preceq$ be a relation which well orders X. Put for every $A \subset X$, $A \neq \varnothing$

$$w(A) = \text{the smallest element in } A.$$

Then $w : (\mathcal{P}(X) \setminus \varnothing) \to X$ is a choice function.   □

1.8   Lemma of Kuratowski-Zorn

In applications the Axiom of Choice is usually used in one of its equivalent forms. We have encountered one of such forms in the preceding section. It was the Zermelo theorem. But perhaps the most famous statement equivalent to the Axiom of Choice is the following

Theorem 1.8.1. [Lemma of Kuratowski-Zorn[7]] *Let* $(X, \leqslant)$ *be an ordered set,* $X \neq \varnothing$, *in which every chain has an upper bound. Then for every* $x_0 \in X$ *there exists in* X *a maximal element* $\overline{x}$ *such that* $\overline{x} \geqslant x_0$.

Proof. Let $\operatorname{card} X = \mathfrak{m}$, $\overline{M(\mathfrak{m})} = \alpha$. For an indirect proof suppose that for every $y \geqslant x_0$ the set $\{z \in X \mid z > y\}$ is non-empty. Define the transfinite sequence $\{y_\beta\}_{\beta < \alpha}$ as

$$y_\beta = \begin{cases} \text{the upper bound of } \{x_\gamma\}_{\gamma < \beta}, \text{if it exists,} \\ x_0 \text{ otherwise ,} \end{cases} \qquad (1.8.1)$$

and define the transfinite sequence $\{x_\beta\}_{\beta < \alpha}$ by

$$x_\beta = w\big(\{z \in X \mid z > y_\beta\}\big), \qquad (1.8.2)$$

where $w : (\mathcal{P}(X) \setminus \varnothing) \to X$ is a choice function. Clearly, by (1.8.1), $y_\beta \geqslant x_0$ for every $\beta < \alpha$ so that the set occurring in (1.8.2) is non-empty. So the sequence $\{x_\beta\}_{\beta < \alpha}$ is well defined.

This sequence is increasing. To show this consider the propositional formula $\Phi(\beta)$ $(\beta < \alpha)$ meaning:

"For $\beta < \alpha$, if $\gamma < \xi \leqslant \beta$, then $x_\gamma < x_\xi$".

If $\Phi(\gamma)$ for $\gamma < \beta$, then $\{x_\delta\}_{\delta < \beta}$ is a chain, and consequently $y_\beta \geqslant x_\delta$, $\delta < \beta$. Hence also $x_\beta > x_\delta$ for $\delta < \beta$, i.e., $\Phi(\beta)$ holds. By Theorem 1.6.1 we have $\Phi(\beta)$ for all $\beta < \alpha$.
Now put

$$B = \bigcup_{\beta < \alpha} \{x_\beta\}.$$

We have $B \subset X$, whence $\operatorname{card} B \leqslant \operatorname{card} X = \mathfrak{m}$, whereas

$$\operatorname{card} B = \operatorname{card} \Gamma(\alpha) = \overline{\alpha} = \operatorname{card} M(\mathfrak{m}) > \mathfrak{m}.$$

This shows that for some $\beta < \alpha$ we must have $\{z \in X \mid z > y_\beta\} = \varnothing$, i.e., y_β is a maximal element in X, and clearly $y_\beta \geqslant x_0$.   □

Theorem 1.8.2. *The Lemma of Kuratowski-Zorn implies the Axiom of Choice.*

Proof. Let $\mathcal{A}$ be any collection of non-empty sets, and put $B = \bigcup \mathcal{A}$. Let X be the family of those sets $F \subset B$ for which the intersection $F \cap A$ contains at most one point for every $A \in \mathcal{A}$. $(X, \subset)$ is an ordered set, $\varnothing \in X$. If $\mathcal{C} \subset X$ is a chain, then $\bigcup \mathcal{C} \in X$. In fact, if $\bigcup \mathcal{C} \cap A$ for an $A \in \mathcal{A}$ contains two different elements, say x, y,

---

[7] Kuratowski [197], Zorn [327].

then there exist sets D_x, $D_y \in \mathcal{C}$ such that $x \in D_x \cap A$ and $y \in D_y \cap A$. But since $\mathcal{C}$ is a chain, one of the sets D_x, D_y is contained in the other, say $D_x \subset D_y$. But then x, $y \in D_y$ and $D_y \cap A$ contains more than one point.

By the Lemma of Kuratowski-Zorn there exists in X a maximal element F_0. We will show that $F_0 \cap A \neq \varnothing$ for every $A \in \mathcal{A}$. If we had $F_0 \cap A_0 = \varnothing$ for an $A_0 \in \mathcal{A}$, then for an $x_0 \in A_0$ we might define a set $F^* = F_0 \cup \{x_0\}$. Clearly $F^* \in X$ and F^* is larger than F_0, which is impossible, since F_0 is maximal in X. Thus $F_0 \cap A$ is a singleton for every $A \in \mathcal{A}$, and we can define a function $w : \mathcal{A} \to B$ by

$$w(A) = A \cap F_0.$$

w is a choice function.   □

Another equivalent formulation of the Axiom of Choice is the following one.

Theorem 1.8.3. *For every non-empty family $\mathcal{A}$ of disjoint non-empty sets, there exists a set which has exactly one element in common with every set $A \in \mathcal{A}$.*

Proof. Let $w : \mathcal{A} \to \bigcup \mathcal{A}$ be a choice function. The set $w(\mathcal{A})$ has the desired properties.   □

We shall show also that Theorem 1.8.3 implies the Axiom of Choice. Let $\mathcal{A}$ be any non-empty family of non-empty sets. Consider the subsets of $(\bigcup \mathcal{A}) \times \mathcal{A}$ of the form $A \times \{A\}$ for $A \in \mathcal{A}$. Let $\mathcal{B}$ be the collection of all such subsets. The sets from $\mathcal{B}$ are disjoint, since for $A_1 \neq A_2$, we have $(A_1 \times \{A_1\}) \cap (A_2 \times \{A_2\}) = \varnothing$. Let F be a set that has exactly one element in common with every $B \in \mathcal{B}$. If $B \cap (A \times \{A\}) = (x, \{A\})$, put $w(A) = x$. This formula defines a choice function $w : \mathcal{A} \to \bigcup \mathcal{A}$.

P. J. Cohen [48] proved that the Axiom of Choice is independent of the remaining Axioms 1–8. A detailed discussion of The Axiom of Choice is found in Jech [154].

Exercises

1. Prove that for any two sets A, B there exists the set $A \cap B$ containing those and only those elements which are common to A and B.

2. The ordered pair $\langle a, b \rangle$ is defined as $\{a, \{a, b\}\}$, where $\{a, b\}$ is the unordered pair consisting of a and b. Prove that $\langle a, b \rangle = \langle c, d \rangle$ if and only if $a = c$ and $b = d$.

3. Prove that every ordered set $(A, \leqslant)$ is similar to a set $(\mathcal{C}, \subset)$, where $\mathcal{C}$ is a collection of sets.

4. Let X, Y be arbitrary sets, and let $\mathcal{F}$ be the family of all functions whose domain is contained in X and range is contained in Y. For $f, g \in \mathcal{F}$ write $f \leqslant g$ iff domain $f \subset$ domain g and $g \mid$ domain $f = f$. Prove that $(\mathcal{F}, \leqslant)$ is an ordered set in which every chain has an upper bound.

5. Let $(A, \leqslant)$ and $(B, \leqslant)$ be well-ordered sets and order the cartesian product $A \times B$ lexicographically. (I.e., $\langle a, b \rangle \leqslant \langle c, d \rangle$ iff $a < b$ or $a = b$ and $c \leqslant d$.) Prove that $(A \times B, \leqslant)$ is well ordered.

6. If $(A, \leqslant)$ is well ordered, then so is also every ordered set similar to $(A, \leqslant)$.

7. Prove that if $(A, \leqslant)$ and $(B, \preccurlyeq)$ are two similar well-ordered sets, then there is exactly one similarity function $f : A \to B$.

8. Prove that if $(X, \leqslant)$ is a well-ordered set, then every subset of X is similar either to X, or to an initial segment of X.

9. A subset A of an ordered set X is called *cofinal* in X iff for every $x \in X$ there exists an $a \in A$ with $a \geqslant x$. Prove that every linearly ordered set has a cofinal well-ordered subset.

10. Let $\{x_\beta\}_{\beta<\alpha}$ be a transfinite sequence of type α, where α is a limit ordinal number, whose terms are ordinal numbers. The smallest ordinal number greater than every x_β, $\beta < \alpha$, is called the *limit* of $\{x_\beta\}_{\beta<\alpha}$ and denoted by $\lim_{\beta<\alpha} x_\beta$. If f and g are two increasing transfinite sequences of ordinal numbers, α is a limit ordinal and $\lim_{\beta<\alpha} f(\beta) = \xi$, then $\lim_{\delta<\xi} g(\delta) = \lim_{\beta<\alpha} g\big(f(\beta)\big)$.

11. An ordinal number ξ is *cofinal* with a limit ordinal number α iff there exists an increasing sequence $\{x_\beta\}_{\beta<\alpha}$ of ordinal numbers such that $\xi = \lim_{\beta<\alpha} x_\beta$. Prove that ξ is cofinal with α if and only if $\Gamma(\xi)$ contains a cofinal subset of type α.

12. Prove the following version of the transfinite induction principle:

    Let $(X, \leqslant)$ be a well-ordered set, x_0 the smallest element in X, and Φ a propositional formula defined on X. If the following conditions are fulfilled

    (i) $\Phi(x_0)$,

    (ii) If $\Phi(y)$ for $y \in P(x)$, then $\Phi(x)$, $x \in X$,

    then $\Phi(x)$ for all $x \in X$.

13. Prove that if $(A, \leqslant)$ is a well-ordered set similar to its subset A_0, then the similarity function $f : A \to A_0$ satisfies $f(x) \geqslant x$ for all $x \in A$.

14. Prove that in every family $\mathcal{E}$ of sets there exists a maximal subfamily $\mathcal{E}_0$ of mutually disjoint sets.

Chapter 2

Topology

2.1 Category

In 2.1–2.2 X is a topological space, so, e.g., X may be a metric space, or, in particular, $\mathbb{R}^N$. A set $A \subset X$ is called *nowhere dense* iff int cl $A = \varnothing$. A set $A \subset X$ is said to be of the *first category* iff A is a countable union of nowhere dense sets:

$$A = \bigcup_{i=1}^{\infty} A_i, \ \text{int cl} \, A_i = \varnothing, \ i \in \mathbb{N}. \tag{2.1.1}$$

A set $A \subset X$ which is not of the first category is said to be of the *second category*. A set $A \subset X$ is called *residual* iff $A' = X \setminus A$ is of the first category.

Theorem 2.1.1. *If $A \subset X$ is of the first category, then there exists a set $B \in \mathcal{F}_\sigma$ such that $A \subset B$, and B is of the first category.*

Proof. A has a representation (2.1.1). Put

$$B = \bigcup_{i=1}^{\infty} \text{cl} \, A_i .$$

The sets cl A_i are nowhere dense, just like A_i:

$$\text{int cl cl} \, A_i = \text{int cl} \, A_i = \varnothing .$$

B is an $\mathcal{F}_\sigma$ set, being a countable union of closed sets, and B is of the first category, being a countable union of nowhere dense sets. Obviously $A \subset B$. $\square$

Theorem 2.1.2. *If $A \subset X$ is residual, then there exists a set $B \in \mathcal{G}_\delta$ such that $B \subset A$, and B is residual.*

Proof. There exists (by Theorem 2.1.1) a set $C \in \mathcal{F}_\sigma$ such that $A' \subset C$, and C is of the first category. The set $B = C' = X \setminus C$ has the desired properties. $\square$

If we assume, moreover, that the space X is complete, then we may assert that the set B in Theorem 2.1.2 is of the second category. Otherwise $X = B \cup C$ would be[1] of the first category contrary to a celebrated Baire theorem that a complete space is of the second category.

The following theorem yields a slight generalization of Theorem 2.1.2:

Theorem 2.1.3. *If $A \subset H \subset X$, $H \in \mathcal{G}_\delta$, and $H \setminus A$ is of the first category, then there exists a set $B \subset A$ such that $B \in \mathcal{G}_\delta$ and $H \setminus B$ is of the first category.*

Proof. [2] Put $A_1 = A \cup (X \setminus H)$. The set A_1 is residual[3] $(X \setminus A_1 \subset H \setminus A)$, so by Theorem 2.1.2 there exists a residual set $B_1 \subset A_1$, $B_1 \in \mathcal{G}_\delta$. Put $B = H \cap B_1$. Since $B_1 \subset A \cup (X \setminus H)$, we have $B \subset H \cap [A \cup (X \setminus H)] = (H \cap A) \cup [H \cap (X \setminus H)] = H \cap A = A$, and clearly $B \in \mathcal{G}_\delta$ as an intersection of $\mathcal{G}_\delta$ sets. It remains to show that $H \setminus B$ is of the first category. Now, $H \setminus B = H \setminus (H \cap B_1) = H \setminus (X \cap B_1) \subset X \setminus B_1$, which is a set of the first category. Consequently also $H \setminus B$ is of the first category.   □

The following lemma will be needed later.

Lemma 2.1.1. *If $F \in \mathcal{F}_\sigma$ and int $F = \varnothing$, then F is of the first category.*

Proof. $F = \bigcup_{i=1}^{\infty} F_i$, where F_i are closed. If, for a certain $i \in \mathbb{N}$, we had int $F_i \neq \varnothing$, then we would have also int $F \neq \varnothing$, since int $F_i \subset$ int F. Consequently int cl $F_i =$ int $F_i = \varnothing$ for all $i \in \mathbb{N}$, and thus the sets F_i are nowhere dense. Hence F is of the first category.   □

We say that a set $A \subset X$ is of the *first category at a point* $x \in X$ iff there exists a neighbourhood U_x of x (in X) such that the set $A \cap U_x$ is of the first category. And we say that a set $A \subset X$ is of the *second category at a point* $x \in X$ iff it is not of the first category at x, i.e., iff $A \cap U_x$ is of the second category for every neighbourhood U_x of x. Put

$$D(A) = \{x \in X \mid A \text{ is of the II category at } x\}.$$

We have the following calculation rules for $D(A)$:

$$D(A \cup B) = D(A) \cup D(B), \tag{2.1.2}$$

$$A \subset B \Rightarrow D(A) \subset D(B). \tag{2.1.3}$$

Simple proofs of formulas (2.1.2) and (2.1.3) are left to the reader (cf. Kuratowski [196]).

---

[1] Cf. Exercise 2.4.
[2] Theorem 2.1.3 follows also from Theorem 2.1.2 on replacing X by H and observing that every $\mathcal{G}_\delta$ set in H is also a $\mathcal{G}_\delta$ set in X.
[3] Cf. Exercise 2.4.

Theorem 2.1.4. *If the space X is separable[4] and $A \cap D(A) = \varnothing$ then A is of the first category.*

Proof. Since X is separable, it has a countable base of neighbourhoods. Let $\{B_n\}_{n \in \mathbb{N}}$ be such a base. For every $x \in A$ there exists a neighbourhood U_x of x such that $A \cap U_x$ is of the first category $\big($since $A \cap D(A) = \varnothing\big)$, and there exists an $n_x \in \mathbb{N}$ such that $x \in B_{n_x} \subset U_x$. The set $A \cap B_{n_x} \subset A \cap U_x$ also is of the first category. Since we have

$$A = \bigcup_{x \in A} A \cap B_{n_x}, \tag{2.1.4}$$

and there are only at most countably many sets B_{n_x}, the union in (2.1.4) is at most countable, and consequently (cf. Exercise 2.4) the set A is of the first category. $\square$

Theorem 2.1.5. *For every set $A \subset X$, the set $D(A)$ is closed.*

Proof. We shall show that the set $X \setminus D(A)$ is open. Take an $x \notin D(A)$. Then there exists an open neighbourhood U_x of x such that $A \cap U_x$ is of the first category. Now, since U_x is a neighbourhood of every point from U_x, we have $U_x \subset X \setminus D(A)$. Consequently x is an inner point of $X \setminus D(A)$, which shows that $X \setminus D(A)$ is open, and so $D(A)$ is closed. $\square$

Theorem 2.1.6. *If the space X is separable, then for every set $A \subset X$, the set $A \setminus D(A)$ is of the first category.*

Proof. By (2.1.3) we have
$$D[A \setminus D(A)] \subset D(A),$$

whence
$$[A \setminus D(A)] \cap D[A \setminus D(A)] \subset [A \setminus D(A)] \cap D(A) = \varnothing.$$

The theorem results now from Theorem 2.1.4. $\square$

Now we are going to investigate category problems in product spaces. Let Y be an arbitrary space. For every set $A \subset X \times Y$ and every $x_0 \in X$ we write

$$A[x_0] = \{y \in Y \mid (x_0, y) \in A\}. \tag{2.1.5}$$

Lemma 2.1.2. *Let Y be a separable[5] topological space, and let $A \subset X \times Y$ be a nowhere dense set. Then there exists a set $P \subset X$ of the first category, such that for every $x \subset P' = X \setminus P$ the set $A[x]$ $\big($defined by (2.1.5)$\big)$ is nowhere dense.*

---

[4] Theorem 2.1.4 (as well as Theorem 2.1.6) is generally valid, without assuming that X is separable, but the proof in the general case is much more difficult (cf., e.g., Kuratowski [196]). Since in the sequel the results of the present chapter will be used for $X = \mathbb{R}^N$, which is separable, we restrict ourselves to separable X only. *Separable* means here having an at most countable neighbourhood base.

[5] Here, as well as in Theorems 2.1.7 and 2.1.8, the assumption that Y is separable is essential.

Proof. Since Y is separable, it has a countable base of neighbourhoods. Let $\{B_n\}_{n \in \mathbb{N}}$ be such a base. Put

$$E_n = \{x \in X \mid \{x\} \times B_n \subset \mathrm{cl}[A \cap (\{x\} \times Y)]\}. \qquad (2.1.6)$$

Hence $E_n \times B_n = \bigcup_{x \in E_n} \{x\} \times B_n \subset \bigcup_{x \in E_n} \mathrm{cl}[A \cap (\{x\} \times Y)] \subset \bigcup_{x \in X} \mathrm{cl}[A \cap (\{x\} \times Y)] \subset$
$\subset \bigcup_{x \in X} [\mathrm{cl}\, A \cap \mathrm{cl}(\{x\} \times Y)] = \mathrm{cl}\, A \cap \bigcup_{x \in X} \mathrm{cl}(\{x\} \times Y) = \mathrm{cl}\, A \cap \bigcup_{x \in X} (\{x\} \times Y) =$
$\mathrm{cl}\, A \cap (X \times Y) = \mathrm{cl}\, A$ so that

$$E_n \times B_n \subset \mathrm{cl}\, A. \qquad (2.1.7)$$

Since A is nowhere dense, it follows from (2.1.7) that $E_n \times B_n$ is nowhere dense. Suppose that E_n is not nowhere dense. Then there exists a non-empty open set $U \subset \mathrm{cl}\, E_n$. Then (cf. Exercise 2.1)

$$U \times B_n \subset \mathrm{cl}\, E_n \times B_n \subset \mathrm{cl}\, E_n \times \mathrm{cl}\, B_n = \mathrm{cl}(E_n \times B_n)$$

so that $U \times B_n \subset \mathrm{int}\,\mathrm{cl}(E_n \times B_n)$, which is impossible, since $E_n \times B_n$ is nowhere dense. Thus, for every $n \in \mathbb{N}$, the set E_n is nowhere dense.

Put

$$P = \bigcup_{n=1}^{\infty} E_n.$$

The set P is of the first category (cf. Exercise 2.4). Suppose that for an $x \in X$ the set $A[x]$ is not nowhere dense. Then $\mathrm{cl}\, A[x]$ contains an open set, and hence a set from the base. So there exists an $n \in \mathbb{N}$ such that $B_n \subset \mathrm{cl}\, A[x]$, whence $\{x\} \times B_n \subset \{x\} \times \mathrm{cl}\, A[x] = \mathrm{cl}\,(\{x\} \times A[x]) = \mathrm{cl}[A \cap (\{x\} \times Y)]$. By (2.1.6) $x \in E_n \subset P$. Hence, for every $x \in P'$, the set $A[x]$ is nowhere dense. $\square$

Theorem 2.1.7. *Let Y be a separable topological space, and let $A \subset X \times Y$ be a set of the first category. Then there exists a set $P \subset X$, of the first category, such that for every $x \in P' = X \setminus P$ the set $A[x]$ (defined by (2.1.5)) is of the first category.*

Proof. We have $A = \bigcup_{n=1}^{\infty} A_n$, where the sets $A_n \subset X \times Y$ are nowhere dense. By Lemma 2.1.2, for every $n \in \mathbb{N}$, there exists a set $P_n \subset X$, of the first category, such that for every $x \in P'_n$ the set $A_n[x]$ is nowhere dense. Put

$$P = \bigcup_{n=1}^{\infty} P_n.$$

So P is a set of the first category, and for every $x \in P' \subset P'_n$ the set $A_n[x]$ is nowhere dense ($n \in \mathbb{N}$), whence

$$A[x] = \bigcup_{n=1}^{\infty} A_n[x] \qquad (2.1.8)$$

is of the first category. To see that (2.1.8) is true observe that $y \in A[x] \Leftrightarrow (x,y) \in A \Leftrightarrow$ there exists an $n \in \mathbb{N}$ such that $(x,y) \in A_n \Leftrightarrow$ there exists an $n \in \mathbb{N}$ such that $y \in A_n[x] \Leftrightarrow y \in \bigcup_{n=1}^{\infty} A_n[x]$. $\square$

Theorem 2.1.8. *Let Y be a separable topological space, and let $A \subset X$, $B \subset Y$. Then the set $A \times B \subset X \times Y$ is of the first category if and only if at least one of the sets A, B is of the first category.*

Proof. Suppose that $A \times B$ is of the first category, and B is of the second category. By Theorem 2.1.7 there exists a set $P \subset X$, of the first category, such that for every $x \in P'$ the set $(A \times B)[x]$ is of the first category. But for $x \in A$ the set $(A \times B)[x] = B$ is of the second category. Hence $A \subset P$, and thus A is of the first category.

To prove the converse implication, we first prove that if A is nowhere dense in X, then $A \times Y$ is nowhere dense in $X \times Y$. We have $\mathrm{cl}(A \times Y) = \mathrm{cl}\, A \times \mathrm{cl}\, Y = \mathrm{cl}\, A \times Y$, whence $\mathrm{int}\,\mathrm{cl}(A \times Y) = \mathrm{int}(\mathrm{cl}\, A \times Y) = \mathrm{int}\,\mathrm{cl}\, A \times \mathrm{int}\, Y = \varnothing \times Y = \varnothing$. Thus $A \times Y$ is nowhere dense.

Now suppose that the set A is of the first category. Then $A = \bigcup_{n=1}^{\infty} A_n$, where A_n are nowhere dense, whence $A \times B \subset A \times Y = \left(\bigcup_{n=1}^{\infty} A_n \right) \times Y = \bigcup_{n=1}^{\infty} (A_n \times Y)$. Every set $A_n \times Y$, $n \in \mathbb{N}$, is nowhere dense, whence $\bigcup_{n=1}^{\infty} (A_n \times Y)$ is of the first category, and so is also $A \times B$.

If the set B is of the first category, then so is $A \times B$. The proof is similar. $\square$

2.2 Baire property

A set $A \subset X$ is said to have the *Baire property* iff

$$A = (G \cup P) \setminus R, \tag{2.2.1}$$

where the set G is open, and the sets P, R are of the first category. It follows directly from the definition that every open set and every set of the first category have the Baire property.

Theorem 2.2.1. *The family of the sets with the Baire property is a σ-algebra.*

Proof. If A_n, $n \in \mathbb{N}$, are sets with the Baire property, then there exist open sets G_n, $n \in \mathbb{N}$, and sets of the first category P_n, R_n, $n \in \mathbb{N}$, such that

$$A_n = (G_n \cup P_n) \setminus R_n, \ n \in \mathbb{N}.$$

Then

$$\bigcup_{n=1}^{\infty} A_n \subset \bigcup_{n=1}^{\infty} G_n \cup \bigcup_{n=1}^{\infty} P_n$$

and

$$\left(\bigcup_{n=1}^{\infty} G_n \cup \bigcup_{n=1}^{\infty} P_n \right) \setminus \bigcup_{n=1}^{\infty} A_n \subset \bigcup_{n=1}^{\infty} R_n$$

which shows that

$$\bigcup_{n=1}^{\infty} A_n = \left(\bigcup_{n=1}^{\infty} G_n \cup \bigcup_{n=1}^{\infty} P_n \right) \setminus \left[\left(\bigcup_{n=1}^{\infty} G_n \cup \bigcup_{n=1}^{\infty} P_n \right) \setminus \bigcup_{n=1}^{\infty} A_n \right]$$

and $\bigcup_{n=1}^{\infty} A_n$ is a set with the Baire property.

Now suppose that A, with a representation (2.2.1), is a set with the Baire property. Then

$$A' = (G' \cap P') \cup R .$$

Put $H = \operatorname{int} G'$, $C = R \cup (G' \setminus H)$, $D = P \setminus R$. Then

$$A' = (H \cup C) \setminus D . \tag{2.2.2}$$

In fact,

$$(H \cup C) \setminus D = [H \cup R \cup (G' \setminus H)] \cap (P \cap R')' = [H \cup R \cup (G' \cap H')] \cap (P' \cup R)$$

$$= (H \cup R \cup G') \cap (H \cup R \cup H') \cap (P' \cup R) .$$

Now, $H \cup R \cup H' = X$, and $H \cup R \cup G' = R \cup G'$, since $H = \operatorname{int} G' \subset G'$. Hence

$$(H \cup C) \setminus D = (R \cup G') \cap (P' \cup R) = (R \cap P') \cup R \cup (G' \cap P') \cup (G' \cap R) = (G' \cap P') \cup R = A'$$

since $R \cap P' \subset R$ and $G' \cap R \subset R$. This proves (2.2.2).

Now, the set H is open, $D \subset P$ is of the first category, and $G' \setminus H$ is a closed set without inner points, and so it is nowhere dense, and hence of the first category. Thus C also is of the first category, and it follows from (2.2.2) that A' has the Baire property. $\square$

Theorem 2.2.2. *If the space X is separable[6], then for every set $A \subset X$ there exists a set B with the Baire property such that $A \subset B$ and for every set Z with the Baire property containing A the set $B \setminus Z$ is of the first category.*

Proof. By Theorem 2.1.6 the set $A \setminus D(A)$ is of the first category, so there exists (Theorem 2.1.1) a set $C \in \mathcal{F}_\sigma$, of the first category such that $A \setminus D(A) \subset C$. Put $B = D(A) \cup C$. The set B has the Baire property, since both, $D(A)$ (being closed; cf. Theorems 2.1.5 and 2.3.1) and C (being of the first category) have the Baire property (now use Theorem 2.2.1), and evidently $A = \big(A \cap D(A)\big) \cup \big(A \setminus D(A)\big) \subset D(A) \cup C = B$. Let $Z \supset A$ be an arbitrary set with the Baire property. Then (cf. (2.1.3))

$$B \setminus Z = \big(D(A) \setminus Z \big) \cup (C \setminus Z) \subset \big(D(Z) \setminus Z \big) \cup (C \setminus Z) \subset \big(D(Z) \setminus Z \big) \cup C$$

is of the first category (cf. Exercise 2.5). $\square$

---

[6] The assumption that X is separable is not essential and can be omitted. Cf. the footnote 4 on p. 21.

Now let Y be a separable topological space. We preserve notation (2.1.5).

Theorem 2.2.3. *Let Y be a separable topological space. If the set $A \subset X \times Y$ has the Baire property, then there exists a set $Q \subset X$, of the first category, such that for every $x \in Q' = X \setminus Q$ the set $A[x]$ has the Baire property.*

Proof. We have $A = (G \cup P) \setminus R$, where G is open, and P, R are of the first category. By Theorem 2.1.7 there exist sets $Q_1 \subset X$ and $Q_2 \subset X$ such that for every $x \in Q_1'$ the set $P[x]$ is of the first category, and for every $x \in Q_2'$ the set $R[x]$ is of the first category. Put $Q = Q_1 \cup Q_2$. For $x \in Q'$ the sets $P[x]$ and $R[x]$ both are of the first category (in Y), and Q itself is of the first category (in X). Now, we have

$$A[x] = \big(G[x] \cup P[x]\big) \setminus R[x].$$

For every $x \in X$ the set $G[x]$ is open (in Y), therefore for every $x \in Q'$ the set $A[x]$ has the Baire property. $\square$

There are far reaching analogies between the sets of the first category and the sets of Lebesgue measure zero (cf. Oxtoby [250]). In this context the sets with the Baire property play the role of measurable sets. These analogies will become even more evident in the sequel of this book.

2.3 Borel sets

Let $\mathcal{B}(X)$ be the family of all Borel sets in a metric space X, i.e., the smallest σ-algebra of subsets of X containing the open sets.

Theorem 2.3.1. *Every Borel set has the Baire property.*

Proof. By Theorem 2.2.1 the family of all sets (in X) with the Baire property is a σ-algebra, and it contains the open sets, so it must contain the smallest σ-algebra containing the open sets, i.e., $\mathcal{B}(X)$. $\square$

Now we describe a construction of $\mathcal{B}(X)$. Let $\mathcal{G}(X)$ and $\mathcal{F}(X)$ be the families of all open and all closed sets in X, respectively. We define (cf. Theorem 1.6.2) two transfinite sequences $\{\mathcal{A}_\alpha\}_{\alpha < \Omega}$ and $\{\mathcal{M}_\alpha\}_{\alpha < \Omega}$ of set classes in X. We put

$$\mathcal{A}_0 = \mathcal{G}(X), \quad \mathcal{M}_0 = \mathcal{F}(X),$$

$$\mathcal{A}_\alpha = \Big(\bigcup_{\xi < \alpha} \mathcal{M}_\xi\Big)_\sigma, \quad \mathcal{M}_\alpha = \Big(\bigcup_{\xi < \alpha} \mathcal{A}_\xi\Big)_\delta, \quad \alpha < \Omega,$$

where, for any collection ζ of sets, ζ_σ denotes the collection of all the countable unions of sets from ζ, and ζ_δ denotes the collection of all the countable intersections of sets from ζ. So $\mathcal{A}_1 = \mathcal{F}_\sigma$, $\mathcal{M}_1 = \mathcal{G}_\delta$.

Lemma 2.3.1. *For every ordinal number $\beta < \alpha < \Omega$*

  (*i*) $\mathcal{A}_\beta \cup \mathcal{M}_\beta \subset \mathcal{A}_\alpha \cap \mathcal{M}_\alpha$;

 (*ii*) *if $Z \in \mathcal{A}_\alpha$, then $Z' \in \mathcal{M}_\alpha$, and conversely;*

(*iii*) $\mathcal{A}_\alpha \cup \mathcal{M}_\alpha \subset \mathcal{B}(X)$.

Proof. (*i*) The proof will be by transfinite induction (Theorem 1.6.1) with respect to α. The smallest possible value of α is $\alpha = 1$, because for $\alpha = 0$ there is no $\beta < \alpha$. If $\alpha = 1$, then the only $\beta < \alpha$ is $\beta = 0$, and (i) becomes

$$\mathcal{G}(X) \cup \mathcal{F}(X) \subset \mathcal{F}_\sigma \cap \mathcal{G}_\delta \,,$$

which is known to be true.

Now take arbitrary ordinal numbers $\beta < \alpha < \Omega$. We must distinguish two cases.
1. $\alpha = \beta + 1$. Then

$$\mathcal{M}_\beta \subset (\mathcal{M}_\beta)_\sigma \subset \Big(\bigcup_{\xi < \alpha} \mathcal{M}_\xi \Big)_\sigma = \mathcal{A}_\alpha \,.$$

Further,

$$\bigcup_{\xi < \beta} \mathcal{A}_\xi \subset \bigcup_{\xi < \alpha} \mathcal{A}_\xi \,,$$

whence also

$$\mathcal{M}_\beta = \Big(\bigcup_{\xi < \beta} \mathcal{A}_\xi \Big)_\delta \subset \Big(\bigcup_{\xi < \alpha} \mathcal{A}_\xi \Big)_\delta = \mathcal{M}_\alpha \,.$$

Hence $\mathcal{M}_\beta \subset \mathcal{A}_\alpha \cap \mathcal{M}_\alpha$. Similarly it is proved that $\mathcal{A}_\beta \subset \mathcal{A}_\alpha \cap \mathcal{M}_\alpha$. Thus (*i*) holds.
2. $\alpha > \beta + 1$. For the inductive proof assume that
(∗) for every $\gamma < \alpha$, if $\eta < \gamma$, then $\mathcal{A}_\eta \cup \mathcal{M}_\eta \subset \mathcal{A}_\gamma \cap \mathcal{M}_\gamma$.
Put $\gamma = \beta + 1$. Then $\beta < \gamma < \alpha$. By (∗)

$$\mathcal{A}_\beta \cup \mathcal{M}_\beta \subset \mathcal{A}_\gamma \cap \mathcal{M}_\gamma \,.$$

Hence

$$\mathcal{A}_\beta \cup \mathcal{M}_\beta \subset \Big(\bigcup_{\xi < \alpha} \mathcal{A}_\xi \Big) \cap \Big(\bigcup_{\xi < \alpha} \mathcal{M}_\xi \Big) \subset \Big(\bigcup_{\xi < \alpha} \mathcal{A}_\xi \Big)_\delta \cap \Big(\bigcup_{\xi < \alpha} \mathcal{M}_\xi \Big)_\sigma = \mathcal{M}_\alpha \cap \mathcal{A}_\alpha \,.$$

Thus (*i*) holds in this case, too.

(*ii*) The proof is again by transfinite induction with respect to α. For $\alpha = 0$ (ii) is true in virtue of the well-known property of the open and closed sets. Suppose that for all $\beta < \alpha < \Omega$ we have
(∗∗) if $Z \in \mathcal{A}_\beta$, then $Z' \in \mathcal{M}_\beta$, and conversely.
Take a $Z \in \mathcal{A}_\alpha$. Then

$$Z = \bigcup_{n=1}^{\infty} E_n \,,$$

where, for every $n \in \mathbb{N}$, $E_n \in \bigcup_{\xi < \alpha} \mathcal{M}_\xi$, i.e., for every $n \in \mathbb{N}$ there exists a $\xi_n < \alpha$ such that $E_n \in \mathcal{M}_{\xi_n}$. Hence, by (∗∗), $E_n' \in \mathcal{A}_{\xi_n}$, $n \in \mathbb{N}$, and

$$Z' = \bigcap_{n=1}^{\infty} E_n' \in \Big(\bigcup_{\xi < \alpha} \mathcal{A}_\xi \Big)_\delta = \mathcal{M}_\alpha \,.$$

The proof of the converse implication is analogous.

(iii) This results from the fact that $\mathcal{B}(X)$ is a σ-algebra (again a proof by transfinite induction is necessary, which is left to the reader). $\square$

Theorem 2.3.2. *We have*

$$\bigcup_{\alpha<\Omega} \mathcal{A}_\alpha = \bigcup_{\alpha<\Omega} \mathcal{M}_\alpha = \mathcal{B}(X).$$

Proof. For every $\alpha < \Omega$ we have $\alpha + 1 < \Omega$, whence by Lemma 2.3.1 (i) $\mathcal{A}_\alpha \subset \mathcal{M}_{\alpha+1} \subset \bigcup_{\alpha<\Omega} \mathcal{M}_\alpha$, and $\bigcup_{\alpha<\Omega} \mathcal{A}_\alpha \subset \bigcup_{\alpha<\Omega} \mathcal{M}_\alpha$. Similarly, for every $\alpha < \Omega$ we have $\mathcal{M}_\alpha \subset \mathcal{A}_{\alpha+1} \subset \bigcup_{\alpha<\Omega} \mathcal{A}_\alpha$, and $\bigcup_{\alpha<\Omega} \mathcal{M}_\alpha \subset \bigcup_{\alpha<\Omega} \mathcal{A}_\alpha$. Consequently

$$\bigcup_{\alpha<\Omega} \mathcal{A}_\alpha = \bigcup_{\alpha<\Omega} \mathcal{M}_\alpha,$$

and by Lemma 2.3.1 (iii) $\bigcup_{\alpha<\Omega} \mathcal{A}_\alpha = \bigcup_{\alpha<\Omega} \mathcal{M}_\alpha \subset \mathcal{B}(X)$. So it is enough to show that $\bigcup_{\alpha<\Omega} \mathcal{A}_\alpha$ is a σ-algebra, for then it must contain the smallest σ-algebra containing $\mathcal{G}(X) = \mathcal{A}_0$, i.e., it must contain $\mathcal{B}(X)$.

Take a sequence of sets $A_n \in \bigcup_{\alpha<\Omega} \mathcal{A}_\alpha$. Then, for every $n \in \mathbb{N}$, there exists an $\alpha_n < \Omega$ such that $A_n \in \mathcal{A}_{\alpha_n}$. By Theorem 1.4.4 there exists an ordinal number α greater than every number α_n. This has been constructed as $\alpha = \overline{B} + 1$, where, in the present case $B = \bigcup_{n=1}^{\infty} \Gamma(\alpha_n)$. We have for every $n \in \mathbb{N}$, $\overline{\Gamma(\alpha_n)} = \alpha_n < \Omega$, whence $\operatorname{card} \Gamma(\alpha_n) = \overline{\alpha}_n \leqslant \aleph_0$. Hence also $\operatorname{card} B \leqslant \aleph_0$, and consequently $\overline{\alpha} \leqslant \aleph_0$. Therefore $\alpha \in M(\aleph_0) = \Gamma(\Omega)$, i.e., $\alpha < \Omega$. Thus we have $\alpha_n < \alpha < \Omega$ for every $n \in \mathbb{N}$. By Lemma 2.3.1 (i) $\mathcal{A}_{\alpha_n} \subset \mathcal{M}_\alpha$ for every $n \in \mathbb{N}$, whence $A_n \in \mathcal{M}_\alpha$ for every $n \in \mathbb{N}$. Hence $\bigcup_{n=1}^{\infty} A_n \in (\mathcal{M}_\alpha)_\sigma \subset \big(\bigcup_{\xi<\alpha+1} \mathcal{M}_\xi \big)_\sigma = \mathcal{A}_{\alpha+1} \subset \bigcup_{\alpha<\Omega} \mathcal{A}_\alpha$.

Now take a set $A \in \bigcup_{\alpha<\Omega} \mathcal{A}_\alpha$. Then $A \in \mathcal{A}_\beta$ for a certain $\beta < \Omega$. By Lemma 2.3.1 (ii) $A' \in \mathcal{M}_\beta \subset \bigcup_{\alpha<\Omega} \mathcal{M}_\alpha = \bigcup_{\alpha<\Omega} \mathcal{A}_\alpha$. Consequently $\bigcup_{\alpha<\Omega} \mathcal{A}_\alpha$ is a σ-algebra. $\square$

Theorem 2.3.3. $\mathcal{B}(X)$ *is the smallest class* $\mathcal{K}$ *of subsets of* X *with the properties:*

(i) $\mathcal{F}(X) \subset \mathcal{K}$,

(ii) *if* $A_n \in \mathcal{K}$ *for* $n \in \mathbb{N}$, *then* $\bigcup_{n=1}^{\infty} A_n \in \mathcal{K}$ *and* $\bigcap_{n=1}^{\infty} A_n \in \mathcal{K}$.

Proof. Clearly $\mathcal{B}(X)$ has properties (i) and (ii), so

$$\mathcal{K} \subset \mathcal{B}(X).$$

To prove the converse inclusion, we will show that for every $\alpha < \Omega$

$$\mathcal{A}_\alpha \cup \mathcal{M}_\alpha \subset \mathcal{K}. \tag{2.3.1}$$

For $\alpha = 0$ (2.3.1) is true by (i) and by the fact that $\mathcal{G}(X) \subset \mathcal{F}_\sigma \subset \mathcal{K}$. Assume that for a certain $\alpha < \Omega$ we have $\mathcal{A}_\xi \cap \mathcal{M}_\xi \subset \mathcal{K}$ for every $\xi < \alpha$. Then also

$$\Big(\bigcup_{\xi < \alpha} \mathcal{A}_\xi \Big) \cup \Big(\bigcup_{\xi < \alpha} \mathcal{M}_\xi \Big) \subset \mathcal{K},$$

and by (ii)

$$\Big(\bigcup_{\xi < \alpha} \mathcal{A}_\xi \Big)_\delta \cup \Big(\bigcup_{\xi < \alpha} \mathcal{M}_\xi \Big)_\sigma \subset \mathcal{K}.$$

Thus (2.3.1) holds.

The inclusion $\mathcal{B}(X) \subset \mathcal{K}$ results now from (2.3.1) in view of Theorem 2.3.2. $\square$

Theorem 2.3.4. *If the space X is separable, then* $\operatorname{card} \mathcal{B}(X) \leqslant \mathfrak{c}$.

Proof. In a separable space X we have $\operatorname{card} \mathcal{G}(X) \leqslant \mathfrak{c}$ (since every open set may be represented as a union of base neighbourhoods, and there exists a countable base), and hence also $\operatorname{card} \mathcal{F}(X) \leqslant \mathfrak{c}$. An easy proof by transfinite induction shows that for every $\alpha < \Omega$ we have

$$\operatorname{card} \mathcal{A}_\alpha \leqslant \mathfrak{c} \quad \text{and} \quad \operatorname{card} \mathcal{M}_\alpha \leqslant \mathfrak{c}.$$

In the union $\bigcup_{\alpha < \Omega} \mathcal{A}_\alpha$ we have $\aleph_1 \leqslant \mathfrak{c}$ summands, so $\operatorname{card} \bigcup_{\alpha < \Omega} \mathcal{A}_\alpha \leqslant \mathfrak{c}$. The theorem results now from Theorem 2.3.2. $\square$

As an easy consequence of Theorem 2.3.4 we obtain that in $\mathbb{R}$ there are sets with the Baire property which are not Borel sets. Take a first category set $C \subset \mathbb{R}$ with $\operatorname{card} C = \mathfrak{c}$ (for instance the Cantor set). Then all subsets of C are of the first category, and consequently they have the Baire property, whereas $\operatorname{card} \mathcal{P}(C) > \operatorname{card} C = \mathfrak{c}$, so among subsets of C there must exist some which are not Borel sets, as $\operatorname{card} \mathcal{B}(\mathbb{R}) = \mathfrak{c}$.

2.4   The space $\mathfrak{z}$

Let X_i (with metrics ϱ_i, respectively), $i \in \mathbb{N}$, be metric spaces and put

$$X = \mathop{\times}_{i=1}^{\infty} X_i.$$

Thus the elements of X are infinite sequences $\{x_i\}$, with $x_i \in X_i$, $i \in \mathbb{N}$.

Define the metric ϱ in X putting for $x = \{x_i\}$, $y = \{y_i\}$, $x, y \in X$,

$$\varrho(x, y) = \sum_{i=1}^{\infty} \frac{1}{2^i} \frac{\varrho_i(x, y)}{1 + \varrho_i(x, y)}. \tag{2.4.1}$$

Lemma 2.4.1. *If $x_n = \{x_{ni}\} \in X$, $n \in \mathbb{N}$, then the sequence $\{x_n\}$ converges:*

$$\lim_{n \to \infty} x_n = x_0 = \{x_{0i}\} \in X \tag{2.4.2}$$

if and only if $\lim_{n \to \infty} x_{ni} = x_{0i}$ for every $i \in \mathbb{N}$.

Proof. If (2.4.2) holds, then

$$\lim_{n\to\infty} \varrho(x_n, x_0) = 0. \tag{2.4.3}$$

By (2.4.1) $\varrho(x_n, x_0) \geqslant \dfrac{1}{2^i} \dfrac{\varrho_i(x_{ni}, x_{0i})}{1 + \varrho_i(x_{ni}, x_{0i})}$, whence

$$\lim_{n\to\infty} \varrho_i(x_{ni}, x_{0i}) = 0 \tag{2.4.4}$$

for every $i \in \mathbb{N}$.

Conversely, suppose that (2.4.4) holds for every $i \in \mathbb{N}$. We have

$$\varrho(x_n, x_0) \leqslant \sum_{i=1}^{m} \frac{1}{2^i} \frac{\varrho_i(x_{ni}, x_{0i})}{1 + \varrho_i(x_{ni}, x_{0i})} + \sum_{i=m+1}^{\infty} \frac{1}{2^i} = \sum_{i=1}^{m} \frac{1}{2^i} \frac{\varrho_i(x_{ni}, x_{0i})}{1 + \varrho_i(x_{ni}, x_{0i})} + \frac{1}{2^m}$$

for every $m \in \mathbb{N}$. By (2.4.4) we get hence $\limsup\limits_{n\to\infty} \varrho(x_n, x_0) \leqslant 2^{-m}$, whence, on letting $m \to \infty$, we obtain (2.4.3). □

Theorem 2.4.1. *If all spaces X_i are separable, then so is also X.*

Proof. For every $i \in \mathbb{N}$ there exists a countable set $Y_i \subset X_i$, dense in X_i. Fix an $\overline{x} = \{\overline{x}_i\} \in X$ and consider the set $Y \subset X$ consisting of the points

$$\begin{aligned} &\{y_1, \overline{x}_2, \overline{x}_3, \ldots\}, \\ &\{y_1, y_2, \overline{x}_3, \ldots\}, \\ &\{y_1, y_2, y_3, \ldots\}, \\ &\qquad \cdots\cdots\cdots\cdots, \end{aligned} \tag{2.4.5}$$

$y_i \in Y_i$. The cardinality of Y equals that of $\bigcup\limits_{n=1}^{\infty} \overset{\infty}{\underset{i=1}{\times}} Y_i$, and thus Y is countable. Take an $x = \{x_i\} \in X$. For every $i \in \mathbb{N}$ there exists a sequence $\{y_{ni}\} \subset Y_i$ of points of Y_i such that

$$\lim_{n\to\infty} y_{ni} = x_i. \tag{2.4.6}$$

Replacing in (2.4.5) in the n-th row y_i by y_{ni} we obtain a sequence of points $y_n \in Y$ and it follows from (2.4.6) and Lemma 2.4.1 that

$$\lim_{n\to\infty} y_n = x.$$

Thus Y is dense in X, and consequently X is separable. □

Theorem 2.4.2. *If all spaces X_i are complete, then so is also X.*

Proof. Let $\{x_n\} \subset X$, $x_n = \{x_{ni}\}$, be an arbitrary Cauchy sequence in X. By (2.4.1)

$$\varrho(x_n, x_{n+k}) \geqslant \frac{1}{2^i} \frac{\varrho_i(x_{ni}, x_{n+k,i})}{1 + \varrho_i(x_{ni}, x_{n+k,i})}$$

for all $n, k, i \in \mathbb{N}$, whence it follows that for every $i \in \mathbb{N}$ the sequence $\{x_{ni}\} \subset X_i$ is a Cauchy sequence in X_i. Since X_i is complete, there exists an $x_i \in X_i$ such that

$$\lim_{n \to \infty} x_{ni} = x_i .$$

Hence, by Lemma 2.4.1,

$$\lim_{n \to \infty} x_n = x ,$$

where $x = \{x_i\} \in X$. Thus the space X is complete. $\qquad \square$

Now take every X_i to be $\mathbb{N}$ with the metric $\varrho_i(n, k) = |n - k|$. In the sequel the resulting product space X will be always denoted by $\mathfrak{z}$. Since the space $\mathbb{N}$ is separable (being countable; $\mathbb{N}$ itself is a countable and dense subset of $\mathbb{N}$) and complete (there are no Cauchy sequences in $\mathbb{N}$ besides stationary ones), from Theorems 2.4.1 and 2.4.2 results

Theorem 2.4.3. *The space $\mathfrak{z}$ is complete and separable.*

The generality of the space $\mathfrak{z}$ may be seen from the following

Theorem 2.4.4. *Every complete and separable metric space X is a continuous image of $\mathfrak{z}$.*

Proof. Let X be a complete and separable metric space. Then

$$X = \bigcup_{x \in X} K\left(x, \frac{1}{2}\right) , \tag{2.4.7}$$

where $K(x, r)$ is the open ball centered at x with the radius r. Denoting by $d(A)$ the diameter of the set $A \subset X$, we have $d\left(K(x, \frac{1}{2})\right) \leqslant 1$ for every $x \in X$.

Since X is separable, by the Lindelöf theorem we can choose from (2.4.7) a countable cover:

$$X = \bigcup_{n=1}^{\infty} K_n^{(1)} , \ d\left(K_n^{(1)}\right) \leqslant 1 ,$$

and also

$$X = \bigcup_{n=1}^{\infty} \operatorname{cl} K_n^{(1)} , \ d\left(\operatorname{cl} K_n^{(1)}\right) \leqslant 1 .$$

We construct a family of sets $\{D_{n_1 \ldots n_m}\}$, $m, n_1, \ldots, n_m \in \mathbb{N}$, as follows. We put $D_n = \operatorname{cl} K_n^{(1)}$. If we have already defined the sets $D_{n_1 \ldots n_m}$ for all $n_1, \ldots, n_m \in \mathbb{N}$ and for a certain $m \in \mathbb{N}$, we have

$$D_{n_1 \ldots n_m} \subset \bigcup_{x \in D_{n_1 \ldots n_m}} K\left(x, \frac{1}{2^m}\right) . \tag{2.4.8}$$

Again we choose from (2.4.8) a countable cover

$$D_{n_1 \ldots n_m} \subset \bigcup_{n=1}^{\infty} K_n^{(m)} , \ d\left(K_n^{(m)}\right) \leqslant \frac{1}{2^{m-1}} ,$$

and we put

$$D_{n_1\ldots n_m, n_{m+1}} = D_{n_1\ldots n_m} \cap \operatorname{cl} K^{(m)}_{n_{m+1}}, \; n_{m+1} \in \mathbb{N}.$$

Thus we have defined inductively a family of sets $\{D_{n_1\ldots n_m}\}$, $m, n_1, \ldots, n_m \in \mathbb{N}$, with the properties

(*i*) Every set $D_{n_1\ldots n_m}$ is closed;

(*ii*) $d(D_{n_1\ldots n_m}) \leqslant \dfrac{1}{2^{m-1}}$;

(*iii*) $D_{n_1\ldots n_m, n_{m+1}} \subset D_{n_1\ldots n_m}, m, n_1, \ldots, n_m, n_{m+1} \in \mathbb{N}.$

Now take a $z = \{n_i\} \in \mathfrak{z}$ and put

$$f(z) = \bigcap_{m=1}^{\infty} D_{n_1\ldots n_m}. \tag{2.4.9}$$

By Cantor's theorem (2.4.9) represents a single point in X, consequently formula (2.4.9) defines a function $f : \mathfrak{z} \to X$. For every $m \in \mathbb{N}$ the sets $D_{n_1\ldots n_m}, n_1 \ldots n_m \in \mathbb{N}$, form a cover of X. Consequently for every $x \in X$ and for every $m \in \mathbb{N}$ there exists a set $D_{n_1\ldots n_m}$ such that $x \in D_{n_1\ldots n_m}$. The sequence $\{n_i\}$ represents a point $z \in \mathfrak{z}$ and we have $f(z) = x$. Thus f is onto:

$$f(\mathfrak{z}) = X.$$

It remains to show that f is continuous. Fix an $\varepsilon > 0$ and choose a $p \in \mathbb{N}$ so that $\dfrac{1}{2^{p-1}} < \varepsilon$. Put $\delta = \dfrac{1}{2^{p+1}}$. Take $z', z'' \in \mathfrak{z}$, $z' = \{n'_i\}$, $z'' = \{n''_i\}$. $\varrho(z', z'') < \delta$ means

$$\sum_{i=1}^{\infty} \frac{1}{2^i} \frac{|n'_i - n''_i|}{1 + |n'_i - n''_i|} < \frac{1}{2^{p+1}}. \tag{2.4.10}$$

Since n'_i and n''_i are integers, $n'_i \neq n''_i$ implies

$$\frac{|n'_i - n''_i|}{1 + |n'_i - n''_i|} \geqslant \frac{1}{2}.$$

Thus it follows from (2.4.10) that $n'_i = n''_i = n_i$ for $i = 1, \ldots, p$. But

$$f(z') = \bigcap_{m=1}^{\infty} D_{n'_1\ldots n'_m}, \; f(z'') = \bigcap_{m=1}^{\infty} D_{n''_1\ldots n''_m}$$

and by (*iii*) above $f(z'), f(z'') \in D_{n_1\ldots n_p}$. By (*ii*) this last set has the diameter at most $\dfrac{1}{2^{p-1}} < \varepsilon$, so the distance between $f(z')$ and $f(z'')$ is less than ε. This proves that f is uniformly continuous in $\mathfrak{z}$. $\qquad\square$

2.5   Analytic sets

Let X be a separable and complete metric space. A set $A \subset X$ is called *analytic* iff there exists a continuous function $f : \mathfrak{z} \to X$ such that $f(\mathfrak{z}) = A$. Moreover, we assume that the empty set is analytic. (It is of the form $f(\mathfrak{z})$ with the empty function f.) The class of all analytic sets in X is denoted by $\mathcal{A}(X)$.

Theorem 2.5.1. *Every closed set in X is analytic, i.e., $\mathcal{F}(X) \subset \mathcal{A}(X)$.*

Proof. If $F \subset X$ is a closed set, then F considered as a metric space (with the same metric as in X) is separable, since the separability is a hereditary property. F is also complete. In fact let $x_n \in F$ be a Cauchy sequence. Then $\{x_n\}$ converges in X, since X is complete:

$$\lim_{n \to \infty} x_n = x \in X.$$

But then $x \in F$, since F is closed. Thus $\{x_n\}$ converges in F:

$$\lim_{n \to \infty} x_n = x \in F.$$

The theorem results now from Theorem 2.4.4. $\square$

Theorem 2.5.2. *Let X, Y be separable and complete metric spaces, $A \subset X$ an analytic set, and $g : A \to Y$ a continuous function. Then the set $g(A) \subset Y$ also is analytic.*

Proof. There exists a continuous function $f : \mathfrak{z} \to X$ such that $f(\mathfrak{z}) = A$. The function $h = g \circ f$ (the composite of g and f) is continuous, $h : \mathfrak{z} \to Y$, and $h(\mathfrak{z}) = g(A)$. $\square$

Theorem 2.5.3. *The cartesian product of a finite or countable number of analytic sets is an analytic set.*

Proof. Let A_i, $i \in I$, where $I = \{1, \ldots, n\}$ or $I = \mathbb{N}$, be in $\mathcal{A}(X_i)$, where X_i are separable complete spaces. Then $\underset{i \in I}{\times} A_i \subset \underset{i \in I}{\times} X_i$, and $\underset{i \in I}{\times} X_i$ is separable and complete. (If $I = \mathbb{N}$, this follows from Theorems 2.4.1 and 2.4.2, and if I is finite, the proof is similar). Thus there exist continuous functions $f_i : \mathfrak{z} \to X_i$ such that $f_i(\mathfrak{z}) = A_i$, $i \in I$. The function $f : \mathfrak{z} \to \underset{i \in I}{\times} X_i$ defined as $f = (f_1, f_2 \ldots)$ is continuous (if $I = \mathbb{N}$, this follows from Lemma 2.4.1, and if I is finite, the proof is similar) and $f(\mathfrak{z}) = \underset{i \in I}{\times} A_i$. Thus $\underset{i \in I}{\times} A_i \in \mathcal{A}(\underset{i \in I}{\times} X_i)$. $\square$

Theorem 2.5.4. *The union of a finite or countable number of analytic sets is an analytic set.*

Proof. We need only consider countable unions, since a finite union can always be written as a countable one. Let X be a separable and complete metric space, and $A_i \subset X$, $i \in \mathbb{N}$, analytic sets. So there exist continuous functions $f_i : \mathfrak{z} \to X$ such that $f_i(\mathfrak{z}) = A_i$, $i \in \mathbb{N}$. We define $f : \mathfrak{z} \times \mathbb{N} \to X$ putting $f(z, n) = f_n(z)$. The function f is continuous. To see this, take a sequence $y_k = (z_k, n_k) \in \mathfrak{z} \times \mathbb{N}$, $k \in \mathbb{N}$, convergent to a $y_0 = (z_0, n_0) \in \mathfrak{z} \times \mathbb{N}$:

$$\lim_{k \to \infty} y_k = y_0 \,.$$

Then $\lim_{k \to \infty} z_k = z_0$ and $\lim_{k \to \infty} n_k = n_0$, i.e., $n_k = n_0$ for k sufficiently large. For such k we have

$$f(y_k) = f(z_k, n_0) = f_{n_0}(z_k) \to f_{n_0}(z_0) = f(z_0, n_0) = f(y_0) \,,$$

since f_{n_0} is continuous. Now, $\mathfrak{z} \times \mathbb{N}$ is complete and separable, being a product of complete and separable spaces, and hence, by Theorem 2.4.4 is an analytic set. Moreover we have

$$f(\mathfrak{z} \times \mathbb{N}) = \bigcup_{i=1}^{\infty} A_i \,.$$

In fact, if $x \in f(\mathfrak{z} \times \mathbb{N})$, then $x = f(z, n)$ for certain $z \in \mathfrak{z}$, $n \in \mathbb{N}$, i.e., $x = f_n(z)$, whence $x \in f_n(\mathfrak{z}) = A_n$ and $x \in \bigcup_{i=1}^{\infty} A_i$. And if $x \in \bigcup_{i=1}^{\infty} A_i$, then there exists an $n \in \mathbb{N}$ such that $x \in A_n = f_n(\mathfrak{z})$, i.e., there exists a $z \in \mathfrak{z}$ such that $x = f_n(z) = f(z, n) \in f(\mathfrak{z} \times \mathbb{N})$. The theorem results now from Theorem 2.5.2. $\qquad \square$

Theorem 2.5.5. *The intersection of a finite or countable number of analytic sets is an analytic set.*

Proof. Again we need only consider countable intersections, and we may assume that the intersection in question is non-empty, for the empty set is always analytic. Let X be a separable and complete metric space, and $A_i \subset X$, $i \in \mathbb{N}$, analytic sets. So there exist continuous functions $f_i : \mathfrak{z} \to X$ such that $f_i(\mathfrak{z}) = A_i$, $i \in \mathbb{N}$. Put $Z = \overset{\infty}{\underset{i=1}{\times}} Z_i$, where $Z_i = \mathfrak{z}$ for every $i \in \mathbb{N}$. By Theorems 2.4.1, 2.4.2 and 2.4.3 Z is a separable and complete space. Further define $Q \subset Z$ as

$$Q = \{z = \{z_i\} \in Z \mid f_i(z_i) \text{ is independent of } i\}$$

(the sequence $f_i(z_i)$ is a constant). First we show that Q is a closed subset of Z, and hence, by Theorem 2.5.1, is analytic.

Take a sequence $z^{(n)} \in Q$, $\lim_{n \to \infty} z^{(n)} = z^{(0)}$. Let $z^{(n)} = \{z_i^{(n)}\}$, $z^{(0)} = \{z_i^{(0)}\}$, $z_i^{(k)} \in \mathfrak{z}$ for $k = 0, 1, 2, \dots$.. By Lemma 2.4.1 $\lim_{n \to \infty} z_i^{(n)} = z_i^{(0)}$ for every $i \in \mathbb{N}$. Thus $f_i(z_i^{(0)}) = \lim_{n \to \infty} f_i(z_i^{(n)})$ is independent of i, since every sequence $f_i(z_i^{(n)})$ is constant. Hence $z^{(0)} \in Q$.

We define a function $f : Q \to X$ as $f(z) = f_i(z_i)$ (independent of i !) for $z \in Q$. Take again sequence $z^{(n)} \in Q$ converging to $z^{(0)} \in Q$, as above. Then $f(z^{(n)}) = f_i(z_i^{(n)}) \to f_i(z_i^{(0)}) = f(z^{(0)})$, since the functions f_i are continuous.

Now,

$$f(Q) = \bigcap_{i=1}^{\infty} A_i \,.$$

For, if $x \in f(Q)$, then $x = f_i(z_i)$ for a z in Z and all $i \in \mathbb{N}$, and hence $x \in f_i(\mathfrak{z}) = A_i$ for all $i \in \mathbb{N}$, i.e., $x \in \bigcap\limits_{i=1}^{\infty} A_i$. And if $x \in \bigcap\limits_{i=1}^{\infty} A_i$, then $x \in A_i = f_i(\mathfrak{z})$ for all $i \in \mathbb{N}$, i.e., there exist $z_i \in \mathfrak{z}$, $i \in \mathbb{N}$, such that $x = f_i(z_i)$, $i \in \mathbb{N}$. Putting $z = \{z_i\} \in Q$ we get $x = f(z) \in f(Q)$.

The theorem results now from Theorem 2.5.2. $\qquad\square$

Theorem 2.5.6. *If X is a complete and separable metric space, then $\mathcal{B}(X) \subset \mathcal{A}(X)$, i.e., every Borel set is analytic.*

Proof. The class $\mathcal{A}(X)$ fulfils conditions (i) and (ii) of Theorem 2.3.3. $((i)$ results from Theorem 2.5.1, and (ii) from Theorems 2.5.4 and 2.5.5$)$, and so it must contain the smallest class fulfilling these conditions, i.e., $\mathcal{B}(X)$. $\qquad\square$

Finally we give here two characterizations of the class $\mathcal{A}(X)$. The first of them will not be proved (cf., e.g., Kuratowski [196], where the condition is taken as the definition of the analytic sets.)

Theorem 2.5.7. *Let X be a complete and separable metric space. A set $A \subset X$ is analytic if and only if there exist a Borel set $B \subset X$ and a continuous function $f : B \to X$ such that $f(B) = A$.*

Let us note that the sufficiency results easily from Theorems 2.5.6 and 2.5.2.

Theorem 2.5.8. *Let X be a complete and separable metric space. The family $\mathcal{A}(X)$ is the smallest class $\mathcal{Z}$ of subsets of X which fulfils the following three conditions*

(i) $\mathcal{F}(X) \subset \mathcal{Z}$;

(ii) *If $A_i \in \mathcal{Z}$ for $i \in \mathbb{N}$, then also $\bigcup\limits_{i=1}^{\infty} A_i$ and $\bigcap\limits_{i=1}^{\infty} A_i \in \mathcal{Z}$;*

(iii) *If $Z \in \mathcal{Z}$, and $f : Z \to X$ is a continuous function, then $f(Z) \in \mathcal{Z}$.*

Proof. The class $\mathcal{A}(X)$ fulfils conditions (i)–(iii) in virtue of Theorems 2.5.1, 2.5.4, 2.5.5 and 2.5.2. So $\mathcal{Z} \subset \mathcal{A}(X)$. Conversely, let $\mathcal{Z} \subset \mathcal{P}(X)$ be a collection of sets such that conditions (i)–(iii) are fulfilled. By Theorem 2.3.3 $\mathcal{B}(X) \subset \mathcal{Z}$ and hence, by (iii), $\mathcal{Z}$ contains also all the continuous images of the Borel subsets of X. But by Theorem 2.5.7 this is exactly the class of analytic subsets of X. So $\mathcal{A}(X) \subset \mathcal{Z}$. $\qquad\square$

However, the class $\mathcal{A}(X)$ is not a σ-algebra. The complement of an analytic set need not be analytic[1]. It can be generally proved that if A and A' both are analytic, then A is a Borel set[2].

The notion of an analytic set is due to M. Souslin and N. Lusin (cf. Lusin [210]). Expositions of the theory of analytic sets may be found, e.g., in Kuratowski [196], Kuratowski-Mostowski [198], Naĭmark [236], Sierpiński [284].

---

[1]Souslin [293], Lusin-Sierpiński [212], Hurewicz [149]. Cf. also Kuratowski-Mostowski [198], p. 443, Kuratowski [196], p. 408.

[2] Souslin [293], Cf. also Kuratowski-Mostowski [198], p. 440, Kuratowski [196] p. 395. Cf. also Exercise E2.10.

2.6 Operation A

Let $\mathcal{S} = \{S_{n_1\ldots n_m}\}$, $m, n_1, \ldots, n_m \in \mathbb{N}$, be a family of sets. (To every $z = \{n_i\} \in \mathfrak{z}$ there corresponds a sequence $\mathcal{S}_z = \{S_{n_1}, S_{n_1 n_2}, S_{n_1 n_2 n_3}, \ldots\}$ of elements of $\mathcal{S}$, and $\mathcal{S} = \bigcup_{z \in \mathfrak{z}} \mathcal{S}_z$). We define the operation A (cf. Lusin-Souslin [213]) for such a family as ($\bigcup_z$ denotes the union extended over all $z = \{n_i\} \in \mathfrak{z}$)

$$A\mathcal{S} = \bigcup_z \bigcap_{m=1}^{\infty} S_{n_1\ldots n_m} .$$

If, for every $m, n_1, \ldots, n_m, n_{m+1} \in \mathbb{N}$, we have

$$S_{n_1\ldots n_m n_{m+1}} \subset S_{n_1\ldots n_m} ,$$

the family $\mathcal{S}$ is called a *regular system*. If $\mathcal{S}$ is any family as described above, then putting

$$S^*_{n_1\ldots n_m} = \bigcap_{i=1}^{m} S_{n_1\ldots n_i} , \quad m, n_1, \ldots, n_m \in \mathbb{N},$$

we obtain a regular system $\mathcal{S}^* = \{S^*_{n_1\ldots n_m}\}$, $m, n_1, \ldots, n_m \in \mathbb{N}$, such that

$$A\mathcal{S}^* = A\mathcal{S} .$$

So when studying the operation A we may always assume, without a loss of generality, that the system $\mathcal{S}$ in question is regular.

Theorem 2.6.1. *Every analytic set is the result of the operation A performed on a regular system of closed sets.*

Proof. Let X be a separable and complete metric space and $A \subset X$ an analytic set. Thus there exists a continuous function $f : \mathfrak{z} \to X$ such that $A = f(\mathfrak{z})$.

Define the sets $\mathfrak{z}_{n_1\ldots n_m} \subset \mathfrak{z}$ by

$$\mathfrak{z}_{n_1\ldots n_m} = \{ z = \{k_i\} \in \mathfrak{z} \mid k_1 = n_1, \ldots, k_m = n_m \}, \quad m, n_1, \ldots, n_m \in \mathbb{N}.$$

It follows from Lemma 2.4.1 that the sets $\mathfrak{z}_{n_1\ldots n_m}$, $m, n_1, \ldots, n_m \in \mathbb{N}$, are closed, and clearly

$$\mathfrak{z}_{n_1\ldots n_m, n_{m+1}} \subset \mathfrak{z}_{n_1\ldots n_m} , \quad m, n_1, \ldots, n_m, n_{m+1} \in \mathbb{N}. \tag{2.6.1}$$

Now put

$$S_{n_1\ldots n_m} = \mathrm{cl}\, f(\mathfrak{z}_{n_1\ldots n_m}), \quad m, n_1, \ldots, n_m \in \mathbb{N}.$$

The family $\mathcal{S} = \{S_{n_1\ldots n_m}\}$, $m, n_1, \ldots, n_m \in \mathbb{N}$, is, by (2.6.1), a regular system of closed sets.

If $z = \{n_i\} \in \mathfrak{z}$ is fixed, then, for every $m \in \mathbb{N}$, $z \in \mathfrak{z}_{n_1\ldots n_m}$, whence, for every $m \in \mathbb{N}$, $f(z) \in S_{n_1\ldots n_m}$ so that

$$f(z) \in \bigcap_{m=1}^{\infty} S_{n_1\ldots n_m} . \tag{2.6.2}$$

But the diameter $d(\mathfrak{z}_{n_1...n_m}) \leqslant \dfrac{1}{2^m}$, whence, by the continuity of f,

$$\lim_{m\to\infty} d(S_{n_1...n_m}) = \lim_{m\to\infty} d\,[\mathrm{cl}\,f(\mathfrak{z}_{n_1...n_m})] = 0\,.$$

So the intersection in (2.6.2), as an intersection of a descending sequence of closed sets with diameters tending to zero in a complete space, represents, by Cantor's theorem, a single point. Thus

$$f(z) = \bigcap_{m=1}^{\infty} S_{n_1...n_m}\,.$$

Hence

$$\mathrm{A}\mathcal{S} = \bigcup_{z} \bigcap_{m=1}^{\infty} S_{n_1...n_m} = \bigcup_{z} f(z) = f(\mathfrak{z}) = A. \qquad \square$$

Theorem 2.6.2. *The result of the operation* A *performed on a family of analytic sets is an analytic set.*

Proof. Let X be a separable and complete metric space and $A_{n_1...n_m} \subset X$, $m, n_1, \ldots,$ $n_m \in \mathbb{N}$ analytic sets. The sets $\mathfrak{z}_{n_1...n_m}$, $m, n_1, \ldots, n_m \in \mathbb{N}$, defined in the proof of Theorem 2.6.1, are analytic in virtue of Theorem 2.5.1. Hence also, by Theorem 2.5.3, the sets $A_{n_1...n_m} \times \mathfrak{z}_{n_1...n_m}$ are analytic subsets of $X \times \mathfrak{z}$, and consequently, by Theorem 2.5.4, so are also the sets (for every fixed $m \in \mathbb{N}$) $\displaystyle\bigcup_{n_1,...,n_m=1}^{\infty} (A_{n_1...n_m} \times \mathfrak{z}_{n_1...n_m})$ and the set (cf. Theorem 2.5.5) $\displaystyle\bigcap_{m=1}^{\infty} \bigcup_{n_1,...,n_m=1}^{\infty} (A_{n_1...n_m} \times \mathfrak{z}_{n_1...n_m})$. The projection $p : X \times \mathfrak{z} \to X$

$$p(x, z) = x$$

is a continuous function, so by Theorem 2.5.2 $p\Big(\displaystyle\bigcap_{m=1}^{\infty} \bigcup_{n_1,...,n_m=1}^{\infty} [A_{n_1...n_m} \times \mathfrak{z}_{n_1...n_m}] \Big)$ is an analytic set. But

$$p\Big(\bigcap_{m=1}^{\infty} \bigcup_{n_1,...,n_m=1}^{\infty} [A_{n_1...n_m} \times \mathfrak{z}_{n_1...n_m}] \Big) = \bigcup_{z} \bigcap_{m=1}^{\infty} A_{n_1...n_m} = \mathrm{A}\{A_{n_1...n_m}\}. \qquad (2.6.3)$$

To prove (2.6.3), take an $x \in p\Big(\displaystyle\bigcap_{m=1}^{\infty} \bigcup_{n_1,...,n_m=1}^{\infty} [A_{n_1...n_m} \times \mathfrak{z}_{n_1...n_m}] \Big)$. Then there exists a $z = \{k_i\} \in \mathfrak{z}$ such that $(x, z) \in \displaystyle\bigcup_{n_1,...,n_m=1}^{\infty} [A_{n_1...n_m} \times \mathfrak{z}_{n_1...n_m}]$ for every $m \in \mathbb{N}$. This means that $x \in A_{n_1...n_m}$ for all $m \in \mathbb{N}$, where $n_i = k_i$ for $i \in \mathbb{N}$ and consequently

$$x \in \bigcap_{m=1}^{\infty} A_{k_1...k_m} \subset \bigcup_{z} \bigcap_{m=1}^{\infty} A_{n_1...n_m}\,.$$

This proves an inclusion in (2.6.3). The proof of the converse inclusion is analogous.

$$\square$$

2.7 Theorem of Marczewski

The theorem of Marczewski (Marczewski [219]) assures that some important regularity properties of sets are preserved by the operation A. We start with two lemmas.

Lemma 2.7.1. *If* $\{S_{n_1...n_m}\}$, $m, n_1, \ldots, n_m \in \mathbb{N}$, *is a regular system, then for every* $k_1, \ldots, k_j \in \mathbb{N}$

$$\bigcup_{l=1}^{\infty} \bigcup_{z} \bigcap_{m=1}^{\infty} S_{k_1...k_j l n_1...n_m} = \bigcup_{z} \bigcap_{m=1}^{\infty} S_{k_1...k_j n_1...n_m} .$$

The easy proof of Lemma 2.7.1 is left to the reader.

Lemma 2.7.2. *If* $\{S_{n_1...n_m}\}, m, n_1, \ldots, n_m \in \mathbb{N}$, *is a regular system, then for every set* A

$$A \setminus \bigcup_{z} \bigcap_{m=1}^{\infty} S_{n_1...n_m} \subset \bigcup_{z} \bigcup_{m=0}^{\infty} \left(S_{n_1...n_m} \setminus \bigcup_{l=1}^{\infty} S_{n_1...n_m l} \right),$$

where for $m = 0$ *we assume* $S_{n_1...n_m} = A$.

Proof. For an indirect proof suppose that there exists an $x \in A \setminus \bigcup_{z} \bigcap_{m=1}^{\infty} S_{n_1...n_m}$ such that $x \notin \bigcup_{z} \bigcup_{m=0}^{\infty} \left(S_{n_1...n_m} \setminus \bigcup_{l=1}^{\infty} S_{n_1...n_m l} \right)$. This means that for every $z \in \mathfrak{z}$ and $m \in \mathbb{N} \cup \{0\}$:

$$\text{if } x \in S_{n_1...n_m}, \text{ then there exists an } l \in \mathbb{N} \text{ such that } x \in S_{n_1...n_m l} . \tag{$*$}$$

We have $x \in A$, so using $(*)$ for $m = 0$ we infer that there exists an $l_1 \in \mathbb{N}$ such that $x \in S_{l_1}$. Using $(*)$ for $m = 1$ and $n_1 = l_1$, we infer that there exists an $l_2 \in \mathbb{N}$ such that $x \in S_{l_1 l_2}$. Proceeding so further we get the existence of a $z_0 = \{l_i\} \in \mathfrak{z}$ such that, for every $m \in \mathbb{N}$, $x \in S_{l_1...l_m}$. Hence $x \in \bigcup_{z} \bigcap_{m=1}^{\infty} S_{n_1...n_m}$, which contradicts our supposition. $\square$

A family Σ of subsets of a set (space) X is called an *M-algebra (Marczewski algebra)* iff it is a σ-algebra and for every set $A \subset X$ there exists a set $B \in \Sigma$, $A \subset B$, such that if $Z \in \Sigma$ and $A \subset Z$, then every set $Y \subset B \setminus Z$ belongs to Σ. By Theorems 2.2.1 and 2.2.2 the class of the sets with the Baire property (in separable metric space X) is an M-algebra. Similarly, as will be shown later, the class of all Lebesgue measurable sets in $\mathbb{R}^N$ is an M-algebra.

Theorem 2.7.1 (Theorem of Marczewski). *If* Σ *is an M-algebra, then the result of the operation* A *performed on a regular system of sets from* Σ *belongs to* Σ.

Proof. Let $S_{n_1...n_m} \in \Sigma, m, n_1, \ldots, n_m \in \mathbb{N}$, form a regular system, and put

$$A = A\{S_{n_1...n_m}\}.$$

For the set A choose a set $B \in \Sigma$ such that $A \subset B$ and if $A \subset Z \in \Sigma$, then every $Y \subset B \setminus Z$ belongs to Σ. Similarly, to every set $\bigcup_{z} \bigcap_{m=1}^{\infty} S_{k_1 \ldots k_j n_1 \ldots n_m}$, $k_1, \ldots, k_j \in \mathbb{N}$, choose a set $B_{k_1 \ldots k_j} \in \Sigma$ according to the same pattern. We may assume that

$$B_{n_1 \ldots n_m} \subset S_{n_1 \ldots n_m}, n_1, \ldots, n_m \in \mathbb{N}, \tag{2.7.1}$$

for otherwise we could replace $B_{n_1 \ldots n_m}$ by $B_{n_1 \ldots n_m} \cap S_{n_1 \ldots n_m}$ which would not spoil the remaining properties of $B_{n_1 \ldots n_m}$.

By Lemma 2.7.2 we have

$$B \setminus A = B \setminus \bigcup_{z} \bigcap_{m=1}^{\infty} S_{n_1 \ldots n_m} \subset B \setminus \bigcup_{z} \bigcap_{m=1}^{\infty} B_{n_1 \ldots n_m} \subset \bigcup_{m=0}^{\infty} \left[B_{n_1 \ldots n_m} \setminus \bigcup_{l=1}^{\infty} B_{n_1 \ldots n_m l} \right],$$

where for $m = 0$ we put $B_{n_1 \ldots n_m} = B$. Now, by Lemma 2.7.1,

$$\bigcup_{z} \bigcap_{m=1}^{\infty} S_{k_1 \ldots k_j, n_1 \ldots n_m} = \bigcup_{l=1}^{\infty} \bigcup_{z} \bigcap_{m=1}^{\infty} S_{k_1 \ldots k_j l n_1 \ldots n_m} \subset \bigcup_{l=1}^{\infty} B_{k_1 \ldots k_j l} \in \Sigma,$$

since Σ is a σ-algebra. Hence, by the choice of $B_{k_1 \ldots k_j}$,

$$(B \setminus A) \cap \left[B_{k_1 \ldots k_j} \setminus \bigcup_{l=1}^{\infty} B_{k_1 \ldots k_j l} \right] \in \Sigma. \tag{2.7.2}$$

Since by (2.7.1) and Lemma 2.7.2

$$B \setminus A = B \setminus \bigcup_{z} \bigcap_{m=1}^{\infty} S_{n_1 \ldots n_m} \subset B \setminus \bigcup_{z} \bigcap_{m=1}^{\infty} B_{n_1 \ldots n_m} \subset \bigcup_{z} \bigcup_{m=0}^{\infty} \left[B_{n_1 \ldots n_m} \setminus \bigcup_{l=1}^{\infty} B_{n_1 \ldots n_m l} \right],$$

we have

$$B \setminus A = (B \setminus A) \cap \bigcup_{z} \bigcup_{m=0}^{\infty} \left[B_{n_1 \ldots n_m} \setminus \bigcup_{l=1}^{\infty} B_{n_1 \ldots n_m l} \right].$$

Let $[\ldots]$ stand for $(B \setminus A) \cap \left[B_{n_1 \ldots n_m} \setminus \bigcup_{l=1}^{\infty} B_{n_1 \ldots n_m l} \right]$. Then

$$B \setminus A = \bigcup_{z} \bigcup_{m=0}^{\infty} [\ldots] = \bigcup_{m=0}^{\infty} \bigcup_{z} [\ldots] = \bigcup_{m=0}^{\infty} \bigcup_{n_1 \ldots n_m=1}^{\infty} [\ldots].$$

The last union contains only countably many summands, therefore, since Σ is a σ-algebra, we get by (2.7.2) $B \setminus A \in \Sigma$. Since $A \subset B$, we have $A = B \setminus (B \setminus A) \in \Sigma$. □

From Theorem 2.7.1, Theorem 2.6.1 and the fact that the class of all sets with the Baire property is an M-algebra, we obtain

Theorem 2.7.2. *Every analytic set has the Baire property.*

In Part II of the present book we will encounter many sets without the Baire property, and hence non-analytic.

2.8 Cantor-Bendixson theorem

The Cantor-Bendixson theorem belongs to standard part of any university course on the topology of metric spaces. But we prove it here because of its application in the theory of analytic sets.

Let X be a metric space and $A \subset X$. A point $x \in X$ is called a *point of accumulation* of A iff for every neighbourhood U of x we have $\operatorname{card}(U \cap A) > 1$. The set of all the accumulation points of A is denoted by A^d. A point $x \in A$ is called a *point of condensation* of A iff for every neighbourhood U of x we have $\operatorname{card}(U \cap A) > \aleph_0$. The set of all the condensation points of A is denoted by $A^\bullet$. A set $A \subset X$ is called *perfect* iff $A = A^d$.

The reader will easily verify the following properties of the operations $^\bullet$ and ^{d}:

$$A^\bullet \subset A^d \subset \operatorname{cl} A,$$

$$(A \cup B)^\bullet = A^\bullet \cup B^\bullet,$$

$$A \subset B \Rightarrow A^\bullet \subset B^\bullet,$$

$$A^\bullet = \operatorname{cl} A^\bullet,$$

$$\text{for every set collection } \mathcal{K}, \ \bigcup_{A \in \mathcal{K}} A^d \subset \Big(\bigcup_{A \in \mathcal{K}} A \Big)^d,$$

$$(A \cup B)^d = A^d \cup B^d,$$

$$A^{dd} \subset A^d,$$

$$\operatorname{cl} A = A \cup A^d.$$

Theorem 2.8.1. *If X is a separable metric space, then for every set A,*

$$\operatorname{card}(A \setminus A^\bullet) \leqslant \aleph_0.$$

Proof. There exists a countable base of neighbourhoods $\{G_n\}_{n \in \mathbb{N}}$ in X. Take an $x \in A \setminus A^\bullet$. Then there exists a neighbourhood U of x such that $\operatorname{card}(U \cap A) \leqslant \aleph_0$. Further, there exists an $n_x \in \mathbb{N}$ such that $x \in G_{n_x} \subset U$, and clearly we have $\operatorname{card}(G_{n_x} \cap A) \leqslant \aleph_0$. Now we have

$$A \setminus A^\bullet \subset \bigcup_{x \in A \setminus A^\bullet} (G_{n_x} \cap A). \qquad (2.8.1)$$

The union in (2.8.1) is at most countable, since there are at most countably many distinct sets G_{n_x}, and hence the set $\bigcup_{x \in A \setminus A^\bullet} (G_{n_x} \cap A)$, and so also $A \setminus A^\bullet$, is at most countable. $\qquad \square$

Corollary 2.8.1. *If X is a separable metric space and $A \subset X$, then $(A \setminus A^\bullet)^\bullet = \varnothing$.*

Corollary 2.8.2. *If X is a separable metric space and $A \subset X$, then $A^{\bullet\bullet} = A^\bullet$.*

Proof. We have, since $A^\bullet$ is closed

$$A^{\bullet\bullet} \subset \operatorname{cl} A^\bullet = A^\bullet.$$

On the other hand $A \subset A^\bullet \cup (A \setminus A^\bullet)$, whence

$$A^\bullet \subset A^{\bullet\bullet} \cup (A \setminus A^\bullet)^\bullet = A^{\bullet\bullet}. \qquad \square$$

Theorem 2.8.2 (Theorem of Cantor-Bendixson). *If X is a separable metric space, then $X = A \cup B$, where $A = A^d$ and $\operatorname{card} B \leqslant \aleph_0$.*

Proof. Let

$$\mathcal{K} = \{\, C \subset X \mid C \subset C^d \,\},$$

and put $A = \bigcup \mathcal{K}$, $B = X \setminus A$. Then $A \subset A^d$, whence $A^d = A \cup A^d = \operatorname{cl} A$ and

$$(\operatorname{cl} A)^d = A^d \cup A^{dd} = A^d = \operatorname{cl} A.$$

Thus $\operatorname{cl} A \in \mathcal{K}$ and $\operatorname{cl} A \subset A$. This means that $A \cup A^d \subset A$, i.e., $A^d \subset A$. Since $A \subset A^d$, we obtain hence

$$A = A^d,$$

i.e., A is perfect.

Now, $B^\bullet = B^{\bullet\bullet} \subset B^{\bullet d}$, i.e., $B^\bullet \in \mathcal{K}$ and $B^\bullet \subset A$. By the definition of B we have $B \cap B^\bullet = \varnothing$ and $B = B \setminus B^\bullet$. Hence by Theorem 2.8.1 $\operatorname{card} B = \operatorname{card}(B \setminus B^\bullet) \leqslant \aleph_0$. $\square$

Any subset D of X may be considered as a space, so $D = A \cup B$ with $A = A^d$ and $\operatorname{card} B \leqslant \aleph_0$. But in the formula $A = A^d$ the accumulation points of A are taken with respect to the relative topology of D. (These are points x such that $\operatorname{card}(D \cap U \cap A) > 1$ for every neighbourhood U of x). But since every accumulation point of A in D is an accumulation point of A in X ($\operatorname{card}(D \cap U \cap A) > 1$ implies $\operatorname{card}(U \cap A) > 1$), A^d in D is contained in A^d in X. Consequently $D = A \cup B$ with $A \subset A^d$ and $\operatorname{card} B \leqslant \aleph_0$.

Theorem 2.8.3. *If X is a complete and separable metric space and $A \in \mathcal{A}(X)$, then either $\operatorname{card} A \leqslant \aleph_0$, or $\operatorname{card} A = \mathfrak{c}$.*

Remark. Theorem 2.8.3 says that the analytic sets realize the continuum hypothesis. Every uncountable analytic set has the power of continuum.

Proof. Let $A \in \mathcal{A}(X)$ and $\operatorname{card} A > \aleph_0$. There exists a continuous function $f : \mathfrak{z} \to X$ such that $A = f(\mathfrak{z})$. Consider the family of sets $f^{-1}(\{y\})$ for $y \in A$. These sets clearly are non-empty and disjoint. By the Axiom of Choice (cf. Theorem 1.8.3) there exists a set $Z \subset \mathfrak{z}$ which has exactly one point in common with every set $f^{-1}(\{y\})$ for $y \in A$. Consequently there is a one-to-one correspondence between the points of A and the points of Z. Thus $\operatorname{card} Z \geqslant \operatorname{card} A > \aleph_0$. Moreover, the function $f \mid Z$ is one-to-one.

By Theorem 2.8.2 $Z = D \cup B$ with

$$D \subset D^d \text{ and } \operatorname{card} B \leqslant \aleph_0. \qquad (2.8.2)$$

It follows that $\operatorname{card} D > \aleph_0$ and the function $f \mid D$ is one-to-one.

Take $x_0, x_2 \in D$, $x_0 \neq x_2$. There exist open balls $K_0, K_2 \subset \mathfrak{z}$ such that $x_i \in K_i$, $d(K_i) < \dfrac{1}{2}$, $i = 0, 2$, and $f(\operatorname{cl} K_0) \cap f(\operatorname{cl} K_2) = \varnothing$, since f is continuous. Let $\mathcal{I}$ be the family of all sequences $I = \{i_n\}$, $i_n = 0, 2$. For every $I = \{i_n\} \in \mathcal{I}$ we shall define a sequence of points $\{x_{i_1}, x_{i_1 i_2}, \ldots x_{i_1 \ldots i_n}, \ldots\}$ and a sequence of open balls $\{K_{i_1}, K_{i_1 i_2}, \ldots K_{i_1 \ldots i_n}, \ldots\}$ in such a manner that

$$x_{i_1 \ldots x_n} \in K_{i_1 \ldots i_n}, \; x_{i_1 \ldots x_n} \in D, \tag{2.8.3}$$

$$d(K_{i_1 \ldots i_n}) < \frac{1}{2^n}, \tag{2.8.4}$$

$$f(\operatorname{cl} K_{i_1 \ldots i_n}) \cap f(\operatorname{cl} K_{j_1 \ldots j_n}) = \varnothing \text{ for } (i_1, \ldots, i_n) \neq (j_1, \ldots, j_n), \tag{2.8.5}$$

where $j_k = 0, 2$. Suppose that we have already defined points $x_{i_1 \ldots i_n}$ and balls $K_{i_1 \ldots i_n}$ for a certain $n \in \mathbb{N}$, so that conditions (2.8.3), (2.8.4), (2.8.5) are fulfilled. Since every neighbourhood U of any $x \in D$ contains points of D different from x (cf. (2.8.2)), we may find points $x_{i_1 \ldots i_n 0}$ and $x_{i_1 \ldots i_n 2}$ in D, different from $x_{i_1 \ldots i_n}$, and open balls $K_{i_1 \ldots i_n 0}$ and $K_{i_1 \ldots i_n 2}$ such that $(i_{n+1} = 0, 2)$

$$x_{i_1 \ldots i_n i_{n+1}} \in K_{i_1 \ldots i_n i_{n+1}}, \; x_{i_1 \ldots i_n i_{n+1}} \in D,$$

$$d(K_{i_1 \ldots i_n i_{n+1}}) < \frac{1}{2^{n+1}},$$

$$f(\operatorname{cl} K_{i_1 \ldots i_n 0}) \cap f(\operatorname{cl} K_{i_1 \ldots i_n 2}) = \varnothing.$$

Moreover, we may choose the balls $K_{i_1 \ldots i_n i_{n+1}} \subset K_{i_1 \ldots i_n}$. In this manner we have inductively defined, for every $I = \{i_n\} \in \mathcal{I}$, a sequence of points $\{x_{i_1 \ldots i_n}\}_{n \in \mathbb{N}}$ and a sequence of balls $\{K_{i_1 \ldots i_n}\}_{n \in \mathbb{N}}$ so that conditions (2.8.3), (2.8.4), (2.8.5) are fulfilled and

$$K_{i_1 \ldots i_{n+1}} \subset K_{i_1 \ldots i_n}. \tag{2.8.6}$$

Thus, by (2.8.4), $\lim_{n \to \infty} d(\operatorname{cl} K_{i_1 \ldots i_n}) = 0$, moreover, (2.8.6) implies that also

$$\operatorname{cl} K_{i_1 \ldots i_{n+1}} \subset \operatorname{cl} K_{i_1 \ldots i_n}.$$

By Cantor's theorem the set $\bigcap\limits_{n=1}^{\infty} \operatorname{cl} K_{i_1 \ldots i_n}$ consists exactly of a single point

$$\bigcap_{n=1}^{\infty} \operatorname{cl} K_{i_1 \ldots i_n} = x_I, \; I = \{i_n\}.$$

Put

$$E = \bigcup_{I \in \mathcal{I}} \{x_I\}.$$

For $I_1 \neq I_2$, $I_1, I_2 \in \mathcal{I}$, we have $x_{I_1} \neq x_{I_2}$, since if $i_1^1 = i_1^2, \ldots, i_{n-1}^1 = i_{n-1}^2, i_n^1 \neq i_n^2$ for an $n \in \mathbb{N}$ $(I_1 = \{i_n^1\}, I_2 = \{i_n^2\})$, then by (2.8.5) $f(\operatorname{cl} K_{i_1^1 \ldots i_n^1}) \cap f(\operatorname{cl} K_{i_1^2 \ldots i_n^2}) = \varnothing$, and also $f(K_{i_1^1 \ldots i_n^1}) \cap f(K_{i_1^2 \ldots i_n^2}) = \varnothing$, whence $f(x_{I_1}) \neq f(x_{I_2})$. Thus $f \mid E$ is one-to-one, and since $\operatorname{card} \mathcal{I} = \mathfrak{c}$, also $\operatorname{card} E = \mathfrak{c}$ and $\operatorname{card} f(E) = \mathfrak{c}$.

Since $E \subset \mathfrak{z}$, $f(E) \subset f(\mathfrak{z}) = A$. Thus $\operatorname{card} A \geqslant \mathfrak{c}$. On the other hand $\operatorname{card} A = \operatorname{card} f(\mathfrak{z}) \leqslant \operatorname{card} \mathfrak{z} = \mathfrak{c}$. So necessarily $\operatorname{card} A = \mathfrak{c}$. $\qquad\square$

We may assign to every $f(x_I) \in f(E)$, $I = \{i_n\}$, the point $\sum\limits_{n=1}^{\infty} \dfrac{i_n}{3^n}$ of the Cantor set (cf. Exercise 2.11), and the correspondence is one-to-one. Actually it can be shown that every uncountable analytic set contains topologically the Cantor set (i.e., contains a subset homeomorphic with the Cantor set). Cf. Exercise 2.12.

2.9   Theorem of S. Piccard

In the year 1942 S. Piccard published a theorem to the effect that if A, $B \subset \mathbb{R}$ are sets of the second category with the Baire property, then the algebraic sum $A + B$ of A and B contains an interval (cf. [256][3]). This theorem was later generalized in various ways (cf., e.g., Kominek [172], Orlicz-Ciesielski [247], Sander [273] and also Smítal-Snoha [291]) Some of these generalizations exceed the scope of this book. Here we will present one of such generalizations based on Dubikajits-Ferens-Ger-Kuczma [72].

Let X be a topological group (not necessary commutative), i.e., a set X endowed with a topology (e.g., a metric space) and with a binary operation $+$ (i.e., $+$ is a function $+ : X^2 \to X$) such that

$$x + (y + z) = (x + y) + z \tag{2.9.1}$$

for arbitrary $x, y, z \in X$, there exists an element $0 \in X$ such that

$$x + 0 = x, \tag{2.9.2}$$

and for every $x \in X$ there exists an element $(-x) \in X$ such that

$$x + (-x) = 0. \tag{2.9.3}$$

$x + (-y)$ is usually written as $x - y$. The commutativity of the operation $+$:

$$x + y = y + x \tag{2.9.4}$$

is, in general, not assumed. Moreover, the functions $f_1 : X^2 \to X$ and $f_2 : X \to X$

$$f_1(x, y) = x + y, \; f_2(x) = -x,$$

are continuous. Then obviously, for every $a \in X$, the functions $T_a : X \to X$ and $S_a : X \to X$

$$T_a(x) = a + x, \; S_a(x) = x + a \tag{2.9.5}$$

are homeomorphisms. For any two sets A, $B \subset X$ we write

$$A + B = \{ x = a + b \mid a \in A, \; b \in B \},$$

and for a set $A \subset X$ and an element $b \in X$ we write

$$A + b = \{ x = a + b \mid a \in A \}, \; b + A = \{ x = b + a \mid a \in A \}.$$

Theorem 2.9.1. *Let $(X, +)$ be a topological group, and A, $B \subset X$ two sets of the second category with the Baire property. Then*

$$\mathrm{int}\,(A + B) \neq \varnothing. \tag{2.9.6}$$

---

[3] Added in the 2nd edition by K. Baron.

Proof. X itself is of the second category, since it contains sets of the second category. Then every non-empty open subset of X is of the second category. In fact, suppose that $G \subset X$ is non-empty, open and of the first category. Fix an $a \in G$. Then, for every $x \in X$, the set $G - a + x$ is of the first category and open (since functions (2.9.5) are homeomorphisms) and $x \in G - a + x$. Thus X is of the first category at every point $x \in X$. By Theorem 2.1.4 X is of the first category, which is impossible.

The sets A, B have the Baire property, i.e., there exist non-empty open sets G, H (if G, H were empty, the sets A, B would be of the first category) and first category sets P, Q, R, S such that

$$A = (G \cup P) \setminus R, \ B = (H \cup Q) \setminus S.$$

Hence it follows that the sets $G \setminus A \subset R$ and $B \setminus H \subset S$ are of the first category. Fix elements $g \in G$ and $h \in H$ and put $G_0 = -g + G$, $H_0 = H - h$, $A_0 = -g + A$, $B_0 = B - h$. Then (since (2.9.5) are homeomorphisms) the set

$$G_0 \setminus A_0 = (-g + G) \setminus (-g + A) = -g + (G \setminus A)$$

is of the first category, and similarly the set $H_0 \setminus B_0 = (H \setminus B) - h$ is of the first category. Put

$$U = G_0 \cap H_0.$$

U is open, since so are G_0 and H_0, and non-empty, since $0 \in U$. Consequently U is of the second category. For every $t \in U$ write

$$U_t = U \cap (t - U).$$

U_t is open and non-empty ($t \in U_t$) for every $t \in U$. The set (for every $t \in U$)

$$U_t \setminus A_0 \subset U \setminus A_0 \subset G_0 \setminus A_0$$

is of the first category. Similarly the set

$$U_t \setminus (t - B_0) \subset (t - U) \setminus (t - B_0) = t - (U \setminus B_0) \subset t - (H_0 \setminus B_0)$$

is of the first category. Since

$$U_t \subset [A_0 \cap (t - B_0)] \cup (U_t \setminus A_0) \cup [U_t \setminus (t - B_0)],$$

the set $A_0 \cap (t - B_0)$ is of the second category (for otherwise U_t would be of the first category) and hence non-empty. So, for every $t \in U$, there exists an $a_t \in A_0 \cap (t - B_0)$, i.e., $a_t \in A_0$ and $t - a_t \in B_0$. This means that there exists a $b_t \in B_0$ such that $a_t + b_t = t$, i.e., $t \in A_0 + B_0$. Now,

$$A_0 + B_0 = (-g + A) + (B - h),$$

so

$$U \subset (-g + A) + (B - h),$$

and

$$g + U + h \subset A + B.$$

Thus the set $A + B$ contains a non-empty open set $g + U + h$, i.e., (2.9.6) holds. $\square$

Exercises

1. Let X, Y be topological spaces. Let $A \subset X$ and $B \subset Y$ be arbitrary sets. Prove that $\operatorname{cl}(A \times B) = \operatorname{cl} A \times \operatorname{cl} B$ and $\operatorname{int}(A \times B) = \operatorname{int} A \times \operatorname{int} B$

2. Let X, Y be topological spaces, and $f : X \to Y$ a homeomorphism. Prove that for any $A \subset X$
$$\operatorname{cl} f(A) = f(\operatorname{cl} A), \quad \operatorname{int} f(A) = f(\operatorname{int} A),$$
and hence the sets A and $f(A)$ are both of the same category.

3. Let X, Y be topological spaces, X compact, and let $f : X \to Y$ be a one-to-one continuous function. Prove that f is a homeomorphism.

4. Prove that any subset of a first category set is of the first category, and the union of a finite or countable collection of first category sets is of the first category.

5. Let X be a separable[4] topological space. Prove that each of the following properties is equivalent to the fact that $A \subset X$ has the Baire property:

   (i) there exists a residual set B such that A is both open and closed in B;

   (ii) there exist sets $B \in \mathcal{G}_\delta$ and R of the first category such that $A = B \cup R$;

   (iii) there exist sets $C \in \mathcal{F}_\sigma$ and Q of the first category such that $A = C \setminus Q$;

   (iv) the set $D(A) \cap D(A')$ is nowhere dense;

   (v) the set $D(A) \setminus A$ is of the first category.

6. Prove that every set E with the Baire property has a decomposition $E = A \cup B$, where A is of the first category, and B is a Borel set.

7. Let X, Y be topological spaces, Y separable. Prove that if a set $A \subset X \times Y$ is residual (with respect to $X \times Y$), then there exists a residual set $C \subset X$ such that for every $x \in C$ the set $A[x]$ is residual.

8. Let X be a topological space. We say that sets A, $B \subset X$ are *Borel separable* iff there exists a set $D \in \mathcal{B}(X)$ such that $A \subset D \subset B'$. Let P_n, $Q_n \subset X$, $n \in \mathbb{N}$, be arbitrary sets, $P = \bigcup_{n=1}^{\infty} P_n$, $Q = \bigcup_{n=1}^{\infty} Q_n$. Show that if for every n, $m \in \mathbb{N}$ the sets P_n and Q_m are Borel separable, then so are also P and Q.
[Hint: Choose $D_{nm} \in \mathcal{B}(X)$ such that $P_n \subset D_{nm} \subset Q'_m$. The set
$$D = \bigcup_{n=1}^{\infty} \bigcap_{m=1}^{\infty} D_{nm} \in \mathcal{B}(X)$$
separates P and Q.]

9. Let X be a separable complete metric space, and let A, $B \subset X$ be disjoint analytic sets. Then A, B are Borel separable. (Lusin separation theorem; Lusin [211]).
[Hint: Choose continuous functions f, $g : \mathfrak{z} \to X$ such that $A = f(\mathfrak{z})$, $B = g(\mathfrak{z})$. For every $m_1, \ldots, m_j \in \mathbb{N}$ put
$$\mathfrak{z}_{m_1 \ldots m_j} = \{ z = \{n_i\} \in \mathfrak{z} \mid n_1 = m_1, \ldots, n_j = m_j \}, \quad j \in \mathbb{N}.$$

---
[4] The assumption that X is separable is not essential, and can be omitted.

Suppose that A, B are not Borel separable, and, using Exercise 2.8, find $m_i \in \mathbb{N}$, $i \in \mathbb{N}$, such that $f(\mathfrak{z}_{m_1 \dots m_j})$ and $g(\mathfrak{z}_{m_1 \dots m_j})$ are not Borel separable, $j \in \mathbb{N}$. Put $z = \{m_i\} \in \mathfrak{z}$. Then $\varrho\big(f(z), g(z)\big) > 0$ (ϱ being the distance in X), and $d\big[f(\mathfrak{z}_{m_1 \dots m_j})\big] + d\big[g(\mathfrak{z}_{m_1 \dots m_j})\big] < \varrho\big(f(z), g(z)\big)$ for every $j \in \mathbb{N}$. Hence $\operatorname{cl} f(\mathfrak{z}_{m_1 \dots m_j}) \cap g(\mathfrak{z}_{m_1 \dots m_j}) = \varnothing$ so that $f(\mathfrak{z}_{m_1 \dots m_j})$ and $g(\mathfrak{z}_{m_1 \dots m_j})$ are Borel separable.]

10. Let X be a separable complete metric space and $A \subset X$ an analytic set. Prove that if A' is analytic, then $A \in \mathcal{B}(X)$.

    [Hint: Use Lusin separation theorem for A and $B = A'$.]

11. The Cantor set $\mathbf{C} \subset \mathbb{R}$ [Cantor discontinuum] is the set

$$\mathbf{C} = \left\{ x \in [0,1] \ \Big| \ x = \sum_{n=1}^{\infty} \frac{i_n}{3^n}, \ i_n = 0, 2; \ n \in \mathbb{N} \right\}.$$

    Prove that $\mathbf{C}$ is a nowhere dense perfect set.

12. Let X be a separable complete metric space, and $A \subset X$ an uncountable analytic set. Put for $\xi = \sum_{n=1}^{\infty} \frac{i_n}{3^n} \in \mathbf{C}$ $h(\xi) = f(x_I)$, where $I = \{i_n\}$ (cf. the proof of Theorem 2.8.3) and $f : \mathfrak{z} \to X$ is a continuous function such that $f(\mathfrak{z}) = A$. Prove that $h : \mathbf{C} \to A$ is a homeomorphism.

13. Prove that $\mathbf{C} + \mathbf{C} = [0, 2]$.

    [Hint: Consider the set

$$D = \left\{ x \in [0,1] \mid x = \sum_{n=1}^{\infty} \frac{j_n}{3^n}, \ j_n = 0, 1; \ n \in \mathbb{N} \right\}$$

    and show that $D + D = [0, 1]$.]

Chapter 3

Measure Theory

3.1 Outer and inner measure

We assume that the reader is acquainted with the theory of Lebesgue measure and integral. We consider sets situated in $\mathbb{R}^N$ with a fixed positive integer N. The Lebesgue measure in $\mathbb{R}^N$ is denoted by m. The σ-algebra of all Lebesgue measurable subsets of $\mathbb{R}^N$ is denoted by $\mathcal{L}$.

There are various possible approaches to the theory of Lebesgue measure (cf., e.g., Halmos [129], Riesz [266], Natanson [237], Carathéodory [40], Łojasiewicz [208], Hartman-Mikusiński [138], etc.). We do not presuppose any of those approaches. At any case the reader should be familiar with the following theorem, which we quote here without proof.

Theorem 3.1.1. *For any set $A \subset \mathbb{R}^N$ the following conditions are equivalent:*

(i) $A \in \mathcal{L}$.

(ii) *For every $\varepsilon > 0$ there exist open sets G and U such that*

$$A \subset G, \ G \setminus A \subset U, \ m(U) < \varepsilon.$$

(iii) *There exist sets $H \in \mathcal{G}_\delta$ and $B \in \mathcal{L}$ such that*

$$A \subset H, \ A = H \setminus B, \ m(B) = 0.$$

(iv) *For every $\varepsilon > 0$ there exist a closed set F and an open set V such that*

$$F \subset A, \ A \setminus F \subset V, \ m(V) < \varepsilon.$$

(v) *There exist sets $E \in \mathcal{F}_\sigma$ and $C \in \mathcal{L}$ such that*

$$E \subset A, \ A = E \cup C, \ m(C) = 0.$$

Theorem 3.1.1 will be our main tool in the sequel of this chapter.

Let $A \subset \mathbb{R}^N$ be an arbitrary set. We define the *outer measure*[1] and the *inner measure* of A (denoted in the sequel by $m_e(A)$ and $m_i(A)$, respectively) as

$$m_e(A) = \inf_{\substack{G \supset A \\ G \text{ open}}} m(G), \ m_i(A) = \sup_{\substack{F \subset A \\ F \text{ closed}}} m(F). \tag{3.1.1}$$

The outer and inner measure are defined for all sets $A \subset \mathbb{R}^N$. It follows directly from definition (3.1.1) that

$$m_e(\varnothing) = m_i(\varnothing) = 0.$$

The connection between the outer measure, inner measure and the measure is expressed by the following theorems:

Theorem 3.1.2. *If $A \in \mathcal{L}$, then*

$$m_e(A) = m_i(A) = m(A). \tag{3.1.2}$$

Proof. If $A \in \mathcal{L}$, then by Theorem 3.1.1 there exist, for every $\varepsilon > 0$, a closed set F and open sets G, U, V such that

$$F \subset A \subset G, \ G \setminus A \subset U, \ A \setminus F \subset V,$$

$$m(U) < \varepsilon, \ m(V) < \varepsilon.$$

Hence $G = A \cup (G \setminus A)$ and $A = F \cup (A \setminus F)$, and since $G \setminus A \in \mathcal{L}$, $A \setminus F \in \mathcal{L}$, $m(G \setminus A) \leqslant m(U) < \varepsilon$, $m(A \setminus F) \leqslant m(V) < \varepsilon$, we have

$$m(G) \leqslant m(A) + \varepsilon, \ m(A) \leqslant m(F) + \varepsilon,$$

i.e.,

$$m(G) - \varepsilon \leqslant m(A) \leqslant m(F) + \varepsilon. \tag{3.1.3}$$

Since $\varepsilon > 0$ has been arbitrary, relation (3.1.3) shows that

$$m(A) = \inf_{\substack{G \supset A \\ G \text{ open}}} m(G) = \sup_{\substack{F \subset A \\ F \text{ closed}}} m(F).$$

which in view of (3.1.1) is equivalent to (3.1.2). $\square$

Theorem 3.1.3. *If*

$$m_i(A) = m_e(A) < \infty, \tag{3.1.4}$$

then $A \in \mathcal{L}$ and (3.1.2) holds.

Proof. Relation (3.1.4) implies that there exist, for every $\varepsilon > 0$, a closed set F and an open set G such that $F \subset A \subset G$ and

$$m(G) - \frac{\varepsilon}{2} \leqslant m_e(A) = m_i(A) \leqslant m(F) + \frac{\varepsilon}{2}.$$

Hence $m(G \setminus F) = m(G) - m(F) \leqslant \varepsilon$, and clearly the set $G \setminus F$ is open, $G \setminus A \subset G \setminus F$, $A \setminus F \subset G \setminus F$. Now the theorem follows from Theorem 3.1.1 if we take $U = V = G \setminus F$. Relation (3.1.2) is a consequence of Theorem 3.1.2. $\square$

---

[1] If the Lebesgue measure is introduced *via* an outer measure, then the reader will easily verify that m_e defined by (3.1.1) coincides with the previous notion.

The next four theorems give the relation between the outer [inner] measure of the union of a countable collection of sets and those of particular sets.

Theorem 3.1.4. *If $A_n \subset \mathbb{R}^N$, $n \in \mathbb{N}$, are arbitrary sets, and $A \subset \bigcup_{n=1}^{\infty} A_n$, then*

$$m_e(A) \leqslant \sum_{n=1}^{\infty} m_e(A_n).$$ (3.1.5)

In particular

$$m_e\left(\bigcup_{n=1}^{\infty} A_n\right) \leqslant \sum_{n=1}^{\infty} m_e(A_n).$$ (3.1.6)

Proof. Fix an $\varepsilon > 0$. According to definition (3.1.1), for every $n \in \mathbb{N}$ we can find an open set G_n such that

$$A_n \subset G_n, \quad m(G_n) \leqslant m_e(A_n) + \frac{\varepsilon}{2^n}.$$ (3.1.7)

$\bigcup_{n=1}^{\infty} G_n$ also is an open set, and hence measurable. For measurable sets (and with m_e replaced by m) formulas (3.1.5) and (3.1.6) are true. Since $A \subset \bigcup_{n=1}^{\infty} G_n$, we have by

(3.1.1) $m_e(A) \leqslant m\left(\bigcup_{n=1}^{\infty} G_n\right)$, and hence

$$m_e(A) \leqslant m\left(\bigcup_{n=1}^{\infty} G_n\right) \leqslant \sum_{n=1}^{\infty} m(G_n) \leqslant \sum_{n=1}^{\infty} m_e(A_n) + \varepsilon,$$ (3.1.8)

by (3.1.7). Letting in (3.1.8) $\varepsilon \to 0$ we obtain (3.1.5). (3.1.6) is a particular case of (3.1.5). $\square$

Relations (3.1.5) and (3.1.6) (as well as similar relations occurring in Theorems 3.1.5–3.1.7) are true also for finite unions. It is enough to put $A_n = \varnothing$ for $n > m$. Taking, in particular, $m = 1$ and $A_1 = B$, we obtain from (3.1.5) the monotonicity of the outer measure: *If $A \subset B$, then*

$$m_e(A) \leqslant m_e(B).$$

The Lebesgue measure, as is well known, is σ-additive: if $A_n \in \mathcal{L}$, $n \in \mathbb{N}$, are pairwise disjoint:

$$A_i \cap A_j = \varnothing \quad \text{for} \quad i \neq j, \; i,j \in \mathbb{N},$$

then

$$m\left(\bigcup_{n=1}^{\infty} A_n\right) = \sum_{n=1}^{\infty} m(A_n).$$

For the outer measure we have the following weaker version of this fact.

Theorem 3.1.5. *If $A_n \in \mathcal{L}$, $n \in \mathbb{N}$, are pairwise disjoint measurable sets, and $A \subset \mathbb{R}^N$ is arbitrary, then*

$$m_e\Big(A \cap \bigcup_{n=1}^{\infty} A_n\Big) = \sum_{n=1}^{\infty} m_e(A \cap A_n). \qquad (3.1.9)$$

Proof. Fix an $\varepsilon > 0$. We can find an open set G such that $A \cap \bigcup_{n=1}^{\infty} A_n \subset G$ and

$$m(G) - \varepsilon \leqslant m_e\Big(A \cap \bigcup_{n=1}^{\infty} A_n\Big). \qquad (3.1.10)$$

The sets $G \cap A_n$, $n \in \mathbb{N}$, are pairwise disjoint measurable sets, $A \cap A_n \subset A \cap A_n \cap \bigcup_{i=1}^{\infty} A_i \subset G \cap A_n$, whence by Theorems 3.1.4 and 3.1.2

$$m_e(A \cap A_n) \leqslant m_e(G \cap A_n) = m(G \cap A_n). \qquad (3.1.11)$$

Since $\bigcup_{n=1}^{\infty}(G \cap A_n) = G \cap \bigcup_{n=1}^{\infty} A_n \subset G$, we get by (3.1.10) and (3.1.11)

$$m_e\Big(A \cap \bigcup_{n=1}^{\infty} A_n\Big) \geqslant m(G) - \varepsilon \geqslant m\Big(\bigcup_{n=1}^{\infty}(G \cap A_n)\Big) - \varepsilon$$

$$= \sum_{n=1}^{\infty} m(G \cap A_n) - \varepsilon \geqslant \sum_{n=1}^{\infty} m_e(A \cap A_n) - \varepsilon. \quad (3.1.12)$$

When $\varepsilon \to 0$ we obtain (3.1.9) from (3.1.12) and Theorem 3.1.4. $\qquad \square$

For $A_2 = A_1' = \mathbb{R}^N \setminus A_1$, $A_n = \varnothing$ for $n \geqslant 3$, we obtain from Theorem 3.1.5 the so-called Carathéodory condition

$$m_e(A) = m_e(A \cap A_1) + m_e(A \cap A_1') \quad \text{for all} \ \ A \subset \mathbb{R}^N,$$

which is equivalent to the measurability of the set A_1, and in the Carathéodory theory of the Lebesgue measure is even taken as the definition of the measurability of A_1.

The following Theorems 3.1.6 and 3.1.7 are counterparts of Theorems 3.1.4 and 3.1.5 for inner measure.

Theorem 3.1.6. *If $A_n \subset \mathbb{R}^N$, $n \in \mathbb{N}$, are pairwise disjoint, and $A \subset \mathbb{R}^N$ is an arbitrary set such that $\bigcup_{n=1}^{\infty} A_n \subset A$, then*

$$m_i(A) \geqslant \sum_{n=1}^{\infty} m_i(A_n). \qquad (3.1.13)$$

In particular,

$$m_i\Big(\bigcup_{n=1}^{\infty} A_n\Big) \geqslant \sum_{n=1}^{\infty} m_i(A_n). \qquad (3.1.14)$$

Proof. For every $n \in \mathbb{N}$ take a real number $a_n < m_i(A_n)$ and a closed set $F_n \subset A_n$ such that

$$m(F_n) > a_n. \tag{3.1.15}$$

Take an arbitrary $k \in \mathbb{N}$. The set $\bigcup_{n=1}^{k} F_n$ is closed and $\bigcup_{n=1}^{k} F_n \subset A$. By the definition of the inner measure we have

$$m_i(A) \geqslant m\Big(\bigcup_{n=1}^{k} F_n\Big) = \sum_{n=1}^{k} m(F_n) > \sum_{n=1}^{k} a_n, \tag{3.1.16}$$

since the sets F_n are disjoint, just like A_n. Now let $a_1 \to m_i(A_1), \ldots, a_k \to m_i(A_k)$. Thus we get from (3.1.16)

$$m_i(A) \geqslant \sum_{n=1}^{k} m_i(A_n).$$

Letting $k \to \infty$ we obtain hence (3.1.13). (3.1.14) is a particular case of (3.1.13). $\square$

Taking in (3.1.13) $A_1 = B$, $A_n = \emptyset$ for $n \geqslant 2$, we obtain the monotonicity of the inner measure: If $B \subset A$, then

$$m_i(B) \leqslant m_i(A).$$

Theorem 3.1.7. *If $A_n \subset \mathbb{R}^N$, $n \in \mathbb{N}$, are pairwise disjoint measurable sets, and $A \subset \mathbb{R}^N$ is arbitrary, then*

$$m_i\Big(A \cap \bigcup_{n=1}^{\infty} A_n\Big) = \sum_{n=1}^{\infty} m_i(A \cap A_n). \tag{3.1.17}$$

Proof. Fix a real number $a < m_i(A \cap \bigcup_{n=1}^{\infty} A_n)$. We can find a closed set $F \subset A \cap \bigcup_{n=1}^{\infty} A_n$ such that

$$a < m(F). \tag{3.1.18}$$

The sets $F \cap A_n$, $n \in \mathbb{N}$, are pairwise disjoint measurable sets. Since $F \subset \bigcup_{n=1}^{\infty} A_n$, also $F \subset F \cap \bigcup_{n=1}^{\infty} A_n = \bigcup_{n=1}^{\infty} (F \cap A_n)$. Moreover, we have $F \cap A_n \subset A \cap A_n$, $n \in \mathbb{N}$. Hence by the monotonicity of the inner measure

$$m(F) \leqslant m\Big(\bigcup_{n=1}^{\infty} (F \cap A_n)\Big) = \sum_{n=1}^{\infty} m(F \cap A_n) = \sum_{n=1}^{\infty} m_i(F \cap A_n) \leqslant \sum_{n=1}^{\infty} m_i(A \cap A_n). \tag{3.1.19}$$

Relations (3.1.18) and (3.1.19) imply

$$a < \sum_{n=1}^{\infty} m_i(A \cap A_n).$$

Letting $a \to m_i(A \cap \bigcup_{n=1}^{\infty} A_n)$, we obtain

$$m_i \left(A \cap \bigcup_{n=1}^{\infty} A_n \right) \leqslant \sum_{n=1}^{\infty} m_i(A \cap A_n),$$

which together with Theorem 3.1.6 yields equality (3.1.17).   □

Before proceeding further let us note that for an arbitrary set $A \subset \mathbb{R}^N$ we have

$$m_i(A) \leqslant m_e(A). \tag{3.1.20}$$

This follows directly from definition (3.1.1), but one can also argue as follows: For an arbitrary $\varepsilon > 0$ there exists a closed set $F \subset A$ such that

$$m_i(A) \leqslant m(F) + \varepsilon.$$

By the monotonicity of the outer measure we have

$$m(F) = m_e(F) \leqslant m_e(A).$$

Hence $m_i(A) \leqslant m_e(A) + \varepsilon$, and on letting $\varepsilon \to 0$ we obtain (3.1.20).

No we prove

Lemma 3.1.1. *For every set $A \subset \mathbb{R}^N$ there exists a measurable set B such that $A \subset B$ and $m_i(B \setminus A) = 0$.*

Proof. Consider first the case where $m_e(A) < \infty$. For every $n \in \mathbb{N}$ there exists an open set G_n such that $A \subset G_n$ and

$$m(G_n) \leqslant m_e(A) + \frac{1}{2^n}. \tag{3.1.21}$$

Put

$$B = \bigcap_{n=1}^{\infty} G_n. \tag{3.1.22}$$

For an indirect proof suppose that $m_i(B \setminus A) = 2\eta > 0$. Then there exists a closed set $F \subset B \setminus A$ such that $m(F) \geqslant \eta$. For every $n \in \mathbb{N}$ the set $G_n \setminus F$ is open, $A \subset G_n \setminus F$, whence

$$m_e(A) \leqslant m(G_n \setminus F) = m(G_n) - m(F) \leqslant m(G_n) - \eta.$$

Hence, by (3.1.21),

$$m(G_n) \leqslant m(G_n) + \frac{1}{2^n} - \eta,$$

or

$$\frac{1}{2^n} - \eta \geqslant 0,$$

whence, on letting $n \to \infty$, we obtain a contradiction.

Now let $m_e(A)$ be arbitrary. Put $A_1 = K(0,1)$, $A_n = K(0,n) \setminus K(0,n-1)$ for $n = 2,3,4,\ldots$, where $K(0,n)$ is the open ball centered at the origin and with radius n. The sets A_n are measurable, pairwise disjoint, and $\bigcup_{n=1}^{\infty} A_n = \mathbb{R}^N$. Moreover, $m(A_n) < \infty$ for $n \in \mathbb{N}$, whence also $m_e(A \cap A_n) \leqslant m(A_n) < \infty$ for $n \in \mathbb{N}$. By the first part of the proof we can find, for every $n \in \mathbb{N}$, a measurable set B_n such that $A \cap A_n \subset B_n$ and $m_i[B_n \setminus (A \cap A_n)] = 0$. Put $B = \bigcup_{n=1}^{\infty} (B_n \cap A_n)$. Then $B \in \mathcal{L}$, $A = A \cap \mathbb{R}^N = A \cap \bigcup_{n=1}^{\infty} A_n = \bigcup_{n=1}^{\infty} (A \cap A_n) \subset \bigcup_{n=1}^{\infty} (B_n \cap A_n) = B$. Moreover,

$$B \setminus A = \bigcup_{n=1}^{\infty} (B_n \cap A_n) \setminus A = \bigcup_{n=1}^{\infty} [(B_n \cap A_n) \setminus A].$$

Now, $(B_n \cap A_n) \setminus A = (B_n \cap A_n) \setminus (A \cap A_n) \subset B_n \setminus (A \cap A_n)$, whence $m_i[(B_n \cap A_n) \setminus A] \leqslant m_i[B_n \setminus (A \cap A_n)] = 0$, and ultimately $m_i[(B_n \cap A_n) \setminus A] = 0$, $n \in \mathbb{N}$. The sets $B_n \cap A_n$ are measurable and pairwise disjoint (since A_n are pairwise disjoint). By Theorem 3.1.7

$$m_i(B \setminus A) = m_i\Big[\bigcup_{n=1}^{\infty} ((B_n \cap A_n) \setminus A)\Big] = m_i\Big[\bigcup_{n=1}^{\infty} ((B_n \cap A_n) \cap A')\Big]$$

$$= \sum_{n=1}^{\infty} m_i[(B_n \cap A_n) \cap A'] = \sum_{n=1}^{\infty} m_i[(B_n \cap A_n) \setminus A] = 0.$$

So the set B has the required properties. $\qquad\square$

Lemma 3.1.2. *For every set $A \subset \mathbb{R}^N$ there exists a measurable set B such that $A \subset B$ and $m(B) = m_e(A)$*

Proof. For every $n \in \mathbb{N}$ there exists an open set G_n such that $A \subset G_n$ and (3.1.21) holds. Define B by (3.1.22). Then $B \subset G_n$, $n \in \mathbb{N}$, whence by (3.1.21) $m(B) \leqslant m(G_n) \leqslant m_e(A) + \dfrac{1}{2^n}$, $n \in \mathbb{N}$. As $n \to \infty$, we obtain hence $m(B) \leqslant m_e(A)$. On the other hand, $A \subset B$, whence $m_e(A) \leqslant m_e(B) = m(B)$. $\qquad\square$

Lemma 3.1.3. *For every set $A \subset \mathbb{R}^N$ there exists a measurable set B such that $A \subset B$ and for every measurable set Z such that $A \subset Z$ we have $m(B \setminus Z) = 0$.*

Proof. Let B be the set occurring in Lemma 3.1.1. For every $Z \in \mathcal{L}$, $A \subset Z$, we have $B \setminus Z \subset B \setminus A$, whence by Lemma 3.1.1 $m_i(B \setminus Z) = 0$. Since $B \setminus Z$ is measurable, this implies in virtue of Theorem 3.1.2 $m(B \setminus Z) = 0$. $\qquad\square$

The Lebesgue measure is complete, i.e., if $Y \subset B \setminus Z$, then $Y \in \mathcal{L}$ and $m(Y) = 0$. Thus Lemma 3.1.3 together with Theorems 2.7.1 and 2.6.1 imply the following

Theorem 3.1.8. *Every analytic set is Lebesgue measurable.*

3.2   Linear transforms

As is well known, the Lebesgue measure in $\mathbb{R}^N$ is invariant under translations. We are going to show that this important property of m is shared also by the outer and inner measure. More exactly, we are going to prove the following (cf. Łojasiewicz [208])

Theorem 3.2.1. *Let* $f : \mathbb{R}^N \to \mathbb{R}^N$ *be linear transform*

$$f(x) = Lx + b, \ x \in \mathbb{R}^N, \tag{3.2.1}$$

where L *is an* $N \times N$ *non-singular square matrix, and* $b \in \mathbb{R}^N$. *Then, for every set* $A \subset \mathbb{R}^N$, *we have*

$$m_e\big(f(A)\big) = |\det L|\, m_e(A), \ \ m_i\big(f(A)\big) = |\det L|\, m_i(A). \tag{3.2.2}$$

Proof. First suppose that the set A is closed[2]. Function (3.2.1) is a homeomorphism, so $f(A)$ also is closed, and thus both A and $f(A)$, are measurable. For such an A formula (3.2.2) takes the form

$$m\big(f(A)\big) = |\det L|\, m(A). \tag{3.2.3}$$

Let χ_A be the characteristic function of the set A:

$$\chi_A(x) = \begin{cases} 1 & \text{if } x \in A, \\ 0 & \text{if } x \notin A. \end{cases}$$

Then $\chi_A \circ f = \chi_{f^{-1}(A)}$ ($\circ$ denotes the functional composition). Hence, by the known rules of the integral calculus,

$$m(A) = \int_{\mathbb{R}^N} \chi_A \, \mathrm{d}x = \int_{\mathbb{R}^N} (\chi_A \circ f)\, |\det L|\, \mathrm{d}x = \int_{\mathbb{R}^N} \chi_{f^{-1}(A)}\, |\det L|\, \mathrm{d}x$$

$$= |\det L| \int_{\mathbb{R}^N} \chi_{f^{-1}(A)}\, \mathrm{d}x = |\det L|\, m\big(f^{-1}(A)\big),$$

whence (3.2.3) follows on replacing A by $f(A)$.

Now let $A \subset \mathbb{R}^N$ be arbitrary. Take an arbitrary closed set $F \subset A$. Then $f(F) \subset f(A)$, whence $m_i\big(f(F)\big) \leqslant m_i\big(f(A)\big)$. By (3.2.3) $m_i\big(f(F)\big) = m\big(f(F)\big) = |\det L|\, m(F)$, whence

$$|\det L|\, m(F) \leqslant m_i\big(f(A)\big). \tag{3.2.4}$$

Taking in (3.2.4) the supremum over all closed sets $F \subset A$, we obtain according to (3.1.1)

$$|\det L|\, m_i(A) \leqslant m_i\big(f(A)\big). \tag{3.2.5}$$

Now take an arbitrary closed set $F_0 \subset f(A)$. Then the set $F = f^{-1}(F_0)$ is also closed, and we have $f(F) = F_0$. By (3.1.3)

$$m(F_0) = |\det L|\, m(F),$$

---

[2] The same argument applies also if A is open.

and $m(F) \leqslant m_i(A)$, since $F \subset A$. Thus

$$m(F_0) \leqslant |\det L| \, m_i(A). \tag{3.2.6}$$

Taking in (3.2.6) the supremum over all closed sets $F_0 \subset f(A)$, we obtain

$$m_i\big(f(A)\big) \leqslant |\det L| \, m_i(A),$$

which together with (3.2.5) proves the second relation in (3.2.2). The proof of the first relation in (3.2.2) is analogous. $\qquad\square$

Let us note that formulas (3.2.2) remain valid also for singular matrices L. Then the set $f(A)$ is contained in an $(N-1)$-dimensional subspace of $\mathbb{R}^N$, whence $m_i\big(f(A)\big) = m_e\big(f(A)\big) = 0$.

Taking in Theorem 3.2.1 $L = E$, the unit matrix of order N, we obtain the invariance of m_e and m_i under translations.

Corollary 3.2.1. *Let $f : \mathbb{R}^N \to \mathbb{R}^N$ be transform (3.2.1), where L is an $N \times N$ non-singular square matrix, and $b \in \mathbb{R}^N$. If a set $A \subset \mathbb{R}^N$ is measurable, then also $f(A)$ is measurable, and relation (3.2.3) holds.*

Proof. Consider first the case where $m(A) < \infty$. By Theorem 3.1.2 $m_e(A) = m_i(A) = m(A)$, whence by Theorem 3.2.1

$$m_e\big(f(A)\big) = |\det L| \, m(A), \quad m_i\big(f(A)\big) = |\det L| \, m(A) \tag{3.2.7}$$

so that $m_e\big(f(A)\big) = m_i\big(f(A)\big) < \infty$. By Theorem 3.1.3 $f(A) \in \mathcal{L}$, and (3.2.3) results from (3.2.7).

If $m(A) = \infty$, then there exist pairwise disjoint measurable sets A_n such that $A = \bigcup\limits_{n=1}^{\infty} A_n$, $m(A_n) < \infty$, $n \subset \mathbb{N}$ (cf. the proof of Lemma 3.1.1). Then $f(A) = \bigcup\limits_{n=1}^{\infty} f(A_n)$, and $f(A_n) \in \mathcal{L}$, $n \in \mathbb{N}$, by the first part of the proof. Hence also $f(A) \in \mathcal{L}$, and by Theorem 3.2.1

$$m\big(f(A)\big) = m_i\big(f(A)\big) = |\det L| \, m_i(A) = |\det L| \, m(A).$$

Thus (3.2.3) holds. $\qquad\square$

Corollary 3.2.2. *Let $f : \mathbb{R}^N \to \mathbb{R}^N$ be the transform*

$$f(x) = ax + b, \quad x \in \mathbb{R}^N, \tag{3.2.8}$$

where $a \in \mathbb{R} \setminus \{0\}$, $b \in \mathbb{R}^N$. Then, for every set $A \subset \mathbb{R}^N$, we have

$$m_e\big(f(A)\big) = m_e(aA + b) = |a|^N \, m_e(A),$$
$$m_i\big(f(A)\big) = m_i(aA + b) = |a|^N \, m_i(A),$$

and, if $A \in \mathcal{L}$, then also $f(A) \in \mathcal{L}$, and

$$m\big(f(A)\big) = m(aA + b) = |a|^N \, m(A).$$

Proof. Take $L = aE$, the diagonal matrix with all the elements of the main diagonal equal a. Then $\det L = a^N \neq 0$, and function (3.2.1) reduces to (3.2.8). Now the corollary results from Theorem 3.2.1 and Corollary 3.2.1. □

3.3   Saturated non-measurable sets

A set $A \subset \mathbb{R}^N$ is called *saturated non-measurable* (Halperin [132]) iff

$$m_i(A) = m_i(A') = 0. \tag{3.3.1}$$

It follows directly from definition (3.3.1) that if a set A is saturated non-measurable, then so is also its complement A'.

To justify the attribute *non-measurable*, let us note that a saturated non-measurable set actually is not Lebesgue measurable. Supposing the contrary, we would get by (3.3.1) $m(A) = m(A') = 0$, whence $m(\mathbb{R}^N) = m(A \cup A') = m(A) + m(A') = 0$, a contradiction.

An insight into the structure of saturated non-measurable sets may be gained from the following

Theorem 3.3.1. *Let $A \subset \mathbb{R}^N$ be an arbitrary set. Then the following conditions are equivalent:*

(i) $m_i(A') = 0$;
(ii) for every measurable set E of positive measure we have $A \cap E \neq \varnothing$;
(iii) for every measurable set E we have $m_e(A \cap E) = m(E)$;
(iv) for every open interval[3] $I \subset \mathbb{R}^N$ we have $m_e(A \cap I) = m(I)$.

Proof. First we prove that condition *(i)* implies *(ii)*. Let a set $A \subset \mathbb{R}^N$ fulfil *(i)*, and let $E \in \mathcal{L}$ be an arbitrary set such that $m(E) > 0$. If we had $A \cap E = \varnothing$, then $E \subset A'$, whence

$$m_i(A') \geqslant m_i(E) = m(E) > 0,$$

which is incompatible with *(i)*.

Now assume that $A \subset \mathbb{R}^N$ fulfils *(ii)*. Let $E \subset \mathbb{R}^N$ be an arbitrary measurable set. Clearly

$$m_e(A \cap E) \leqslant m_e(E) = m(E).$$

It remains to exclude the strict inequality. For an indirect proof suppose that for a certain $E \in \mathcal{L}$ we have

$$m_e(A \cap E) < m(E).$$

By Lemma 3.1.2 there exists a set $B \in \mathcal{L}$ such that $A \cap E \subset B$ and $m(B) = m_e(A \cap E) < m(E)$. Hence also

$$m(B \cap E) \leqslant m(B) < m(E). \tag{3.3.2}$$

Now, by (3.3.2) $m(B \cap E) < \infty$, whence, again by (3.3.2)

$$m[E \setminus (B \cap E)] = m(E) - m(B \cap E) > 0.$$

---

[3] An open interval in $\mathbb{R}^N$ is a cartesian product of N one-dimensional open intervals.

Since $A \cap E \subset B \cap E$, we have $E \setminus (B \cap E) \subset E \setminus (A \cap E) \subset A'$. Thus we have constructed a measurable set of positive measure $\left(\text{viz. } E \setminus (B \cap E)\right)$ disjoint with A. This contradicts (ii).

Condition (iv) is a particular case of condition (iii), so $(iii) \Rightarrow (iv)$. It remains to show that $(v) \Rightarrow (i)$.

Let condition (iv) be fulfilled, and put $J_n = (-n, n)^N$, $n \in \mathbb{N}$. Thus $J_n \subset \mathbb{R}^N$ are finite intervals, $J_n \subset J_{n+1}$, $n \in \mathbb{N}$, $\bigcup_{n=1}^{\infty} J_n = \mathbb{R}^N$. Suppose that $m_i(A') > 0$. Then there exists a closed set $F \subset A'$ such that $m(F) > 0$. We have

$$F = F \cap \mathbb{R}^N = F \cap \bigcup_{n=1}^{\infty} J_n = \bigcup_{n=1}^{\infty} (F \cap J_n),$$

whence

$$m(F) = \lim_{n \to \infty} m(F \cap J_n).$$

Thus $m(F \cap J_n) > 0$ for n sufficiently large. Let J denote an arbitrary interval from the sequence $\{J_n\}$ such that $m(F \cap J) > 0$. We have

$$A \cap J = (A \cap J \cap F) \cup (A \cap J \cap F') = A \cap J \cap F' \subset J \cap F',$$

since $A \cap J \cap F = \varnothing$. Hence by (iv)

$$m(J) = m_e(A \cap J) \leqslant m_e(J \cap F') = m(J \cap F')$$
$$= m\left[J \setminus (J \cap F)\right] = m(J) - m(J \cap F) < m(J),$$

a contradiction.   □

It is clear from the proof that in condition (iv) open intervals may be replaced by closed intervals.

Let us note that conditions of Theorem 3.3.1 do not imply that the set A is non-measurable. They are fulfilled, e.g., for $A = \mathbb{R}^N$. But for saturated non-measurable sets we can derive from Theorem 3.3.1 the following result, which to a certain extent justifies the attribute *saturated*.

Theorem 3.3.2. *Let $A \subset \mathbb{R}^N$ be a saturated non-measurable set. Then, for every measurable set B of positive measure, the set $A \cap B$ is non-measurable.*

Proof. We have $A \cap B \subset A$, whence $m_i(A \cap B) \leqslant m_i(A) = 0$. Thus $m_i(A \cap B) = 0$, whereas by Theorem 3.3.1 $m_e(A \cap B) = m(B) > 0$. The theorem now follows from Theorem 3.1.2.   □

We defer the construction of a saturated non-measurable set $A \subset \mathbb{R}$ to the last section. Many examples of saturated non-measurable sets $A \subset \mathbb{R}^N$ will occur in the second part of this book. However, in the sequel we assume that there exist saturated non-measurable sets, the assumption which will be justified only later. Let us note that if A is a saturated non-measurable set, then by Theorem 3.3.1

$$m_e(A) = m_e(A \cap \mathbb{R}^N) = m(\mathbb{R}^N) = \infty.$$

Theorem 3.3.3. *Every set $B \subset \mathbb{R}^N$ of positive outer measure contains a non-measurable set.*

Proof. If B is non-measurable, then B itself is good. Now suppose that $B \in \mathcal{L}$ so that $m(B) = m_e(B) > 0$, and let A be a saturated non-measurable set. Then the set $A \cap B \subset B$ is non-measurable in virtue of Theorem 3.3.2. $\square$

The saturated non-measurable set A constructed in 3.8 will also have the property:

(∗) If $B \subset A$ or $B \subset A'$, and B has the Baire property, then B is of the first category.

Property (∗) is a topological analogue of the saturated non-measurability (Kuczma [189]). Making use of (∗) we will prove topological analogues of Theorems 3.3.1, 3.3.2, 3.3.3.

Theorem 3.3.4. *If $A \subset \mathbb{R}^N$ has property (∗), and $B \subset \mathbb{R}^N$ is a set of the second category and with the Baire property, then the set $A \cap B$ is of the second category and without the Baire property.*

Proof. $B = (A \cap B) \cup (A' \cap B)$. If $A \cap B$ had the Baire property, then also $A' \cap B = B \setminus (A \cap B)$ would have the Baire property, and since $A' \cap B \subset A'$, we would get by (∗) that $A' \cap B$ is of the first category. Similarly, $A \cap B \subset A$ would be of the first category, and hence also B would be of the first category. If $A \cap B$ were of the first category, then it would have the Baire property, and as we have just seen, this leads to a contradiction. Consequently $A \cap B$ is of the second category and without the Baire property. $\square$

Corollary 3.3.1. *If a set $A \subset \mathbb{R}^N$ has property (∗), then for every set B of the second category and with the Baire property we have $A \cap B \neq \varnothing$.*

Theorem 3.3.5. *If $A \subset \mathbb{R}^N$ has property (∗), and $B \subset \mathbb{R}^N$ is of the second category and with the Baire property, then $D(A \cap B) = D(B)$.*

Proof. The inclusion $D(A \cap B) \subset D(B)$ follows from $A \cap B \subset B$. Now let $x \in D(B)$. For every open neighbourhood U of x the set $U \cap B$ is of the second category. The set $U \cap B$ has the Baire property, since B has it, and U, being open, also has the Baire property. By Theorem 3.3.4 the set $U \cap B \cap A$ is of the second category, which proves that $x \in D(A \cap B)$. Consequently $D(B) \subset D(A \cap B)$, which completes the proof. $\square$

Taking in Theorem 3.3.5 $B = \mathbb{R}^N$, we obtain

Corollary 3.3.2. *If a set $A \subset \mathbb{R}^N$ has property (∗), then $D(A) = \mathbb{R}^N$.*

Theorem 3.3.6. *Every set $B \subset \mathbb{R}^N$ of the second category contains a subset without the Baire property.*

Proof. If B is without the Baire property, then B itself is good. Otherwise, let $A \subset \mathbb{R}^N$ be a set with property (∗). By Theorem 3.3.4 $A \cap B \subset B$ is a set without the Baire property. $\square$

3.4 Lusin sets

An uncountable set $A \subset \mathbb{R}^N$ is called a *Lusin set* (Lusin [210]) iff it intersects every nowhere dense perfect set in $\mathbb{R}^N$ at most countably many points. In other words, A is a Lusin set iff card $A > \aleph_0$ and for every nowhere dense perfect set $F \subset \mathbb{R}^N$ we have

$$\operatorname{card}(A \cap F) \leqslant \aleph_0 \, .$$

It follows directly from the definition that an uncountable subset of a Lusin set is a Lusin set. The existence of Lusin sets was shown, under the assumption of the continuum hypothesis, by Lusin [210]. This will follow also from theorems in 11.6.

We will need the following

Lemma 3.4.1. *Let $E \subset \mathbb{R}^N$ be an arbitrary set, and let $f : E \to \mathbb{R}^M$, $M \in \mathbb{N}$, be a continuous function. There exists a decomposition*

$$E = A \cup B \, , \tag{3.4.1}$$

where $A \cap B = \varnothing$, A is of the first category, and $f(B)$ is measurable and of measure zero.

Proof. If E is finite, the lemma is obvious. So in the sequel we assume that E is infinite. Let $\{p_k\}_{k \in \mathbb{N}}$ be a sequence of points of E, dense in E. By the continuity of f we may choose positive numbers δ_{nk}, $n, k \in \mathbb{N}$, such that for $x \in K(p_k, \delta_{nk})$ (the open ball centered at p_k and with the radius δ_{nk}) we have

$$f(x) \in K\left(f(p_k), \frac{1}{2^k n} \right) . \tag{3.4.2}$$

Put $E_n = E \setminus \bigcup_{k=1}^{\infty} K(p_k, \delta_{nk})$, $A = \bigcup_{n=1}^{\infty} E_n$, $B = E \setminus A$. Then (3.4.1) holds, and $A \cap B = \varnothing$. Suppose that $y \in \operatorname{int} \operatorname{cl} E_n$. So there exists an open ball $K \subset \mathbb{R}^N$ such that $y \in K \subset \operatorname{cl} E_n$. There exists a $k \in \mathbb{N}$ such that $p_k \in K$, since the sequence $\{p_k\}$ is dense in E. Hence $K(p_k, \delta_{nk}) \cap K \neq \varnothing$. On the other hand, $K(p_k, \delta_{nk}) \cap E_n = \varnothing$, so $K(p_k, \delta_{nk}) \cap K$ cannot contain any point of E_n. But this contradicts the fact that $K \subset \operatorname{cl} E_n$.

Thus $\operatorname{int} \operatorname{cl} E_n = \varnothing$, which means that E_n are nowhere dense, and A is of the first category. It remains to show that $m\bigl(f(B)\bigr) = 0$.

Let $U_n = \bigcup_{k=1}^{\infty} K\left(f(p_k), \frac{1}{2^k n} \right) \subset \mathbb{R}^M$, and let c denote the M-dimensional measure of the unit ball in $\mathbb{R}^M$. By Corollary 3.2.2

$$m\left(K\left(f(p_k), \frac{1}{2^k n} \right) \right) = m\left(\frac{1}{2^k n} K(0,1) + f(p_k) \right) = \frac{1}{(2^k n)^M} m\bigl(K(0,1)\bigr) = \frac{c}{(2^k n)^M} \, ,$$

whence by Theorem 3.1.4

$$m(U_n) \leqslant \sum_{k=1}^{\infty} m\left(K\left(f(p_k), \frac{1}{2^k n} \right) \right) = \sum_{k=1}^{\infty} \frac{c}{(2^k n)^M} = \frac{c}{n^M} \frac{1}{2^M - 1} \, .$$

If an $x \in B$, then $x \notin E_n$ for $n \in \mathbb{N}$, i.e., $x \in K(p_k, \delta_{nk})$ for a certain $k_n \in \mathbb{N}$. Hence by (3.4.2) $f(x) \in U_n$ for every $n \in \mathbb{N}$, which means that $f(B) \subset U_n$ for every $n \in \mathbb{N}$. Hence

$$m_e\big(f(B)\big) \leqslant m(U_n) \leqslant \frac{c}{n^M} \frac{1}{2^M - 1}$$

for all $n \in \mathbb{N}$. Letting $n \to \infty$ we obtain hence $m_e\big(f(B)\big) = 0$, i.e., $f(B)$ is measurable and of measure zero. $\qquad\square$

An interesting property of Lusin sets is expressed by the following theorem of Sierpiński [281].

Theorem 3.4.1. *If $E \subset \mathbb{R}^N$ is a Lusin set, and $f : E \to \mathbb{R}^M$ a continuous function, then $f(E)$ is Lebesgue measurable and of measure zero.*

Proof. By Lemma 3.4.1 the set E has a decomposition (3.4.1), where A is of the first category, and $m\big(f(B)\big) = 0$. Thus $A = \bigcup_{n=1}^{\infty} E_n$, where $\operatorname{int} \operatorname{cl} E_n = \varnothing$ for $n \in \mathbb{N}$. By Theorem 2.8.2 $\operatorname{cl} E_n = F_n \cup P_n$, where F_n is perfect in $\operatorname{cl} E_n$, and $\operatorname{card} P_n \leqslant \aleph_0$, $n \in \mathbb{N}$. But $\operatorname{cl} E_n$ is closed, so F_n is perfect. Moreover, $F_n \subset \operatorname{cl} E_n$, and $\operatorname{cl} \operatorname{cl} E_n = \operatorname{cl} E_n$ so that $\operatorname{int} \operatorname{cl} \operatorname{cl} E_n = \operatorname{int} \operatorname{cl} E_n = \varnothing$. This means that $\operatorname{cl} E_n$ is nowhere dense, and hence also F_n is nowhere dense. Since $A \subset E$, A itself is a Lusin set[4], whence $\operatorname{card}(A \cap F_n) \leqslant \aleph_0$ for $n \in \mathbb{N}$. We have

$$A \subset \bigcup_{n=1}^{\infty} \operatorname{cl} E_n = \bigcup_{n=1}^{\infty} (F_n \cup P_n),$$

so $A = \bigcup_{n=1}^{\infty} [(A \cap F_n) \cup (A \cap P_n)]$. The sets occurring under the union sign are at most countable, so also $\operatorname{card} A \leqslant \aleph_0$, and $\operatorname{card} f(A) \leqslant \operatorname{card} A \leqslant \aleph_0$, whence $m\big(f(A)\big) = 0$. Now $f(E) = f(A) \cup f(B)$, and being a union of measurable sets of measure zero, $f(E)$ itself is a measurable set of measure zero. $\qquad\square$

Taking $M = N$ and $f(x) = x$ for $x \in E$, we obtain from Theorem 3.4.1

Corollary 3.4.1. *Every Lusin set is Lebesgue measurable and of measure zero.*

Theorem 3.4.1 and Corollary 3.4.1 express the fact that from the point of view of measure theory Lusin sets are very small. The situation is different from the point of view of topology (Sierpiński [282]).

Theorem 3.4.2. *If $E \subset \mathbb{R}^N$ is a Lusin set, then E is of the second category and without the Baire property.*

Proof. Suppose that E is of the first category. Then $E = \bigcup_{n=1}^{\infty} E_n$, where the sets E_n are nowhere dense. Since E, being a Lusin set, is uncountable, at least one of the sets E_n, say E_m, is uncountable. By Theorem 2.8.2 $E_m = A \cup B$, where A is perfect in

---

[4] If A is uncountable. If $\operatorname{card} A \leqslant \aleph_0$, then so the more $\operatorname{card}(A \cap F_n) \leqslant \aleph_0$.

E_m and card $B \leqslant \aleph_0$. The fact that A is perfect in E_m means that every point of A is its accumulation point (this fact is independent of E_m), in other words

$$A \subset A^d, \qquad (3.4.3)$$

and every accumulation point of A, belonging to E_m, belongs also to A: $A^d \cap E_m \subset A$. Put $F = \operatorname{cl} A = A \cup A^d$. Hence $F^d = A^d \cup A^{dd} = A^d$, and by (3.4.3) $F = A \cup A^d \subset A^d \cup A^d = A^d = F^d$. On the other hand, $F^d \subset F$, since F is closed. Consequently $F = F^d$, and F is perfect.

Further, since $A \subset E_m$, A is nowhere dense, and also $F = \operatorname{cl} A$ is nowhere dense. $E_m \subset E$ is a Lusin set, so $\operatorname{card}(E_m \cap F) \leqslant \aleph_0$. Hence also $\operatorname{card}(E_m \cap A) \leqslant \aleph_0$, since $A \subset F$, and since $E_m \cap A = A$, $\operatorname{card} A \leqslant \aleph_0$. Therefore also $E_m = A \cup B$ is at most countable. This contradiction shows that E cannot be of the first category.

So E is of the second category, and suppose that E has the Baire property:

$$E = (G \cup P) \setminus R,$$

where G is open, and P, R are of the first category. $G \neq \varnothing$, since otherwise E would be of the first category. By Theorem 2.1.1 there exists a set $B \in \mathcal{F}_\sigma$, of the first category, such that $R \subset B$. Put $A = G \setminus B$. The set A is analytic, since it is a Borel set, and we have $A \subset E$. Clearly A is of the second category, and hence uncountable. There exists (cf. Exercise 2.13) a homeomorphism $f : \mathbf{C} \to A$, where $\mathbf{C}$ is the Cantor set. $\mathbf{C}$ is perfect and nowhere dense (Exercise 2.11), so also $F = f(\mathbf{C}) \subset A \subset E$ is perfect and nowhere dense, since f is a homeomorphism. Therefore $\operatorname{card}(E \cap F) \leqslant \aleph_0$. But $E \cap F = F$, so $\operatorname{card} F \leqslant \aleph_0$. But f, being a homeomorphism, is one-to-one, so $\operatorname{card} F = \operatorname{card} f(\mathbf{C}) = \operatorname{card} \mathbf{C} = \mathfrak{c}$. This contradiction shows that E cannot have the Baire property. $\qquad\square$

3.5 Outer density

Let $A \subset \mathbb{R}^N$ be an arbitrary set. A point $x \in \mathbb{R}^N$ is called a point of *outer density* of A iff

$$\lim_{r \to 0} \frac{m_e(A \cap Q^r)}{m(Q^r)} = 1,$$

where Q^r denotes the closed cube with the edge of length r centered at x:

$$Q^r = \mathop{\times}_{i=1}^{N} \left[x_i - \frac{r}{2}, x_i + \frac{r}{2} \right], \quad x = (x_1, \ldots, x_N).$$

If $A \in \mathcal{L}$, we omit the attribute outer and say simply that x is a *density point* of A.

The famous Lebesgue density theorem says that if $A \in \mathcal{L}$, then almost every point $x \in A$ is a density point of A. For an arbitrary set A, let A^* denote the set of the points of outer density of A, which reduces to the set of density points of A whenever A is measurable. The Lebesgue density theorem may be expressed as follows: If $A \in \mathcal{L}$, then

$$m(A \setminus A^*) = 0. \qquad (3.5.1)$$

Formula (3.5.1) can be extended to arbitrary sets $A \subset \mathbb{R}^N$. We will prove this below making use of the Vitali covering theorem, with which, as we assume, the reader is familiar. The proof given here is due to Z. Opial (cf. Lojasiewicz [208]).

Theorem 3.5.1. *Almost every point of an arbitrary set $A \subset \mathbb{R}^N$ is a point of outer density of A.*

Proof. Let $E = A \setminus A^*$ be the set of those $x \in A$ which are not points of outer density of A. The thing to prove is that $m(E) = 0$. First assume that A is bounded. For every cube Q^r we have $A \cap Q^r \subset Q^r$, whence $m_e(A \cap Q^r) \leqslant m(Q^r)$. Thus for $x \in E$ we have

$$\liminf_{r \to 0} \frac{m_e(A \cap Q^r)}{m(Q^r)} < 1.$$

Let $H(\alpha)$, $\alpha \in (0,1)$, be the set of those $x \in A$ for which

$$\liminf_{r \to 0} \frac{m_e(A \cap Q^r)}{m(Q^r)} < \alpha.$$

Since $E = \bigcup_{k=2}^{\infty} H\left(1 - \frac{1}{k}\right)$, it is enough to prove that $m\big(H(\alpha)\big) = 0$ for every $\alpha \in (0,1)$.

Fix an $\alpha \in (0,1)$, and write $H = H(\alpha)$. Take an $\varepsilon > 0$. There exists an open set G such that $H \subset G$ and

$$m(G) \leqslant m_e(H) + \varepsilon. \tag{3.5.2}$$

Moreover, since $H \subset A$ is bounded, we have $m_e(H) < \infty$.

Let $\mathcal{Q}$ be the family of all closed cubes $Q \subset G$ such that

$$\frac{m_e(A \cap Q)}{m(Q)} < \alpha. \tag{3.5.3}$$

By definition $\mathcal{Q}$ is a Vitali cover of H. By the Vitali covering theorem there exists a finite or countably infinite sequence of pairwise disjoint cubes $Q_n \in \mathcal{Q}$ such that

$$m\Big(H \setminus \bigcup_n Q_n\Big) = 0. \tag{3.5.4}$$

Since

$$H = \bigcup_n (H \cap Q_n) \cup \Big(H \setminus \bigcup_n Q_n\Big) \subset \bigcup_n (A \cap Q_n) \cup \Big(H \setminus \bigcup_n Q_n\Big),$$

we have by (3.5.4) and by Theorem 3.1.5

$$m_e(H) \leqslant m_e\Big[\bigcup_n (A \cap Q_n)\Big] = \sum_n m_e(A \cap Q_n).$$

But all $Q_n \in \mathcal{Q}$, and so fulfil (3.5.3). Moreover, $Q_n \subset G$, whence also $\bigcup_n Q_n \subset G$. Thus by (3.5.2)

$$m_e(H) \leqslant \alpha \sum_n m(Q_n) = \alpha \, m\Big(\bigcup_n Q_n\Big) \leqslant \alpha \, m(G) \leqslant \alpha \, \big(m_e(H) + \varepsilon\big),$$

whence, on letting $\varepsilon \to 0$, we obtain $m_e(H) \leqslant \alpha\, m_e(H)$, whence $m_e(H) = 0$. In view of the inequalities $0 \leqslant m_i(H) \leqslant m_e(H)$ this implies by Theorem 3.1.3 that $H \in \mathcal{L}$ and $m(H) = 0$.

Now let A be unbounded. Then A can be represented as a union $A = \bigcup\limits_{n=1}^{\infty} A_n$ of disjoint bounded sets. Since $A_n \subset A$, we have $m_e(A_n \cap Q^r) \leqslant m_e(A \cap Q^r) \leqslant m(Q^r)$ for every $n \in \mathbb{N}$ and $r > 0$. Thus, if an x is a point of outer density of an A_n, it must be also a point of outer density of A. In other words, $A_n^* \subset A^*$, $n \in \mathbb{N}$. Hence, for every $n \in \mathbb{N}$, $A_n \setminus A^* \subset A_n \setminus A_n^*$, and

$$E = A \setminus A^* = \Big(\bigcup_{n=1}^{\infty} A_n \Big) \setminus A^* = \bigcup_{n=1}^{\infty} (A_n \setminus A^*) \subset \bigcup_{n=1}^{\infty} (A_n \setminus A_n^*) . \qquad (3.5.5)$$

By the first part of the proof $m(A_n \setminus A_n^*) = 0$ for $n \in \mathbb{N}$. Thus all the sets in (3.5.5) are of measure zero, and, in particular, $m(E) = 0$. $\qquad \square$

The classical Lebesgue density theorem is a particular case of Theorem 3.5.1 above.

3.6 Some lemmas

In this section we prove some unrelated theorems which will be useful in later parts of this book. Theorem 3.6.1 is known as Smítal's lemma (Kuczma-Smítal [190], Kuczma [185]), although similar ideas are found also in earlier works (Ostrowski [248], Halmos [129], Erdős [76]; R. A. Rosenbaum [269] mentions an equivalent theorem). Lemma 3.6.1 is Zygmund's lemma (cf. Łojasiewicz [208]). Theorem 3.6.4 is found in Borel [36].

Theorem 3.6.1. *Let* B, $D \subset \mathbb{R}^N$ *be such that* $m_e(B) > 0$ *and* D *is dense in* $\mathbb{R}^N$. *Put* $A = B + D$. *Then* $m_i(A') = 0$.

Proof. By Theorem 3.5.1 there exists an $x_0 \in B$ which is a point of outer density of B. Thus, for every $c \in (0,1)$, there exists a $\delta > 0$ such that, if Q^r is the cube centered at x_0 and with the edge r, then $r < \delta$ implies

$$m_e(B \cap Q^r) > c\, m(Q^r) .$$

Put $D_0 = x_0 + D$. Then every $x \in D_0$ may be written as $x = x_0 + y$, $y \in D$. By Corollary 3.2.2 $m_e\left[(B \cap Q^r) + y\right] = m_e(B \cap Q^r)$ and $m(Q^r + y) = m(Q^r)$. Thus for any $x = x_0 + y \in D_0$ and for any cube Q centered at x and with the edge $r < \delta$ we have $m_e\left[(B + y) \cap Q\right] > c\, m(Q)$. Since $A = B + D = \bigcup\limits_{y \in D} (B + y)$, we have $m_e(A \cap Q) \geqslant m_e\left[(B + y) \cap Q\right]$, whence

$$m_e(A \cap Q) > c\, m(Q) \qquad (3.6.1)$$

for every cube Q centered at a point of D_0 and with the edge $r < \delta$.

Take an arbitrary open interval[5] $I \subset \mathbb{R}^N$. Let $\mathcal{Q}$ be the family of cubes $Q \subset I$ centered at points of D_0 and with edges less than δ. The family $\mathcal{Q}$ is a Vitali cover of I, since the set D_0 is, like D, dense in $\mathbb{R}^N$. By the Vitali covering theorem[6] there exists a (finite of infinite) sequence of pairwise disjoint cubes $Q_n \in \mathcal{Q}$ such that

$$m\Big(I \setminus \bigcup_n Q_n\Big) = 0 . \tag{3.6.2}$$

Then we have[7] by Theorem 3.1.5 and by (3.6.1)

$$m(I) \geqslant m_e(I \cap A) = m_e\Big(A \cap \bigcup_n Q_n\Big) + m_e\Big[A \cap \Big(I \setminus \bigcup_n Q_n\Big)\Big]$$

$$= m_e\Big(A \cap \bigcup_n Q_n\Big) = \sum_n m_e(A \cap Q_n) \geqslant c \sum_n m(Q_n) = c\, m(I) ,$$

since $m(I) = m(\bigcup_n Q_n) + m(I \setminus \bigcup_n Q_n)$, and by (3.6.2). Hence, letting $c \to 1$, we get

$$m(I) = m(I \cap A) . \tag{3.6.3}$$

Relation (3.6.3) is valid for an arbitrary open interval $I \subset \mathbb{R}^N$. The theorem now follows from Theorem 3.3.1. $\square$

The following theorem (Kuczma [189]) is a topological analogue of Theorem 3.6.1.

Theorem 3.6.2. *Let X be a separeble[8] topological group, and let $B, D \subset X$ be such that B is of the second category, and D is dense in X. Put $A = B + D$. Then the set $A' = X \setminus A$ does not contain any subset of the second category and with the Baire property.*

Proof. It is enough to show that A intersects every set of the second category and with the Baire property in X. Let $E \subset X$ be such a set. Then $E = (G \cup P) \setminus R$, where G is open, and P, R are of the first category. Since E is of the second category, we must have $G \neq \varnothing$.

Since B is of the second category, $D(B) \neq \varnothing$ by Theorem 2.1.4. Let $x_0 \in D(B)$. There exists a $d \in D$ such that $x_0 + d \in G$. Hence the set $G - d$ is a neighbourhood of x_0. Consequently $B \cap (G - d)$ is of the second category, and so is also its translation $[B \cap (G - d)] + d$. (Translation is a homeomorphism). But $[B \cap (G - d)] + d = (B + d) \cap G \subset (B + D) \cap G = A \cap G$. Thus $A \cap G$ is of the second category, and so is also $(A \cap G) \setminus R$. On the other hand, $(A \cap G) \setminus R = A \cap (G \setminus R) \subset A \cap [(G \cup P) \setminus R] = A \cap E$. Consequently $A \cap E$ also is of the second category, and, in particular, $A \cap E \neq \varnothing$. $\square$

---

[5] Cf. 3.3.

[6] The proof given in Kuczma-Smítal [190] does not use the Vitali theorem

[7] Since $Q_n \subset I$ for all n, we have $I \cap A \cap \bigcup_n Q_n = A \cap \bigcup_n Q_n$.

[8] The assumption that X is separable is not essential and may be omitted.

Now let $f : (a, b) \to \mathbb{R}$ be a function of a single real variable, and let $D^+ f(x)$ denote the upper right Dini derivative of f at x:

$$D^+ f(x) = \limsup_{h \to 0+0} \frac{f(x+h) - f(x)}{h}.$$

Lemma 3.6.1. *Let $f : (a, b) \to \mathbb{R}$ be a continuous function, and let $Z \subset (a, b)$ be a set such that $\operatorname{int} f(Z) = \varnothing$. If $D^+ f(x) > 0$ for $x \in (a, b) \setminus Z$, then f is increasing*[9].

Proof. For an indirect proof suppose that there exist $c, d \in (a, b)$ such that $c < d$, and

$$f(d) < f(c).$$

Take an arbitrary $y \in \big(f(d), f(c)\big)$. There exist $x \in (c, d)$ such that $f(x) = y$. Put

$$A = \{\, x \in (c, d) \mid f(x) = y \,\}.$$

By the continuity of f the set A is closed[10], and since $A \subset (c, d)$, A is bounded. Thus A is compact. Hence

$$x_0 = \sup A \in A \subset (c, d).$$

For $x \in (x_0, d)$ we have $f(x) < y = f(x_0)$. Thus for $x \in (x_0, d)$

$$\frac{f(x) - f(x_0)}{x - x_0} = \frac{f(x) - y}{x - x_0} < 0.$$

Hence $D^+ f(x_0) \leqslant 0$, whence $x_0 \in Z$ and $y \in f(Z)$. Since y was arbitrary in $\big(f(d), f(c)\big)$ this shows that $\big(f(d), f(c)\big) \subset f(Z)$, contrary to the assumption that $\operatorname{int} f(Z) = \varnothing$. Thus f must be increasing in (a, b). $\qquad\square$

Theorem 3.6.3. *Let $f : (a, b) \to \mathbb{R}$ be a continuous function, and let $Z \subset (a, b)$ be a countable set. If $D^+ f(x) \geqslant 0$ for $x \in (a, b) \setminus Z$, then f is increasing.*

Proof. For every $n \in \mathbb{N}$ put $g_n(x) = f(x) + \dfrac{x}{n}$. Then

$$D^+ g_n(x) = D^+ f(x) + \frac{1}{n} > 0 \text{ for } x \in (a, b) \setminus Z.$$

Since $\operatorname{card} g_n(Z) \leqslant \operatorname{card} Z \leqslant \aleph_0$, we have $\operatorname{int} g_n(Z) = \varnothing$ for $n \in \mathbb{N}$. By Lemma 3.6.1 all functions g_n are increasing, and hence also

$$f(x) = \lim_{n \to \infty} g_n(x)$$

is increasing. $\qquad\square$

---

[9] In this book increasing resp. decreasing means weakly increasing resp. weakly decreasing, i.e., $x < y$ implies $f(x) \leqslant f(y)$ resp. $f(x) \geqslant f(y)$.
[10] The set A is closed in (c, d). But since $f(d) < y < f(c)$,

$$A = \{\, x \in [c, d] \mid f(x) = y \,\},$$

and so A is also closed in $[c, d]$, and hence closed.

Now fix integers k, p, q, $p \geqslant 1$, $q \geqslant 2$, $0 \leqslant k \leqslant q - 1$, and put

$$C = \{ c \in \mathbb{N} \mid 0 \leqslant c \leqslant q - 1, \ c \neq k \},\tag{3.6.4}$$

$$A_p = \left\{ x \in [0,1] \mid x = \sum_{i=1}^{\infty} \frac{c_i}{q^i}, \ c_i \in C \ \text{ for } i = 1, \ldots, p-1, \ c_p = k, \right.$$

$$\left. 0 \leqslant c_i \leqslant q - 1 \ \text{ for } i > p \right\}.\tag{3.6.5}$$

Lemma 3.6.2. *Set* (3.6.5) *is measurable and*

$$m(A_p) = (q - 1)^{p-1} q^{-p}.\tag{3.6.6}$$

Proof. For every $c_1 \ldots, c_{p-1} \in C$ we define the interval $P(c_1, \ldots, c_{p-1})$ as

$$P(c_1, \ldots, c_{p-1}) = \begin{cases} \left[\displaystyle\sum_{i=1}^{p-1} \frac{c_i}{q^i} + \frac{k}{q^p}, \ \sum_{i=1}^{p-1} \frac{c_i}{q^i} + \frac{k+1}{q^p} \right] & \text{if } k < q-1, \\[4mm] \left[\displaystyle\sum_{i=1}^{p-1} \frac{c_i}{q^i} + \frac{k}{q^p}, \ \sum_{i=1}^{p-2} \frac{c_i}{q^i} + \frac{c_{p-1}+1}{q^{p-1}} \right] & \text{if } k = q-1. \end{cases}$$

Then

$$A_p = \bigcup_{c_1, \ldots, c_{p-1} \in C} P(c_1, \ldots, c_{p-1}).\tag{3.6.7}$$

The length of each interval $P(c_1, \ldots, c_{p-1})$ is q^{-p}, their number is $(q-1)^{p-1}$, and the different intervals are disjoint. Thus (3.6.6) is a direct consequence of (3.6.7). $\qquad\square$

Theorem 3.6.4. *Let* $q \geqslant 3$ *and let*

$$A = \left\{ x \in [0,1] \mid x = \sum_{i=1}^{\infty} \frac{c_i}{q^i}, \ c_i \in C \right\}.$$

Then $A \in \mathcal{L}$ *and* $m(A) = 0$.

Proof. Put $A' = [0,1] \setminus A$. Then

$$A' = \bigcup_{p=1}^{\infty} A_p,$$

where A_p are given by (3.6.5). The sets A_p are pairwise disjoint and measurable, thus $A' \in \mathcal{L}$ and by Lemma 3.6.2

$$m(A') = \sum_{p=1}^{\infty} m(A_p) = \sum_{p=1}^{\infty} \frac{(q-1)^{p-1}}{q^p} = 1.$$

Hence also $A \in \mathcal{L}$ and $m(A) = 1 - m(A') = 0$. $\qquad\square$

Corollary 3.6.1. *The Cantor set* **C** *(cf. Exercise 2.11) has measure zero.*

3.7 Theorem of Steinhaus

The celebrated theorem of Steinhaus is the measure theoretical analogue of the theorem of S. Piccard (Theorem 2.9.1). Originally Steinhaus proved it (Steinhaus [297]) for measurable subsets of $\mathbb{R}$, but then it was extended to more general settings (cf., e.g., Kurepa [200], Kuczma-Kuczma [194], Sander [274], M. E. Kuczma [192], [193], Paganoni [251]). Here we will present two different proofs of Theorem 3.7.1. The first one (Kuczma-Kuczma [194]) is based on an idea of H. Kestelman [169] and will use the following

Lemma 3.7.1. *If A, B, $C \subset \mathbb{R}^N$ are measurable sets, $m(A) < \infty$, then*

$$|m(A \cap B) - m(A \cap C)| \leqslant m(B \div C),$$

where $B \div C = (B \setminus C) \cup (C \setminus B)$ is the symmetric difference of the sets B and C.

Proof. We have

$$m(A \cap B) - m(A \cap C) = m(A \cap B \cap C) + m(A \cap B \cap C') - m(A \cap B \cap C) - m(A \cap B' \cap C),$$

whence

$$|m(A \cap B) - m(A \cap C)| \leqslant m(A \cap B \cap C') + m(A \cap B' \cap C)$$
$$\leqslant m(B \cap C') + m(B' \cap C) = m(B \div C). \qquad \square$$

Theorem 3.7.1 (Theorem of Steinhaus). *Let A, $B \subset \mathbb{R}^N$ be arbitrary sets such that $m_i(A) > 0$, $m_i(B) > 0$. Then $\operatorname{int}(A + B) \neq \varnothing$.*

Proof. Every set A such that $m_i(A) > 0$ contains a compact subset A_0 such that $m(A_0) > 0$. So there exist compact sets $A_0 \subset A$ and $B_0 \subset B$, $m(A_0) > 0$, $m(B_0) > 0$. Since $A_0 + B_0 \subset A + B$, it is enough to prove the theorem for compact A, B of positive measure.

Assume that A, $B \subset \mathbb{R}^N$ are compact, $m(A) > 0$, $m(B) > 0$. For every $t \in \mathbb{R}^N$ put $B_t = t - B$. We define a function $\omega : \mathbb{R}^N \to \mathbb{R}$ as

$$\omega(t) = m(A \cap B_t) = \int_{\mathbb{R}^N} \chi_A(x) \chi_{B_t}(x) \, dx, \qquad (3.7.1)$$

where χ denotes the characteristic function (cf. 3.2). First we prove that the function ω is continuous.

Fix a $t \in \mathbb{R}^N$ and take an $\varepsilon > 0$. By Theorem 3.1.1 there exists an open set G_t such that $B_t \subset G_t$ and $m(G_t \setminus B_t) < \frac{1}{2}\varepsilon$. Thus $B_t \cap G_t' = \varnothing$, and both these sets being closed, their distance δ is a positive number. So for $h \in \mathbb{R}^N$ with[11] $|h| < \delta$ we have $B_{t+h} \subset G_t$, whence

$$m(B_{t+h} \setminus B_t) \leqslant m(G_t \setminus B_t) < \frac{1}{2}\varepsilon.$$

Similarly, since $B_{t+h} = B_t + h$, we have by Corollary 3.2.2 $m(B_{t+h}) = m(B_t)$, and

$$m(B_t \setminus B_{t+h}) \leqslant m(G_t \setminus B_{t+h}) = m(G_t) - m(B_{t+h})$$
$$= m(G_t) - m(B_t) = m(G_t \setminus B_t) < \frac{1}{2}\varepsilon.$$

Hence

$$m(B_{t+h} \div B_t) = m(B_{t+h} \setminus B_t) + m(B_t \setminus B_{t+h}) < \varepsilon \quad \text{for} \quad |h| < \delta.$$

For such h we have by Lemma 3.7.1

$$|\omega(t+h) - \omega(t)| = |m(A \cap B_{t+h}) - m(A \cap B_t)| \leqslant m(B_{t+h} \div B_t) < \varepsilon.$$

Thus ω is continuous at t, and $t \in \mathbb{R}^N$ being arbitrary, ω is continuous in $\mathbb{R}^N$.

Next we have by (3.7.1)

$$\int_{\mathbb{R}^N} \omega(t)\,\mathrm{d}t = \int_{\mathbb{R}^N} \chi_A(x)\Big(\int_{\mathbb{R}^N} \chi_{B_t}(x)\,\mathrm{d}t\Big)\,\mathrm{d}x = \int_{\mathbb{R}^N} \chi_A(x)\Big(\int_{\mathbb{R}^N} \chi_B(t-x)\,\mathrm{d}t\Big)\,\mathrm{d}x.$$

Making in the inner integral the change of variables $u = t - x$, we get hence

$$\int_{\mathbb{R}^N} \omega(t)\,\mathrm{d}t = \int_{\mathbb{R}^N} \chi_A(x)\Big(\int_{\mathbb{R}^N} \chi_B(u)\,\mathrm{d}u\Big)\,\mathrm{d}x$$
$$= \Big(\int_{\mathbb{R}^N} \chi_B(u)\,\mathrm{d}u\Big)\Big(\int_{\mathbb{R}^N} \chi_A(x)\,\mathrm{d}x\Big) = m(B)\,m(A) > 0.$$

This shows that $\omega \not\equiv 0$ in $\mathbb{R}^N$, and by (3.7.1) $\omega(t) \geqslant 0$ in $\mathbb{R}^N$. Consequently there exists a $t_0 \in \mathbb{R}^N$ such that $\omega(t_0) > 0$, and, by the continuity of ω, there exists an open neighbourhood U of t_0 such that $\omega(t) > 0$ for $t \in U$. Thus (cf. (3.7.1)) $m(A \cap B_t) > 0$ for $t \in U$, in particular, $A \cap B_t \neq \emptyset$ for $t \in U$. Thus for every $t \in U$ there exists an $a \in A$ such that $a \in B_t$, i.e., $a = t - b$ with a $b \in B$. In other words, $t = a + b \in A + B$, whence $U \subset A + B$ and $\mathrm{int}(A + B) \neq \emptyset$. $\qquad\square$

The other proof is due to J. H. B. Kemperman [165]. It is based on the following

Lemma 3.7.2. *Let A, $B \subset \mathbb{R}^N$ be measurable sets, and let A^*, B^* be the sets of their density points. If $a \in A^*$ and $b \in B^*$, then there exist positive numbers δ and η such that for $x, y \in \mathbb{R}^N$ such that $|x - a| < \delta$, $|y - b| < \delta$ we have*

$$m\left[(A - x) \cap (y - B)\right] \geqslant \eta.\tag{3.7.2}$$

---

[11] $|h|$ denotes the Euclidean norm of h. If $h = (h_1, \ldots, h_N)$, then $|h| = (\sum_{i=1}^{N} h_i^2)^{1/2}$.

Proof. The reader will easily verify that 0 is a density point of the sets $A - a$ and $B - b$. Fix an ε, $0 < \varepsilon < \frac{1}{2}$. There exists a cube Q centered at the origin and such that

$$m(C) \geqslant (1 - \varepsilon)\, m(Q)\,, \quad m(D) \geqslant (1 - \varepsilon)\, m(Q)\,, \qquad (3.7.3)$$

where $C = Q \cap (A - a)$, $D = Q \cap (B - b)$. We have $C \subset A - a$, $D \subset B - b$, whence $C + a \subset A$, $D + b \subset B$ and $C + a - x \subset A - x$, $y - (D + b) \subset y - B$. Hence

$$Q \cap (C + a - x) \cap (y - D - b) \subset (C + a - x) \cap (y - D - b) \subset (A - x) \cap (y - B)\,,$$

and

$$m\left[(A - x) \cap (y - B)\right] \geqslant m\left[Q \cap (C + a - x) \cap (y - D - b)\right]\,. \qquad (3.7.4)$$

Put

$$C_x = C + a - x\,, \quad D_y = y - D - b\,.$$

Since $(C_x \cap D_y) \cup C_x' \cup D_y' = \mathbb{R}^N$, we have

$$m(Q) \leqslant m(Q \cap C_x \cap D_y) + m(Q \cap C_x') + m(Q \cap D_y')\,,$$

and by (3.7.4)

$$m\left[(A - x) \cap (y - B)\right] \geqslant m(Q \cap C_x \cap D_y) \geqslant m(Q) - m(Q \cap C_x') - m(Q \cap D_y')\,. \quad (3.7.5)$$

Further, $(Q \cap C_x') \cup C_x = Q \cup C_x = Q \cup (Q' \cap C_x)$, and similarly $(Q \cap D_y') \cup D_y = Q \cup (Q' \cap D_y)$, whence

$$m(Q \cap C_x') + m(C_x) - m(Q) + m(Q' \cap C_x)\,, \quad m(Q \cap D_y') + m(D_y) = m(Q) + m(Q' \cap D_y)\,,$$

the sets in question being disjoint and measurable, whence by (3.7.3), since by Corollary 3.2.2 $m(C_x) = m(C)$ and $m(D_y) = m(D)$,

$$m(Q \cap C_x') = m(Q) - m(C_x) + m(Q' \cap C_x)$$
$$= m(Q) - m(C) + m(Q' \cap C_x) \leqslant \varepsilon m(Q) + m(Q' \cap C_x)\,,$$
$$m(Q \cap D_y') = m(Q) - m(D_y) + m(Q' \cap D_y)$$
$$= m(Q) - m(D) + m(Q' \cap D_y) \leqslant \varepsilon m(Q) + m(Q' \cap D_y)\,.$$

The cube Q is symmetric with respect to the origin, i.e., $-Q = Q$. Thus we have $C \subset Q$, $D \subset Q$, $-D \subset -Q \subset Q$. Let Q_x and Q_y be the cubes concentric with Q and such that $\operatorname{edge} Q_x = \operatorname{edge} Q + 2\,|a - x|$, $\operatorname{edge} Q_y = \operatorname{edge} Q + 2\,|b - y|$. Then $Q' \cap C_x \subset Q_x \setminus Q$, $Q' \cap D_y \subset Q_y \setminus Q$, whence $m(Q' \cap C_x) \leqslant m(Q_x \setminus Q) = m(Q_x) - m(Q)$, $m(Q' \cap D_y) \leqslant m(Q_y \setminus Q) = m(Q_y) - m(Q)$, and

$$m(Q \cap C_x') \leqslant \varepsilon m(Q) + m(Q_x) - m(Q)\,,$$

$$m(Q \cap D_y') \leqslant \varepsilon m(Q) + m(Q_y) - m(Q)\,.$$

The functions $\varphi(x) = m(Q_x) - m(Q)$, $\psi(y) = m(Q_y) - m(Q)$ approach zero as $x \to a$, $y \to b$. We have by (3.7.5)

$$m\left[(A - x) \cap (y - B)\right] \geqslant (1 - 2\varepsilon)m(Q) - \varphi(x) - \psi(y). \qquad (3.7.6)$$

Put $\eta = \dfrac{1}{2}(1 - 2\varepsilon)m(Q)$. There exists a $\delta > 0$ such that $\varphi(x) \leqslant \dfrac{1}{2}\eta$ and $\psi(y) \leqslant \dfrac{1}{2}\eta$ whenever $|x - a| < \delta$ and $|y - b| < \delta$. For such x, y (3.7.2) follows from (3.7.6). $\qquad \square$

Theorem 3.7.2. *Let A, $B \subset \mathbb{R}^N$ be measurable sets of positive measure, and let A^*, B^* be sets of their density points. Then the set $(A \cap A^*) + (B \cap B^*)$ is open and non-empty.*

Proof. By the Lebesgue density theorem $A \cap A^* \neq \varnothing$, $B \cap B^* \neq \varnothing$, so also $(A \cap A^*) + (B \cap B^*) \neq \varnothing$. Take a $d \in (A \cap A^*) + (B \cap B^*)$. Thus $d = a + b$, where, in particular, $a \in A^*$, $b \in B^*$. Choose η and δ according to Lemma 3.7.2 and take a $t \in \mathbb{R}^N$ such that $|d - t| < \delta$. Put

$$x = a, \quad y = t - a.$$

Then $|x - a| = 0 < \delta$, and $|y - b| = |t - a - b| = |t - d| < \delta$. By Lemma 3.7.2

$$m\left[(A - x) \cap (y - B)\right] \geqslant \eta. \qquad (3.7.7)$$

By Lebesgue density theorem the sets A and $A \cap A^*$ differ only by a set of measure zero, and similarly the sets B and $B \cap B^*$ differ only by a set of measure zero. Therefore also the sets $A \cap (t - B)$ and $(A \cap A^*) \cap [t - (B \cap B^*)]$ differ only by a set of measure zero, and hence they have equal measure. Thus (3.7.7) implies in virtue of Corollary 3.2.2

$$m\left((A \cap A^*) \cap [t - (B \cap B^*)]\right) = m\left[A \cap (t - B)\right] = m\left[(A - a) \cap (t - B - a)\right]$$
$$= m\left[(A - x) \cap (y - B)\right] \geqslant \eta.$$

Hence $(A \cap A^*) \cap [t - (B \cap B^*)] \neq \varnothing$, i.e., there exists an $a^* \in A \cap A^*$ such that $a^* \in t - (B \cap B^*)$, or $a^* = t - b^*$ with a $b^* \in B \cap B^*$. Thus $t = a^* + b^* \in (A \cap A^*) + (B \cap B^*)$.

The set

$$\left\{ t \in \mathbb{R}^N \mid |d - t| < \delta \right\} \qquad (3.7.8)$$

is a neighbourhood of d. The argument presented above shows that the set (3.7.8) is contained in $(A \cap A^*) + (B \cap B^*)$. Hence $(A \cap A^*) + (B \cap B^*)$ is open. $\qquad \square$

Theorem 3.7.1 is an easy consequence of Theorem 3.7.2. In fact, let A, $B \subset \mathbb{R}^N$ be arbitrary sets such that $m_i(A) > 0$, $m_i(B) > 0$. Similarly as previously we may assume that A, B are measurable and of positive measure. By Theorem 3.7.2 the set $(A \cap A^*) + (B \cap B^*)$ is open and non-empty, and clearly $(A \cap A^*) + (B \cap B^*) \subset A + B$, and hence $(A \cap A^*) + (B \cap B^*) \subset \mathrm{int}(A + B)$. Thus $\mathrm{int}(A + B) \neq \varnothing$.

3.8 Non-measurable sets

We conclude this chapter with a construction, following P. R. Halmos [129], of an example of a non-measurable set and of a saturated non-measurable set.

Lemma 3.8.1. *Let $\xi \in \mathbb{R}$ be an irrational number. Then the sets*

$$D = \{\, x \in \mathbb{R} \mid x = p + q\xi, \ p, q \in \mathbb{Z} \,\},$$
$$D_1 = \{\, x \in \mathbb{R} \mid x = p + q\xi, q \in \mathbb{Z}, \ p \in 2\mathbb{Z}, \},$$
$$D_2 = \{\, x \in \mathbb{R} \mid x = p + q\xi, q \in \mathbb{Z}, \ p \in 2\mathbb{Z} + 1 \}$$

are dense in $\mathbb{R}$.

Proof. For every $n \in \mathbb{N}$ there exists a unique $p_n \in \mathbb{Z}$ such that $p_n + n\xi \in (0,1)$. Fix a non-empty open interval $I \subset \mathbb{R}$. There exists a $k \in \mathbb{N}$ such that $m(I) > \dfrac{1}{k}$. Choose $k + 1$ distinct integers $n_1, \ldots, n_{k+1} \in \mathbb{N}$ and corresponding $p_{n_1}, \ldots, p_{n_{k+1}} \in \mathbb{Z}$ such that $x_i = p_{n_i} + n_i \xi \in (0, 1)$, $i = 1, \ldots, k + 1$. The points x_i are distinct, since ξ is irrational. So among the points x_i there are at least two, say x_r and x_s, such that $|x_r - x_s| < \dfrac{1}{k}$. Thus for some $l \in \mathbb{Z}$ we have

$$l(x_r - x_s) \in I .$$

But $l(x_r - x_s) = l(p_{n_r} - p_{n_s}) + l(n_r - n_s)\xi \in D$. Thus $D \cap I \neq \varnothing$ for every non-empty open interval $I \subset \mathbb{R}$, i.e., D is dense $\mathbb{R}$.

The proof for D_1 is similar. It is enough to replace the interval $(0, 1)$ by $(0, 2)$. The proof for D_2 follows from the equality $D_2 = D_1 + 1$. $\qquad\square$

Now we proceed to construct a non-measurable set. For any $x, y \in \mathbb{R}$ we write $x \sim y$ iff $x - y \in D$. The relation $\sim$ is an equivalence relation, so $\mathbb{R}$ may be split into a family of mutually disjoint equivalence classes; two numbers x, y belong to the same class if and only if $x \sim y$. By Theorem 1.8.3 there exists a set $B \subset \mathbb{R}$ which has exactly one point in common with every such class. So any two different members b_1, b_2 of B are in different classes, whence $b_1 - b_2 \notin D$. In other words $b_1 - b_2 \in D$, $b_1, b_2 \in B$, means $b_1 = b_2$. Hence $(B - B) \cap D = \{0\}$.

Suppose that $m_i(B) > 0$. By Corollary 3.2.2 $m_i(-B) = m_i(B) > 0$. By Theorem 3.7.1 $\operatorname{int}(B - B) = \operatorname{int}[B + (-B)] \neq \varnothing$. But then the set $B - B$ must contain points of D other than 0, since by Lemma 3.7.1 D is dense in $\mathbb{R}$. But as we have just seen, this is not the case. Consequently

$$m_i(B) = 0 . \tag{3.8.1}$$

Now suppose that $m_e(B) = 0$. By Corollary 3.2.2 $m_e(B+d) = 0$ for every $d \in D$. Since $0 \leqslant m_i(B + d) \leqslant m_e(B + d) = 0$, this actually means (cf. Theorem 3.1.3) that all the sets $B + d$, $d \in D$, are measurable and of measure zero.

Every $x \in \mathbb{R}$ belongs to an equivalence class with respect to $\sim$. If b is the only element of B lying in the same class, then $x \sim b$, i.e., $d = x - b \in D$. Hence

$x = b + d \in B + d$. Consequently

$$\bigcup_{d \in D} (B + d) = \mathbb{R}. \tag{3.8.2}$$

The union in (3.8.2) is countable (since D is countable), and thus $\mathbb{R}$, as a union of countably many sets of measure zero, would be of measure zero, which is not the case. Thus we must have

$$m_e(B) > 0. \tag{3.8.3}$$

Let us note that the same argument shows that the set B cannot be of the first category, and consequently it is of the second category.

By (3.8.1), (3.8.3), and Theorem 3.1.2, the set B is not Lebesgue measurable.

Now we define a set $A \subset \mathbb{R}$ as

$$A = B + D_1, \tag{3.8.4}$$

where B is the non-measurable set constructed above, and D_1 is defined in Lemma 3.7.1. Let $x \in A - A$ and suppose that $x \in D_2$. Since $x \in A - A$, there exist $b_1, b_2 \in B$ and $d_1, d_2 \in D_1$ such that $x = (b_1 + d_1) - (b_2 + d_2) = (b_1 - b_2) + (d_1 - d_2)$. Obviously $D_1 \subset D$ and $D_2 \subset D$, whence $d_1 - d_2 \in D$, and, as we have just supposed, $x \in D_2$, whence $x - (d_1 - d_2) \in D$ and $b_1 - b_2 = x - (d_1 - d_2) \in D$. But, as pointed out above, this implies that $b_1 = b_2$, whence $x = d_1 - d_2$. Now, $d_1 - d_2 \in D_1$, whereas $x \in D_2$, and evidently $D_1 \cap D_2 = \varnothing$. This contradiction shows that

$$(A - A) \cap D_2 = \varnothing. \tag{3.8.5}$$

Theorem 3.8.1. *The set A defined by (3.8.4) is saturated non-measurable.*

Proof. Relation (3.8.5) and Lemma 3.8.1 imply that $\text{int}(A - A) = \varnothing$, whence it follows from Theorem 3.7.1 that $m_i(A) = 0$. The relation $m_i(A') = 0$ results from Theorem 3.6.1 in view of (3.8.3) and of Lemma 3.8.1. □

Theorem 3.8.2. *The set A defined by (3.8.4) has property $(*)$ from 3.3.*

Proof. Suppose that a set $E \subset A$ is of the second category and with the Baire property. Then also $-E$ is of the second category and with the Baire property, since the function $f(x) = -x$ is a homeomorphism. We have $E - E = E + (-E)$, and so by Theorem 2.9.1 $\text{int}(E - E) \neq \varnothing$. Since $E \subset A$, we have $E - E \subset A - A$, whence also $\text{int}(A - A) \neq \varnothing$, which contradicts (3.8.5). Thus A cannot contain any set of the second category and with the Baire property. Neither A' can contain such a set, as results from Theorem 3.6.2, in view of (3.8.4), of the fact that (as pointed above) B is of the second category, and of Lemma 3.8.1. □

Exercises

1. Show that if $A \subset \mathbb{R}^N$ is a set such that $m_i(A) > 0$, then there exists a compact set $F \subset A$ such that $0 < m(F) < \infty$.

2. Let $A_1 \subset A_2 \subset \mathbb{R}^N$ be arbitrary sets, $m_i(A_2) < \infty$. Show that there exist measurable sets $K_1 \subset A_1$, $K_2 \subset A_2$ such that $K_1 \subset K_2$, $m(K_1) = m_i(A_1)$, $m(K_2) = m_i(A_2)$.

3. Let $A_n \subset \mathbb{R}^N$, $n \in \mathbb{N}$, be arbitrary sets such that $A_n \subset A_{n+1}$, $n \in \mathbb{N}$. Show that

$$m_e\left(\bigcup_{n=1}^{\infty} A_n\right) = \lim_{n \to \infty} m_e(A_n).$$

4. Let $A_n \subset \mathbb{R}^N$, $n \in \mathbb{N}$, be arbitrary sets such that $A_{n+1} \subset A_n$, $n \in \mathbb{N}$, $m_i(A_1) < \infty$. Show that

$$m_i\left(\bigcap_{n=1}^{\infty} A_n\right) = \lim_{n \to \infty} m_i(A_n).$$

5. Show by an example that there exist sets A_1, $A_2 \subset \mathbb{R}^N$ such that $A_1 \cap A_2 = \varnothing$ and $m_i(A_1) + m_i(A_2) < m_i(A_1 \cup A_2) \leqslant m_e(A_1 \cup A_2) < m_e(A_1) + m_e(A_2)$.

6. For every set $A \subset \mathbb{R}^N$ let A^* denote the set of all points of outer density of A. Show that if $A \subset \mathbb{R}^N$ is a saturated non-measurable set, and $E \subset \mathbb{R}^N$ is a measurable set, then $(E \cap A)^* = E^*$.

7. Show that there exist disjoint measurable sets A, $B \subset \mathbb{R}^N$ which are not Borel separable (cf. Exercise 2.8).
   [Hint: Take A to be a non-Borel subset of the Cantor set $\mathbf{C}$, and $B = \mathbf{C} \setminus A$].

8. Show that there exists a decomposition $\mathbb{R}^N = A \cup B$ such that A, $B \in \mathcal{L}$, $A \cap B = \varnothing$, $m(A) = 0$, and B is of the first category.
   [Hint: Let D be a countable dense subset of $\mathbb{R}^N$. Thus $m(D) = 0$. For every $n \in \mathbb{N}$ there exists an open set G_n such that $D \subset G_n$, $m(G_n) < 2^{-n}$. Take $A = \bigcap_{n=1}^{\infty} G_n$].

9. Show that if $f : (a,b) \to \mathbb{R}$ is a continuous real function of a single real variable, and $f'(x) = 0$ in (a,b) except for at most countably many points, then f is constant in (a,b).

10. Let $A \subset \mathbb{R}$ be symmetric with respect to zero (i.e., $-A = A$) and such that $A + A = A$. Show that either $A = \mathbb{R}$, or $A \in \mathcal{L}$ and $m(A) = 0$, or A is saturated non-measurable.

11. Let $\mathcal{A}$ be a collection of mutually disjoint measurable subsets of $\mathbb{R}^N$ of positive measure. Show that $\operatorname{card} \mathcal{A} \leqslant \aleph_0$.
    [Hint: Let $K_r = K(0,r)$ denote the open ball centered at the origin and with radius r. Given a $c > 0$, let $\mathcal{A}(c) = \{A \in \mathcal{A} \mid m(A) \geqslant c\}$, and let $\mathcal{A}_r = \{A \in \mathcal{A} \mid m(A \cap K_r) > 0\}$. Consider first the case where there exists an $r > 0$ such that $\bigcup \mathcal{A} \subset K_r$. Show that, for every $c > 0$, $\operatorname{card} \mathcal{A}(c) < \aleph_0$ and $\mathcal{A} = \bigcup_{n=1}^{\infty} \mathcal{A}(1/n)$. In the general case show that, for every $r > 0$, $\operatorname{card} \mathcal{A}_r \leqslant \aleph_0$ and $\mathcal{A} = \bigcup_{r=1}^{\infty} \mathcal{A}_r$].

Chapter 4

Algebra

4.1 Linear independence and dependence

Let F be a field (cf. 4.7), and let L be a set endowed with two operations: the addition of elements of L, and the multiplication of elements of L by elements of F such that $(L, +)$ is a commutative group (i.e., fulfils conditions (2.9.1)–(2.9.4); cf. 4.5), and moreover

$$1x = x$$

for every $x \in L$,

$$\alpha(\beta x) = (\alpha\beta)x$$

for every $\alpha, \beta \in F$, $x \in L$,

$$(\alpha + \beta)x = \alpha x + \beta x, \quad \alpha(x + y) = \alpha x + \alpha y$$

for every $\alpha, \beta \in F$, $x, y \in L$.

Then the quadruple $(L; F; +; \cdot)$ is called a *linear space* over the field F. Often we say that L is a linear space (over F), without making specific reference to the operations $+$ and $\cdot$. It follows from the group property of L that there exists an element $0 \in L$ such that $x + 0 = 0 + x = x$ for every $x \in L$, and for every $x \in L$ there exists an element $-x \in L$ such that $(-x) + x = x + (-x) = 0$. Instead of writing $x + (-y)$, we write simply $x - y$.

Elements $x_1, \ldots, x_n \in L$ are called *linearly dependent* (over F) iff there exist $\alpha_1, \ldots, \alpha_n \in F$, not all zero, such that[1] $\alpha_1 x_1 + \cdots + \alpha_n x_n = 0$. Elements $x_1, \ldots, x_n \in L$ are called *linearly independent* (over F) iff they are not linearly dependent, i.e., iff the equality $\alpha_1 x_1 + \cdots + \alpha_n x_n = 0$ implies that $\alpha_1 = \cdots = \alpha_n = 0$.

It follows that a singleton $x \in L$ is linearly dependent if and only if $x = 0$, and if a system $x_1, \ldots, x_n \in L$ is linearly dependent, then so is also every system containing $x_1, \ldots, x_n$. (In particular, any linearly independent system cannot contain zero.) For if there exist $\alpha_1, \ldots, \alpha_n \in F$, not all zero, such that

$$\alpha_1 x_1 + \cdots + \alpha_n x_n = 0,$$

---

[1] Every expression $\alpha_1 x_1 + \cdots + \alpha_n x_n$, $x_i \in L$, $\alpha_i \in F$, $i = 1, \ldots, n$, is called the linear combination of $x_1, \ldots, x_n$ with the coefficients $\alpha_1, \ldots, \alpha_n$.

then for any $y_1, \ldots, y_k \in L$ we have

$$\alpha_1 x_1 + \cdots + \alpha_n x_n + 0 y_1 + \cdots + 0 y_k = 0,$$

and $\alpha_1, \ldots, \alpha_n, 0, \ldots, 0 \in F$ are not all zeros.

This last property gives rise to the following definition. A set $B \subset L$ is called *linearly independent* (over F) iff the elements of every finite subset of B are linearly independent. A set $B \subset L$ is called *linearly dependent* (over F) iff it is not linearly independent, i.e., iff B contains a finite subset whose elements are linearly dependent.

An important example of a linear space is $(\mathbb{R}^N; \mathbb{R}; +; \cdot)$ with customary addition of elements of $\mathbb{R}^N$ and customary multiplication of elements of $\mathbb{R}^N$ by real numbers. Actually, for every field $F \subset \mathbb{R}$, the quadruple $(\mathbb{R}^N; F; +; \cdot)$ is a linear space. In the sequel of this book an important role will be played by the linear space $(\mathbb{R}^N; \mathbb{Q}; +; \cdot)$.

If $(L; F; +; \cdot)$ is a linear space, and $L_0 \subset L$ is such that $(L_0; F; +; \cdot)$ (with the same addition and multiplication by scalars) itself is a linear space, then L_0 is called a subspace of L. If $A \subset L$ is an arbitrary non-empty set, then by $E(A)$ we denote the linear subspace of L spanned by A, i.e., the set[2]

$$E(A) = \{x \in L \mid x = \alpha_1 a_1 + \cdots + \alpha_n a_n, \ \alpha_1, \ldots, \alpha_n \in F, \ a_1, \ldots, a_n \in A, \ n \in \mathbb{N}\}$$

of all the finite linear combinations of elements of A (with coefficients from F). In other sources the set $E(A)$ is also denoted by $\operatorname{Span} A$, $\operatorname{Lin} A$.

The reader will easily check the following properties of the operation E. Here $(L; F; +; \cdot)$ is a linear space.

(*i*) For every $A \subset L$ we have $A \subset E(A) \subset L$.

(*ii*) If $A \subset B \subset L$, then $E(A) \subset E(B)$.

(*iii*) $E(L) = L$.

Further properties are contained in the following lemmas.

Lemma 4.1.1. *If $(L; F; +; \cdot)$ is a linear space and $A \subset L$ is a linearly independent set, then every $x \in E(A)$ has a representation, unique up to the terms with coefficients zero,*

$$x = \alpha_1 a_1 + \cdots + \alpha_n a_n, \ \ \alpha_i \in F, \ a_i \in A, \ i = 1, \ldots, n. \qquad (4.1.1)$$

Proof. It follows from the definition of $E(A)$ that every $x \in E(A)$ has a representation (4.1.1). Suppose that x has also another representation

$$x = \beta_1 b_1 + \cdots + \beta_m b_m, \ \ \beta_i \in F, \ b_i \in A, \ i = 1, \ldots, m. \qquad (4.1.2)$$

We may assume that $m = n$ and $a_i = b_i$, $i = 1, \ldots, n$, adding in (4.1.1) and (4.1.2) the lacking terms with coefficients zero, and changing, if necessary, the numeration of a_i's and b_i's. Hence

$$\alpha_1 a_1 + \cdots + \alpha_n a_n = \beta_1 a_1 + \cdots + \beta_n a_n,$$

---

[2] We will write $E_F(A)$ instead of $E(A)$ if a question may arise which field is involved. From Chapter 5 on $E(A)$ will always mean $E_{\mathbb{Q}}(A)$.

or

$$(\alpha_1 - \beta_1)a_1 + \cdots + (\alpha_n - \beta_n)a_n = 0. \tag{4.1.3}$$

Since A is linearly independent and $\{a_1, \ldots, a_n\} \subset A$, (4.1.3) implies that $\alpha_i - \beta_i = 0$, i.e., $\alpha_i = \beta_i$, $i = 1, \ldots, n$. Thus representations (4.1.1) and (4.1.2) may differ only by the added terms with coefficients zero. $\square$

Lemma 4.1.2. *Let $(L; F; +; \cdot)$ be a linear space. If $A \subset L$ is linearly independent, and $a \in L \backslash A$, then the set $A \cup \{a\}$ is linearly independent if and only if $a \in L \backslash E(A)$.*

Proof. Suppose that $A \cup \{a\}$ is linearly dependent. Then there exists a finite set $\{a_1, \ldots, a_n\} \subset A \cup \{a\}$ and $\alpha_i \in F$, $i = 1, \ldots, n$, such that

$$\alpha_1 a_1 + \cdots + \alpha_n a_n = 0.$$

If we had $a_i \in A$ for $i = 1, \ldots, n$, then A itself would be linearly dependent. So one of a_i's, say a_1, equals a. If we had $\alpha_1 = 0$, then we are back in the previous situation. So $\alpha_1 \neq 0$ and

$$a = a_1 = -\frac{\alpha_2}{\alpha_1} a_2 - \cdots - \frac{\alpha_n}{\alpha_1} a_n,$$

which means that $a \in E(A)$.

Now, if $a \in E(A)$, then

$$a = \alpha_1 a_1 + \cdots + \alpha_n a_n, \ \alpha_i \in F, \ a_i \in A, \ i = 1, \ldots, n,$$

and

$$a - \alpha_1 a_1 - \cdots - \alpha_n a_n = 0.$$

Not all coefficients in the above linear combination are zeros (the coefficient of a is 1). Consequently the elements of the finite set $\{a, a_1, \ldots, a_n\} \subset A \cup \{a\}$ are linearly dependent, and hence also $A \cup \{a\}$ is linearly dependent. $\square$

Lemma 4.1.3. *Let $A \subset \mathbb{R}^N$ be a finite or countable set. Then the set $E(A) = E_{\mathbb{Q}}(A)$ is countable.*

Proof. For every $n \in \mathbb{N}$, let $\mathcal{H}_n$ be the collection of all finite subsets K of A such that $\operatorname{card} K = n$. If $n > \operatorname{card} A$, we assume $\mathcal{H}_n = \varnothing$. For $K \in \mathcal{H}_n$ the set $E(K)$ may be identified with $\mathbb{Q}^n$, and hence $\operatorname{card} E(K) = \aleph_0$. On the other hand, we have $\operatorname{card} \mathcal{H}_n \leqslant (\operatorname{card} A)^n$, or $\operatorname{card} \mathcal{H}_n = 0$ whenever $\mathcal{H}_n = \varnothing$, so every $\mathcal{H}_n$ is finite or countable, and so is also $\mathcal{H} = \bigcup_{n=1}^{\infty} \mathcal{H}_n$. For every set $K \in \mathcal{H}$ there exists an $n \in \mathbb{N}$ such that $K \subset \mathcal{H}_n$, whence $\operatorname{card} E(K) \leqslant \aleph_0$, as observed above. The reader will easily check the formula

$$E(A) = \bigcup_{K \in \mathcal{H}} E(K). \tag{4.1.4}$$

Thus (4.1.4) is a finite or countable union of countable sets, and therefore $E(A)$ is countable.

$\square$

Let $(L; F; +; \cdot)$ be a linear space over a field $F \subset \mathbb{R}$. A set $C \subset L$ is called a *cone* iff $x + y \in C$ for all $x, y \in C$, and $\alpha x \in C$ for all $x \in C$, $\alpha \in F$, $\alpha \geqslant 0$. For an arbitrary set $A \subset L$ the symbol $E^+(A)$ will denote the cone spanned by A, i.e., the set

$$E^+(A) = \{\, x \in L \mid x = \alpha_1 a_1 + \cdots + \alpha_n a_n,\ a_1, \ldots, a_n \in A,\ \alpha_1, \ldots, \alpha_n \in F,$$
$$\alpha_1 \geqslant 0, \ldots, \alpha_n \geqslant 0,\ n \in \mathbb{N}\}$$

of all the finite linear combinations of elements of A with non-negative coefficients from F.

4.2 Bases

Let $(L; F; +; \cdot)$ be a linear space. A set $B \subset L$ is called a *base* of L iff

(i) B is linearly independent;

(ii) $E(B) = L$.

Similarly, if $C \subset L$ is a cone, then a set $B \subset C$ is called a *cone-base* of C if B fulfils (i) and

(iii) $E^+(B) = C$.

As we shall see later, only very few cones have a base, whereas we have the following

Lemma 4.2.1. *Let $(L; F; +; \cdot)$ be a linear space, and let $A \subset L$ be a linearly independent set. Let $C \subset L$ be an arbitrary set such that $A \subset C$. Then there exists a linearly independent set B such that $A \subset B \subset C$ and $E(B) = E(C)$.*

Proof. Let $\mathcal{R}$ be the family of sets

$$\mathcal{R} = \{\, D \subset L \mid A \subset D \subset C \text{ and } D \text{ is linearly independent}\,\}.$$

$A \in \mathcal{R}$, so $\mathcal{R} \neq \varnothing$. $(\mathcal{R}, \subset)$ is an ordered set. Let $\mathcal{Z} \subset \mathcal{R}$ be any chain, and put $E = \bigcup \mathcal{Z}$. Then E is linearly independent. In fact, let $\{a_1, \ldots, a_n\} \subset E$ be an arbitrary finite set. Then, since $\mathcal{Z}$ is a chain, there exists a set $D_0 \in \mathcal{Z}$ such that $\{a_1, \ldots, a_n\} \subset D_0$, and consequently $a_1, \ldots, a_n$ are linearly independent. Clearly $A \subset E \subset C$ so that $E \in \mathcal{R}$. Thus E is an upper bound of $\mathcal{Z}$ in $\mathcal{R}$.

By Theorem 1.8.1 there exists in $\mathcal{R}$ a maximal element B. Thus B is linearly independent, $A \subset B \subset C$. Hence $E(B) \subset E(C)$. Suppose that $E(B) \neq E(C)$. If we had $C \subset E(B)$, then $E(C) \subset E\big(E(B)\big) = E(B)$, and $E(C) = E(B)$, which we have supposed untrue. Consequently there exists an $a \in C \backslash E(B) \subset L \backslash B$. By Lemma 4.1.2 the set $B \cup \{a\}$ is linearly independent, and, since $a \in C$, we have $A \subset B \cup \{a\} \subset C$. Thus $B \cup \{a\} \in \mathcal{R}$ and is strictly larger than B, which contradicts the fact that B is maximal in $\mathcal{R}$. Consequently we must have $E(B) = E(C)$. $\qquad\square$

Taking $C = L$ we have hence

Theorem 4.2.1. *If $(L; F; +; \cdot)$ is a linear space, and $A \subset L$ is a linearly independent set, then there exists a base B of L such that $A \subset B$.*

Corollary 4.2.1. *Every linear space $L \neq \{0\}$ has a base.*

To see this take in Theorem 4.2.1 $A = \{a\}$ with an arbitrary $a \in L$, $a \neq 0$.
Two further important corollaries of Lemma 4.2.1 run as follows.

Corollary 4.2.2. *Let $L \neq \{0\}$ be a linear space. Then every set $C \subset L$ such that $E(C) = L$ contains a base of L.*

Corollary 4.2.3. *If $(L; F; +; \cdot)$ is a linear space, and $L_0 \subset L$ is a linear subspace of L, then for every base B_0 of L_0 there exists a base B of L such that $B_0 \subset B$.*

In other words, every base of a subspace can be extended to a base of the entire space.

In proving Lemma 4.2.1 we have used Theorem 1.8.1, and so the Axiom of Choice. As has been shown by J. D. Halpern [133], the statement in Lemma 4.2.1 is actually equivalent to the Axiom of Choice (cf. also Läuchli [205]).

There is no uniqueness attached to bases of a linear space. One linear space has many different bases. But there is an important relation between all those bases.

Theorem 4.2.2. *Let $(L; F; +; \cdot)$ be a linear space. Then any two bases of L are equipollent.*

Proof. Let $B \subset L$ and $C \subset L$ be two bases of L. Consider the family Φ of all functions φ whose domain D_φ is contained in B, range R_φ is contained in C, the set $R_\varphi \cup (B \setminus D_\varphi)$ is linearly independent[3], and φ is one-to-one. For $\varphi_1, \varphi_2 \in \Phi$ we introduce the relation $\leqslant$: $\varphi_1 \leqslant \varphi_2$ iff $D_{\varphi_1} \subset D_{\varphi_2}$ and $\varphi_2 \mid D_{\varphi_1} = \varphi_1$ (cf. Exercise 1.4). $(\Phi, \leqslant)$ is an ordered set. Fix an $x_0 \in B$. We have $C \setminus E(B \setminus \{x_0\}) \neq \varnothing$. In fact, if we had $C \subset E(B \setminus \{x_0\})$, then $L = E(C) \subset E(E(B \setminus \{x_0\})) = E(B \setminus \{x_0\}) \subset E(B) = L$, whence $E(B \setminus \{x_0\}) = L$. This implies that $x_0 \in E(B \setminus \{x_0\})$, whence by Lemma 4.1.2 the set $B = (B \setminus \{x_0\}) \cup \{x_0\}$ would be linearly dependent.

Consequently there exists a $y_0 \in C \setminus E(B \setminus \{x_0\})$. We define a transform φ : $\{x_0\} \rightarrow \{y_0\}$ by putting $\varphi(x_0) = y_0$. Then $D_\varphi = \{x_0\} \subset B$, $R_\varphi = \{y_0\} \subset C$, and $R_\varphi \cup (B \setminus D_\varphi) = \{y_0\} \cup (B \setminus \{x_0\})$ is linearly independent in virtue of Lemma 4.1.2. Clearly φ is one-to-one. Thus $\varphi \in \Phi$, i.e., $\Phi \neq \varnothing$.

Now let $\mathcal{Z} \subset \Phi$ be any chain. Put $D = \bigcup_{\varphi \in \mathcal{Z}} D_\varphi$ and define a function $\varphi_0 : D \rightarrow L$ by putting $\varphi_0(x) = \varphi(x)$ if $x \in D_\varphi$, $\varphi \in \mathcal{Z}$. This definition is correct, since if $x \in D_{\varphi_1} \cap D_{\varphi_2}$, $\varphi_1, \varphi_2 \in \mathcal{Z}$, then either $\varphi_1 \leqslant \varphi_2$, or $\varphi_2 \leqslant \varphi_1$. Let, e.g., $\varphi_1 \leqslant \varphi_2$. This means that $D_{\varphi_1} \subset D_{\varphi_2}$ and $\varphi_2 \mid D_{\varphi_1} = \varphi_1$ so that $\varphi_1(x) = \varphi_2(x)$.

Clearly $\varphi \leqslant \varphi_0$ for every $\varphi \in \mathcal{Z}$. Further we have $R_{\varphi_0} = \bigcup_{\varphi \subset \mathcal{Z}} R_\varphi$. Since for $\varphi \in \mathcal{Z}$ we have $D_\varphi \subset B$, $R_\varphi \subset C$, also $D_{\varphi_0} \subset B$, $R_{\varphi_0} \subset C$. We are going to show that the set $R_{\varphi_0} \cup (B \setminus D_{\varphi_0})$ is linearly independent.

Supposing the contrary, let $\{a_1, \ldots, a_n\} \subset R_{\varphi_0} \cup (B \setminus D_{\varphi_0})$ be linearly dependent. There exists a k, $0 \leqslant k \leqslant n$, such that $\{a_1, \ldots, a_k\} \subset R_{\varphi_0}$, $\{a_{k+1}, \ldots, a_n\} \subset B \setminus D_{\varphi_0}$. (If necessary, we may change the numeration of $a_1, \ldots, a_n$). Note that $\varphi_1 \leqslant \varphi_2$ implies

---

[3] In the present proof we understand under this fact also that $R_\varphi \cap (B \setminus D_\varphi) = \varnothing$.

$R_{\varphi_1} \subset R_{\varphi_2}$. Since $\mathcal{Z}$ is a chain, there exists a $\varphi \in \mathcal{Z}$ such that $a_1, \ldots, a_k \in R_\varphi$. Since $D_\varphi \subset D_{\varphi_0}$, we have $B \setminus D_{\varphi_0} \subset B \setminus D_\varphi$, and $a_{k+1}, \ldots, a_n \in B \setminus D_\varphi$. Hence $\{a_1, \ldots, a_n\} \subset R_\varphi \cup (B \setminus D_\varphi)$, which is incompatible with the linear independence of the latter. It is easy to check that φ_0 is one-to-one.

Consequently $\varphi_0 \in \Phi$ and is an upper bound of $\mathcal{Z}$ in Φ. By Theorem 1.8.1 there exists in Φ a maximal element $\hat\varphi$. We will show that $D_{\hat\varphi} = B$. Suppose that $D_{\hat\varphi} \neq B$. If we had $R_{\hat\varphi} = C$, then taking an $x_0 \in B \setminus D_{\hat\varphi} \neq \varnothing$ we would get that $C \cup \{x_0\} = R_{\hat\varphi} \cup \{x_0\}$ would be linearly independent, which, according to Lemma 4.1.2, is impossible, since $x_0 \in E(C) = L$. So $R_{\hat\varphi} \neq C$. There exists a $y_0 \in C \setminus R_{\hat\varphi}$. We must distinguish two cases.

1. $y_0 \notin E\big(R_{\hat\varphi} \cup (B \setminus D_{\hat\varphi})\big)$.

Take an $x_0 \in B \setminus D_{\hat\varphi}$. Define a function $\psi : D_{\hat\varphi} \cup \{x_0\} \to L$ by putting $\psi(x) = \hat\varphi(x)$ for $x \in D_{\hat\varphi}$, $\psi(x_0) = y_0$. Then $D_\psi = D_{\hat\varphi} \cup \{x_0\} \subset B$, $R_\psi = R_{\hat\varphi} \cup \{y_0\} \subset C$. Further, since $D_{\hat\varphi} \subset D_\psi$,

$$R_\psi \cup (B \setminus D_\psi) \subset R_{\hat\varphi} \cup \{y_0\} \cup (B \setminus D_{\hat\varphi}) = [R_{\hat\varphi} \cup (B \setminus D_{\hat\varphi})] \cup \{y_0\}.$$

By Lemma 4.1.2 the set $[R_{\hat\varphi} \cup (B \setminus D_{\hat\varphi})] \cup \{y_0\}$ is linearly independent, and hence also its subset $R_\psi \cup (B \setminus D_\psi)$ is linearly independent. Since $y_0 \notin R_{\hat\varphi}$, the function ψ is one-to-one. Hence $\psi \in \Phi$ and is strictly larger than $\hat\varphi$, which contradicts the maximality of $\hat\varphi$.

2. $y_0 \in E\big(R_{\hat\varphi} \cup (B \setminus D_{\hat\varphi})\big)$.

Then y_0 has a representation

$$y_0 = \sum_{i=1}^{k} \alpha_i t_i + \sum_{i=k+1}^{n} \alpha_i s_i, \tag{4.2.1}$$

$\alpha_i \in F$, $i = 1, \ldots, n$; $t_i \in R_{\hat\varphi}$, $i = 1, \ldots, k$; $s_i \in B \setminus D_{\hat\varphi}$, $i = k+1, \ldots, n$.

If we had $\alpha_{k+1} = \cdots = \alpha_n = 0$, then we would get by (4.2.1)

$$y_0 - \sum_{i=1}^{k} \alpha_i t_i = 0, \ y_0 \in C, \ t_i \in R_{\hat\varphi} \subset C, \ i = 1, \ldots, k,$$

i.e., C would be linearly dependent. So there must exist an i_0, $k + 1 \leqslant i_0 \leqslant n$, such that $\alpha_{i_0} \neq 0$. Put $x_0 = s_{i_0} \in B \setminus D_{\hat\varphi}$, and define a function $\psi : D_{\hat\varphi} \cup \{x_0\} \to L$ as above

$$\psi(x) = \hat\varphi(x) \text{ for } x \in D_{\hat\varphi}, \ \psi(x_0) = y_0.$$

As in case 1 $D_\psi \subset B$, $R_\psi \subset C$, ψ is one-to-one. Suppose that the set $R_\psi \cup (B \setminus D_\psi)$ is linearly dependent. But $R_{\hat\varphi} \cup (B \setminus D_\psi) \subset R_{\hat\varphi} \cup (B \setminus D_{\hat\varphi})$, and since the latter set is linearly independent, also its subset $R_{\hat\varphi} \cup (B \setminus D_\psi)$ is linearly independent. Since $R_\psi = R_{\hat\varphi} \cup \{y_0\}$, we have

$$R_\psi \cup (B \setminus D_\psi) = [R_{\hat\varphi} \cup (B \setminus D_\psi)] \cup \{y_0\},$$

whence by Lemma 4.1.2

$$y_0 \in E\big(R_{\hat{\varphi}} \cup (B \setminus D_\psi)\big). \tag{4.2.2}$$

But $x_0 = s_{i_0} \in D_\psi$ and $x_0 \notin R_{\hat{\varphi}}$ (in representation (4.2.1) only t_i's belong to $R_{\hat{\varphi}}$), so $x_0 \notin R_{\hat{\varphi}} \cup (B \setminus D_\psi)$. By (4.2.2) y_0 has a representation

$$y_0 = \sum_{i=1}^m \gamma_i b_i, \tag{4.2.3}$$

where $b_i \in R_{\hat{\varphi}} \cup (B \setminus D_\psi)$, and so $b_i \neq x_0$ for $i = 1, \ldots, m$. Thus the term $\alpha_{i_0} x_0$ occurs in representation (4.2.1), but not in representation (4.2.3), which contradicts Lemma 4.1.1, since $\alpha_{i_0} \neq 0$.

Consequently the set $R_\psi \cup (B \setminus D_\psi)$ is linearly independent, and since $y_0 \notin R_{\hat{\varphi}}$, the function ψ is one-to-one. Hence $\psi \in \Phi$ and is strictly larger that $\hat{\varphi}$, which contradicts the maximality of $\hat{\varphi}$.

Thus we have proved that $D_{\hat{\varphi}} = B$. Thus there exists a one-to-one function $\hat{\varphi} : B \to C$, which proves that $\operatorname{card} B \leqslant \operatorname{card} C$. Since the role of B and C is symmetric, we have also $\operatorname{card} C \leqslant \operatorname{card} B$. Hence

$$\operatorname{card} B = \operatorname{card} C$$

which means that B and C are equipollent. $\qquad\square$

Let $(L; F; +; \cdot)$ be a linear space, $L \neq \{0\}$. We define the dimension of L as

$$\dim L = \operatorname{card} B, \tag{4.2.4}$$

where B is an arbitrary base of L. Theorem 4.2.2 shows that definition (4.2.4) is correct (is independent of the choice of the base B). We assume that if $L = \{0\}$, then $\dim L = 0$ (for every field Γ).

The following lemma is an analogue of Lemma 4.1.3 for sets of higher cardinality.

Lemma 4.2.2. [4] *Let $A \subset \mathbb{R}^N$ be a set such that $\operatorname{card} A > \aleph_0$. Then $\operatorname{card} E_{\mathbb{Q}}(A) = \operatorname{card} A$.*

Proof. Write $E(A) = E_{\mathbb{Q}}(A)$, and $\operatorname{card} A = \mathfrak{n}$. Let $B \subset A$ be a base of $E(A)$ so that $E(B) = E(A)$ (cf. Corollary 4.2.2). Put $\mathfrak{m} = \operatorname{card} B$. We have $\mathfrak{m} > \aleph_0$, for $\mathfrak{m} \leqslant \aleph_0$ would imply by Lemma 4.1.3 $\operatorname{card} A \leqslant \operatorname{card} E(A) = \operatorname{card} E(B) = \aleph_0$. For every $n \in \mathbb{N}$ let $\mathcal{H}_n$ be the collection of all subsets K of B such that $\operatorname{card} K = n$. Put $\mathcal{H} = \bigcup_{n=1}^\infty \mathcal{H}_n$. For every $n \in \mathbb{N}$ we have $\operatorname{card} \mathcal{H}_n \leqslant \mathfrak{m}^n = \mathfrak{m}$, whence $\operatorname{card} \mathcal{H} \leqslant \aleph_0 \mathfrak{m} = \mathfrak{m}$. On the other hand, for every $K \in \mathcal{H}$ we have $\operatorname{card} E(K) = \aleph_0$. Since

$$E(B) = \bigcup_{K \in \mathcal{H}} E(K),$$

we have $\operatorname{card} E(B) \leqslant \mathfrak{m} \aleph_0 = \mathfrak{m}$, and $\operatorname{card} E(A) = \operatorname{card} E(B) \leqslant \mathfrak{m}$. Since $B \subset A \subset E(A)$, we get hence $\mathfrak{m} \leqslant \mathfrak{n} \leqslant \mathfrak{m}$, i.e., $\mathfrak{m} = \mathfrak{n}$. Hence $\operatorname{card} E(A) = \mathfrak{m} = \mathfrak{n} = \operatorname{card} A$. $\qquad\square$

---

[4] Tarski [314]

We will also need the following

Lemma 4.2.3. *Let $\{h_1, \ldots, h_N\} \subset \mathbb{R}^N$ be a base of the space $(\mathbb{R}^N; \mathbb{R}; +; \cdot)$. Then the set $E_{\mathbb{Q}}(\{h_1, \ldots, h_N\})$ is dense in $\mathbb{R}^N$.*

Proof. Take an $x \in \mathbb{R}^N$ and an $\varepsilon > 0$. x may be written as

$$x = x_1 h_1 + \cdots + x_N h_N, \; x_i \in \mathbb{R}, \; i = 1, \ldots, N.$$

To every x_i, $i = 1, \ldots, N$, there exists an $\alpha_i \in \mathbb{Q}$ such that $|x_i - \alpha_i| < \varepsilon/N \, |h_i|$. Put

$$y = \alpha_1 h_1 + \cdots + \alpha_N h_N \in E_{\mathbb{Q}}(\{h_1, \ldots, h_N\}).$$

Then

$$x - y = (x_1 - \alpha_1)h_1 + \cdots + (x_N - \alpha_N)h_N,$$

whence

$$|x - y| \leqslant \sum_{i=1}^{N} |x_i - \alpha_i| \, |h_i| < \varepsilon.$$

Thus arbitrarily close to every $x \in \mathbb{R}^N$ there is a $y \in E_{\mathbb{Q}}(\{h_1, \ldots, h_N\})$, which means that the latter set is dense in $\mathbb{R}^N$. $\qquad\square$

Lemma 4.2.4. *Let $L \subset \mathbb{R}^N$ be such that $(L; \mathbb{Q}; +; \cdot)$ is a linear subspace of $(\mathbb{R}^N; \mathbb{Q}; +; \cdot)$. If $m_e(L) > 0$, or L is of the second category, then L is dense in $\mathbb{R}^N$.*

Proof. Put $M = E_{\mathbb{R}}(L)$. Then $m_e(M) > 0$ or M is of the second category, since $L \subset M$. Hence $M = \mathbb{R}^N$, for lower dimensional subspaces of $\mathbb{R}^N$ over $\mathbb{R}$ (k-dimensional hyperplanes, $k < N$) are of measure zero and nowhere dense, and hence of the first category. By Corollary 4.2.2 L contains a base B of $\mathbb{R}^N$ over $\mathbb{R}$. By Lemma 4.2.3 $E_{\mathbb{Q}}(B)$ is dense in $\mathbb{R}^N$, and so is also L, since $E_{\mathbb{Q}}(B) \subset E_{\mathbb{Q}}(L) = L$. $\qquad\square$

Every base of the space $(\mathbb{R}^N; \mathbb{Q}; +; \cdot)$ will be referred to as a *Hamel*[5] *basis*[6].

Theorem 4.2.3. *Every Hamel basis has the power of continuum.*

Proof. Let $H \subset \mathbb{R}^N$ be a Hamel basis, and let $\operatorname{card} H = \mathfrak{n}$. By Lemma 4.2.2 and Lemma 4.1.3

$$\mathfrak{n} = \operatorname{card} H = \operatorname{card} E_{\mathbb{Q}}(H) = \operatorname{card} \mathbb{R}^N = \mathfrak{c}$$

Thus $\operatorname{card} H = \mathfrak{c}$. $\qquad\square$

Consequently the dimension of the space $(\mathbb{R}^N; \mathbb{Q}; +; \cdot)$ is $\mathfrak{c}$, whereas, as is well known, the space $(\mathbb{R}^N; \mathbb{R}; +; \cdot)$ has dimension N.

---

[5] After G. Hamel, who in a celebrated paper Hamel [134] first proved the existence of a base of the space $(\mathbb{R}; \mathbb{Q}; +; \cdot)$.

[6] Some authors use the term Hamel basis to denote any base of a linear space, as defined above, in distinction from other types of bases (e.g., a Schauder basis, etc.), where topology is involved. This use will not be followed in the present book.

4.3 Homomorphisms

Let $(L; F; +; \cdot)$ and $(M; F; +; \cdot)$ be linear spaces (over the same field F). A mapping $f : L \to M$ is called a *homomorphism* iff

$$f(x + y) = f(x) + f(y) \tag{4.3.1}$$

for arbitrary $x, y \in L$, and

$$f(\alpha x) = \alpha f(x) \tag{4.3.2}$$

for arbitrary $x \in L$, $\alpha \in F$.

It follows from (4.3.1) by induction that

$$f\left(\sum_{i=1}^{n} x_i\right) = \sum_{i=1}^{n} f(x_i) \tag{4.3.3}$$

for arbitrary $x_i \in L$, $i = 1, \ldots, n$, and $n \in \mathbb{N}$.

The general construction of homomorphisms $f : L \to M$ is described by the following

Theorem 4.3.1. *Let $(L; F; +; \cdot)$ and $(M; F; +; \cdot)$ be linear spaces, and let $B \subset L$ be a base of L. Then for every function $g : B \to M$ there exists a unique homomorphism $f : L \to M$ such that $f \mid B = g$.*

Proof. Take an arbitrary $x \in L$. By Lemma 4.1.1 x has a representation (unique up to terms with coefficients zero)

$$x = \sum_{i=1}^{n} \alpha_i b_i, \ \alpha_i \in F, \ b_i \in B, \ i = 1, \ldots, n. \tag{4.3.4}$$

For such an x we define $f(x)$ as[7]

$$f(x) = \sum_{i=1}^{n} \alpha_i g(b_i). \tag{4.3.5}$$

Thus the function f has been unambiguously defined in the whole of L.

If $x \in B$, then representation (4.3.4) takes the form

$$x = x,$$

whence by (4.3.5)

$$f(x) = g(x).$$

Thus $f \mid B = g$.

Now, besides point (4.3.4) take also a $y \in L$. y has a representation

$$y = \sum_{i=1}^{n} \beta_i b_i, \ \beta_i \in F, \ i = 1, \ldots, n, \tag{4.3.6}$$

---

[7] Note that adding in (4.3.4) terms with coefficients zero does not change x, nor $f(x)$.

because, adding in representations (4.3.4) and (4.3.6) terms with coefficients zero, we may assume that the number of summands is the same in (4.3.4) and (4.3.6), and (4.3.4) and (4.3.6) contain the same b_i's. Hence

$$x + y = \sum_{i=1}^{n} (\alpha_i + \beta_i) b_i.$$

By (4.3.5)

$$f(x + y) = \sum_{i=1}^{n} (\alpha_i + \beta_i) g(b_i) = \sum_{i=1}^{n} \alpha_i g(b_i) + \sum_{i=1}^{n} \beta_i g(b_i) = f(x) + f(y).$$

Thus f satisfies (4.3.1).

Take an arbitrary $\alpha \in F$. By (4.3.4)

$$\alpha x = \sum_{i=1}^{n} \alpha \alpha_i b_i,$$

where $\alpha \alpha_i \in F$. According to definition (4.3.5)

$$f(\alpha x) = \sum_{i=1}^{n} \alpha \alpha_i g(b_i) = \alpha \sum_{i=1}^{n} \alpha_i g(b_i) = \alpha f(x).$$

Thus f satisfies also (4.3.2) and, consequently, is a homomorphism.

Thus the desired homomorphism exists. To prove the uniqueness, let $f : L \to M$ be an arbitrary homomorphism such that

$$f(b) = g(b) \text{ for } b \in B. \tag{4.3.7}$$

Take an $x \in L$ with a representation (4.3.4). We have by (4.3.3), (4.3.2) and (4.3.7)

$$f(x) = f\Big(\sum_{i=1}^{n} \alpha_i b_i \Big) = \sum_{i=1}^{n} f(\alpha_i b_i) = \sum_{i=1}^{n} \alpha_i f(b_i) = \sum_{i=1}^{n} \alpha_i g(b_i).$$

Thus f must be given by (4.3.5), and so the function defined by (4.3.5) is the unique homomorphism $f : L \to M$ such that $f \mid B = g$. $\square$

Theorem 4.3.1 says that every function $g : B \to M$ can be extended onto L to a homomorphism $f : L \to M$, and the extension is unique. We prove also the following theorem about the extensions of homomorphisms.

Theorem 4.3.2. *Let $(L; F; +; \cdot)$ and $(M; F; +; \cdot)$ be linear spaces, and let L_0 be a linear subspace of L. Then every homomorphism $f_0 : L_0 \to M$ can be extended onto L to a homomorphism $f : L \to M$.*

Proof. Let $f_0 : L_0 \to M$ be a homomorphism, and let $B_0 \subset L_0$ be a base of L_0. By Corollary 4.2.3 there exists a base B of L such that $B_0 \subset B$. Put

$$g(b) = f_0(b) \text{ for } b \in B_0, \tag{4.3.8}$$

and define the function $g : B \to M$ arbitrarily on $B \setminus B_0$. By Theorem 4.3.1 g can be extended onto L to a homomorphism $f : L \to M$. Take an $x \in L_0$. Then x has a representation

$$x = \sum_{i=1}^{n} \alpha_i b_i, \ \alpha_i \in F, \ b_i \in B_0.$$

We get by (4.3.8), since f and f_0 are homomorphisms,

$$f(x) = f\left(\sum_{i=1}^{n} \alpha_i b_i\right) = \sum_{i=1}^{n} \alpha_i f(b_i) = \sum_{i=1}^{n} \alpha_i g(b_i) = \sum_{i=1}^{n} \alpha_i f_0(b_i) = f_0\left(\sum_{i=1}^{n} \alpha_i b_i\right) = f_0(x).$$

Thus $f \mid L_0 = f_0$, i.e., f is an extension of f_0. $\qquad\square$

The next theorem is slightly more sophisticated. If $f : L \to M$ is a homomorphism, the *kernel* of f is defined as

$$\operatorname{Ker} f = f^{-1}(0) = \{ x \in L \mid f(x) = 0 \}.$$

Lemma 4.3.1. *Let $(L; F; +; \cdot)$ and $(M; F; +; \cdot)$ be linear spaces, and let $f : L \to M$ be a homomorphism. Then $\operatorname{Ker} f$ is a linear subspace of L.*

Proof. The thing to show is that for $x, y \in \operatorname{Ker} f$ and $\alpha, \beta \in F$ we have $\alpha x + \beta y \in \operatorname{Ker} f$. Now, $x, y \in \operatorname{Ker} f$ means $f(x) = f(y) = 0$, whence by (4.3.1) and (4.3.2)

$$f(\alpha x + \beta y) = f(\alpha x) + f(\beta y) = \alpha f(x) + \beta f(y) = 0,$$

i.e., $\alpha x + \beta y \in \operatorname{Ker} f$. $\qquad\square$

Theorem 4.3.3. *Let $(L; F; +; \cdot)$ and $(M; F; +; \cdot)$ be linear spaces, and let L_0 be a linear subspace of L. Let B_0 be a base of L_0 and B a base of L such that $B_0 \subset B$. If $\dim M \geqslant \operatorname{card}(B \setminus B_0)$, then there exists a homomorphism $f : L \to M$ such that $\operatorname{Ker} f = L_0$.*

Proof. Let $C \subset M$ be a base of M. We have

$$\operatorname{card}(B \setminus B_0) \leqslant \dim M = \operatorname{card} C,$$

so there exists a one-to-one mapping $g : B \setminus B_0 \to C$. Put

$$g(b) = 0 \text{ for } b \in B_0. \tag{4.3.9}$$

Thus g is a mapping from B into M, and by Theorem 4.3.1 there exists a homomorphism $f : L \to M$ such that $f \mid B = g$.

Now take an $x \in \operatorname{Ker} f \subset L$, i.e., $f(x) = 0$. Let x have a representation

$$x = \sum_{i=1}^{k} \alpha_i t_i + \sum_{i=k+1}^{n} \alpha_i s_i, \tag{4.3.10}$$

where $\alpha_i \in F$, $i = 1, \ldots, n$; $t_i \in B_0$, $i = 1, \ldots, k$; $s_i \in B \setminus B_0$, $i = k+1, \ldots, n$. Then

$$0 = f(x) = \sum_{i=1}^{k} \alpha_i g(t_i) + \sum_{i=k+1}^{n} \alpha_i g(s_i).$$

By (4.3.9)

$$\sum_{i=k+1}^{n} \alpha_i g(s_i) = 0. \tag{4.3.11}$$

But since $g \mid B \setminus B_0$ is one-to-one, $g(s_i)$ are different members of C, and since the latter is linearly independent, elements $g(s_{k+1}), \ldots, g(s_n)$ also are linearly independent. Thus (4.3.11) implies

$$\alpha_i = 0, \quad i = k+1, \ldots, n,$$

whence (4.3.10) reduces to

$$x = \sum_{i=1}^{k} \alpha_i t_i, \quad \alpha_i \in F, \; t_i \in B_0, \; i = 1, \ldots, k. \tag{4.3.12}$$

Hence $x \in E(B_0) = L_0$, i.e., $\operatorname{Ker} f \subset L_0$.

On the other hand, let $x \in L_0$, and let x have a representation (4.3.12). Then

$$f(x) = \sum_{i=1}^{k} \alpha_i f(t_i) = \sum_{i=1}^{k} \alpha_i g(t_i) = 0$$

by (4.3.9). Thus $x \in \operatorname{Ker} f$, whence $L_0 \subset \operatorname{Ker} f$. Consequently $L_0 = \operatorname{Ker} f$. $\square$

Theorem 4.3.4. *Let $(L; F; +; \cdot)$ and $(M; F; +; \cdot)$ be linear spaces, and let L_0 be a linear subspace of L. Let B_0 be a base of L_0 and B a base of L such that $B_0 \subset B$. If $\dim M = \operatorname{card}(B \setminus B_0)$, then there exists a homomorphism $f : L \to M$ such that $\operatorname{Ker} f = L_0$, $f(L) = M$.*

Proof. We argue as in the proof of Theorem 4.3.3. Let $C \subset M$ be a base of M. Because of the equality

$$\operatorname{card}(B \setminus B_0) = \dim M = \operatorname{card} C$$

we may find a $g : B \setminus B_0 \to C$ which is one-to-one and onto. If we define g on B_0 by (4.3.9), then we can extend it onto L to a homomorphism $f : L \to M$.

The relation $\mathrm{Ker}\, f = L_0$ is shown in the proof of Theorem 4.3.3. Clearly $f(L) \subset M$. Now take a $y \in M$. It has a representation

$$y = \sum_{i=1}^{n} \beta_i c_i, \ \beta_i \in F, \ c_i \in C, \ i = 1, \ldots, n. \tag{4.3.13}$$

Since $g \mid B \setminus B_0$ is onto C, there exist $b_i \in B \setminus B_0$, $i = 1, \ldots, n$, such that

$$g(b_i) = c_i, \ i = 1, \ldots, n. \tag{4.3.14}$$

Put

$$x = \sum_{i=1}^{n} \beta_i b_i \in E(B \setminus B_0) \subset E(B) = L.$$

Then by (4.3.14) and (4.3.13), since f is a homomorphism,

$$f(x) = f\Big(\sum_{i=1}^{n} \beta_i b_i\Big) = \sum_{i=1}^{n} \beta_i f(b_i) = \sum_{i=1}^{n} \beta_i g(b_i) = \sum_{i=1}^{n} \beta_i c_i = y.$$

Thus $y \in f(L)$, whence $M \subset f(L)$, and ultimately $f(L) = M$. Thus f has the desired properties. $\qquad\square$

4.4 Cones

Let $(L; F; +; \cdot)$ and $(M; F; +; \cdot)$ be linear spaces (over the same field $F \subset \mathbb{R}$), and let $C \subset L$ be a cone. Let a function $f : C \to M$ fulfil condition (4.3.1) for arbitrary $x, y \in C$ $\big($and hence also (4.3.3) for arbitrary $x_1, \ldots, x_n \in C\big)$ and (4.3.2) for arbitrary $x \in C$, $\alpha \in F$, $\alpha \geqslant 0$. Such a function is called *additive and positively homogenous*. In the present section we are going to investigate the possibility of extending an additive and positively homogenous function onto L to a homomorphism from L into M (Kuczma [179]).

Lemma 4.4.1. *Let $(L; F; +; \cdot)$ and $(M; F; +; \cdot)$ be linear spaces over a field $F \subset \mathbb{R}$, and let $C \subset L$ be a cone such that*

$$E(C) = L. \tag{4.4.1}$$

Further, let $B \subset C$ be a base of L. Then any two additive and positively homogenous functions from C into M which coincide on B are identical on C.

Proof. Let $f : C \to M$ and $g : C \to M$ be additive and positively homogenous functions such that

$$f(b) = g(b) \ \text{for} \ b \in B. \tag{4.4.2}$$

Take an arbitrary $x \in C$. Then x has a representation

$$x = \sum_{i=1}^{n} \alpha_i b_i, \ \alpha_i \in F, \ b_i \in B, \ i = 1, \ldots, n. \tag{4.4.3}$$

Setting in (4.3.1) $x = y = 0$ we get $f(0) = 2f(0)$, whence $f(0) = 0$, and similarly $g(0) = 0$. So $f(0) = g(0)$. If $x \neq 0$, then some α_i's in (4.4.3) must be different from zero, and hence $\alpha_0 = \max_i |\alpha_i| > 0$. We put $\varrho = 1/2\alpha_0$. Then $1 + \varrho\alpha_i \in F$ and $1 + \varrho\alpha_i > 0$ for $i = 1, \ldots, n$, and consequently

$$\sum_{i=1}^{n} b_i + \varrho x = \sum_{i=1}^{n}(1 + \varrho\alpha_i)b_i \in C,$$

since $b_i \in C$, $i = 1, \ldots, n$. Further

$$f\left(\sum_{i=1}^{n} b_i + \varrho x\right) = \sum_{i=1}^{n} f(b_i) + \varrho f(x), \tag{4.4.4}$$

$$f\left(\sum_{i=1}^{n}(1 + \varrho\alpha_i)b_i\right) = \sum_{i=1}^{n}(1 + \varrho\alpha_i)f(b_i). \tag{4.4.5}$$

Relations (4.4.4) and (4.4.5) imply

$$f(x) = \sum_{i=1}^{n} \alpha_i f(b_i).$$

Similarly,

$$g(x) = \sum_{i=1}^{n} \alpha_i g(b_i).$$

Condition (4.4.2) now implies that $f(x) = g(x)$. $\square$

Theorem 4.4.1. *Let $(L; F; +; \cdot)$ and $(M; F; +; \cdot)$ be linear spaces over a field $F \subset \mathbb{R}$, and let $C \subset L$ be a cone fulfilling condition (4.4.1). Then every additive and positively homogenous function $f : C \to M$ can be uniquely extended onto L to a homomorphism $f_0 : L \to M$.*

Proof. In virtue of Corollary 4.2.2 and of condition (4.4.1) there exists a base B of L such that $B \subset C$. By Theorem 4.3.1 the function $f \mid B$ can be uniquely extended onto L to a homomorphism $f_0 : L \to M$. Since f and $f_0 \mid C$ both are additive and positively homogenous, and they coincide on B, we have $f_0 \mid C = f$ by Lemma 4.4.1. The uniqueness of the extension also results from Lemma 4.4.1. $\square$

Condition (4.4.1) can be released, if we renounce from the uniqueness of the extension.

Theorem 4.4.2. *Let $(L; F; +; \cdot)$ and $(M; F; +; \cdot)$ be linear spaces over a field $F \subset \mathbb{R}$, and let $C \subset L$ be a cone. Then every additive and positively homogenous function $f : C \to M$ can be extended onto L to a homomorphism $f_0 : L \to M$.*

Proof. By Theorem 4.4.1 there exists a unique homomorphism $f_1 : E(C) \to M$ such that $f_1 \mid C = f$. By Theorem 4.3.2 f_1 can be extended onto L to a homomorphism $f_0 : L \to M$. f_0 is the required extension of f. $\square$

4.5 Groups and semigroups

As was pointed out in 2.9, a *group* $(G, +)$ is a non-empty set G endowed with a binary inner operation[8] $+$ satisfying condition (2.9.1), (2.9.2) and (2.9.3). If, moreover, $(G, +)$ satisfies also condition (2.9.4), then it is said to be *commutative*, or *abelian*. A *semigroup* $(S, +)$ is a non-empty set S endowed with a binary inner operation $+$ satisfying conditions (2.9.1). Of course, every group is, in particular, a semigroup. If $(G, +)$ is a group, and $\varnothing \neq H \subset G$, and if $(H, +)$ (with the same operation $+$) is a group resp. semigroup, then $(H, +)$ is called a *subgroup* resp. *subsemigroup* of $(G, +)$. If $(G, +)$ is a group resp. semigroup, then the n-fold sum $(n \in \mathbb{N})$ $x + \cdots + x$ is usually denoted by nx. Often, when speaking about a group $(G, +)$, we omit the reference to the operation in G, and say simply: the group G.

Lemma 4.5.1. *Let $(G, +)$ be a group, and $\varnothing \neq H \subset G$. In order that $(H, +)$ be a group (a subgroup of G) it is necessary and sufficient that*

$$x - y \in H \tag{4.5.1}$$

for every $x, y \in H$.

Proof. Suppose that condition (4.5.1) is satisfied for every $x, y \in H$. The associativity of the operation $+$ in H (i.e., condition (2.9.1) for all $x, y, z \in H$) results from $H \subset G$ and from the fact that (2.9.1) holds for all $x, y, z \in G$. Taking an $x \in H$ and $y = x$, we get from (4.5.1) $0 \in H$, i.e., (2.9.2) is fulfilled in H. Finally, taking in (4.5.1) $x = 0$, we obtain that $-y \in H$ for every $y \in H$. This means that every $x \in H$ has in H the inverse element $-x$, i.e., condition (2.9.3) is fulfilled, too. Now, for arbitrary $u, v \in H$, also $-v \in H$, as has been just pointed out, and taking in (4.5.1) $x = u$, $y = -v$, we obtain $u + v \in H$, which means that $+$ restricted to H is an inner operation.

Conversely, if $(H, +)$ is a subgroup of $(G, +)$, then (4.5.1) results from (2.9.3) and from the fact that $+$ is an inner operation in H. $\square$

Lemma 4.5.2. *Let $(G, +)$ be a semigroup, and $\varnothing \neq S \subset G$. In order that $(S, +)$ be a semigroup (a subsemigroup of G) it is necessary and sufficient that*[9]

$$x + y \in S \tag{4.5.2}$$

for every $x, y \in S$.

Proof. Condition (4.5.2) guarantees that $+$ is an inner operation in S. The associativity of $+$ in S results from that in G, as in the proof of Lemma 4.5.1. Conversely, if $(S, +)$ is a semigroup, then (4.5.2) results from the fact that $+$ is an inner operation in S. $\square$

---

[8] Usually the operation in a group G is denoted by $+$ only in the case of commutative groups. For non-commutative groups the multiplicative notation is employed. However, because of the connection with additive functions, in the present book we use the additive notation also in non-commutative cases.
An operation in a set A is called *binary* iff it is defined for pairs of elements from A, and a binary operation in a set A is called *inner*, if the result of the operation performed on two elements from A is again an element of A.
[9] Condition (4.5.2) can equivalently be written as $S + S \subset S$.

Lemma 4.5.3. *Let* $(G, +)$ *be a group, and* $(H, +)$ *a subgroup of* $(G, +)$. *Then* $(H' = G \setminus H)$

$$H = -H, \ H' = -H', \ H + H' = H' + H = H'.$$

Proof. The relation $H = -H$ results from the fact that $(H, +)$ is a group. Take an $x \in H'$. If we had $-x \in H$, then we would have $x = -(-x) \in -H = H$, a contradiction. Thus $-x \in H'$ and $-H' \subset H'$. Hence $H' = -(-H') \subset -H'$ and so $-H' = H'$.

If $x \in H$ and $y \in H'$, then $x + y \in H'$, since otherwise we would have $y = (-x) + (x + y) \in -H + H = H + H = H$. Similarly, $y + x \in H'$, since otherwise we would have $y = (y + x) - x \in H - H = H + H = H$. Thus $H + H' \subset H'$ and $H' + H \subset H'$, whence[10] $H' \subset -H + H' = H + H'$ and $H' \subset H' - H = H' + H$. So $H + H' = H' + H = H'$.   □

Let $(G, +)$ be an arbitrary group, and $(H, +)$ a subgroup of $(G, +)$, $H \neq G$. We say that the *index* of H is 2 iff $x - y \in H$ for every $x, y \in H'$ (i.e., iff $H' - H' \subset H$).

Lemma 4.5.4. *Let* $(G, +)$ *be a group, and* $(H, +)$ *a subgroup of* $(G, +)$, $H \neq G$. *Then the index of* H *is* 2 *if and only if*

$$H' + H' \subset H. \qquad\qquad (4.5.3)$$

Proof. Let the index of H be 2. Take arbitrary $x, y \in H'$, and suppose that $x + y \in H'$. Then $x = (x + y) - y \in H' - H' \subset H$, since the index of H is 2. This contradiction shows that $x + y \in H$, i.e., (4.5.3) holds.

Now assume that (4.5.3) holds. By Lemma 4.5.3 $-H' = H'$ so that by (4.5.3) $H' - H' = H' + H' \subset H$ and the index of H is 2.   □

Corollary 4.5.1. *Let* $(G, +)$ *be a group, and* $(H, +)$ *a subgroup of* $(G, +)$, $H \neq G$. *Then the index of* H *is* 2 *if and only if* $H' + H' = H$.

Proof. By Lemma 4.5.3, $H + H' \subset H'$, whence $H \subset H' - H' = H' + H'$. Hence $H' + H' = H$ if and only if (4.5.3) holds.   □

As an example we mention the additive groups of integers $(\mathbb{Z}, +)$ and of even integers $(2\mathbb{Z}, +)$, where $+$ denotes the usual addition of numbers. $(2\mathbb{Z}, +)$ is a subgroup of $(\mathbb{Z}, +)$, and its index is 2. On the other hand, $(\mathbb{Q}, +)$ and $(\mathbb{R}, +)$ have no subgroups of index 2 (cf. Corollary 4.5.2 below).

Let $(G, +)$ be a group, and $n \in \mathbb{N}$. We say that the *division by n is performable* in G iff for every $y \in G$ there exists a unique $x \in G$ such that $nx = y$. In such a case this unique x is denoted by y/n.

If the division by n is performable in G for every $n \in \mathbb{N}$, then the group $(G, +)$ is called *divisible*. The additive groups $(\mathbb{Q}, +)$, $(\mathbb{R}, +)$, $(\mathbb{C}, +)$, $(\mathbb{R}^N, +)$ (where $+$ denotes the usual addition, as defined in the corresponding sets) are all divisible.

---

[10] If $A, B, C \subset G$ are arbitrary sets such that $A + B \subset C$, then $A \subset A + 0 \subset A + B - B \subset C - B$, and, similarly, $B \subset 0 + B \subset -A + A + B \subset -A + C$. On the other hand, in general it is not true that if $A + B = C$, then $A = C - B$.

Lemma 4.5.5. *Let $(G, +)$ be a group in which the division by 2 is performable. Then G has no subgroup of index 2.*

Proof. Every $x \in G$ can be written as $x = 2(x/2) \in 2G$. So $G \subset 2G$. The converse inclusion is evident so that $G = 2G$.

Now suppose that $H \neq G$ is a subgroup of index 2. Then $2H \subset H$, since $(H, +)$ is a group, and $2H' \subset H' + H' \subset H$ by (4.5.3). Now,

$$G = 2G = 2(H \cup H') = 2H \cup 2H' \subset H,$$

whereas $H \subset G$, since $(H, +)$ is a subgroup of $(G, +)$. Thus $H = G$, contrary to the supposition. □

Corollary 4.5.2. *No divisible group has a subgroup of index 2.*

Corollary 4.5.3. *The group $(\mathbb{R}^N, +)$ has no subgroup of index 2.*

Let $(G, +)$ be a group, and $(H, +)$ a subgroup of $(G, +)$. We say that the subgroup $(H, +)$ of $(G, +)$ is *normal* iff $-y + x + y \in H$ for every $x \in H$, $y \in G$, i.e., iff

$$-y + H + y \subset H \tag{4.5.4}$$

for every $y \in G$. Condition (4.5.4) can also be formulated as

$$-y + H + y = H \tag{4.5.5}$$

for every $y \in G$. In fact, since in (4.5.4) $y \in G$ is arbitrary, we may replace y by $-y$, arriving at $y + H - y \subset H$, i.e., $H \subset -y + H + y$ (for every y in G), which together with (4.5.4) yields (4.5.5).

Let $(H, +)$ be a group, and $S \subset H$ an arbitrary set. We say that a subgroup $(G, +)$ of $(H, +)$ is *generated* by S iff $(G, +)$ is the smallest subgroup of $(H, +)$ containing S, i.e., for every subgroup $(K, +)$ of $(H, +)$ such that $S \subset K$ we have

$$G \subset K. \tag{4.5.6}$$

(In this connection cf. Exercise 4.8.) Note that a set $S \subset H$ can generate only one subgroup of $(H, +)$. If $(\tilde{G}, +)$ is another subgroup of $(H, +)$ generated by S, then we must have, according to (4.5.6), $G \subset \tilde{G}$ and $\tilde{G} \subset G$, which yields $\tilde{G} = G$.

Theorem 4.5.1. *Let $(H, +)$ be a commutative group, and $(S, +)$ a subsemigroup of $(H, +)$. Then the group G generated by S is*

$$G = S - S. \tag{4.5.7}$$

Proof. Suppose that (4.5.7) holds. For every $x, y \in G$ there exist $u, v, w, z \in S$ such that $x = u - v$, $y = w - z$. Hence, since $S \subset H$, and $(H, +)$ is commutative

$$x - y = (u - v) - (w - z) = (u + z) - (v + w) \in S - S = G,$$

since by Lemma 4.5.2 $u + z \in S$, $v + w \in S$. It follows by Lemma 4.5.1 that $(G, +)$ is a subgroup of $(H, +)$. Now take an arbitrary $x \in S$. By Lemma 4.5.2 $2x \in S$, and hence, by (4.5.7),

$$x = 2x - x \in S - S = G.$$

Thus $S \subset G$. Let $(K, +)$ be an arbitrary subgroup of $(H, +)$ such that $S \subset K$. For every $x, y \in S$ we have $x - y \in K$, i.e., $S - S \subset K$. By (4.5.7) we get hence (4.5.6). Consequently $(G, +)$ is the group generated by S. $\square$

Lemma 4.5.6. *Let $(G, +)$ be a group, and let $(S, +)$ be a subsemigroup of $(G, +)$ such that for every $x \in G$, $x \neq 0$, either $x \in S$, or $-x \in S$ (or both). Then S generates G and (4.5.7) holds.*

Proof. Take an arbitrary $x \in G$, $x \neq 0$. If $x \in S$, then also $2x \in S$ (since S is a semigroup), and

$$x = 2x - x \in S - S. \tag{4.5.8}$$

If $x \notin S$, then $-x \in S$ and $-2x = 2(-x) \in S$. Hence

$$x = (-x) - (-2x) \in S - S. \tag{4.5.9}$$

Finally, with an arbitrary $y \in S$

$$0 = y - y \in S - S. \tag{4.5.10}$$

Relations (4.5.8)–(4.5.10) imply that $G \subset S - S$. On the other hand $S - S \subset G$, since $S \subset G$ and $(G, +)$ is a group. Hence we get (4.5.7).

If $(K, +)$ is an arbitrary subgroup of $(H, +)$ such that $S \subset K$, then $G = S - S \subset K$. Thus S generates G. $\square$

Let $(S, +)$ be a semigroup. We say that the *left cancellation law* holds in S iff

$$z + x = z + y \text{ implies } x = y$$

for arbitrary $x, y, z \in S$. We say that the *right cancellation law* holds in S iff

$$x + z = y + z \text{ implies } x = y$$

for arbitrary $x, y, z \in S$. We say that S is *cancellative* iff both, left and right cancellation laws hold in S. We say that S is *left reversible* iff

$$(x + S) \cap (y + S) \neq \varnothing$$

for arbitrary $x, y \in S$.

We quote here without proof the following Theorem of Øre [246] (cf. also Rees [265], Dubreil [73], and also Clifford-Preston [47]).

Theorem 4.5.2. *Let $(S, +)$ be a semigroup. There exists a group $(G, +)$ such that $(S, +)$ is a subsemigroup of $(G, +)$, S generates G, and (4.5.7) holds, if and only if S is left reversible and cancellative.*

Let $(G, +)$ be a group, and let $(H, +)$ be a subgroup of $(G, +)$. We may introduce in G the equivalence relation[11]

$$x \sim y \text{ iff } -x + y \in H. \tag{4.5.11}$$

---

[11] The verification of the fact that (4.5.11) actually is an equivalence relation is left to the reader.

The equivalence classes with respect to relation (4.5.11) are called the left cosets[12] with respect to H, and the coset generated by an $x \in G$ is denoted by $[x]$:

$$[x] = \{\, y \in G \mid x \sim y \,\} = \{\, y \in G \mid -x + y \in H \,\} = \{\, y \in G \mid y \in x + H \,\} = x + H.$$
$$(4.5.12)$$

In particular, we have

$$[0] = H. \qquad\qquad (4.5.13)$$

The set of all left cosets in G with respect to H will be denoted by G/H.

If $(H, +)$ is a normal subgroup of $(G, +)$, then in the set G/H we may define a binary operation as follows:

$$[x] + [y] = [x + y]. \qquad\qquad (4.5.14)$$

We must show that definition (4.5.14) does not depend on the choice of x and y, i.e., that if $[x] = [u]$ and $[y] = [v]$, then $[x+y] = [u+v]$. But $[x] = [u]$ and $[y] = [v]$ means $x \sim u$ and $y \sim v$, i.e., $-x + u \in H$, $-y + v \in H$. Put $a = -x + u \in H$. Thus $u = a + x$, and $-(x+y) + (u+v) = -y - x + u + v = -y + a + y + (-y + v) \in -y + H + y + H = H + H \subset H$. This means that $x + y \sim u + v$, i.e., $[x+y] = [u+v]$, which was to be shown.

Theorem 4.5.3. *Let $(G, +)$ be a group, and $(H, +)$ a normal subgroup of $(G, +)$. With the operation $+$ defined by (4.5.14), $(G/H, +)$ is a group.*

Proof. If $a, b, c \in G/H$, then there exist $x, y, z \in G$ such that $a = [x]$, $b = [y]$, $c = [z]$. Then

$$(a + b) + c = ([x] + [y]) + [z] = [x + y] + [z] = [(x + y) + z]$$
$$= [x + (y + z)] = [x] + [y + z] = [x] + ([y] + [z]) = a + (b + c).$$

Consequently condition (2.9.1) is satisfied.

Let $a \in G/H$ be arbitrary, and $x \in G$ be such that $a = [x]$. Then

$$a + [0] = [x] + [0] = [x + 0] = [x] = a.$$

Thus $[0] = H$ $\big($cf. (4.5.13)$\big)$ is the neutral element of $(G/H, +)$, which shows that condition (2.9.2) is satisfied.

Finally, for every $a \in G/H$, we have (with $x \in G$ such that $a = [x]$)

$$a + [-x] = [x] + [-x] = [x - x] = [0],$$

which shows that $-a = [-x]$ is the inverse element to a. Consequently, also condition (2.9.3) is satisfied. □

---

[12] The right cosets are the equivalence classes with respect to the relation

$$x \approx y \text{ iff } x - y \in H.$$

The left cosets and right cosets coincide if and only if the subgroup H is normal.

Let $(G, +)$ and $(H, +)$ be two arbitrary groups.[13] A mapping $f : G \to H$ is called a *homomorphism*[14] iff

$$f(x + y) = f(x) + f(y) \tag{4.5.15}$$

holds for all $x, y \in G$. The set

$$\operatorname{Ker} f = f^{-1}(0) = \{\, x \in G \mid f(x) = 0 \,\}$$

is called the *kernel* of f .

Lemma 4.5.7. *Let $(G, +)$ and $(H, +)$ be groups, and let $f : G \to H$ be a homomorphism. Put $K = \operatorname{Ker} f$. Then $(K, +)$ is a normal subgroup of $(G, +)$.*

Proof. Putting in (4.5.15) $x = y = 0$ we get $f(0) = 0$, whence (putting in (4.5.15) $y = -x$) $f(-x) = -f(x)$. Hence, replacing in (4.5.15) y by $-y$,

$$f(x - y) = f(x) - f(y). \tag{4.5.16}$$

Now, if $x, y \in K$, then by (4.5.16)

$$f(x - y) = f(x) - f(y) = 0 - 0 = 0,$$

i.e., $x - y \in K$. In virtue of Lemma 4.5.1, $(K, +)$ is a subgroup of $(G, +)$.
    Take arbitrary $x \in K$, $y \in G$. Then

$$f(-y + x + y) = f(-y) + f(x) + f(y) = f(-y) + f(y) = -f(y) + f(y) = 0.$$

Thus $-y + x + y \in K$, which means that K is normal.   □

Lemma 4.5.8. *Let $(G, +)$ and $(H, +)$ be groups, and let $f : G \to H$ be a homomorphism. Let $K = \operatorname{Ker} f$. Then for every $x \in G$*

$$[x] = \{\, y \in G \mid f(y) = f(x) \,\}, \tag{4.5.17}$$

where $[x]$ denotes the elements of G/K containing x.

Proof. Let $y \in [x]$. This means that $x \sim y$, i.e., $-x + y \in K$. In other words $f(-x + y) = 0$. Now, since f is a homomorphism

$$f(y) = f\bigl(x + (-x + y)\bigr) = f(x) + f(-x + y) = f(x). \tag{4.5.18}$$

Thus $y \in \{\, y \in G \mid f(y) = f(x) \,\}$. Conversely, let $f(y) = f(x)$. Then from (4.5.18) we get $f(-x + y) = 0$, whence $-x + y \in K$, i.e., $x \sim y$ and $y \in [x]$. Hence we get (4.5.17).   □

---

[13] In particular, $(H, +)$ need not be a subgroup of $(G, +)$, and so $+$ in G and in H may denote two quite unrelated operations.

[14] The word *homomorphism* is used to denote several distinct notions: homomorphism of linear spaces, homomorphism of groups (and later of rings and fields). In fact, all these notions are particular cases of one more general notion: homomorphism of algebraic structures. We presume that the reader is familiar with this situation, and we give here the definitions only for the sake of completeness. No ambiguities can possibly arise.

Corollary 4.5.4. *Let* $(G, +)$ *and* $(H, +)$ *be groups, and let* $f : G \to H$ *be a homomorphism. Let* $K = \operatorname{Ker} f$. *Then, for arbitrary* $[x]$, $[y] \in G/K$, $[x] = [y]$ *if and only if* $f(x) = f(y)$.

Let $(G, +)$ be a group, and let $(H, +)$ be an arbitrary subgroup of $(G, +)$. Let the mapping $\pi : G \to G/H$ be defined by

$$\pi(x) = [x], \quad x \in G \tag{4.5.19}$$

Mapping (4.5.19) is called the *canonical projection*, or (in the case, where the subgroup H is normal), the *canonical homomorphism* from G onto G/H. This last name is due to the following fact.

Lemma 4.5.9. *Let* $(G, +)$ *be a group, and let* $(H, +)$ *be a normal subgroup of* $(G, +)$. *Then the mapping* $\pi : G \to G/H$, *defined by* (4.5.19), *is a homomorphism.*

Proof. Take arbitrary $x, y \in G$. By (4.5.19) and (4.5.14)

$$\pi(x + y) = [x + y] = [x] + [y] = \pi(x) + \pi(y),$$

which shows that π is a homomorphism. $\qquad\qquad\qquad\qquad\qquad\qquad\qquad\qquad\square$

Let $(G, +)$ be a group, and let $(H, +)$ be an arbitrary subgroup of it. Every mapping $\xi : G/H \to G$ such that $\pi\big(\xi(a)\big) = a$ for every $a \in G/H$ is called a *lifting*[15] from G/H into G. If we take an arbitrary set $T \subset G$ which has exactly one element in common with every class in G/H, then the mapping $\xi : G/H \to G$ defined by

$$\xi(a) = T \cap a, \quad a \in G/H, \tag{4.5.20}$$

is a lifting from G/H into G. Conversely, if $\xi : G/H \to G$ is a lifting, then the set $T = \xi(G/H)$ has exactly one element in common with every class in G/H, and has property (4.5.20); the verification of these facts is left to the reader.

4.6 Partitions of groups

Let A and T be arbitrary sets. Every family $\{A_t\}_{t \in T}$ of subsets of A satisfying the conditions

$$\bigcup_{t \in T} A_t = A \tag{4.6.1}$$

and

$$A_{t'} \cap A_{t''} = \varnothing \text{ for } t' \neq t'', \ t', t'' \in T, \tag{4.6.2}$$

is called a *partition* of A. A partition $\{B_s\}_{s \in S}$ of a set $B \subset A$ is called a *subpartition* of a partition $\{A_t\}_{t \in T}$ of A iff the following condition is fulfilled:

(i) For every $s \in S$ there exists a $t \in T$ such that $B_s \subset A_t$.

It is easily seen that, by (4.6.2), this $t \in T$ is unique. Hence we have the following

Lemma 4.6.1. *Let* $B \subset A$ *be arbitrary non-empty sets, and let a partition* $\{B_s\}_{s \in S}$ *of B be a subpartition of a partition* $\{A_t\}_{t \in T}$ *of* A. *If, for any* $s \in S$ *and* $t \in T$, *we have* $B_s \cap A_t \neq \varnothing$, *then actually* $B_s \subset A_t$.

---

[15] There is no analogue of Lemma 4.5.9 for liftings.

Proof. By (i) there exists a $t' \in T$ such that $B_s \subset A_{t'}$. Then evidently $A_t \cap A_{t'} \neq \varnothing$, whence by (4.6.2) $t = t'$ and $A_t = A_{t'}$. Consequently $B_s \subset A_t$.   □

Let $\varnothing \neq B \subset A$. A partition $\{B_s\}_{s \in S}$ of B which is a subpartition of a partition $\{A_t\}_{t \in T}$ of A is called *semiselective* iff the following condition is fulfilled:

(ii) For every $t \in T$ and s, $s' \in S$, if $B_s \subset A_t$ and $B_{s'} \subset A_t$, then $B_s = B_{s'}$.

A partition $\{B_s\}_{s \in S}$ of B which is a subpartition of a partition $\{A_t\}_{t \in T}$ of A is called *selective* iff it is semiselective and

(iii) For every $t \in T$ there exists an $s \in S$ such that $B_s \subset A_t$.

Again, as a result of (ii), this $s \in S$ is unique.

Let $(G, +)$ be an arbitrary group, and let $(H, +)$ be a subgroup of $(G, +)$. Then clearly G/H (the family of the left cosets with respect to H) is a partition[16] of G. The following three theorems are due to A. Grząślewicz [123].

Theorem 4.6.1. *Let $(G, +)$ be a group, and $(H, +)$ its subgroup. Let $(G_0, +)$ be another subgroup of $(G, +)$, and let $(H_0, +)$ be a subgroup of $(H, +)$. Then H/H_0 is a subpartition of G/G_0 if and only if $H_0 \subset G_0$.*

Proof. Let $H_0 \subset G_0$. Let $[x]$ be a member of H/H_0, and $\langle x \rangle$ a member of G/G_0 generated by the same $x \in H \subset G$. Thus $u \in [x]$ means $-x+u \in H_0$. But then also $-x+u \in G_0$, which means that $u \in \langle x \rangle$. Consequently $[x] \subset \langle x \rangle$, i.e., condition (i) is satisfied.

Now suppose that for every $x \in H$ there exists a $y \in G$ such that $[x] \subset \langle y \rangle$. In particular, $[0] \subset \langle y \rangle$ for a certain $y \in G$. But then 0, which clearly belongs to $[0]$, belongs also to $\langle y \rangle$, which means that 0 and y are equivalent (with respect to G_0), or $\langle y \rangle = \langle 0 \rangle$. Hence $H_0 = [0] \subset \langle 0 \rangle = G_0$, which ends the proof.   □

Theorem 4.6.2. *Let $(G, +)$ be a group, and $(H, +)$ its subgroup. Let $(G_0, +)$ be another subgroup of $(G, +)$, and let $(H_0, +)$ be a subgroup of $(H, +)$. Then H/H_0 is a semiselective subpartition of G/G_0 if and only if*

$$H_0 = H \cap G_0. \tag{4.6.3}$$

Proof. Suppose that H/H_0 is a semiselective subpartition of G/G_0. Similarly as in the proof of Theorem 4.6.1, the elements of H/H_0 will be denoted by $[\,\cdot\,]$, whereas the elements of G/G_0 will be denoted by $\langle\,\cdot\,\rangle$. We clearly have $H_0 \subset H$, and by Theorem 4.6.1 $H_0 \subset G_0$, whence

$$H_0 \subset H \cap G_0. \tag{4.6.4}$$

Now take an arbitrary $x \in H \cap G_0$. Then, in particular, $x \in H$, so $x \in [x]$, and, on the other hand $x \in G_0 = \langle 0 \rangle$. Thus $[x] \cap \langle 0 \rangle \neq \varnothing$, and, by Lemma 4.6.1, $[x] \subset \langle 0 \rangle$. Further, $[0] = H_0 \subset H \cap G_0 \subset G_0 = \langle 0 \rangle$ (cf. (4.6.4)). Condition (ii) then implies that $[x] = [0] = H_0$, i.e., $x \in H_0$. Hence

$$H \cap G_0 \subset H_0,$$

which together with (4.6.4) implies (4.6.3).

---

[16] However, G/H cannot be indexed as $\{[x]\}_{x \in G}$, for then it could happen that $[x] = [x']$ for $x \neq x'$. We may take as T an arbitrary set which has exactly one element in common with every class contained in G/H, and writing $A_t = [t]$ consider G/H as $\{A_t\}_{t \in T}$.

Now suppose that condition (4.6.3) holds. Hence $H_0 \subset G_0$, and according to Theorem 4.6.1 H/H_0 is a subpartition of G/G_0. Now take arbitrary $x, y \in H$ and $z \in G$ such that $[x] \subset \langle z \rangle$ and $[y] \subset \langle z \rangle$. In order to prove (ii) we need to show that then $[x] = [y]$. But $[x] = x + H_0$ and $[y] = y + H_0$ (cf. (4.5.12)). Thus $x + H_0 \subset \langle z \rangle$ and $y + H_0 \subset \langle z \rangle$. Since $(H_0, +)$ is a group, $0 \in H_0$, and so, in particular, $x \in \langle z \rangle$, and $y \in \langle z \rangle$, which means that x and y are equivalent with respect to G_0, i.e., $-x + y \in G_0$. Since $x, y \in H$, also $-x + y \in H$, and we get by (4.6.3) $-x + y \in H_0$. Thus x and y are equivalent with respect to H_0, i.e., $[x] = [y]$. □

Theorem 4.6.3. *Let $(G, +)$ be a group, and $(H, +)$ its subgroup. Let $(G_0, +)$ be another subgroup of $(G, +)$, and let $(H_0, +)$ be a subgroup of $(H, +)$. Then H/H_0 is a selective subpartition of G/G_0 if and only if relation (4.6.3) and the relation*

$$G = H + G_0 \tag{4.6.5}$$

hold.

Proof. Let H/H_0 be a selective subpartition of G/G_0. Then, in particular, H/H_0 is a semiselective subpartition of G/G_0, and in virtue of Theorem 4.6.2 condition (4.6.3) holds. Since $H \subset G$, $G_0 \subset G$, and $(G, +)$ is a group, we have

$$H + G_0 \subset G + G = G. \tag{4.6.6}$$

In order to prove the converse inclusion, let z be an arbitrary element of G. We continue to employ the notation $[\,\cdot\,]$, $\langle\,\cdot\,\rangle$, to denote elements of H/H_0 and G/G_0, respectively. Thus we have $z \in \langle z \rangle$. H/H_0 is a selective subpartition of G/G_0, therefore by (iii) there exists an $x \in H$ such that $[x] \subset \langle z \rangle$. Since $x \in [x]$, we have $x \in \langle z \rangle$, i.e., x and z are equivalent with respect to G_0, which means that $-x + z \in G_0$. Put $y = -x + z$ so that $y \in G_0$. We have $z = x + y \in H + G_0$. This shows that $G \subset H + G_0$, which together with (4.6.6) implies (4.6.5).

Conversely, suppose that conditions (4.6.3) and (4.6.5) are fulfilled. By Theorem 4.6.2 H/H_0 is a semiselective subpartition of G/G_0. Take an arbitrary class $\langle z \rangle \in G/G_0$. Then $z \in G$, and by (4.6.5) there exist $x \in H$ and $y \in G_0$ such that $z = x + y$, whence $-x + z = y \in G_0$, or $-z + x \in -G_0 = G_0$ (since G_0 is a group). Thus $x \in z + G_0 = \langle z \rangle$ (cf. (4.5.12)). Hence $x \in [x] \cap \langle z \rangle$, whence $[x] \cap \langle z \rangle \neq \varnothing$. By Lemma 4.6.1 $[x] \subset \langle z \rangle$, which means that for an arbitrary class $\langle z \rangle$ from G/G_0 there exists a class $[x]$ from H/H_0 such that $[x] \subset \langle z \rangle$. Consequently condition (iii) is fulfilled and H/H_0 is a selective subpartition of G/G_0. □

Example 4.6.1. Let $G = \mathbb{Z} + 2i\mathbb{Z}$ be the set of all complex numbers $p + iq$ such that p is an integer ($p \in \mathbb{Z}$), and q is an even integer ($q \in 2\mathbb{Z}$). Let $G_0 = 2\mathbb{Z} + 2i\mathbb{Z}$ be the set of all complex numbers $p + iq$ such that both, p and q, are even integers ($p, q \in 2\mathbb{Z}$). Let $H = \mathbb{Z}$ be the set of integers and let $H_0 = 2\mathbb{Z}$ be the set of even integers. Endowed with the operation $+$ of the usual addition of complex or real numbers, all these sets become groups. These groups fulfil conditions (4.6.3) and (4.6.5).

4.7 Rings and fields

A set $R \neq \varnothing$ endowed with two inner binary operations $+$ and $\cdot$ is called a *ring* iff $(R, +)$ is a commutative group (i.e., fulfils conditions (2.9.1)–(2.9.4)), and, moreover,

$$(xy)z = x(yz) \tag{4.7.1}$$

for arbitrary $x, y, z \in R$, and

$$(x + y)z = xz + yz, \ z(x + y) = zx + zy \tag{4.7.2}$$

for arbitrary $x, y, z \in R$. If, moreover,

$$xy = yx \tag{4.7.3}$$

for arbitrary $x, y \in R$, then the ring $(R; +, \cdot)$ is called *commutative*. A commutative[17] ring $(F; +, \cdot)$ is called a *field* iff $(F \setminus \{0\}, \cdot)$ is a (commutative) group. In every field[18] F there exists an element $1 \in F$ such that

$$1x = x1 = x \tag{4.7.4}$$

for every $x \in F$, and to every $x \in F$, $x \neq 0$, there exists an element $x^{-1} \in F \setminus \{0\}$ such that

$$xx^{-1} = x^{-1}x = 1. \tag{4.7.5}$$

Let $(R; +, \cdot)$ be a ring. Every expression

$$p(x_1, \ldots, x_m) = \sum_{k=0}^{n} \sum_{\substack{i_1 + \cdots + i_m = k \\ i_j \in \mathbb{N} \cup \{0\}}} a_{i_1 \ldots i_m} x_1^{i_1} \ldots x_m^{i_m}, \ a_{i_1 \ldots i_m} \in R, \tag{4.7.6}$$

is called a *polynomial* in variables $x_1, \ldots, x_m$. Every such polynomial can be considered as a function from R^m into R, for if we replace $x_1, \ldots, x_m$ by concrete elements of R, and then calculate expression (4.7.6), then $p(x_1, \ldots, x_m)$ will become a concrete element of R.

The collection of all polynomials (4.7.6) is denoted by $R[x_1, \ldots, x_m]$. For $m = 1$ (4.7.6) becomes simply

$$p(x) = \sum_{k=0}^{n} a_k x^k, \tag{4.7.7}$$

and the collection of all such polynomials is denoted by $R[x]$.

Let R be a subring of a ring P and let $a \in P$. We denote by $R[a]$ the set of the values of all polynomials (4.7.7) for $x = a$, i.e.,

$$R[a] = \{ y \in P \mid y = p(a) = \sum_{k=0}^{n} a_k a^k, \ p \in R[x] \}, \tag{4.7.8}$$

---

[17] Sometimes the commutativity of the operation $\cdot$ in a field is not required, but in the present book we are going to consider such fields only in which the operation $\cdot$ is commutative.
[18] Sometimes, where there is no need to underline the operations in R resp. F, we say simply *a ring R*, or *a field F*, instead of *a ring* $(R; +, \cdot)$, or *a field* $(F; +, \cdot)$.

and similarly for polynomials in more variables. It is easily seen that if $(R; +, \cdot)$ is a ring, then also $(R[x]; +, \cdot)$ and $(R[x_1, \ldots, x_m]; +, \cdot)$, with the addition and multiplication of polynomials defined in the usual way, are rings. Similarly, if a is an element of P, then[19] $(R[a]; +, \cdot)$ is a ring.

If all a_k in (4.7.7) (or, more generally, all $a_{i_1 \ldots i_m}$ in (4.7.6)) are zeros, we write $p = 0$. The condition $p \neq 0$ means that at least one of the coefficients in (4.7.7) (resp. (4.7.6)) is different from zero.

If p is a polynomial (4.7.7), then the polynomial

$$p'(x) = \sum_{k=1}^{n} k a_k x^{k-1}$$

is called the *derivative* of the polynomial p.

If in (4.7.7) $a_n \neq 0$, then the number n is called the *degree* of p. Polynomials of degree zero $(p(x) = a_0 \neq 0)$ are elements of R. For $p = 0$ we assume that its degree is -1.

A structure intermediate between a ring and a field is an integral domain. An *integral domain* $(R; +, \cdot)$ is a commutative ring with an element $1 \in R$ fulfilling (4.7.4) for all $x \in R$, and without divisors of zero, i.e., fulfilling, for arbitrary $x, y \in R$, the condition

$$xy = 0 \text{ implies that either } x = 0, \text{ or } y = 0. \tag{4.7.9}$$

Every field is an integral domain. (If $xy = 0$, but $x \neq 0$, then $x^{-1}xy = 0$, whence $y = 0$).

However, in a field, or in an integral domain, or even in a group, it may happen that for an element $u \neq 0$ we have $u + u = 0$, or, more generally, for a certain $k \in \mathbb{N}$

$$ku = 0. \tag{4.7.10}$$

If (4.7.10) holds, but $lu \neq 0$ for $l = 1, \ldots, k - 1$, then we say that u is an element *of order k*. If such a k does not exist, we say that u is an element of order zero. In an integral domain (and so also in a field) all elements $u \neq 0$ have the same order. In fact, if u fulfils (4.7.10), and $v \neq 0$ is arbitrary, then

$$u(kv) = (ku)v = 0,$$

whence by (4.7.9), since $u \neq 0$, we get $kv = 0$. The same argument shows also that if $lu \neq 0$ for an $l \in \mathbb{N}$, then also necessarily $lv \neq 0$ (for otherwise we would have $lu = 0$), and so u and v have the same order.

The common order k of all $u \neq 0$ in an integral domain (or in a field) $(R; +, \cdot)$ is called the *characteristic* of R:

$$\text{char } R = k.$$

If all elements $u \neq 0$ in R have order zero, then we say that R is of characteristic zero.

---

[19] The operations in $R[a]$ are those existing in P, since by (4.7.8) $R[a] \subset P$. So $(R[a]; +, \cdot)$ is a subring of $(P; +, \cdot)$.

If $(R; +, \cdot)$ is an integral domain, then we may associate with R a field $(F; +, \cdot)$, the so called *field of fractions* of R. The elements of F are the expressions $\dfrac{x}{y}, x, y \in R$, $y \neq 0$, where it is to be understood that $\dfrac{x}{y} = \dfrac{u}{v}$ iff $xv = yu$. The operations (the addition and multiplication) for such expressions are defined as for usual fractions (elements of $\mathbb{Q}$, which is the field of fractions of the integral domain $(\mathbb{Z}; +, \cdot)$), and the rules of calculations are the same. We do not go into details here.

Let $(F; +, \cdot)$ be a field. It is not difficult to check that then, for every $m \in \mathbb{N}$, the ring $F[x_1, \ldots, x_m]$ is an integral domain. The field of fractions of $F[x_1, \ldots, x_m]$ is denoted by $F(x_1, \ldots, x_m)$, and its elements are called *rational functions* (over F). Thus a rational function is an expression of the form

$$w(x_1, \ldots, x_m) = \frac{p(x_1, \ldots, x_m)}{q(x_1, \ldots, x_m)}, \; p, q \in F[x_1, \ldots, x_m], \; q \neq 0. \tag{4.7.11}$$

(Of course, the polynomials p and q need not depend effectively on all variables $x_1, \ldots, x_m$). Again, if we substitute for $x_1, \ldots, x_m$ concrete elements $a_1, \ldots, a_m \in F$ such that $q(a_1, \ldots, a_m) \neq 0$, and calculate expression (4.7.11) with such a substitution, then we obtain an element of F:

$$w(a_1, \ldots, a_m) = \frac{p(a_1, \ldots, a_m)}{q(a_1, \ldots, a_m)} \in F.$$

Thus a rational function w may be considered as a function from a subset of F^m into F.

Lemma 4.7.1. *Let $(K; +, \cdot)$ be a field, and $\varnothing \neq F \subset K$. Then $(F; +, \cdot)$ is a field if and only if for every $x, y \in F$*

$$x - y \in F, \tag{4.7.12}$$

and, if $y \neq 0$,

$$xy^{-1} \in F. \tag{4.7.13}$$

Proof. Suppose that conditions (4.7.12) and (4.7.13) are fulfilled. By Lemma 4.5.1 $(F, +)$ is a group, and, of course, $(F, +)$ is commutative, since $(K, +)$ is a commutative group. Conditions (4.7.1)–(4.7.3) hold for all $x, y \in F$, since they hold in K and $F \subset K$. It remains to check that $(F \setminus \{0\}, \cdot)$ is a group. But this follows again from Lemma 4.5.1, in virtue of condition (4.7.13).

Conversely, if $(F; +, \cdot)$ is a field, then conditions (4.7.12) and (4.7.13) are evidently fulfilled. (They result also from Lemma 4.5.1.) $\qquad \square$

Theorem 4.7.1. *Let $\mathcal{K} \neq \varnothing$ be an arbitrary non-empty collection of subfields of a field. Then $K_0 = \bigcap \mathcal{K}$ is a field.*

Proof. If $x, y \in K_0$, then $x, y \in K$ for every $K \in \mathcal{K}$. Hence also $x - y \in K$ and, if $y \neq 0$, $xy^{-1} \in K$, for every $K \in \mathcal{K}$. Consequently $x - y \in K_0$ and $xy^{-1} \in K_0$. The theorem results now from Lemma 4.7.1. $\qquad \square$

Let $(K; +, \cdot)$ be a field, and let $(F; +, \cdot)$, $F \subset K$, be a subfield of $(K; +, \cdot)$. Let $A \subset K$ be an arbitrary set. Let $\mathcal{K}$ be the collection of all fields contained in K (subfields of K) and containing $F \cup A$. By Theorem 4.7.1 the intersection K_0 of all fields in $\mathcal{K}$ ($\mathcal{K} \neq \varnothing$, since $K \in \mathcal{K}$) is a field containing $F \cup A$, and clearly is the smallest field with this property. In the sequel this field K_0 will be denoted[20] by $F(A)$ and called the *extension* of F by the set A. If $A = \{a\}$ is a singleton, then instead of $F(\{a\})$ we write simply $F(a)$ and call the extension $F(a)$ *simple extension*. Similarly, the extension of F by a finite set $\{a_1, \ldots, a_m\}$ will be denoted by $F(a_1, \ldots, a_m)$. As is easily seen we have

$$F(a_1, \ldots, a_m) = F(a_1)(a_2) \ldots (a_m). \tag{4.7.14}$$

We assume that the reader is familiar with the main properties of rings, fields, and polynomials, and here we present only not-so-known properties and definitions, and also we recall some known facts and notions in order to refresh the reader's memory and fix the terminology and notation.

4.8 Algebraic independence and dependence

Let F, K be fields, $F \subset K$. The elements $a_1, \ldots, a_m \in K$ are called *algebraically dependent* (over F) iff there exists a polynomial $p \in F[x_1, \ldots, x_m]$, $p \neq 0$, such that

$$p(a_1, \ldots, a_m) = 0.$$

If $a_1, \ldots, a_m$ are not algebraically dependent, they are called *algebraically independent* (over F). Then $p(a_1, \ldots, a_m) \neq 0$ for every polynomial $p \in F[x_1, \ldots, x_m]$, $p \neq 0$.

If $a_1, \ldots, a_m \in F$, then they are algebraically dependent over F. One can take the polynomial

$$p(x_1, \ldots, x_m) = \sum_{i=1}^{m} (x_i - a_i) \in F[x_1, \ldots, x_m].$$

Of course, $p \neq 0$, and $p(a_1, \ldots, a_m) = 0$.

The reader will verify that every part (every sub-system) of an algebraically independent system is algebraically independent. This gives rise to the following definition.

An arbitrary set $A \subset K$ is called *algebraically independent* (over F) iff every finite system $\{a_1, \ldots, a_m\} \subset A$ is algebraically independent, i.e., iff $p(a_1, \ldots, a_m) \neq 0$ for every $a_1, \ldots, a_m \in A$, $m \in \mathbb{N}$, and for every $p \in F[x_1, \ldots, x_m]$, $p \neq 0$. A set $A \subset K$ is called *algebraically dependent* (over F) iff it is not algebraically independent (over F), i.e., iff there exist $a_1, \ldots, a_m \in A$ and $p \in F[x_1, \ldots, x_m]$, $p \neq 0$, such that $p(a_1, \ldots, a_m) = 0$. The notion of the algebraic independence [dependence] extends

---

[20] The operations in $F(A)$ are those originally existing in K. Also in Theorem 4.7.1 it is understood that the operations in all the fields $K \in \mathcal{K}$, and in K_0, are the same. In the sequel of this book, whenever two fields F and K occur, and $F \subset K$, it is understood that $(F; +, \cdot)$ is a subfield of $(K; +, \cdot)$, i.e., the operations in F are the same as in K.

that of the linear independence [dependence], which is the special case, where the polynomial p is linear homogenous:

$$p(x_1,\ldots,x_m) = \sum_{i=1}^{m} \alpha_i x_i, \ \alpha_i \in F, \ i = 1,\ldots,m.$$

Let $F \subset K$ be fields, and let $S \subset K$ be an arbitrary non-empty set. Consider the set[21]

$$\tilde{F} = \bigcup_{m=1}^{\infty} \bigcup_{w \in F(x_1,\ldots,x_m)} \bigcup_{t_i} \{w(t_1,\ldots,t_m)\}, \tag{4.8.1}$$

where the third union extends over all $t_1,\ldots,t_m \in S$ such that $w(t_1,\ldots,t_m)$ has sense[22].

Lemma 4.8.1. *Let $F \subset K$ be fields, and let $\varnothing \neq S \subset K$. Then with notation (4.8.1), $\tilde{F} = F(S)$.*

Proof. Let $x \in \tilde{F}$. Then there exist an $m \in \mathbb{N}$, a $w \in F(x_1,\ldots,x_m)$ and $t_1,\ldots,t_m \in S$ such that

$$x = w(t_1,\ldots,t_m). \tag{4.8.2}$$

But since the coefficients in w belong to $F \subset F(S)$, and $t_1,\ldots,t_m \in S \subset F(S)$, and rational operations[23] do not lead us out of a field, this implies that $x \in F(S)$. Thus $\tilde{F} \subset F(S)$.

Now observe that $(\tilde{F}; +, \cdot)$ is a field. Indeed, if $x, y \in \tilde{F}$, then x is given by (4.8.2) and

$$y = u(s_1,\ldots,s_n),$$

with some $m, n \in \mathbb{N}$, $w \in F(x_1,\ldots,x_m)$, $u \in F(x_1,\ldots,x_n)$, and $t_1,\ldots,t_m$, $s_1,\ldots,s_n \in S$. Moreover, if $y \neq 0$, then $u \neq 0$. Define rational functions $v, z \in F(x_1,\ldots,x_m, x_{m+1},\ldots,x_{m+n})$ as

$$v(x_1,\ldots,x_{m+n}) = w(x_1,\ldots,x_m) - u(x_{m+1},\ldots,x_{m+n}),$$

and (if $u \neq 0$)

$$z(x_1,\ldots,x_{m+n}) = \frac{w(x_1,\ldots,x_m)}{u(x_{m+1},\ldots,x_{m+n})}.$$

Then

$$x - y = v(t_1,\ldots,t_m,s_1,\ldots,s_n) \in \tilde{F},$$

and (if $y \neq 0$)

$$xy^{-1} = z(t_1,\ldots,t_m,s_1,\ldots,s_n) \in \tilde{F}.$$

Since $\tilde{F} \subset F(S)$, this implies in virtue of Lemma 4.7.1 that $\tilde{F}$ is a field.

[21] Of course, since $F \subset K$, we have $F[x_1,\ldots,x_m] \subset K[x_1,\ldots,x_m]$ and $F(x_1,\ldots,x_m) \subset K(x_1,\ldots,x_m)$.
[22] I.e., the denominator of $w(t_1,\ldots,t_m)$ does not vanish. If the set S is algebraically independent over F, then $w(t_1,\ldots,t_m)$ has sense for every $t_1,\ldots,t_m \in S$, and the third union in (4.8.1) extends over all $t_1,\ldots,t_m \in S$.
[23] Addition, subtraction, multiplication, and division (by an element different from zero).

The constant polynomials $p_a(x) = a$, and $q(x) = 1$, as well as the polynomial $r(x) = x$, belong to $F[x]$ for every $a \in F$. Hence the rational functions

$$w_a(x) = \frac{p_a(x)}{q(x)} \text{ and } u(x) = \frac{r(x)}{q(x)}$$

belong to $F(x)$. Now, for every $a \in F$ and arbitrary $t \in S$ we have

$$a = w_a(t) \in \tilde{F}, \ t = u(t) \in \tilde{F},$$

i.e., $F \subset \tilde{F}$, $S \subset \tilde{F}$. Thus $\tilde{F}$ is a field containing $F \cup S$, and consequently it must contain the smallest field with these properties, i.e., $F(S)$. Thus $F(S) \subset \tilde{F}$, and consequently $\tilde{F} = F(S)$. $\qquad \square$

Corollary 4.8.1. *Let $F \subset K$ be fields, and let $a_1, \ldots, a_n \in K$. Then for every $b \in F(a_1, \ldots, a_n)$ there exists a $w \in F(x_1, \ldots, x_n)$ such that $b = w(a_1, \ldots, a_n)$.*

Proof. By Lemma 4.8.1 $b = w_0(t_1, \ldots, t_k)$, where $w_0 \in F(x_1, \ldots, x_k)$, and $t_1, \ldots, t_k \in \{a_1, \ldots, a_n\}$. Every such w_0 can be considered as a rational function $w \in F(x_1, \ldots, x_n)$ such that $w(a_1, \ldots, a_n) = w_0(t_1, \ldots, t_k)$. $\qquad \square$

Lemma 4.8.2. *Let $F \subset K$ be fields, and let $\varnothing \neq S \subset K$. Let $a_1, \ldots, a_n \in F(S)$. Then there exists a finite set $S_1 \subset S$ such that $a_1, \ldots, a_n \in F(S_1)$.*

Proof. By Lemma 4.8.1 there exist rational functions

$$w_1 \in F(x_1, \ldots, x_{m_1}), \ldots, w_n \in F(x_1, \ldots, x_{m_n}), \ m_1, \ldots, m_n \in \mathbb{N},$$

and points $t_1^1, \ldots, t_{m_1}^1, \ldots, t_1^n, \ldots, t_{m_n}^n \in S$ such that

$$a_i = w_i(t_1^i, \ldots, t_{m_i}^i), \ i = 1, \ldots, n.$$

Then, again by Lemma 4.8.1, $a_1, \ldots, a_n \in F(S_1)$, where

$$S_1 = \{ t_1^1, \ldots, t_{m_1}^1, \ldots, t_1^n, \ldots, t_{m_n}^n \}$$

is a finite subset of S. $\qquad \square$

4.9 Algebraic and transcendental elements

Let $F \subset K$ be fields. An element $t \in K$ is called *algebraic* (over F) iff the singleton $\{t\}$ forms an algebraically dependent system over F. If a $t \in K$ is not algebraic (over F), then it is called *transcendental* (over F).

Alternatively, a $t \in K$ is algebraic (over F) iff there exists a non-trivial ($\neq 0$) polynomial $p \in F[x]$ such that $p(t) = 0$. This follows directly from the definition of the algebraic dependence.

The *algebraic closure* of F (in K) is the set

$$\operatorname{algcl} F = \{ t \in K \mid t \text{ is algebraic over } F \}.$$

It can be shown that $\operatorname{algcl} F$ is a field (a subfield of K; cf. Exercise 4.15). The algebraic closure depends not only on F, but also on K. For example, $\operatorname{algcl} \mathbb{Q}$ in $\mathbb{C}$ is larger than $\operatorname{algcl} \mathbb{Q}$ in $\mathbb{R}$. E.g., the imaginary unit i is an element of $\operatorname{algcl} \mathbb{Q}$ in $\mathbb{C}$, but not of $\operatorname{algcl} \mathbb{Q}$ in $\mathbb{R}$. As is well known, we have $\operatorname{card} \operatorname{algcl} \mathbb{Q} = \aleph_0$ so that $\mathbb{R} \setminus \operatorname{algcl} \mathbb{Q} \neq \varnothing$.

If $t \in K$ is algebraic over F, then among all the polynomials $p \in F[x]$, $p \neq 0$, such that $p(t) = 0$ there are polynomials of the smallest degree. (This is a consequence of the fact that $\mathbb{N}$ is well ordered.) If p_0 is such a polynomial and

$$p_0(x) = \sum_{k=0}^{n} a_k x^k, \ a_n \neq 0,$$

then $\dfrac{1}{a_n} p_0(x)$ also is such a polynomial, and has the coefficient of x^n equal one. For a given algebraic $t \in K$ such a polynomial (of the smallest possible degree, with the coefficient of x in the highest power equal one and such that its value at the point $x = t$ is zero) is unique, and is called the *minimal polynomial* of t. The degree of the minimal polynomial of t is called the *degree* of t. Algebraic elements of degree 1 are simply all the elements of F. In fact, if $t \in F$, then the polynomial $p(x) = x - t$ belongs to $F[x]$, $p(t) = 0$, and clearly p is the minimal polynomial of t. On the other hand, if t is algebraic of degree 1, then there exists a polynomial $p(x) = x - c \in F[x]$ such that $p(t) = 0$. Thus $t = c$, and $p \in F[x]$ implies that $c \in F$. Consequently $t \in F$.

It may be shown that a $p \in F[x]$ is the minimal polynomial of an algebraic element $t \in K$ if and only if $p(t) = 0$, p is irreducible[24] (in $F[x]$), and the coefficient of x in the highest power in p is one (cf. Exercise 4.11).

Lemma 4.9.1. *Let $F \subset K$ be fields, and let a set $S \subset K$ be algebraically independent over F. Let $a \in K \setminus S$. Then $S \cup \{a\}$ is algebraically independent if and only if $a \notin \operatorname{algcl} F(S)$.*

Proof. Suppose that $a \notin \operatorname{algcl} F(S)$. Take an arbitrary finite system $\{a_1, \ldots, a_m\} \subset S \cup \{a\}$, and for an indirect proof suppose that the system $\{a_1, \ldots, a_m\}$ is algebraically dependent. Note that the system $\{a_1 \ldots, a_m\}$ must contain a, for otherwise S would be algebraically dependent. Let, e.g., $a_m = a$. There exists a polynomial $p \in F[x_1, \ldots, x_m]$, $p \neq 0$, such that

$$p(a_1, \ldots, a_m) = 0.$$

Arrange the polynomial p according to the increasing powers of x_m:

$$p(x_1, \ldots, x_m) = \sum_{i=0}^{n} p_i(x_1, \ldots, x_{m-1}) x_m^i, \ p_i \in F[x_1, \ldots, x_{m-1}], \ i = 0, \ldots, n.$$

Thus

$$\sum_{i=0}^{n} p_i(a_1, \ldots, a_{m-1}) a^i = 0. \tag{4.9.1}$$

---

[24] A polynomial $p \in F[x]$ is called *reducible* (in $F[x]$, or over F) iff there exist polynomials $u, v \in F[x]$, of degrees smaller than that of p, and such that $p = uv$. If p is not reducible, it is said to be *irreducible*.

Since a is transcendental over $F(S)$, (4.9.1) implies that $p_i(a_1, \ldots, a_{m-1}) = 0$, $i = 0, \ldots, n$, and at least one of p_i's is non-trivial, for otherwise we would have $p = 0$. So the system $\{a_1, \ldots, a_{m-1}\} \subset S$ is algebraically dependent, which is incompatible with the fact that S is algebraically independent.

Now assume that $S \cup \{a\}$ is algebraically independent, and for an indirect proof suppose that $a \in \text{algcl}\, F(S)$. Consequently there exists a polynomial $p \in F(S)[x]$, $p \neq 0$, such that $p(a) = 0$. Let

$$p(x) = \sum_{i=0}^{n} a_i x^i, \ a_i \in F(S), \ i = 0, \ldots, n. \tag{4.9.2}$$

By Lemma 4.8.2 there exist $t_1, \ldots, t_m \in S$ such that $a_i \in F(t_1, \ldots, t_m)$, $i = 0, \ldots, n$. According to Lemma 4.8.1 there exist rational functions $w_0, \ldots, w_n \in F(x_1, \ldots, x_m)$ such that[25]

$$a_i = w_i(t_1, \ldots, t_m), \ i = 0, \ldots, n. \tag{4.9.3}$$

Write every w_i as

$$w_i = \frac{L_i}{M}, \ M, L_i \in F[x_1, \ldots, x_m], \ i = 0, \ldots, n. \tag{4.9.4}$$

(We can find a common denominator M for all w_i, $i = 0, \ldots, n$.) Not all L_i are zero, for otherwise, by (4.9.2), (4.9.3) and (4.9.4), we would have $p = 0$. Inserting (4.9.3) and (4.9.4) into (4.9.2) we obtain

$$\sum_{i=0}^{n} L_i(t_1, \ldots, t_m) a^i = 0. \tag{4.9.5}$$

Define a polynomial $P \in F[x_1, \ldots, x_m, x_{m+1}]$ as

$$P(x_1, \ldots, x_m, x_{m+1}) = \sum_{i=0}^{n} L_i(x_1, \ldots, x_m) x_{m+1}^i .$$

Then by (4.9.5) $P(t_1, \ldots, t_m, a) = 0$, i.e., the system $\{t_1, \ldots, t_m, a\} \subset S \cup \{a\}$ is algebraically dependent, which is incompatible with the algebraic independence of $S \cup \{a\}$. $\qquad \square$

4.10 Algebraic bases

Let $F \subset K$ be fields. A set $S \subset K$ is called an *algebraic base* of K over F iff it fulfils the two conditions:

(i) S is algebraically independent over F;

(ii) $\text{algcl}\, F(S) = K$.

---

[25] We may assume all w_i belong to $F(x_1, \ldots, x_m)$ (with the same m), because a rational function of a smaller number of variables may always be regarded as a rational function of larger number of variables, which does not depend effectively on all the variables.

In order to prove the existence of algebraic bases, we prove first the following

Lemma 4.10.1. *Let $F \subset K$ be fields. A set $S \subset K$ is an algebraic base of K over F if and only if it is a maximal[26] algebraically independent subset of K.*

Proof. Let $S \subset K$ be an algebraic base of K over F. Suppose that there exists an algebraically independent set $S_1 \subset K$ such that $S \subset S_1$ and $S \neq S_1$. Then there exists an $a \in S_1 \setminus S$. The set $S \cup \{a\} \subset S_1$ is algebraically independent (as a subset of an algebraically independent set), and so, by Lemma 4.9.1, $a \notin \operatorname{algcl} F(S) = K$, which is impossible.

Conversely, suppose that S is a maximal algebraically independent subset of K, but is not an algebraic base of K over F. Being algebraically independent, S satisfies (i) above, so necessarily we must have $\operatorname{algcl} F(S) \neq K$. Thus there exists an $a \in K \setminus \operatorname{algcl} F(S)$, and then, by Lemma 4.9.1, the set $S_1 = S \cup \{a\}$ is algebraically independent, $S \subset S_1$, $S \neq S_1$. Thus S cannot be maximal. $\qquad\square$

Theorem 4.10.1. *Let $F \subset K$ be fields, and let $K \neq \operatorname{algcl} F$. Then there exists an algebraic base of K over F.*

Proof. Let $\mathcal{R}$ be the family

$$\mathcal{R} = \{\, S \subset K \mid S \text{ is algebraically independent over } F \,\}.$$

Any singleton $\{t\}$, where $t \in K \setminus \operatorname{algcl} F$ (so that t is transcendental over F) belongs to $\mathcal{R}$, and so $\mathcal{R} \neq \varnothing$. $(\mathcal{R}, \subset)$ is an ordered set. If $\mathcal{Z} \subset \mathcal{R}$ is a chain, then, as the reader will easily verify, $S_0 = \bigcup \mathcal{Z} \in \mathcal{R}$ is an upper bound for $\mathcal{Z}$. By Theorem 1.8.1 $\mathcal{R}$ contains a maximal element $S_{\max}$. By Lemma 4.10.1, $S_{\max}$ is an algebraic base of K over F. $\qquad\square$

4.11 Simple extensions of fields

In this section we prove some lemmas concerning simple extensions of a field. Let $F \subset K$ be fields.

Lemma 4.11.1. *If $F \subset K$ are fields, and $a \in K$, then*

$$F(a) = \bigcup \{w(a)\},$$

where the union extends over $w \in F(x)$, whose denominator does not vanish at the point a.

This is a particular case of Lemma 4.8.1. It is enough to observe that if $w \in F(x_1 \ldots, x_m)$, then $w(x, \ldots, x) \in F(x)$.

Corollary 4.11.1. *If $F \subset K$ are fields, and $a \in K$ is transcendental over F, then*

$$F(a) = \bigcup_{w \in F(x)} \{w(a)\}.$$

---

[26] I.e., there does not exist an algebraically independent set $S_1 \subset K$ such that $S \subset S_1$ and $S \neq S_1$.

Lemma 4.11.2. *If $F \subset K$ are fields, and $a \in K$ is algebraic over F, then*

$$F(a) = F[a] = \bigcup_{p \in F[x]} \{p(a)\}.$$

Proof. The inclusion $F[a] \subset F(a)$ is clear. Now we show that $F[a]$ is a field. Of course, $F[a] \subset K$. If $b, c \in F[a]$, $b = p(a)$, $c = q(a)$, $p, q \in F[x]$, then also $r = p - q \in F[x]$, and $b - c = p(a) - q(a) = r(a) \in F[a]$. If, moreover, $c \neq 0$ and $m \in F[x]$ is the minimal polynomial of a, then[27] $(q, m) = 1$. It follows that there exist polynomials $r, s \in F[x]$ such that (cf. Exercise 4.10)

$$r(x)\,q(x) + s(x)\,m(x) = 1. \tag{4.11.1}$$

Putting in (4.11.1) $x = a$ we obtain $cr(a) = 1$, i.e., $c^{-1} = r(a) \in F[a]$, whence also

$$bc^{-1} = p(a)\,r(a) \in F[a].$$

By Lemma 4.7.1 $F[a]$ is a field. Clearly $F \subset F[a]$ $\big($any $d \in F$ is the value at a of the constant polynomial $p(x) = d\big)$, and $a \in F[a]$ $\big($a is the value at a of the polynomial $p(x) = x\big)$. $F[a]$ must contain the smallest field with these properties, i.e., $F(a) \subset F[a]$. Hence $F[a] = F(a)$. $\qquad\square$

If $F \subset K$ are fields, then we can, in particular, add elements of K, and multiply elements of K by elements of F. Thus we may consider K as a linear space[28] over F:

$$(K; F; +; \cdot).$$

Lemma 4.11.3. *If $F \subset K$ are fields, and $a \in K$ is algebraic over F, of a degree k, then the system $\{1, a, a^2, \ldots, a^{k-1}\}$ forms a base of the linear space $\big(F(a); F; +; \cdot\big)$.*

Proof. If the system $\{1, a, \ldots, a^{k-1}\}$ were linearly dependent, then there would exist $\alpha_0, \alpha_1, \ldots, \alpha_{k-1} \in F$, not all zero, and such that

$$\alpha_0 + \alpha_1 a + \cdots + \alpha_{k-1} a^{k-1} = 0. \tag{4.11.2}$$

Put $p(x) = \alpha_0 + \alpha_1 x + \cdots + \alpha_{k-1} x^{k-1}$. We have $p \in F[x]$, $p \neq 0$. Relation (4.11.2) means that $p(a) = 0$, and since the degree of p is at most $k - 1$, the degree of a cannot exceed $k - 1$. This shows that the set $B = \{1, a, \ldots, a^{k-1}\}$ is linearly independent.

Now take an arbitrary $b \in F(a)$, $b \neq 0$. By Lemma 4.11.2 $b \in F[a]$, i.e., there exists a $p \in F[x]$ such that $b = p(a)$. Let $q \in F[x]$ be the minimal polynomial of a. There exist polynomials $u, r \in F[x]$ such that

$$p(x) = u(x)\,q(x) + r(x),$$

---

[27] (q, m) denotes the greatest common divisor of q and m, i.e., the polynomial $u \in F[x]$, normalized so that the coefficient of x in the highest power is 1, and such that u divides both, q and m, and if a polynomial $v \in F[x]$ divides q and m, then v divides also u. If we had $u \neq 1$, then we would have $q = ut$ and $m = uv$ with certain $t, v \in F[x]$. Since $q(a) = c \neq 0$, we get $u(a) \neq 0$, and hence, since $m(a) = 0$, we get $v(a) = 0$. But since $u \neq 1$, the degree of v is smaller than that of m, and m would not be minimal.

[28] The reader will verify that all the conditions from the definition of a linear space (cf. 4.1) are fulfilled.

and the degree of $r \leqslant k-1$, or else $r = 0$. The latter case is impossible, since otherwise we would have for $x = a$

$$b = p(a) = u(a)\, q(a) = 0.$$

Thus $r(x) = \alpha_0 + \alpha_1 x + \cdots + \alpha_n x^n$, $\alpha_n \neq 0$, $n \leqslant k - 1$, and

$$b = p(a) = r(a) = \alpha_0 + \alpha_1 a + \cdots + \alpha_n a^n \in E(B).$$

Of course, $0 \in E(B)$ (take all the coefficients 0), and so $F(a) \subset E(B)$. The converse inclusion is obvious, whence $E(B) = F(a)$, and, as has already been shown, B is linearly independent. This means that B is a base of $\big(F(a);\, F;\, +;\, \cdot\big)$. $\square$

Lemma 4.11.4. *Let $F \subset K$ be fields, let $a \in K$ be algebraic over F, and let $q \in F[x]$ be its minimal polynomial. If p_1, $p_2 \in F[x]$ and $p_1(a) = p_2(a)$, then q divides $p_1 - p_2$.*

Proof. Let $r = p_1 - p_2$ (so that $r(a) = 0$), and let $u = (q, r)$. Since q is a minimal polynomial (and hence irreducible), we must have either $u = 1$, or $u = q$. $u = 1$ is impossible, for otherwise there would exist s, $t \in F[x]$ such that $s(x)\, q(x) + t(x)\, r(x) = 1$, and for $x = a$ the left-hand side becomes zero. So $u = q$, and hence q divides r. $\square$

4.12   Isomorphism of fields and rings

Let $(R_1;\, +,\, \cdot)$ and $(R_2;\, +,\, \cdot)$ be rings. A mapping $\varphi : R_1 \to R_2$ is called a *homomorphism* (of R_1 into R_2) iff

$$\varphi(x + y) = \varphi(x) + \varphi(y) \tag{4.12.1}$$

for every x, $y \in R_1$, and

$$\varphi(xy) = \varphi(x)\, \varphi(y) \tag{4.12.2}$$

for every x, $y \in R_1$.

If, moreover, φ is one-to-one, then φ is called a *monomorphism*. If φ is onto $\big(\varphi(R_1) = R_2\big)$, then φ is called an *epimorphism*. A homomorphism which is a monomorphism and an epimorphism is called an *isomorphism*.

A homomorphism $\varphi : R \to R$ $(R_2 = R_1 = R)$ is called an *endomorphism*. An isomorphism $\varphi : R \to R$ is called an *automorphism*.

R_1 and R_2 may be in particular (and same terminology is applied) fields. An isomorphism between two algebraic structures states their complete similarity (from the algebraic point of view). All algebraic properties of one structure are also shared by the other.

Note that if in (4.12.1) we put $x = y = 0$, we get

$$\varphi(0) = 0, \tag{4.12.3}$$

and if in (4.12.2) we put $x = y$, and apply the mathematical induction, then we get

$$\varphi(x^i) = \varphi(x)^i,\ i \in \mathbb{N}, \tag{4.12.4}$$

for every homomorphism φ. Further, if φ is a monomorphism, and thus one-to-one, then by (4.12.3) $\varphi(x) \neq 0$ for $x \neq 0$, and setting in (4.12.2) $x = y = 1$ we obtain

$$\varphi(1) = 1. \tag{4.12.5}$$

Let F_1, F_2 be fields, and $\varphi : F_1 \to F_2$ a homomorphism. We define a mapping $I_\varphi : F_1[x] \to F_2[x]$. If

$$p(x) = \sum_{i=0}^{n} \alpha_i x^i, \ \alpha_i \in F_1, \ i = 0, \ldots, n, \tag{4.12.6}$$

then we put

$$I_\varphi(p)(x) = \sum_{i=0}^{n} \varphi(\alpha_i) x^i. \tag{4.12.7}$$

Thus $I_\varphi(p) \in F_2[x]$. We have the following

Lemma 4.12.1. *Let F_1, F_2 be fields, and $\varphi : F_1 \to F_2$ a homomorphism. Then the mapping $I_\varphi : F_1[x] \to F_2[x]$ defined by (4.12.6) and (4.12.7) also is a homomorphism. If φ is an isomorphism, then I_φ also is an isomorphism.*

Proof. Let $q \in F_1[x]$,

$$q(x) = \sum_{j=0}^{m} \beta_j x^j, \ \beta_j \in F_1, \ j = 0, \ldots, m,$$

be an arbitrary polynomial from $F_1[x]$. Consider polynomial (4.12.6). We may assume that, e.g., $m \leqslant n$. Then

$$(p + q)(x) = \sum_{j=0}^{m} (\alpha_j + \beta_j) x^j + \sum_{j=m+1}^{n} \alpha_j x^j,$$

whence

$$\begin{aligned}
I_\varphi(p+q)(x) &= \sum_{j=0}^{m} \varphi(\alpha_j + \beta_j) x^j + \sum_{j=m+1}^{n} \varphi(\alpha_j) x^j \\
&= \sum_{j=0}^{m} \varphi(\alpha_j) x^j + \sum_{j=0}^{m} \varphi(\beta_j) x^j + \sum_{j=m+1}^{n} \varphi(\alpha_j) x^j \\
&= \sum_{j=0}^{n} \varphi(\alpha_j) x^j + \sum_{j=0}^{m} \varphi(\beta_j) x^j = I_\varphi(p)(x) + I_\varphi(q)(x).
\end{aligned}$$

Similarly

$$(pq)(x) = \sum_{k=0}^{m+n} \Big(\sum_{\substack{i+j=k \\ 0 \leqslant i \leqslant n \\ 0 \leqslant j \leqslant m}} \alpha_i \beta_j \Big) x^k,$$

whence

$$I_\varphi(pq)(x) = \sum_{k=0}^{m+n} \varphi\Big(\sum_{\substack{i+j=k \\ 0\leqslant i\leqslant n \\ 0\leqslant j\leqslant m}} \alpha_i\beta_j\Big)\, x^k = \sum_{k=0}^{m+n}\Big(\sum_{\substack{i+j=k \\ 0\leqslant i\leqslant n \\ 0\leqslant j\leqslant m}} \varphi(\alpha_i)\varphi(\beta_j)\Big)\, x^k$$

$$= \Big(\sum_{i=0}^{n}\varphi(\alpha_i)\, x^i\Big)\Big(\sum_{j=0}^{m}\varphi(\beta_j)\, x^j\Big) = I_\varphi(p)(x)I_\varphi(q)(x).$$

Thus I_φ is a homomorphism.

Now let φ be an isomorphism. If $r \in F_2[x]$,

$$r(x) = \sum_{i=0}^{k}\gamma_i x^i,\ \gamma_i \in F_2,\ i = 0,\dots,k,$$

then there is a unique $p \in F_1[x]$ such that $I_\varphi(p) = r$, viz.

$$p(x) = \sum_{i=0}^{k}\varphi^{-1}(\gamma_i)x^i.$$

It follows that I_φ is one-to-one and onto. Consequently I_φ is an isomorphism. □

Lemma 4.12.2. *Let F_1, F_2 be fields, and $\varphi : F_1 \to F_2$ a homomorphism. Then either $\varphi = 0$ in F_1, or φ is one-to-one, and hence a monomorphism.*

Proof. Suppose that φ is not one-to-one. Then there exist x_1, $x_2 \in F_1$, $x_1 \neq x_2$, such that $\varphi(x_1) = \varphi(x_2)$. Put $y = x_1 - x_2 \neq 0$. Then $x_1 = y + x_2$, whence by (4.12.1) $\varphi(x_1) = \varphi(y) + \varphi(x_2)$, and $\varphi(y) = 0$. Thus we have by (4.12.2) for every $x \in F_1$

$$\varphi(x) = \varphi\Big(y\,\frac{x}{y}\Big) = \varphi(y)\,\varphi\Big(\frac{x}{y}\Big) = 0.$$

Consequently $\varphi = 0$ in F_1. □

Now we are going to investigate the possibility of extending an isomorphism of fields to an isomorphism of their simple extensions. Let F_1, F_2, K_1, K_2 be fields, $F_1 \subset K_1$, $F_2 \subset K_2$, and let $\varphi : F_1 \to F_2$ be an isomorphism. We say that elements $a \in K_1$ and $A \in K_2$ are φ-conjugate iff either they are both transcendental over F_1 and F_2, respectively, or they are both algebraic, and their minimal polynomials q and Q, respectively, are related in the following fashion

$$Q = I_\varphi(q). \tag{4.12.8}$$

If $F \subset K$ are fields, then the identity map

$$e(x) = x \text{ for } x \in F \tag{4.12.9}$$

is an automorphism of F. If elements a, $A \in K$ are e-conjugate, then we say shortly that they are *conjugate*. Thus a, $A \in K$ are conjugate iff either they are both transcendental over F, or they are both algebraic over F and have the same minimal polynomial. Two elements of F are conjugate if and only if they are equal. The "if" part is trivial. Now assume that a, $A \in F$ are conjugate, and let q be their common minimal polynomial. Then q must have the form $q(x) = x - c$, whence, since $q(a) = q(A) = 0$, we get $a = c$ and $A = c$. Thus $a = A$.

Now we prove the following

Theorem 4.12.1. *Let F_1, F_2, K_1, K_2 be fields, $F_1 \subset K_1$, $F_2 \subset K_2$, and let $\varphi : F_1 \to F_2$ be an isomorphism. Let $a \in K_1$, $A \in K_2$, $a \neq 0$, $A \neq 0$. Then there exists an isomorphism $\Phi : F_1(a) \to F_2(A)$ such that $\Phi(a) = A$, $\Phi \mid F_1 = \varphi$, if and only if a and A are φ-conjugate.*

Proof. First we assume that the required isomorphism Φ exists. Let a be algebraic over F_1, and let q be its minimal polynomial. Then q is irreducible, and since, by Lemma 4.12.1, I_φ is an isomorphism, the polynomial $Q = I_\varphi(q)$ also is irreducible (cf. Exercise 4.16). By (4.12.5) the coefficient of x in the highest power in Q is 1 (just like in q). If we show that $Q(A) = 0$, then it follows that A is algebraic over F_2, and Q is its minimal polynomial. (4.12.8) holds by the definition of Q.

If we assume A to be algebraic, we argue in the same way (replacing I_φ by I_φ^{-1}), since by Lemma 4.12.1 I_φ is an isomorphism.

Let

$$q(x) = \sum_{i=0}^{n} \alpha_i x^i.$$

Then

$$Q(x) = I_\varphi(q)(x) = \sum_{i=0}^{n} \varphi(\alpha_i) x^i,$$

whence, using (4.12.1), (4.12.2), (4.12.3), (4.12.4) and the facts that $\Phi(a) = A$, $\Phi \mid F_1 = \varphi$, we obtain

$$Q(A) = \sum_{i=0}^{n} \varphi(\alpha_i) A^i = \sum_{i=0}^{n} \Phi(\alpha_i)\Phi(a)^i = \Phi\Big(\sum_{i=0}^{n} \alpha_i a^i\Big) = \Phi\big(q(a)\big) = \Phi(0) = 0.$$

Now assume that a and A are φ-conjugate, and let a and A be algebraic over F_1 and F_2, respectively, their minimal polynomials being q and Q, respectively. Since by Lemma 4.11.2 $F_1(a) = F_1[a]$, for every $b \in F_1(a)$ there exists a polynomial $p \in F_1[x]$ such that

$$b = p(a). \tag{4.12.10}$$

For such a b we put

$$\Phi(b) = I_\varphi(p)(A). \tag{4.12.11}$$

To see that this definition is correct (i.e., it is independent of the choice of p fulfilling (4.12.10)) suppose that $p_1(a) = p_2(a)$ for some $p_1, p_2 \in F_1[x]$. Then, by Lemma 4.11.4,

$p_1 - p_2 = uq$ with a polynomial $u \in F_1[x]$, whence, by Lemma 4.12.1 and (4.12.8),

$$I_\varphi(p_1)(A) - I_\varphi(p_2)(A) = I_\varphi(p_1 - p_2)(A) = I_\varphi(u)(A)I_\varphi(q)(A) = I_\varphi(u)(A)Q(A) = 0.$$

Thus Φ is unambiguously defined.

Take arbitrary u, $v \in F_1(a) = F_1[a]$. There exist polynomials r, $s \in F_1[x]$ such that $u = r(a)$, $v = s(a)$, whence $u + v = r(a) + s(a) = (r + s)(a)$, $uv = r(a)\,s(a) = (rs)(a)$. By Lemma 4.12.1

$$\Phi(u + v) = I_\varphi(r + s)(A) = I_\varphi(r)(A) + I_\varphi(s)(A) = \Phi(u) + \Phi(v),$$

and
$$\Phi(uv) = I_\varphi(rs)(A) = I_\varphi(r)(A)I_\varphi(s)(A) = \Phi(u)\Phi(v).$$

Thus Φ is a homomorphism. Take an arbitrary $C \in F_2(A)$. By Lemma 4.11.2 there exists a polynomial $P \in F_2[x]$ such that $C = P(A)$. Put $p = I_\varphi^{-1}(P)$ and $c = p(a) \in F_1[a] = F_1(a)$. Then

$$\Phi(c) = I_\varphi(p)(A) = P(A) = C.$$

Consequently Φ is an epimorphism, and so we cannot have $\Phi = 0$ in $F_1(a)$. By Lemma 4.12.2 Φ is a monomorphism, and hence an isomorphism.

For polynomial (4.12.9) we have $e(a) = a$, whence

$$\Phi(a) = I_\varphi(e)(A) = A,$$

whereas for the polynomial $p(x) = \alpha \in F_1$ we have $I_\varphi(p) = \varphi(\alpha)$, whence

$$\Phi(\alpha) = I_\varphi(p)(A) = \varphi(\alpha),$$

i.e., $\Phi \mid F_1 = \varphi$. Thus Φ has all the required properties.

If a and A are φ-conjugate and transcendental, then by Corollary 4.11.1 $F_1(a) = \bigcup_{w \in F_1(x)} \{w(a)\}$, and for every $b \in F_1(a)$ there exist polynomials r, $s \in F_1[x]$ such that

$$b = \frac{r(a)}{s(a)}. \tag{4.12.12}$$

For such a b we put

$$\Phi(b) = \frac{I_\varphi(r)(A)}{I_\varphi(s)(A)}. \tag{4.12.13}$$

Again we must check that this definition is correct. Suppose that also $b = u(a)/v(a)$ with u, $v \in F_1[x]$. Hence $(rv - su)(a) = 0$, $rv - su \in F_1[x]$, and since a is a transcendental over F_1, we must have $rv - su = 0$, whence by Lemma 4.12.1 $I_\varphi(r)I_\varphi(v) = I_\varphi(s)I_\varphi(u)$, and

$$\frac{I_\varphi(u)(A)}{I_\varphi(v)(A)} = \frac{I_\varphi(r)(A)}{I_\varphi(s)(A)}.$$

Thus Φ is well defined by (4.12.13). Similarly as in the case, where a and A are algebraic, it can be checked that Φ is an isomorphism.

We have $a = \dfrac{e(a)}{1}$, whence

$$\Phi(a) = \frac{I_\varphi(e)(A)}{I_\varphi(1)(A)} = A,$$

and for every $\alpha \in F_1$, taking $p(x) = \alpha$, we have $\alpha = \dfrac{p(a)}{1}$, whence

$$\Phi(\alpha) = \frac{I_\varphi(p)(A)}{I_\varphi(1)(A)} = \varphi(\alpha).$$

Thus Φ has all the required properties. $\square$

Hence we derive the following result, which will be needed later.

Theorem 4.12.2. *Let $F \subset K$ be fields, and let a, b, A, $B \in K$, $abAB \neq 0$. There exists an isomorphism $\varphi : F(a,b) \to F(A,B)$ such that $\varphi(a) = A$, $\varphi(b) = B$, $\varphi \mid F = e$ if and only if a and A are conjugate and for a certain isomorphism $\hat{\varphi} : F(a) \to F(A)$ such that $\hat{\varphi}(a) = A$, $\hat{\varphi} \mid F = e$, the elements b and B are $\hat{\varphi}$-conjugate.*

Proof. The theorem follows by applying twice Theorem 4.12.1 according to (4.7.14). In the proof of the necessity one has to put $\hat{\varphi} = \varphi \mid F(a)$. $\square$

Exercises

1. Let $(L; F; +; \cdot)$ be a linear space. Show that $L_0 \subset L$ is a linear subspace if and only if $E(L_0) = L_0$. Let, moreover, $F \subset \mathbb{R}$. Show that $C \subset L$ is a cone if and only if $E^+(C) = C$.

2. Let $(L; F; +; \cdot)$ be a linear space, let $F \subset \mathbb{R}$, and let $C \subset L$ be a cone such that $C \cap (-C) = \{0\}$. Show that the relation $\leqslant$ defined as

$$x \leqslant y \text{ iff } y - x \in C$$

   is an order in L.

3. Let $(L; F; +; \cdot)$ be a linear space, and let a set $B \subset L$ be linearly independent. Suppose that $B = B_1 \cup B_2$, $B_1 \cap B_2 = \varnothing$. Show that $E(B_1) \cap E(B_2) = \{0\}$.

4. Let $(L; F; +; \cdot)$ and $(M; F; +; \cdot)$ be linear spaces, and let $f : L \to M$ be a homomorphism. Show that for every set $A \subset L$ we have $f\big(E(A)\big) = E\big(f(A)\big)$.

5. Let $(L; F; +; \cdot)$ and $(M; F; +; \cdot)$ be linear spaces, and let $f : L \to M$ be a homomorphism. Show that $f(L)$ is a linear subspace of M, and $\dim \operatorname{Ker} f + \dim f(L) = \dim L$.
   [Hint: Let B_0 be a base of $\operatorname{Ker} f$, and B a base of L such that $B_0 \subset B$. Prove that $f \mid B \setminus B_0$ is one-to-one and that $f(B \setminus B_0)$ is a base of $f(L)$.]

6. Let $(G, +)$ be a group, and $(S, +)$ its subsemigroup. Show that if S is finite, then $(S, +)$ is a subgroup of $(G, +)$.
   [Hint: Let $S = \{a_1, \ldots, a_n\}$. Take arbitrary a, $b \in S$. Then $\{a_1 + a, \ldots, a_n + a\}$ is a set of n distinct elements of S, so one of them must equal b. Therefore the equation $x + a = b$ has a unique solution $x \in S$.]

7. Let $(G, +)$ be a group, and $(H, +)$ its subgroup. Show that the index of H is 2 if and only if there exists a $y_0 \in G \setminus H$ such that $x + y_0 \in H$ for every $x \in G \setminus H$ (Grząślewicz-Powązka-Tabor [124]).
   [Hint: If $u, v \in G \setminus H$, then $u + y_0 \in H$, $v + y_0 \in H$, and $u - v = (u + y_0) - (v + y_0) \in H$.]

8. Let $(H, +)$ be a group. Show that for every family of subgroups of $(H, +)$ their intersection is a subgroup of $(H, +)$, and hence for every set $S \subset H$ there exists the subgroup of $(H, +)$ generated by S.

9. Let $(R; +, \cdot)$ be a ring. A *principal ideal* of R is every set of the form $I_a = \{ x \in R \mid x = pa, \ p \in R \}$, where $a \in R$. Show that[29] $I_a + I_b = I_{(a,b)}$ for arbitrary $a, b \in R$ such that (a, b) exists.

10. Let $(R; +, \cdot)$ be a ring, and let $a, b \in R$. Show that if (a, b) exists, then there exist $r, s \in R$ such that $(a, b) = ra + sb$.
    [Hint: Use Exercise 4.9.]

11. Let $F \subset K$ be fields, and let $a \in K$ be algebraic over F. Show that $p \in F[x]$ is the minimal polynomial of a if and only if p is irreducible (over F), $p(a) = 0$, and the coefficient of x in the highest power in p is one.
    [Hint: If p is irreducible, and there exists a $q \in F[x]$, of a degree smaller than that of p, such that $q(a) = 0$, then consider $(p, q) \in F[x]$. It follows that p must be reducible.]

12. Show that if $F_1 \subset F_2 \subset K$ are fields, and $(F_2; F_1; +; \cdot)$ has a finite dimension, then $F_2 \subset \operatorname{algcl} F_1$.
    [Hint: Let $\dim F_2 = k \in \mathbb{N}$. For every $a \in F_2$ the elements $1, a, \ldots, a^k$ are linearly dependent.]

13. Show that if $F_1 \subset F_2 \subset F_3$ are fields and the dimensions of both the spaces $(F_2; F_1; +; \cdot)$ and $(F_3; F_2; +; \cdot)$ are finite, then also the dimension of the space $(F_3; F_1; +; \cdot)$ is finite.

14. Show that if $F \subset K$ are fields, and $a, b \in K$ are algebraic over F, then $F(a, b) \subset \operatorname{algcl} F$.

15. Show that if $F \subset K$ are fields, then $\operatorname{algcl} F$ is a field.
    [Hint: For every $a, b \in \operatorname{algcl} F$, consider the field $F(a, b) \subset \operatorname{algcl} F$.]

16. Let F_1 and F_2 be fields, and let $\varphi : F_1 \to F_2$ be an isomorphism. Show that a polynomial $p \in F_1[x]$ is reducible if and only if $I_\varphi(p) \in F_2[x]$ is reducible.

17. A field F is called *algebraically closed* iff every polynomial $p \in F[x]$ of a degree greater than one is reducible (over F). Show that if a field F is algebraically closed, then for every field $K \supset F$ we have $\operatorname{algcl} F = F$.

18. Let F_1 and F_2 be fields, and let $\varphi : F_1 \to F_2$ be an isomorphism. Show that F_1 is algebraically closed if and only if F_2 is algebraically closed.

19. Let $(K; +, \cdot)$ be a field, let $(F; +, \cdot)$ be a subfield of $(K; +, \cdot)$, and let $S \subset K$ be algebraically independent over F. Show that if $S_1 \subset S$ and $a \in S \setminus S_1$, then a is transcendental over $F(S_1)$.

---

[29] (a, b) denotes the greatest common divisor of a and b.

Part II

Cauchy's Functional Equation and Jensen's Inequality

Chapter 5

Additive Functions and Convex Functions

5.1 Convex sets

Let $C \subset \mathbb{R}$ be an arbitrary set. A non-empty set $A \subset \mathbb{R}^N$ is called *C-convex* iff

$$\lambda x + (1 - \lambda) y \in A \text{ for all } x, y \in A \text{ and } \lambda \in C \cap [0, 1]. \tag{5.1.1}$$

In order to avoid the tiresome necessity of considering exceptional cases, we note the convention that the empty set $\varnothing$ is *not* C-convex for any C (cf. also Eggleston [74]). C-convex sets have been thoroughly investigated by J. W. Green–W. Gustin [120].

Some particular sets C will play a prominent role in the sequel, and we shall apply a special terminology in such cases. If $C = \mathbb{R}$, then a set A fulfilling (5.1.1) is called shortly *convex*[1]. If $C = \mathbb{Q}$, then A is called *Q-convex*. If $C = \left\{ \dfrac{1}{2} \right\}$, then A is called *J-convex*.

Let $\mathbb{D}$ be the set of diadic numbers:

$$\mathbb{D} = \left\{ x \in \mathbb{R} \mid x = \frac{k}{2^n}, k \in \mathbb{Z}, n \in \mathbb{N} \cup \{0\} \right\}.$$

Lemma 5.1.1. *A set $A \subset \mathbb{R}^N$ is J-convex if and only if it is $\mathbb{D}$-convex.*

Proof. Both conditions of convexity imply that $A \neq \varnothing$. The "if" part of the lemma is trivial, since $\dfrac{1}{2} \in \mathbb{D}$. Now suppose that A is J-convex. We have to show that

$$\lambda x + (1 - \lambda) y \in A \text{ for all } x, y \in A \text{ and for every } \lambda \text{ of the form } \lambda = \frac{k}{2^n}, 0 \leqslant k \leqslant$$

---

[1] The theory of convex sets is well developed and there exist many monographs of this subject. Cf., e.g., Bonnesen–Fenchel [35], Eggleston [74]. When λ runs over $[0, 1]$, the point $\lambda x + (1 - \lambda) y$ runs over the segment joining x and y. Thus the geometrical meaning of convexity is: a non-empty set $A \subset \mathbb{R}^N$ is convex iff together with arbitrary two its points it contains also the segment joining these points.

2^n, $n \in \mathbb{N} \cup \{0\}$. This is certainly true for $n = 0$ and $n = 1$. Now assume it to be true for an $n \geqslant 1$ and consider a $\lambda = \dfrac{k}{2^{n+1}}$. If $k = 2^n$, then $\lambda = \dfrac{1}{2}$, and we are back in the case $n = 1$. So let $k < 2^n$ (if $k > 2^n$ we need only interchange the roles of x and y). Then

$$\lambda x + (1 - \lambda) y = \frac{k}{2^{n+1}} x + \frac{2^n - k}{2^{n+1}} y + \frac{1}{2} y. \tag{5.1.2}$$

By the induction hypothesis the point $\dfrac{k}{2^n} x + \dfrac{2^n - k}{2^n} y$ belongs to A. But then (5.1.2) may be written as

$$\lambda x + (1 - \lambda) y = \frac{1}{2} z + \frac{1}{2} y,$$

and this is a point of A by the J-convexity of A.   □

Lemma 5.1.2. *Suppose that we are given $n + 1$ non-negative real numbers $\alpha_1, \ldots, \alpha_{n+1}$, $n \geqslant 2$, such that*

$$\sum_{i=1}^{n+1} \alpha_i = 1 \tag{5.1.3}$$

and

$$\alpha_{n+1} \geqslant \alpha_i, \ i = 1, \ldots, n. \tag{5.1.4}$$

Then the numbers $\alpha_1, \ldots, \alpha_n$ can be partitioned into two non-empty disjoint classes such that the sum of numbers in each class does not exceed $\dfrac{1}{2}$.

Proof. Let us divide the numbers $\alpha_1, \ldots, \alpha_n$ into two non-empty disjoint classes A and B in an arbitrary fashion. Suppose that the sum of numbers in one class, say B, is greater than $\dfrac{1}{2}$. Let

$$s = \sum_{\alpha_i \in A} \alpha_i, \alpha_j = \max_{\alpha_i \in B} \alpha_i.$$

Observe that α_j cannot be the only member of B, for otherwise we would have $\alpha_j > \dfrac{1}{2}$ and by (5.1.4) $\alpha_{n+1} > \dfrac{1}{2}$, which is incompatible with (5.1.3). The sum of the elements in B is, according to (5.1.3), $1 - \alpha_{n+1} - s > \dfrac{1}{2}$, whence $s + \alpha_{n+1} < \dfrac{1}{2}$, and by (5.1.4), $s + \alpha_j < \dfrac{1}{2}$. Thus we may transfer α_j from class B to class A, keeping still the sum of the elements in A under $\dfrac{1}{2}$. Continuing this procedure, after a finite number of steps we arrive at the desired decomposition.   □

Lemma 5.1.3. *Let $C \subset \mathbb{R}$ be a ring[2] such that $\dfrac{1}{2} \in C$, and let $A \subset \mathbb{R}^N$ be a C-convex set. Then for every $n \in \mathbb{N}$, every $a_1, \ldots, a_n \in A$, and every $\alpha_1, \ldots, \alpha_n \in C \cap [0, 1]$*

---

[2] It would be sufficient to assume that $(C, +)$ is a group. But in Theorem 5.1.3 the assumption that $(C; +, \cdot)$ is a ring is essential, and anyhow the only important cases are $C = \mathbb{R}, C = \mathbb{Q}$, and $C = D$.

such that $\sum\limits_{i=1}^{n} \alpha_i = 1$ *we have*

$$\alpha_1 a_1 + \cdots + \alpha_n a_n \in A. \tag{5.1.5}$$

Proof. For $n = 1$ (5.1.5) is trivial, for $n = 2$ (5.1.5) is equivalent to (5.1.1). Assume (5.1.5) to be true for an $n \in \mathbb{N}, n \geqslant 2$, and consider a combination

$$x_0 = \alpha_1 a_1 + \cdots + \alpha_{n+1} a_{n+1},$$

where $a_1, \ldots, a_{n+1} \in A$, $\alpha_1, \ldots, \alpha_{n+1} \in C \cap [0, 1]$, and (5.1.3) holds. Taking into account Lemma 5.1.2, we can change the numeration of $a_1, \ldots, a_{n+1}$ (and, correspondingly, that of $\alpha_1, \ldots, \alpha_{n+1}$) in such a manner that (5.1.4) holds, and, for a certain $k \in \mathbb{N}, 1 \leqslant k \leqslant n - 1$, we have

$$s_1 = \sum_{i=1}^{k} \alpha_i \leqslant \frac{1}{2}, \quad s_2 = \sum_{i=k+1}^{n} \alpha_i \leqslant \frac{1}{2}.$$

Put

$$x = \alpha_1 a_1 + \cdots + \alpha_k a_k + 0 \cdot a_{k+1} + \cdots + 0 \cdot a_{n-1} + \left(\frac{1}{2} - s_1\right) a_{n+1},$$

$$y = 0 \cdot a_2 + \cdots + 0 \cdot a_k + \alpha_{k+1} a_{k+1} + \cdots + \alpha_n a_n + \left(\frac{1}{2} - s_2\right) a_{n+1}.$$

By the induction hypothesis we have $2x \in A, 2y \in A$, and hence, since $\frac{1}{2} \in C$, $x_0 = \frac{1}{2}(2x + 2y) \in A$. $\qquad\square$

The following two results are almost self-evident, so we leave the task of proving them to the reader.

Theorem 5.1.1. *Let $a \in \mathbb{R}, b \in \mathbb{R}^N$, and let $C \subset \mathbb{R}$ be an arbitrary set. If a set $A \subset \mathbb{R}^N$ is C-convex, then also the set $aA + b$ is C-convex.*

Lemma 5.1.4. *Let $H \subset \mathbb{R}^N$ be a k-dimensional hyperplane, $1 \leqslant k \leqslant N - 1$. If a set $A \subset \mathbb{R}^N$ is convex, then also the perpendicular projection of A onto H is convex.*

The next result, although equally simple, has a fundamental importance.

Theorem 5.1.2. *Let $C \subset \mathbb{R}$ be an arbitrary set, and let $\mathcal{A}$ be a non-empty collection of C-convex subsets of $\mathbb{R}^N$. Then the set $A_0 = \bigcap \mathcal{A} = \bigcap\limits_{A \in \mathcal{A}} A$ is either empty or C-convex.*

Proof. Suppose that $A_0 \neq \varnothing$, and take arbitrary $x, y \in A_0$, $\lambda \in C \cap [0, 1]$. Then $x, y \in A$ for every $A \in \mathcal{A}$, whence also $\lambda x + (1 - \lambda) y \in A$ for every $A \in \mathcal{A}$. Therefore $\lambda x + (1 - \lambda) y \in A_0$, which shows that A_0 is C-convex. $\qquad\square$

Let $A \subset \mathbb{R}^N$ and $C \subset \mathbb{R}$ be arbitrary sets, $A \neq \varnothing$. Taking as $\mathcal{A}$ the family of all C-convex sets $B \subset \mathbb{R}^N$ such that $A \subset B$ ($\mathcal{A} \neq \varnothing$, since $\mathbb{R}^N \in \mathcal{A}$), we infer from Theorem 5.1.2 that there exists the smallest C-convex subset of $\mathbb{R}^N$ containing A ($\bigcap \mathcal{A} \neq \varnothing$, since $A \subset \bigcap \mathcal{A}$). This set is called the C-*convex hull* of A.

For any non-empty set $A \subset \mathbb{R}^N$ the convex hull ($C = \mathbb{R}$) of A will be denoted by $\operatorname{conv} A$, the Q-convex hull ($C = \mathbb{Q}$) of A will be denoted by $Q(A)$, the J-convex hull ($C = \left\{ \frac{1}{2} \right\}$) of A will be denoted by $J(A)$.

Theorem 5.1.3. *Let $A \subset \mathbb{R}^N$ be an arbitrary set, $A \neq \varnothing$, and let $C \subset \mathbb{R}$ be a ring. Then the C-convex hull of A is the set B given by*

$$B = \left\{ x \in \mathbb{R}^N \mid x = \sum_{i=1}^n \alpha_i a_i, \alpha_i \in C \cap [0,1], a_i \in A, i = 1, \ldots, n, \sum_{i=1}^n \alpha_i = 1; n \in \mathbb{N} \right\}.$$

$$(5.1.6)$$

Proof. Let $C(A)$ denote the C-convex hull of A. It is easily checked that the set B given by (5.1.6) is C-convex, whence $C(A) \subset B$. The inclusion $B \subset C(A)$ results from Lemma 5.1.3. Hence $C(A) = B$. $\qquad \square$

Hence we derive a number of results, which will be needed only later.

Lemma 5.1.5. *For an arbitrary non-empty set $A \subset \mathbb{R}^N$ and for an arbitrary $a \in \mathbb{R}$ we have*

$$Q(aA) = aQ(A). \qquad (5.1.7)$$

Proof. If $x \in Q(aA)$, then by Theorem 5.1.3 there exist $n \in \mathbb{N}$, $\alpha_1, \ldots, \alpha_n \in \mathbb{Q} \cap [0,1]$ and $a_1, \ldots, a_n \in A$ such that $\sum_{i=1}^n \alpha_i = 1$ and

$$x = \alpha_1(aa_1) + \cdots + \alpha_n(aa_n) = a(\alpha_1 a_1 + \cdots + \alpha_n a_n) \in aQ(A).$$

Conversely, if $x \in aQ(A)$, then (with suitable $\alpha_i \in \mathbb{Q} \cap [0,1]$ and $a_i \in A, i = 1, \ldots, n$)

$$x = a(\alpha_1 a_1 + \cdots + \alpha_n a_n) = \alpha_1(aa_1) + \cdots + \alpha_n(aa_n) \in Q(aA).$$

Consequently (5.1.7) holds. $\qquad \square$

Corollary 5.1.1. *If a set $A \subset \mathbb{R}^N, A \neq \varnothing$, is symmetric with respect to zero: $A = -A$, then so is also $Q(A)$.*

Proof. By Lemma 5.1.5 $-Q(A) = Q(-A) = Q(A)$. $\qquad \square$

Lemma 5.1.6. *For arbitrary non-empty sets $A, B \subset \mathbb{R}^N$ we have*

$$Q(A + B) = Q(A) + Q(B). \qquad (5.1.8)$$

Proof. By Theorem 5.1.3 if an $x \in Q(A+B)$ then there exist $n \in \mathbb{N}, \alpha_1, \ldots, \alpha_n \in \mathbb{Q} \cap [0,1]$, $\sum_{i=1}^{n} \alpha_i = 1$, and $a_1, \ldots, a_n \in A$, $b_1, \ldots, b_n \in B$ such that

$$x = \sum_{i=1}^{n} \alpha_i (a_i + b_i) = \sum_{i=1}^{n} \alpha_i a_i + \sum_{i=1}^{n} \alpha_i b_i \in Q(A) + Q(B).$$

On the other hand, if $x \in Q(A) + Q(B)$, then

$$x = \sum_{i=1}^{n} \alpha_i a_i + \sum_{j=1}^{m} \beta_j b_j,$$

with $\alpha_i, \beta_j \in \mathbb{Q}$, $a_i \in A, b_j \in B, i = 1, \ldots, n; j = 1, \ldots, m$, $\sum_{i=1}^{n} \alpha_i = 1, \sum_{j=1}^{m} \beta_i = 1$. Let $d \in \mathbb{N}$ be a common denominator of all α_i's and β_j's. Then x may be written as

$$x = \sum_{k=1}^{d} \gamma_k a'_k + \sum_{k=1}^{d} \gamma_k b'_k = \sum_{k=1}^{d} \gamma_k (a'_k + b'_k) \in Q(A+B),$$

where $\gamma_k = 1/d, k = 1, \ldots, d$, and a'_k, b'_k equal to suitable a_i, b_j. $\qquad\square$

If $a, b \in \mathbb{R}^N$ are arbitrary points, we write $Q(a,b)$ instead of $Q(\{a,b\})$, and call $Q(a,b)$ the *rational segment* joining the points a, b. It is easily seen from Theorem 5.1.3 that

$$Q(a,b) = \left\{ x \in \mathbb{R}^N \mid x = \lambda a + (1-\lambda) b, \lambda \in \mathbb{Q} \cap [0,1] \right\}, \qquad (5.1.9)$$

for if in the combination $x = \alpha_1 a_1 + \cdots + \alpha_n a_n$ several a_i's equal a, and the remaining equal b, we can make the suitable reduction, denoting the sum of the coefficients of $a_i = a$ by λ. And if one of a, b is lacking in the combination, we can always add it with the coefficient zero.

The set $(a \neq b)$

$$l(a,b) = \left\{ x \in \mathbb{R}^N \mid x = \lambda a + (1-\lambda) b, \lambda \in \mathbb{Q} \right\} \qquad (5.1.10)$$

is called the *rational line* passing through the points a and b. If we take $b = 0$, (5.1.10) becomes

$$l(a,0) = \left\{ x \in \mathbb{R}^N \mid x = \lambda a, \lambda \in \mathbb{Q} \right\}.$$

This is the general form of a rational line passing through the origin. Note that $a \neq 0$ (i.e., a is linearly independent over $\mathbb{Q}$) so that we have $l(a,0) = E(\{a\})$. Similarly, if a, b are linearly independent over $\mathbb{Q}$, then the set $E(\{a,b\})$ is called the rational plane passing through the origin, a, and b. Generally, for arbitrary $k \in \mathbb{N}$, if the points $a_1, \ldots, a_k \in \mathbb{R}^N$ are linearly independent over $\mathbb{Q}$, then the set $E(\{a_1, \ldots, a_k\})$ is called a *k-dimensional rational hyperplane* through the origin and the points $a_1, \ldots, a_k$. Note that for any $k \in \mathbb{N}$ it may happen that a k-dimensional rational hyperplane is entirely contained in a less dimensional (real) hyperplane. In particular, in $\mathbb{R}$ there exist k-dimensional rational hyperplanes for every $k \in \mathbb{N}$.

One of the simplest examples of a convex set is a simplex. Suppose that we are given $k + 1$ $(k \leqslant N)$ points $p_1, \ldots, p_{k+1} \in \mathbb{R}^N$ which do not lie entirely on a $(k-1)$-dimensional (real) hyperplane. The convex hull S of the set $\{p_1, \ldots, p_{k+1}\}$ is called the *simplex* with *vertices* $p_1, \ldots, p_{k+1}$, and we say, that the dimension of S is k. By Theorem 5.1.3 (cf. the remarks following formula (5.1.9))

$$S = \operatorname{conv}\{p_1, \ldots, p_{k+1}\}$$
$$= \left\{ x \in \mathbb{R}^N \mid x = \sum_{i=1}^{k+1} \alpha_i p_i, \alpha_i \in [0, 1], i = 1, \ldots, k+1; \sum_{i=1}^{k+1} \alpha_i = 1 \right\}.$$

Thus one-dimensional simplex is the segment joining the points p_1 and p_2, the two dimensional simplex is the triangle (including its interior) with the vertices p_1, p_2, p_3, and so on.

Lemma 5.1.7. *Suppose that the points* $p_1, \ldots, p_{N+1} \in \mathbb{R}^N$ *do not lie entirely on an* $(N-1)$*-dimensional hyperplane. Then the set*

$$S^\circ = \left\{ x \in \mathbb{R}^N \mid x = \sum_{i=1}^{N+1} \alpha_i p_i, \alpha_i \in (0, 1), i = 1, \ldots, N+1; \sum_{i=1}^{N+1} \alpha_i = 1 \right\}$$

is open.

Proof. Take an $x_0 \in S^\circ$,

$$x_0 = \sum_{i=1}^{N+1} \alpha_i^0 p_i, \alpha_i^0 \in (0, 1), i = 1, \ldots, N+1; \sum_{i=1}^{N+1} \alpha_i^0 = 1.$$

Consider the equalities

$$\begin{cases} x = \alpha_1 p_1 + \cdots + \alpha_{N+1} p_{N+1}, \\ 1 = \alpha_1 + \cdots + \alpha_{N+1}. \end{cases} \tag{5.1.11}$$

We may consider (5.1.11) as a system of $N+1$ linear equations with $N+1$ unknowns $\alpha_1, \ldots, \alpha_{N+1}$. Since the points $p_1, \ldots, p_{N+1}$ do not lie on an $(N-1)$-dimensional hyperplane, the determinant of the system is different from zero, and thus (5.1.11) is a Cramer system. Thus for every $x \in \mathbb{R}^N$ (5.1.11) has a unique solution $(\alpha_1, \ldots, \alpha_{N+1})$, and this solution depends on x in a continuous manner. For $x = x_0$ the solution is $(\alpha_1^0, \ldots, \alpha_{N+1}^0)$, $\alpha_i^0 \in (0, 1)$, $i = 1, \ldots, N+1$. Therefore if x remains in a small neighbourhood U of x_0, the resulting α's remain in $(0, 1)$, and thus $x \in S^\circ$. Consequently $U \subset S^\circ$, which shows that S° is open. $\qquad\square$

Corollary 5.1.2. *If* $S \subset \mathbb{R}^N$ *is an* N*-dimensional simplex, then* $\operatorname{int} S \neq \varnothing$.

Proof. S being N-dimensional means that $S = \operatorname{conv}\{p_1, \ldots, p_{N+1}\}$, where the points $p_1, \ldots, p_{N+1}$ do not lie entirely on an $(N-1)$-dimensional hyperplane. By Lemma 5.1.7 the set S° is open, and clearly $S^\circ \subset S$, whence $S^\circ \subset \operatorname{int} S$. Thus $\operatorname{int} S \neq \varnothing$. $\quad\square$

Remark 5.1.1. If $G \subset \mathbb{R}^N$ is a non-empty open set, then for every $x \in G$ there exists an N-dimensional simplex $S \subset G$ such that $x \in \operatorname{int} S$. In fact, let $\{e_1, \ldots, e_N\}$ be the usual orthonormal base of $\mathbb{R}^N$ (over $\mathbb{R}$). Consider the points

$$p_i = x + re_i, \quad i = 1, \ldots, N, \quad p_{N+1} = x - r \sum_{i=1}^N e_i, \tag{5.1.12}$$

where $r > 0$ is so small that all points (5.1.12) lie in an open ball K centered at x and contained in G. Since K is convex, also the set $S = \operatorname{conv} \{p_1, \ldots, p_{N+1}\}$ is contained in K, and hence in G. It is easy to check that points (5.1.12) do not lie on an $(N-1)$-dimensional hyperplane, and thus S is an N-dimensional simplex. Since $x = \sum_{i=1}^{N+1} \alpha p_i$, where $\alpha = 1/(N+1)$, $x \in S^\circ \subset \operatorname{int} S$.

Theorem 5.1.4. *Let $A \subset \mathbb{R}^N$ be a convex set, which does not lie entirely on an $(N-1)$-dimensional hyperplane. Then $\operatorname{int} A \neq \varnothing$.*

Proof. Since A does not lie on an $(N-1)$-dimensional hyperplane, it must contain $N+1$ points $p_1, \ldots, p_{N+1}$ which do not lie on an $(N-1)$-dimensional hyperplane. Let S be the N-dimensional simplex with the vertices $p_1, \ldots, p_{N+1}$. Since A is convex, $S \subset A$, and by Corollary 5.1.2 we obtain $\operatorname{int} A \neq \varnothing$. $\square$

Lemma 5.1.8. *Let $A \subset \mathbb{R}^N$ be a convex set, and let $a \in \operatorname{int} A, b \in A$. Then, for every $\lambda \in (0, 1)$,*

$$x = \lambda a + (1 - \lambda) b \in \operatorname{int} A. \tag{5.1.13}$$

Proof. Since $a \in \operatorname{int} A$, there exists an $r > 0$ such that the open ball $K = K(a, r)$ centered at a and with the radius r is contained in A:

$$K \subset A. \tag{5.1.14}$$

Fix a $\lambda \in (0, 1)$, and consider x given by (5.1.13). Let $U = K(x, \lambda r)$ be the open ball centered at x and with the radius λr. Take a $y \in U$. Then $|y - x| < \lambda r$, i.e., $|y - \lambda a - (1 - \lambda) b| < \lambda r$, or

$$\left| \frac{y - (1 - \lambda) b}{\lambda} - a \right| < r.$$

By (5.1.14)

$$\frac{y - (1 - \lambda) b}{\lambda} \in A. \tag{5.1.15}$$

Now we have

$$y = \lambda \frac{y - (1 - \lambda) b}{\lambda} + (1 - \lambda) b,$$

which, in virtue of (5.1.15) and of the convexity of A, implies that $y \in A$. Hence $U \subset A$, which means that x is n interior point of A. $\square$

Corollary 5.1.3. *Let $A \subset \mathbb{R}^N$ be a convex set, and let $a \in \operatorname{int} A, b \in \operatorname{cl} A$. Then, for every $\lambda \in (0,1)$, (5.1.13) holds.*

Proof. Choose $r > 0$ so that $K = K(a, r)$ fulfils (5.1.14). Fix a $\lambda \in (0,1)$, and consider points (5.1.13). We may find a $y \in A$ such that

$$|y - b| < r \frac{|b - x|}{|a - x|}. \tag{5.1.16}$$

Put

$$z = a - \frac{|a - x|}{|b - x|}(y - b). \tag{5.1.17}$$

Then by (5.1.16)

$$|z - a| = \frac{|a - x|}{|b - x|}|y - b| < r,$$

whence $z \in K$, and by (5.1.14) $z \in A$ so that actually $z \in \operatorname{int} A$. Now, since $x = \lambda a + (1 - \lambda) b = \lambda(a - b) + b = (1 - \lambda)(b - a) + a$, we have

$$\lambda = \frac{|b - x|}{|b - a|}, \quad 1 - \lambda = \frac{|a - x|}{|b - a|}.$$

Thus (5.1.17) may be written as

$$z = a - \frac{1 - \lambda}{\lambda}(y - b),$$

whence $\lambda a + (1 - \lambda) b = \lambda z + (1 - \lambda) y$, i.e.,

$$x = \lambda z + (1 - \lambda) y.$$

By Lemma 5.1.8 $x \in \operatorname{int} A$. $\qquad \square$

Hence we derive some consequences pointing at a certain regularity of convex sets.

Theorem 5.1.5. *Let $A \subset \mathbb{R}^N$ be a convex set such that $\operatorname{int} A \neq \varnothing$. Then the set $\operatorname{int} A$ is convex[3].*

Proof. This is an immediate consequence of Lemma 5.1.8. $\qquad \square$

Theorem 5.1.6. *Let $A \subset \mathbb{R}^N$ be a convex set such that $\operatorname{int} A \neq \varnothing$. Then $\operatorname{cl}\operatorname{int} A = \operatorname{cl} A$.*

Proof. Since $\operatorname{int} A \subset A$, we have $\operatorname{cl}\operatorname{int} A \subset \operatorname{cl} A$. Now let $b \in \operatorname{cl} A$, and choose an arbitrary $a \in \operatorname{int} A$. By Corollary 5.1.3 we have (5.1.13) for every $\lambda \in (0,1)$. As $\lambda \to 0$, the point x tends to b, whence $b \in \operatorname{cl}\operatorname{int} A$. Consequently $\operatorname{cl} A \subset \operatorname{cl}\operatorname{int} A$, whence $\operatorname{cl}\operatorname{int} A = \operatorname{cl} A$. $\qquad \square$

---

[3] Cf. also Exercise 5.4 for a dual theorem.

A dual theorem is also true (cf. Exercise 5.5).

Let $H \subset \mathbb{R}^N$ be an $(N-1)$-dimensional hyperplane. Then H separates $\mathbb{R}^N$ into two open half-spaces. These half-spaces will be denoted by π_H^1 and π_H^2. So $\mathbb{R}^N = \pi_H^1 \cup \pi_H^2 \cup H$, and the sets occurring in this decomposition are pairwise disjoint.

Let $A \subset \mathbb{R}^N, A \neq \varnothing$, be an arbitrary set. An $(N-1)$-dimensional hyperplane H is called a *support hyperplane* of A iff

(i) $H \cap \mathrm{cl}\, A \neq \varnothing$;

(ii) $A \cap \pi_H^1 = \varnothing$, or $A \cap \pi_H^2 = \varnothing$.

In other words, H is a support hyperplane of A iff $H \cap \mathrm{cl}\, A \neq \varnothing$ and A lies entirely on one side of H.

Lemma 5.1.9. *Let $A \subset \mathbb{R}^N$ be a convex set such that $\mathrm{int}\, A \neq \varnothing$. An $(N-1)$-dimensional hyperplane H is a support hyperplane of A if and only if $H \cap \mathrm{cl}\, A \neq \varnothing$ and $H \cap \mathrm{int}\, A = \varnothing$.*

Proof. Suppose that H is a support hyperplane of A. Then H satisfies conditions (i) and (ii). Let, e.g., $A \cap \pi_H^1 = \varnothing$. Then $A \subset \pi_H^2 \cup H$, whence $\mathrm{int}\, A \subset \mathrm{int}\, \left(\pi_H^2 \cup H\right) = \pi_H^2$, and $H \cap \mathrm{int}\, A = \varnothing$.

Now suppose that H fulfils (i) and $H \cap \mathrm{int}\, A = \varnothing$. Put $B = \mathrm{int}\, A$, and suppose that $B \cap \pi_H^1 \neq \varnothing$ and $B \cap \pi_H^2 \neq \varnothing$. Then there exist points $a \in B \cap \pi_H^1$ and $b \in B \cap \pi_H^2$. By Theorem 5.1.5 the set B is convex, so the whole segment joining a and b lies in B. But this segment must intersect H, whence we would get $H \cap B \neq \varnothing$. Consequently, either $\pi_H^1 \cap \mathrm{int}\, A = \varnothing$, or $\pi_H^2 \cap \mathrm{int}\, A = \varnothing$. Suppose that, e.g., $\pi_H^1 \cap \mathrm{int}\, A = \varnothing$. Then $\mathrm{int}\, A \subset \pi_H^2 \cup H$, and since $H \cap \mathrm{int}\, A = \varnothing$, $\mathrm{int}\, A \subset \pi_H^2$. Hence, using Theorem 5.1.6, we get $A \subset \mathrm{cl}\, A = \mathrm{cl\,int}\, A \subset \mathrm{cl}\, \pi_H^2 = \pi_H^2 \cup H$. Hence $A \cap \pi_H^1 = \varnothing$, and H fulfils (ii). Consequently H is support hyperplane of A. $\square$

Lemma 5.1.10. *Let $D \subset \mathbb{R}^N$ be a convex and open set, and let L_0 be a linear subspace of $\mathbb{R}^N$ (over $\mathbb{R}$) of a dimension r, $0 \leqslant r < N$, such that $L_0 \cap D = \varnothing$. Then there exists a linear subspace L of $\mathbb{R}^N$, of the dimension $N-1$, and such that $L_0 \subset L$ and $L \cap D = \varnothing$.*

Proof. Let L be a linear subspace of $\mathbb{R}^N$, of the greatest possible dimension, such that $L_0 \subset L$ and $L \cap D = \varnothing$. Let $\dim L = s$ so that $r \leqslant s < N$. The thing to prove is that $s = N-1$.

Let M be a linear subspace of $\mathbb{R}^N$ perpendicular to L of the dimension $N-s$. Let B be the perpendicular projection of D onto M. The set B is convex (cf. Lemma 5.1.4) and is open in the relative topology of M. The perpendicular projection of L onto M is the point 0. Let $l \subset M$ be a straight line passing through 0. If $l \cap B = \varnothing$, then the linear subspace L_1 of $\mathbb{R}^N$ spanned by $L \cup l$ is disjoint with D, contains L_0, and $\dim L_1 = s+1$, which contradicts the fact that the dimension of L is maximal. Thus $l \cap B \neq \varnothing$ for every line l in M passing through the origin. Of course, $0 \notin B$, since $L \cap D = \varnothing$.

If $N - s = 1$, the lemma is proved. So suppose that $N - s \geqslant 2$. Let $P \subset M$ be a plane through the origin (a two-dimensional linear subspace of $\mathbb{R}^N$). The set $A = B \cap P$ is not empty and hence, by Theorem 5.1.2, A is convex. A is also open in the relative topology of P. We have $l \cap A \neq \varnothing$ for every line $l \subset P$ passing through the origin. Moreover,

$$0 \neq A. \tag{5.1.18}$$

All rays starting from the origin and passing through the points of A form a sector C in P with vertex at 0. We have

$$A \subset C. \tag{5.1.19}$$

Let α be the angle of C at the vertex. If $\alpha < \pi$, then we may draw a line l in P such that $l \cap C = \{0\}$, whence $l \cap A = \varnothing$, which is impossible. Suppose that $\alpha = \pi$. Then C is a half-plane determined by a line $l \subset P$ passing through zero. But $A \cap l \neq \varnothing$, so A must have a point on l, and being open in P, must have points on both the sides of l, which is incompatible with (5.1.19). Consequently $\alpha > \pi$. Then C contains a half-plane P_1. Let P_2 be the other half-plane of P. Since $\alpha > \pi$, $C \cap P_2 \neq \varnothing$, and on every ray (from 0) contained in $C \cap P_2$ there are points of A. Choose an $a \in A \cap P_2$. Draw a line l in P through a and 0. Then $l \cap P_1$ is a ray in C, whence $A \cap (l \cap P_1) \neq \varnothing$. Choose a $b \in A \cap P_1$. Then the whole segment joining a and b is contained in A, since A is convex, hence $0 \in A$, which contradicts (5.1.18). Consequently we must have $N - s = 1$, i.e., $s = N - 1$. $\square$

Now we are able to prove one of the fundamental results in the theory of convex sets.

Theorem 5.1.7. *Let $A \subset \mathbb{R}^N$ be a convex set. Through every point of the frontier of A there passes a support hyperplane of A.*

Proof. Suppose that A lies on an $(N-1)$-dimensional hyperplane H: $A \subset H$. Then H fulfils conditions (i) and (ii), and hence it is a support hyperplane of A.

Now suppose that A does not lie on any $(N-1)$-dimensional hyperplane. Then by Theorem 5.1.4 the set $\operatorname{int} A$ is non-empty, and hence, by Theorem 5.1.5, $\operatorname{int} A$ is a convex open set. Let p be a frontier point of A so that $p \in \operatorname{cl} A$, but $p \notin \operatorname{int} A$. Put $D = \operatorname{int} A - p$. According to Theorem 5.1.1 D is a convex and open set, $0 \notin D$. The set $L_0 = \{0\}$ is a 0-dimensional linear subspace of $\mathbb{R}^N$, and $L_0 \cap D = \varnothing$, since $0 \notin D$. By Lemma 5.1.10 there exists an $(N-1)$-dimensional linear subspace L of $\mathbb{R}^N$ such that $0 \in L$ (which is true about every subspace of $\mathbb{R}^N$), and $L \cap D = \varnothing$. Then $H = L + p$ is an $(N-1)$-dimensional hyperplane, $p \in H$ so that $H \cap \operatorname{cl} A \neq \varnothing$, and $H \cap \operatorname{int} A = H \cap (D + p) = \varnothing$. By Lemma 5.1.9 H is a support hyperplane of A. $\square$

There is no uniqueness attached to the constructions in Lemma 5.1.10 and Theorem 5.1.7, and there is no uniqueness attached to support hyperplanes. A convex set may have many support hyperplanes passing through the same point. E.g., in the case $N = 2$ consider the set

$$A = \left\{ (x, y) \in \mathbb{R}^2 \mid y > |x| \right\}. \tag{5.1.20}$$

It is easy to check that the set (5.1.20) is convex (cf. also Exercise 7.1), the origin is its frontier point, and every line $y = ax$ with $|a| \leqslant 1$ is a support line[4] of A passing through the origin.

---

[4] In the case where $N = 2$ (and thus $(N-1)$-dimensional hyperplanes are simply lines) we say *support line* instead of *support hyperplane*.

The existence of support hyperplanes is a very characteristic feature of convex sets. It may be proved that if $A \subset \mathbb{R}^N$ is a closed and bounded set such that $\operatorname{int} A \neq \varnothing$, and A has a support hyperplane at every frontier point, then A is convex (Eggleston [74]).

Lemma 5.1.11. *Let $A \subset \mathbb{R}^N$ be convex and closed set, and let $x \notin A$. Then there exists a support hyperplane H of A which separates x from A.*

Proof. Let $r = d(x, A) > 0$ be the distance from x to A, and let K be a closed ball centered at x and with a radius greater than r. Then

$$d(x, A) = \inf_{y \in A} |x - y| = \inf_{y \in A \cap K} |x - y|,$$

and since the set $A \cap K$ is compact (being closed and bounded), there is a point $z \in A$ such that

$$d(x, A) = |x - z|. \tag{5.1.21}$$

Let H be the $(N-1)$-dimensional hyperplane passing through z and perpendicular to the segment $\overline{xz}$. Let π denote that of the two open half-spaces into which H divides $\mathbb{R}^N$ which contains x. Suppose that there exists a $u \in A \cap \pi$. Then the angle $\Theta = \angle xzu$ is acute so that

$$\cos \Theta > 0. \tag{5.1.22}$$

Consider the point

$$v = z + \lambda(u - z) = \lambda u + (1 - \lambda) z,$$

with a $\lambda \in (0, 1)$. Since z and u belong to A, and A is convex, also v belongs to A. Now, we have $v - z = \lambda(u - z)$, whence $x - v = (x - z) - (v - z) = (x - z) - \lambda(u - z)$, and

$$
\begin{aligned}
|x - v|^2 &= |x - z|^2 + \lambda^2 |u - z|^2 - 2\lambda(x - z)(u - z) \\
&= |x - z|^2 + \lambda^2 |u - z|^2 - 2\lambda |x - z| |u - z| \cos \Theta \\
&= |x - z|^2 + \lambda |u - z| (\lambda |u - z| - 2|x - z| \cos \Theta).
\end{aligned}
$$

If $\lambda \in (0, 1)$ is small enough, we have by (5.1.22) $\lambda |u - z| < 2|x - z| \cos \Theta$, whence $|x - v| < |x - z|$. But this contradicts (5.1.21), since $v \in A$.

Consequently we must have $A \cap \pi = \varnothing$, which means that H is a support hyperplane of A and that H separates x from A. $\qquad\square$

Theorem 5.1.8. *Let $A \subset \mathbb{R}^N, A \neq \mathbb{R}^N$, be a convex and closed set[5]. Then A is equal to the intersection of all the closed half-spaces containing A, determined by all the support hyperplanes of A.*

Proof. Let $\mathcal{H}$ be the collection of all support hyperplanes of A. Since $A \neq \mathbb{R}^N$ the frontier of A is non-empty, and in virtue of Theorem 5.1.7 also $\mathcal{H} \neq \varnothing$. For every

---

[5] Theorem remains true also for $A = \mathbb{R}^N$ if we agree that $\bigcap \varnothing = \mathbb{R}^N$.

$H \in \mathcal{H}$ let Π_H denote that of the two closed subspaces into which H divides $\mathbb{R}^N$ which contains A:

$$A \subset \Pi_H \text{ for } H \in \mathcal{H}. \tag{5.1.23}$$

The thing to prove is that

$$A = \bigcap_{H \in \mathcal{H}} \Pi_H. \tag{5.1.24}$$

By (5.1.23) we have $A \subset \bigcap_{H \in \mathcal{H}} \Pi_H$. Take an $x \notin A$. By Lemma 5.1.11 there exists an $H_0 \in \mathcal{H}$ such that $x \notin \Pi_{H_0}$, whence $x \notin \bigcap_{H \in \mathcal{H}} \Pi_H$. This shows that $\bigcap_{H \in \mathcal{H}} \Pi_H \subset A$. Hence (5.1.24) follows. $\square$

5.2 Additive functions

A function[6] $f : \mathbb{R}^N \to \mathbb{R}$ is called *additive* iff it satisfies Cauchy's functional equation

$$f(x + y) = f(x) + f(y) \tag{5.2.1}$$

for all $x, y \in \mathbb{R}^N$. For $N = 1$ equation (5.2.1) was first treated by A. M. Legendre [206] and C. F. Gauss [96], but A. L. Cauchy [41] first found its general continuous solution, and the equation has been named after him. Concerning equation (5.2.1) and the vast literature of the subject, cf. also Aczél [5].

It follows from (5.2.1) by induction that if $f : \mathbb{R}^N \to \mathbb{R}$ is additive, then

$$f\left(\sum_{i=1}^{n} x_i\right) = \sum_{i=1}^{n} f(x_i) \tag{5.2.2}$$

for every $n \in \mathbb{N}$ and for every $x_1, \ldots, x_n \in \mathbb{R}^N$.

Lemma 5.2.1. *If $f_1 : \mathbb{R}^N \to \mathbb{R}$ and $f_2 : \mathbb{R}^N \to \mathbb{R}$ are additive functions, then, for every $a, b \in \mathbb{R}$, the function $f = af_1 + bf_2$ is additive.*

This results immediately from (5.2.1).

The following theorem gives a fundamental property of solutions of equation (5.2.1).

Theorem 5.2.1. *If $f : \mathbb{R}^N \to \mathbb{R}$ satisfies equation (5.2.1), then*

$$f(\lambda x) = \lambda f(x) \tag{5.2.3}$$

for every $x \in \mathbb{R}^N$ and $\lambda \in \mathbb{Q}$.

Proof. For $x = y = 0$ we get from (5.2.1)

$$f(0) = 0, \tag{5.2.4}$$

---

[6] So in the present book we consider only finite-valued additive functions. Additive functions with infinite values have been considered by I. Halperin [131]. He showed that the only such functions are $f = +\infty$ and $f = -\infty$ (cf. also 16.6).

whence, setting in (5.2.1) $y = -x$, we obtain

$$0 = f(0) = f(x - x) = f(x) + f(-x),$$

i.e.,

$$f(-x) = -f(x). \tag{5.2.5}$$

Thus the function f is odd.

Taking in (5.2.2) $x_1 = \cdots = x_n = x$ we obtain (5.2.3) for all $\lambda \in \mathbb{N}$. According to (5.2.4) and (5.2.5), (5.2.3) holds for all $\lambda \in \mathbb{Z}$.

Now, an arbitrary $\lambda \in \mathbb{Q}$ may be written as $\lambda = \dfrac{k}{m}$, where $k \in \mathbb{Z}, m \in \mathbb{N}$. Hence $kx = m(\lambda x)$ and so, by what already has been proved,

$$kf(x) = f(kx) = f(m(\lambda x)) = mf(\lambda x),$$

whence (5.2.3) follows.   □

If $N = 1$, then we get from (5.2.3) with $c = f(1)$

$$f(\lambda x) = c\lambda \text{ for } \lambda \in \mathbb{Q},$$

whence, if f is continuous,

$$f(x) = cx \text{ for } \lambda \in \mathbb{R}. \tag{5.2.6}$$

Formula (5.2.6) gives the general continuous additive functions $f : \mathbb{R} \to \mathbb{R}$. For continuous additive functions $f : \mathbb{R}^N \to \mathbb{R}$, $N > 1$, the corresponding formula will be found in 5.5.

Theorem 5.2.1 says that every additive function $f : \mathbb{R}^N \to \mathbb{R}$ is a homomorphism from the space $(\mathbb{R}^N; \mathbb{Q}; +; \cdot)$ into $(\mathbb{R}; \mathbb{Q}; +; \cdot)$(cf. 4.3). Thus as an immediate consequence of Theorem 4.3.1 we obtain the following

Theorem 5.2.2. *Let H be an arbitrary Hamel basis of the space $(\mathbb{R}^N; \mathbb{Q}; +; \cdot)$. Then for every function $g : H \to \mathbb{R}$ there exists a unique additive function $f : \mathbb{R}^N \to \mathbb{R}$ such that $f \mid H = g$.*

Theorem 5.2.2 gives the general construction of all the additive functions $f : \mathbb{R}^N \to \mathbb{R}$. In fact, every additive f may be obtained as the unique additive extension of a certain function $g : H \to \mathbb{R}$, viz. $g = f \mid H$.

For many years the existence of discontinuous additive functions was an open problem. Mathematicians could neither prove that every additive function is continuous (which is false), nor exhibit an example of a discontinuous additive function (as we will see later, it was a hopeless task, since there do not exist effective examples of discontinuous additive functions). It was G. Hamel who first succeeded in proving that there exist discontinuous additive functions. In Hamel [134] he proved Theorem 5.2.2 above (for $N = 1$), whence the existence of discontinuous additive functions follows easily. In fact, we obtain from Theorem 5.2.2

Corollary 5.2.1. *Let H be an arbitrary Hamel basis of the space $(\mathbb{R}; \mathbb{Q}; +; \cdot)$, and let $g : H \to \mathbb{R}$ be an arbitrary function. Let $f : \mathbb{R} \to \mathbb{R}$ be the unique additive extension of g. The function f is continuous if and only if $g(x)/x = $ const for $x \in H$.*

Proof. If f is continuous, then it has form (5.2.6), and in particular, for $x \in H$, we must have $g(x) = f(x) = cx$, i.e., $g(x)/x = c = $ const. If $g(x) = cx$ for $x \in H$, then the function cx is an additive extension of g, and by the uniqueness of such an extension f must be given by (5.2.6). Thus f is continuous. $\square$

Now, it is enough to take any function $g : H \to \mathbb{R}$ such that $g(x)/x \neq$ const on H, and as an additive extension of g we obtain a discontinuous additive function.

This argument would have to be modified for $N > 1$, but we can also argue otherwise. It follows from the proof of Theorem 4.3.1 that the values of an additive extension $f : \mathbb{R}^N \to \mathbb{R}$ of a function $g : H \to \mathbb{R}$ are given as linear combinations of the values of g (at suitable points of H) with rational coefficients. Hence, if g takes on only rational values, then also all values of f will be rational, and thus f cannot be continuous. We shall often make use of this observation, therefore we will formulate it as

Corollary 5.2.2. *Let H be an arbitrary Hamel basis of the space $(\mathbb{R}^N; \mathbb{Q}; +; \cdot)$, and let $g : H \to \mathbb{R}$ be a function with rational values only $(g(H) \subset \mathbb{Q})$, $g \neq 0$. Then the additive extension $f : \mathbb{R}^N \to \mathbb{R}$ of g is discontinuous.*

Hence we infer the existence of discontinuous additive functions $f : \mathbb{R}^N \to \mathbb{R}$ for arbitrary $N \in \mathbb{N}$.

Discontinuous additive functions are sometimes called *Hamel functions*. They exhibit many pathological properties, as will be seen later in this book. Since such functions are undesirable in applications, we will try to find possibly weak conditions which assure the continuity of an additive function.

5.3 Convex functions

Let $D \subset \mathbb{R}^N$ be a convex and open[7] set. A function $f : D \to \mathbb{R}$ is called *convex*[8] iff it satisfies Jensen's functional inequality

$$f\left(\frac{x+y}{2}\right) \leqslant \frac{f(x) + f(y)}{2} \tag{5.3.1}$$

for all $x, y \in D$. If the inequality in (5.3.1) for $x \neq y$ is sharp, f is called *strictly convex*.

If a function $f : D \to \mathbb{R}$ (where $D \subset \mathbb{R}^N$ is a convex and open set) satisfies the inequality

$$f\left(\frac{x+y}{2}\right) \geqslant \frac{f(x) + f(y)}{2} \tag{5.3.2}$$

---

[7] In this book we consider convex functions only on convex and open subsets of $\mathbb{R}^N$, and we consider only finite-valued functions. Continuous convex functions defined on subsets of $\mathbb{R}^N$ which are convex, but not necessarily open, and also infinite-valued convex functions, are discussed in detail in Rockafellar [268]. Adjoining to D also some of its frontier points would seriously complicate the presentation, but the essentials would remain the same. (Let us note also that convex functions are often considered in more general spaces than $\mathbb{R}^N$; cf., e.g., Roberts–Varberg [267]).

[8] Some authors call functions fulfilling (5.3.1) *J-convex* or *Jensen-convex*, reserving the name convex for functions fulfilling (5.3.7) for all real $\lambda \in [0,1]$. Since in the present book we are primarily interested in functions fulfilling Jensen's inequality (5.3.1), we call such functions simply convex, as many authors do. Functions fulfilling (5.3.7) for all real $\lambda \in [0,1]$ will be referred to as *continuous and convex*. As we shall see later, they are necessarily continuous.

for all $x, y \in D$, then f is called *concave*. If the inequality for $x \neq y$ is sharp, f is called *strictly concave*. It is easily seen that a function $f : D \to \mathbb{R}$ is concave if and only if the function $-f$ is convex. So the properties of concave functions can easily be obtained from those of convex functions. Every theorem about convex functions can be "translated" into a theorem about concave functions, and *vice versa*. Therefore concave functions will not be separately treated in this book.

Convex functions were introduced (for $N = 1$) by J. L. W. V. Jensen [155], [156], although functions satisfying similar conditions were already treated by O. Hölder [143], J. Hadamard [127] and O. Stolz [298]. Basic properties of convex functions in the one-dimensional case were proved by Jensen himself and by F. Bernstein - G. Doetsch [30]. Generalizations to higher dimensions were made by H. Blumberg [33] and E. Mohr [227].

In his fundamental paper J. L. W. V. Jensen [156] wrote: "Il semble que la notion de fonction convexe est a peu près aussi fondamentale que celles-ci fonction positive, fonction croissante. Si je ne me trompe pas en ceci, la notion devra trouver sa place dans les expositions élémentaires de la théorie des fonctions réelles." (It seems to me that the notion of convex function is just as fundamental as positive function or increasing function. If I am not mistaken in this, the notion ought to find its place in elementary expositions of the theory of real functions). And he was certainly not mistaken. Convex functions are more or less extensively treated in various textbooks on calculus (cf., e.g., Bourbaki [37], Haupt–Aumann [139]). The whole monographs[9] (e.g., Rockafellar [268], Roberts–Varberg [267], Popoviciou [260]; cf. also Krasnosel'skiĭ–Rutickiĭ [175]) and expositions(Beckenbach [22], Moldovan [230], Kuczma [181]) have been devoted to them. They have been given many generalizations (Beckenbach [21], Beckenbach–Bing [25], Popoviciou [259], Moldovan [228], [229], [230], Kemperman [167], Ger [104], Deák [65], Czerwik [51], Ponstein [258], Vertgeĭn–Rubinšteĭn [317], Mitrinović [226], Guerraggio–Paganoni [126], Sander [275]; cf. also Chapter 15). They play extremely important role in many branches of mathematics.

In the present book we deal with seemingly less important aspects of the theory of convex functions. However, the importance of considering pathological examples of non-measurable functions lies not in investigating their properties, but in establishing that such examples may occur and finding what conditions are essential and cannot be released when we want to avoid such functions. Continuous convex functions have very beautiful properties (cf. Chapter 7). But in order to enjoy them we must know, what (possibly weak) conditions should be imposed on the functions considered in order to guarantee that they are actually continuous.

Properties of convex functions are strikingly similar to those of additive functions. Actually we have by Theorem 5.2.1 for any additive function $f : \mathbb{R}^N \to \mathbb{R}$

$$f\left(\frac{x+y}{2}\right) = \frac{1}{2} f(x+y) = \frac{f(x)+f(y)}{2},$$

i.e., *every additive function is convex*[10].

---

[9] These monographs deal mainly with continuous convex functions.
[10] As well as concave. The converse is not true: if a function $f : \mathbb{R}^N \to \mathbb{R}$ is at the same time convex and concave, then it differs from an additive function by a constant. Cf. 13.2 for details.

Many other convex functions are well known; for $N = 1$ the functions $f(x) = x^2$, $f(x) = \exp x$, $f(x) = |x|$ are all convex on any open interval $I \subset \mathbb{R}$ (also on $I = \mathbb{R}$). In Chapter 7 we will see how to find examples of convex functions in $\mathbb{R}^N$. Some examples of convex functions will also be given at the end of this section.

We start with some trivial results.

Theorem 5.3.1. *The linear combination, with non-negative real coefficients, of convex functions is a convex function.*

Theorem 5.3.1[11] is quite obvious and requires no proof. Let us note that, as a result of Theorem 5.3.1, the sum of a finite number of convex functions is a convex function, as is also the product of a convex function by a non-negative constant.

On the other hand, the product of (even non-negative) convex functions need not be convex (the function $f(x) = x^2 \exp x$ on $\mathbb{R}$ is a counter-example). But we have (cf. also Corollary 7.2.1)

Theorem 5.3.2. *The square of a non-negative convex function is a convex function.*

Also Theorem 5.3.2 is almost self-evident and will be given no proof, similarly as the following

Theorem 5.3.3. *The limit of a convergent sequence of convex functions is a convex function.*

Corollary 5.3.1. *The sum of a convergent series of convex functions is a convex function.*

Theorem 5.3.4. *Let $D \subset \mathbb{R}^N$ be a convex and open set, and let $f_1, f_2 : D \to \mathbb{R}$ be convex functions. Then the function f given by $f(x) = \max\big(f_1(x), f_2(\dot{x})\big)$, $x \in D$, is convex.*

Proof. Take arbitrary $x, y \in D$. We have

$$f_1\left(\frac{x+y}{2}\right) \leqslant \frac{f_1(x) + f_1(y)}{2} \leqslant \frac{f(x) + f(y)}{2},$$

and

$$f_2\left(\frac{x+y}{2}\right) \leqslant \frac{f_2(x) + f_2(y)}{2} \leqslant \frac{f(x) + f(y)}{2}.$$

Whence also

$$f\left(\frac{x+y}{2}\right) = \max\left(f_1\left(\frac{x+y}{2}\right), f_2\left(\frac{x+y}{2}\right)\right) \leqslant \frac{f(x) + f(y)}{2},$$

i.e., f is convex. $\square$

---

[11] Theorem 5.2.1 says that the family of all convex functions $f : D \to \mathbb{R}$ forms a cone in the space of all functions $f : D \to \mathbb{R}$. Cf. Kemperman [168] concerning further research in this direction.

Lemma 5.3.1. *Let $D \subset \mathbb{R}^N$ be a convex and open set. If $f : D \to \mathbb{R}$ is a convex function, then for every $n \in \mathbb{N}$ and for every $x_1, \ldots, x_n \in D$*

$$f\left(\frac{x_1 + \cdots + x_n}{n}\right) \leqslant \frac{f(x_1) + \cdots + f(x_n)}{n}. \qquad (5.3.3)$$

Proof. It follows from (5.3.1) by induction that for every $p \in \mathbb{N}$ and for every $x_1, \ldots, x_n \in D$

$$f\left(\frac{1}{2^p} \sum_{i=1}^{2^p} x_i\right) \leqslant \frac{1}{2^p} \sum_{i=1}^{2^p} f(x_i). \qquad (5.3.4)$$

Now fix an $n \in \mathbb{N}$, and choose a $p \in \mathbb{N}$ such that $n < 2^p$. Take arbitrary $x_1, \ldots, x_n \in D$, and put

$$x_k = \frac{1}{n} \sum_{i=1}^{n} x_i \quad \text{for} \quad k = n+1, \ldots, 2^p. \qquad (5.3.5)$$

Since D is convex, points (5.3.5) belong to D in virtue of Lemma 5.1.3. We have

$$\frac{1}{2^p} \sum_{i=1}^{2^p} x_i = \frac{1}{n} \sum_{i=1}^{n} x_i,$$

whence by (5.3.4)

$$f\left(\frac{1}{n} \sum_{i=1}^{n} x_i\right) = f\left(\frac{1}{2^p} \sum_{i=1}^{2^p} x_i\right) \leqslant \frac{1}{2^p} \sum_{i=1}^{2^p} f(x_i)$$

$$= \frac{1}{2^p}\left[\sum_{i=1}^{n} f(x_i) + (2^p - n) f\left(\frac{1}{n} \sum_{i=1}^{n} x_i\right)\right]. \qquad (5.3.6)$$

From (5.3.6) we obtain

$$n f\left(\frac{1}{n} \sum_{i=1}^{n} x_i\right) \leqslant \frac{1}{n} \sum_{i=1}^{n} f(x_i),$$

which yields (5.3.3). $\qquad \square$

Theorem 5.3.5. *Let $D \subset \mathbb{R}^N$ be a convex and open set. If $f : D \to \mathbb{R}$ is a convex function, then for every $x, y \in D$ and for every $\lambda \in \mathbb{Q} \cap [0, 1]$ we have*

$$f(\lambda x + (1 - \lambda)y) \leqslant \lambda f(x) + (1 - \lambda)f(y). \qquad (5.3.7)$$

Proof. Let $\lambda = \dfrac{k}{n}, n \in \mathbb{N}, 0 < k < n$. Put $x_1 = \cdots = x_k = x$, $x_{k+1} = \cdots = x_n = y$. By (5.3.3)

$$f\left(\frac{kx + (n-k)y}{n}\right) \leqslant \frac{kf(x) + (n-k) f(y)}{n},$$

which is the same as (5.3.7). If $\lambda = 0$ or 1, (5.3.7) is trivial. $\qquad \square$

If $f : D \to \mathbb{R}$ is convex and *continuous*, then (5.3.7) holds for all real $\lambda \in [0, 1]$. The converse theorem is also true (cf. 7.1).

Lemma 5.3.2. *Let $D \subset \mathbb{R}^N$ be a convex set, and let $F \subset \mathbb{R}$ be a field. If a function*[12] *$f : D \to [-\infty, \infty)$ satisfies (5.3.7) for all $\lambda \in F \cap [0,1]$, then*

$$f\left(\sum_{i=1}^{n} \lambda_i x_i\right) \leqslant \sum_{i=1}^{n} \lambda_i f(x_i) \tag{5.3.8}$$

for every $n \in \mathbb{N}$, $x_1, \ldots, x_n \in D$, and $\lambda_1, \ldots, \lambda_n \in F \cap [0,1]$ such that $\sum_{i=1}^{n} \lambda_i = 1$.

Proof. The proof runs by induction. For $n = 2$ (5.3.8) is identical with (5.3.7). Now suppose (5.3.8) to be true for an $n \in \mathbb{N}$. Take arbitrary $x_1, \ldots, x_{n+1} \in D$ and $\lambda_1, \ldots, \lambda_{n+1} \in F \cap [0,1]$ such that $\lambda_1 + \cdots + \lambda_{n+1} = 1$. If $\lambda_1 = \cdots = \lambda_n = 0, \lambda_{n+1} = 1$, then (5.3.8) is trivial. If $\lambda_1 + \cdots + \lambda_n \neq 0$, then we get by (5.3.7)

$$f\left(\sum_{i=1}^{n+1} \lambda_i x_i\right) = f\left(\left(\sum_{i=1}^{n} \lambda_i\right) \frac{\lambda_1 x_1 + \cdots + \lambda_n x_n}{\lambda_1 + \cdots + \lambda_n} + \lambda_{n+1} x_{n+1}\right)$$

$$\leqslant \left(\sum_{i=1}^{n} \lambda_i\right) f\left(\frac{\lambda_1}{\lambda_1 + \cdots + \lambda_n} x_1 + \cdots + \frac{\lambda_n}{\lambda_1 + \cdots + \lambda_n} x_n\right) + \lambda_{n+1} f(x_{n+1}). \tag{5.3.9}$$

By the induction hypothesis

$$f\left(\frac{\lambda_1}{\lambda_1 + \cdots + \lambda_n} x_1 + \cdots + \frac{\lambda_n}{\lambda_1 + \cdots + \lambda_n} x_n\right)$$

$$\leqslant \frac{\lambda_1}{\lambda_1 + \cdots + \lambda_n} f(x_1) + \cdots + \frac{\lambda_n}{\lambda_1 + \cdots + \lambda_n} f(x_n),$$

and we obtain from (5.3.9)

$$f\left(\sum_{i=1}^{n+1} \lambda_i x_i\right) \leqslant \sum_{i=1}^{n+1} \lambda_i f(x_i),$$

i.e., (5.3.8) for $n + 1$. Thus (5.3.8) is generally true. $\square$

Thus inequality (5.3.8) with arbitrary $x_1, \ldots, x_n \in D$ is valid for every convex function $f : D \to \mathbb{R}$ with arbitrary $\lambda_1, \ldots, \lambda_n \in \mathbb{Q} \cap [0,1]$, adding up to 1, and for every continuous convex function $f : D \to \mathbb{R}$ with arbitrary $\lambda_1, \ldots, \lambda_n \in [0,1]$, adding up to 1. (Cf. Theorem 5.3.5 and the subsequent remark).

Theorem 5.3.5 is an analogue of Theorem 5.2.1. Theorem 5.2.2 has no analogue for convex functions. We do not know a way to construct all convex functions $f : D \to \mathbb{R}$. Below we give a few examples of convex functions which are not additive.

Example 5.3.1. Let $g : \mathbb{R} \to \mathbb{R}$ be convex function (possibly continuous), and let $a : \mathbb{R}^N \to \mathbb{R}$ be an additive function. Then the function $f : \mathbb{R}^N \to \mathbb{R}$:

$$f(x) = g\big(a(x)\big) \tag{5.3.10}$$

---

[12] Thus we admit that f may assume the value $-\infty$. The lemma is also true for $f : D \to (-\infty, \infty]$.

is convex. In fact,

$$f\left(\frac{x+y}{2}\right) = g\left(a\left(\frac{x+y}{2}\right)\right) = g\left(\frac{a(x)+a(y)}{2}\right) \leqslant \frac{g(a(x))+g(a(y))}{2} = \frac{f(x)+f(y)}{2}.$$

It would be tempting to conjecture that every convex function $f : \mathbb{R}^N \to \mathbb{R}$ can be written in form (5.3.10), where a is additive, and g is a continuous convex function. But this is not the case, as will become apparent from the following

Example 5.3.2. Let H be a Hamel basis for $\mathbb{R}^N$, and let, for every $h \in H$, a convex function (possibly continuous) $g_h : \mathbb{R} \to \mathbb{R}$ be given such that

$$g_h(0) = 0. \tag{5.3.11}$$

We define a function $f : \mathbb{R}^N \to \mathbb{R}$ as follows. If $x \in \mathbb{R}^N$, then x has an expansion

$$x = \sum_{i=1}^{n} \beta_i h_i, \ \beta_i \in \mathbb{Q}, \ h_i \in H, i = 1, \ldots, n. \tag{5.3.12}$$

For such an x we put

$$f(x) = \sum_{i=1}^{n} g_{h_i}(\beta_i). \tag{5.3.13}$$

The function f thus defined is convex. In fact, take an arbitrary $y \in \mathbb{R}^N$. y has an expansion

$$y = \sum_{i=1}^{n} \gamma_i h_i, \ \gamma_i \in \mathbb{Q}, \ i = 1, \ldots, n. \tag{5.3.14}$$

We may assume that the same h_i's occur in (5.3.12) and (5.3.14), adding in both expressions the lacking h_i's with the coefficients 0. Due to condition (5.3.11) this does not affect formula (5.3.13).

According to the definition of f

$$f(y) = \sum_{i=1}^{n} g_{h_i}(\gamma_i),$$

and, since by (5.3.12) and (5.3.14)

$$\frac{x+y}{2} = \sum_{i=1}^{n} \frac{\beta_i + \gamma_i}{2} h_i,$$

we have

$$f\left(\frac{x+y}{2}\right) = \sum_{i=1}^{n} g_{h_i}\left(\frac{\beta_i + \gamma_i}{2}\right) \leqslant \sum_{i=1}^{n} \frac{g_{h_i}(\beta_i) + g_{h_i}(\gamma_i)}{2} = \frac{f(x)+f(y)}{2}.$$

Now fix $h_1, h_2 \in H$, and take $g_{h_1}(x) = x^2$, $g_{h_2}(x) = x^4$, $g_h(x) = 0$ for $h \in H \setminus \{h_1, h_2\}$. Suppose that function (5.3.13) can be written in form (5.3.10) with a continuous g and additive a. Then for $x = \lambda h_1, \lambda \in \mathbb{Q}$, we have according to Theorem 5.2.1 $\left(a_1 = a\left(h_1\right)\right)$

$$f\left(x\right) = f\left(\lambda h_1\right) = g\left(a\left(\lambda h_1\right)\right) = g\left(\lambda a\left(h_1\right)\right) = g\left(\lambda a_1\right),$$

whereas from (5.3.13)

$$f\left(x\right) = f\left(\lambda h_1\right) = g_{h_1}\left(\lambda\right) = \lambda^2.$$

Hence $g\left(\lambda a_1\right) = \lambda^2$ for all $\lambda \in \mathbb{Q}$. Note that a_1 cannot be zero, for otherwise we would obtain $\lambda^2 = g\left(0\right)$ for all $\lambda \in \mathbb{Q}$, which is clearly impossible. Therefore $g\left(x\right) = x^2/a_1^2$ for all $x = \lambda a_1, \lambda \in \mathbb{Q}$, and since g is continuous, this implies that $g\left(x\right) = x^2/a_1^2$ for all real x.

Similarly, we have for $x = \lambda h_2, \lambda \in \mathbb{Q}$ $\left(a_2 = a\left(h_2\right)\right)$:

$$f(x) = f(\lambda h_2) = g\left(a(\lambda h_2)\right) = g\left(\lambda a(h_2)\right) = g(\lambda a_2),$$

and by (5.3.13)

$$f(x) = f(\lambda h_2) = g_{h_2}(\lambda) = \lambda^4,$$

whence it follows, as in the preceding case, $g\left(x\right) = x^4/a_2^4$. Consequently

$$\frac{x^2}{a_1^2} = \frac{x^4}{a_2^4}$$

for all real x, which is impossible.

But also formula (5.3.13) with continuous g_h does not yield the general form of convex functions $f : \mathbb{R}^N \to \mathbb{R}$. To show this, we consider

Example 5.3.3. Let H be a Hamel basis for $\mathbb{R}^N$. For $x \in \mathbb{R}^N$ with expansion (5.3.12) we define $f\left(x\right)$ as[13]

$$f\left(x\right) = \exp \sum_{i=1}^{n} \beta_i. \tag{5.3.15}$$

Adding to (5.3.12) further h_i's with the coefficients 0 does not affect expression (5.3.15). Thus if we take a $y \in \mathbb{R}^N$, we may assume that y has an expansion (5.3.14), and we have, since the function $\exp$ is convex,

$$f\left(\frac{x+y}{2}\right) = \exp \sum_{i=1}^{n} \frac{\beta_i + \gamma_i}{2} \leqslant \frac{1}{2}\left(\exp \sum_{i=1}^{n} \beta_i + \exp \sum_{i=1}^{n} \gamma_i\right) = \frac{f\left(x\right) + f\left(y\right)}{2}.$$

Consequently our function f is convex. If we suppose that f can be written in form (5.3.13) with continuous g_h's, then[14] for $x = \lambda h, \lambda \in \mathbb{Q}, h \in H$, we get by (5.3.13) $f(x) = g_h(\lambda)$, whereas by (5.3.15) $f(x) = e^{\lambda}$. Hence $g_h(\lambda) = e^{\lambda}$ for $\lambda \in \mathbb{Q}$, and for all $h \in H$. Hence, for $x = \lambda h_1 + \mu h_2, \lambda, \mu \in \mathbb{Q}, h_1, h_2 \in H$, we get by (5.3.13) $f(x) = e^{\lambda} + e^{\mu}$, whereas by (5.3.15) $f(x) = e^{\lambda+\mu}$. Thus $e^{\lambda+\mu} = e^{\lambda} + e^{\mu}$ for all $\lambda, \mu \in \mathbb{Q}$, which is impossible.

---

[13] It is easily seen that expression (5.3.15) is a particular case of (5.3.10).

[14] We could argue simpler: (5.3.15) yields $f\left(0\right) = 1$, whereas from (5.3.13) $f(0) = 0$. But this argument fails if instead of (5.3.13) we consider the formula $f(x) = c + \sum_{i=1}^{n} g_{h_i}(\beta_i)$, where $c \in \mathbb{R}$ is a constant.

5.4 Homogeneity fields

Let $f : \mathbb{R}^N \to \mathbb{R}$ be an additive function. Then, by Theorem 5.2.1, f satisfies relation (5.2.3) for all $x \in \mathbb{R}^N$ and all $\lambda \in \mathbb{Q}$. But it may happen that f satisfies (5.2.3) (for all $x \in \mathbb{R}^N$) also for some $\lambda \notin \mathbb{Q}$. Let

$$H_f = \left\{ \lambda \in \mathbb{R} \mid f(\lambda x) = \lambda f(x) \text{ for all } x \in \mathbb{R}^N \right\}.$$

The set H_f is called the *homogeneity field* of the function f. The name is well motivated, as may be seen from the following theorem, due to J. Rätz [263].

Theorem 5.4.1. *Let $f : \mathbb{R}^N \to \mathbb{R}$ be an additive function. Then the set H_f is a field.*

Proof. Let $\lambda, \mu \in H_f$. Then, for an arbitrary $x \in \mathbb{R}^N$,

$$f((\lambda - \mu)x) = f(\lambda x - \mu x) = f(\lambda x) - f(\mu x) = \lambda f(x) - \mu f(x) = (\lambda - \mu)f(x).$$

Consequently $\lambda - \mu \in H_f$. Similarly, if, moreover, $\mu \neq 0$

$$\lambda f(x) = f(\lambda x) = f\left(\mu \frac{\lambda}{\mu} x\right) = \mu f\left(\frac{\lambda}{\mu} x\right),$$

whence

$$f\left(\frac{\lambda}{\mu} x\right) = \frac{\lambda}{\mu} f(x).$$

Thus $\lambda/\mu \in H_f$, and in virtue of Lemma 4.7.1 H_f is a field. $\qquad\square$

The following result (Rätz [263]) is in a certain sense converse to Theorem 5.4.1.

Theorem 5.4.2. *Let $F \subset \mathbb{R}$ be an arbitrary field. Then there exists an additive function $f : \mathbb{R}^N \to \mathbb{R}$ such that $H_f = F$.*

Proof. Consider the linear space

$$(\mathbb{R}^N; F; +; \cdot), \tag{5.4.1}$$

and let B be a base of (5.4.1) (cf. Corollary 4.2.1). Fix an arbitrary $c \in \mathbb{R}, c \neq 0$, and let $g : B \to \mathbb{R}$ be the constant function

$$g(x) = c \text{ for } x \in B. \tag{5.4.2}$$

According to Theorem 4.3.1 there exists a homomorphism f from (5.4.1) into $(\mathbb{R}; F; +; \cdot)$ such that $f|B = g$. Since f is homomorphism, f is an additive function, and satisfies

$$f(\lambda x) = \lambda f(x) \tag{5.4.3}$$

for all $x \in \mathbb{R}^N$ and $\lambda \in F$. Consequently $F \subset H_f$.

Now take an arbitrary $x \in \mathbb{R}^N$. x can be written in the form

$$x = \sum_{i=1}^{n} \lambda_i b_i, \ \lambda_i \in F, \ b_i \in B, \ i = 1, \ldots, n.$$

By (5.4.3) and (5.4.2), since f is additive and $f \mid B = g$,

$$f(x) = \sum_{i=1}^{n} \lambda_i f(b_i) = \sum_{i=1}^{n} \lambda_i g(b_i) = c \sum_{i=1}^{n} \lambda_i \in cF.$$

Thus

$$f\left(\mathbb{R}^N\right) \subset cF. \tag{5.4.4}$$

Take arbitrary $\lambda \in H_f$ and $b_0 \in B$. We have

$$f(\lambda b_0) = \lambda f(b_0) = \lambda g(b_0) = \lambda c. \tag{5.4.5}$$

On the other hand, since $\lambda b_0 \in \mathbb{R}^N$, we infer from (5.4.4) that there exists an $\alpha \in F$ such that

$$f(\lambda b_0) = \alpha c. \tag{5.4.6}$$

Since $c \neq 0$, relations (5.4.5) and (5.4.6) yield $\lambda = \alpha \in F$. Hence $H_f \subset F$, and ultimately we obtain that $H_f = F$. $\qquad\qquad\qquad\qquad\qquad\qquad\qquad\qquad\square$

Similar investigations for convex functions have been carried out by R. Ger [113]. However, the results are not so complete as those for additive functions. Therefore we shall not go into details here.

5.5 Additive functions on product spaces

Consider $\mathbb{R}^N$ as a product of lower dimensional spaces:

$$\mathbb{R}^N = \mathbb{R}^p \times \mathbb{R}^q \tag{5.5.1}$$

with $p \in \mathbb{N}, q \in \mathbb{N}, p + q = N$. Every $x \in \mathbb{R}^N$ can be represented as $x = (x_p, x_q)$, with $x_p \in \mathbb{R}^p$, $x_q \in \mathbb{R}^q$, and if $y \in \mathbb{R}^N$, $y = (y_p, y_q)$, $y_p \in \mathbb{R}^p, y_q \in \mathbb{R}^q$, then

$$x + y = (x_p, x_q) + (y_p, y_q) = (x_p + y_p, x_q + y_q). \tag{5.5.2}$$

With these notations we have the following (Aczél [5])

Theorem 5.5.1. *If $f : \mathbb{R}^N \to \mathbb{R}$ is an additive function, and $\mathbb{R}^N$ has decomposition (5.5.1), then there exist additive functions $f_p : \mathbb{R}^p \to \mathbb{R}$ and $f_q : \mathbb{R}^q \to \mathbb{R}$ such that*

$$f(x) = f(x_p, x_q) = f_p(x_p) + f_q(x_q). \tag{5.5.3}$$

Proof. Put

$$f_p(x_p) = f(x_p, 0), \quad f_q(x_q) = f(0, x_q).$$

Then by (5.5.2) $x = (x_p, x_q) = (x_p, 0) + (0, x_q)$, and

$$f(x) = f(x_p, x_q) = f(x_p, 0) + f(0, x_q) = f_p(x_p) + f_q(x_q),$$

i.e., we obtain formula (5.5.3). Further, we have for arbitrary $x_p, y_p \in \mathbb{R}^p$ and x_q, $y_q \in \mathbb{R}^q$

$$f_p(x_p + y_p) = f(x_p + y_p, 0) = f\big((x_p, 0) + (y_p, 0)\big) = f(x_p, 0) + f(y_p, 0) = f_p(x_p) + f_p(y_p),$$

and

$$f_q(x_q+y_q) = f(0, x_q+y_q) = f\big((0,x_0)+(0,y_q)\big) = f(0,x_q)+f(0,y_q) = f_q(x_q)+f_q(y_q),$$

i.e., the functions f_p and f_q are additive. This completes the proof[15]. $\square$

Applying Theorem 5.5.1 several times we arrive at the representation

$$f(x) = f(x_1,\ldots,x_N) = \sum_{i=1}^{N} f_i(x_i), \tag{5.5.4}$$

valid for additive functions $f : \mathbb{R}^N \to \mathbb{R}$, where $f_i : \mathbb{R} \to \mathbb{R}, i = 1,\ldots,N$, are additive functions of a single variable. Hence we derive the following result.

Theorem 5.5.2. *If $f : \mathbb{R}^N \to \mathbb{R}$ is continuous additive function, then there exists a $c \in \mathbb{R}^N$ such that*

$$f(x) = cx, \tag{5.5.5}$$

where $cx = \sum_{i=1}^{N} c_i x_i \left(c = (c_1,\ldots,c_N), x = (x_1,\ldots,x_N)\right)$ denotes the scalar product.

Proof. As we saw in 5.2, the theorem is true for $N = 1$. If $N > 1$, we apply formula (5.5.4). Fix an i, $1 \leqslant i \leqslant N$, and put in (5.5.4) $x_1 = \cdots = x_{i-1} = x_{i+1} = \cdots = x_N = 0$. Since $f_j(0) = 0$, $j = 1,\ldots,N$ (cf. Theorem 5.2.1), we get

$$f_i(x_i) = f(0,\ldots,0,x_i,0,\ldots,0),$$

whence it follows that the function f_i is continuous. Consequently all f_i in (5.5.4) are continuous additive functions of a single variable, and thus there exist constants $c_i \in \mathbb{R}, i = 1,\ldots,N$, such that

$$f_i(x_i) = c_i x_i, \ i = 1,\ldots,N. \tag{5.5.6}$$

Put $c = (c_1,\ldots,c_N)$. Then (5.5.5) is a consequence of (5.5.4) and (5.5.6). $\square$

Thus it has been rather easy to find the general form of continuous additive functions $f : \mathbb{R}^N \to \mathbb{R}$. For discontinuous additive functions no such formula exists. As we shall see later (Chapter 9), such functions are non-measurable, and consequently they cannot be effectively displayed. They can be obtained only by a use of a Hamel basis, and thus, implicitly, the Axiom of Choice.

5.6 Additive functions on $\mathbb{C}$

In this section we consider functions $f : \mathbb{C} \to \mathbb{C}$ which are additive, i.e., satisfy the functional equation

$$f(z_1 + z_2) = f(z_1) + f(z_2) \tag{5.6.1}$$

for all $z_1, z_2 \in \mathbb{C}$.

---

[15] The same argument works for homomorphisms on products of groups. Concerning homomorphisms on products of semigroups, cf. Kuczma [182], Szymiczek [311], Martin [221].

If we abstract from the analytic structure of $\mathbb{C}$, $\mathbb{C}$ may be considered as $\mathbb{R}^2$:

$$\mathbb{C} = \mathbb{R} \times \mathbb{R},$$

and $z \in \mathbb{C}$ may be considered as $z = (\operatorname{Re} z, \operatorname{Im} z)$. An arbitrary function $f : \mathbb{C} \to \mathbb{C}$ may be written as

$$f(z) = f_1(z) + i f_2(z), \tag{5.6.2}$$

where $f_1 : \mathbb{C} \to \mathbb{R}$ and $f_2 : \mathbb{C} \to \mathbb{R}$ are the functions

$$f_1(z) = \operatorname{Re} f(z), \, f_2(z) = \operatorname{Im} f(z). \tag{5.6.3}$$

If f is additive, then by (5.6.1) and (5.6.3), for arbitrary $z_1, z_2 \in \mathbb{C}$

$$f_1(z_1+z_2) = \operatorname{Re} f(z_1+z_2) = \operatorname{Re}[f(z_1)+f(z_2)] = \operatorname{Re} f(z_1) + \operatorname{Re} f(z_2) = f_1(z_1) + f_1(z_2),$$

and

$$f_2(z_1+z_2) = \operatorname{Im} f(z_1+z_2) = \operatorname{Im}[f(z_1)+f(z_2)] = \operatorname{Im} f(z_1) + \operatorname{Im} f(z_2) = f_2(z_1) + f_2(z_2),$$

i.e., the functions f_1 and f_2 (which may be considered as functions from $\mathbb{R}^2$ into $\mathbb{R}$) are additive. Thus we obtain from Theorem 5.5.1 according to (5.6.2)

Theorem 5.6.1. *If $f : \mathbb{C} \to \mathbb{C}$ is additive, then there exist additive functions $f_{kj} : \mathbb{R} \to \mathbb{R}$, $k, j = 1, 2$, such that*

$$f(z) = f_{11}(\operatorname{Re} z) + f_{12}(\operatorname{Im} z) + i f_{21}(\operatorname{Re} z) + i f_{22}(\operatorname{Im} z). \tag{5.6.4}$$

Similar is the situation with the continuous solutions of (5.6.1).

Theorem 5.6.2. *If $f : \mathbb{C} \to \mathbb{C}$ is a continuous additive function, then there exist complex constants c_1, c_2 such that*

$$f(z) = c_1 z + c_2 \bar{z}, \tag{5.6.5}$$

where $\bar{z}$ denotes the complex conjugate of z.

Proof. The continuity of f implies the continuity of each function f_{kj} occurring in (5.6.4), and so there exist real constants c_{kj}, $k, j = 1, 2$, such that

$$f_{kj}(x) = c_{kj} x, \, k, j = 1, 2.$$

Hence

$$f(z) = c_{11} \operatorname{Re} z + c_{12} \operatorname{Im} z + i c_{21} \operatorname{Re} z + i c_{22} \operatorname{Im} z = (c_{11} + i c_{21}) \operatorname{Re} z + (c_{12} + i c_{22}) \operatorname{Im} z.$$

Put $a = c_{11} + i c_{21} \in \mathbb{C}$, $b = c_{12} + i c_{22} \in \mathbb{C}$. Then

$$f(z) = a \operatorname{Re} z + b \operatorname{Im} z. \tag{5.6.6}$$

Since $i^2 = -1$, relation (5.6.6) can be written as

$$f(z) = a\operatorname{Re} z - (bi)\,i\operatorname{Im} z = \frac{a+bi}{2}\operatorname{Re} z + \frac{a-bi}{2}\operatorname{Re} z - \frac{a+bi}{2}i\operatorname{Im} z + \frac{a-bi}{2}i\operatorname{Im} z,$$

and with $c_1 = \dfrac{1}{2}(a-bi)$, $c_2 = \dfrac{1}{2}(a+bi)$, we obtain

$$f(z) = c_1\left(\operatorname{Re} z + i\operatorname{Im} z\right) + c_2\left(\operatorname{Re} z - i\operatorname{Im} z\right),$$

i.e., (5.6.5). $\square$

Functions (5.6.5) are not analytic, since $\bar{z}$ is not an analytic function of z. For analytic additive functions we have the following result.

Theorem 5.6.3. *If $f : \mathbb{C} \to \mathbb{C}$ is an analytic additive function, then there exists a complex constant c such that*

$$f(z) = cz. \tag{5.6.7}$$

Proof. f, being analytic, is differentiable. Differentiating relation (5.6.1) with respect to z_1 we obtain

$$f'(z_1 + z_2) = f'(z_1). \tag{5.6.8}$$

Relation (5.6.8), valid for arbitrary $z_1, z_2 \in \mathbb{C}$, says that the function f' is constant, i.e.,

$$f'(z) = c,\ z \in \mathbb{C},$$

with a $c \in \mathbb{C}$. Hence

$$f(z) = cz + b, \tag{5.6.9}$$

with $a, b \in \mathbb{C}$. Inserting (5.6.9) into (5.6.1) we obtain $b = 0$, whence (5.6.7) follows. $\square$

Exercises

1. Let $C \subset \mathbb{R}$ be a set such that $C \cap (0,1) \neq \varnothing$. Show that if a set $A \subset \mathbb{R}^N$ is C-convex, then there exists a set $D \subset \mathbb{R}$ dense in $(0,1)$ and such that A is D-convex.
   [Hint: Let $D = \{\lambda \in [0,1] \mid \lambda x + (1-\lambda)y \in A$ for every $x, y \in A\}$. If $\lambda \in C \cap (0,1)$, then $\alpha_{nk} = \left(1 - \lambda^k\right)^n \in D$ for every $k, n \in \mathbb{N}$. Moreover, $0 < \alpha_{nk} - \alpha_{n+1,k} < \lambda^k$ for $n, k \in \mathbb{N}$. Let $(a,b) \subset (0,1)$ be a non-empty interval. If $k \in \mathbb{N}$ is such that $\lambda^k < \min(1-b, b-a)$, then there exists an $n \in \mathbb{N}$, such that $\alpha_{nk} \in (a,b)$.]

2. Let $C \subset \mathbb{R}$ be a set such $C \cap (0,1) \neq \varnothing$. Show that if $A \subset \mathbb{R}^N$ is C-convex, then $\operatorname{cl} A$ is convex.

3. For arbitrary $a \in \mathbb{R}^N$ and $A \subset \mathbb{R}^N$ let $d(a, A)$ denote the Euclidean distance of a from A, and, given an $\varepsilon > 0$, let

$$A_\varepsilon = \left\{x \in \mathbb{R}^N \mid d(x, A) < \varepsilon\right\}.$$

Show that if $A \subset \mathbb{R}^N$ is convex, then so is also A_ε for every $\varepsilon > 0$.

4. Show that if $A \subset \mathbb{R}^N$ is convex, then so is also cl A.
   [Hint: Use Exercise 4.3.]

5. Let $A \subset \mathbb{R}^N$ be a convex set such that int $A \neq \varnothing$. Show that int cl $A = $ int A.
   [Hint: Let $y \in$ int $A, x \in$ cl $A \setminus$ int A. Using Corollary 5.1.3 show that $y + \lambda(x - y) \notin$ cl A for $\lambda > 1$, whence $x \notin$ int cl A.]

6. Let $f : \mathbb{R} \to \mathbb{R}$ be an additive function, which is not one-to-one. Show that, for every $y \in f(\mathbb{R})$, the set $f^{-1}(y)$ is dense in $\mathbb{R}$. (Smítal [286].)

7. Let $f : \mathbb{R} \to \mathbb{R}$ be an additive function such that $f(\mathbb{R}) = \mathbb{R}$. Show that f either is one-to-one, or has the Darboux property (the intermediate value property; Smítal [286]).

8. Let $f : \mathbb{R} \to \mathbb{R}$, $f \neq 0$ be an additive function. Show that, for every $y \in f(\mathbb{R})$, $m_i\big(f^{-1}(y)\big) = 0$. (Smítal [286]).

9. Let $D \subset \mathbb{R}^N$ be a convex and open set. Show that a function $f : D \to \mathbb{R}$ is convex if and only if the set

$$\{(x, y) \in D \times \mathbb{R} \mid y > f(x)\}$$

   is J-convex.

10. Show that if an additive function $f : \mathbb{R}^2 \to \mathbb{R}$ is continuous with respect to either variable, then it is jointly continuous in $\mathbb{R}^2$.

11. Show that if an additive function $f : \mathbb{R}^2 \to \mathbb{R}$ is unbounded on a square $[a, b] \times [a, b] \subset \mathbb{R}^2$, then for every fixed $y \in \mathbb{R}$, either $f(x, y)$, or $f(y, x)$ is unbounded on $[a, b]$.

12. Let $f : \mathbb{R}^2 \to \mathbb{R}$ be an additive function. Show that if $f(x, y)$ (as a function of x) is unbounded on an interval $I \subset \mathbb{R}$ for a $y \in \mathbb{R}$, then it is unbounded on I for every $y \in \mathbb{R}$.

Chapter 6

Elementary Properties of Convex Functions

6.1 Convex functions on rational lines

In this chapter we discuss some properties of convex functions connected with their boundedness and continuity. We start with the following

Lemma 6.1.1. *Let $D \subset \mathbb{R}^N$ be a convex and open set, and let $f : D \to \mathbb{R}$ be a convex function. Then*

$$\frac{f(x) - f(x - nd)}{n} \leqslant \frac{f(x) - f(x - md)}{m} \leqslant \frac{f(x + md) - f(x)}{m} \leqslant \frac{f(x + nd) - f(x)}{n}$$
(6.1.1)

for every $x \in D, d \in \mathbb{R}^N$ and $m, n \in \mathbb{N}$ such that $0 < m < n$ and $x \pm nd \in D$.

Proof. We use Lemma 5.3.1. Take in (5.3.3) $x_1 = \cdots = x_m = x + nd$, $x_{m+1} = \cdots = x_n = x$. We obtain

$$f\left(\frac{m(x + nd) + (n - m)x}{n}\right) \leqslant \frac{mf(x + nd) + (n - m)f(x)}{n},$$

i.e.,

$$nf(x + md) \leqslant mf(x + nd) + (n - m)f(x),$$

and

$$n[f(x + md) - f(x)] \leqslant m[f(x + nd) - f(x)]. \tag{6.1.2}$$

Hence the last inequality results. The first inequality in (6.1.1) results from (6.1.2) if we replace d by $-d$. For the proof of the middle inequality in (6.1.1) use (5.3.1) with x, y replaced by $x + md, x - md$, respectively. We obtain

$$f(x) \leqslant \frac{f(x + md) + f(x - md)}{2},$$

whence

$$f(x) - f(x - md) \leqslant f(x + md) - f(x),$$

and the middle inequality in (6.1.1) results on dividing both the sides by m. $\qquad \square$

Lemma 6.1.2. *Let $D \subset \mathbb{R}^N$ be a convex and open set, and let $f : D \to \mathbb{R}$ be a convex function. If f is bounded above on a set $A \subset D$, then it is also bounded above (by the same constant) on $Q(A)$.*

Proof. Suppose that for a certain constant $M \in \mathbb{R}$ we have

$$f(t) \leqslant M \text{ for } t \in A. \tag{6.1.3}$$

Take an arbitrary $x \in Q(A)$. By Theorem 5.1.3 there exist an $n \in \mathbb{N}, t_1, \ldots, t_n \in A$, and $\lambda_1, \ldots, \lambda_n \in \mathbb{Q} \cap [0, 1]$ such that

$$\sum_{i=1}^n \lambda_i = 1, \tag{6.1.4}$$

and

$$x = \sum_{i=1}^n \lambda_i t_i. \tag{6.1.5}$$

By Lemma 5.3.2 and Theorem 5.3.5 we obtain from (6.1.5), (6.1.3), and (6.1.4)

$$f(x) \leqslant \sum_{i=1}^n \lambda_i f(t_i) \leqslant \sum_{i=1}^n \lambda_i M = M \sum_{i=1}^n \lambda_i = M.$$

This shows that f is bounded above by M on $Q(A)$. $\square$

Corollary 6.1.1. *Let $D \subset \mathbb{R}^N$ be a convex and open set, and let $f : D \to \mathbb{R}$ be a convex function. If f is bounded above on a set $A \subset D$, then it is also bounded above (by the same constant) on $J(A)$.*

Proof. This could be proved by the same argument as Lemma 6.1.2, using Lemma 5.1.1, but we will derive this directly from Lemma 6.1.2. Since $\dfrac{1}{2} \in \mathbb{Q}$, the set $Q(A)$ is J-convex, and hence $J(A) \subset Q(A)$. So our corollary is an immediate consequence of Lemma 6.1.2. $\square$

Corollary 6.1.2. *Let $D \subset \mathbb{R}^N$ be a convex and open set, and let $f : D \to \mathbb{R}$ be a convex function. For arbitrary $x, y \in D$, the function f is bounded above on the rational segment $Q(x, y)$.*

Proof. This follows from Lemma 6.1.2 in view of the fact that every function is bounded on every finite set. $\square$

Lemma 6.1.3. *Let $D \subset \mathbb{R}^N$ be a convex and open set, and let $f : D \to \mathbb{R}$ be a convex function. For arbitrary $x, y \in D$, the function f is bounded below on the rational segment $Q(x, y)$.*

Proof. Take arbitrary $x, y \in D$, and $t \in Q(x, y)$. According to (5.1.9) there exists a $\lambda \in \mathbb{Q} \cap [0, 1]$ such that

$$t = \lambda x + (1 - \lambda) y. \tag{6.1.6}$$

Moreover, by Corollary 6.1.2, there exists a constant M such that

$$f(s) \leqslant M \text{ for } s \in Q(x, y). \tag{6.1.7}$$

Put $u = \dfrac{1}{2}(x + y)$, and $v = 2u - t$. By (6.1.6)

$$v = x + y - \lambda x - (1 - \lambda) y = \lambda y + (1 - \lambda) x \in Q(x, y),$$

whence by (6.1.7)

$$f(v) \leqslant M. \tag{6.1.8}$$

Since $u = \dfrac{1}{2}(v + t)$, and f is convex, we have

$$f(u) \leqslant \frac{1}{2}[f(v) + f(t)],$$

whence by (6.1.8)

$$f(t) \geqslant 2f(u) - f(v) \geqslant 2f(u) - M.$$

Thus f is bounded below on $Q(x, y)$ by the constant[1] $2f(u) - M$. $\qquad \square$

Theorem 6.1.1. *Let $D \subset \mathbb{R}^N$ be a convex and open set, and let $f : D \to \mathbb{R}$ be a convex function. For arbitrary $a, b \in D$ the function $f \,|\, Q(a, b)$ is uniformly continuous.*

Proof. Fix $a, b \in D$. We may assume that $a \neq b$. Since D is open, there exists a positive number $\rho \in \mathbb{Q}$ such that

$$a' = a - \rho(b - a) \in D \text{ and } b' = b + \rho(b - a) \in D. \tag{6.1.9}$$

By Corollary 6.1.2 and Lemma 6.1.3 there exist real constants $K \leqslant M$ such that

$$K \leqslant f(x) \leqslant M \text{ for } x \in Q(a', b'). \tag{6.1.10}$$

First we show that

$$Q(a, b) \subset Q(a', b'). \tag{6.1.11}$$

In fact, if $t \in Q(a, b)$, then, according to (5.1.9), there exists a $\lambda \in \mathbb{Q} \cap [0, 1]$ such that $t = \lambda a + (1 - \lambda) b$. Put $\mu = \dfrac{\rho + \lambda}{1 + 2\rho}$. It is easily seen that $\mu \in \mathbb{Q} \cap (0, 1)$. Moreover, by (6.1.9),

$$\mu a' + (1 - \mu) b' = \frac{\rho + \lambda}{1 + 2\rho}[a - \rho(b - a)] + \frac{1 + \rho - \lambda}{1 + 2\rho}[b + \rho(b - a)] = \lambda a + (1 - \lambda) b = t.$$

Consequently $t \in Q(a', b')$, which proves inclusion (6.1.11).

---

[1] This constant depends on x and y, but not on a particular $t \in Q(x, y)$.

Given an arbitrary $\varepsilon > 0$, put $n = \left[\dfrac{M-K}{\varepsilon} + 1\right]$ (the integral part of $\dfrac{M-K}{\varepsilon} +$

1) so that $n \in \mathbb{N}$ and

$$\frac{M-K}{\varepsilon} < n. \tag{6.1.12}$$

Further put $\delta = \dfrac{\rho}{n}\,|b-a|$. Take arbitrary $x', x'' \in Q\,(a, b)$ such that

$$|x' - x''| < \delta. \tag{6.1.13}$$

Then there exist $\alpha, \beta \in \mathbb{Q} \cap [0, 1]$ such that

$$x' = \alpha a + (1-\alpha)\,b, \ \ x'' = \beta a + (1-\beta)\,b,$$

whence

$$x' - x'' = (\beta - \alpha)\,(b - a)\,,$$

and by (6.1.13) $|\beta - \alpha|\,|b-a| < \delta = \dfrac{\rho}{n}\,|b-a|$, and finally

$$n\,|\beta - \alpha| < \rho. \tag{6.1.14}$$

Consider the numbers

$$\mu_{1,2} = \frac{\rho + \alpha \pm n\,(\beta - \alpha)}{1 + 2\rho} \in \mathbb{Q}.$$

By (6.1.14)

$$0 \leqslant \frac{\alpha}{1 + 2\rho} < \mu_{1,2} < \frac{\alpha + 2\rho}{1 + 2\rho} \leqslant 1$$

so that $\mu_{1,2} \in \mathbb{Q} \cap [0, 1]$. Now,

$$\mu_{1,2} a' + (1 - \mu_{1,2})\,b' = [\alpha \pm n\,(\beta - \alpha)]\,a + [1 - \alpha \mp n\,(\beta - \alpha)]\,b$$
$$= \alpha a + (1-\alpha)\,b \pm n\,(\beta - \alpha)\,(b-a) = x' \pm n\,(x' - x'')\,.$$

Consequently $x' \pm n\,(x' - x'') \in Q\,(a', b') \subset D$.

Taking in Lemma 6.1.1 $x = x'$, $d = x' - x''$, and $m = 1$, we get by (6.1.10) and (6.1.11)

$$\frac{K-M}{n} \leqslant \frac{f\,(x') - f\,(x' - nd)}{n} \leqslant f\,(x') - f\,(x'') \leqslant \frac{f\,(x' + nd) - f\,(x')}{n} \leqslant \frac{M-K}{n},$$

and by (6.1.12)

$$|f\,(x') - f\,(x'')| \leqslant \frac{M-K}{n} < \varepsilon.$$

This shows that f is uniformly continuous on $Q\,(a, b)$. $\square$

Every uniformly continuous function on a set $A \subset \mathbb{R}^N$ can be uniquely extended onto $\mathrm{cl}\,A$ to a continuous function (cf. Exercise 6.1), so we obtain from Theorem 6.1.1 the following

Theorem 6.1.2. *Let $D \subset \mathbb{R}^N$ be a convex and open set, and let $f : D \to \mathbb{R}$ be a convex function. For arbitrary $a, b \in D$ there exists a unique continuous function[2] $g_{ab} : \overline{ab} \to \mathbb{R}$ such that $g_{ab} \mid Q(a, b) = f \mid Q(a, b)$. The function g_{ab} satisfies*

$$g_{ab}\left(\frac{x+y}{2}\right) \leqslant \frac{g_{ab}(x) + g_{ab}(y)}{2} \qquad (6.1.15)$$

for every $x, y \in \overline{ab}$.

Proof. The set $\mathrm{cl}\, Q(a, b)$ is convex (cf. Exercise 5.2), so $\overline{ab} \subset \mathrm{cl}\, Q(a, b)$. On the other hand, $Q(a, b) \subset \overline{ab}$, whence $\mathrm{cl}\, Q(a, b) \subset \mathrm{cl}\, \overline{ab} = \overline{ab}$. Thus $\overline{ab} = \mathrm{cl}\, Q(a, b)$.

The existence of a unique continuous extension g_{ab} of $f \mid Q(a, b)$ onto $\overline{ab}$ results from Theorem 6.1.1. It remains to prove formula (6.1.15).

Take arbitrary $x, y \in \overline{ab} = \mathrm{cl}\, Q(a, b)$. There exist two sequences of points $x_n, y_n \in Q(a, b)$, $n \in \mathbb{N}$, such that $\lim_{n \to \infty} x_n = x$, $\lim_{n \to \infty} y_n = y$. Then $\lim_{n \to \infty} \frac{1}{2}(x_n + y_n) = \frac{1}{2}(x + y)$, and, since f is convex, g_{ab} is continuous, and $g_{ab} \mid Q(a, b) = f \mid Q(a, b)$,

$$g_{ab}\left(\frac{x+y}{2}\right) = \lim_{n \to \infty} g_{ab}\left(\frac{x_n + y_n}{2}\right) = \lim_{n \to \infty} f\left(\frac{x_n + y_n}{2}\right) \leqslant \lim_{n \to \infty} \frac{f(x_n) + f(y_n)}{2}$$

$$= \lim_{n \to \infty} \frac{g_{ab}(x_n) + g_{ab}(y_n)}{2} = \frac{g_{ab}(x) + g_{ab}(y)}{2}.$$

This proves relation (6.1.15). $\qquad\square$

Remark 6.1.1. We do not say that g_{ab} is convex, since g_{ab} is not defined on an open subset of $\mathbb{R}^N$.

Let $l = l(a, b)$ be the rational line passing through a and b (cf. 5.1), and put $l_0 = l \cap D$. Similarly, let L be the (real) line passing through a and b, and let $L_0 = L \cap D$. With this notation we have

Theorem 6.1.3. *Let $D \subset \mathbb{R}^N$ be a convex and open set, and let $f : D \to \mathbb{R}$ be a convex function. For every $a, b \in D$ there exists a unique continuous function $G_{ab} : L_0 \to \mathbb{R}$ such that $G_{ab} \mid l_0 = f \mid l_0$. The function G_{ab} satisfies inequality (6.1.15) for every $x, y \in L_0$.*

Proof. Let $a_n, b_n \in l_0$, $n \in \mathbb{N}$, be two monotonic sequences of points converging to the two ends of L_0. As is easy to check, we have $Q(a_n, b_n) \subset l_0$, $\overline{a_n b_n} \subset L_0$ for every $n \in \mathbb{N}$, and $\bigcup_{n=1}^{\infty} Q(a_n, b_n) = l_0$, $\bigcup_{n=1}^{\infty} \overline{a_n, b_n} = L_0$. Put $g_n = g_{a_n b_n}$ (Theorem 6.1.2). If for some $m, n \in \mathbb{N}$ we have $\overline{a_m b_m} \subset \overline{a_n b_n}$, then $g \mid \overline{a_m b_m}$ is a continuous function on $\overline{a_m b_m}$ which restricted to $Q(a_m, b_m)$ coincides with f. But g_m is the unique function with these properties, whence $g_n \mid \overline{a_m b_m} = g_m$. Consequently we my define a function G_{ab} on L_0 in the following way:

$$G_{ab}(x) = g_n(x) \text{ if } x \in \overline{a_n b_n}, n \in \mathbb{N}. \qquad (6.1.16)$$

---

[2] $\overline{ab} = \mathrm{cl}\, Q(a, b) = \mathrm{conv}(a, b)$ denotes the closed segment joining the points a and b.

Inequality (6.1.15) for G_{ab} results from Theorem 6.1.2 in virtue of the fact that for every $x, y \in L_0$ there exists an $n \in \mathbb{N}$ such that $x, y \in \overline{a_n b_n}$. The continuity of G_{ab} results from that of g_n, $n \in \mathbb{N}$. Finally, if $G : L_0 \to \mathbb{R}$ is an arbitrary continuous function such that $G \mid l_0 = f \mid l_0$, then $G_n = G \mid \overline{a_n b_n}$ is continuous, and $G_n \mid Q(a_n, b_n) = f \mid Q(a_n, b_n)$ for every $n \in \mathbb{N}$, whence it follows that $G_n = g_n$, $n \in \mathbb{N}$. Thus G must be given by (6.1.16), which proves the uniqueness of G_{ab}.   □

With the same notation we have

Theorem 6.1.4. *Let $D \subset \mathbb{R}^N$ be a convex and open set, and let $f : D \to \mathbb{R}$ be a convex function. Then, for arbitrary $a, b \in D$, the function $f \mid l_0$ is continuous.*

This is an immediate consequence of Theorem 6.1.3.

However, f itself need not be continuous in D. In order to find conditions ensuring the continuity of f in D, we are going to investigate, in the next section, the problem of the boundedness of f on open subsets of D.

6.2   Local boundedness of convex functions

Let $D \subset \mathbb{R}^N$. A function $f : D \to \mathbb{R}$ is called *locally bounded* [*locally bounded above, locally bounded below*] at a point $x_0 \in D$ iff there exists a neighbourhood $U \subset D$ of x_0 such that the function f is bounded [bounded above, bounded below] on U. The following three theorems refer to the local boundedness of convex functions.

Theorem 6.2.1. *Let $D \subset \mathbb{R}^N$ be a convex and open set, and let $f : D \to \mathbb{R}$ be a convex function. If f is locally bounded above at a point $x_0 \in D$, then it is locally bounded above at every point $x \in D$.*

Proof. Take an arbitrary $x \in D$. We may assume that $x \neq x_0$, for otherwise f is locally bounded above at x by hypothesis. There exists a positive number $\rho \in \mathbb{Q}$ such that the point

$$y = x - \rho(x_0 - x) \tag{6.2.1}$$

belongs to D. Let $\lambda = \dfrac{\rho}{1 + \rho} \in \mathbb{Q} \cap (0, 1)$. We have by (6.2.1)

$$x = \lambda x_0 + (1 - \lambda) y. \tag{6.2.2}$$

By hypothesis there exists an $r > 0$ such that the open ball $K = K(x_0, r)$ centered at x_0 and with the radius r, is contained in D, and f is bounded above on K

$$f(t) \leqslant M \text{ for } t \in K, \tag{6.2.3}$$

where M is a real constant. Let $U = K(x, \lambda r)$ be the open ball centered at x and with the radius λr. Take an arbitrary $u \in U$, and put

$$t = \frac{u - (1 - \lambda) y}{\lambda}. \tag{6.2.4}$$

Since $u \in U$, we have $|u - x| < \lambda r$, whence by (6.2.2) $|u - \lambda x_0 - (1 - \lambda) y| < \lambda r$, and

$$\left| \frac{u - (1 - \lambda) y}{\lambda} - x_0 \right| < r,$$

i.e., $t - x_0 < r$, which means that $t \in K$. Since by (6.2.4)

$$u = \lambda t + (1 - \lambda) y, \qquad (6.2.5)$$

and $y \in D$, $t \in K \subset D$, and the set D is convex, we have $u \in D$, whence $U \subset D$. Moreover, by (6.2.5), Theorem 5.3.5, and (6.2.3)

$$f(u) \leqslant \lambda f(t) + (1 - \lambda) f(y) \leqslant \lambda M + (1 - \lambda) f(y) \leqslant \max\left(M, f(y)\right).$$

Thus f is bounded above on U by the constant $\max\left(M, f(y)\right)$.   □

Theorem 6.2.2. *Let $D \subset \mathbb{R}^N$ be a convex and open set, and let $f : D \to \mathbb{R}$ be a convex function. If f is locally bounded below at a point $x_0 \in D$, then it is locally bounded below at every point $x \in D$.*

Proof. Take an arbitrary $x \in D$, $x \neq x_0$. There exists a positive number $\rho \in \mathbb{Q}$ such that the point

$$y = x_0 - \rho(x - x_0)$$

belongs to D. With $\lambda = \dfrac{\rho}{1 + \rho} \in \mathbb{Q} \cap (0, 1)$ we have

$$x_0 = \lambda x + (1 - \lambda) y. \qquad (6.2.6)$$

By hypothesis there exists an $r > 0$, such that the open ball $K = K(x_0, r)$ centered at x_0 and with the radius r, is contained in D and f is bounded below on K

$$f(t) \geqslant M \quad \text{for} \quad t \in K, \qquad (6.2.7)$$

where M is a real constant. Moreover, we may assume that r is so small that the open ball $U = K\left(x, \dfrac{r}{\lambda}\right)$ is contained in D.

Now take an arbitrary $u \in U$, and put

$$t = \lambda u + (1 - \lambda) y. \qquad (6.2.8)$$

Then by (6.2.6) and (6.2.8) $|t - x_0| = \lambda |u - x| < \lambda \dfrac{r}{\lambda} = r$, which means that $t \in K$. By Theorem 5.3.5 we have

$$f(t) \leqslant \lambda f(u) + (1 - \lambda) f(y),$$

whence by (6.2.7)

$$f(u) \geqslant \frac{1}{\lambda} f(t) - \frac{1 - \lambda}{\lambda} f(y) \geqslant \frac{1}{\lambda} M - \frac{1 - \lambda}{\lambda} f(y) = \left(1 + \frac{1}{\rho}\right) M - \frac{1}{\rho} f(y).$$

Thus f is bounded below on U by constant, which depends on the constants M, $f(y)$, ρ, but not on a particular $u \in U$. Hence f is locally bounded below at x.   □

Theorem 6.2.3. *Let $D \subset \mathbb{R}^N$ be a convex and open set, and let $f : D \to \mathbb{R}$ be a convex function. If f is locally bounded above at a point $x_0 \in D$, then it is locally bounded at every point $x \in D$.*

Proof. By Theorem 6.2.1 f is locally bounded above at every point $x \in D$, and so if we show that f is locally bounded below at x_0, then it will follow by Theorem 6.2.2 that it is locally bounded below, and hence locally bounded, at every point $x \in D$.

Let $K \subset D$ be an open ball around x_0 such that f is bounded above on K, i.e., (6.2.3) holds with a real constant M.

Take an arbitrary $u \in K$, and put $t = 2x_0 - u$ so that $x_0 = \dfrac{1}{2}(u+t)$. We have $t - x_0 = -(u - x_0)$, whence $|t - x_0| = |u - x_0|$, which shows that $t \in K$. We have $f(x_0) \leqslant \dfrac{1}{2}[f(u) + f(t)]$, whence by (6.2.3)

$$f(u) \geqslant 2f(x_0) - f(t) \geqslant 2f(x_0) - M.$$

Thus f is bounded below on K by the constant $2f(x_0) - M$. $\square$

6.3   The lower hull of a convex functions

Throughout this section the symbol $K(x, r)$ denotes the open ball in $\mathbb{R}^N$ centered at x and with the radius r.

Let $D \subset \mathbb{R}^N$ be an open set, and let $f : D \to \mathbb{R}$ be function. For $x \in D$ and $r > 0$ sufficiently small (such that $K(x, r) \subset D$) we define $\varphi_x(r)$ as

$$\varphi_x(r) = \inf_{K(x,r)} f.$$

If $r_1 < r_2$ and $K(x, r_2) \subset D$, we have $K(x, r_1) \subset K(x, r_2)$, whence

$$\varphi_x(r_1) = \inf_{K(x,r_1)} f \geqslant \inf_{K(x,r_2)} f = \varphi_x(r_2).$$

Thus the function φ_x is decreasing. Consequently there exists the limit

$$m_f(x) = \lim_{r \to 0+} \varphi_x(r) = \lim_{r \to 0+} \inf_{K(x,r)} f. \qquad (6.3.1)$$

Function $m_f : D \to [-\infty, +\infty)$ is called the *lower hull* of f, and its value at an $x \in D$ is called the *infimum of f at x*. Similarly we can define the function $M_f : D \to (-\infty, +\infty]$

$$M_f(x) = \lim_{r \to 0+} \sup_{K(x,r)} f, \qquad (6.3.2)$$

which is called an *upper hull* of f, and its value at an $x \in D$ is called the *supremum of f at x*.

If the set D and the function f are convex, and f is locally bounded below t a point $x_0 \in D$, then, by Theorem 6.2.2, f is locally bounded below at every point $x \in D$ and the infimum of f at every $x \in D$ is finite. Thus, in such a case, the function m_f is finite: $m_f : D \to \mathbb{R}$. In the other case f is locally unbounded below at every point $x \in D$, which means that $m_f(x) = -\infty$ for every $x \in D$. Similarly, it follows from Theorem 6.2.1 that either $M_f(x)$ is finite for every $x \in D$: $M_f : D \to \mathbb{R}$, or $M_f(x) = +\infty$ for every $x \in D$.

Theorem 6.3.1. *Let $D \subset \mathbb{R}^N$ be a convex and open set, and let $f : D \to \mathbb{R}$ be a convex function. If $m_f \neq -\infty$, then the function $m_f : D \to \mathbb{R}$ is continuous and convex.*

Proof. For every $z \in D$ and $\varepsilon > 0$ there exists a $\delta > 0$ such that

$$\varphi_z(r) \geqslant \lim_{\rho \to 0+} \varphi_z(\rho) - \varepsilon = m_f(z) - \varepsilon \text{ for } r < \delta, \tag{6.3.3}$$

whence for $t \in K(z,r)$, $r < \delta$,

$$f(t) \geqslant \inf_{K(z,r)} f = \varphi_z(r) \geqslant m_f(z) - \varepsilon. \tag{6.3.4}$$

Now take arbitrary $x, y \in D$, put $z = \dfrac{1}{2}(x+y)$, fix an $\varepsilon > 0$, and choose a $\delta > 0$ according to (6.3.3). Fix a positive $r < \delta$. Then (6.3.4) holds for $t \in K(z,r)$. Further, we can find points $u \in K(x,r)$ and $v \in K(y,r)$ such that (we assume that r has been chosen so small that $K(x,r), K(y,r) \subset D$)

$$f(u) \leqslant \inf_{K(x,r)} f + \varepsilon = \varphi_x(r) + \varepsilon \leqslant \lim_{\rho \to 0+} \varphi_x(\rho) + \varepsilon = m_f(x) + \varepsilon, \tag{6.3.5}$$

and

$$f(v) \leqslant \inf_{K(y,r)} f + \varepsilon = \varphi_y(r) + \varepsilon \leqslant \lim_{\rho \to 0+} \varphi_y(\rho) + \varepsilon = m_f(y) + \varepsilon, \tag{6.3.6}$$

since the functions φ_x, φ_y are decreasing. We have

$$\left| \frac{u+v}{2} - z \right| = \left| \frac{u+v}{2} - \frac{x+y}{2} \right| \leqslant \frac{1}{2}|u-x| + \frac{1}{2}|v-y| < \frac{1}{2}r + \frac{1}{2}r = r,$$

which means that $\dfrac{1}{2}(u+v) \in K(z,r)$. Thus we get by (6.3.4)

$$f\left(\frac{u+v}{2} \right) \geqslant m_f\left(\frac{x+y}{2} \right) - \varepsilon. \tag{6.3.7}$$

From (6.3.7), (6.3.5) and (6.3.6) we obtain, since the function f is convex,

$$m_f\left(\frac{x+y}{2} \right) - \varepsilon \leqslant f\left(\frac{u+v}{2} \right) \leqslant \frac{f(u) + f(v)}{2} \leqslant \frac{m_f(x) + m_f(y)}{2} + \varepsilon,$$

i.e.,

$$m_f\left(\frac{x+y}{2} \right) \leqslant \frac{m_f(x) + m_f(y)}{2} + 2\varepsilon.$$

Letting $\varepsilon \to 0$, we obtain hence the convexity of m_f.

Now we turn to the proof of the continuity of m_f. Fix an $x_0 \in D$. Let $\{e_1, \ldots, e_N\}$ be an orthonormal base of $\mathbb{R}^N$ over $\mathbb{R}$ so that we have, in particular,

$$|e_i| = 1, \ i = 1, \ldots, N. \tag{6.3.8}$$

For every $x \in \mathbb{R}^N$ we can write $x - x_0$ uniquely in the form

$$x - x_0 = \lambda_1 e_1 + \cdots + \lambda_N e_N, \tag{6.3.9}$$

where $\lambda_i \in \mathbb{R}$, $i = 1, \ldots, N$. For every k, $1 \leqslant k \leqslant N$, let M_k denote the set of those points $x \in D$ for which in expansion (6.3.9) at least $N - k$ of the coefficients λ_i are zeros.

By induction on k we will prove that $m_f \mid M_k$ is continuous at x_0. Let $k = 1$. Put

$$L_i = \{ x \in D \mid x = x_0 + \lambda e_i, \lambda \in \mathbb{R} \}, \ i = 1, \ldots, N.$$

We have

$$M_1 = \bigcup_{i=1}^{N} L_i,$$

so in order to prove that $m_f \mid M_1$ is continuous at x_0 it is enough to prove that $m_f \mid L_i$ is continuous at x_0 for every $i = 1, \ldots, N$.

Fix an i, $1 \leqslant i \leqslant N$. Since $x_0 \in L_i$, we can find a $\lambda \in \mathbb{R}$, $\lambda \neq 0$ such that $u = x_0 + \lambda e_i$ and $v = x_0 - \lambda e_i$ both belong to D. Hence $x_0 = \frac{1}{2}u + \frac{1}{2}v \in Q(u, v)$. By Theorem 6.1.1 $m_f \mid Q(u, v)$ is continuous, so given an $\varepsilon > 0$ we can find a $\delta > 0$ such that

$$|m_f(t) - m_f(x_0)| < \frac{\varepsilon}{2} \ \text{for } t \in Q(u, v) \cap K(x_0, \delta). \tag{6.3.10}$$

Moreover, we may assume that $\delta < |\lambda|$.

On the other hand, we have by (6.3.4) for $y \in K(x_0, \delta)$, provided that δ has been chosen small enough,

$$f(y) \geqslant m_f(x_0) - \frac{\varepsilon}{2}.$$

If $z \in K(x_0, \delta)$, and $r > 0$ is so small that $K(z, r) \subset K(x_0, \delta)$, we obtain hence

$$\inf_{K(z,r)} f \geqslant m_f(x_0) - \frac{\varepsilon}{2},$$

and on letting $r \to 0+$ we get (for arbitrary $z \in K(x_0, \delta)$)

$$m_f(z) \geqslant m_f(x_0) - \frac{\varepsilon}{2} > m_f(x_0) - \varepsilon. \tag{6.3.11}$$

Now suppose that $z \in K(x_0, \delta) \cap L_i$, and take an arbitrary $r > 0$ such that $K(z, r) \subset K(x_0, \delta)$. Since $z \in L_i$, there exists a $\kappa \in \mathbb{R}$ such that $z = x_0 + \kappa e_i$, and since $z \in K(x_0, \delta)$, we have $|\kappa| |e_i| = |z - x_0| < \delta$, i.e., by (6.3.8), $|\kappa| < \delta < |\lambda|$. Hence

$$\frac{\kappa + \lambda}{2\lambda} = \frac{1}{2} + \frac{1}{2}\frac{\kappa}{\lambda} \in (0, 1).$$

Consequently there exists a $\mu \in \mathbb{Q} \cap (0,1)$ such that

$$\left| \mu - \frac{\kappa + \lambda}{2\lambda} \right| < \frac{r}{2|\lambda|}. \tag{6.3.12}$$

Put $t = \mu u + (1 - \mu) v = x_0 + (2\mu\lambda - \lambda) e_i$. Thus $t \in Q(u,v)$. Moreover by (6.3.8) and (6.3.12)

$$|t - z| = |(2\mu\lambda - \lambda - \kappa) e_i| = |2\mu\lambda - \lambda - \kappa| = 2|\lambda| \left| \mu - \frac{\kappa + \lambda}{2\lambda} \right| < r,$$

whence $t \in K(z,r) \subset K(x_0, \delta)$. By (6.3.10)

$$m_f(t) < m_f(x_0) + \frac{\varepsilon}{2}. \tag{6.3.13}$$

Let U_t be an arbitrary ball centered at t and such that $U_t \subset K(z,r)$. There exists an $s \in U_t$ such that $f(s) \leqslant m_f(t) + \frac{\varepsilon}{2}$, or by (6.3.13)

$$f(s) < m_f(x_0) + \varepsilon. \tag{6.3.14}$$

Consequently every ball $K(z,r) \subset K(x_0, \delta)$ contains a point s such that (6.3.14) holds. Hence

$$\inf_{K(z,r)} f \leqslant m_f(x_0) + \varepsilon,$$

and on letting $r \to 0+$ we obtain

$$m_f(z) \leqslant m_f(x_0) + \varepsilon. \tag{6.3.15}$$

Relations (6.3.11) and (6.3.15) imply that

$$|m_f(z) - m_f(x_0)| \leqslant \varepsilon$$

for $z \in K(x_0, \delta) \cap L_i$. Consequently the function $m_f \mid L_i$ is continuous at x_0, and hence also the function $m_f \mid M_1$ is continuous at x_0.

Now assume that for a $k \in \mathbb{N}$, $2 \leqslant k \leqslant N$, $m_f \mid M_{k-1}$ is continuous at x_0. Thus, given an $\varepsilon > 0$, there exists a $\delta > 0$ such that for all $y \in M_{k-1} \cap K(x_0, \delta)$ we have

$$|m_f(y) - m_f(x_0)| < \frac{1}{3}\varepsilon. \tag{6.3.16}$$

Now take an $x \in M_k \cap K\left(x_0, \frac{1}{2}\delta \right)$. Then, possibly after a suitable renumbering of e_i's,

$$x - x_0 = \lambda_1 e_1 + \cdots + \lambda_k e_k$$

with some $\lambda_i \in \mathbb{R}$, $i = 1, \ldots, k$. Put

$$u = x_0 - \lambda_1 e_1 + \lambda_3 e_3 + \cdots + \lambda_k e_k, \quad v = x_0 + \frac{1}{2}\lambda_2 e_2 + \lambda_3 e_3 + \cdots + \lambda_k e_k.$$

Then $u, v \in M_{k-1}$ and $v = \dfrac{1}{2}(x + u)$. Moreover,

$$|u - x_0| = \left(\lambda_1^2 + \lambda_3^2 + \cdots + \lambda_k^2\right)^{\frac{1}{2}} \leqslant \left(\lambda_1^2 + \lambda_2^2 + \lambda_3^2 + \cdots + \lambda_k^2\right)^{\frac{1}{2}} = |x - x_0| < \frac{1}{2}\delta < \delta,$$

$$|v - x_0| = \left(\frac{1}{4}\lambda_2^2 + \lambda_3^2 + \cdots + \lambda_k^2\right)^{\frac{1}{2}} \leqslant \left(\lambda_1^2 + \lambda_2^2 + \lambda_3^2 + \cdots + \lambda_k^2\right)^{\frac{1}{2}} = |x - x_0| < \frac{1}{2}\delta < \delta,$$

which means that $u, v \in K(x_0, \delta)$. By (6.3.16)

$$|m_f(u) - m_f(x_0)| < \frac{1}{3}\varepsilon, \quad |m_f(v) - m_f(x_0)| < \frac{1}{3}\varepsilon, \qquad (6.3.17)$$

and, since the function m_f is convex,

$$m_f(v) = m_f\left(\frac{u + x}{2}\right) \leqslant \frac{1}{2}[m_f(u) + m_f(x)],$$

i.e.,

$$m_f(x) \geqslant 2 m_f(v) - m_f(u).$$

Hence we obtain in view of (6.3.17)

$$m_f(x) > 2\left[m_f(x_0) - \frac{1}{3}\varepsilon\right] - \left[m_f(x_0) + \frac{1}{3}\varepsilon\right] = m_f(x_0) - \varepsilon. \qquad (6.3.18)$$

Now put

$$w = x_0 + 2\lambda_2 e_2 + \lambda_3 e_3 + \cdots + \lambda_k e_k, \quad z = x_0 + 2\lambda_1 e_1 + \lambda_3 e_3 + \cdots + \lambda_k e_k.$$

Then $w, z \in M_{k-1}$, $x = \dfrac{1}{2}(w + z)$. Moreover, $|w - x_0| \leqslant 2|x - x_0| < \delta$, $|z - x_0| \leqslant 2|x - x_0| < \delta$ so that $w, z \in K(x_0, \delta)$. By (6.3.16)

$$|m_f(w) - m_f(x_0)| < \frac{1}{3}\varepsilon, \quad |m_f(z) - m_f(x_0)| < \frac{1}{3}\varepsilon, \qquad (6.3.19)$$

and, since the function m_f is convex,

$$m_f(x) = m_f\left(\frac{z + w}{2}\right) \leqslant \frac{1}{2}[m_f(z) + m_f(w)].$$

Hence we obtain in view of (6.3.19)

$$m_f(x) < \frac{1}{2}\left[m_f(x_0) + \frac{1}{3}\varepsilon + m_f(x_0) + \frac{1}{3}\varepsilon\right] = m_f(x_0) + \frac{1}{3}\varepsilon < m_f(x_0) + \varepsilon. \quad (6.3.20)$$

Relations (6.3.18) and (6.3.20) imply that

$$|m_f(x) - m_f(x_0)| < \varepsilon. \qquad (6.3.21)$$

Relation (6.3.21) holds for every $x \in M_k \cap K(x_0, \delta)$, which means that the function $m_f \mid M_k$ is continuous at x_0.

Induction shows that $m_f \mid M_k$ is continuous at x_0 for $k = 1, \ldots, N$. But $M_N = \mathbb{R}^N$, so $m_f \mid M_N = m_f$. Thus the function m_f is continuous at x_0. Since $x_0 \in D$ has been arbitrary, the function m_f is continuous in D. $\qquad \square$

6.4 Theorem of Bernstein-Doetsch

Let $D \subset \mathbb{R}^N$ be an open set, and let $f : D \to \mathbb{R}$ be a function. It follows directly from the definition of the lower hull of f that

$$m_f(x) \leqslant f(x) \text{ for } x \in D. \tag{6.4.1}$$

But if the function f and the set D are convex, we have more precise informations.

Theorem 6.4.1. *Let $D \subset \mathbb{R}^N$ be a convex and open set, and let $f : D \to \mathbb{R}$ be a convex function. If at a point $\xi \in D$ we have $f(\xi) \neq m_f(\xi)$, then f is not locally bounded at ξ.*

Proof. If $m_f(\xi) = -\infty$, then f is not locally bounded below at ξ, as results from the definition of m_f. So let $m_f(\xi) > -\infty$. Let $U \subset D$ be an arbitrary neighbourhood of ξ. Take an arbitrary $M > 0$. According to (6.4.1)

$$p = f(\xi) - m_f(\xi) > 0. \tag{6.4.2}$$

Thus there exists an $n \in \mathbb{N}$ such that

$$\frac{1}{2}(n-1)p + m_f(\xi) > M. \tag{6.4.3}$$

Further, there exists an $h \in \mathbb{R}^N$ such that $\xi \pm ih \in U$, $i = 1, \ldots, n$, and

$$m_f(\xi) - \frac{1}{2}p < f(\xi - h) < m_f(\xi) + \frac{1}{2}p. \tag{6.4.4}$$

Take in Lemma 5.3.1 $x_1 = \cdots = x_{n-1} = \xi - h$, $x_n = \xi + (n-1)h$. We obtain

$$f(\xi) = f\left(\frac{(n-1)(\xi-h) + \xi + (n-1)h}{n}\right) \leqslant \frac{(n-1)f(\xi-h) + f(\xi+(n-1)h)}{n}.$$

Hence

$$f(\xi + (n-1)h) \geqslant nf(\xi) - (n-1)f(\xi-h) = n[f(\xi) - f(\xi-h)] + f(\xi-h). \tag{6.4.5}$$

By (6.4.2) and (6.4.4) $f(\xi - h) < f(\xi) - \frac{1}{2}p$, so we obtain by (6.4.4), (6.4.5) and (6.4.3)

$$f(\xi + (n-1)h) > \frac{n}{2}p + m_f(\xi) - \frac{1}{2}p = \frac{1}{2}(n-1)p + m_f(\xi) > M.$$

Thus for every neighbourhood U of ξ and for every $M > 0$ there exists a point $t \in U$ [viz. $t = \xi + (n-1)h$] such that $f(t) > M$. It follows that f is not locally bounded above at ξ. $\qquad\square$

Hence we derive the following Theorem of Bernstein-Doetsch [30].

Theorem 6.4.2. *Let $D \subset \mathbb{R}^N$ be a convex and open set, and let $f : D \to \mathbb{R}$ be a convex function. If f is locally bounded above at a point of D, then it is continuous in D.*

Proof. Let f be locally bounded above at a point of D. By Theorem 6.2.3 f is locally bounded at every point of D, whence it follows in virtue of Theorem 6.4.1 that $f(x) = m_f(x)$ for $x \in D$. Since f is locally bounded, we have $m_f \neq -\infty$, whence by Theorem 6.3.1 m_f is continuous in D. Consequently f is continuous in D.   □

Because of the basic importance of the Theorem of Bernstein-Doetsch in the theory of convex functions, we present below two other proofs, which do not rely on the difficult Theorem 6.3.1.

Second proof of Theorem 6.4.2. If f is locally bounded above at a point of D, then, by Theorem 6.2.3, it is locally bounded at every point of D. Consequently formulas (6.3.1) and (6.3.2) define finite numbers for every $x \in D$. Evidently

$$m_f(x) \leqslant f(x) \leqslant M_f(x) \tag{6.4.6}$$

for every $x \in D$.

Now take an arbitrary $x \in D$. There exists a sequence of points $x_n \in D$, $n \in \mathbb{N}$, such that

$$\lim_{n \to \infty} x_n = x, \quad \lim_{n \to \infty} f(x_n) = m_f(x), \tag{6.4.7}$$

and the sequence of points $z_n \in D$, $n \in \mathbb{N}$, such that

$$\lim_{n \to \infty} z_n = x, \quad \lim_{n \to \infty} f(z_n) = M_f(x). \tag{6.4.8}$$

Put $y_n = 2z_n - x_n$, $n \in \mathbb{N}$. We have by (6.4.7) and (6.4.8)

$$\lim_{n \to \infty} y_n = x. \tag{6.4.9}$$

Moreover $z_n = \dfrac{1}{2}(x_n + y_n)$, $n \in \mathbb{N}$, whence by the convexity of f

$$f(z_n) \leqslant \frac{1}{2}[f(x_n) + f(y_n)],$$

or,

$$f(y_n) \geqslant 2f(z_n) - f(x_n).$$

Hence, on letting $n \to \infty$ we obtain in virtue of (6.4.7) and (6.4.8)

$$\liminf_{n \to \infty} f(y_n) \geqslant 2M_f(x) - m_f(x), \tag{6.4.10}$$

whereas by (6.4.9)

$$\liminf_{n \to \infty} f(y_n) \leqslant M_f(x). \tag{6.4.11}$$

Relations (6.4.10) and (6.4.11) yield the inequality $M_f(x) \leqslant m_f(x)$, which together with (6.4.6) implies that $M_f(x) = m_f(x)$. Thus (cf. Exercise 6.8) f is continuous at x, and so, since $x \in D$ has been arbitrary, f is continuous in D.   □

Third proof of Theorem 6.4.2. Let f be locally bounded above at a point of D. Take an arbitrary $x_0 \in D$. By Theorem 6.2.3 f is locally bounded at x_0, thus there exists positive constants M and δ such that $K(x_0, \delta) \subset D$ and

$$|f(t)| \leqslant M \text{ for } t \in K(x_0, \delta). \tag{6.4.12}$$

For an arbitrary $x \in K(x_0, \delta)$ we have $|x - x_0| < \delta$, and consequently we can find a $\lambda \in \mathbb{Q} \cap (0, 1)$ such that

$$\frac{|x - x_0|}{\delta} < \lambda < 2 \frac{|x - x_0|}{\delta}. \tag{6.4.13}$$

Further put

$$y = x_0 + \frac{x - x_0}{\lambda}, \quad z = x_0 - \frac{x - x_0}{\lambda}. \tag{6.4.14}$$

Hence, in view of (6.4.13),

$$|y - x_0| = \frac{|x - x_0|}{\lambda} < \delta, \quad |z - x_0| = \frac{|x - x_0|}{\lambda} < \delta,$$

which means that $y, z \in K(x_0, \delta)$. Moreover, by (6.4.14)

$$x = \lambda y + (1 - \lambda) x_0, \quad x_0 = \frac{1}{\lambda + 1} x + \frac{\lambda}{\lambda + 1} z.$$

By Theorem 5.3.5, since f is convex

$$f(x) \leqslant \lambda f(y) + (1 - \lambda) f(x_0), \quad f(x_0) \leqslant \frac{1}{\lambda + 1} f(x) + \frac{\lambda}{\lambda + 1} f(z),$$

i.e.,

$$f(x) - f(x_0) \leqslant \lambda [f(y) - f(x_0)], \quad f(x) - f(x_0) \geqslant \lambda [f(x_0) - f(z)].$$

Hence we obtain by (6.4.12), since $x_0, y, z \in K(x_0, \delta)$,

$$-2M\lambda \leqslant f(x) - f(x_0) \leqslant 2M\lambda,$$

and taking into account (6.4.13) we get

$$-\frac{4M}{\delta} |x - x_0| < f(x) - f(x_0) < \frac{4M}{\delta} |x - x_0|,$$

or

$$|f(x) - f(x_0)| < \frac{4M}{\delta} |x - x_0|. \tag{6.4.15}$$

Relation (6.4.15) proves the continuity of f at x_0, whence, since $x_0 \in D$ has been arbitrary, the continuity of f in D follows. $\square$

Since every open set is a neighbourhood of any of its points, we get hence

Corollary 6.4.1. *Let $D \subset \mathbb{R}^N$ be a convex and open set, and let $f : D \to \mathbb{R}$ be a convex function. If f is bounded above on a non-empty open set $U \subset D$, then it is continuous in D.*

Let us observe, that the local (or even global) boundedness *below* does not imply the continuity of a convex function. For example, the function exp is continuous and convex on $\mathbb{R}$, and let $a : \mathbb{R}^N \to \mathbb{R}$ be a discontinuous additive function. Then the function (cf. Example 5.3.1) $f : \mathbb{R}^N \to \mathbb{R}$:

$$f(x) = \exp a(x)$$

is convex, discontinuous, and

$$f(x) \geqslant 0 \text{ for } x \in \mathbb{R}^N.$$

If a function is continuous at a point, then it is locally bounded at this point. Thus the Theorem of Bernstein-Doetsch implies the following

Theorem 6.4.3. *Let $D \subset \mathbb{R}^N$ be a convex and open set, and let $f : D \to \mathbb{R}$ be a convex function. Then either f is continuous in D, or f is totally discontinuous[3] in D.*

In the next chapter we are going to discuss the main properties of continuous convex functions.

Exercises

1. Let $A \subset \mathbb{R}^N$ be an arbitrary set, and let $f : A \to \mathbb{R}$ be a function uniformly continuous on A. Prove that there exists a unique continuous function $F : \operatorname{cl} A \to \mathbb{R}$ such that $F \,|\, A = f$.

2. Let $D \subset \mathbb{R}^N$ be a convex and open set, and let $f : D \to \mathbb{R}$ be a convex function. Let $x \in D$ and $y \in \mathbb{R}^N$ be such that $x + y \in D$. Prove that

$$\lim_{n \to \infty} f\left(x + \frac{y}{2^n}\right) = f(x).$$

3. Let $D \subset \mathbb{R}^N$ be a convex and open set, and let $f : D \to \mathbb{R}$ be a convex function. Let $0, a, b, a + b \in D$. Prove that the set

$$A = \left\{x \in \mathbb{R}^N \mid x = \lambda_1 a + \lambda_2 b, \lambda_1, \lambda_2 \in \mathbb{Q} \cap [0, 1]\right\}$$

is contained in D and that the function f is bounded on A.
   [Hint: Clearly f is bounded above on A, and bounded below on $Q\,(0, a + b)$. For any $x \in A$, $x = \lambda_1 a + \lambda_2 b$, $\lambda_1, \lambda_2 \in \mathbb{Q} \cap (0, 1)$ consider the points $y = \lambda_2 a + \lambda_1 b$ and $z = \dfrac{1}{2}\,(x + y)$.]

4. Let $D \subset \mathbb{R}^N$ be a convex and open set, and let $f : D \to \mathbb{R}$ be a convex function. Let $0, a, b, a+b \in D$, and let $\lambda_n, \mu_n \in \mathbb{Q} \cap [0, 1]$, $n \in \mathbb{N}$, be such that $\lim\limits_{n \to \infty} \lambda_n = \lambda_0$, $\lim\limits_{n \to \infty} \mu_n = \mu_0$. Prove that if $\lambda_0, \mu_0 \in \mathbb{Q} \cap [0, 1]$, then

$$\lim_{n \to \infty} f\,(\lambda_n a + \mu_n b) = f\,(\lambda_0 a + \mu_0 b). \tag{$*$}$$

---

[3] I.e., discontinuous at every point of D.

5. Let $D \subset \mathbb{R}^N$ be a convex and open set, and let $f : D \to \mathbb{R}$ be a convex function. Let $0, a, b \in D$, and let $\lambda_n, \mu_n \in \mathbb{Q}$, $n \in \mathbb{N}$, be such that $\lim\limits_{n\to\infty} \lambda_n = \lambda_0$, $\lim\limits_{n\to\infty} \mu_n = \mu_0$, and $\lambda_0 a + \mu_0 b \in D$, $\lambda_n a + \mu_n b \in D$, $n \in \mathbb{N}$. Prove that if $\lambda_0, \mu_0 \in \mathbb{Q}$, then $(*)$ holds.

6. Let $D \subset \mathbb{R}^N$ be a convex and open set, and let $f : D \to \mathbb{R}$ be a convex function. Let $0, a, b \in D$. Prove that if a, b are linearly independent over $\mathbb{R}$, then the function $f \, | \, E(a, b) \cap D$ is continuous.

7. Show that the conclusion of Exercise 6.6 may be invalid if a, b are linearly dependent over $\mathbb{R}$.

   [Hint: Take incommensurable $a, b \in \mathbb{R}$, and let $f : \mathbb{R} \to \mathbb{R}$ be an additive function such that $f(a) = 1$, $f(b) = 0$. Consider a sequence $x_n = p_n a + q_n b$, $p_n, q_n \in \mathbb{Z}$, $n \in \mathbb{N}$, such that $\lim\limits_{n\to\infty} x_n = 0$].

8. Let $D \subset \mathbb{R}^N$ be an open set, and let $f : D \to \mathbb{R}$ be a function. Prove that the condition $M_f(x) = m_f(x)$ (cf. (6.3.1) and (6.3.2)) is necessary and sufficient for the continuity of f at x, $x \in D$.

Chapter 7

Continuous Convex Functions

7.1 The basic theorem

Let $D \subset \mathbb{R}^N$ be a convex and open set. In 5.3 we saw that a convex function $f : D \to \mathbb{R}$ fulfills the inequality

$$f\big(\lambda x + (1 - \lambda)y\big) \leqslant \lambda f(x) + (1 - \lambda)f(y) \tag{7.1.1}$$

for all $x, y \in D$ and all $\lambda \in \mathbb{Q} \cap [0, 1]$. It was also pointed out that if, moreover, f is continuous, then inequality (7.1.1) holds actually for all real $\lambda \in [0, 1]$.

The following lemma is analogous to Lemma 6.1.2

Lemma 7.1.1. *Let $D \subset \mathbb{R}^N$ be a convex and open set, and let $f : D \to \mathbb{R}$ be a function fulfilling inequality (7.1.1) for all $\lambda \in [0, 1]$. If the function f is bounded above on a set $A \subset D$, then it is also bounded above (by the same constant) on the set $\mathrm{conv}(A)$.*

Proof. Suppose that

$$f(t) \leqslant M \quad \text{for } t \in A, \tag{7.1.2}$$

with a certain real constant M. Take an arbitrary $x \in \mathrm{conv}(A)$. By Theorem 5.1.3 there exist an $n \in \mathbb{N}$, $t_1, \ldots, t_n \in A$, and $\lambda_1, \ldots \lambda_n \in [0, 1]$ such that $\sum_{i=1}^{n} \lambda_i = 1$ and

$$x = \sum_{i=1}^{n} \lambda_i t_i. \tag{7.1.3}$$

Relation (7.1.3) implies in virtue of Lemma 5.3.2 that

$$f(x) \leqslant \sum_{n}^{i=1} \lambda_i f(t_i) \leqslant \sum_{n}^{i=1} \lambda_i M = M \sum_{n}^{i=1} \lambda_i = M$$

where we have used also (7.1.2). Consequently f is also bounded above by M at an arbitrary point of $\mathrm{conv}(A)$. $\qquad\square$

Theorem 7.1.1. *Let $D \subset \mathbb{R}^N$ be a convex and open set, and let $f : D \to \mathbb{R}$ be a function. The function f is convex and continuous if and only if it satisfies inequality* (7.1.1) *for all $x, y \in D$ and all $\lambda \in [0, 1]$.*

Proof. If the function f is convex and continuous, then clearly it satisfies (7.1.1) for all $x, y \in D$ and all $\lambda \in [0, 1]$. Now suppose that a function $f : D \to \mathbb{R}$ satisfies inequality (7.1.1) for all $x, y \in D$ and $\lambda \in [0, 1]$. Putting $\lambda = \dfrac{1}{2}$ we see that f is convex. Take an arbitrary $x_0 \in D$. According to the Remark in 5.1, theres exist an N-dimensional simplex $S \subset D$ such that $x_0 \in \mathrm{int}(S)$. Let $p_1, \ldots, p_{N+1}$ be the vertices of S:

$$S = \mathrm{conv} \{p_1, \ldots, p_{N+1}\}$$

Since the set $A = \{p_1, \ldots, p_{N+1}\}$ is finite, f is bounded above on A. By Lemma 7.1.1 f is bounded above on S, and thus on a neighborhood of x_0. In other words, f is locally bounded above at x_0. By Theorem 6.4.2 f is continuous. $\square$

Condition (7.1.1) has a very simple geometric interpretation. When λ runs over the interval $[0, 1]$, the points $\lambda x + (1 - \lambda) y$ fill the segment joining x and y, and thus the points $\big(\lambda x + (1 - \lambda)y, \, \lambda f(x) + (1 - \lambda)f(y)\big) = \lambda\big(x, f(x)\big) + (1 - \lambda)\big(y, f(y)\big)$ fill the segment joining the points $\big(x, f(x)\big)$ and $\big(y, f(y)\big)$ $\big($in $\mathbb{R}^{N+1}\big)$, i.e., the chord joining the corresponding points of the graph of f. Condition (7.1.1) says that the points of the graph of f between $\big(x, f(x)\big)$ and $\big(y, f(y)\big)$, for arbitrary $x, y \in D$, lie below the points of the corresponding chord.

In many sources (where the authors are interested exclusively in continuous convex functions) condition (7.1.1) is taken as the definition of convexity. Then discontinuous convex functions are automatically eliminated. In the present book, however, a *convex function* is any function satisfying the Jensen-inequality (5.3.1). If we want the function in question to be continuous, we shall always distinctly say so.

Contrary to discontinuous convex functions, which behave in a rather pathological manner, continuous convex functions display a nice behaviour. Continuous convex functions are very thoroughly treated, e.g., in Rockafellar [268], and Roberts-Varberg [267]. In the sequel of this chapter we are going to present only a few properties of continuous convex functions. Our exposition is based on Rockafellar [268].

7.2–7.5 are devoted particularly to continuous convex functions of a single real variable. The results presented there have no simple extension to the case of several variables. In the rest if this chapter we deal with continuous convex functions of an arbitrary number of variables.

7.2 Compositions and inverses

In the present section we investigate the convexity of compositions of, and of inverse functions to, convex and concave functions. Note that continuity of the functions in question is never explicitly used, however, a monotonic function an interval is locally bounded at every point of this interval, and hence a monotonicity condition on a convex or concave function authomatically implies its continuity. Only the functions f and φ in Theorem 7.2.1 below need not be continuous.

Theorem 7.2.1. *Let $d \subset \mathbb{R}^N$ be a convex and open set, and let $J \subset \mathbb{R}$ be an open interval[1]. Let $f : D \to J$, and $g : J \to \mathbb{R}$, be functions, and let $\varphi : D \to \mathbb{R}$ be the composition $\varphi = g \circ f = g(f)$. Then:*

1. *if f is convex, and g is convex and increasing, then φ is convex;*
2. *if f is concave, and g is concave and increasing, then φ is concave;*
3. *if f is convex, and g is concave and decreasing, then φ is concave;*
4. *if f is concave, and g is convex and decreasing, then φ is convex.*

Proof. Let f be convex and g convex and increasing. We have for arbitrary $x, y \in D$

$$f\left(\frac{x+y}{2}\right) \leqslant \frac{f(x)+f(y)}{2}$$

whence

$$g\left(f\left(\frac{x+y}{2}\right)\right) \leqslant g\left(\frac{f(x)+f(y)}{2}\right) \leqslant \frac{g(f(x))+g(f(y))}{2},$$

i.e.,

$$\varphi\left(\frac{x+y}{2}\right) \leqslant \frac{\varphi(x)+\varphi(y)}{2}$$

Thus φ is convex. The remaining cases are dealt with similarly. $\qquad\square$

Note that if f is additive, then we do not need the monotonicity condition on g (cf. Example 5.3.1)

The function $g : \mathbb{R} \to \mathbb{R}$ given by

$$g(x) = \begin{cases} x^p & \text{for } x \geqslant 0, \\ 0 & \text{for } x < 0, \end{cases} \tag{7.2.1}$$

is convex and increasing whenever $p > 1$ (cf. Theorem 7.5.2). Hence we obtain from Theorem 7.2.1

Corollary 7.2.1. *Let $D \subset \mathbb{R}^N$ be a convex and open set, and let $f : D \to \mathbb{R}$ be a non-negative convex function. If $p > 1$, then the function $\varphi : D \to \mathbb{R}$, $\varphi(x) = [f(x)]^p$, $x \in D$, is convex.*

Theorem 7.2.2. *Let $J \subset \mathbb{R}$ be an open interval, and let $f : J \to \mathbb{R}$ be a strictly monotonic function. Let $f^{-1} : f(J) \to J$ be the function inverse to f. Then:*

1. *if f is convex and increasing, then f^{-1} is concave;*
2. *if f is convex and decreasing, then f^{-1} is convex;*
3. *if f is concave and increasing, then f^{-1} is convex;*
4. *if f is concave and decreasing, then f^{-1} is concave.*

---

[1] By an open interval we understand in this book every interval of the form (a, b), where $-\infty \leqslant a < b \leqslant +\infty$

Proof. Let f be convex and increasing. Take arbitrary $x, y \in f(J)$, and put $u = f^{-1}(x)$, $v = f^{-1}(y)$. By the convexity of f

$$f\left(\frac{u+v}{2}\right) \leqslant \frac{f(u)+f(v)}{2} = \frac{x+y}{2}$$

whence, since f^{-1} is increasing, just like f,

$$\frac{u+v}{2} \leqslant f^{-1}\left(\frac{x+y}{2}\right),$$

or

$$\frac{f^{-1}(x)+f^{-1}(y)}{2} \leqslant f^{-1}\left(\frac{x+y}{2}\right).$$

Thus f^{-1} is concave. The remaining cases are dealt similarly. □

7.3 Differences quotients

We start with the following result.

Theorem 7.3.1. *Let $J \subset \mathbb{R}$ be an open interval, and let $f : D \to \mathbb{R}$ be a function. Each of the following conditions (postulated for every $x_1, x_2, x_3 \in J$, $x_1 < x_2 < x_3$) is necessary and sufficient for the function f to be continuous and convex:*

$$(x_3 - x_1) f(x_2) \leqslant (x_2 - x_1) f(x_3) + (x_3 - x_2) f(x_1); \qquad (7.3.1)$$

$$\frac{f(x_2) - f(x_1)}{x_2 - x_1} \leqslant \frac{f(x_3) - f(x_1)}{x_3 - x_1}; \qquad (7.3.2)$$

$$\frac{f(x_3) - f(x_1)}{x_3 - x_1} \leqslant \frac{f(x_3) - f(x_2)}{x_3 - x_2}. \qquad (7.3.3)$$

Proof. First we show that the three conditions (7.3.1), (7.3.2), (7.3.3) are equivalent to each other. Take arbitrary $x_1, x_2, x_3 \in J$, $x_1 < x_2 < x_3$. Suppose that a function $f : J \to \mathbb{R}$ fulfills (7.3.1). Then we get subtracting the term $(x_3 - x_1) f(x_1)$ from both the sides of (7.3.1)

$$(x_3 - x_1) [f(x_2) - f(x_1)] \leqslant (x_2 - x_1) [f(x_3) - f(x_1)] \qquad (7.3.4)$$

whence (7.3.2) follows on dividing by $(x_2 - x_1)(x_3 - x_1) > 0$.

Conversely, (7.3.4) results from (7.3.2), and adding the term $(x_3 - x_1)f(x_1)$ to both sides we obtain (7.3.1). Thus (7.3.1) and (7.3.2) are equivalent.

Similarly, adding to both sides of (7.3.1) the term $(x_3 - x_2)f(x_3)$ we obtain

$$(x_3 - x_1) f(x_2) + (x_3 - x_2) f(x_3) \leqslant (x_3 - x_1) f(x_3) + (x_3 - x_2) f(x_1), \qquad (7.3.5)$$

whence

$$(x_3 - x_2) [f(x_3) - f(x_1)] \leqslant (x_3 - x_1) [f(x_3) - f(x_2)], \qquad (7.3.6)$$

and (7.3.3) results on dividing by $(x_3 - x_2)(x_3 - x_1) > 0$. Conversely, (7.3.3) implies (7.3.6) and (7.3.5), and hence (7.3.1) follows by subtracting the term $(x_3 - x_2) f(x_3)$ from both the sides of (7.3.5). Thus (7.3.1) and (7.3.3) are equivalent.

Hence it follows that all the three conditions (7.3.1), (7.3.2), (7.3.3) are equivalent to each other.

Now let $f : J \to \mathbb{R}$ be a continuous and convex function. By Theorem 7.1.1

$$f\big(\lambda x + (1 - \lambda)y\big) \leqslant \lambda f(x) + (1 - \lambda)f(y) \tag{7.3.7}$$

holds for arbitrary $x, y \in J$ and $\lambda \in [0, 1]$. Take arbitrary $x_1, x_2, x_3 \in J$, $x_1 < x_2 < x_3$, and put in (7.3.7) $x = x_1$, $y = x_3$, $\lambda = \dfrac{x_2 - x_1}{x_3 - x_1}$. Since $x_1 < x_2 < x_3$, we have $0 < x_3 - x_2 < x_3 - x_1$, and consequently $\lambda \in (0, 1)$. Then $1 - \lambda = \dfrac{x_2 - x_1}{x_3 - x_1}$, and

$$\lambda x + (1 - \lambda)y = \frac{x_3 - x_2}{x_3 - x_1}x_1 + \frac{x_2 - x_1}{x_3 - x_1}x_3 = x_2.$$

Thus (7.3.7) yields

$$f(x_2) \leqslant \frac{x_3 - x_2}{x_3 - x_1} f(x_1) + \frac{x_2 - x_1}{x_3 - x_1} f(x_3),$$

and (7.3.1) follows on multiplying both the sides by $x_3 - x_1 > 0$.

Now assume that a function $f : J \to \mathbb{R}$ fulfills (7.3.1) for arbitrary $x_1, x_2, x_3 \in J$, $x_1 < x_2 < x_3$. Take arbitrary $x, y \in J$, $x \neq y$, and an arbitrary $\lambda \in (0, 1)$. One of x, y is smaller than the other, e.g., $x < y$. Put in (7.3.1) $x_1 = x$, $x_3 = y$, $x_2 = \lambda x + (1 - \lambda)y$. We have

$$x_2 - x_1 = \lambda x + (1 - \lambda)y - x = (1 - \lambda)(y - x) > 0,$$

and

$$x_3 - x_2 = y - \lambda x - (1 - \lambda)y = \lambda(y - x) > 0$$

so that $x_1 < x_2 < x_3$. (7.3.1) yields

$$(y - x)f\big(\lambda x + (1 - \lambda)y\big) \leqslant (1 - \lambda)(y - x)f(y) + \lambda(y - x)f(x).$$

Dividing both the sides by $y - x > 0$ we obtain (7.3.7).

If $x = y$, or if $\lambda = 0$, or $\lambda = 1$, then (7.3.7) is trivially fulfilled. So f satisfies (7.3.7) for all $x, y \in J$ and all $\lambda \in [0, 1]$. By Theorem 7.1.1 f is continuous and convex. $\square$

Now let $J \subset \mathbb{R}$ be an open interval, and let $f : J \to \mathbb{R}$ be a function. We define a function $I(x, h)$ for $x \in J$ and $h \in \mathbb{R}$ such that $h \neq 0$, $x + h \in J$:

$$I(x, h) = \frac{f(x + h) - f(x)}{h}. \tag{7.3.8}$$

Theorem 7.3.2. *Let $J \subset \mathbb{R}$ be an open interval, and let $f : J \to \mathbb{R}$ be a continuous and convex function. Then the corresponding function I defined by (7.3.8) is increasing with respect to either variable.*

Proof. Fix an arbitrary $x \in J$, and take $h_1, h_2 \in \mathbb{R}$ such that $h_1 < h_2$, $x + h_1 \in J$, $x + h_2 \in J$. If $0 < h_1 < h_2$, put in (7.3.2) $x_1 = x$, $x_2 = x + h_1$, $x_3 = x + h_2$. Then we get $I(x, h_1) \leqslant I(x, h_2)$. If $h_1 < h_2 < 0$, then we put in (7.3.3) $x_1 = x + h_1$, $x_2 = x + h_2$, $x_3 = x$. We obtain

$$\frac{f(x) - f(x + h_1)}{-h_1} \leqslant \frac{f(x) - f(x + h_2)}{-h_2},$$

i.e.,

$$\frac{f(x + h_1) - f(x)}{h_1} \leqslant \frac{f(x + h_2) - f(x)}{h_2},$$

and so $I(x, h_1) \leqslant I(x, h_2)$. Finally, if $h_1 < 0 < h_2$, then in the inequality

$$\frac{f(x_2) - f(x_1)}{x_2 - x_1} \leqslant \frac{f(x_3) - f(x_2)}{x_3 - x_2},$$

resulting from (7.3.2) and (7.3.3), we put $x_1 = x + h_1$, $x_2 = x$, $x_3 = x + h_2$, and obtain again $I(x, h_1) \leqslant I(x, h_2)$. Consequently the function I is increasing with respect to h.

Now take arbitrary $x_1, x_2 \in J$, $x_1 < x_2$, and an $h \neq 0$ such that $x_1 + h, x_2 + h \in J$. Using the monotonicity (already proved) of I with respect to the second variable, we have

$$I(x_1, h) = \frac{f(x_1 + h) - f(x_1)}{h} \leqslant \frac{f(x_2 + h) - f(x_1)}{h + (x_2 - x_1)},$$

and

$$\begin{aligned}
\frac{f(x_2 + h) - f(x_1)}{h + (x_2 - x_1)} &= \frac{f\big((x_2 + h) + \big(-h - (x_2 - x_1)\big)\big) - f(x_2 + h)}{-h - (x_2 - x_1)} \\
&\leqslant \frac{f\big((x_2 + h) + (-h)\big) - f(x_2 + h)}{-h} \\
&= \frac{f(x_2 + h) - f(x_2)}{h} = I(x_2, h).
\end{aligned}$$

This means that I is increasing also with respect to the first variable. $\qquad\square$

The converse is only partially true.

Theorem 7.3.3. *Let $J \subset \mathbb{R}$ be an open interval, and let $f : J \to \mathbb{R}$ be a function. Let I be defined by (7.3.8). If for $h > 0$ the function I is increasing with respect to h, then f is continuous and convex. If I is increasing with respect to x, then the function is convex, but need not be continuous.*

Proof. Suppose that I is increasing with respect to h. Take arbitrary $x_1, x_2, x_3 \in J$, $x_1 < x_2 < x_3$, and put $x = x_1$, $h_1 = x_2 - x_1$, $h_2 = x_3 - x_1 > h_1$. Then $I(x, h_1) \leqslant I(x, h_2)$, i.e.,

$$\frac{f(x_2) - f(x_1)}{x_2 - x_1} \leqslant \frac{f(x_3) - f(x_1)}{x_3 - x_1}.$$

Consequently f satisfies condition (7.3.2) of Theorem 7.3.1, and thus it is continuous and convex.

Now let I be increasing with respect to x. Take arbitrary $x, y \in J$, $x \neq y$. Thus one of x, y is smaller than the other; let, e.g., $x < y$. Put $h = \frac{1}{2}(y - x)$, $x_1 = x$, $x_2 = x + \frac{1}{2}(y - x) = x_1 + h > x_1$. Hence $I(x_1, h) \leqslant I(x_2, h)$, i.e.,

$$\frac{f\left(\frac{x+y}{2}\right) - f(x)}{\frac{1}{2}(y-x)} \leqslant \frac{f(y) - f\left(\frac{x+y}{2}\right)}{\frac{1}{2}(y-x)},$$

or, after multiplication by $\frac{1}{2}(y - x) > 0$, and rearranging the remaining terms,

$$2f\left(\frac{x+y}{2}\right) \leqslant f(x) + f(y). \tag{7.3.9}$$

If $x = y$, (7.3.9) is trivially fulfilled. (7.3.9) is equivalent to the Jensen inequality (5.3.1). Thus the function f is convex.

The monotonicity of I with respect to x does not imply the continuity of f. For example, if $f : \mathbb{R} \to \mathbb{R}$ is a discontinuous additive function, then the function

$$I(x, h) = \frac{f(x+h) - f(x)}{h} = \frac{f(h)}{h}$$

does not depend on x, and thus, as a function of x, it is constant, and hence increasing. But f is discontinuous. $\qquad\square$

On the other hand, we may observe that the convexity of f alone does not imply the monotonicity if I with respect to x. Let $g : \mathbb{R} \to \mathbb{R}$ be a discontinuous additive function, and let $f : \mathbb{R} \to \mathbb{R}$ be defined by $f(x) = [g(x)]^2$, $x \in \mathbb{R}$. Thus f is convex (cf. Example 5.3.1). Fix an h such that $g(h) \neq 0$. We have

$$I(x, h) = \frac{2}{h}g(x)g(h) + \frac{1}{h}[g(h)]^2.$$

Hence

$$g(x) = \frac{h}{2g(h)}\left(I(x, h) - \frac{1}{h}[g(h)]^2\right).$$

If I were increasing with respect to x, then the function g would be monotonic, and hence continuous. Consequently I cannot be increasing with respect to x.

As particular case of Theorem 7.3.2 we get

Theorem 7.3.4. *Let $J \subset \mathbb{R}$ be an open interval, and let $f : J \to \mathbb{R}$ be a continuous and convex function. Then, for every fixed $h > 0$, the function $\Delta_h f(x) = f(x + h) - f(x)$ (defined for $x \in J$ such that $x + h \in J$) is increasing.*

Proof. This follows from Theorem 7.3.2 and the equality $\Delta_h f(x) = hI(x, h)$ $\qquad\square$

Theorem 7.3.5. *Let $J \subset \mathbb{R}$ be an open interval, $J = (a, b)$, and let $f : J \to \mathbb{R}$ be a continuous and convex function. Then either of f is monotonic in J, or there exist a point $x_0 \in J$ such that f is decreasing in (a, x_0), and increasing in (x_0, b).*

Proof. If f is decreasing in J, there is nothing more to prove. Otherwise there exist $u, v \in J$ such that $u < v$ and $f(u) < f(v)$. Then we have for every $x \in (u, v)$

$$x = \lambda u + (1 - \lambda) v$$

with certain $\lambda \in (0, 1)$, and hence, by Theorem 7.1.1,

$$f(x) \leqslant \lambda f(u) + (1 - \lambda) f(v) < \lambda f(v) + (1 - \lambda) f(v).$$

Thus there exist arbitrary close points $u, v \in J$ such that $u < v$ and $f(u) < f(v)$. Define a set $A \subset J$:

$$A = \{u \in J \mid \text{For every } \varepsilon > 0 \text{ there exist a } v \in J \text{ such that}$$
$$u < v < u + \varepsilon \text{ and } f(u) < f(v)\}.$$

As we have already seen, $A \neq \varnothing$. Put[2] $x_0 = \inf A$. By what has already been established, f is decreasing in (a, x_0). Now take arbitrary $x, y \in J$, $x_0 \leqslant x \leqslant y$. By the definition of x_0 there exist a $u \in A$ such that $x_0 \leqslant u \leqslant x$, and by the definition of A there exist a $v \in J$ such that $u < v < y$ and $f(u) < f(v)$. By Theorem 7.3.2

$$0 < \frac{f(v) - f(u)}{v - u} \leqslant \frac{f(y) - f(u)}{y - u} \leqslant \frac{f(y) - f(x)}{y - x}.$$

Consequently $f(x) < f(y)$, which proves that f is increasing in (x_0, b). $\square$

7.4   Differentiation

According to Theorem 7.3.2, if $f : J \to \mathbb{R}$ (J – an open interval) is convex and continuous, then for every fixed $x \in J$ the differences quotient $I(x, h)$ is an increasing function of h. Consequently it has finite one-sided limits as h tends to zero from the right, and from the left. But these limits are one-sided derivatives of f at x:

$$f'_+(x) = \lim_{h \to 0+} I(x, h), \ f'_-(x) = \lim_{h \to 0-} I(x, h) \tag{7.4.1}$$

Thus we have

Theorem 7.4.1. *Let $J \subset \mathbb{R}$ be an open interval, and let $f : J \to \mathbb{R}$ be a continuous and convex function. Then at every point $x \in J$ there exist the right derivative $f'_+(x)$, and the right derivative $f'_-(x)$, and we have for every $x, y \in J$, $x < y$,*

$$f'_-(x) \leqslant f'_+(x) \leqslant f'_-(y) \leqslant f'_+(y). \tag{7.4.2}$$

---

[2] It may happen that $x_0 = a$, but, as we prove further, the function f is increasing in (x_0, b). So, if $x_0 = a$, the function f is increasing in $(x_0, b) = (a, b) = J$, and the theorem is true also in this case.

Moreover,

$$\lim_{t \to x+} f'_+ (t) = \lim_{t \to x+} f'_- (t) = f'_+ (x), \tag{7.4.3}$$

$$\lim_{t \to x-} f'_+ (t) = \lim_{t \to x-} f'_- (t) = f'_- (x). \tag{7.4.4}$$

Proof. The existence of derivatives (7.4.1) results from the monotonicity of I as a function of h, as has been pointed out above. Moreover, we have for every $x, y \in J$, $x < y$, and for every h such that $x - h \in J$, $x + h \in J$, $0 < h < y - x$,

$$I (x, -h) \leqslant I (x, h) \leqslant I (y, -h) \leqslant I (y, h). \tag{7.4.5}$$

Only the inequality $I (x, h) \leqslant I (y, -h)$ requires motivation, the remaining ones result from the inequality $-h < h$ and from Theorem 7.3.2. Now,

$$I (y, -h) = \frac{f (y - h) - f (y)}{-h} = \frac{f ((y - h) + h) - f (y - h)}{h}$$
$$\geqslant \frac{f (x + h) - f (x)}{h} = I (x, h)$$

since $y - h > x$, and the function I is monotonic with respect to the first variable (Theorem 7.3.2). Letting in (7.4.5) $h \to 0+$ we obtain inequalities (7.4.2).

Inequalities (7.4.2) show that the functions f'_+ and f'_- are increasing, and hence at every point $x \in J$ they have one-sided limits. Since f is continuous, the function I is continuous with respect to x. Moreover, because of the monotonicity of I with respect to h we have for every $t \in J$ and every fixed $h > 0$ such that $t + x \in J$

$$f'_+ (t) \leqslant I (t, h). \tag{7.4.6}$$

Now fix an $x \in J$ and an $h > 0$ such that $x + h \in J$. By (7.4.2) and (7.4.6) we have for every $t > x$ such that $t + x \in J$

$$f'_+ (x) \leqslant f'_- (t) \leqslant f'_+ (t) \leqslant I (t, h). \tag{7.4.7}$$

Letting in (7.4.7) $t \to x+$ we obtain

$$f'_+ (x) \leqslant \lim_{t \to x+} f'_- (t) \leqslant \lim_{t \to x+} f'_+ (t) \leqslant I (x, h),$$

whence, as $h \to 0+$,

$$f'_+ (x) \leqslant \lim_{t \to x+} f'_- (t) \leqslant \lim_{t \to x+} f'_+ (t) \leqslant f'_+ (x). \tag{7.4.8}$$

Inequalities (7.4.8) yield (7.4.3). Relation (7.4.4) may be established in a similar manner. $\square$

Corollary 7.4.1. *Let $J \subset \mathbb{R}$ be an open interval, and let $f : J \to \mathbb{R}$ be a continuous and convex function. Then the functions $f'_+, f'_- : J \to \mathbb{R}$ are increasing.*

Theorem 7.4.2. *Let $J \subset \mathbb{R}$ be an open interval, and let $f : J \to \mathbb{R}$ be a continuous and convex function. Then the following three conditions are equivalent for every $x \in J$:*

(i) f is differentiable at x;

(ii) f'_+ is continuous at x;

(iii) f'_- is continuous at x.

Proof. Condition (i) means that $f'_+(x) = f'_-(x)$, condition (ii) that $\lim_{t \to x+} f'_+(t) = \lim_{t \to x-} f'_-(t)$, and condition (iii) that $\lim_{t \to x-} f'_-(t) = \lim_{t \to x-} f'_-(t)$. Relations (7.4.3) and (7.4.4) show that all the three conditions are equivalent. $\square$

Theorem 7.4.3. *Let $J \subset \mathbb{R}$ be an open interval, and let $f : J \to \mathbb{R}$ be a continuous and convex function. Then f is differentiable in J except at most countably many points. If $J_0 \subset J$ is the set of the points of the differentiability of f, then the function $f' : J_0 \to \mathbb{R}$ is increasing and continuous.*

Proof. The function f'_+, being monotonic, may be discontinuous at at most countably many points. In view of Theorem 7.4.2 we obtain hence the first part of our assertion. The second results from the fact that for $x \in J_0$ we have $f'(x) = f'_+(x)$, and by the Theorem 7.4.2 f'_+ is continuous at every point of J_0. $\square$

In sequel we say that f is twice differentiable at a point $x_0 \in J$ iff $x_0 \in J_0$ (the set of the points of differentiability of f) and the limit

$$f''(x_0) = \lim_{\substack{y \to x \\ y \in J_0}} \frac{f'(y) - f'(x_0)}{y - x_0} \tag{7.4.9}$$

exists. Limit (7.4.9) is equal to

$$f''(x_0) = \lim_{y \to x_0} \frac{f'_+(y) - f'_+(x_0)}{y - x_0}, \tag{7.4.10}$$

whenever the latter limit exists. But since the function f'_+ is monotonic, it follows from the famous Theorem of Lebesgue (cf., e.g., Łojasiewicz [208]) that limit (7.4.10) exists almost everywhere in J. Moreover, the difference quotient in (7.4.10) is non-negative (since f'_+ is increasing), and consequently so is also $f''(x_0)$ if it exist. Thus we have

Theorem 7.4.4. *Let $J \subset \mathbb{R}$ be an open interval, and let $f : J \to \mathbb{R}$ be a continuous and convex function. Then f is twice differentiable almost everywhere in J. Whenever it exist, $f''(x) > 0$.*

Since f' (defined almost everywhere in J) is monotonic, it is measurable, and bounded on every compact subinterval of J. Thus the expression[3]

$$\int_{x_0}^{x} f'(t)dt \qquad (7.4.11)$$

is meaningful for every $x_0, x \in J$. We have the following

Theorem 7.4.5. *Let $J \subset \mathbb{R}$ be an open interval, and let $f : J \to \mathbb{R}$ be a continuous and convex function. Then, for every $x_0, x \in J$, we have*

$$f(x) = f(x_0) + \int_{x_0}^{x} f'(t)dt. \qquad (7.4.12)$$

Proof. Put

$$F(x) = f(x_0) + \int_{x_0}^{x} f'(t)dt.$$

The function F is differentiable whenever f' is continuous, and thus everywhere in J with the exception of an at most countable set. If $x \in J$ is a point of the continuity of f', then we have $F'(x) = f'(x)$, i.e., $(F - f)'(x) = 0$. By Theorem 3.6.3 (cf. also Exercise 3.9) $F - f = $ const. Since at the point x_0 we have $F - f = 0$, the constant must be zero, and $F = f$. Hence we obtain (7.4.12).   □

As an immediate consequence of Theorem 7.4.5 we obtain

Theorem 7.4.6. *Let $J \subset \mathbb{R}$ be an open interval, and let $f : J \to \mathbb{R}$ be a continuous and convex function. The function f is absolutely continuous in J.*

7.5 Differential conditions of convexity

The convexity of differentiable function can be inferred from the behaviour of its derivatives. The theorems in this section are in some respect converse to those of 7.4.

Theorem 7.5.1. *Let $J \subset \mathbb{R}$ be an open interval, and let $f : J \to \mathbb{R}$ be a differentiable function. The function f is convex if and only if the function f' is increasing in J.*

Proof. The "only if" part results from Corollary 7.4.1. now assume that f' is increasing in J, and take arbitrary $x, y \in J$, $x < y$. We have by the mean-value theorem

$$f(y) - f\left(\frac{x+y}{2}\right) = f'(v)\left(y - \frac{x+y}{2}\right) = f'(v)\frac{y-x}{2},$$

$$f\left(\frac{x+y}{2}\right) - f(x) = f'(u)\left(\frac{x+y}{2} - x\right) = f'(u)\frac{y-x}{2},$$

---

[3] The integral in (7.4.11) and (7.4.12) are Lebesgue integrals, but everything remains unchanged if we replace them by the Riemann integral $\int_{x_0}^{x} f'_+(t)dt$

where $u, v \in J$ are points such that

$$x < u < \frac{x+y}{2} < v < y.$$

Hence $f'(u) \leqslant f'(v)$, i.e.,

$$f\left(\frac{x+y}{2}\right) - f(x) \leqslant f(y) - f\left(\frac{x+y}{2}\right),$$

and

$$f\left(\frac{x+y}{2}\right) \leqslant \frac{f(x) + f(y)}{2} \tag{7.5.1}$$

Relation (7.5.1) has been so far established for $x < y$, but due to the symmetry of (7.5.1) with respect to x and y we actually obtain (7.5.1) for $x \neq y$. For $x = y$ (7.5.1) is obvious. $\qquad\square$

Theorem 7.5.2. *Let $J \subset \mathbb{R}$ be an open interval, and let $f : J \to \mathbb{R}$ be a twice differentiable function. The function f is convex if and only if the function f'' is non-negative in J.*

Proof. This is a direct consequence of Theorem 7.5.1 above. $\qquad\square$

It follows from Theorem 7.5.2 that the function (7.2.1) ($p > 1$), and the function exp are convex in $\mathbb{R}$, the fact that has already been used earlier in this book.

For every function $f : J \to \mathbb{R}$ and for every $x \in J$ the expressions (finite or not)

$$D^+ f(x) = \limsup_{h \to 0+} I(x, h), \quad D^- f(x) = \limsup_{h \to 0-} I(x, h),$$

$$d^+ f(x) = \liminf_{h \to 0+} I(x, h), \quad d^- f(x) = \liminf_{h \to 0-} I(x, h),$$

are meaningful. They are called the *Dini derivatives* of f at x. Thus $D^+ f$, $D^- f$, $d^+ f$ and $d^- f$ are functions from J into $[-\infty, \infty]$.

Lemma 7.5.1. *Let $J \subset \mathbb{R}$ be an open interval, and let $f : J \to \mathbb{R}$ be a continuous function. If one of the Dini derivatives of f is finite and increasing in J, then f is absolutely continuous in J.*

Proof. Assume that the upper right Dini derivative $D^+ f$ of f is finite and increasing in J. Thus $D^+ f$ is measurable and bounded on every compact subinterval of J. Consequently for every $x, y \in J$ the integral $\int_x^y |D^+ f(t)|\, dt$ is meaningful. In order to prove the lemma it is enough to show that for every $x, y \in J$, $x < y$, we have

$$|f(y) - f(x)| \leqslant \int_x^y |D^+ f(t)|\, dt. \tag{7.5.2}$$

Fix an $x \in J$, and put

$$\varphi(y) = |f(y) - f(x)|, \quad \psi(y) = \int_x^y |D^+f(t)| \, dt.$$

We have

$$\psi'(y) = |D^+f(y)| \tag{7.5.3}$$

at every point $y \in J$ at which D^+f, and hence also $|D^+f|$, is continuous. Take arbitrary $y \in J$ such that D^+f is continuous at y, and take an arbitrary sequence of points $y_n \in J$, $y_n > y$, $n \in \mathbb{N}$, such that

$$\lim_{n\to\infty} y_n = y \quad \text{and} \quad \lim_{n\to\infty} \frac{f(y_n) - f(y)}{y_n - y} = D^+f(y).$$

Since

$$\varphi(y_n) - \varphi(y) = |f(y_n) - f(y)| - |f(y) - f(x)| \leqslant |f(y_n) - f(y)|,$$

we get hence

$$d^+\varphi(y) \leqslant \liminf_{n\to\infty} \frac{\varphi(y_n) - \varphi(y)}{y_n - y} \leqslant |D^+f(y)|. \tag{7.5.4}$$

By (7.5.3) and (7.5.4)

$$D^+(\psi - \varphi)(y) = \psi'(y) - d^+\varphi(y) \geqslant |D^+f(y)| - d^+\varphi(y) \geqslant 0.$$

Thus $D^+(\psi - \varphi) \geqslant 0$ at all points of continuity of D^+f, and since the latter function is increasing, $D^+(\psi - \varphi) \geqslant$ in J except at at most countably many points. By Theorem 3.6.3 the function $\psi - \varphi$ is increasing. For $x = y$ we have $\varphi(x) = \psi(x) = 0$, and so $\psi(y) \geqslant \varphi(y)$ for $y > x$. But this is just relation (7.5.2). $\square$

Theorem 7.5.3. *Let $J \subset \mathbb{R}$ be an open interval, and let $f : J \to \mathbb{R}$ be a continuous function. If one of the Dini derivatives of f is finite and increasing in J, then f is convex.*

Proof. Assume that the upper right Dini derivative D^+f of f is finite and increasing in J. By Lemma 7.5.1 f is absolutely continuous in J, and hence it is differentiable almost everywhere in J, and we have

$$f(x) = f(x_0) + \int_{x_0}^x f'(t) \, dt = f(x_0) + \int_{x_0}^x D^+f(t) \, dt, \tag{7.5.5}$$

where $x_0 \in J$ is an arbitrary fixed point. Take arbitrary $x, y \in J$, $x < y$. By (7.5.5)

$$f(y) - f\left(\frac{x+y}{2}\right) = \int_{\frac{1}{2}(x+y)}^y D^+f(t)dt, \quad f\left(\frac{x+y}{2}\right) - f(x) = \int_x^{\frac{1}{2}(x+y)} D^+f(t)dt.$$

$$\tag{7.5.6}$$

Since D^+f is increasing, and $x < \dfrac{x+y}{2} < y$, we have

$$\int\limits_{\frac{1}{2}(x+y)}^{y} D^+f(t)dt \geqslant \int\limits_{x}^{\frac{1}{2}(x+y)} D^+f(t)dt,$$

whence by (7.5.6)

$$f(y) - f\left(\frac{x+y}{2}\right) \geqslant f\left(\frac{x+y}{2}\right) - f(x).$$

Hence we obtain (7.5.1), and in the same way as in the proof of Theorem 7.5.1 it follows that f is convex. □

7.6   Functions of several variables

In the present section we consider arbitrary families $\{f_\alpha\}_{\alpha \in A}$ of continuous and convex functions.

Lemma 7.6.1. *Let $D \subset \mathbb{R}^N$ be a convex and open set, and let $A \neq \varnothing$ be an arbitrary set. Suppose that for every $\alpha \in A$ we are given a continuous and convex function $f_\alpha : D \to \mathbb{R}$. Then the function $f : D \to (-\infty, \infty)$, $f(x) = \sup\limits_{\alpha} f_\alpha(x)$ satisfies the inequality*

$$f\big(\lambda x + (1-\lambda)y\big) \leqslant \lambda f(x) + (1-\lambda)f(y) \tag{7.6.1}$$

for all $x, y \in D$ and $\lambda \in [0, 1]$.

Proof. Take arbitrary $x, y \in D$, $\lambda \in [0, 1]$, and $a \in \mathbb{R}$ such that $a < f\big(\lambda x + (1-\lambda)y\big)$. There exist an $\alpha \in A$ such that

$$a < f_\alpha\big(\lambda x + (1-\lambda)y\big)$$

Hence by Theorem 7.1.1

$$a < f_\alpha\big(\lambda x + (1-\lambda)y\big) \leqslant \lambda f_\alpha(x) + (1-\lambda)f_\alpha(y) \leqslant \lambda f(x) + (1-\lambda)f(y).$$

Letting a tend to $f\big(\lambda x + (1-\lambda)y\big)$, we obtain (7.6.1). □

Lemma 7.6.2. *Let $D \subset \mathbb{R}^N$ be a convex and open set, and let $\Delta \subset D$ be a dense subset of D, and let $A \neq \varnothing$ be an arbitrary set. Suppose that for every $\alpha \in A$ the function $f_\alpha : D \to \mathbb{R}$ is continuous and convex, and*

$$\sup_{\alpha} f_\alpha(x) < \infty \quad \text{for } x \in \Delta, \tag{7.6.2}$$

$$\inf_{\alpha} f_\alpha(x_0) > -\infty \tag{7.6.3}$$

for an $x_0 \in D$. Then the functions f_α, $\alpha \in A$ are jointly bounded on every compact set contained in D.

Proof. Put

$$f(x) = \sup_\alpha f_\alpha(x), \quad x \in D.$$

By (7.6.2) $f(x) < \infty$ for $x \in \Delta$ and by Lemma 7.6.1 f satisfies (7.6.1) for all $x, y \in D$, $\lambda \in [0, 1]$. Take an arbitrary $x \in D$. According to the Remark in 5.1 x is an interior point of an N-dimensional simplex contained in D. Very small changes of the vertices of this simplex do not affect this fact, so we may assume that the vertices of the simplex belong to Δ. Consequently there exist points $p_1, \ldots, p_{N+1} \in \Delta$, and numbers $\lambda_1, \ldots, \lambda_{N+1} \in [0, 1]$ such that

$$x = \sum_{i=1}^{N+1} \lambda_i p_i, \quad \sum_{i=1}^{N+1} \lambda_i = 1.$$

Hence by Lemma 5.3.2

$$f(x) \leqslant \sum_{i=1}^{N+1} \lambda_i f(p_i) < \infty.$$

Consequently $f(x) < \infty$ for all $x \in D$, and in virtue of Lemma 7.6.1, and Theorem 7.1.1 f is continuous and convex function. Being continuous, f is bounded above on every compact set contained in D, and so is also every function $f_\alpha \leqslant f$, $\alpha \in A$.

Now write

$$L = \inf_\alpha f_\alpha(x_0) > -\infty \qquad (7.6.4)$$

(cf. (7.6.3)), and let K be a closed ball centered at x_0 and with a radius $r > 0$ such that $K \subset D$. The ball K is a compact set, so by the first part of the proof theres exist a constant M such that

$$f_\alpha(x) \leqslant M \quad \text{for } x \in K \text{ and } \alpha \in A. \qquad (7.6.5)$$

Take an arbitrary $x \in D$, $x \neq x_0$, and put

$$y = x_0 - \frac{r}{|x - x_0|}(x - x_0)$$

Then $|y - x_0| = r$ so that $y \in K$. Moreover,

$$x_0 = \lambda x + (1 - \lambda) y \qquad (7.6.6)$$

with

$$\lambda = \frac{r}{r + |x - x_0|} \in (0, 1). \qquad (7.6.7)$$

For every $\alpha \in A$ we have by (7.6.6)

$$f_\alpha(x) \leqslant \lambda f_\alpha(x) + (1 - \lambda) f_\alpha(y),$$

whence by (7.6.4) and (7.6.5)

$$f_\alpha(x) \geqslant \frac{1}{\lambda} f_\alpha(x_0) - \frac{1 - \lambda}{\lambda} f_\alpha(y) \geqslant \frac{1}{\lambda} L - \frac{1 - \lambda}{\lambda} M \geqslant \frac{1}{\lambda}(L - M),$$

and by (7.6.7)

$$f_\alpha(x) \geqslant \frac{L-M}{r}(r + |x - x_0|), \quad \alpha \in A. \qquad (7.6.8)$$

So far estimation (7.6.8) has been derived for all $x \in D$, $x \neq x_0$. For x_0 we have according to (7.6.4), for all $\alpha \in A$,

$$f_\alpha(x_0) \geqslant L \geqslant L - M = \frac{L-M}{r}(r + |x - x_0|).$$

Thus (7.6.8) is valid for all $x \in D$. The expression on the right-hand side of (7.6.8) is a continuous function of x, and thus it is bounded below on every compact set contained in D, and in virtue of (7.6.8) so is also every function f_α, $\alpha \in A$. □

Theorem 7.6.1. *Let $D \subset \mathbb{R}^N$ be a convex and open set, and let $\Delta \subset D$ be a dense subset of D and let $A \neq \varnothing$ be an arbitrary set. Suppose that for every $\alpha \in A$ the function $f_\alpha : D \to \mathbb{R}$ is continuous and convex and that relations (7.6.2) and (7.6.3) hold with an $x_0 \in D$. Then all functions f_α, $\alpha \in A$, fulfil a Lipschitz condition with a common constant on every compact set contained in D.*

Proof. Take a compact $C \subset D$, $C \neq \varnothing$, and let $d(C, D')$ be the distance of the sets C and D' (we assume $d(C, D') = \infty$ if $D' = \varnothing$). Since C and D' are disjoint closed sets, $d(C, D') > 0$, and so we can find an r,

$$0 < r < d(C, D'). \qquad (7.6.9)$$

Further, let $d(x, C)$ denote the distance of the point x from C and put

$$K = \{x \in \mathbb{R}^N \mid d(x, C) \leqslant r\}.$$

K is a compact set, and by (7.6.9) $K \subset D$. By Lemma 7.6.2 there exist constant Q and M such that

$$-\infty < Q \leqslant f_\alpha(x) \leqslant M < \infty \quad \text{for all } x \in K, \alpha \in A. \qquad (7.6.10)$$

Now take arbitrary $x, y \in C$, $x \neq y$, and put

$$z = y - \frac{r}{|x - y|}(x - y).$$

We have $|z - y| = r$, whence $z \in K$; moreover $x \in C \subset K$. Since

$$y = \lambda x + (1 - \lambda) z$$

with

$$\lambda = \frac{r}{r + |x - y|}, \qquad (7.6.11)$$

we have by (7.6.10) for every $\alpha \in A$

$$f_\alpha(y) \leqslant \lambda f_\alpha(x) + (1 - \lambda) f_\alpha(z) = f_\alpha(x) - (1 - \lambda)[f_\alpha(x) - f_\alpha(z)]$$
$$\leqslant f_\alpha(x) - (1 - \lambda)(Q - M) = f_\alpha(x) + (1 - \lambda)(M - Q),$$

and by (7.6.11)

$$f_\alpha(y) - f_\alpha(x) \leqslant (1 - \lambda)(Q - M) = \frac{|x - y|}{r + |x - y|}(M - Q) \leqslant \frac{M - Q}{r}|x - y|.$$

$$(7.6.12)$$

Since the role of x and y has been symmetric, we may interchange x and y arriving at

$$f_\alpha(x) - f_\alpha(y) \leqslant \frac{M - Q}{r}|x - y|. \qquad (7.6.13)$$

Relations (7.6.12) and (7.6.13) together yield

$$|f_\alpha(x) - f_\alpha(y)| \leqslant \frac{M - Q}{r}|x - y|. \qquad (7.6.14)$$

Relation (7.6.14), so far obtained for $x \neq y$, $y \in C$ and all $\alpha \in A$, is evidently valid also for $x = y$. Condition (7.6.14) means that all functions f_α, $\alpha \in A$ fulfill on C the Lipschitz condition with the same constant $(M - Q)/r$. □

Taking as A a singleton, we obtain as a particular case of Theorem 7.6.1

Theorem 7.6.2. *Let $D \subset \mathbb{R}^N$ be a convex and open set, and let $f : D \to \mathbb{R}$ be a continuous and convex function. The function fulfils a Lipschitz condition on every compact set contained in D.*

Let us note that Theorem 7.4.6 results also from Theorem 7.6.2 above.

7.7 Derivatives of a function

For functions of several variables we have three different notions of the differentiability at a point, in general not equivalent to each other[4]. The simplest fact is that the function in question, say f, has at a point x all the partial derivatives $\partial f/\partial \xi_i$. This fact refers only to the behaviour of f may be quite wild[5]. Therefore we need stronger conditions.

A function $f : \mathbb{R}^N \to \mathbb{R}$ is called *linear* if the function f_0 defined by $f_0(x) = f(x) - f(0)$ is additive and homogeneous:

$$f_0(\lambda x) = \lambda f_0(x) \qquad (7.7.1)$$

for all $\lambda \in \mathbb{R}$. If (7.7.1) holds only for positive $\lambda \in \mathbb{R}$, the function f_0 (independently of whether it is additive, or not) is called *positively homogeneous*.

If $f : \mathbb{R}^N \to \mathbb{R}$ is linear, then for the corresponding function f_0 we have by (7.7.1)

$$f_0(\lambda x + (1 - \lambda)y) = f_0(\lambda x) + f_0((1 - \lambda)y) = \lambda f_0(x) + (1 - \lambda)f_0(y)$$

---

[4] In the one-dimensional space ($N = 1$) the difference between the three notions disappears and they are all equivalent to each other. But since for N=1 the problem of the differentiability of continuous convex functions are related questions have been thoroughly discussed in 4–5, here we may restrict ourselves to $N > 1$.

[5] However, if we know that the partial derivatives exist at every point of an open set D, and are continuous in D, then it follows already that the function in question has a Stolz differential (cf. below) at every point of D.

for every $x, y \in \mathbb{R}^N$, $\lambda \in [0,1]$. Thus f_0 we have by (7.7.1), and by Theorem 7.1.1 it is continuous. Now it follows from Theorem 5.5.2 that $f_0(x) = cx$ with a certain $c \in \mathbb{R}^N$ (cx is the scalar product), and

$$f(x) = cx + b, \quad x \in \mathbb{R}^N, \tag{7.7.2}$$

where $b = f(0)$. Conversely, every function $f : \mathbb{R}^N \to \mathbb{R}$ of form (7.7.2) is linear.

Let $D \subset \mathbb{R}^N$ be an open set, and let $f : D \to \mathbb{R}$ be a function. We say that f is *differentiable* at a point $x \in D$ iff for every $y \in \mathbb{R}^N$ there exist the limit

$$f'_y(x) = \lim_{\lambda \to 0+} \frac{f(x + \lambda y) - f(x)}{\lambda}, \tag{7.7.3}$$

and $f'_y(x)$ is a linear function of y.

It follows from (7.7.3) that we always have

$$f'_0(x) = 0 \tag{7.7.4}$$

Lemma 7.7.1. *Let $D \subset \mathbb{R}^N$ be an open set, and let $f : D \to \mathbb{R}$ be a function. If f is differentiable at a point $x \in D$, then f has all partial derivatives at x.*

Proof. Let $e_i = (0, \ldots, 0, 1, 0, \ldots, 0)$ (1 at i-th place), $i = 1, \ldots, N$, be the usual orthonormal base of $\mathbb{R}^N$ (over $\mathbb{R}$). There exist the limits ($i = 1, \ldots, N$)

$$\lim_{\lambda \to 0+} \frac{f(x + \lambda e_i) - f(x)}{\lambda} = f'_{e_i}(x)$$

$$\lim_{\lambda \to 0-} \frac{f(x + \lambda e_i) - f(x)}{\lambda} = - \lim_{\lambda \to 0+} \frac{f(x - \lambda e_i) - f(x)}{\lambda} = -f'_{-e_i}(x).$$

It follows from the linearity of the function $f'(x)$ and from condition (7.7.4) that $f'(x)$ is homogeneous so that in particular $-f'_{-e_i}(x) = f'_{e_i}(x)$. Consequently there exists the limit

$$\lim_{\lambda \to 0} \frac{f(x + \lambda e_i) - f(x)}{\lambda} = \frac{\partial f}{\partial \xi_i}(x),$$

$i = 1, \ldots, N$. Thus f has all its partial derivatives at x. $\qquad\square$

Conversely, the function $f : \mathbb{R}^2 \to \mathbb{R}$ which is one whenever $\xi_1, \xi_2 > 0$ ($x = (\xi_1, \xi_2)$), and 0 otherwise, has at zero the partial derivatives equal to zero, but is not differentiable at zero, since limit (7.7.3) does not exist for $x = 0$ and y such that $f(y) = 1$.

But also this notion of differentiability is not sufficient in order to rend the function in question sufficiently smooth around the point at which it is differentiable. For example, the function $f : \mathbb{R}^2 \to \mathbb{R}$ defined as

$$f(x) = f(\xi_1, \xi_2) = \begin{cases} 1 & \text{if } \xi_2 = \xi_1^2, \, \xi_1 > 0, \\ 0 & \text{otherwise,} \end{cases} \tag{7.7.5}$$

is differentiable at zero $\left(f_y'(0) = 0 \text{ for all } y \in \mathbb{R}^N \right)$ but is not continuous at the point $x = 0$. So we introduce one more notion.

Let $D \subset \mathbb{R}^N$ be an open set, and let $f : D \to \mathbb{R}$ be a function. We say that f has a *Stolz differential* at the point $x \in D$ iff there exists an $x^* \in \mathbb{R}^N$ such that

$$f(z) = f(x) + x^*(z - x) + r(x, z) \quad \text{for} z \in D, \tag{7.7.6}$$

where the function $r : \mathbb{R}^{2N} \to \mathbb{R}$ fulfills the condition

$$\lim_{z \to x} \frac{r(x, z)}{|z - x|} = 0.^6 \tag{7.7.7}$$

If it exists, x^* in (7.7.6) is called the *Stolz differential* or the *gradient* of f at x, and is denoted by $\nabla f(x)$.

In the above example function (7.7.5) has no Stolz differential at zero. In the opposite direction we have the following

Lemma 7.7.2. *Let $D \subset \mathbb{R}^N$ be an open set, and $f : D \to \mathbb{R}$ be a function. If f has a Stolz differential $\nabla f(x)$ at a point $x \in D$, then it is differentiable at x, and we have*

$$\nabla f(x) = \left(\frac{\partial f}{\partial \xi_1(x)}, \ldots, \frac{\partial f}{\partial \xi_N(x)} \right), \tag{7.7.8}$$

and

$$f_y'(x) = \big(\nabla f(x) \big) y. \tag{7.7.9}$$

Proof. Take an arbitrary $y \in \mathbb{R}^N$, $y \neq 0$. Putting in (7.7.6) $z = x + \lambda y$, $\lambda > 0$, we have by (7.7.7)

$$\lim_{\lambda \to 0+} \frac{f(x + \lambda y) - f(x) - \big(\nabla f(x) \big) \lambda y}{\lambda |y|} = 0,$$

i.e.,

$$\lim_{\lambda \to 0+} \frac{1}{|y|} \left[\frac{f(x + \lambda y) - f(x)}{\lambda} - \big(\nabla f(x) \big) \lambda y \right] = 0.$$

Hence we obtain (7.7.9). According to (7.7.4), (7.7.9) holds for $y = 0$. It follows from (7.7.9) that $f_y'(x)$ is a linear function of y. (7.7.8) results from (7.7.9) on putting $y = e_i$, $i = 1, \ldots, N$ (compare the proof of Lemma 7.7.1). $\square$

In the sequel we will need one more notion. Let $D \subset \mathbb{R}^N$ be an open set, and let $f : D \to \mathbb{R}$ be a function. Any $x^* \in \mathbb{R}^N$ such that

$$f(z) \geqslant f(x) + x^*(z - x) \quad \text{for } z \in D \tag{7.7.10}$$

is called a *subgradient* of f at x. The set of all subgradients of f at x will be denoted by $\partial f(x)$. Of course, it may happen that $\partial f(x) = \varnothing$.

---

[6] Of course, every function f can be written in form (7.7.6) with every $x^* \in \mathbb{R}^N$. It is enough to put $r(x, z) = f(z) - f(x) - x^*(z - x)$. Here the problem lies in that x^* should be chosen in such a manner that the corresponding function r should satisfy (7.7.7).

The simple examples given above show that a function of several variables may have all partial derivatives at a point, but be non-differentiable at this point, and may be differentiable at a point, but have no Stolz differential at this point. However, the considerations of next section aim at showing that in the case of a continuous convex function all the three conditions (the existence of partial derivatives, the differentiability, the existence of a Stolz differential) are equivalent to each other.

7.8   Derivatives of convex functions

In this section we will be concerned with continuous and convex functions.

Lemma 7.8.1. *Let $D \subset \mathbb{R}^N$ be a convex and open set, and let $f : D \to \mathbb{R}$ be a convex and continuous function. At every point $x \in D$ the derivative $f'_y(x)$ exist for every $y \in \mathbb{R}^N$. Moreover, $f'(x)$ is a convex, continuous and positively homogeneous function and, for every $y \in \mathbb{R}^N$,*

$$-f'_y(x) \leqslant f'_{-y}(x). \tag{7.8.1}$$

Proof. Fix an $x \in D$ and $y \in \mathbb{R}^N$, and put

$$J = \{\lambda \in \mathbb{R} \mid x + \lambda y \in D\}.$$

J is an open interval containing zero. Further put

$$g(\lambda) = f(x + \lambda y) - f(x), \quad \lambda \in J. \tag{7.8.2}$$

The function $g : J \to \mathbb{R}$ defined by (7.8.2) clearly is continuous, and we have for arbitrary $\lambda, \mu \in J$

$$g\left(\frac{\lambda + \mu}{2}\right) = f\left(x + \frac{\lambda + \mu}{2}y\right) - f(x) = f\left(\frac{(x + \lambda y) + (x + \mu y)}{2}\right) - f(x)$$

$$\leqslant \frac{f(x + \lambda y) + f(x + \mu y)}{2} - f(x)$$

$$= \frac{[f(x + \lambda y) - f(x)] + [f(x + \mu y) - f(x)]}{2} = \frac{g(\lambda) + g(\mu)}{2}$$

so that g is convex. By Theorem 7.4.1 there exist the derivative

$$g'_+(0) = \lim_{\lambda \to 0+} \frac{g(\lambda)}{\lambda} = \lim_{\lambda \to 0+} \frac{f(x + \lambda y) - f(x)}{\lambda} = f'_y(x).$$

Now, we have for arbitrary $u, v \in \mathbb{R}^N$, $\mu \in [0, 1]$ and $\lambda > 0$ sufficiently small

$$\frac{1}{\lambda}[f(x + \lambda[\mu u + (1 - \mu)v]) - f(x)]$$

$$= \frac{1}{\lambda}\left[f\big(\mu(x + \lambda u) + (1 - \mu)(x + \lambda v)\big) - f(x)\right]$$

$$\leqslant \frac{1}{\lambda}[\mu f(x + \lambda u) + (1 - \mu)(x + \lambda v) - f(x)]$$

$$= \mu \frac{f(x + \lambda u) - f(x)}{\lambda} + (1 - \mu)\frac{f(x + \lambda v) - f(x)}{\lambda},$$

and as $\lambda \to 0+$ we obtain in the limit

$$f'_{\mu u + (1-\mu)v}(x) \leqslant \mu f'_u(x) + (1-\mu) f'_v(x).$$

By Theorem 7.1.1 the function $f'(x)$ is convex and continuous. Further, we have for arbitrary positive $\mu \in \mathbb{R}$ and $\lambda > 0$ sufficiently small

$$\frac{f(x + \lambda \mu y) - f(x)}{\lambda} = \mu \frac{f(x + \lambda \mu y) - f(x)}{\lambda \mu},$$

hence, letting $\lambda \to 0+$, we obtain

$$f'_{\mu y}(x) = \mu f'_y(x),$$

i.e., $f'(x)$ is positively homogeneous. Finally we have for arbitrary $y \in \mathbb{R}^N$ by (7.7.4) and by the convexity of $f'(x)$

$$0 = f'_0(x) = f'_{\frac{y-y}{2}}(x) \leqslant \frac{1}{2}\left[f'_y(x) + f'_{-y}(x)\right],$$

whence (7.8.1) follows. $\qquad\square$

Lemma 7.8.2. [7] *Let $D \subset \mathbb{R}^N$ be a convex and open set, and let $f : D \to \mathbb{R}$ be a convex and continuous function. For every $x \in D$, we have $\partial f(x) \neq \varnothing$*

Proof. Consider the set

$$A = \left\{(x,y) \in \mathbb{R}^{N+1} \mid x \in D, y > f(x)\right\}$$

The set A is open and convex (cf. Exercise 7.1) and the points $(x, f(x)) \in \mathbb{R}^{N+1}$ are its frontier points. Take an $x \in D$. According to Theorem 5.1.7 there exist a support hyperplane H of A passing through the point $(x, f(x))$. The hyperplane H has an equation of the form

$$c(z - x) + \alpha(y - f(x)) = 0, \qquad (7.8.3)$$

where $c \in \mathbb{R}^N$, $\alpha \in \mathbb{R}$, and (z, y) is the current point of $\mathbb{R}^{N+1}$. If we had $\alpha = 0$, then every point (x, y) with an arbitrary $y \in \mathbb{R}$, would satisfy (7.8.3), and taking $y > f(x)$ we would get that $H \cap A \neq \varnothing$, and so A being open, would have points on both sides of H, which is impossible. Consequently $\alpha \neq 0$. So we may write (7.8.3) in the form

$$y = f(x) - \frac{c}{\alpha}(z - x).$$

The two halfspaces into which H divides $\mathbb{R}^{N+1}$ are determined by the inequalities

$$y > f(x) - \frac{c}{\alpha}(z - x) \quad \text{and} \quad y < f(x) - \frac{c}{\alpha}(z - x).$$

---

[7] The converse result is also true. Cf. Exercise 7.8.

For arbitrary $y > f(x)$ the point $(x, y) \in A$ and clearly we have $y > f(x) - \dfrac{c}{\alpha}(x - x)$ so that the set A must be contained in the half-space determined by the inequality

$$y > f(x) - \frac{c}{\alpha}(z - x).$$

Take an arbitrary $z \in D$, and an arbitrary $t > f(z)$. Then $(z, t) \in A$, whence

$$t > f(x) - \frac{c}{\alpha}(z - x) \ .$$

Letting $t \to f(z)$, we obtain hence

$$f(z) \geqslant f(x) - \frac{c}{\alpha}(z - x) \ ,$$

and this is valid for all $z \in D$, which shows that $-\dfrac{c}{\alpha} \in \partial f(x)$, and $\partial f(x) \neq \varnothing$.   □

Lemma 7.8.3. *Let $D \subset \mathbb{R}^N$ be a convex and open set, and let $f : D \to \mathbb{R}$ be a convex and continuous function. For every $x^* \in \mathbb{R}^N$ we have $x^* \in \partial f(x)$ if and only if*

$$f'_y \geqslant x^* y \tag{7.8.4}$$

for all $y \in \mathbb{R}^N$.

Proof. By the definition of a subgradient, $x^* \in \partial f(x)$ iff (7.7.10) holds. Fix a $\lambda > 0$ and put $y = \dfrac{1}{\lambda}(z - x)$. Then (7.7.10) goes into

$$f(x + \lambda y) \geqslant f(x) + x^* \lambda y. \tag{7.8.5}$$

Conversely (7.7.10) results from (7.8.5) on putting $z = x + \lambda y$. thus $x^* \in \partial f(x)$ iff (7.8.5) holds for all $y \in \mathbb{R}^N$ and $\lambda > 0$ such that $x + \lambda y \in D$.

Now, if $x^* \in \partial f(x)$, then we have by (7.8.5)

$$\frac{f(x + \lambda y) - f(x)}{\lambda} \geqslant x^* y$$

for all $y \in \mathbb{R}^N$ and $\lambda > 0$ such that $x + \lambda y \in D$, whence on passing to the limit as $\lambda \to 0+$ we obtain (7.8.4).

Conversely, let (7.8.4) hold. Function (7.8.2) is continuous and convex, whence by Theorem 7.3.2

$$\frac{g(\lambda)}{\lambda} \geqslant g'_+(0)$$

for $\lambda > 0$, i.e., by (7.8.4)

$$\frac{f(x + \lambda y) - f(x)}{\lambda} \geqslant f'_y(x) \geqslant x^* y.$$

Hence we obtain (7.8.5), i.e., $x^* \in \partial f(x)$.   □

Lemma 7.8.4. *Let $D \subset \mathbb{R}^N$ be a convex and open set, and let $f : D \to \mathbb{R}$ be a convex function. Then, for every $x \in D$ and $y \in \mathbb{R}^N$,*

$$f'_y(x) = \sup_{x^* \in \partial f(x)} x^* y.$$

Proof. In virtue of Lemma 7.8.3 it is enough to show that there exists an $x^* \in \partial f(x)$ such that

$$f'_y(x) = x^* y \qquad (7.8.6)$$

($x \in D$ and $y \in \mathbb{R}^N$ fixed). If $y = 0$, (7.8.6) results from (7.7.4) and Lemma 7.8.2. So let $y \neq 0$. By a simple transformation of the coordinate system we can make $y = e_1$. We define a function $\varphi : \mathbb{R}^N \to \mathbb{R}$ putting

$$\varphi(z) = f'_z(x).$$

By Lemma 7.8.1 φ is convex, continuous and positively homogeneous. Next we define a function $\psi : \mathbb{R}^{N-1} \to \mathbb{R}$ putting

$$\psi(\zeta_2, \ldots, \zeta_N) = \varphi(1, \zeta_2, \ldots, \zeta_N)$$

It is easily seen that also the function ψ is convex and continuous.

For every $z = (\zeta_1, \ldots, \zeta_N) \in \mathbb{R}^N$ we write $\tilde{z} = (\zeta_2, \ldots, \zeta_N) \in \mathbb{R}^{N-1}$. By Lemma 7.8.2 there exists a $z_0^* \in \partial f(0)$ (so that, in particular, $z_0^* \in \mathbb{R}^{N-1}$). By the definition of a subgradient we have

$$\psi(\tilde{z}) - \psi(0) \geqslant z_0^* \tilde{z} \qquad (7.8.7)$$

for every $z \in \mathbb{R}^N$.

Now let $H \subset \mathbb{R}^N$ be the hyperplane

$$H = \left\{ z = (\zeta_1, \ldots, \zeta_N) \in \mathbb{R}^N \mid \zeta_1 = 1 \right\}.$$

For $z \in H$ we have $\varphi(z) = \psi(\tilde{z})$ whence by (7.8.7)

$$\varphi(z) - \varphi(e_1) = \psi(\tilde{z}) - \psi(0) \geqslant z_0^* \tilde{z}, \quad z \in H. \qquad (7.8.8)$$

Put $z^* = (\varphi(e_1), z_0^*) \in \mathbb{R}^n$. For $z \in H$ we have by (7.8.8)

$$\varphi(z) \geqslant z^* z, \quad z \in H. \qquad (7.8.9)$$

In particular, according to the definition of z^*,

$$\varphi(e_1) = z^* e_1. \qquad (7.8.10)$$

Let P be the half-space

$$P = \left\{ z = (\zeta_1, \ldots, \zeta_N) \, \mathbb{R}^N \mid \zeta_1 > 0 \right\}.$$

For every $z \in P$ there exist a $\hat{z} \in H$ and a $\lambda > 0$ such that $z = \lambda \hat{z}$. Hence we get by (7.8.9) and by the positive homogeneity of φ

$$\varphi(z) = \varphi(\lambda \hat{z}) = \lambda \varphi(\hat{z}) \geqslant \lambda z^* \hat{z} = z^* z, \quad z \in P. \qquad (7.8.11)$$

Now take an arbitrary $z \in \mathbb{R}^N$ and consider the point

$$\bar{z} = \lambda z + (1 - \lambda)e_1 \qquad (7.8.12)$$

with $\lambda \in \mathbb{R}$. When $\lambda \to 0$, point (7.8.12) approaches $e_1 \in P = \text{int } P$ so we may find a $\lambda \in (0, 1)$ such that for point (7.8.12) we have $\bar{z} \in P$. Suppose that we have fixed a suitable $\lambda \in (0, 1)$ and $\bar{z} \in P$ given by (7.8.12) and that

$$\varphi(z) < z^* z. \qquad (7.8.13)$$

Then we have by (7.8.12), (7.8.10) and by the convexity of φ

$$\varphi(\bar{z}) \leqslant \lambda \varphi(z) + (1 - \lambda)\varphi(e_1) < \lambda z^* z + (1 - \lambda)z^* e_1 = z^*(\lambda z + (1 - \lambda)e_1) = z^* \bar{z},$$

which contradicts (7.8.11), since $\bar{z} \in P$. The contradiction obtained shows that supposition (7.8.13) was false and that we actually have

$$\varphi(z) \geqslant z^* z$$

for all $z \in \mathbb{R}^N$. By Lemma 7.8.3 $z^* \in \partial f(x)$. (7.8.10) means that we have (7.8.6) with $x^* = z^*$ and $y = e_1$. $\qquad \square$

Lemma 7.8.5. *Let $f : \mathbb{R}^N \to \mathbb{R}$ and $g : \mathbb{R}^N \to \mathbb{R}$ be additive functions. If $f(x) \leqslant g(x)$ for all $x \in \mathbb{R}^N$, then $f = g$.*

Proof. By Theorem 5.2.1 f and g are odd. Suppose that for an $x \in \mathbb{R}^N$ we have $f(x) < g(x)$. Then

$$g(-x) - f(-x) = -g(x) + f(x) = -[g(x) - f(x)] < 0$$

i.e., $g(-x) < f(-x)$, contrary to the assumption. Thus we must have $f(x) = g(x)$ for all $x \in \mathbb{R}^N$. $\qquad \square$

Theorem 7.8.1. *Let $D \subset \mathbb{R}^N$ be a convex and open set, and let $f : D \to \mathbb{R}$ be a convex and continuous function. If at a point $x \in D$ the function f has Stolz differential $\nabla f(x)$, then $\partial f(x) = \{\nabla f(x)\}$. Conversely, if $\partial f(x)$ is a singleton, $\partial f(x) = \{x^*\}$, then f has at x Stolz differential $\nabla f(x)$, and $\nabla f(x) = x^*$.*

Proof. Assume that at a certain point $x \in D$ the function f has Stolz differential $\nabla f(x)$. By Lemma 7.7.2 $(\nabla f(x))y = f_y'(x)$. Take any $x^* \in \partial f(x)$. By Lemma 7.8.3 $f_y'(x) \geqslant x^* y$, or

$$(\nabla f(x))y \geqslant x^* y \qquad (7.8.14)$$

for all $y \in \mathbb{R}^N$. The expressions on both the sides of (7.8.14) are additive functions of y, whence by Lemma 7.8.5 $(\nabla f(x))y = x^* y$ for all $y \in \mathbb{R}^N$. But this is only possible if $x^* = \nabla f(x)$. Consequently $\partial f(x) = \{\nabla f(x)\}$.

Conversely, assume $\partial f(x)$ to be a singleton, $\partial f(x) = \{x^*\}$. We define a function $p : (D - x) \to \mathbb{R}$ by

$$p(y) = f(x + y) - f(x) - x^* y.$$

Since x^* is a subgradient, $p \geqslant 0$ in $D - x$, and moreover, as may be easily checked, we have for every $u, v \in D - x$ and $\mu \in [0, 1]$

$$p(\mu u + (1 - \mu) v) \leqslant \mu p(u) + (1 - \mu) p(v),$$

whence it follows in virtue of Theorem 7.1.1 that the function p is convex and continuous. Also the function $g : \mathbb{R} \to \mathbb{R}$ defined for a fixed, but arbitrary, $y \in \mathbb{R}^N$ by (7.8.2), is convex and continuous. Thus we have by Theorem 7.3.2

$$\frac{f(x + \lambda y) - f(x)}{\lambda} = \frac{g(\lambda)}{\lambda} \geqslant \lim_{\lambda \to 0+} \frac{g(\lambda)}{\lambda} = f'_y(x)$$

for every $y \in \mathbb{R}^N$ and $\lambda > 0$ such that $x + \lambda y \in D$. Taking $\lambda = 1$ we obtain hence by Lemma 7.8.3

$$p(y) = f(x + y) - f(x) - x^* y \geqslant f'_y(x) - x^* \geqslant 0 \qquad (7.8.15)$$

for every $y \in D - x$. Since $p(0) = 0$, (7.8.15) can be written as

$$p(y) \geqslant p(0) + 0(y - 0),$$

which shows that $0 \in \partial p(0)$. And if a $y^* \in \partial p(0)$, then we have for all $y \in D - x$

$$p(y) \geqslant p(0) + y^*(y - 0) = y^* y,$$

or

$$f(x + y) \geqslant f(x) + (x^* + y^*) y.$$

This means that $x^* + y^* \in \partial f(x) = \{x^*\}$, whence $x^* + y^* = x^*$, and $y^* = 0$. Consequently $\partial p(0) = \{0\}$, and by Lemma 7.8.4 $p'_u(0) = 0$ for every $u \in \mathbb{R}^N$.

Now write

$$h(\lambda; u) = \frac{p(\lambda u)}{\lambda}, \qquad \lambda > 0.$$

Since $p(0) = 0$, we have

$$h(\lambda; u) = \frac{p(\lambda u) - p(0)}{\lambda} = \frac{p(0 + \lambda u) - p(0)}{\lambda},$$

and

$$\lim_{\lambda \to 0+} h(\lambda; u) = p'_u(0) = 0 \qquad (7.8.16)$$

The set $D - x$ is open, and $0 \in D - x$, so by the Remark in 5.1 there exists a simplex

$$S = \text{conv}\{p_1, \ldots, p_{N+1}\}$$

such that $S \subset D - x$ and $0 \in \text{int } S$. Consequently there exists a closed ball K centered at the origin and with a radius $r > 0$, contained in S. Take an arbitrary $u \in K$. Then by Theorem 5.1.3 there exist $\lambda_1, \ldots, \lambda_{N+1} \in [0, 1]$ such that $\sum_{i=1}^{N+1} \lambda_i = 1$ and

$$u = \sum_{i=1}^{N+1} \lambda_i p_i.$$

Hence we have, using the convexity and continuity if p, and Lemma 5.3.2, for $\lambda \in (0,1)$,

$$h(\lambda; u) = \frac{1}{\lambda} p(\lambda u) = \frac{1}{\lambda} p \left(\sum_{i=1}^{N+1} \lambda_i (\lambda p_i) \right) \leqslant \sum_{i=1}^{N+1} \lambda_i \frac{p(\lambda p_i)}{\lambda} \leqslant \sum_{i=1}^{N+1} h(\lambda; p_i), \quad (7.8.17)$$

since for $\lambda \in (0,1)$ we have $\lambda p_i = \lambda p_i + (1 - \lambda) 0 \in S \subset D - x$, and since $\lambda_i \in [0,1]$, $i = 1, \ldots, N + 1$. Since evidently $h \geqslant 0$, relations (7.8.16) and (7.8.17) show that $\lim_{\lambda \to 0} h(\lambda; u) = 0$ uniformly on K.

Now take an arbitrary $y \in D - x$. There exist a $u \in K$ and a $\lambda > 0$ such that $|u| = r$, and $y = \lambda u$. Hence we have for arbitrary $\varepsilon > 0$

$$\frac{p(y)}{|y|} = \frac{p(\lambda u)}{\lambda r} = \frac{1}{r} h(\lambda; u) < \varepsilon$$

provided that λ is small enough. In other words

$$\frac{f(x + y) - f(x) - x^* y}{|y|} = \frac{p(y)}{|y|} < \varepsilon$$

provided that $|y|$ is small enough, which means that x^* is Stolz differential of f at x. $\square$

Lemma 7.8.6. *Let $f : \mathbb{R}^N \to \mathbb{R}$ be a convex and positively homogeneous function, and let $b_1, \ldots, b_N$ be a base of $\mathbb{R}^N$ over $\mathbb{R}$. If*

$$f(-b_i) = -f(b_i), \quad i = 1, \ldots, N, \quad (7.8.18)$$

then f is linear.

Proof. Since f is positively homogeneous, we have $f(0) = f(2 \cdot 0) = 2f(0)$, whence

$$f(0) = 0 \quad (7.8.19)$$

Further, for arbitrary $\lambda < 0$ we have in view of (7.8.18) for $i = 1, \ldots, N$

$$f(\lambda b_i) = f(-|\lambda| b_i) = |\lambda| f(-b_i) = -|\lambda| f(b_i) = \lambda f(b_i),$$

which together with (7.8.19) and the positive homogeneity of f shows that

$$f(\lambda b_i) = \lambda f(b_i) \quad \text{for } \lambda \in \mathbb{R} \text{ and } i = 1, \ldots, N. \quad (7.8.20)$$

We also have, since f is positively homogeneous and convex,

$$f(x + y) = 2f \left(\frac{x + y}{2} \right) \leqslant 2 \frac{f(x) + f(y)}{2} = f(x) + f(y),$$

whence by introduction

$$f \left(\sum_{i=1}^{n} x_i \right) \leqslant \sum_{i=1}^{n} f(x_i) \quad (7.8.21)$$

for every $n \in \mathbb{N}$ and $x_1, \cdots, x_n \in \mathbb{R}^N$.

It follows from (7.8.19) and (7.8.21) that for $x \in \mathbb{R}^N$

$$0 = f(0) = f(x - x) \leqslant f(x) + f(-x)$$

and so

$$-f(-x) \leqslant f(x) \quad \text{for } x \in \mathbb{R}^N. \tag{7.8.22}$$

Now take an arbitrary $x \in \mathbb{R}^N$. There exist $\lambda_1, \cdots, \lambda_N \in \mathbb{R}$ such that

$$x = \sum_{i=1}^{N} \lambda_i b_i.$$

By (7.8.20), (7.8.21) and (7.8.22)

$$\sum_{i=1}^{N} \lambda_i f(b_i) = \sum_{i=1}^{N} f(\lambda_i b_i) \geqslant f\left(\sum_{i=1}^{N} \lambda_i b_i\right) = f(x) \geqslant -f(-x)$$

$$= -f\left(\sum_{i=1}^{N} (-\lambda_i) b_i\right) \geqslant -\sum_{i=1}^{N} f(-\lambda_i b_i) = \sum_{i=1}^{N} \lambda_i f(b_i),$$

whence it follows, in particular, that f is odd:

$$f(-x) = -f(x) \quad \text{for } x \in \mathbb{R}^N.$$

Thus, for arbitrary $x, y \in \mathbb{R}^N$,

$$f(x + y) = -f(-x - y) \geqslant -[f(x) + f(-y)] = f(x) + f(y),$$

which together with (7.8.21) yields that f is additive. Similarly, the argument leading to (7.8.20) shows that f is homogeneous. Thus f is linear. $\quad\square$

Theorem 7.8.2. *Let $D \subset \mathbb{R}^N$ be a convex and open set, and let $f : D \to \mathbb{R}$ be a convex and continuous function. For every $x \in D$ the following three conditions are equivalent:*

(i) f has a Stolz differential at x;
(ii) f is differentiable at x;
(iii) f has at x all partial derivatives.

Proof. The implications (i) $\Rightarrow$ (ii) $\Rightarrow$ (iii) result from Lemmas 7.7.1 and 7.7.2. Now we are going to prove the converse implications.

Assume that f satisfies (iii). By Lemma 7.8.1 the derivatives $f'_y(x)$ exist for every $y \in \mathbb{R}^N$, moreover the function $f'(x)$ is convex and positively homogeneous. Let $e_1, \ldots, e_N$ be the usual orthonormal base of $\mathbb{R}^N$ over $\mathbb{R}$. We have for $i = 1, \ldots, N$

$$f'_{e_i}(x) = \lim_{\lambda \to 0+} \frac{f(x + \lambda e_i) - f(x)}{\lambda} = \frac{\partial f}{\partial \xi_i}(x),$$

and
$$f'_{-e_i}(x) = \lim_{\lambda \to 0+} \frac{f(x - \lambda e_i) - f(x)}{\lambda} = -\frac{\partial f}{\partial \xi_i}(x),$$

Hence $f'_{-e_i}(x) = -f'_{e_i}(x)$, $i = 1, \ldots, N$. By Lemma 7.8.6 the function $f'(x)$ is linear, which means that f is differentiable at x.

Now assume that f satisfies (ii). In view of (7.7.4) there exist a $y^* \in \mathbb{R}^N$ such that
$$f'_y(x) = y^* y. \tag{7.8.23}$$

Take an arbitrary $x^* \in \partial f(x)$. By (7.8.23) and Lemma 7.8.3 $y^* y \geqslant x^* y$ for every $y \in \mathbb{R}^N$, whence by Lemma 7.8.5 $y^* y = x^* y$ for all $y \in \mathbb{R}^N$, i.e., $y^* = x^*$. Thus $\partial f(x) = \{y^*\}$ is a singleton, and by Theorem 7.8.1 f satisfies (i). $\qquad \square$

7.9 Differentiability of convex functions

In this section we extend Theorem 7.4.3 to the case of functions of several variables. We start with some lemmas.

Lemma 7.9.1. *Let $D \subset \mathbb{R}^N$ be a convex and open set, and let $f : D \to \mathbb{R}$ be a convex and continuous function. Let $x_n \in D$ and $y_n \in \mathbb{R}^N$, $n \in \mathbb{N}$, be convergent sequences: $\lim_{n \to \infty} x_n = x_0 \in D$, $\lim_{n \to \infty} y_n = y_0 \in \mathbb{R}^N$. Then*

$$\limsup_{n \to \infty} f'_{y_n}(x_n) \leqslant f'_{y_0}(x_0). \tag{7.9.1}$$

Proof. Since f is continuous and convex, $f'_{y_0}(x_0)$ exist. Take an arbitrary $\mu \geqslant f'_{y_0}(x_0)$. Then
$$\frac{f(x_0 + \lambda y_0) - f(x_0)}{\lambda} < \mu \tag{7.9.2}$$

for $\lambda > 0$ sufficiently small. On the other hand, for every fixed $n \in \mathbb{N}$, the differences quotient $[f(x_n + \lambda y_n) - f(x_n)]/\lambda$ is an increasing function of λ (cf. Theorem 7.3.2), whence for $\lambda > 0$

$$\frac{f(x_n + \lambda y_n) - f(x_n)}{\lambda} \geqslant f'_{y_n}(x_n). \tag{7.9.3}$$

Moreover,
$$\lim_{n \to \infty} [f(x_n + \lambda y_n) - f(x_n)] = f(x_0 + \lambda y_0) - f(x_0), \tag{7.9.4}$$

since f is continuous. Let us find a $\lambda > 0$ such that (7.9.2) holds. We have by (7.9.3), (7.9.4), and (7.9.2)

$$\limsup_{n \to \infty} f'_{y_n}(x_n) \leqslant \lim_{n \to \infty} \frac{f(x_n + \lambda y_n) - f(x_n)}{\lambda} = \frac{f(x_0 + \lambda y_0) - f(x_0)}{\lambda} < \mu.$$

Letting $\mu \to f'_{y_0}(x_0)$, we obtain hence (7.9.1). $\qquad \square$

Corollary 7.9.1. *Let $D \subset \mathbb{R}^N$ be a convex and open set, and let $f : D \to \mathbb{R}$ be a convex and continuous function. Then the function $f' : D \times \mathbb{R}^N \to \mathbb{R}$ is upper semicontinuous.*

Lemma 7.9.2. *Let $D \subset \mathbb{R}^N$ be a convex and open set, and let $f : D \to \mathbb{R}$ be a convex and continuous function. Then, for every $x \in D$ and $y \in \mathbb{R}^N$,*

$$\liminf_{z \to x} f'_y(z) = f'_{-y}(x). \tag{7.9.5}$$

Proof. By Lemma 7.8.1 $-f'_{-y} \leqslant f'_y$, whence we obtain according to Lemma 7.9.1

$$\liminf_{z \to x} f'_y(z) \geqslant \liminf_{z \to x} (-f'_{-y})(z) = -\limsup_{z \to x} f'_{-y}(z) \geqslant -f'_{-y}(x). \tag{7.9.6}$$

Fix $x \in D$, $y \in \mathbb{R}^N$, and let

$$J = \{\lambda \in \mathbb{R} \mid x + \lambda y \in D\}. \tag{7.9.7}$$

J is an open interval, and we define a function $g : J \to \mathbb{R}$ by

$$g(\lambda) = f(x + \lambda y). \tag{7.9.8}$$

g is a convex and continuous function. For arbitrary $\lambda \in J$ we have

$$f'_y(x + \lambda y) = \lim_{\mu \to 0+} \frac{f(x + \lambda y + \mu y) - f(x + \lambda y)}{\mu} \tag{7.9.9}$$

$$= \lim_{\mu \to 0+} \frac{g(\lambda + \mu) - g(\lambda)}{\mu} = g'_+(\lambda).$$

and

$$f'_{-y}(x + \lambda y) = \lim_{\mu \to 0+} \frac{f(x + \lambda y - \mu y) - f(x + \lambda y)}{\mu} \tag{7.9.10}$$

$$= \lim_{\mu \to 0+} \frac{g(\lambda - \mu) - g(\lambda)}{\mu} = -g'_-(\lambda).$$

By Theorem 7.4.1 we obtain in virtue of (7.9.9) and (7.9.10)

$$\lim_{\lambda \to 0-} f'_y(x + \lambda y) = \lim_{\lambda \to 0-} g'_+(\lambda) = g'_-(0) = -f'_{-y}(x). \tag{7.9.11}$$

Relations (7.9.6) and (7.9.11) imply (7.9.5). $\qquad\square$

Lemma 7.9.3. *Let $D \subset \mathbb{R}^N$ be a convex and open set, and let $f : D \to \mathbb{R}$ be a convex and continuous function. Let a $y \in \mathbb{R}^N$ be fixed, and let*

$$Y_0 = \{x \in D \mid f'_{-y}(x) = -f'_y(x)\}. \tag{7.9.12}$$

Then the set Y_0 is the set of the points of continuity of the function f'_{-y}. Y_0 is dense in D, $D \setminus Y_0$ is of measure zero and of the first category in $\mathbb{R}^N$. More exactly

$$D \setminus Y_0 = \bigcup_{k=1}^{\infty} S_k, \tag{7.9.13}$$

where the sets S_k, $k \in \mathbb{N}$, are closed in D, and at most countable on every straight line with the direction y.

Proof. First assume that $y \neq 0$. Take an $x \in Y_0$. Then by Lemmas 7.9.2 and 7.9.1

$$\liminf_{z \to x} f'_y(z) = -f'_{-y}(x) = f'_y(x) \geqslant \limsup_{z \to x} f'_y(z)$$

whence it follows that the limit $\lim_{z \to x} f'_y(z)$ exist, and

$$\lim_{z \to x} f'_y(z) = f'_y(x).$$

Thus the function f'_y is continuous at x.

Conversely, let f'_y be continuous at an $x \in D$. We have by Lemma 7.8.1 and Lemma 7.9.2

$$f'_y(x) \geqslant -f'_{-y}(x) = \liminf_{z \to x} f'_y(z) = \lim_{z \to x} f'_y(z) = f'_y(x),$$

whence $f'_{-y}(x) = -f'_y(x)$, and $x \in Y_0$. Consequently Y_0 is the set of points of continuity of the function f'_y.

Now define a function $h : D \to \mathbb{R}$ putting

$$h(x) = f'_y(x) + f'_{-y}(x), \quad x \in D.$$

By Lemma 7.8.1, $h \geqslant 0$ in D, and we have $Y_0 = \{x \in D | h(x) = 0\}$. Thus relation (7.9.13) holds with

$$S_k = \left\{ x \in D \mid h(x) \geqslant \frac{1}{k} \right\}, \quad k \in \mathbb{N}. \tag{7.9.14}$$

Since by Corollary 7.9.1 h is upper semicontinuous in D, the sets (7.9.14) are closed in D. The straight line L_x passing through a point $x \in \mathbb{R}^N$ and with the direction y can be written as

$$L_x = \{z \in \mathbb{R}^N \mid z = x + \lambda y, \lambda \in \mathbb{R}\}.$$

Define the interval J by (7.9.7), and the function $g : J \to \mathbb{R}$ by (7.9.8). Take an $x \in \mathbb{R}^N$ and an arbitrary $z \in D \cap L_x$ (if $D \cap L_x \neq \varnothing$). Then $z = x + \lambda y$ with $\lambda \in J$, and we have by (7.9.9) and (7.9.10)

$$f'_y(z) = g'_+(\lambda), \quad f'_{-y}(z) = -g'_-(\lambda)$$

Hence $h(z) = g'_+(\lambda) - g'_-(\lambda)$, and

$$S_k \cap L_x = \left\{ z \in D \cap L_x \mid h(z) \geqslant \frac{1}{k} \right\} = \left\{ z = x + \lambda y \in D \mid g'_+(\lambda) - g'_-(\lambda) \geqslant \frac{1}{k} \right\}. \tag{7.9.15}$$

Set (7.9.15) is at most countable, since it is contained in the set of the points of non-diffentiability (cf. Theorem (7.4.3) of the continuous and convex function $g : J \to \mathbb{R}$.

Sets (7.9.14) being closed in D, are Lebesgue measurable. Let H be an $(N-1)$-dimensional hyperplane perpendicular to y. Then

$$m(S_k) = \int_H m(S_k \cap L_x)\, dx = 0, \quad k \in \mathbb{N}, \tag{7.9.16}$$

since $m(S_k \cap L_x) = 0$ for every $x \in \mathbb{R}^N$, the set (7.9.15) being at most countable. It follows by (7.9.13) that $m(D \setminus Y_0) = 0$. This last property implies, in particular, that the set Y_0 is dense in D.

Similarly, (7.9.16) implies that $\operatorname{int} S_k = \varnothing$, $k \in \mathbb{N}$. The fact that S_k is closed in D means that $S_k = D \cap F_k$, where F_k is closed $\left(\text{in } \mathbb{R}^N\right)$. Thus both sets, D and F_k, belong to the class $\mathcal{F}_\sigma$, and consequently so does also S_k. By Lemma 2.1.1 S_k is of the first category, $k \in \mathbb{N}$, and by (7.9.13) also $D \setminus Y_0$ is of the first category.

If $y = 0$, then by (7.7.4) $Y_0 = D$ and in this case the theorem is evident. $\square$

Theorem 7.9.1. *Let $D \subset \mathbb{R}^N$ be a convex and open set, and let $f : D \to \mathbb{R}$ be a convex and continuous function. Let*

$$D_0 = \{x \in D \mid f \text{ is differentiable at } x\}.$$

Then the set D_0 is dense in D, $D \setminus D_0$ is of measure zero and of first category, and the function $\nabla f : D_0 \to \mathbb{R}^N$ is continuous in D_0.

Proof. Put

$$Y_j = \left\{ x \in D \mid \frac{\partial f}{\partial \xi_j}(x) \text{ exists} \right\} = \left\{ x \in D \mid f'_{e_j}(x) = -f'_{-e_j}(x) \right\}, \quad j = 1, \ldots, N,$$

where $e_1, \ldots, e_N$ is the usual orthonormal base of $\mathbb{R}^N$ (over $\mathbb{R}$). By Theorem 7.8.2

$$D_0 = \bigcap_{j=1}^N Y_j. \qquad \text{Hence} \qquad D \setminus D_0 = \bigcup_{j=1}^N (D \setminus Y_j).$$

The properties of $D \setminus D_0$ result now from Lemma 7.9.3. Also by Lemma 7.9.3 it follows that all the partial derivatives of f are continuous in D_0, and hence, in view of Theorem 7.8.2 and Lemma 7.7.2 $\left(\text{cf., in particular, formula (7.7.8)}\right)$, also ∇f is continuous in D_0. $\square$

Corollary 7.9.2. *Let $D \subset \mathbb{R}^N$ be a convex and open set, and let $f : D \to \mathbb{R}$ be a convex and continuous function. Then f is differentiable almost everywhere in D.*

Corollary 7.9.3. *Let $D \subset \mathbb{R}^N$ be a convex and open set, and let $f : D \to \mathbb{R}$ be a convex and differentiable function. Then f is of class C^1 in D.*

We prove yet the following generalization of Theorem 7.5.2.

Theorem 7.9.2. *Let $D \subset \mathbb{R}^N$ be a convex and open set, and let $f : D \to \mathbb{R}$ be a twice differentiable[8] function. The function f is convex if and only if the matrix*

$$\left(\frac{\partial^2 f}{\partial \xi_i \partial \xi_j}(x) \right) \tag{7.9.17}$$

is positive semi-definite for every $x \in D$.

---

[8] I.e., f is differentiable in D and f'_y is differentiable in D for every $y \in \mathbb{R}^N$ (in the sense of the definition in 7.7).

Proof. For every $x \in D$ and $y \in \mathbb{R}^N$ define the interval $J = J_{xy}$ by (7.9.7) and the function $g = g_{xy} : J \to \mathbb{R}$ by (7.9.8). Observe that g is twice differentiable in J, and f is convex in D if and only if $g = g_{xy}$ is convex in J_{xy} for every choice of $x \in D$ and $y \in \mathbb{R}^N$. In fact, if f is convex, then it is easy to check that g_{xy} is convex for every $x \in D$, $y \in \mathbb{R}^N$. Conversely, let g_{xy} be convex for every $x \in D$, $y \in \mathbb{R}^N$, and take an arbitrary $u, v \in D$. Put $x = u$, $y = v - u$. Then

$$f\left(\frac{u+v}{2}\right) = f\left(u + \frac{1}{2}(v - u)\right) = g_{xy}\left(\frac{1}{2}\right) \leqslant \frac{g_{xy}(0) + g_{xy}(1)}{2} = \frac{f(u) + f(v)}{2}.$$

So f is convex.

By Theorem 7.5.2 the function g is convex if and only if $g'' \geqslant 0$ in J. Therefore f is convex if and only if $\dfrac{d^2}{d\lambda^2} f(x + \lambda y) \geqslant 0$ for every choice of $x \in D$ and $y \in \mathbb{R}^N$, and for every λ from the corresponding interval $J_x y$. In other words, f is convex if and only if $\dfrac{d^2}{d\lambda^2} f(x + \lambda y) |_{x+\lambda y=u} \geqslant 0$ for every $u \in D$ and $y \in \mathbb{R}^N$. But

$$\frac{d^2}{d\lambda^2} f(x + \lambda y) |_{x+\lambda y=u} = \sum_{i=1}^{N} \sum_{j=1}^{N} \frac{\partial^2 f}{\partial \xi_i \partial \xi_j}(u)\eta_i\eta_j, \qquad (7.9.18)$$

where $y = (\eta_1, \dots, \eta_N)$. But the condition that (7.9.18) is non-negative for every $u \in D$ and $y \in \mathbb{R}^N$ is equivalent to the fact that the matrix (7.9.17) is positive semi-definite. $\square$

7.10   Sequences of convex functions

First let us note the following variant of Theorem 5.3.3.

Theorem 7.10.1. *Let $D \subset \mathbb{R}^N$ be a convex and open set, and let $f_n : D \to \mathbb{R}$, $n \in \mathbb{N}$, be a sequence of convex and continuous functions. If the sequence $\{f_n\}$ converges in D to a finite function f, then f is convex and continuous.*

Proof. Fix an arbitrary $x, y \in D$ and $\lambda \in [0, 1]$. Then by Theorem 7.1.1

$$f_n(\lambda x + (1 - \lambda)y) \leqslant \lambda f_n(x) + (1 - \lambda) f_n(y) \quad \text{for all } n \in \mathbb{N}.$$

As $n \to \infty$, we obtain hence

$$f(\lambda x + (1 - \lambda)y) \leqslant \lambda f(x) + (1 - \lambda)f(y). \qquad (7.10.1)$$

Relation (7.10.1) holds for arbitrary $x, y \in D$ and $\lambda \in [0, 1]$, whence, again by Theorem 7.1.1, the function f is convex and continuous. $\square$

Theorem 7.10.2. *Let $D \subset \mathbb{R}^N$ be a convex and open set, and let $\Delta \subset D$ be a dense subset of D. Let $f_n : D \to \mathbb{R}$, $n \in \mathbb{N}$, be a sequence of convex and continuous functions. If the sequence $\{f_n(x)\}$ converges (to a finite limit) for every $x \in \Delta$, then the sequence $\{f_n\}$ converges uniformly on every compact set contained in D.*

Proof. Fix a compact $C \subset D$, and an $\varepsilon > 0$. By Theorem 7.6.1 there exist a constant $L > 0$ such that

$$|f_n(x) - f_n(y)| \leqslant L|x - y| \qquad (7.10.2)$$

for all $x, y \in C$ and all $n \in \mathbb{N}$. For every $x \in \Delta$, let B_x be the open ball centered at x and with the radius $\varepsilon/3L$. Since Δ is dense in D, we have

$$C \subset D \subset \bigcup_{x \in \Delta} B_x.$$

Since C is compact, there exist $x_1, \ldots, x_m \in \Delta$ such that

$$C \subset \bigcup_{i=1}^{m} B_{x_i}. \qquad (7.10.3)$$

Since the set $X = \{x_1, \ldots, x_m\}$ is finite, the sequence f_n converges uniformly on X. Therefore there exist an $n_0 \in \mathbb{N}$ such that for every indices $p, q > n_0$ we have

$$|f_p(x) - f_q(x)| < \frac{\varepsilon}{3} \quad \text{for } x \in X. \qquad (7.10.4)$$

Take an arbitrary $u \in C$. By (7.10.3) there exists an $x_j \in X$ such that

$$|u - x_j| < \frac{\varepsilon}{3L}. \qquad (7.10.5)$$

Thus we have by (7.10.2), (7.10.5) and (7.10.4) for $p, q > n_0$

$$|f_p(u) - f_q(u)| \leqslant |f_p(u) - f_p(x_j)| + |f_p(x_j) - f_q(x_j)| + |f_q(x_j) - f_q(u)|$$
$$\leqslant L|u - x_j| + |f_p(x_j) - f_q(x_j)| + L|u - x_j| < \varepsilon.$$

Thus the sequence $\{f_n\}$ satisfies on C the uniform Cauchy condition, and therefore converges uniformly on C. $\square$

Corollary 7.10.1. *Let $D \subset \mathbb{R}^N$ be a convex and open set, and let $f, f_n : D \to \mathbb{R}$, $n \in \mathbb{N}$, be convex and continuous functions. If*

$$\limsup_{n \to \infty} f_n(x) \leqslant f(x) \quad \text{for } x \in D, \qquad (7.10.6)$$

then to every compact $C \subset D$ and every $\varepsilon > 0$ there exists an index $n_0 \in \mathbb{N}$ such that for every index $n > n_0$ and every $x \in C$

$$f_n(x) \leqslant f(x) + \varepsilon. \qquad (7.10.7)$$

Proof. Put $g_n(x) = \max(f_n(x), f(x))$, $x \in D$, $n \in \mathbb{N}$. The functions $g_n : D \to \mathbb{R}$ are continuous, and by Theorem 5.3.4 they are convex. Relation (7.10.6) implies that

$$\lim_{n \to \infty} g_n(x) = f(x) \quad \text{for } x \in D.$$

Consequently, by Theorem 7.10.2, given a compact $C \subset D$ and an $\varepsilon > 0$, we can find an $n_0 \in \mathbb{N}$ such that for $n > n_o$, $n \in \mathbb{N}$,

$$g_n(x) \leqslant f(x) + \varepsilon.$$

Since $f_n(x) \leqslant \max(f_n(x), f(x)) = g_n(x)$, we obtain hence (7.10.7). $\square$

Corollary 7.10.2. *Let $D \subset \mathbb{R}^N$ be a convex and open set, and let $\Delta \subset D$ be a dense subset of D. Let $f_n : D \to \mathbb{R}$, $n \in \mathbb{N}$, be a sequence of convex and continuous functions. If the sequence $\{f_n(x)\}$ is bounded for every $x \in \Delta$, then it is possible to choose from the sequence $\{f_n\}$ a subsequence uniformly convergent on every compact set contained in D.*

Proof. Since the space $\mathbb{R}^N$ is separable, there exist a countable set $\Delta_0 \subset \Delta$ dense in D. Using the diagonal method of choice (cf., e.g., Łojasiewicz [208]), we can choose from $\{f_n\}$ a subsequence convergent at every point of Δ_0. The corollary results now from Theorem 7.10.2.   □

Theorem 7.10.3. *Let $D \subset \mathbb{R}^N$ be a convex and open set, and let $f_n : D \to \mathbb{R}$, $n \in \mathbb{N}$, be a sequence of convex and differentiable functions, convergent in D:*

$$\lim_{n \to \infty} f_n = f.$$

If the function $f : D \to \mathbb{R}$ is differentiable in D, then

$$\lim_{n \to \infty} \nabla f_n = \nabla f$$

uniformly on every compact set contained in D.

Proof. It is enough to show that for every $y \in \mathbb{R}^N$, $y \neq 0$,

$$\lim_{n \to \infty} (\nabla f_n(x))y = (\nabla f(x))y$$

uniformly on every compact set contained in D. (For $y = e_i, i = 1, \ldots, N$, we obtain hence the suitable convergence of partial derivatives of f, and thus, by formula (7.7.8), of ∇f_n). For an indirect proof suppose that there exist a compact $C \subset D$, a $y \in \mathbb{R}^N \setminus \{0\}$, and an $\varepsilon > 0$ such that for infinitely many $n \in \mathbb{N}$ there exist points $x_n \in C$ such that

$$|(\nabla f_n(x_n))y - (\nabla f(x_n))y| \geqslant \varepsilon,$$

i.e., either

$$(\nabla f_n(x_n))y \geqslant (\nabla f(x_n))y + \varepsilon, \tag{7.10.8}$$

or

$$(\nabla f_n(x_n))y \leqslant (\nabla f(x_n))y - \varepsilon, \tag{7.10.9}$$

which is equivalent to

$$(\nabla f_n(x_n))(-y) \geqslant (\nabla f(x_n))(-y) + \varepsilon,$$

Of the two possibilities, (7.10.8) and (7.10.9), at least one must occur for infinitely many n. We may assume that (7.10.8) occurs for infinitely many n, for (7.10.9) may be reduced to (7.10.8) on replacing y by $-y$. Since C is compact, we may assume that the sequence $\{x_n\}$ converges to an $x_0 \in C$:

$$\lim_{n \to \infty} x_n = x_0$$

and that(7.10.8) holds for all $n \in \mathbb{N}$, for otherwise we replace $\{f_n\}$ and $\{x_n\}$ by suitable subsequences. The point $x_0 + \lambda y$ belongs to D for small $\lambda > 0$, and hence $x_n + \lambda y \in D$ for small $\lambda > 0$ and large $n \in \mathbb{N}$. By Theorem 7.10.2

$$\lim_{n \to \infty} f(x_n) = f(x_0), \tag{7.10.10}$$

and

$$\lim_{n \to \infty} f(x_n + \lambda y) = f(x_0 + \lambda y) \tag{7.10.11}$$

(cf., e.g., Łojasiewicz [208]). Thus we have by Lemma 7.7.2 and Theorem 7.3.2 for $\lambda > 0$ and $n \in \mathbb{N}$ such that $x_n + \lambda y \in D$

$$(\nabla f_n(x_n))y = (f_n)'_y(x_n) \leqslant \frac{f_n(x_n + \lambda y) - f_n(x_n)}{\lambda},$$

and letting $n \to \infty$, we obtain hence in view of (7.10.10) and (7.10.11)

$$\liminf_{n \to \infty} (\nabla f_n(x_n))y \leqslant \frac{f(x_0 + \lambda y) - f(x_0)}{\lambda}. \tag{7.10.12}$$

On the other hand, we have by Corollary 7.9.3 and by (7.10.8)

$$(\nabla f(x_0))y + \varepsilon = \lim_{n \to \infty} (\nabla f(x_n))y + \varepsilon \leqslant \liminf_{n \to \infty} (\nabla f_n(x_n))y. \tag{7.10.13}$$

Relations (7.10.12) and (7.10.13) yield

$$(\nabla f(x_0)) + \varepsilon \leqslant \frac{f(x_0 + \lambda y) - f(x_0)}{\lambda}. \tag{7.10.14}$$

for sufficiently small $\lambda > 0$. When $\lambda \to 0+$, (7.10.14) goes into (cf. 7.7.9)

$$(\nabla f(x_0))y + \varepsilon \leqslant f'_y(x_0) = (\nabla f(x_0))y$$

which is a contradiction with $\varepsilon > 0$. $\qquad \square$

Exercises

1. Let $D \subset \mathbb{R}^N$ be a convex and open set. Show that a function $f : D \to \mathbb{R}$ is convex and continuous if and only if the set

$$\{(x, y) \in \mathbb{R}^{N+1} \mid x \in D, \, y > f(x)\}$$

is convex.

2. Let $f : \mathbb{R}^N \to \mathbb{R}$ be a convex function. Show that if f is bounded above on $\mathbb{R}^N$, then it is constant.

3. Let $f : \mathbb{R}^N \to \mathbb{R}$ be a convex and positively homogeneous function. Show that f is continuous.

4. Let $J \subset \mathbb{R}$ be an open interval, and let $f : J \to \mathbb{R}$ be a continuous function. Suppose that the derivative $f'(x)$ exists for $x \in J_0 \subset J$ such that the set $J \setminus J_0$ is at most countable. Show that f is convex if and only if f' is an increasing function on J_0.

5. Let $J \subset \mathbb{R}$ be an open interval, and let $J_0 \subset J$ be a set such that the set $J \setminus J_0$ is at most countable. Show that there exists a convex and continuous function $f : J \to \mathbb{R}$ such that the derivative f' exists if and only if $x \in J_0$.

6. Let $J \subset \mathbb{R}$ be an open interval, and let $f : J \to \mathbb{R}$ and $g : J \to \mathbb{R}$ be non-negative, increasing, convex functions. Show that the product fg is a convex function.

7. Let $J \subset \mathbb{R}$ be an open interval, and let $f : J \to \mathbb{R}$ be a convex and continuous function. Prove that, for every $x \in J$, we have $\partial f(x) = \left[f'_-(x), f'_+(x) \right]$.

8. Let $D \subset \mathbb{R}^N$ be a convex and open set, and let $f : D \to \mathbb{R}$ be a function. Show that if $\partial f(x) \neq \varnothing$ for every $x \in D$, then f is convex and continuous.
   [Hint: Take arbitrary $u, v \in D$, $\lambda \in [0,1]$, and let $x = \lambda u + (1 - \lambda) v$, $x^* \in \partial f(x)$. Then $f(x) = \lambda [f(x) + x^*(u - x)] + (1 - \lambda)[f(x) + x^*(v - x)] \leqslant \lambda f(u) + (1 - \lambda) f(v)$.]

9. Prove the following extension of Lemma 7.9.1:
   Let $D \subset \mathbb{R}^N$ be a convex and open set, and let $f_n : D \to \mathbb{R}$, $n \in \mathbb{N}$, be a sequence of convex and continuous functions. Further, let $x_n \in D$ and $y \in \mathbb{R}^N$, $n \in \mathbb{N}$, be convergent sequences:

   $$\lim_{n \to \infty} x_n = x_0 \in D, \quad \lim_{n \to \infty} y_n = y_0 \in \mathbb{R}^N.$$

   If the sequence $\{f_n\}$ converges in D to a finite function $f : D \to \mathbb{R}$

   $$\lim_{n \to \infty} f_n = f,$$

   then

   $$\limsup_{n \to \infty} (f'_n)_{y_n}(x_n) \leqslant f'_{y_0}(x_0).$$

Chapter 8

Inequalities

8.1 Jensen inequality

Since the convex functions are defined by a functional inequality, it is not surprising that this notion will lead to a number of interesting and important inequalities. Some inequalities connected with the notion of convexity will be presented in this chapter.

Nowadays the theory of inequalities is a well-developed mathematical subject with scores of results which have been gathered in various books and monographs devoted to this topic (e.g., Hardy-Littlewood-Pólya [136], Mitrinović [226], Beckenbach-Bellman [24]). A number of interesting results about inequalities (a.o. about inequalities connected with the notion of convexity) can be found in various articles published in Publikacije Elektrotehničkog Fakulteta Univerzita u Beogradu. An ample bibliography of the subject can also be found in those sources. Here we will not endeavour to give complete references related to every particular inequality considered. The reader is referred to the above mentioned books and articles. Also, we do not claim to have exhausted the topic (cf. also Roberts-Varberg [267]). Simply we present here a number of inequalities connected with the notion of convexity which seem to us particulary interesting and important.

Jensen's inequality is inequality (5.3.1). Also direct consequences of (5.3.1), viz. (5.3.3), (5.3.7) and (5.3.13) are referred to as Jensen's inequalities. Now we are going to derive further inequalities of this type, all known under the name of Jensen's inequalities.

Theorem 8.1.1. *Let $D \subset \mathbb{R}^N$ be a convex and open set, and let $f : D \to \mathbb{R}$ be a continuous and convex function. Then for every $n \in \mathbb{N}$, $x_1, \ldots, x_n \in D$, and for every non-negative real numbers $q_1, \ldots, q_n$ with $\sum_{i=1}^{n} q_i > 0$, we have*

$$f\left(\frac{\sum_{i=1}^{n} q_i x_i}{\sum_{i=1}^{n} q_i}\right) \leqslant \frac{\sum_{i=1}^{n} q_i f(x_i)}{\sum_{i=1}^{n} q_i}. \tag{8.1.1}$$

Proof. Inequality (8.1.1) results from (5.3.13) on setting $\lambda_i = q_i / \sum\limits_{j=1}^{n} q_j$.   □

Theorem 8.1.2. *Let $(X, \mathfrak{M}, \mu)$ be a measure space such that $\mu(X) = 1$, and let $J \subset \mathbb{R}$ be an open interval. Let $p : X \to J$ be an integrable function, and let $f : J \to \mathbb{R}$ be a continuous and convex function. Then*

$$f\left(\int\limits_X p d\mu\right) \leqslant \int\limits_X (f \circ p) \, d\mu. \tag{8.1.2}$$

Proof. (Zygmund [328]). Put $t = \int\limits_X p d\mu$. Clearly $t \in J$. Take a $k \in \partial f(t)$. Then

$$f(y) - f(t) \geqslant k(y - t) \tag{8.1.3}$$

for all $y \in J$. Replacing y by $p(x)$, and integrating over $x \in X$, we get from (8.1.3)

$$\int\limits_X (f \circ p) \, d\mu - f(t) \geqslant k\left(\int\limits_X p d\mu - t\right) = 0,$$

which is equivalent to (8.1.2).   □

Theorem 8.1.3. *Let $(X, \mathfrak{M}, \mu)$ be a measure space, and let $J \subset \mathbb{R}$ be an open interval. Let $p : X \to J$ and $q : X \to [0, \infty]$ be integrable functions such that the product pq is integrable and $\int\limits_X q d\mu > 0$, and let $f : J \to \mathbb{R}$ be a continuous and convex function. Then*

$$f\left(\frac{\int\limits_X p q d\mu}{\int\limits_X q d\mu}\right) \leqslant \frac{\int\limits_X q(f \circ p) \, d\mu}{\int\limits_X q d\mu}. \tag{8.1.4}$$

Proof. Define the new measure $\nu : \mathfrak{M} \to [0, \infty]$ putting for $A \in \mathfrak{M}$

$$\nu(A) = \frac{\int\limits_A q d\mu}{\int\limits_X q d\mu}.$$

In particular, we have $\nu(X) = 1$, and

$$d\nu = \frac{q}{\int\limits_X q d\mu} d\mu. \tag{8.1.5}$$

By Theorem 8.1.2

$$f\left(\int\limits_X p d\nu\right) \leqslant \int\limits_X (f \circ p) \, d\nu,$$

whence by (8.1.5) we obtain (8.1.4).   □

Theorem 8.1.4. *Let $X \subset \mathbb{R}^N$ be an open set, and let $\Lambda \geqslant 0$ be an additive function of intervals in X, $0 < \int\limits_X d\Lambda < \infty$. Let $J \subset \mathbb{R}$ be an open interval, let $p : X \to J$ be a function integrable (in the sense of Lebesgue-Stieltjes) with respect to Λ, and let $f : J \to \mathbb{R}$ be a continuous and convex function. Then*

$$f\left(\frac{\int\limits_X p \, d\Lambda}{\int\limits_X d\Lambda}\right) \leqslant \frac{\int\limits_X (f \circ p) \, d\Lambda}{\int\limits_X d\Lambda}. \tag{8.1.6}$$

Proof. Inequality (8.1.6) is a particular case of (8.1.2), where $\mu(A) = \int\limits_A d\Lambda / \int\limits_X d\Lambda$. $\square$

From the above theorems we can derive some classical inequalities.

Corollary 8.1.1. *Let $n \in \mathbb{N}$, and let $a_1, \ldots, a_n$, $b_1, \ldots, b_n$ be arbitrary real numbers. Then*

$$\left(\sum_{i=1}^{n} a_i b_i\right)^2 \leqslant \left(\sum_{i=1}^{n} a_i^2\right)\left(\sum_{i=1}^{n} b_i^2\right). \tag{8.1.7}$$

Proof. Assume first that $a_i \neq 0$ for $i = 1, \ldots, n$. Put in Theorem 8.1.1 $D = \mathbb{R}$, $f(x) = x^2$, $x_i = \frac{b_i}{a_i}$, $q_i = a_i^2$, $i = 1, \ldots, n$. Then (8.1.1) becomes (8.1.7). If all a_i's are zeros, (8.1.7) is trivial. If some a_i's are zeros, we can rearrange sequences $\{a_i\}$ and $\{b_i\}$ so that $a_i \neq 0$ for $i = 1, \ldots, m$, $a_i = 0$ for $i = m+1, \ldots, n$, $0 < m < n$. Then, on account of what has already been proved

$$\left(\sum_{i=1}^{n} a_i b_i\right)^2 = \left(\sum_{i=1}^{m} a_i b_i\right)^2 \leqslant \left(\sum_{i=1}^{m} a_i^2\right)\left(\sum_{i=1}^{m} b_i^2\right)$$
$$= \left(\sum_{i=1}^{n} a_i^2\right)\left(\sum_{i=1}^{m} b_i^2\right) \leqslant \left(\sum_{i=1}^{n} a_i^2\right)\left(\sum_{i=1}^{n} b_i^2\right),$$

and (8.1.7) holds true. $\square$

Inequality (8.1.7) is the famous Cauchy-Buniakowski-Schwarz inequality.

Corollary 8.1.2. *Let $(X, \mathfrak{M}, \mu)$ be a measure space, and let $a, b : X \to [-\infty, \infty]$ be integrable functions. Then*

$$\left(\int\limits_X ab \, d\mu\right)^2 \leqslant \left(\int\limits_X a^2 \, d\mu\right)\left(\int\limits_X b^2 \, d\mu\right). \tag{8.1.8}$$

Proof. Since b is integrable, it is equivalent to a finite function (i.e., $b = b^*$ a.e. in X, where b^* is finite), and thus we may assume that b itself is finite: $b : X \to \mathbb{R}$. Assume first that $a \neq 0$ in X. Set in Theorem 8.1.3 $J = \mathbb{R}$, $f(x) = x^2$, $p(x) = b(x)/a(x)$ and $q(x) = a(x)^2$. Then (8.1.4) becomes (8.1.8).

If $a = 0$ a.e. in X, inequality (8.1.8) is trivial. Otherwise put

$$Y = \{\, x \in X \mid a(x) \neq 0 \,\}.$$

By what has already been proved

$$\left(\int_X abd\mu \right)^2 = \left(\int_Y abd\mu \right)^2 \leqslant \left(\int_Y a^2 d\mu \right) \left(\int_Y b^2 d\mu \right)$$

$$= \left(\int_X a^2 d\mu \right) \left(\int_Y b^2 d\mu \right) \leqslant \left(\int_X a^2 d\mu \right) \left(\int_X b^2 d\mu \right),$$

and (8.1.8) holds true.   □

Corollary 8.1.3. *Let $n \in \mathbb{N}$, and let $x_1, \ldots, x_n, q_1, \ldots, q_n$ be non-negative real numbers. Then*

$$\left(\sum_{i=1}^n q_i x_i \right)^p \leqslant \left(\sum_{i=1}^n q_i \right)^{p-1} \sum_{i=1}^n q_i x_i^p \quad \text{for } p > 1 \text{ and } p < 0, \tag{8.1.9}$$

and if, moreover, $x_i > 0$ for $i = 1, \ldots, n$, then

$$\left(\sum_{i=1}^n q_i x_i \right)^p \geqslant \left(\sum_{i=1}^n q_i \right)^{p-1} \sum_{i=1}^n q_i x_i^p \quad \text{for } 0 < p < 1. \tag{8.1.10}$$

Proof. If all q_i's are zeros, (8.1.9) and (8.1.10) are trivial, so we assume that $\sum_{i=1}^n q_i > 0$. Let $p > 1$ or $p < 0$, and assume that $x_i > 0$ for $i = 1, \ldots, n$. Set in Theorem 8.1.1 $N = 1$, $D = (0, \infty)$, $f(x) = x^p$, and (8.1.1) becomes (8.1.9). If some x_i's are zeros, we argue as in the proof of Corollary 8.1.1.

If $0 < p < 1$ and all x_i's are positive, we take in Theorem 8.1.1 $N = 1$, $D = (0, \infty)$, and $f(x) = -x^p$, and then we multiply the resulting inequality by (-1), arriving thus at (8.1.10).   □

Corollary 8.1.4. *Let $n \in \mathbb{N}$, and let $x_1, \ldots, x_n$ be non-negative real numbers. Then*

$$\left(\sum_{i=1}^n x_i \right)^p \leqslant n^{p-1} \sum_{i=1}^n x_i^p \quad \text{for } p > 1 \text{ and } p < 0, \tag{8.1.11}$$

and if, moreover, $x_i > 0$ for $i = 1, \ldots, n$, then

$$\left(\sum_{i=1}^n x_i \right)^p \geqslant n^{p-1} \sum_{i=1}^n x_i^p \quad \text{for } 0 < p < 1. \tag{8.1.12}$$

Proof. We obtain (8.1.11) and (8.1.12) from (8.1.9) and (8.1.10) taking $q_i = 1$, $i = 1, \ldots, n$.   □

Corollary 8.1.5. *Let* $n \in \mathbb{N}$, *let* $a_1, \ldots, a_n$, $b_1, \ldots, b_n$ *be arbitrary non-negative numbers. Let* $p > 1$ *and let* q *be given by* $\dfrac{1}{p} + \dfrac{1}{q} = 1$. *Then*

$$\sum_{i=1}^{n} a_i b_i \leqslant \left(\sum_{i=1}^{n} a_i^p \right)^{1/p} \left(\sum_{i=1}^{n} b_i^q \right)^{1/q}. \tag{8.1.13}$$

Proof. Assume first that $a_i > 0$, $b_i > 0$ for $i = 1, \ldots, n$. Taking in Theorem 8.1.1 $N = 1$, $D = (0, \infty)$, $f(x) = x^p$, $q_i = b_i^q$, $x_i = (a_i^p / b_i^q)^{1/p}$, $i = 1, \ldots, n$, we obtain (8.1.13). If some a_i's and/or some b_i's are zeros, we argue as in the proof of Corollary 8.1.1. □

Inequality (8.1.13) is the famous Hölder inequality.

Corollary 8.1.6. *Let* $(X, \mathfrak{M}, \mu)$ *be a measure space, and let* $a, b : X \to [-\infty, \infty]$ *be functions integrable in p-th and q-th power, respectively, where* $p > 1$, $\frac{1}{p} + \frac{1}{q} = 1$. *Then the function ab is integrable, and*

$$\int_X |ab| \, d\mu \leqslant \left(\int_X |a|^p \, d\mu \right)^{1/p} \left(\int_X |b|^q \, d\mu \right)^{1/q}. \tag{8.1.14}$$

Proof. The functions a and b are equivalent to finite functions, so we may assume that a and b are themselves finite. Assume first that $a \neq 0$ and $b \neq 0$ in X. Setting in Theorem 8.1.3 $J = (0, \infty)$, $p(x) = \left(|a(x)|^p / |b(x)|^q \right)^{1/p}$, $q(x) = |b(x)|^q$, $f(x) = x^p$, we obtain inequality (8.1.14). The integrability of the product ab results from (8.1.14). If a, b are allowed to assume the value zero, we argue as in the proof of Corollary 8.1.2. Here again ab is integrable, since by the above the function ab restricted to the set $\{ x \in X \mid a(x) b(x) \neq 0 \}$ is integrable. □

8.2 Jensen-Steffensen inequalities

Now we are going to generalize the result of the preceding section (cf., e.g., Mitrinović [225], [226], Boas [34]). We start with a lemma.

Lemma 8.2.1. *Let* $J = [a, b]$ *be a closed real interval, and let* $p, \lambda : J \to \mathbb{R}$ *be functions such that* p *is non-negative and monotonic in* J *and the Riemann-Stieltjes integral* $\int_a^b p \, d\lambda$ *exists*[1]. *Then, if* p *is decreasing*

$$p(a) \inf_{a \leqslant c \leqslant b} [\lambda(c) - \lambda(a)] \leqslant \int_a^b p \, d\lambda \leqslant p(a) \sup_{a \leqslant c \leqslant b} [\lambda(c) - \lambda(a)], \tag{8.2.1}$$

---

[1] The Stieltjes integrals in this section denote Riemann-Stieltjes integrals. The Riemann-Stieltjes integral $\int_a^b p \, d\lambda$ certainly exists if one of the functions p, λ is continuous in J, and the other has a finite variation in J. In the case of Lemma 8.2.1 the function p (being monotonic) is of the finite variation, so for the existence of the integral in question it is sufficient (but not necessary) that the function λ is continuous. Note that if the function λ is increasing, Lemma 8.2.1 is trivial, and Theorem 8.2.1 is a particular case of Theorem 8.1.4. The main interest of inequality (8.2.3) below lies in the fact that λ need not be a monotonic function. Similarly, inequality (8.2.6) below is more general than (8.1.1) in that now q_i need not be non-negative.

and if p is increasing,

$$p\left(b\right)\inf_{a\leqslant c\leqslant b}\left[\lambda\left(b\right)-\lambda\left(c\right)\right]\leqslant\int_{a}^{b}pd\lambda\leqslant p\left(b\right)\sup_{a\leqslant c\leqslant b}\left[\lambda\left(b\right)-\lambda\left(c\right)\right].\qquad(8.2.2)$$

Proof. Assume p to be decreasing. Then

$$\int_{a}^{b}p\left(x\right)d\lambda\left(x\right)=\int_{a}^{b}p\left(x\right)d\left[\lambda\left(x\right)-\lambda\left(a\right)\right]$$

$$=\left[\lambda\left(b\right)-\lambda\left(a\right)\right]p\left(b\right)-\int_{a}^{b}\left[\lambda\left(x\right)-\lambda\left(a\right)\right]dp\left(x\right)$$

$$\leqslant p\left(b\right)\sup_{a\leqslant c\leqslant b}\left[\lambda\left(c\right)-\lambda\left(a\right)\right]+\left[p\left(a\right)-p\left(b\right)\right]\sup_{a\leqslant c\leqslant b}\left[\lambda\left(c\right)-\lambda\left(a\right)\right]$$

$$=p\left(a\right)\sup_{a\leqslant c\leqslant b}\left[\lambda\left(c\right)-\lambda\left(a\right)\right].$$

This is the right inequality in (8.2.1). The left inequality in (8.2.1), and, in the case where p is increasing, inequalities (8.2.2) are proved similarly. □

Theorem 8.2.1. *Let $J=[a,b]$ be a closed real interval, and let $p,\lambda:J\to\mathbb{R}$ be functions such that $\lambda\left(a\right)\leqslant\lambda\left(x\right)\leqslant\lambda\left(b\right)$ in J, $\lambda\left(a\right)<\lambda\left(b\right)$, p is monotonic in J, and $\int_{a}^{b}pd\lambda$ exists. Let $D\subset\mathbb{R}$ be an open interval such that $p\left([a,b]\right)\subset D$, and let $f:D\to\mathbb{R}$ be a continuous and convex function. Then*

$$f\left(\frac{\int_{a}^{b}pd\lambda}{\lambda\left(b\right)-\lambda\left(a\right)}\right)\leqslant\frac{\int_{a}^{b}\left(f\circ p\right)d\lambda}{\lambda\left(b\right)-\lambda\left(a\right)}.\qquad(8.2.3)$$

Proof. We assume that the function p is decreasing; in the case of an increasing p the proof is similar. Put

$$t=\frac{\int_{a}^{b}pd\lambda}{\lambda\left(b\right)-\lambda\left(a\right)}.\qquad(8.2.4)$$

We have

$$\int_{a}^{b}pd\lambda-p\left(b\right)\left[\lambda\left(b\right)-\lambda\left(a\right)\right]=\int_{a}^{b}pd\lambda-p\left(b\right)\int_{a}^{b}d\lambda=\int_{a}^{b}\left[p\left(x\right)-p\left(b\right)\right]d\lambda\left(x\right).$$

By Lemma 8.2.1

$$\int_a^b [p(x) - p(b)] \, d\lambda(x) \leqslant [p(a) - p(b)] \sup_{a \leqslant c \leqslant b} [\lambda(c) - \lambda(a)]$$

$$= [p(a) - p(b)][\lambda(b) - \lambda(a)].$$

Hence

$$\int_a^b p \, d\lambda \leqslant p(a)[\lambda(b) - \lambda(a)].$$

Similarly

$$\int_a^b p \, d\lambda \geqslant p(b)[\lambda(b) - \lambda(a)].$$

The above inequalities show that $p(b) \leqslant t \leqslant p(a)$.

Take a $k \in \partial f(t)$. Then

$$f(y) - f(t) \geqslant k(y - t)$$

for all $y \in D$. Put

$$r(y) = f(y) - f(t) - k(y - t), \quad y \in D.$$

It is easily seen that $r : D \to \mathbb{R}$ is a non-negative convex function, and $r(t) = 0$. It follows from Theorem 7.3.5 that r decreases in $[p(b), t]$ and increases in $[t, p(a)]$. Next put $R(x) = r(p(x))$, $T \in p^{-1}(t)$. Then $R \geqslant 0$, and R decreases in $[a, T]$, R increases in $[T, b]$. By Lemma 8.2.1

$$\int_a^b R \, d\lambda = \int_a^T R \, d\lambda + \int_T^b R \, d\lambda \geqslant R(a) \inf_{a \leqslant c \leqslant b} [\lambda(c) - \lambda(a)] + R(b) \inf_{a \leqslant c \leqslant b} [\lambda(b) - \lambda(c)] = 0.$$

Next, by (8.2.4),

$$0 \leqslant \int_a^b R \, d\lambda = \int_a^b (r \circ p) \, d\lambda = \int_a^b [f(p(x)) - f(t) - kp(x) + kt] \, d\lambda$$

$$= \int_a^b (f \circ p) \, d\lambda - f(t)[\lambda(b) - \lambda(a)] - k \left[\int_a^b p \, d\lambda - t(\lambda(b) - \lambda(a)) \right]$$

$$= \int_a^b (f \circ p) \, d\lambda - f(t)[\lambda(b) - \lambda(a)].$$

Hence (8.2.3) results. $\qquad\qquad\qquad\qquad\qquad\qquad\qquad\qquad\qquad\qquad\qquad\quad\square$

Theorem 8.2.2. *Let $J = [a, b]$ be a closed real interval, and let $p : J \to \mathbb{R}$ and $q : J \to \mathbb{R}$ be a monotonic and an integrable function, respectively, such that*

$$0 \leqslant \int\limits_a^x q \, dt \leqslant \int\limits_a^b q \, dt \quad for \quad x \in J, \; \int\limits_a^b q \, dt > 0.$$

Let $D \subset \mathbb{R}$ be an open interval such that $p([a, b]) \subset D$, and let $f : D \to \mathbb{R}$ be a continuous and convex function. Then

$$f \left(\frac{\int\limits_a^b pq \, dt}{\int\limits_a^b q \, dt} \right) \leqslant \frac{\int\limits_a^b (f \circ p) \, q \, dt}{\int\limits_a^b q \, dt}. \tag{8.2.5}$$

Proof. We take in Theorem 8.2.1 $\lambda(x) = \int\limits_a^x q(t) \, dt$, and (8.2.5) results from (8.2.3).   □

Theorem 8.2.3. *Let $n \in \mathbb{N}$, and let $a < x_1 \leqslant \cdots \leqslant x_n < b$ be real numbers. Further, let $q_1, \ldots, q_n$ be real numbers such that for every $k = 1, \ldots, n$*

$$0 \leqslant \sum_{i=k}^n q_i \leqslant \sum_{i=1}^n q_i, \quad \sum_{i=1}^n q_i > 0.$$

If $f : (a, b) \to \mathbb{R}$ is a continuous and convex function, then

$$f \left(\frac{\sum\limits_{i=1}^n q_i x_i}{\sum\limits_{i=1}^n q_i} \right) \leqslant \frac{\sum\limits_{i=1}^n q_i f(x_i)}{\sum\limits_{i=1}^n q_i}. \tag{8.2.6}$$

Proof. Take arbitrary points $\xi_1, \ldots, \xi_n$ such that $0 < \xi_1 < \cdots < \xi_n < 1$, and let p be an arbitrary continuous increasing function from $[0, 1]$ onto $[a, b]$ such that $p(\xi_i) = x_i$, $i = 1, \ldots, n$. Define the function $\lambda : [0, 1] \to \mathbb{R}$ by

$$\lambda(x) = \sum_{\xi_i < x} q_i.$$

Then inequality (8.2.3) goes into (8.2.6).   □

Originally inequalities (8.2.3) and (8.2.5) were obtained (Steffensen [296]) from the Steffensen inequality (Steffensen [295])

$$\int\limits_{b-\lambda}^b f(t) \, dt \leqslant \int\limits_a^b f(t) \, g(t) \, dt \leqslant \int\limits_a^{a+\lambda} f(t) \, dt, \quad where \; \lambda = \int\limits_a^b g(t) \, dt,$$

valid for arbitrary integrable functions $f, g : (a, b) \to \mathbb{R}$ such that f is decreasing and $0 \leqslant g \leqslant 1$ in (a, b). The generalizations presented here are due to R. P. Boas [34]. Theorem 8.2.5 and 8.2.6 are also related to some results of Z. Ciesielski [45].

Theorem 8.2.4. *Let $J = [a, b]$ be a closed real interval, let $p : J \to \mathbb{R}$ be a continuous, non-negative, decreasing function, let $\lambda : J \to \mathbb{R}$ be a function such that $\int_a^b p d\lambda$ exists, and let $\mu : J \to \mathbb{R}$ be a function. Assume that the following conditions are fulfilled:*

$$\lambda(a) \leqslant \lambda(x) \leqslant \lambda(b) + \lambda^* \quad \text{for } x \in J,$$

where $\lambda^* \geqslant 0$, and

$$0 < \mu(b) - \mu(a), \quad \lambda(b) - \lambda(a) + \lambda^* \leqslant \mu(b) - \mu(a).$$

Let $D \subset \mathbb{R}$ be an open interval such that $0, p(a) \in D$, and let $f : D \to \mathbb{R}$ be a continuous and convex function such that $f(0) \leqslant 0$. Then

$$f\left(\frac{\int_a^b p d\lambda}{\mu(b) - \mu(a)} \right) \leqslant \frac{\int_a^b (f \circ p) d\lambda}{\mu(b) - \mu(a)}. \tag{8.2.7}$$

Proof. We shall distinguish a few cases.

I. $\lambda^* = 0$. Put $A = \int_a^b p d\lambda$. For $y_0 = \lambda(b) - \lambda(a)$ we have by Theorem 8.2.1

$$\int_a^b (f \circ p) d\lambda \geqslant y_0 f\left(\frac{A}{y_0} \right). \tag{8.2.8}$$

We will prove that the function $F(y) = y f\left(\frac{A}{y} \right)$ is decreasing. With the exception of at most countably many points we have (cf. Theorem 7.4.3)

$$F'(y) = f\left(\frac{A}{y} \right) - \frac{A}{y} f'\left(\frac{A}{y} \right).$$

Since f is continuous and convex and $f(0) \leqslant 0$, we have by Theorem 7.3.2 for $u > 0$, $u \in D$,

$$\frac{f(u)}{u} \leqslant \frac{f(u) - f(0)}{u - 0} \leqslant f'_-(u).$$

Thus $F'(y) \leqslant 0$ (for $y > 0$ such that $A/y \in D$ and $F'(y)$ exists). By Theorem 3.6.3 F is decreasing. Consequently for $y_1 = \mu(b) - \mu(a) \geqslant \lambda(b) - \lambda(a) = y_0$ we have by (8.2.8)

$$\int_a^b (f \circ p) d\lambda \geqslant y_0 f\left(\frac{A}{y_0} \right) \geqslant y_1 f\left(\frac{A}{y_1} \right),$$

which is equivalent to (8.2.7).

II. $\lambda^* > 0$ and $p(b) = 0$. We put

$$\delta(x) = \begin{cases} 0 & \text{for } x \in [a, b), \\ 1 & \text{for } x = b, \end{cases} \qquad \text{and} \qquad \omega(x) = \lambda(x) + \lambda^* \delta(x), \quad x \in J.$$

Then

$$\omega(a) \leqslant \omega(x) \leqslant \omega(b) \quad \text{for } x \in J,$$

and

$$\omega(b) - \omega(a) = \lambda(b) + \lambda^* - \lambda(a) \leqslant \mu(b) - \mu(a).$$

By the first part of the proof

$$f\left(\frac{\int\limits_a^b p\,d\omega}{\mu(b) - \mu(a)} \right) \leqslant \frac{\int\limits_a^b (f \circ p)\,d\omega}{\mu(b) - \mu(a)}. \tag{8.2.9}$$

But

$$\int\limits_a^b p\,d\omega = \int\limits_a^b p\,d\lambda + \lambda^* \int\limits_a^b p\,d\delta = \int\limits_a^b p\,d\lambda + \lambda^* p(b) = \int\limits_a^b p\,d\lambda,$$

and

$$\int\limits_a^b (f \circ p)\,d\omega = \int\limits_a^b (f \circ p)\,d\lambda + \lambda^* \int\limits_a^b (f \circ p)\,d\delta = \int\limits_a^b (f \circ p)\,d\lambda + \lambda^* f(0) \leqslant \int\limits_a^b (f \circ p)\,d\lambda,$$

and so (8.2.7) results from (8.2.9).

III. $\lambda^* > 0$ and $p(b) > 0$. Then we extend the functions p, λ, μ onto an interval $[a, c]$, where $c > b$, so that p is continuous, non-negative, decreasing in $[a, c]$, $p(c) = 0$, and

$$\lambda(x) = \lambda(b), \quad \mu(x) = \mu(b) \quad \text{for } x \in [b, c].$$

By the second part of the proof

$$f\left(\frac{\int\limits_a^c p\,d\lambda}{\mu(c) - \mu(a)} \right) \leqslant \frac{\int\limits_a^c (f \circ p)\,d\lambda}{\mu(c) - \mu(a)}. \tag{8.2.10}$$

But $\mu(c) - \mu(a) = \mu(b) - \mu(a)$,

$$\int\limits_a^c p\,d\lambda = \int\limits_a^b p\,d\lambda, \qquad \int\limits_a^c (f \circ p)\,d\lambda = \int\limits_a^b (f \circ p)\,d\lambda.$$

So (8.2.7) results from (8.2.10). $\square$

Theorem 8.2.5. *Let $J = [a, b]$ be a closed real interval, let $p : J \to \mathbb{R}$ be a continuous, non-negative, decreasing function, and let $q : J \to \mathbb{R}$ be an integrable function such that*

$$\int_a^x q \, dt \geqslant 0 \quad \text{for } x \in J, \quad \int_a^b |q| \, dt > 0.$$

Let $D \subset \mathbb{R}$ be an open interval such that $0, p(a) \in D$, and let $f : D \to \mathbb{R}$ be a continuous and convex function such that $f(0) \leqslant 0$. Then

$$f\left(\frac{\int_a^b pq \, dt}{\int_a^b |q| \, dt} \right) \leqslant \frac{\int_a^b (f \circ p) q \, dt}{\int_a^b |q| \, dt}. \tag{8.2.11}$$

Proof. We take in Theorem 8.2.4

$$\lambda x = \int_a^x q \, dt, \quad \mu x = \int_a^x |q| \, dt, \quad \lambda^* = \int_a^b (|q| - q) \, dt,$$

and (8.2.7) becomes (8.2.11). $\qquad \square$

Theorem 8.2.6. *Let $n \in \mathbb{N}$, let $x_1, \ldots, x_n, a, b$ be real numbers such that $a < 0 \leqslant x_1 \leqslant \cdots \leqslant x_n < b$, and let $q_1, \ldots, q_n$ be real numbers such that for $k = 1, \ldots, n$*

$$\sum_{i=k}^n q_i \geqslant 0, \quad \sum_{i=1}^n |q_i| > 0.$$

If $f : (a, b) \to \mathbb{R}$ is a continuous and convex function such that $f(0) \leqslant 0$, then

$$f\left(\frac{\sum_{i=1}^n q_i x_i}{\sum_{i=1}^n |q_i|} \right) \leqslant \frac{\sum_{i=1}^n q_i f(x_i)}{\sum_{i=1}^n |q_i|}. \tag{8.2.12}$$

Proof. We take arbitrary $\xi_1, \ldots, \xi_n$ such that $0 < \xi_n < \cdots < \xi_1 < 1$, and let p be an arbitrary continuous, non-negative, decreasing function from $[0, 1]$ onto $[0, b]$ such that $p(\xi_i) = x_i$, $i = 1, \ldots, n$. Define the function $\lambda, \mu : [0, 1] \to \mathbb{R}$ and the number λ^* by

$$\lambda(x) = \sum_{\xi_i < x} q_i, \quad \mu(x) = \sum_{\xi_i < x} |q_i|, \quad \lambda^* = \sum_{i=1}^n (|q_i| - q_i).$$

Then inequality (8.2.7) goes over into (8.2.12). $\qquad \square$

8.3 Inequalities for means

Let F be a strictly monotonic function on an interval $J \subset \mathbb{R}$, let $w_1, \ldots, w_n$ be non-negative real numbers $\left(\sum\limits_{i=1}^{n} w_i > 0 \right)$, and let $a_1, \ldots, a_n \in F(J)$ be arbitrary points. The expression

$$M_n(F; \mathbf{a}; \mathbf{w}) = F \left(\frac{\sum\limits_{i=1}^{n} w_i F^{-1}(a_i)}{\sum\limits_{i=1}^{n} w_i} \right) \tag{8.3.1}$$

is called a *weighted quasiarithmetic mean* of $a_1, \ldots, a_n$ with the weights $w_1, \ldots, w_n$. In (8.3.1) and in the sequel $\mathbf{a}$ stands for $(a_1, \ldots, a_n)$, and $\mathbf{w}$ stands for $(w_1, \ldots, w_n)$.

Theorem 8.3.1. *Let $J \subset \mathbb{R}$ be an open interval, and let the functions $F, G : J \to \mathbb{R}$ be strictly monotonic, $F(J) = G(J)$. Let the function $G^{-1} \circ F$ be convex[2]. Then, for every $n \in \mathbb{N}$, every $a_1, \ldots, a_n \in F(J)$, and every non-negative $w_1, \ldots, w_n$ such that $\sum\limits_{i=1}^{n} w_i > 0$, we have*

$$M_n(F; \mathbf{a}; \mathbf{w}) \leqslant M_n(G; \mathbf{a}; \mathbf{w}) \tag{8.3.2}$$

whenever G is increasing. If G is decreasing, inequality (8.3.2) gets reversed.

Proof. We take in (8.1.1) $f = G^{-1} \circ F$, $q_i = w_i$, $x_i = F^{-1}(a_i)$, $i = 1, \ldots, n$. We get

$$G^{-1} \circ F \left(\frac{\sum\limits_{i=1}^{n} w_i F^{-1}(a_i)}{\sum\limits_{i=1}^{n} w_i} \right) \leqslant \frac{\sum\limits_{i=1}^{n} w_i G^{-1}(a_i)}{\sum\limits_{i=1}^{n} w_i}. \tag{8.3.3}$$

Applying to both the sides of (8.3.3) the function G we obtain (8.3.2) if G is increasing, and the reversed inequality if G is decreasing. $\square$

Remark. If the function $G^{-1} \circ F$ is concave, then inequality (8.3.2) holds whenever G is decreasing, and gets reversed whenever G is increasing.

The well-known inequality $H_n \leqslant G_n \leqslant A_n$ between the arithmetic mean

$$A_n = \frac{1}{n}(a_1 + \cdots + a_n), \tag{8.3.4}$$

the geometric mean

$$G_n = (a_1 \cdots a_n)^{1/n}, \tag{8.3.5}$$

and the harmonic mean

$$H_n = n \left(a_1^{-1} + \cdots + a_n^{-1} \right)^{-1}, \tag{8.3.6}$$

$(a_1, \ldots, a_n > 0)$ can be obtained from Theorem 8.3.1, since each mean (8.3.4), (8.3.5), (8.3.6) is a quasiarithmetic mean. Actually, we can prove a more general result.

---

[2] Since the function $G^{-1} \circ F$ is monotonic its convexity implies the continuity.

For every positive $a_1, \ldots, a_n$ and non-negative $w_1, \ldots, w_n$ such that $\sum\limits_{i=1}^{n} w_i > 0$, and for every $r \in \mathbb{R}$, $r \neq 0$, we put

$$M_n^r [\mathbf{a}; \mathbf{w}] = \left(\sum_{i=1}^{n} w_i a_i^r / \sum_{i=1}^{n} w_i \right)^{1/r}. \qquad (8.3.7)$$

For $r = 0$ we assume

$$M_n^0 [\mathbf{a}; \mathbf{w}] = \exp \left(\sum_{i=1}^{n} w_i \log a_i / \sum_{i=1}^{n} w_i \right) = \prod_{i=1}^{n} a_i^{w_i / \sum_{j=1}^{n} w_j}. \qquad (8.3.8)$$

We define also M_n^r for $r = +\infty$ and $r = -\infty$:

$$M_n^\infty [\mathbf{a}; \mathbf{w}] = \max_i a_i, \quad M_n^{-\infty} [\mathbf{a}; \mathbf{w}] = \min_i a_i. \qquad (8.3.9)$$

In particular, we have for means (8.3.4), (8.3.5), (8.3.6)

$$A_n = M_n^1 [\mathbf{a}; \mathbf{1}], \quad G_n = M_n^0 [\mathbf{a}; \mathbf{1}], \quad H_n = M_n^{-1} [\mathbf{a}; \mathbf{1}],$$

where $\mathbf{1}$ means $(1, \ldots, 1)$.

Theorem 8.3.2. *With fixed $n \in \mathbb{N}$, $\mathbf{a} = (a_1, \ldots, a_n)$ and $\mathbf{w} = (w_1, \ldots, w_n)$ ($a_i > 0, w_i \geq 0$, $i = 1, \ldots, n$; $\sum\limits_{i=1}^{n} w_i > 0$) the function $\varphi(r) = M_n^r [\mathbf{a}; \mathbf{w}]$ is increasing on $[-\infty, \infty]$.*

Proof. Inequalities $\varphi(-\infty) \leq \varphi(r) \leq \varphi(\infty)$ for $-\infty < r < \infty$ are obvious. Now the theorem results from Theorem 8.3.1. For $r \neq 0$ (8.3.7) coincides with (8.3.1) for $F(x) = x^{1/r}$, whereas (8.3.8) coincides with (8.3.1) for $F(x) = e^x$. Taking $r, s \in \mathbb{R}$, $r < s$, we need only use Theorem 8.3.1 for F and G as follows.

1. If $r \neq 0$, $s \neq 0$, then $F(x) = x^{1/r}$, $G(x) = x^{1/s}$, whence $G^{-1} \circ F(x) = x^{s/r}$. If $0 < r < s$ or $r < 0 < s$, then G is increasing and $G^{-1} \circ F$ is convex, and if $r < s < 0$, then G is decreasing and $G^{-1} \circ F$ is concave, so at any case $\varphi(r) \leq \varphi(s)$.
2. $r = 0 < s$. Then $F(x) = e^x$, $G(x) = x^{1/s}$. The function $G^{-1} \circ F(x) = e^{sx}$ is convex and G is increasing, so $\varphi(0) \leq \varphi(s)$.
3. $r < 0 = s$. Then $F(x) = x^{1/r}$, $G(x) = e^x$. The function $G^{-1} \circ F(x) = \frac{1}{r} \log x$ is convex since $r < 0$, and the function G is increasing, so $\varphi(r) \leq \varphi(0)$. $\square$

We will prove one more generalization of the inequality between means (8.3.4) and (8.3.5). Let F be a strictly monotonic function on an interval $D \subset \mathbb{R}$, let $n \in \mathbb{N}$, let I be a subset of $\{1, \ldots, n\}$, and let $a_1, \ldots, a_n \in F(D)$, $w_1, \ldots, w_n \geq 0$, $\sum\limits_{i \in I} w_i > 0$.
Let $H : F(D) \to \mathbb{R}$ be a continuous function. Put

$$\alpha(H; F; \mathbf{a}; \mathbf{w}; I) = \left(\sum_{i \in I} w_i \right) H \circ F \left(\frac{\sum\limits_{i \in I} w_i F^{-1}(a_i)}{\sum\limits_{i \in I} w_i} \right).$$

With this notation, and under the conditions specified, we have

Theorem 8.3.3. *If the function F is continuous, the function $H \circ F$ is convex, and I, J are disjoint subsets of $\{1, \ldots, n\}$, then for fixed $\mathbf{a}$ and $\mathbf{w}$ the function $\alpha(F; I) = \alpha(H; F; \mathbf{a}; \mathbf{w}; I)$ fulfils*

$$\alpha(F; I \cup J) \leqslant \alpha(F; I) + \alpha(F; J). \tag{8.3.10}$$

If the function $H \circ F$ is concave, the inequality gets reversed.

Proof. Take in (8.1.1) $n = 2$, $f = H \circ F$, $q_1 = \sum\limits_{i \in I} w_i$, $q_2 = \sum\limits_{i \in J} w_i$, $x_1 = \sum\limits_{i \in I} w_i F^{-1}(a_i) /$ $/ \sum\limits_{i \in I} w_i$, $x_2 = \sum\limits_{i \in J} w_i F^{-1}(a_i) / \sum\limits_{i \in J} w_i$. Then

$$H \circ F \left(\frac{\sum\limits_{i \in I \cup J} w_i F^{-1}(a_i)}{\sum\limits_{i \in I \cup J} w_i} \right) \leqslant \frac{q_1 H \circ F(x_1) + q_2 H \circ F(x_2)}{\sum\limits_{i \in I \cup J} w_i}, \tag{8.3.11}$$

whence, after multiplying by $\sum\limits_{i \in I \cup J} w_i$ we obtain (8.3.10). If $H \circ F$ is concave, then we obtain (8.3.11) with $H \circ F$ replaced by $-H \circ F$, and the reversed inequality (8.3.10) results on multiplying by $-\sum\limits_{i \in I \cup J} w_i$. $\qquad \square$

Now suppose that $F, G : D \to \mathbb{R}$ are strictly monotonic continuous functions such that $F(D) \cap G(D) \neq \varnothing$, $H : F(D) \cup G(D) \to \mathbb{R}$ is continuous, $H \circ F$ is convex, $H \circ G$ is concave. By Theorem 8.3.3

$$\alpha(F; I \cup J) - \alpha(G; I \cup J) \leqslant \alpha(F; I) - \alpha(G; I) + \alpha(F; J) - \alpha(G; J). \tag{8.3.12}$$

Take in particular $I = \{1, \ldots, n-1\}$, $J = \{n\}$. Then $\alpha(F; J) = \alpha(G; J) = w_n H(a_n)$, and with notation (8.3.1) and $W_k = \sum\limits_{i=1}^{k} w_i$ we get from (8.3.12)

$$W_n \{ H(M_n[F; \mathbf{a}; \mathbf{w}]) - H(M_n[G; \mathbf{a}; \mathbf{w}]) \}$$
$$\leqslant W_{n-1} \{ H(M_{n-1}[F; \mathbf{a}; \mathbf{w}]) - H(M_{n-1}[G; \mathbf{a}; \mathbf{w}]) \}.$$

Specifying further $\mathbf{w} = \mathbf{1}$, $H(x) = G(x) = x$, $F(x) = e^x$, we get hence with notation (8.3.4), (8.3.5) $n(G_n - A_n) \leqslant (n-1)(G_{n-1} - A_{n-1})$, or

$$n(A_n - G_n) \geqslant (n-1)(A_{n-1} - G_{n-1}). \tag{8.3.13}$$

Since $A_1 = G_1$ for every $a_1 > 0$, inequality (8.3.13) implies that $A_n \geqslant G_n$ for $n \in \mathbb{N}$. From (8.3.13) one can obtain the inequality

$$A_n - G_n \geqslant \frac{1}{n} \max_{i,j=1,\ldots,n} \left(\sqrt{a_i} - \sqrt{a_j} \right)^2. \tag{8.3.14}$$

Inequalities (8.3.13), (8.3.14) are due to R. Rado[3]. Similarly one can derive the following inequalities of T. Popoviciu[4]

$$\left(\frac{A_n}{G_n}\right)^n \geqslant \left(\frac{A_{n-1}}{G_{n-1}}\right)^{n-1}, \quad \text{and} \quad \left(\frac{A_n}{G_n}\right) \geqslant \max_{i,j=1,\ldots,n}\left\{\frac{1}{2}\left(\sqrt{\frac{a_j}{a_i}}+\sqrt{\frac{a_i}{a_j}}\right)\right\}^{2/n}.$$

One can also show that if $\lim\limits_{n\to\infty} a_n = \infty$, then $\lim\limits_{n\to\infty} n\left(A_n - G_n\right) = \infty$; and if $\lim\limits_{n\to\infty} a_n = \alpha < \infty$, then $\lim\limits_{n\to\infty} n\left(A_n - G_n\right) < \infty$ if and only if $\sum\limits_{n=1}^{\infty} \left(a_n - \alpha\right)^2 < \infty$.

8.4 Hardy-Littlewood-Pólya majorization principle

The Hardy-Littlewood-Pólya majorization principle [5] reads as follows.

Theorem 8.4.1. *Let $J \subset \mathbb{R}$ be an interval[6], let $x_i, y_i \in J$, $i = 1, \ldots, n$, be real numbers such that*

$$x_1 \geqslant \cdots \geqslant x_n, \quad y_1 \geqslant \cdots \geqslant y_n, \tag{8.4.1}$$

$$\sum_{i=1}^{k} x_i \leqslant \sum_{i=1}^{k} y_i, \quad k = 1, \ldots, n-1, \quad \sum_{i=1}^{n} x_i = \sum_{i=1}^{n} y_i. \tag{8.4.2}$$

If $f : J \to \mathbb{R}$ is a continuous and convex function, then

$$\sum_{i=1}^{n} f\left(x_i\right) \leqslant \sum_{i=1}^{n} f\left(y_i\right). \tag{8.4.3}$$

Conversely, if for some $x_i, y_i \in J$, $i = 1, \ldots, n$, such that (8.4.1) holds, inequality (8.4.3) is fulfilled for every continuous and convex function, then relations (8.4.2) hold.

Theorem 8.4.1 is a particular case of the following theorem of L. Fuchs [93].

Theorem 8.4.2. *Let $J \subset \mathbb{R}$ be an interval, let $x_i, y_i \in J$, $i = 1, \ldots, n$, be real numbers fulfilling (8.4.1), and let $p_1, \ldots, p_n \in \mathbb{R}$ be such that*

$$\sum_{i=1}^{k} p_i x_i \leqslant \sum_{i=1}^{k} p_i y_i, \quad k = 1, \ldots, n-1, \quad \sum_{i=1}^{n} p_i x_i = \sum_{i=1}^{n} p_i y_i. \tag{8.4.4}$$

If $f : J \to \mathbb{R}$ is a continuous and convex function, then

$$\sum_{i=1}^{n} p_i f\left(x_i\right) \leqslant \sum_{i=1}^{n} p_i f\left(y_i\right). \tag{8.4.5}$$

---

[3] Cf. Mitrinović [226]
[4] Cf. Mitrinović [226]
[5] Hardy-Littlewood-Pólya [135]. Sometimes Theorem 8.4.1 is inscribed to J. Karamata. (Karamata [164]. Cf., e.g., Beckenbach-Bellman [24]).
[6] Here and in the sequel, if the interval J is not open, the expression "$f : J \to \mathbb{R}$ is a continuous and convex function" should be understood: "$f : J \to \mathbb{R}$ is continuous, and $f_0 = f \mid \mathrm{int}\, J$, $f_0 : \mathrm{int}\, J \to \mathbb{R}$ is a convex function".

Conversely, if $p_i \geqslant 0$ for $i = 1, \ldots, n$, and if for some $x_i, y_i \in J$, $i = 1, \ldots, n$, such that (8.4.1) holds, inequality (8.4.5) is fulfilled for every continuous and convex function $f : J \to \mathbb{R}$, then relations (8.4.4) hold.

The following theorem of M. Tomić [315] is very similar to Theorem 8.4.2.

Theorem 8.4.3. *Let $J \subset \mathbb{R}$ be an interval, let $x_i, y_i \in J$, $i = 1, \ldots, n$, be real numbers fulfilling (8.4.1), and let $p_1, \ldots, p_n \in \mathbb{R}$ be such that*

$$\sum_{i=1}^{k} p_i x_i \leqslant \sum_{i=1}^{k} p_i y_i, \quad k = 1, \ldots, n. \tag{8.4.6}$$

If $f : J \to \mathbb{R}$ is a continuous, convex and increasing function, then inequality (8.4.5) holds. Conversely, if $p_i \geqslant 0$ for $i = 1, \ldots, n$, and if for some $x_i, y_i \in J$, $i = 1, \ldots, n$, such that (8.4.1) holds inequality (8.4.5) is fulfilled for every continuous convex, and increasing function $f : J \to \mathbb{R}$, then relations (8.4.6) hold.

We will prove Theorems 8.4.2 and 8.4.3 simultaneously.

Proof. I. Sufficiency. Adding the same term to both sides of (8.4.5) does not affect the inequality. Therefore we may assume that $x_i \neq y_i$ for $i = 1, \ldots, n$.

Put
$$D_i = \frac{f(x_i) - f(y_i)}{x_i - y_i}, \quad i = 1, \ldots, n.$$

We have by Theorem 7.3.2 and by (8.4.1)

$$D_i = \frac{f(x_i) - f(y_i)}{x_i - y_i} \geqslant \frac{f(x_{i+1}) - f(y_i)}{x_{i+1} - y_i} \geqslant \frac{f(x_{i+1}) - f(y_{i+1})}{x_{i+1} - y_{i+1}} = D_{i+1}. \tag{8.4.7}$$

Further, put

$$X_0 = Y_0 = 0, \quad X_k = \sum_{i=1}^{k} p_i x_i, \quad Y_k = \sum_{i=1}^{k} p_i y_i, \quad k = 1, \ldots, n.$$

By (8.4.4) and (8.4.7), or by (8.4.6), (8.4.7) and the monotonicity of f,

$$\sum_{k=1}^{n-1} (X_k - Y_k)(D_k - D_{k+1}) + (X_n - Y_n) D_n \leqslant 0,$$

or

$$\sum_{k=1}^{n-1} X_k (D_k - D_{k+1}) + X_n D_n \leqslant \sum_{k=1}^{n-1} Y_k (D_k - D_{k+1}) + Y_n D_n. \tag{8.4.8}$$

But

$$\sum_{k=1}^{n-1} X_k (D_k - D_{k+1}) + X_n D_n = \sum_{k=1}^{n} X_k D_k - \sum_{k=1}^{n-1} X_k D_{k+1} = \sum_{k=1}^{n} X_k D_k - \sum_{k=1}^{n} X_{k-1} D_k$$

$$= \sum_{k=1}^{n} (X_k - X_{k-1}) D_k = \sum_{k=1}^{n} p_k x_k D_k, \tag{8.4.9}$$

and similarly

$$\sum_{k=1}^{n-1} Y_k (D_k - D_{k+1}) + Y_n D_n = \sum_{k=1}^{n} p_k y_k D_k. \tag{8.4.10}$$

Thus we get from (8.4.8), (8.4.9) and (8.4.10)

$$\sum_{k=1}^{n} p_k x_k D_k \leqslant \sum_{k=1}^{n} p_k y_k D_k, \quad \text{i.e.,} \quad \sum_{k=1}^{n} p_k (x_k - y_k) D_k \leqslant 0,$$

which is equivalent to (8.4.5).

II. Necessity. Suppose that $p_i \geqslant 0$ for $i = 1, \ldots, n$, and that (8.4.5) holds for some $x_i, y_i \in J$, $i = 1, \ldots, n$, and for every continuous and convex [and increasing] function $f : J \to \mathbb{R}$. Taking $f(x) = x$ we obtain from (8.4.5)

$$\sum_{i=1}^{n} p_i x_i \leqslant \sum_{i=1}^{n} p_i y_i, \tag{8.4.11}$$

i.e., (8.4.6) for $k = n$. Moreover, if there is no restriction on the monotonicity of f, we take $f(x) = -x$ to obtain

$$-\sum_{i=1}^{n} p_i x_i \leqslant -\sum_{i=1}^{n} p_i y_i. \tag{8.4.12}$$

Relations (8.4.11) and (8.4.12) together yield the last equality in (8.4.4).

Now take an arbitrary k, $1 \leqslant k < n$, and $f(x) = \max(0, x - y_k)$. Thus $f(x) \geqslant 0$, $f(x) \geqslant x - y_k$, and, since $p_i \geqslant 0$ $(i = 1, \ldots, n)$, we have

$$p_i f(x) \geqslant p_i (x - y_k). \tag{8.4.13}$$

Putting in (8.4.13) $x = x_i$, $i = 1, \ldots, k$, and summing up over $i = 1, \ldots, k$, we get by (8.4.5)

$$\sum_{i=1}^{k} p_i (x_i - y_k) \leqslant \sum_{i=1}^{k} p_i f(x_i) \leqslant \sum_{i=1}^{n} p_i f(x_i) \leqslant \sum_{i=1}^{n} p_i f(y_i). \tag{8.4.14}$$

Now, by (8.4.1), since $f(x) = 0$ for $x \leqslant y_k$,

$$\sum_{i=1}^{n} p_i f(y_i) = \sum_{i=1}^{k} p_i f(y_i) = \sum_{i=1}^{k} p_i (y_i - y_k). \tag{8.4.15}$$

Relations (8.4.14) and (8.4.15) yield

$$\sum_{i=1}^{k} p_i (x_i - y_k) \leqslant \sum_{i=1}^{k} p_i (y_i - y_k), \quad k = 1, \ldots, n-1,$$

whence the first $n - 1$ inequalities in (8.4.4) follow. $\qquad\square$

8.5 Lim's inequality

From Theorem 8.4.1 we derive the following result:

Theorem 8.5.1. *Let $a \geqslant 0, b \geqslant 0, c \geqslant a + b$ be real numbers, and let $f : [0, \infty) \to \mathbb{R}$ be a continuous and convex function. Then*

$$f(a) + f(b + c) \geqslant f(a + b) + f(c). \tag{8.5.1}$$

Proof. We use Theorem 8.4.1 with $n = 2$, $x_1 = c$, $x_2 = a + b$, $y_1 = b + c$, $y_2 = a$, and (8.5.1) follows. $\square$

Lim's inequality (Lim [207]) is the special case $f(x) = x^r$, $r \geqslant 1$.

Lim's inequality can also be generalized as follows (cf. Stanković-Lacković [294]).

Theorem 8.5.2. *Let $m \in \mathbb{N}$, and let $Q_i \geqslant 0$, $i = 1, \dots, m$, be real numbers. Let $Q \geqslant \max\limits_{i=1,\dots,m} Q_i$ and $Q_0 = \sum\limits_{i=1}^{m} Q_i - nQ$ for a certain non-negative integer n. If $0 \leqslant Q_0 \leqslant Q$, and $f : [0, \infty) \to \mathbb{R}$ is a continuous and convex function such that $f(0) = 0$, then*

$$nf(Q) + f(Q_0) \geqslant \sum_{i=1}^{m} f(Q_i). \tag{8.5.2}$$

Proof. We assume that $Q_0 > 0$. The case, where $Q_0 = 0$, will be covered by Theorem 8.5.3. Then

$$nQ < nQ + Q_0 = \sum_{i=1}^{m} Q_i \leqslant mQ,$$

whence $n < m$. We may assume that Q_i's are numbered in such a way that

$$Q_1 \geqslant \cdots \geqslant Q_m,$$

and we put in Theorem 8.4.1 (with n replaced by m) $y_1 = \cdots = y_n = Q$, $y_{n+1} = Q_0$, $y_i = 0$ for $i = n + 2, \dots, m$; $x_i = Q_i$, $i = 1, \dots, m$. Then

$$nf(Q) + f(Q_0) = \sum_{i=1}^{m} f(y_i) \geqslant \sum_{i=1}^{m} f(x_i) = \sum_{i=1}^{m} f(Q_i),$$

i.e., we obtain (8.5.2). $\square$

In order to obtain (8.5.1) from (8.5.2) (in such an argument f must be assumed to vanish at the origin) we put $m = 2$, $n = 1$, $Q_1 = a + b$, $Q_2 = c$, $Q = b + c$, $Q_0 = a$.

Remark. If n is not an integer, Theorem 8.5.2 fails to hold. Take, for example, $m = 2$, $n = \frac{3}{2}$, $Q_1 = Q_2 = Q = 1$, $Q_0 = \frac{1}{2}$, $f(x) = x^2$. Then all the remaining assumptions of Theorem 8.5.2 are fulfilled, and (8.5.2) turns into the false inequality

$$\frac{3}{2} + \frac{1}{4} \geqslant 2.$$

But for such a case we have the following

Theorem 8.5.3. *Let $m \in \mathbb{N}$, and let $Q_i \geqslant 0$, $i = 1, \ldots, m$, be real numbers. Let $Q \geqslant \max\limits_{i=1,\ldots,m} Q_i$, and let $s \in \mathbb{R}$ be such that $sQ = \sum\limits_{i=1}^{m} Q_i$. If $f : [0,\infty) \to \mathbb{R}$ is a continuous and convex function such that $f(0) = 0$, then*

$$sf(Q) \geqslant \sum_{i=1}^{m} f(Q_i). \tag{8.5.3}$$

Proof. If $Q = 0$, then all Q_i's are zeros, and (8.5.3) becomes $0 \geqslant 0$, which is true. Now assume that $Q > 0$. For every $\lambda \in [0, 1]$ we have

$$f(\lambda x) = f(\lambda x + (1 - \lambda) 0) \leqslant \lambda f(x) + (1 - \lambda) f(0) = \lambda f(x).$$

Hence

$$f(Q_i) = f\left(\frac{Q_i}{Q} Q\right) \leqslant \frac{Q_i}{Q} f(Q), \quad i = 1, \ldots, m,$$

and

$$\sum_{i=1}^{m} f(Q_i) \leqslant \frac{1}{Q} \left(\sum_{i=1}^{m} Q_i\right) f(Q) = sf(Q),$$

i.e., we obtain (8.5.3). □

8.6 Hadamard inequality

We start with the following, more general theorem of Lacković-Stanković [203].

Theorem 8.6.1. *Let $J = [a, b]$ be a closed real interval, let $p : J \to \mathbb{R}$ be a non-negative integrable function such that $\int_a^b p\,dx > 0$, let $f : J \to \mathbb{R}$ be a continuous and convex function, let $c \in \mathrm{int}\, J$, and let $k \in \partial f(c)$. Then*

$$f(c) + k\frac{\int_a^b xp(x)\,dx}{\int_a^b p(x)\,dx} - ck \leqslant \frac{\int_a^b p(x) f(x)\,dx}{\int_a^b p(x)\,dx}$$

$$\leqslant \frac{f(a)\int_a^c p(x)\,dx + f(b)\int_c^b p(x)\,dx + k\int_a^b xp(x)\,dx - k\left[a\int_a^c p(x)\,dx + b\int_c^b p(x)\,dx\right]}{\int_a^b p(x)\,dx}.$$

$$\tag{8.6.1}$$

Proof. Put

$$g(x) = f(x) - kx, \quad x \in J. \tag{8.6.2}$$

Then $g : J \to \mathbb{R}$ is a continuous and convex function, $0 \in \partial g\left(c\right)$, and thus $g\left(x\right) - g\left(c\right) \geqslant 0$ in J. Since $p \geqslant 0$ we obtain hence

$$\int_a^b p\left(x\right) g\left(x\right) dx \geqslant \int_a^b p\left(x\right) g\left(c\right) dx. \tag{8.6.3}$$

Relations (8.6.3) and (8.6.2) yield

$$\int_a^b p\left(x\right) f\left(x\right) dx - k \int_a^b x p\left(x\right) dx \geqslant f\left(c\right) \int_a^b p\left(x\right) dx - kc \int_a^b p\left(x\right) dx.$$

Dividing by $\int_a^b p\left(x\right) dx > 0$ we obtain hence the left inequality in (8.6.1).

It follows from Theorem 7.3.5 that $g\left(x\right)$ decreases in $[a, c]$ and increases in $[c, b]$ so that $g\left(a\right) - g\left(x\right) \geqslant 0$ in $[a, c]$, and $g\left(b\right) - g\left(x\right) \geqslant 0$ in $[c, b]$. Hence

$$\int_a^b p\left(x\right) g\left(a\right) dx \geqslant \int_a^c p\left(x\right) g\left(x\right) dx, \quad \int_c^b p\left(x\right) g\left(b\right) dx \geqslant \int_c^b p\left(x\right) g\left(x\right) dx,$$

and thus

$$\int_a^c p\left(x\right) g\left(a\right) dx + \int_c^b p\left(x\right) g\left(b\right) dx \geqslant \int_a^b p\left(x\right) g\left(x\right) dx.$$

By (8.6.2) we get hence

$$f\left(a\right) \int_a^c p\left(x\right) dx - ka \int_a^c p\left(x\right) dx + f\left(b\right) \int_c^b p\left(x\right) dx - kb \int_c^b p\left(x\right) dx$$

$$\geqslant \int_a^b p\left(x\right) f\left(x\right) dx - k \int_a^b x p\left(x\right) dx.$$

Dividing by $\int_a^b p\left(x\right) dx$ we obtain hence the right inequality in (8.6.1). □

Setting $c = \dfrac{1}{2}\left(a + b\right)$, $p\left(x\right) = 1$ for $x \in J$, we obtain from (8.6.1) the Hadamard inequality (Hadamard [127]; cf. also Vasić-Lacković [316]).

Theorem 8.6.2. *Let* $J = [a, b]$ *be a closed real interval, and let* $f : J \to \mathbb{R}$ *be a continuous and convex function. Then*

$$f\left(\frac{a + b}{2}\right) \leqslant \frac{1}{b - a} \int_a^b f\left(x\right) dx \leqslant \frac{f\left(a\right) + f\left(b\right)}{2}.$$

8.7 Petrović inequality

The following inequality is due to M. Petrović [253].

Theorem 8.7.1. *Let $[0, a) \subset \mathbb{R}$, $0 < a \leqslant \infty$, and let $f : [0, a) \to \mathbb{R}$ be a continuous and convex function. Then for every $n \in \mathbb{N}$ and every $x_1, \ldots, x_n \in [0, a)$ such that $x_1 + \cdots + x_n \in [0, a)$ we have*

$$f(x_1 + \cdots + x_n) \geqslant f(x_1) + \cdots + f(x_n) - (n-1)f(0). \qquad (8.7.1)$$

Proof I. For $n = 2$ we obtain from (8.1.1)

$$f\left(\frac{px + qy}{p+q}\right) \leqslant \frac{p}{p+q}f(x) + \frac{q}{p+q}f(y) \qquad (8.7.2)$$

for every $p, q \geqslant 0$ such that $p + q > 0$. Assume that $x_1, x_2 \in [0, a)$ are such that $x_1 + x_2 \in (0, a)$, and set in (8.7.2) $x = x_1 + x_2$, $y = 0$, $p = x_1$, $q = x_2$. Then (8.7.2) goes into

$$f(x_1) \leqslant \frac{x_1}{x_1 + x_2}f(x_1 + x_2) + \frac{x_2}{x_1 + x_2}f(0). \qquad (8.7.3)$$

Interchanging in (8.7.3) x_1 and x_2, we obtain

$$f(x_2) \leqslant \frac{x_2}{x_1 + x_2}f(x_1 + x_2) + \frac{x_1}{x_1 + x_2}f(0). \qquad (8.7.4)$$

Adding (8.7.3) and (8.7.4) we get

$$f(x_1) + f(x_2) \leqslant f(x_1 + x_2) + f(0), \qquad (8.7.5)$$

which is (8.7.1) for $n = 2$. If $x_1 = x_2 = 0$, (8.7.5) is trivial.

Further the proof runs by induction. $\qquad \square$

We will give also another proof of Theorem 8.7.1.

Proof II. Put

$$g(x_1, \ldots, x_n) = f(x_1 + \cdots + x_n) - f(x_1) - \cdots - f(x_n). \qquad (8.7.6)$$

Fix a j, $1 \leqslant j \leqslant n$. Differentiating (8.7.6) with respect to x_j, we get by Theorem 7.4.3

$$\frac{\partial g}{\partial x_j}(x_1, \ldots, x_n) = f'(x_1 + \cdots + x_n) - f'(x_j) \geqslant 0$$

except for at most countably many values of x_j. By Theorem 3.6.3 g is increasing with respect to each variable. Hence, in particular, for $x_1, \ldots, x_n \in [0, a)$ such that $x_1 + \cdots + x_n \in [0, a)$

$$g(x_1, \ldots, x_n) \geqslant g(0, \ldots, 0) = (1 - n)f(0). \qquad (8.7.7)$$

Relations (8.7.6) and (8.7.7) yield inequality (8.7.1). $\qquad \square$

8.8   Mulholland's inequality

The results in this section are due to H. P. Mulholland [234].

Consider the two sets of conditions:

(i) The function $\varphi : [0, \infty) \to [0, \infty)$ is continuous, strictly increasing, convex, $\varphi(0) = 0$, and $\log \varphi$ is a convex function of $\log x$ (i.e., the function $F : \mathbb{R} \to \mathbb{R}$, $F(t) = \log \varphi(e^t)$ is convex).

(ii) There exists a function $\gamma : \mathbb{R} \to \mathbb{R}$, continuous, convex and increasing and such that
$$\varphi(x) = x \exp\left[\gamma(\log x)\right] \text{ for } x > 0, \quad \varphi(0) = 0. \tag{8.8.1}$$

Lemma 8.8.1. *Conditions (i) and (ii) are equivalent.*

Proof. I. Assume (i) to be fulfilled. Put
$$\gamma(t) = \log \frac{\varphi(e^t)}{e^t} = \log \varphi(e^t) - t. \tag{8.8.2}$$

It is clear that γ is defined by (8.8.2) for $t \in \mathbb{R}$, and it is continuous and convex. For $x > 0$ the function
$$\frac{\varphi(x)}{x} = \frac{\varphi(x) - \varphi(0)}{x - 0}$$

is increasing in virtue of Theorem 7.3.2, and thus $\varphi(e^t)/e^t$, and hence also $\gamma(t)$ is an increasing function of t. Formula (8.8.1) results from (8.8.2).

II. Assume (ii) to be fulfilled. Then we have (8.8.1) with γ given by (8.8.2). Hence
$$F(t) = \log \varphi(e^t) = \gamma(t) + t$$

is a convex function. Moreover, since γ is increasing, we have for $0 < x < 1$ the inequality $\gamma(\log x) \leqslant \gamma(0)$, whence by (8.8.1)
$$0 < \varphi(x) \leqslant x \exp \gamma(0),$$

whence $\lim_{x \to 0} \varphi(x) = 0$. Thus if we put $\varphi(0) = 0$, φ will be defined and continuous in $[0, \infty)$, since γ is continuous in $\mathbb{R}$. It follows from (8.8.1) that φ is strictly increasing. Further, we have for $x > 0$
$$D^+\varphi(x) = e^{\gamma(\log x)} + x e^{\gamma(\log x)} \gamma'_+(\log x) \frac{1}{x},$$

and thus $D^+\varphi$ is an increasing function. By Theorem 7.5.3 φ is convex.

$\square$

Lemma 8.8.2. *Let φ satisfy condition (i), and put $f(x) = \varphi(x)/x$, $x > 0$. Let $n \in \mathbb{N}$, and let $s_1, \ldots, s_n$ be positive real numbers. Put*
$$Q = \left\{ x = (x_1, \ldots, x_n) \in \mathbb{R}^N \mid 0 \leqslant x_i \leqslant s_i, i = 1, \ldots, n \right\} \setminus \{(0, \ldots, 0)\},$$

and define the function $\Phi : Q \to \mathbb{R}$ by

$$\Phi(x_1, \ldots, x_n) = \sum_{i=1}^{n} x_i f(s_i) \big/ \varphi^{-1}\left(\sum_{i=1}^{n} \varphi(x_i)\right).$$

Then Φ admits the maximum at the point $(s_1, \ldots, s_n)$.

Proof. By Theorem 7.3.2 f is an increasing function. We have on suitable subsets of Q (where the derivative in question exists), and for every $j = 1, \ldots, n$,

$$\frac{\partial}{\partial x_j} \log \Phi = \frac{\partial}{\partial x_j} \log \sum_{i=1}^{n} x_i f(s_i) - \frac{\partial}{\partial x_j} \log \psi\left(\sum_{i=1}^{n} \varphi(x_i)\right)$$

$$= \frac{f(s_j)}{\sum\limits_{i=1}^{n} x_i f(s_i)} - \frac{\partial}{\partial x_j} \log \psi\left(\sum_{i=1}^{n} \varphi(x_i)\right),$$

where $\psi = \varphi^{-1}$ is concave in virtue of Theorem 7.2.2. For the function $F(t) = \log \varphi(e^t)$ we have $F^{-1}(t) = \log \psi(e^t)$, and again by Theorem 7.2.2, since F is convex and increasing, F^{-1} is concave, and thus its derivative $(F^{-1})'$ is decreasing on the set where it exists. But

$$(F^{-1})'(\log u) = \frac{\psi'(u)}{\psi(u)} u.$$

Consequently

$$\frac{\partial}{\partial x_j} \log \psi\left(\sum_{i=1}^{n} \varphi(x_i)\right) = \frac{\psi'\left(\sum\limits_{i=1}^{n} \varphi(x_i)\right)}{\psi\left(\sum\limits_{i=1}^{n} \varphi(x_i)\right)} \varphi'(x_j)$$

$$= \frac{\psi'\left(\sum\limits_{i=1}^{n} \varphi(x_i)\right)}{\psi\left(\sum\limits_{i=1}^{n} \varphi(x_i)\right)} \left(\sum_{i=1}^{n} \varphi(x_i)\right) \frac{\varphi'(x_j)}{\sum\limits_{i=1}^{n} \varphi(x_i)}$$

$$\leqslant \frac{\psi'(\varphi(x_j))}{\psi(\varphi(x_j))} \varphi(x_j) \frac{\varphi'(x_j)}{\sum\limits_{i=1}^{n} \varphi(x_i)} = \frac{\varphi(x_j)}{x_j \sum\limits_{i=1}^{n} \varphi(x_i)} = \frac{f(x_j)}{\sum\limits_{i=1}^{n} x_i f(x_i)}.$$

Thus

$$\frac{\partial}{\partial x_j} \log \Phi \geqslant \frac{f(s_j)}{\sum\limits_{i=1}^{n} x_i f(s_i)} - \frac{f(x_j)}{\sum\limits_{i=1}^{n} x_i f(x_i)} = \frac{\sum\limits_{i=1}^{n} x_i f(x_i) f(s_j) - \sum\limits_{i=1}^{n} x_i f(s_i) f(x_j)}{\left(\sum\limits_{i=1}^{n} x_i f(s_i)\right)\left(\sum\limits_{i=1}^{n} x_i f(x_i)\right)}.$$

Let

$$x_i = x_i(t), \quad i = 1, \ldots, n, \tag{8.8.3}$$

be a regular curve passing inside Q. Then

$$\frac{d}{dt} \log \Phi \left(x_1 \left(t\right), \ldots, x_n \left(t\right)\right) = \sum_{j=1}^{n} \left(\frac{\partial}{\partial x_j} \log \Phi\right) x_j'$$

$$\geqslant \frac{\displaystyle\sum_{j=1}^{n} \sum_{i=1}^{n} \left(x_i f \left(x_i\right) f \left(s_j\right) - x_i f \left(s_i\right) f \left(x_j\right)\right) x_j'}{\left(\displaystyle\sum_{i=1}^{n} x_i f \left(s_i\right)\right) \left(\displaystyle\sum_{i=1}^{n} x_i f \left(x_i\right)\right)}.$$

We have

$$\sum_{j=1}^{n} \sum_{i=1}^{n} \left(x_i f \left(x_i\right) f \left(s_j\right) - x_i f \left(s_i\right) f \left(x_j\right)\right) x_j'$$

$$= \sum_{1 \leqslant i \leqslant j \leqslant n} \left(x_i f \left(x_i\right) f \left(s_j\right) x_j' - x_i f \left(s_i\right) f \left(x_j\right) x_j' + x_j f \left(x_j\right) f \left(s_i\right) x_i' - x_j f \left(s_j\right) f \left(x_i\right) x_i'\right)$$

$$= \sum_{1 \leqslant i \leqslant j \leqslant n} \left(f \left(x_i\right) f \left(s_j\right) - f \left(x_j\right) f \left(s_i\right)\right) \left(x_i x_j' - x_j x_i'\right) = \sum_{1 \leqslant i \leqslant j \leqslant n} A_{ij},$$

where

$$A_{ij} = \left(f \left(x_i\right) f \left(s_j\right) - f \left(x_j\right) f \left(s_i\right)\right) \left(x_i x_j' - x_j x_i'\right), \quad i, j = 1, \ldots, n.$$

Now take the functions $x_i \left(t\right)$ of the special form $x_i \left(t\right) = k_i t$, $k_i \geqslant 0$, or $x_i \left(t\right) = s_i$. In order to evaluate the sign of A_{ij}, we must distinguish four cases.

1. $x_i \left(t\right) = k_i t$, $x_j \left(t\right) = k_j t$. Then $x_i x_j' = x_j x_i' = k_i k_j t$, whence $A_{ij} = 0$.
2. $x_i \left(t\right) = s_i$, $x_j \left(t\right) = s_j$. Then $x_i' = x_j' = 0$, and $A_{ij} = 0$.
3. $x_i \left(t\right) = k_i t$, $x_j \left(t\right) = s_j$. Then

$$A_{ij} = f \left(s_j\right) \left(f \left(x_i\right) - f \left(s_i\right)\right) \left(-k_i s_j\right) \geqslant 0.$$

(Note that $x_i \leqslant s_i$ and f is increasing.)

4. $x_i \left(t\right) = s_i$, $x_j \left(t\right) = k_j t$. Then

$$A_{ij} = f \left(s_i\right) \left(f \left(s_j\right) - f \left(x_j\right)\right) s_i k_j \geqslant 0.$$

Ultimately $\frac{d}{dt} \log \Phi\left(x_1(t), \ldots, x_n(t)\right)$ is non-negative whenever it exists, and we could carry out this argument for $D^+ \log \Phi\left(x_1(t), \ldots, x_n(t)\right)$. By Theorem 3.6.3 $\log \Phi$, and hence also Φ, is increasing along every line (8.8.3), where every $x_i(t)$ is either $k_i t$, $k_i \geqslant 0$, or s_i. Since every point in Q different from $(s_1, \ldots, s_n)$ can be joint with $(s_1, \ldots, s_n)$ by segments of lines of this form, the value of Φ at $(s_1, \ldots, s_n)$ exceeds or equals any other value of Φ in Q. $\square$

Theorem 8.8.1. *Let a function $\varphi : [0, \infty) \rightarrow [0, \infty)$ fulfil condition (i), let $n \in \mathbb{N}$, let a_i, b_i, $i = 1, \ldots, n$, be non-negative real numbers. Then*

$$\varphi^{-1} \left(\sum_{i=1}^{n} \varphi \left(a_i + b_i \right) \right) \leqslant \varphi^{-1} \left(\sum_{i=1}^{n} \varphi \left(a_i \right) \right) + \varphi^{-1} \left(\sum_{i=1}^{n} \varphi \left(b_i \right) \right). \tag{8.8.4}$$

Proof. It is enough to prove (8.8.4) for $a_i > 0$, $b_i > 0$, $i = 1, \ldots, n$. By the continuity of φ (8.8.4) then results for arbitrary non-negative a_i, b_i, $i = 1, \ldots, n$.

Put in Lemma 8.8.2 $x_i = a_i$, $s_i = a_i + b_i$, $i = 1, \ldots, n$. Then

$$\frac{\sum_{i=1}^{n} a_i f \left(a_i + b_i \right)}{\varphi^{-1} \left(\sum_{i=1}^{n} \varphi \left(a_i \right) \right)} \leqslant \frac{\sum_{i=1}^{n} \left(a_i + b_i \right) f \left(a_i + b_i \right)}{\varphi^{-1} \left(\sum_{i=1}^{n} \varphi \left(a_i + b_i \right) \right)}. \tag{8.8.5}$$

Similarly, putting $x_i = b_i$, $s_i = a_i + b_i$, $i = 1, \ldots, n$, we get

$$\frac{\sum_{i=1}^{n} b_i f \left(a_i + b_i \right)}{\varphi^{-1} \left(\sum_{i=1}^{n} \varphi \left(b_i \right) \right)} \leqslant \frac{\sum_{i=1}^{n} \left(a_i + b_i \right) f \left(a_i + b_i \right)}{\varphi^{-1} \left(\sum_{i=1}^{n} \varphi \left(a_i + b_i \right) \right)}. \tag{8.8.6}$$

Multiplying (8.8.5) by $\varphi^{-1} \left(\sum_{i=1}^{n} \varphi \left(a_i \right) \right)$ and (8.8.6) by $\varphi^{-1} \left(\sum_{i=1}^{n} \varphi \left(b_i \right) \right)$, and adding together the resulting expressions, we obtain

$$\sum_{i=1}^{n} \left(a_i + b_i \right) f \left(a_i + b_i \right) \leqslant \frac{\sum_{i=1}^{n} \left(a_i + b_i \right) f \left(a_i + b_i \right)}{\varphi^{-1} \left(\sum_{i=1}^{n} \varphi \left(a_i + b_i \right) \right)} \left[\varphi^{-1} \left(\sum_{i=1}^{n} \varphi \left(a_i \right) \right) + \varphi^{-1} \left(\sum_{i=1}^{n} \varphi \left(b_i \right) \right) \right],$$

whence (8.8.4) results. $\square$

Taking in Theorem 8.8.1 $\varphi \left(x \right) = x^r$, $r \geqslant 1$, we obtain

Corollary 8.8.1. *Let $n \in \mathbb{N}$, let a_i, b_i, $i = 1, \ldots, n$, be non-negative real numbers, and let $r \geqslant 1$. Then*

$$\sqrt[r]{\sum_{i=1}^{n} \left(a_i + b_i \right)^r} \leqslant \sqrt[r]{\sum_{i=1}^{n} a_i^r} + \sqrt[r]{\sum_{i=1}^{n} b_i^r}. \tag{8.8.7}$$

Inequality (8.8.7) is the famous Minkowski inequality.

Theorem 8.8.2. *Let a function $\varphi : [0, \infty) \rightarrow [0, \infty)$ fulfil condition (i), and let the function $f \left(x \right) = \varphi \left(x \right) / x$ be strictly increasing. Put $g = f^{-1}$ and $\psi \left(x \right) = x g \left(x \right)$ for $x > 0$, $\psi \left(0 \right) = 0$. Let $n \in \mathbb{N}$, and let a_i, b_i, $i = 1, \ldots, n$, be non-negative real numbers such that*

$$0 \leqslant \varphi \left(a_i \right) \leqslant \psi \left(b_i \right), \quad i = 1, \ldots, n. \tag{8.8.8}$$

Then

$$\sum_{i=1}^{n} a_i b_i \leqslant \varphi^{-1}\left(\sum_{i=1}^{n} \varphi(a_i)\right) \psi^{-1}\left(\sum_{i=1}^{n} \psi(b_i)\right). \tag{8.8.9}$$

Proof. Taking in Lemma 8.8.2 $x_i = a_i$, $s_i = \varphi^{-1}\big(\psi(b_i)\big)$, $i = 1, \ldots, n$ (cf. (8.8.8)), we have

$$\frac{\sum_{i=1}^{n} a_i f(s_i)}{\varphi^{-1}\left(\sum_{i=1}^{n} \varphi(a_i)\right)} \leqslant \frac{\sum_{i=1}^{n} s_i f(s_i)}{\varphi^{-1}\left(\sum_{i=1}^{n} \varphi(s_i)\right)}. \tag{8.8.10}$$

Put

$$A = \sum_{i=1}^{n} \varphi(s_i) = \sum_{i=1}^{n} \psi(b_i), \quad B = \frac{\sum_{i=1}^{n} s_i f(s_i)}{\varphi^{-1}\left(\sum_{i=1}^{n} \varphi(s_i)\right)} = \frac{\sum_{i=1}^{n} \varphi(s_i)}{\varphi^{-1}\left(\sum_{i=1}^{n} \varphi(s_i)\right)} = \frac{A}{\varphi^{-1}(A)}. \tag{8.8.11}$$

Then

$$\varphi^{-1}(A) = A/B,$$

whence

$$A = \varphi\left(\frac{A}{B}\right) = \frac{A}{B} f\left(\frac{A}{B}\right), \quad \text{and} \quad f\left(\frac{A}{B}\right) = B,$$

i.e.,

$$\frac{A}{B} = f^{-1}(B) = g(B), \quad \text{and} \quad \psi(B) = Bg(B) = A.$$

Hence

$$B = \psi^{-1}(A) = \psi^{-1}\left(\sum_{i=1}^{n} \psi(b_i)\right). \tag{8.8.12}$$

Relations (8.8.10), (8.8.11) and (8.8.12) yield

$$\sum_{i=1}^{n} a_i f(s_i) \leqslant \varphi^{-1}\left(\sum_{i=1}^{n} \varphi(a_i)\right) \psi^{-1}\left(\sum_{i=1}^{n} \psi(b_i)\right). \tag{8.8.13}$$

Now,

$$\varphi(s_i) = \psi(b_i), \quad i = 1, \ldots, n. \tag{8.8.14}$$

On the other hand,

$$\varphi(s_i) = s_i f(s_i) = g[f(s_i)] f(s_i) = \psi[f(s_i)], \quad i = 1, \ldots, n. \tag{8.8.15}$$

Relations (8.8.14) and (8.8.15) yield

$$\psi(b_i) = \psi[f(s_i)], \quad i = 1, \ldots, n,$$

and, since ψ is one-to-one,

$$b_i = f(s_i), \quad i = 1, \ldots, n. \tag{8.8.16}$$

(8.8.16) inserted into (8.8.13) yield (8.8.9).   □

8.9 The general inequality of convexity

Let E be an arbitrary set, and let P be a family of functions $f : E \to [0, \infty)$ such that $f_1 + f_2 \in P$ and $\lambda f \in P$ whenever $f, f_1, f_2 \in P$ and $\lambda \geqslant 0$ (thus P is a cone in the space of all functions $f : E \to \mathbb{R}$). Let $M : P \to [0, \infty]$ be a function such that $M(0) = 0$ and

$$M(\lambda f) = \lambda M(f) \quad \text{for } \lambda > 0, f \in P, \tag{8.9.1}$$

$$M(f + g) \leqslant M(f) + M(g) \quad \text{for } f, g \in P, \tag{8.9.2}$$

$$f \leqslant g \Rightarrow M(f) \leqslant M(g) \quad \text{for } f, g \in P, \tag{8.9.3}$$

where $f \leqslant g$ means $f(x) \leqslant g(x)$ for all $x \in E$. Thus M is positively homogeneous, subadditive, and increasing. E.g., if $(E, \mathfrak{M}, \mu)$ is a measure space, we may take P to be the family of all integrable functions $f : E \to [0, \infty)$, and $M(f) = \int_E f d\mu$. Another example: $E = \{1, \ldots, N\}$ a finite set, $P = \mathbb{R}_+^N$, where

$$\mathbb{R}_+^N = \left\{ x = (x_1, \ldots, x_N) \in \mathbb{R}^N \mid x_1 \geqslant 0, \ldots, x_N \geqslant 0 \right\},$$

and M any standard norm in $\mathbb{R}^N$.

Theorem 8.9.1. Let $\varphi : \mathbb{R}_+^N \to [0, \infty)$ be a continuous and positively homogeneous function such that $\varphi(x_1, \ldots, x_N) > 0$ whenever $x_1 > 0, \ldots, x_N > 0$, and the set

$$K = \left\{ x = (x_1, \ldots, x_N) \in \mathbb{R}_+^N \mid \varphi(x_1, \ldots, x_N) \geqslant 1 \right\}$$

is convex. If $f_1, \ldots, f_N \in P$ are such that $M(f_i) < \infty$ for $i = 1, \ldots, N$, and $\varphi(f_1, \ldots, f_N) \in P$, then

$$M(\varphi(f_1, \ldots, f_N)) \leqslant \varphi(M(f_1), \ldots, M(f_N)). \tag{8.9.4}$$

Proof. By Theorem 5.1.8 the set K is equal to the intersection of all the closed half-spaces containing K, determined by all the support hyperplanes of K. Thus K is the set of points $(x_1, \ldots, x_N) \in \mathbb{R}^N$ such that

$$\alpha_1^\tau x_1 + \cdots + \alpha_N^\tau x_N - \beta_\tau \geqslant 0$$

for all τ from a certain set T of indices. If $x_1 > 0, \ldots, x_N > 0$, then

$$\varphi(\lambda x_1, \ldots, \lambda x_N) = \lambda \varphi(x_1, \ldots, x_N) \geqslant 1$$

for λ sufficiently large, and thus $(\lambda x_1, \ldots, \lambda x_N) \in K$ for λ sufficiently large. In other words,

$$\lambda(\alpha_1^\tau x_1 + \cdots + \alpha_N^\tau x_N) - \beta_\tau \geqslant 0$$

for λ sufficiently large, whence $\alpha_i^\tau \geqslant 0$ for $i = 1, \ldots, N$ and all $\tau \in T$. If for a $\tau \in T$ we had $\beta_\tau \leqslant 0$, then the point $0 = (0, \ldots, 0)$ would lie on the suitable support hyperplane of K. If we draw a line from 0, passing through a point $x = (x_1, \ldots, x_N)$ such that $x_1 > 0, \ldots, x_N > 0$ (note that $K \subset \mathbb{R}_+^N$) and crossing the support hyperplane in question, then for large λ the points λx would lie on the other side of the hyperplane, and thus do not belong to K, which is impossible. Consequently $\beta_\tau > 0$ for all $\tau \in T$.

Let

$$C = \left\{ (x_1, \ldots, x_N, x_{N+1}) \in \mathbb{R}_+^{N+1} \mid x_{N+1} \leqslant \varphi(x_1, \ldots, x_N) \right\}.$$

If $x_{N+1} = 0$, then $(x_1, \ldots, x_N, x_{N+1}) \in C$ for every $(x_1, \ldots, x_N) \in \mathbb{R}_+^N$, and also

$$0 = \beta_\tau x_{N+1} \leqslant \alpha_1^\tau x_1 + \cdots + \alpha_N^\tau x_N$$

for every $(x_1, \ldots, x_N) \in \mathbb{R}_+^N$. If $x_{N+1} > 0$, then, by the positive homogeneity of φ, the inequality $x_{N+1} \leqslant \varphi(x_1, \ldots, x_N)$ is equivalent to $1 \leqslant \varphi\left(\dfrac{x_1}{x_{N+1}}, \ldots, \dfrac{x_N}{x_{N+1}}\right)$, which, in turn, is equivalent to

$$\beta_\tau x_{N+1} \leqslant \alpha_1^\tau x_1 + \cdots + \alpha_N^\tau x_N \quad \text{for all } \tau \in T. \tag{8.9.5}$$

Thus condition (8.9.5) is necessary and sufficient for $(x_1, \ldots, x_N, x_{N+1}) \in C$.

Now take arbitrary $f_1, \ldots, f_N \in P$ such that $M(f_i) < \infty$ for $i = 1, \ldots, N$, and let $x \in E$ be arbitrary. We have

$$\left(f_1(x), \ldots, f_N(x), \varphi\big(f_1(x), \ldots, f_N(x)\big)\right) \in C,$$

whence

$$\beta_\tau \varphi\big(f_1(x), \ldots, f_N(x)\big) \leqslant \alpha_1^\tau f_1(x) + \cdots + \alpha_N^\tau f_N(x) \quad \text{for all } \tau \in T. \tag{8.9.6}$$

Relation (8.9.6) holds for all $x \in E$, consequently $\beta_\tau \varphi(f_1, \ldots, f_N) \leqslant \alpha_1^\tau f_1 + \cdots + \alpha_N^\tau f_N$ (for all $\tau \in T$), and by (8.9.1), (8.9.2) and (8.9.3)

$$\begin{aligned}
\beta_\tau M\big(\varphi(f_1, \ldots, f_N)\big) &= M\big(\beta_\tau \varphi(f_1, \ldots, f_N)\big) \\
&\leqslant M\left(\alpha_1^\tau f_1 + \cdots + \alpha_N^\tau f_N\right) \leqslant \alpha_1^\tau M(f_1) + \cdots + \alpha_N^\tau M(f_N)
\end{aligned}$$

for all $\tau \in T$. But this means that

$$\big(M(f_1), \ldots, M(f_N), M(\varphi(f_1, \ldots, f_N))\big) \in C,$$

which is equivalent to (8.9.4). $\qquad \square$

Example 8.9.1. Let $N = 2$, and let $\varphi(x_1, x_2) = x_1^\alpha x_2^\beta$, where $\alpha > 0$, $\beta > 0$, $\alpha + \beta = 1$. Then $\varphi(x_1, x_2) = 1$ means $x_2 = x_1^{-\alpha/\beta}$, and consequently the set K is the part of the first quadrant lying above the hiperbola $x_2 = x_1^{-\alpha/\beta}$, i.e., it is a convex set. From inequality (8.9.4) we obtain a generalization of the Hölder inequality (8.1.13)

$$M\left(f_1^\alpha f_2^\beta\right) \leqslant M(f_1)^\alpha M(f_2)^\beta. \tag{8.9.7}$$

Example 8.9.2. Let $N = 2$, and let $\varphi(x_1, x_2) = \left(x_1^{1/r} + x_2^{1/r}\right)^r$, where $r \geqslant 1$. Then $\varphi(x_1, x_2) = 1$ means $x_2 = \left(1 - x_1^{1/r}\right)^r$, and consequently the set K is the part of the first quadrant lying above the curve $x_2 = \left(1 - x_1^{1/r}\right)^r$, i.e., it is a convex set. From inequality (8.9.4) we obtain

$$M\left[\left(f_1^{1/r} + f_2^{1/r}\right)^r\right] \leqslant \left[M(f_1)^{1/r} + M(f_2)^{1/r}\right]^r. \tag{8.9.8}$$

Replacing in (8.9.8) f_j by f_j^r, $j = 1, 2$, we obtain a generalization of the Minkowski inequality (8.8.7)

$$M\left[(f_1 + f_2)^r\right]^{1/r} \leqslant M(f_1^r)^{1/r} + M(f_2^r)^{1/r}.$$

Exercises

1. Let $D \subset \mathbb{R}^N$ be a convex and open set, and let $f : D \to (0, \infty)$ be a positive function. Show that if the function $\log f$ is convex, then f also is convex.

2. Let $f : [0, \infty) \to \mathbb{R}$ be a continuous and concave function such that $f(0) \leqslant 0$. Show that for every $n \in \mathbb{N}$ and $x_1, \ldots, x_n \in [0, \infty)$ we have $f(x_1 + \cdots + x_n) \leqslant f(x_1) + \cdots + f(x_n)$.

3. Let $f : [0, \infty) \to \mathbb{R}$ be a continuous and convex function. Derive inequality (8.7.1) from Theorem 8.5.3.

4. Derive inequality (8.1.13) from (8.9.7).

5. Prove that if $n \in \mathbb{N}$, $x_i \geqslant 0$, $q_i > 0$ for $i = 1, \ldots, n$, and $\sum_{i=1}^{n} q_i = 1$, then

$$\prod_{i=1}^{n} x_i^{q_i} \leqslant \sum_{i=1}^{n} q_i x_i.$$

6. Under conditions of Exercise 8.5, show that, for every $p \geqslant 1$,

$$\left(\sum_{i=1}^{n} q_i x_i\right)^p \leqslant \sum_{i=1}^{n} q_i x_i^p.$$

7. Let $n \in \mathbb{N}$, $x_i \geqslant 0$, $y_i \geqslant 0$, $i = 1, \ldots, n$. Show that

$$\left(\prod_{i=1}^{n}(x_i + y_i)\right)^{1/n} \geqslant \left(\prod_{i=1}^{n} x_i\right)^{1/n} + \left(\prod_{i=1}^{n} y_i\right)^{1/n}$$

(cf. Roberts-Varberg [267]).

[Hint: Note that

$$\frac{\left(\prod_{i=1}^{n} x_i\right)^{1/n} + \left(\prod_{i=1}^{n} y_i\right)^{1/n}}{\left(\prod_{i=1}^{n}(x_i + y_i)\right)^{1/n}} = \left(\prod_{i=1}^{n} \frac{x_i}{x_i + y_i}\right)^{1/n} + \left(\prod_{i=1}^{n} \frac{y_i}{x_i + y_i}\right)^{1/n},$$

and use Exercise 8.5.]

8. Let $J \subset \mathbb{R}$ be an open interval, and let $f : J \to \mathbb{R}$ be a continuous function. Prove that f is convex if and only if

$$f\left(x\right) \leqslant \frac{1}{2h} \int\limits_{-h}^{h} f\left(x+t\right) dt$$

for every $x \in J$ and for every $h > 0$ such that $x - h, x + h \in J$. (Beckenbach [23], [22].)

Chapter 9

Boundedness and Continuity of Convex Functions and Additive Functions

9.1 The classes $\mathfrak{A}, \mathfrak{B}, \mathfrak{C}$

The Theorem of Bernstein–Doetsch (cf., in particular, Corollary 6.4.1) says that if $D \subset \mathbb{R}^N$ is a convex and open set, $f : D \to \mathbb{R}$ is a convex function, $T \subset D$ is open and non-empty, and f is bounded above on T, then f is continuous in D. Are there other sets T with this property? What are possibly weak conditions which assure the continuity of a convex function, or of an additive function? In this and in the next chapter we will deal with such questions.

In order to simplify the notation we introduce the following classes of sets (Ger-Kuczma [115]).

$$
\begin{aligned}
\mathfrak{A} &= \{\, T \subset \mathbb{R}^N \mid \text{ every convex function } f : D \to \mathbb{R}, \text{ where } T \subset D \subset \mathbb{R}^N \\
&\qquad \text{and } D \text{ is convex and open, bounded above on } T \text{ is} \\
&\qquad \text{continuous in } D\}, \\
\mathfrak{B} &= \{\, T \subset \mathbb{R}^N \mid \text{ every additive function } f : \mathbb{R}^N \to \mathbb{R} \text{ bounded above on } T \\
&\qquad \text{is continuous}\}, \\
\mathfrak{C} &= \{\, T \subset \mathbb{R}^N \mid \text{ every additive function } f : \mathbb{R}^N \to \mathbb{R} \text{ bounded on } T \\
&\qquad \text{is continuous}\}.
\end{aligned}
$$

If we want to stress that the classes $\mathfrak{A}, \mathfrak{B}, \mathfrak{C}$ refer to a particular space $\mathbb{R}^N$, we will write $\mathfrak{A}_N, \mathfrak{B}_N, \mathfrak{C}_N$ instead of $\mathfrak{A}, \mathfrak{B}, \mathfrak{C}$.

Since every additive function is convex (cf. 5.3), it follows directly from the definition that

$$\mathfrak{A} \subset \mathfrak{B} \subset \mathfrak{C}. \tag{9.1.1}$$

In the next chapter it will be proved (Theorem 10.2.2) that actually

$$\mathfrak{A} = \mathfrak{B}. \tag{9.1.2}$$

Relation (9.1.2) implies that the class $\mathfrak{A}$ is independent of the choice of the convex and open set D occurring in the definition. Relation (9.1.2) implies also that there are no special conditions guaranteeing that a set $T \in \mathfrak{B}$ which would not guarantee at the same time that also $T \in \mathfrak{A}$. In the sequel of this chapter we will freely use relation (9.1.2), deferring its proof to 10.2.

On the other hand, the inclusion $\mathfrak{B} \subset \mathfrak{C}$ is strict, i.e., the class $\mathfrak{C}$ is strictly larger than $\mathfrak{B}$. Namely we have the following

Theorem 9.1.1. $\mathfrak{C} \setminus \mathfrak{B} \neq \varnothing$.

Proof. Let $f_0 : \mathbb{R}^N \to \mathbb{R}$ be a discontinuous additive function, and put

$$T = \{\, x \in \mathbb{R}^N \mid f_0(x) \leqslant 0 \,\}. \tag{9.1.3}$$

Then $T \notin \mathfrak{B}$, since there exists a discontinuous additive function (viz. f_0) bounded above on T. We will show that $T \in \mathfrak{C}$.

Let $f : \mathbb{R}^N \to \mathbb{R}$ be an arbitrary additive function which is bounded on T. Thus there exists a constant $M > 0$ such that

$$|f(x)| \leqslant M \qquad \text{for } x \in T. \tag{9.1.4}$$

Take an arbitrary $x_0 \in \mathbb{R}^N$. By Theorem 5.2.1

$$f(kx_0) = kf(x_0), \tag{9.1.5}$$

and

$$f_0(kx_0) = kf_0(x_0) \tag{9.1.6}$$

for every $k \in \mathbb{Z}$. By (9.1.6), depending on the sign of $f_0(x_0)$, $kx_0 \in T$ for $k = 1, 2, 3, \ldots$, or for $k = -1, -2, -3, \ldots$. Making use of (9.1.5) and (9.1.4) we get hence

$$|kf(x_0)| \leqslant M$$

for $k = 1, 2, 3, \ldots$, or for $k = -1, -2, -3, \ldots$. This is possible if and only if $f(x_0) = 0$.

Consequently $f(x_0) = 0$ for arbitrary $x_0 \in \mathbb{R}^N$, i.e., $f = 0$, and hence it is continuous. Thus every additive function $f : \mathbb{R}^N \to \mathbb{R}$ bounded on T is continuous, i.e., $T \in \mathfrak{C}$. Hence $T \in \mathfrak{C} \setminus \mathfrak{B}$. $\qquad \square$

Let us also note the following

Theorem 9.1.2. *Let $f : \mathbb{R}^N \to \mathbb{R}$ be an additive function bounded below on a set $T \in \mathfrak{B}$. Then f is continuous.*

Proof. If f is an additive function bounded below on T, then $-f$ is an additive function bounded above on T. Since $T \in \mathfrak{B}$, this implies that $-f$ is continuous, and hence also f is continuous. $\qquad \square$

9.2   Conservative operations

We recall that $\mathcal{P}(X)$ denotes the collection of all subsets of X (cf. Axiom 1.1.4).

A function $F : \mathcal{P}(\mathbb{R}^N) \to \mathcal{P}(\mathbb{R}^N)$ is called an $\mathfrak{A}$-*conservative operation* iff, for every set $T \subset \mathbb{R}^N$ and every additive function $f : \mathbb{R}^N \to \mathbb{R}$, if f is bounded above on T, then f is bounded above also on $F(T)$. A function $F : \mathcal{P}(\mathbb{R}^N) \to \mathcal{P}(\mathbb{R}^N)$ is called a $\mathfrak{C}$-*conservative operation* iff, for every set $T \subset \mathbb{R}^N$ and every additive function $f : \mathbb{R}^N \to \mathbb{R}$, if f is bounded on T, then f is bounded also on $F(T)$.

Lemma 9.2.1. *Every $\mathfrak{A}$-conservative operation is also $\mathfrak{C}$-conservative.*

Proof. Let $F : \mathcal{P}(\mathbb{R}^N) \to \mathcal{P}(\mathbb{R}^N)$ be an $\mathfrak{A}$-conservative operation, let $T \subset \mathbb{R}^N$, and let $f : \mathbb{R}^N \to \mathbb{R}$ be an additive function bounded on the set T. Then both the additive functions, f and $-f$, are bounded above on T, and hence also on $F(T)$. Consequently f is bounded on $F(T)$, which proves that F is $\mathfrak{C}$-conservative. $\square$

But, as we shall see later, there exist $\mathfrak{C}$-conservative operations which are not $\mathfrak{A}$-conservative.

Theorem 9.2.1. *If $F : \mathcal{P}(\mathbb{R}^N) \to \mathcal{P}(\mathbb{R}^N)$ is an $\mathfrak{A}$-conservative operation and, for a certain $T \subset \mathbb{R}^N$, $F(T) \in \mathfrak{A}$, then also $T \in \mathfrak{A}$.*

Proof. Let $f : \mathbb{R}^N \to \mathbb{R}$ be an arbitrary additive function bounded above on T. Then f is also bounded above on $F(T)$, and since $F(T) \in \mathfrak{A} = \mathfrak{B}$ (cf. (9.1.2)), f is continuous. Consequently $T \in \mathfrak{B} = \mathfrak{A}$. $\square$

Theorem 9.2.2. *If $F : \mathcal{P}(\mathbb{R}^N) \to \mathcal{P}(\mathbb{R}^N)$ is a $\mathfrak{C}$-conservative operation and, for a certain $T \subset \mathbb{R}^N$, $F(T) \in \mathfrak{C}$, then also $T \in \mathfrak{C}$.*

Proof. Let $f : \mathbb{R}^N \to \mathbb{R}$ be an arbitrary additive function bounded on T. Then f is bounded on $F(T)$, and since $F(T) \in \mathfrak{C}$, f is continuous. Consequently $T \in \mathfrak{C}$. $\square$

Theorem 9.2.3. *The composition of $\mathfrak{A}$-conservative [$\mathfrak{C}$-conservative] operations is $\mathfrak{A}$-conservative [$\mathfrak{C}$-conservative].*

Proof. We will prove Theorem 9.2.3 for $\mathfrak{A}$-conservative operations only. The proof for $\mathfrak{C}$-conservative operations is similar.

Let $F_1, F_2 : \mathcal{P}(\mathbb{R}^N) \to \mathcal{P}(\mathbb{R}^N)$ be $\mathfrak{A}$-conservative operations, let $T \subset \mathbb{R}^N$, and let $f : \mathbb{R}^N \to \mathbb{R}$ be an arbitrary additive function bounded above on T. Then f is bounded above on $F_1(T)$, and hence also on $F_2(F_1(T))$. Thus $F_2 \circ F_1$ is $\mathfrak{A}$-conservative. $\square$

By induction this generalizes to an arbitrary finite number of operations $F_1, \dots, F_n$. Trivially we have

Theorem 9.2.4. *Any operation $F : \mathcal{P}(\mathbb{R}^N) \to \mathcal{P}(\mathbb{R}^N)$ such that $F(T) \subset T$ for every set $T \subset \mathbb{R}^N$ is $\mathfrak{A}$-conservative.*

Corollary 9.2.1. *If $A \subset B \subset \mathbb{R}^N$ and $A \in \mathfrak{A}$ [$A \in \mathfrak{C}$], then also $B \in \mathfrak{A}$ [$B \in \mathfrak{C}$].*

Proof. By Theorem 9.2.4 and Lemma 9.2.1 operation $F(T) = A \cap T$ is $\mathfrak{A}$-conservative and $\mathfrak{C}$-conservative, and $F(B) = A$. Theorems 9.2.1 and 9.2.2 complete the proof. $\square$

But a real interest lies in operations F such that $F(T)$ is larger than T. Below we exhibit a number of such conservative operations. For every $n \in \mathbb{N}$ we define $S_n : \mathcal{P}(\mathbb{R}^N) \to \mathcal{P}(\mathbb{R}^N)$ by

$$S_n(A) = A + \cdots + A$$

(n summands). The operations J and Q are defined in 5.1.

Theorem 9.2.5. *The operations J, Q and S_n for every $n \in \mathbb{N}$ are $\mathfrak{A}$-conservative.*

Proof. Let $T \subset \mathbb{R}^N$, and let $f : \mathbb{R}^N \to \mathbb{R}$ be an additive (and hence convex) function bounded above on T

$$f(x) \leqslant M \quad \text{for} \quad x \in T. \tag{9.2.1}$$

By Lemma 6.1.2 and Corollary 6.1.1 f is bounded above on $Q(T)$ and $J(T)$, which shows that the operations J and Q are $\mathfrak{A}$-conservative.

Take a $y \in S_n(T)$. Then $y = t_1 + \cdots + t_n$, where $t_i \in T$ for $i = 1, \ldots, n$. Hence

$$f(y) = f(t_1) + \cdots + f(t_n) \leqslant M + \cdots + M = nM.$$

Thus f is bounded above on $S_n(T)$ by the constant nM, which shows that the operation S_n is $\mathfrak{A}$-conservative. $\qquad\square$

Now we define the operations R and U

$$R(T) = T - T \tag{9.2.2}$$

$$U(T) = \{\, x \in \mathbb{R}^N \mid x = t_1 + (t_2 - t_3), t_1, t_2, t_3 \in T \,\} = T + (T - T) = T + R(T). \tag{9.2.3}$$

Theorem 9.2.6. *The operations R and U are $\mathfrak{C}$-conservative.*

Proof. Let $T \subset \mathbb{R}^N$, and let $f : \mathbb{R}^N \to \mathbb{R}$ be an arbitrary additive function bounded on T

$$|f(x)| \leqslant M \quad \text{for} \quad x \in T. \tag{9.2.4}$$

If $y \in R(T)$, then there exist $t_1, t_2 \in T$ such that $y = t_1 - t_2$, whence

$$|f(y)| = |f(t_1) - f(t_2)| \leqslant |f(t_1)| + |f(t_2)| \leqslant 2M.$$

If $y \in U(T)$, then there exist $t_1, t_2, t_3 \in T$ such that $y = t_1 + (t_2 - t_3)$, whence

$$|f(y)| = |f(t_1) + f(t_2) - f(t_3)| \leqslant |f(t_1)| + |f(t_2)| + |f(t_3)| \leqslant 3M.$$

Thus f is bounded on $R(T)$ and on $U(T)$, which shows that the operations R and U are $\mathfrak{C}$-conservative. $\qquad\square$

On the other hand, R and U are not $\mathfrak{A}$-conservative. Let $f_0 : \mathbb{R}^N \to \mathbb{R}$ be a discontinuous additive function, and define the set $T \subset \mathbb{R}^N$ by (9.1.3). Then f_0 is bounded above on T. Let $x_0 \in \mathbb{R}^N$ be such that $f_0(x_0) < 0$. Then we have (9.1.6) for all $k \in \mathbb{Z}$, whence $kx_0 \in T$ for $k \in \mathbb{N}$. Hence

$$-kx_0 = x_0 - (k+1)x_0 \in T - T = R(T)$$

for every $k \in \mathbb{N}$, and

$$\lim_{k \to \infty} f_0(-kx_0) = \lim_{k \to \infty} \left(- k f_0(x_0) \right) = \infty.$$

Consequently f_0 is not bounded above on $R(T)$. Similarly it can be shown that f_0 is not bounded above on $U(T)$. It follows that the operations R and U are not $\mathfrak{A}$-conservative.

Let us note also the following two simple results.

Theorem 9.2.7. *The operation* $F : \mathcal{P}(\mathbb{R}^N) \to \mathcal{P}(\mathbb{R}^N)$ *defined by*

$$F(T) = \alpha T, \tag{9.2.5}$$

where α *is a fixed rational number, is* $\mathfrak{C}$-*conservative. If* $\alpha \geqslant 0$, *then* F *is* $\mathfrak{A}$-*conservative.*

Proof. Let $T \subset \mathbb{R}^N$, and let $f : \mathbb{R}^N \to \mathbb{R}$ be an arbitrary additive function fulfilling (9.2.4). If $y \in \alpha T$, then $y = \alpha t$ for a certain $t \in T$. Hence by Theorem 5.2.1

$$|f(y)| = |f(\alpha t)| = |\alpha f(t)| = |\alpha|\,|f(t)| \leqslant |\alpha| M.$$

Thus f is bounded on αT, which means that operation (9.2.5) is $\mathfrak{C}$-conservative.

If f fulfils (9.2.1) and $\alpha \geqslant 0$, then

$$f(y) = \alpha f(t) \leqslant \alpha M,$$

and f is bounded above on αT. Consequently in this case operation (9.2.5) is $\mathfrak{A}$-conservative. $\square$

Theorem 9.2.8. *The operation* $F : \mathcal{P}(\mathbb{R}^N) \to \mathcal{P}(\mathbb{R}^N)$ *defined by*

$$F(T) = T + t, \tag{9.2.6}$$

where $t \in \mathbb{R}^N$ *is fixed, is* $\mathfrak{A}$-*conservative.*

Proof. Let $T \subset \mathbb{R}^N$, and let $f : \mathbb{R}^N \to \mathbb{R}$ be an arbitrary additive function fulfilling (9.2.1). If $y \in T + t$, then $y = x + t$, where $x \in T$. Thus by (9.2.1)

$$f(y) = f(x) + f(t) \leqslant M + f(t).$$

This means that f is bounded above on $T + t$, and operation (9.2.6) is $\mathfrak{A}$-conservative. $\square$

9.3   Simple conditions

In this section we give some sufficient conditions and some necessary conditions for a set $T \subset \mathbb{R}^N$ to belong to the class $\mathfrak{A}$ or $\mathfrak{C}$.

Lemma 9.3.1. *If* $T \subset \mathbb{R}^N$ *and* $\mathrm{int}\, T \neq \varnothing$, *then* $T \in \mathfrak{A}$.

Proof. Corollary 6.4.1 says that $\operatorname{int} T \in \mathfrak{A}$. The operation $F(T) = \operatorname{int} T$ fulfils the condition of Theorem 9.2.4, and hence is $\mathfrak{A}$-conservative. In virtue of Theorem 9.2.1 we have $T \in \mathfrak{A}$. $\qquad\qquad\square$

Theorem 9.3.1. [Theorem of Ostrowski]. *If $T \subset \mathbb{R}^N$ and $m_i(T) > 0$, then $T \in \mathfrak{A}$.*

Proof. By the Theorem of Steinhaus (Theorem 3.7.1) $\operatorname{int} S_2(T) \neq \varnothing$, whence by Lemma 9.3.1 $S_2(T) \in \mathfrak{A}$. Theorems 9.2.5 and 9.2.1 now imply that $T \in \mathfrak{A}$. $\qquad\square$

For $N = 1$ Theorem 9.3.1 was proved by A. Ostrowski [248]. The generalizations to higher dimensions are due to A. Császár [50] and S. Marcus [218].

The topological analogue of the Theorem of Ostrowski is the following theorem of M. R. Mehdi [224] (cf. also Kominek [172]).

Theorem 9.3.2. [Theorem of Mehdi]. *If $T \subset \mathbb{R}^N$ contains a second category set with the Baire property, then $T \in \mathfrak{A}$.*

Proof. By the Theorem of S. Piccard (Theorem 2.9.1) $\operatorname{int} S_2(T) \neq \varnothing$, and the proof runs as that of Theorem 9.3.1. $\qquad\qquad\square$

From Theorems 9.3.1 and 9.3.2 and from Theorems 9.2.5 and 9.2.1 we obtain the following

Theorem 9.3.3. *Let $T \subset \mathbb{R}^N$. Each of the following conditions is sufficient for T to belong to $\mathfrak{A}$:*

$(i)_n \quad m_i\big(S_n(T)\big) > 0$.

$(ii)_n \quad S_n(T)$ *contains a set of the second category and with the Baire property.*

$(iii) \quad m_i\big(J(T)\big) > 0$.

$(iv) \quad J(T)$ *contains a set of the second category and with the Baire property.*

$(v) \quad m_i\big(Q(T)\big) > 0$.

$(vi) \quad Q(T)$ *contains a set of the second category and with the Baire property.*

Condition $(i)_2$ is due to S. Kurepa [199]; $(i)_n$ for arbitrary n to R. Ger [97] (cf. also Kemperman [165], Marcus [218]). Condition (iii) is found with R. Ger–M. Kuczma [115] (cf. also Kuczma [176]), condition (v) with R. Ger [97].

From Theorem 9.3.3 we obtain immediately (cf. Exercise 2.11 and Exercise 2.13)

Corollary 9.3.1. $\mathbf{C} \in \mathfrak{A} = \mathfrak{A}_1$.

Concerning the mutual relations of the conditions occurring in Theorem 9.3.3, we have the following

Lemma 9.3.2. *With the notation of Theorem 9.3.3,*

$$(i)_n \Rightarrow (i)_{n+1}, \quad n \in \mathbb{N},$$
$$(i)_n \Rightarrow (iii), \qquad n \in \mathbb{N},$$
$$(iii) \Rightarrow (v).$$

Similarly

$$(ii)_n \Rightarrow (ii)_{n+1}, \quad n \in \mathbb{N},$$
$$(ii)_n \Rightarrow (iv), \qquad n \in \mathbb{N},$$
$$(iv) \Rightarrow (vi).$$

None of these implications can be reversed.

Proof. Fix an $n \in \mathbb{N}, T \subset \mathbb{R}^N$, and take a $t_0 \in T$. Clearly we have

$$S_n(T) + t_0 \subset S_{n+1}(T). \tag{9.3.1}$$

Relation (9.3.1) proves the implication $(i)_n \Rightarrow (i)_{n+1}$ and $(ii)_n \Rightarrow (ii)_{n+1}$.

Now, to a given $n \in \mathbb{N}$, choose a p so that $m = 2^p > n$. Take an $x \in \dfrac{1}{m} S_m(T)$.
There exist $t_1, \ldots, t_m \in T$ such that

$$x = \frac{1}{m} \sum_{i=1}^{m} t_i = \sum_{i=1}^{m} \frac{1}{m} t_i.$$

By Theorem 5.1.3 $x \in J(T)$, which shows that

$$\frac{1}{m} S_m(T) \subset J(T). \tag{9.3.2}$$

By what we have already proved, $(i)_n \Rightarrow (i)_m$ and $(ii)_n \Rightarrow (ii)_m$. Inclusion (9.3.2) proves now the implication $(i)_n \Rightarrow (iii)$ and $(ii)_n \Rightarrow (iv)$.

Since every Q-convex set is, in particular, J-convex, we have $J(T) \subset Q(T)$ for every $T \subset \mathbb{R}^N$. Hence we obtain the implications $(iii) \Rightarrow (v)$ and $(iv) \Rightarrow (vi)$.

Now we are going to show, by suitable counterexamples, that none of the implications in Theorem 9.3.3 can be reversed. These counterexamples are due to R. Ger [97].

Fix an $n \in \mathbb{N}$, and consider the set $T_0 \subset \mathbb{R}$:

$$T_0 = \left\{ x \in [0,1] \mid x = \sum_{i=1}^{\infty} \frac{\varepsilon_i}{(n+2)^i}, \quad \varepsilon_i = 0 \text{ or } 1, i \in \mathbb{N} \right\}.$$

Then for every $x \in S_n(T_0)$ the expansion of x with the base $(n+2)$ does not contain the digit $(n+1)$. By Theorem 3.6.4 the set $S_n(T_0)$ has measure zero, and it is easy to check that $S_n(T_0)$ is nowhere dense. On the other hand, $S_{n+1}(T_0)$ contains all the numbers from the interval $[0,1]$. The set $T = T_0 \times \mathbb{R}^{N-1} \subset \mathbb{R}^N$ has similar properties: $S_n(T)$ is nowhere dense and of measure zero, whereas $S_{n+1}(T)$ contains the strip $[0,1] \times \mathbb{R}^{N-1}$. Consequently conditions $(i)_{n+1}$ and $(ii)_{n+1}$ are fulfilled, whereas $(i)_n$ and $(ii)_n$ are not.

Now let $H \subset \mathbb{R}^N$ be a Hamel basis, and put $T = \mathbb{Q}H = \{x \in \mathbb{R}^N \mid x = \lambda h, \lambda \in \mathbb{Q}, h \in H\}$. Suppose that for a certain $n \in \mathbb{N}$ condition $(i)_n$ or $(ii)_n$ is fulfilled. By Theorem 3.7.1 or 2.9.1

$$\text{int } S_{2n}(T) = \text{int } \big[S_n(T) + S_n(T) \big] \neq \varnothing.$$

Every $x \in \operatorname{int} S_{2n}(T)$ (except, at most, $x = 0$) may be written in the form

$$x = \lambda_1(x)h_1(x) + \cdots + \lambda_{k(x)}(x)h_{k(x)}(x), \qquad (9.3.3)$$

where $\lambda_i(x) \in \mathbb{Q}$, $\lambda_i(x) \neq 0$, $h_i(x) \in H$, $i = 1, \ldots, k(x)$, $k(x) \in \mathbb{N}$. Moreover,

$$k(x) \leqslant 2n, \qquad (9.3.4)$$

since $x \in S_{2n}(T)$, and due to the uniqueness of expansion (9.3.3) (Lemma 4.1.1). Write $m = \sup_{x \in \operatorname{int} S_{2n}(T)} k(x)$. By (9.3.4) there exists an $x_0 \in \operatorname{int} S_{2n}(T)$ such that $k(x_0) = m$. Let

$$x_0 = \lambda_1(x_0)h_1(x_0) + \cdots + \lambda_m(x_0)h_m(x_0), \qquad \lambda_i(x_0) \neq 0, \ i = 1, \ldots, m,$$

be the Hamel expansion of x_0, and take an $h_0 \in H$, $h_0 \neq h_i(x_0)$, $i = 1, \ldots, m$. We have $\bar{x} = x_0 + \lambda h_0 \in \operatorname{int} S_{2n}(T)$ for sufficiently small positive $\lambda \in \mathbb{Q}$. For such a λ we have

$$\bar{x} = \lambda_1(x_0)h_1(x_0) + \cdots + \lambda_m(x_0)h_m(x_0) + \lambda h_0,$$

and $\bar{x} \in \operatorname{int} S_{2n}(T)$, whence $k(\bar{x}) = m + 1$, which contradicts the maximality of m. Consequently none of conditions $(i)_n$, $(ii)_n$, $n \in \mathbb{N}$, can be fulfilled.

In order to show that conditions (iii) and (iv) are fulfilled, we will prove that

$$J(T) = \mathbb{R}^N. \qquad (9.3.5)$$

Take an arbitrary $x \in \mathbb{R}^N$. x has an expansion (9.3.3), where $\lambda_i(x) \in \mathbb{Q}$, $h_i(x) \in H$, $i = 1, \ldots, k(x)$. Let $p \in \mathbb{N}$ be chosen so that $m = 2^p > n$, take arbitrary $h_j \in H$, $j = k(x) + 1, \ldots, m$, so that $h_j \neq h_i(x)$, $i = 1, \ldots, k(x)$; $j = k(x) + 1, \ldots, m$, and put $\lambda_j = 0$, $j = k(x) + 1, \ldots, m$. Then

$$mx = m\lambda_1(x)h_1(x) + m\lambda_{k(x)}(x)h_{k(x)}(x) + \lambda_{k(x)+1}h_{k(x)+1} + \cdots + \lambda_m h_m \in S_m(T),$$

whence $x \in \dfrac{1}{m} S_m(T)$, and by (9.3.2) $x \in J(T)$. Consequently $\mathbb{R}^N \subset J(T)$, and since the converse inclusion is obvious, we obtain (9.3.5).

Now, again, let $H \subset \mathbb{R}^N$ be an arbitrary Hamel basis, and put

$$T = \Big\{ x \in \mathbb{R}^N \mid x = \sum_{i=1}^{n} \lambda_i h_i, \ \lambda_i \in \mathbb{Z}, \ h_i \in H, \ i = 1, \ldots, n; \ n \in \mathbb{N} \Big\}.$$

The set $J(T)$ contains those $x \in \mathbb{R}^N$ whose coefficients in the Hamel expansions are diadic numbers. A sum of such x's again has such a form, whence $J(T) + J(T) \subset J(T)$. Consequently if either of conditions (iii), (iv) were fulfilled, we would have $\operatorname{int} J(T) \neq \varnothing$. Let $x \in \operatorname{int} J(T)$ have a Hamel expansion

$$x = \lambda_1 h_1 + \cdots + \lambda_k h_k, \qquad (9.3.6)$$

$\lambda_i \in \mathbf{D}$, $h_i \in H$, $i = 1, \ldots, k$, and take an $h_0 \in H$, $h_0 \neq h_i$, $i = 1, \ldots, k$. The point

$$\bar{x} = \lambda_1 h_1 + \cdots + \lambda_k h_k + 3^{-n} h_0$$

does not belong to $J(T)$ (since $3^{-n} \notin \mathbf{D}$), and, if n is large enough, belongs to int $J(T)$. The contradiction obtained shows that neither (iii), nor (iv) is fulfilled.

On the other hand, we have clearly $Q(T) = \mathbb{R}^N$. Consequently conditions (v) and (vi) are fulfilled. $\qquad\square$

We are going to show yet that neither of the conditions

$$(vii) \quad m_i \Big(\bigcup_{n=1}^{\infty} S_n(T) \Big) > 0,$$

$$(viii) \quad \bigcup_{n=1}^{\infty} S_n(T) \text{ contains a set of the second category and with the Baire property,}$$

does imply that $T \in \mathfrak{C}$. Let $H \subset \mathbb{R}^N$ be a Hamel basis, and put

$$T = \{ x \in \mathbb{R}^N \mid x = \lambda h, \ \lambda \in \mathbb{Q} \cap [-1, 1], \ h \in H \}.$$

First we show that $T \notin \mathfrak{C}$. Let $g : H \to \mathbb{R}$ be given by $g(h) = 1$, $h \in H$, and let $f : \mathbb{R}^N \to \mathbb{R}$ be the additive extension of g (Theorem 5.2.2). By Corollary 5.2.2 f is discontinuous, and by Theorem 5.2.1 we have for $x \in T$, $x = \lambda h$,

$$|f(x)| = |f(\lambda h)| = |\lambda f(h)| = |\lambda g(h)| = |\lambda| \leqslant 1.$$

Consequently f is bounded on T, and hence $T \notin \mathfrak{C}$.

Now take an arbitrary $x \in \mathbb{R}^N$. x has a Hamel expansion (9.3.6), which may be written as

$$x = \lambda_{11} h_1 + \cdots + \lambda_{1p_1} h_1 + \cdots + \lambda_{k1} h_k + \cdots + \lambda_{kp_k} h_k,$$

where $\lambda_{ij} \in \mathbb{Q} \cap [-1, 1]$, $\lambda_{i1} + \cdots + \lambda_{ip_i} = \lambda_i$, $i = 1, \ldots, k$. Consequently $x \in S_m(T)$, where $m = p_1 + \cdots + p_k$. Hence $\mathbb{R}^N \subset \bigcup_{n=1}^{\infty} S_n$, i.e., $\bigcup_{n=1}^{\infty} S_n = \mathbb{R}^N$, which shows that conditions (vii) and $(viii)$ are fulfilled.

Also this example is due to R. Ger [97].

Theorem 9.3.4. *None of the conditions occurring in Theorem 9.3.3 is necessary for the set T to belong to $\mathfrak{A}$.*

Proof. None of conditions $(i)_n$, $(ii)_n$, $n \in \mathbb{N}$, (iii), (iv) is necessary, as results from Lemma 9.3.2. It remains to exhibit an example of a set $T \in \mathfrak{A}$ which does not fulfil (v) nor (vi). Because the set we are going to construct is of a more general interest, we will employ a special letter (V_0) to denote it.

Let $H \subset \mathbb{R}^N$ be a Hamel basis. By Theorem 1.7.1 H can be well ordered, and so we can arrange the elements of H into a transfinite sequence of the type γ (where γ is the smallest ordinal with $\bar{\gamma} = \mathfrak{c}$):

$$H = \{h_\alpha\}_{\alpha < \gamma}.$$

Now every element $x \in \mathbb{R}^N$, except for $x = 0$, can be uniquely written in the form

$$x = \lambda_1 h_{\alpha_1} + \cdots + \lambda_n h_{\alpha_n}, \tag{9.3.7}$$

where $n \in \mathbb{N}$, $\lambda_i \in \mathbb{Q}$, $\lambda_i \neq 0$, $i = 1, \ldots, n$, $\alpha_1 < \cdots < \alpha_n$. We define V_0 as the set of those $x \in \mathbb{R}^N$ which in expansion (9.3.7) have $\lambda_n > 0$ (the last non-zero coefficient is positive). Additionally we assume that $0 \in V_0$.

It is easily seen from this definition that V_0 is a cone in the linear space $(\mathbb{R}^N; \mathbb{Q}; +; \cdot)$; in particular,

$$V_0 + V_0 \subset V_0. \tag{9.3.8}$$

Moreover, V_0 is Q-convex, and so $Q(V_0) = V_0$.

If the set $T = V_0$ fulfilled either of conditions (v), (vi), then we would have by (9.3.8) int $V_0 \neq \varnothing$. Suppose that an x with expansion (9.3.7) belongs to int V_0. Then, for sufficiently small positive $\lambda \in \mathbb{Q}$, we have $x - \lambda h_{\alpha_n+1} \in$ int V_0. On the other hand for such λ we have $x - \lambda h_{\alpha_n+1} \notin V_0$, since the coefficient of h_{α_n+1} is negative. This contradiction shows that $T = V_0$ fulfils neither (v), nor (vi).

In order to prove that $V_0 \in \mathfrak{A}$ consider an arbitrary additive function $f : \mathbb{R}^N \to \mathbb{R}$ bounded above on V_0. Take an arbitrary $h \in H$. There exists an $\alpha < \gamma$ such that $h = h_\alpha$. The points $\lambda h_\alpha + h_{\alpha+1}$ belong to V_0 for every $\lambda \in \mathbb{Q}$, and $f(\lambda h_\alpha + h_{\alpha+1}) = \lambda f(h_\alpha) + f(h_{\alpha+1})$. When λ ranges over $\mathbb{Q}$ this expression should remain bounded above, which is possible only if $f(h_\alpha) = 0$. Consequently $f \mid H = 0$, whence by the uniqueness of the extension $f = 0$, i.e., f is continuous. This shows that $V_0 \in \mathfrak{A}$. $\square$

For the class $\mathfrak{C}$ we have the following

Theorem 9.3.5. *Let $T \subset \mathbb{R}^N$. Each of the following conditions is sufficient for T to belong to $\mathfrak{C}$:*

(ix) $m_i(R(T)) > 0$.

(x) $R(T)$ *contains a set of the second category and with the Baire property.*

(xi) $measurable_i(U(T)) > 0$.

(xii) $U(T)$ *contains a set of the second category and with the Baire property.*

Theorem 9.3.5 results from Theorems 9.3.1 and 9.3.2 and from Theorems 9.2.6 and 9.2.2.

Again, none of the above conditions is necessary for a set T to belong to $\mathfrak{C}$, as may be seen from the example of the set $T = \mathbb{Q} \cdot H$ considered in the proof of Lemma 9.3.2. Note that for this particular T we have $-T = T$ so that $R(T) = S_2(T)$ and $U(T) = S_3(T)$.

Now we pass to some necessary conditions for a set to belong to $\mathfrak{C}$.

Theorem 9.3.6. *Let $T \subset \mathbb{R}^N$. If $T \in \mathfrak{C}$, then T contains a Hamel basis.*

Proof. Suppose that T does not contain a Hamel basis. It follows from Corollary 4.2.2 that then

$$E(T) \neq \mathbb{R}^N. \tag{9.3.9}$$

Let $B \subset T$ be a base of $E(T)$ (Corollary 4.2.2), and let H be a Hamel basis of $\mathbb{R}^N$ such that $B \subset H$ (Corollary 4.2.3). By (9.3.9) we have $H \setminus B \neq \emptyset$. Define a function $g : H \to \mathbb{Q}$ putting

$$g(x) = \begin{cases} 0 & \text{for } x \in B, \\ 1 & \text{for } x \in H \setminus B, \end{cases}$$

and let $f : \mathbb{R}^N \to \mathbb{R}$ be the additive extension of g (Theorem 5.2.2). By Corollary 5.2.2 f is discontinuous ($f \neq 0$, since $f \mid H \setminus B = g \mid H \setminus B = 1$), and, as is easy to see, f vanishes on $E(B) = E(T)$. Thus, in particular, f is bounded on $T \subset E(T)$, which shows that $T \notin \mathfrak{C}$. $\square$

From relation (9.1.1) and Theorem 9.3.6 we obtain

Corollary 9.3.2. *Let $T \subset \mathbb{R}^N$. If $T \in \mathfrak{A}$, then T contains a Hamel basis.*

The converse to Theorem 9.3.6 or to Corollary 9.3.2 is not true. Any Hamel basis yields an example of a set containing a Hamel basis, but belonging neither to $\mathfrak{A}$, nor to $\mathfrak{C}$ (cf. 11.1). But we have the following (Kominek [173])

Theorem 9.3.7. *Let $T \subset \mathbb{R}^N$. If T is analytic and contains a Hamel basis, then $T \in \mathfrak{C}$.*

Proof. For every $\lambda_1, \ldots, \lambda_n \in \mathbb{R}$ the function $g : \mathbb{R}^{nN} \to \mathbb{R}^N$

$$g(x_1, \ldots, x_n) = \lambda_1 x_1 + \cdots + \lambda_n x_n$$

is continuous. By Theorem 2.5.3 the set $T^n = T \times \cdots \times T$ (n-fold product) is analytic, and hence, by Theorem 2.5.2 the set

$$g(T^n) = \lambda_1 T + \cdots + \lambda_n T$$

is analytic, and hence, by Theorem 3.1.8, Lebesgue measurable.

Let $H \subset T$ be a Hamel basis. We have

$$\mathbb{R}^N = \bigcup_{n=1}^{\infty} \bigcup_{\lambda_1, \ldots, \lambda_n \in \mathbb{Q}} (\lambda_1 T + \cdots + \lambda_n T). \qquad (9.3.10)$$

In fact, if $x \in \mathbb{R}^N$, then

$$x = \alpha_1 h_1 + \cdots + \alpha_k h_k,$$

where $\alpha_i \in \mathbb{Q}$, $h_i \in H \subset T$, $i = 1, \ldots, k$. Hence $x \in \alpha_1 T + \cdots + \alpha_k T$, which is one of the summands in (9.3.10). Consequently (9.3.10) holds.

As pointed out above, the summands in (9.3.10) are measurable. If they all had measure zero, also $\mathbb{R}^N$ would be of measure zero, since the union in (9.3.10) is countable. Consequently there exist $\lambda_1, \ldots, \lambda_n \in \mathbb{Q}$ such that

$$T_0 = \lambda_1 T + \cdots + \lambda_n T$$

has a positive measure: $m(T_0) = m_i(T_0) > 0$.

Let $f : \mathbb{R}^N \to \mathbb{R}$ be an arbitrary additive function bounded on T:

$$|f(x)| \leqslant M \qquad \text{for} \ \ x \in T.$$

Then we have for $x \in T_0$

$$|f(x)| \leqslant |\lambda_1| M + \cdots + |\lambda_n| M \leqslant (\max_i |\lambda_i|) M.$$

Thus f is bounded on T_0. By the Theorem of Ostrowski (Theorem 9.3.1) and by (9.1.1) $T_0 \in \mathfrak{C}$, which implies that f is continuous. Consequently $T \in \mathfrak{C}$. $\qquad\square$

From Theorem 9.3.6 and Corollary 9.3.1 we obtain yet

Corollary 9.3.3. *The Cantor set* **C** *contains a Hamel basis for* $\mathbb{R}$.

Next we prove

Lemma 9.3.3. *Let* $T \subset \mathbb{R}^N$. *If* $m(J(T)) = 0$, *then* $m(E^+(T)) = 0$.

Proof. We will distinguish two cases.

I. $0 \in J(T)$. Fix arbitrarily $\lambda_1, \ldots, \lambda_n \in \mathbb{Q} \cap [0, \infty)$. We may write

$$\lambda_i = a_i/b, \qquad i = 1, \ldots, n, \tag{9.3.11}$$

where $a_i \in \mathbb{N} \cup \{0\}$, $b \in \mathbb{N}$. (We may assume that all λ_i's have been written with a common denominator). There exists a $p \in \mathbb{N}$ such that $a_1 + \cdots + a_n < 2^p$. Take an arbitrary x such that

$$x \in \lambda_1 J(T) + \cdots + \lambda_n J(T). \tag{9.3.12}$$

Then

$$x = \lambda_1 t_1 + \cdots + \lambda_n t_n = \frac{1}{b}(a_1 t_1 + \cdots + a_n t_n), \qquad t_i \in J(T), \qquad i = 1, \ldots, n,$$

or

$$x = \frac{1}{b}[(t_1 + \cdots + t_1) + \cdots + (t_n + \cdots + t_n) + 0 + \cdots + 0] = \frac{2^p}{b} \frac{s_1 + \cdots + s_{2^p}}{2^p},$$

where

$$s_i = t_1 \quad \text{for} \ \ i = 1, \ldots, a_1,$$

$$\cdots\cdots\cdots\cdots\cdots\cdots\cdots\cdots\cdots\cdots\cdots\cdots$$

$$s_i = t_n \quad \text{for} \ \ i = \sum_{k=1}^{n-1} a_k + 1, \ldots, \sum_{k=1}^{n} a_k,$$

$$s_i = 0 \quad \text{for} \ \ i = \sum_{k=1}^{n} a_k + 1, \ldots, 2^p.$$

Since $J(T)$ is J-convex, we have by Lemma 5.1.3 $2^{-p}(s_1 + \cdots + s_{2^p}) \in J(T)$, and hence

$$\lambda_1 J(T) + \cdots + \lambda_n J(T) \subset \frac{2^p}{b} J(T).$$

Since $m(J(T)) = 0$, we get hence

$$m[\lambda_1 J(T) + \cdots + \lambda_n J(T)] = 0. \tag{9.3.13}$$

Relation (9.3.13) is valid for arbitrary $\lambda_1, \ldots, \lambda_n \in \mathbb{Q} \cap [0, \infty)$ $(n \in \mathbb{N})$. We have

$$E^+(J(T)) = \bigcup_{n=1}^{\infty} \bigcup_{\lambda_1, \ldots, \lambda_n \in \mathbb{Q} \cap [0, \infty)} [\lambda_1 J(T) + \cdots + \lambda_n J(T)], \qquad (9.3.14)$$

whence $m(E^+(J(T))) = 0$. (The union in (9.3.14) is countable.) Since $T \subset J(T)$, we have $E^+(T) \subset E^+(J(T))$, whence $m(E^+(T)) = 0$.

II. $0 \notin J(T)$. Take an $a \in J(T)$ and put $T_0 = J(T) - a$. The set T_0 is J-convex: if $x, y \in T_0$, then $x = x' - a$, $y = y' - a$ with $x', y' \in J(T)$, whence $\frac{1}{2}(x+y) = \frac{1}{2}(x'+y') - a \in J(T) - a = T_0$. Hence $J(T_0) = T_0$. By Corollary 3.2.2 $m(J(T_0)) = m(J(T)) = 0$. Clearly $0 \in T_0 = J(T_0)$. By Part I of the proof $m(E^+(T_0)) = 0$. We will show that

$$E^+(J(T)) \subset \bigcup_{\lambda \in \mathbb{Q} \cap [0, \infty)} (E^+(T_0) + \lambda a). \qquad (9.3.15)$$

Let $x \in E^+(J(T))$. Then x may be written as

$$x = \lambda_1 t_1 + \cdots + \lambda_n t_n + \lambda_0 a,$$

where $n \in \mathbb{N}$, $\lambda_0, \ldots, \lambda_n \in \mathbb{Q} \cap [0, \infty)$, $t_1, \ldots, t_n \in J(T)$. Hence

$$x = \lambda_1 (t_1 - a) + \cdots + \lambda_n (t_n - a) + (\lambda_0 + \lambda_1 + \cdots + \lambda_n) a.$$

Now, $\lambda_i \in \mathbb{Q} \cap [0, \infty)$, $i = 1, \ldots, n$; $\lambda = \lambda_0 + \lambda_1 + \cdots + \lambda_n \in \mathbb{Q} \cap [0, \infty)$; $t_i - a \in J(T) - a = T_0$, $i = 1, \ldots, n$. Hence

$$x \in E^+(T_0) + \lambda a \subset \bigcup_{\lambda \in \mathbb{Q} \cap [0, \infty)} (E^+(T_0) + \lambda a),$$

which proves (9.3.15).

Relation (9.3.15) yields $m(E^+(J(T))) = 0$, whence also $m(E^+(T)) = 0$. $\square$

Corollary 9.3.4. *Let $T \subset \mathbb{R}^N$. If $m(J(T)) = 0$, then also $m(Q(T)) = 0$.*

This follows from the inclusion $J(T) \subset Q(T) \subset E^+(T)$ and from Lemma 9.3.3.

Lemma 9.3.4. *Let $T \subset \mathbb{R}^N$. If $m(S_n(T)) = 0$ for all $n \in \mathbb{N}$, then also $m(J(T)) = 0$.*

Proof. Take an $x \in J(T)$. Then, by Theorem 5.1.3

$$x = \lambda_1 t_1 + \cdots + \lambda_n t_n,$$

where $n \in \mathbb{N}$, $t_i \in T$, $\lambda_i \in \mathbf{D}$, $i = 1, \ldots, n$, $\sum_{i=1}^{n} \lambda_i = 1$. λ_i's may be written in form (9.3.11), where $b = 2^p$ for a certain $p \in \mathbb{N} \cup \{0\}$, and $a_i \in \mathbb{N}$, $i = 1, \ldots, n$, $\sum_{i=1}^{n} a_i = b$. By a similar argument as in the proof of Lemma 9.3.3 we get $x \in 2^{-p} S_{2^p}(T)$, whence

$$J(T) \subset \bigcup_{p=0}^{\infty} 2^{-p} S_{2^p}(T),$$

and the lemma follows. $\square$

Lemma 9.3.5. *Let $T \subset \mathbb{R}^N$. If $m\big(R\big(J(T)\big)\big) = 0$, then also $m\big(E(T)\big) = 0$.*

Proof. Again we distinguish two cases.

I. $0 \in J(T)$. Fix arbitrarily $\lambda_1, \ldots, \lambda_n \in \mathbb{Q}$. They may be written in form (9.3.11), with $a_i \in \mathbb{Z}$, $i = 1, \ldots, n$; $b \in \mathbb{N}$. Moreover, they may be numbered in such a way that $a_i \geqslant 0$ for $i = 1, \ldots, q$; $a_i < 0$ for $i = q+1, \ldots, n$. There exists a $p \in \mathbb{N}$ such that

$$2^p > \max\Big(\sum_{i=1}^{q} a_i,\ \sum_{i=q+1}^{n} |a_i| \Big).$$

Take an arbitrary x fulfilling (9.3.12). Then $x = y_1 - y_2$, where

$$y_1 = \frac{1}{b}\big[(t_1 + \cdots + t_1) + \cdots + (t_q + \cdots + t_q) + 0 + \cdots + 0\big],$$

$$y_2 = \frac{1}{b}\big[(t_{q+1} + \cdots + t_{q+1}) + \cdots + (t_n + \cdots + t_n) + 0 + \cdots + 0\big],$$

whence (similarly as in the proof of Lemma 9.3.3) $y_1, y_2 \in \dfrac{2^p}{b} J(T)$ and $x \in \dfrac{2^p}{b}\big[J(T) - J(T)\big] = \dfrac{2^p}{b} R\big(J(T)\big)$. Hence it follows that

$$\lambda_1 J(T) + \cdots + \lambda_n J(T) \subset \frac{2^p}{b} R\big(J(T)\big),$$

and since

$$E\big(J(T)\big) = \bigcup_{n=1}^{\infty}\ \bigcup_{\lambda_1, \ldots, \lambda_n \in \mathbb{Q}} \big[\lambda_1 J(T) + \cdots + \lambda_n J(T)\big],$$

we obtain hence $m\big(E(T)\big) = 0$.

II. $0 \notin J(T)$. Take an $a \in J(T)$ and put $T_0 = J(T) - a$. Then $J(T_0) = T_0$, and

$$R\big(J(T)\big) = J(T) - J(T) = [J(T) - a] - [J(T) - a] = R[J(T) - a] = R\big(J(T_0)\big).$$

By the first part of the proof $m\big(E(T_0)\big) = 0$. As in the proof of Lemma 9.3.3 we can show that

$$E\big(J(T)\big) \subset \bigcup_{\lambda \in \mathbb{Q}} \big(E(T_0) + \lambda a\big).$$

Hence $m\big(E\big(J(T)\big)\big) = 0$, whence also $m\big(E(T)\big) = 0$. □

Theorem 9.3.8. *Let $T \subset \mathbb{R}^N$. If $T \in \mathfrak{C}$, then $m_e\big(R\big(J(T)\big)\big) > 0$.*

Proof. Suppose the contrary. Then $m\big(R\big(J(T)\big)\big) = 0$, and by Lemma 9.3.5 $m\big(E(T)\big) = 0$. Hence $E(T) \neq \mathbb{R}^N$, and T cannot contain a Hamel basis.

By Theorem 9.3.6 $T \notin \mathfrak{C}$. □

A full characterization of the classes $\mathfrak{A}$ and $\mathfrak{C}$ (necessary and sufficient conditions for a set $T \subset \mathbb{R}^N$ to belong to $\mathfrak{A}$ resp. $\mathfrak{C}$) will be given in the next chapter.

9.4 Measurability of convex functions

We start with the following result.

Theorem 9.4.1. *Let $D \subset \mathbb{R}^N$ be a convex and open set, and let $f : D \to \mathbb{R}$ be a convex function. Further, let $T \subset D$ be a Lebesgue measurable set of positive measure, and let $g : T \to \mathbb{R}$ be a measurable function. If*

$$f(x) \leqslant g(x) \qquad for \quad x \in T, \tag{9.4.1}$$

then the function f is continuous.

Proof. Put

$$T_k = \{x \in T \mid k - 1 \leqslant g(x) < k\}, \qquad k \in \mathbb{Z}. \tag{9.4.2}$$

Since the function g is measurable, all sets T_k are measurable. Moreover, the sets T_k are pairwise disjoint and $\bigcup\limits_{k=-\infty}^{\infty} T_k = T$. Hence

$$\sum_{k=-\infty}^{\infty} m(T_k) = m(T) > 0.$$

Thus there exists a $k_0 \in \mathbb{Z}$ such that

$$m(T_{k_0}) > 0. \tag{9.4.3}$$

By (9.4.1) and (9.4.2)

$$f(x) \leqslant g(x) < k_0 \qquad \text{for } x \subset T_{k_0},$$

and thus f is bounded above on a set of positive measure (cf. (9.4.3)). By the Theorem of Ostrowski (Theorem 9.3.1) f is continuous. $\qquad\square$

Taking in Theorem 9.4.1 $T = D$ and $g = f$, we obtain the following (Sierpiński [277], Blumberg [33], Bonnesen-Fenchel [35]).

Theorem 9.4.2. [Theorem of Sierpiński]. *Let $D \subset \mathbb{R}^N$ be a convex and open set, and let $f : D \to \mathbb{R}$ be a convex function. If f is measurable, then it is continuous.*

Since every additive function is convex, we obtain as a particular case of Theorem 9.4.2 the following result, which we will formulate as a separate theorem.

Theorem 9.4.3. [Theorem of Fréchet]. *Every measurable additive function $f : \mathbb{R}^N \to \mathbb{R}$ is continuous.*

In the case $N = 1$, Theorem 9.4.3 was first proved by M. Fréchet [90], [91]. Various other proofs were then supplied by W. Sierpiński [279], [280], S. Banach [19], M. Kac [160], A. Alexiewicz-W. Orlicz [13] and T. Figiel [80].

Concerning other measures than the Lebesgue measure cf. Kuczma-Smítal [190], Paganoni [251]; cf. also Fischer-Słodkowski [85].

Theorem 9.4.2 explains why all examples of discontinuous convex functions are non-effective. In fact, all such functions are necessarily non-measurable, and non-measurable functions exist only under the assumption of the Axiom of Choice. Such functions are highly pathological (cf. also Chapter 12), and have no practical importance. Therefore, when dealing with convex functions, it is of importance to know possibly weak conditions which would guarantee that the functions dealt with are continuous. The theorems in the present chapter furnish such conditions.

9.5 Plane curves

We have seen that in $\mathbb{R}$ there exist sets of measure zero and of the first category which nevertheless belong to the class $\mathfrak{C}_1$. (The Cantor set $\mathbf{C}$ is such a set; cf. Corollary 9.3.1). Do there exist sets with similar properties in spaces of higher dimensions? In the present section we answer this question in the affirmative for $N = 2$. The case of an arbitrary $N > 1$ will be dealt with in 9.6. The results in the present section are due to M. Kuczma [184].

Suppose that $N > 1$, let $D \subset \mathbb{R}^{N-1}$ be an arbitrary set, and let $\varphi : D \to \mathbb{R}$ be an arbitrary function. By $\mathrm{Gr}(\varphi)$ we denote the graph of φ, i.e., the set

$$\mathrm{Gr}(\varphi) = \{\, (x, y) \in \mathbb{R}^N \mid x \in D,\ y = \varphi(x) \,\}.$$

If the function φ is continuous, then the set $\mathrm{Gr}(\varphi) \subset \mathbb{R}^N$ is of measure zero and nowhere dense.

Theorem 9.5.1. *Let $D \subset \mathbb{R}^{N-1}$ be an arbitrary set, and let $\varphi : D \to \mathbb{R}$ be given as*

$$\varphi(x) = c + f(x) \qquad \text{for } x \in D, \tag{9.5.1}$$

where $f : \mathbb{R}^{N-1} \to \mathbb{R}$ is an additive function, and $c \in \mathbb{R}$ is a constant. Then $\mathrm{Gr}(\varphi) \notin \mathfrak{C}_N$.

Proof. Assume first that $D = \mathbb{R}^{N-1}$. Suppose that $\mathrm{Gr}(\varphi) \in \mathfrak{C}_N$. Let $\mathbf{c} = (0, \dots, 0, c) \in \mathbb{R}^N$. We have $\mathrm{Gr}(f) = \mathrm{Gr}(\varphi) - \mathbf{c}$. By Theorems 9.2.8 and 9.2.2 we get hence $\mathrm{Gr}(f) \in \mathfrak{C}_N$.

Take arbitrary $p_1, \dots, p_n \in \mathrm{Gr}(f)$, and arbitrary $\lambda_1, \dots, \lambda_n \in \mathbb{Q}$. We may write $p_i = (x_i, y_i)$, where $x_i \in \mathbb{R}^{N-1}$, $y_i = f(x_i)$, $i = 1, \dots, n$. Hence by Theorem 5.2.1

$$\lambda_1 y_1 + \dots + \lambda_n y_n = \lambda_1 f(x_1) + \dots + \lambda_n f(x_n) = f(\lambda_1 x_1 + \dots + \lambda_n x_n). \tag{9.5.2}$$

Relation (9.5.2) means that the point $\lambda_1 p_1 + \dots + \lambda_n p_n \in \mathrm{Gr}(f)$. Hence[1] $E\big(\mathrm{Gr}(f)\big) \subset \mathrm{Gr}(f) \neq \mathbb{R}^N$. By Theorem 9.3.6 $\mathrm{Gr}(f) \notin \mathfrak{C}_N$. This contradiction shows that also $\mathrm{Gr}(\varphi) \notin \mathfrak{C}_N$.

Now, if $D \neq \mathbb{R}^{N-1}$, then take the extension φ_0 of φ onto $\mathbb{R}^{N-1}$:

$$\varphi_0(x) = c + f(x) \qquad \text{for } x \in \mathbb{R}^{N-1}.$$

We have $\mathrm{Gr}(\varphi) \subset \mathrm{Gr}(\varphi_0)$, and, as we have already shown, $\mathrm{Gr}(\varphi_0) \notin \mathfrak{C}_N$. By Corollary 9.2.1 also $\mathrm{Gr}(\varphi) \notin \mathfrak{C}_N$. ∎

---

[1] Actually we have the equality, since the converse inclusion is trivial.

Now we restrict ourselves to the case $N = 2$, so x is a single real variable. If $D \subset \mathbb{R}$ is an interval, and a function φ of form (9.5.1) is continuous, then also f is continuous on D, and hence bounded on every compact subinterval of D. By Lemma 9.3.1 f is continuous in $\mathbb{R}$, and hence, by Theorem 5.5.2, of the form $f(x) = ax$ with a certain $a \in \mathbb{R}$. Thus (9.5.1) reduces to

$$\varphi(x) = c + ax \qquad \text{for } x \in D. \tag{9.5.3}$$

Theorem 9.5.2. *Let $D \subset \mathbb{R}$ be an interval, and let $\varphi : D \to \mathbb{R}$ be a continuous function which is not of form (9.5.3) for any constants $a, c \in \mathbb{R}$. Then $\mathrm{Gr}(\varphi) \in \mathfrak{A}_2$.*

Proof. There exists an interval $[\alpha, \beta] \subset D$ such that φ is not of form (9.5.3) for all $x \in [\alpha, \beta]$. Put $\widehat{\varphi}(x) = \varphi(x + \alpha) - \varphi(\alpha)$ for $x \in [0, \ \beta - \alpha]$. We have $\mathrm{Gr}(\varphi) = \mathrm{Gr}(\widehat{\varphi}) - (\alpha, \varphi(\alpha))$, and thus, by Theorems 9.2.8 and 9.2.2, it is enough to prove that $\mathrm{Gr}(\widehat{\varphi}) \in \mathfrak{A}_2$. Note that also the function $\widehat{\varphi}$ is not of form (9.5.3) for $x \in I = [0, \beta - \alpha]$.

Put $\phi(x, t) = \widehat{\varphi}(t) + \widehat{\varphi}(x - t)$ for $x \in I$, $t \in [0, x]$. Suppose that for every $x \in I$ and for all $t, s \in [0, x]$ we have

$$\phi(x, t) = \phi(x, s),$$

i.e., $\widehat{\varphi}(t) + \widehat{\varphi}(x - t) = \widehat{\varphi}(s) + \widehat{\varphi}(x - s)$. Setting, in particular, $t = x$, we obtain $\widehat{\varphi}(x) + \widehat{\varphi}(0) = \widehat{\varphi}(s) + \widehat{\varphi}(x - s)$, or, with $u = s, v = x - s$, (since $\widehat{\varphi}(0) = 0$),

$$\widehat{\varphi}(u + v) = \widehat{\varphi}(u) + \widehat{\varphi}(v). \tag{9.5.4}$$

Relation (9.5.4) holds, in particular, for all $u, v \in \frac{1}{2}I$. In 13.5 we will show that then $\widehat{\varphi}$ can be extended onto $\mathbb{R}$ to an additive function, and thus, being continuous, is of the form $\widehat{\varphi}(x) = ax$ for $x \in \frac{1}{2}I$. Now take an arbitrary $x \in I$. Then $\frac{1}{2}x \in \frac{1}{2}I$, whence $\widehat{\varphi}\left(\frac{1}{2}x\right) = \frac{1}{2}ax$. Setting in (9.5.4) $u = v = \frac{1}{2}x$ we get

$$\widehat{\varphi}(x) = \widehat{\varphi}\left(\frac{1}{2}x\right) + \widehat{\varphi}\left(\frac{1}{2}x\right) = \frac{1}{2}ax + \frac{1}{2}ax = ax.$$

Thus $\widehat{\varphi}$ has form (9.5.3) (with $c = 0$) for all $x \in I$, contrary to the supposition.

Consequently we may assume that there exists an $x_0 \in I$ such that

$$M = \sup_{t \in [0, \ x_0]} \phi(x_0, \ t) > \inf_{t \in [0, \ x_0]} \phi(x_0, \ t) = m,$$

and due to the continuity of $\widehat{\varphi}$ we may even assume that $x_0 \in \mathrm{int}\, I$. Since $\phi(x_0, \ t)$ is a continuous function of t, there exist in the interval $[0, \ x_0]$ points t_0 and T_0 such that

$$\phi(x_0, \ t_0) = m, \qquad \phi(x_0, \ T_0) = M.$$

Put $d = M - m$. Again by the continuity of ϕ (now with respect to x) we can find an $x_1 \in (x_0, \ \beta - \alpha)$ such that

$$\phi(x, \ t_0) < m + \frac{1}{3}d, \qquad \phi(x, \ T_0) > M - \frac{1}{3}d \qquad \text{for } x \in [x_0, \ x_1]. \tag{9.5.5}$$

Now take a $p = (x, y) \in R = (x_0, x_1) \times \left(m + \frac{1}{3}d, M - \frac{1}{3}d \right) \neq \varnothing$. It follows from (9.5.5) and from the Darboux property (intermediate value property) of $\phi(x, t)$ (as a function of t) that there exists a t between t_0 and T_0 (and so in $[0, x_0]$) such that

$$\phi(x, t) = y. \tag{9.5.6}$$

Put $s = x - t$. Then, by (9.5.6) and by the definition of ϕ,

$$x = t + s, \qquad y = \widehat{\varphi}(t) + \widehat{\varphi}(s). \tag{9.5.7}$$

Relation (9.5.7) shows that $p \in \mathrm{Gr}(\widehat{\varphi}) + \mathrm{Gr}(\widehat{\varphi})$. Consequently $R \subset \mathrm{Gr}(\widehat{\varphi}) + \mathrm{Gr}(\widehat{\varphi})$ so that $\mathrm{int}[\mathrm{Gr}(\widehat{\varphi}) + \mathrm{Gr}(\widehat{\varphi})] \neq \varnothing$. By Theorem 9.3.3 $\mathrm{Gr}(\widehat{\varphi}) \in \mathfrak{A}_2$, whence also $\mathrm{Gr}(\varphi) \in \mathfrak{A}_2$. $\qquad\square$

Theorem 9.5.2 has been extended by R. Ger [102], [103] to higher dimensions, but he had to make stronger assumptions concerning the regularity of φ. In the next section we will present one of these generalizations (Ger [103]). At any case, this result implies, in particular, that in the space $\mathbb{R}^N$ ($N > 1$) there exist sets belonging to $\mathfrak{A}_N$ which are nowhere dense and of the N-dimensional Lebesgue measure zero (cf. also Exercise 10.6).

9.6   Skew curves

Also in this section we assume that $N > 1$. Let $D \subset \mathbb{R}$ be an interval, and let $\varphi : D \to \mathbb{R}^N$ be a C^1 function such that[2] $\varphi(0) = 0$. The function φ can be written as $\varphi = (\varphi_1, \ldots, \varphi_N)$, and the functions $\varphi_i : D \to \mathbb{R}$, $i = 1, \ldots, N$, are of class C^1 in D. The function φ describes a curve in $\mathbb{R}^N$. We assume that this curve does not lie entirely on an $(N-1)$-dimensional hyperplane. Analytically this means that there do not exist constants $c_1, \ldots, c_N \in \mathbb{R}$ such that $|c_1| + \cdots + |c_N| > 0$ and

$$c_1\varphi_1 + \cdots + c_N\varphi_N = 0 \qquad \text{in } D. \tag{9.6.1}$$

In other words, the functions $\varphi_1, \ldots, \varphi_N$ are linearly independent over $\mathbb{R}$. Then also the functions $\varphi'_1, \ldots, \varphi'_N$ are linearly independent (over $\mathbb{R}$), for if we had

$$c_1\varphi'_1 + \cdots + c_N\varphi'_N = 0 \qquad \text{in } D \tag{9.6.2}$$

with certain $c_1, \ldots, c_n \in \mathbb{R}$, then, integrating relation (9.6.2), we would get (9.6.1). We may form the determinant $d : D^N \to \mathbb{R}$:

$$d(x_1, \ldots, x_N) = \begin{vmatrix} \varphi'_1(x_1), & \ldots, & \varphi'_1(x_N) \\ \vdots & & \vdots \\ \varphi'_N(x_1), & \ldots, & \varphi'_N(x_N) \end{vmatrix} \tag{9.6.3}$$

---

[2] This assumption is not essential, since it can always be realized by a suitable shift.

and the linear independence of $\varphi'_1, \ldots, \varphi'_N$ implies that $d(\widehat{x}_1, \ldots, \widehat{x}_N) \neq 0$ for some $\widehat{x}_1, \ldots, \widehat{x}_N \in D$. Write $\widehat{p} = (\widehat{x}_1, \ldots, \widehat{x}_N)$, then $d(\widehat{p}) \neq 0$. The continuity of the derivatives of φ implies that there exists a neighbourhood U of $\widehat{p}$ such that

$$d(p) \neq 0 \qquad \text{for } p \in U$$

$(p = (x_1, \ldots, x_N))$. Define the function $\phi : U \to \mathbb{R}^N$ by

$$\phi(x_1, \ldots, x_N) = \Big(\sum_{i=1}^{N} \varphi_1(x_i), \ldots, \sum_{i=1}^{N} \varphi_N(x_i) \Big).$$

The derivative of ϕ is an $N \times N$ square matrix with the entries $\dfrac{\partial \phi_i}{\partial x_j} = \varphi'_i(x_j)$, $i, j = 1, \ldots, N$. Hence

$$\det \phi'(x_1, \ldots, x_N) = d(x_1, \ldots, x_N),$$

whence $\det \phi'(p) \neq 0$ in U. Consequently there exist intervals $(\alpha_i, \ \beta_i) \subset D$, $i = 1, \ldots, N$, such that $\widehat{x}_i \in (\alpha_i, \ \beta_i)$, $i = 1, \ldots, N$, and ϕ is a diffeomorphism of $V = \overset{N}{\underset{i=1}{\times}} (\alpha_i, \ \beta_i)$ onto a suitable domain in $\mathbb{R}^N$, and $V \subset U$.

For any $A \subset D$ put

$$H_A = \{ p \in \mathbb{R}^N \mid p = \varphi(x), \ x \in A \}.$$

Theorem 9.6.1. *With the above notation, assume that $Z \subset D$ is a set of positive one-dimensional Lebesgue measure and that $\widehat{x}_i$, $i = 1, \ldots, N$, are density points of Z. Then $H_Z \in \mathfrak{A}_N$.*

Proof. Put

$$Z_i = Z \cap (\alpha_i, \ \beta_i), \qquad i = 1, \ldots, N.$$

Since $\widehat{x}_i$, $i = 1, \ldots, N$, are density points of Z, the one dimensional Lebesgue measure of Z_i is positive for $i = 1, \ldots, N$, whence also $m\big(\overset{N}{\underset{i=1}{\times}} Z_i \big) > 0$ (m stands here for the N-dimensional Lebesgue measure). Further, we have

$$m(H_{Z_1} + \cdots + H_{Z_N}) = \int\limits_{\overset{N}{\underset{i=1}{\times}} Z_i} | \det \phi'(p)| \, dp > 0,$$

since $\det \phi' \neq 0$ in $\overset{N}{\underset{i=1}{\times}} Z_i \subset V \subset U$. On the other hand,

$$H_{Z_1} + \cdots + H_{Z_N} \subset H_Z + \cdots + H_Z = S_N(H_Z)$$

so that $m\big(S_N(H_Z)\big) > 0$. By Theorem 9.3.3 $H_Z \in \mathfrak{A}_N$. $\qquad\square$

The condition that $\widehat{x}_i$, $i = 1, \ldots, N$, are density points of Z can be replaced by the condition, which is less general but easier to verify that the one-dimensional measure of $Z \cap (\alpha, \beta)$ is positive for every interval $(\alpha, \beta) \subset D$. Another possible variant is that $d(x_1, \ldots, x_N) \neq 0$ for almost every $(x_1, \ldots, x_N) \in D^N$ such that $x_i \neq x_j$ for $i \neq j$. Then $\widehat{x}_1, \ldots, \widehat{x}_N$ can always be chosen so that $d(\widehat{x}_1, \ldots, \widehat{x}_N) \neq 0$ and $\widehat{x}_1, \ldots, \widehat{x}_N$ are density points of Z. Taking these remarks into account, we may formulate the following corollary to Theorem 9.6.1.

Theorem 9.6.2. *Let $D \subset \mathbb{R}$ be an interval, and let $\varphi : D \to \mathbb{R}^N$ be a function of class C^1 defining in $\mathbb{R}^N$ a curve which does not lie entirely on an $(N-1)$-dimensional hyperplane. Let $Z \subset D$ be a measurable set of positive one-dimensional measure. If either of the conditions is fulfilled:*

(i) *The one-dimensional measure of $Z \cap (\alpha, \beta)$ is positive for every interval $(\alpha, \beta) \subset D$;*

(ii) *The determinant (9.6.3) is non-zero for almost every $(x_1, \ldots, x_N) \in D^N$ such that $x_i \neq x_j$ for $i \neq j$;*

then $H_Z \in \mathfrak{A}_N$.

The set H_Z is very small. It may even be much smaller than the curve $y = \varphi(x)$, which is of N-dimensional measure zero and nowhere dense. Thus in every space $\mathbb{R}^N$ there exist very small sets belonging to $\mathfrak{A}_N$.

As an example consider the curve

$$\varphi_1(x) = x, \ldots, \varphi_i(x) = x^i, \ldots, \varphi_N(x) = x^N,$$

$x \in D$, an interval. We have $\big(\text{cf. } (9.6.3)\big)$ for $x_1, \ldots, x_N \in D$ such that $x_i \neq x_j$ for $i \neq j$

$$d(x_1, \ldots, x_N) = \begin{vmatrix} 1 & , \ldots, & 1 \\ 2x_1 & , \ldots, & 2x_N \\ \vdots & & \vdots \\ Nx_1^{N-1} & , \ldots, & Nx_N^{N-1} \end{vmatrix} = N! \begin{vmatrix} 1 & , \ldots, & 1 \\ x_1 & , \ldots, & x_N \\ \vdots & & \vdots \\ x_1^{N-1} & , \ldots, & x_N^{N-1} \end{vmatrix} \neq 0,$$

since the last determinant is Van der Monde's determinant. The condition (ii) of Theorem 9.6.2 is fulfilled. For any set $Z \subset D$ of positive one-dimensional measure we have $H_Z \in \mathfrak{A}_N$.

9.7 Boundedness below

In the definition of the class $\mathfrak{B}$ the boundedness above may be replaced by the boundedness below. In fact, if $f : \mathbb{R}^N \to \mathbb{R}$ is an additive function bounded below on a set $T \in \mathfrak{B}$, then $-f$ is an additive function bounded above on T. Thus $-f$, and hence also f, is continuous.

But in the definition of the class $\mathfrak{A}$ the boundedness above cannot be replaced by the boundedness below. There exist discontinuous convex functions even globally bounded below (cf. the end of 6.4). But Theorem 6.2.2 suggests that one might look for conditions that would imply that a convex function is locally bounded below at every point of its domain of definition. In this connection J. Smítal [289] introduced the classes $\mathfrak{D}(D)$ which may be defined as follows.

Let $D \subset \mathbb{R}^N$ be a convex and open set. $\mathfrak{D}(D)$ is the class of all sets $T \subset D$ with the property that every convex function $f : D \to \mathbb{R}$ which is bounded below on T is locally bounded below at every point of D:

$$\mathfrak{D}(D) = \{\, T \subset D \mid \text{ every convex function } f : D \to \mathbb{R} \text{ bounded below on } T \text{ is} $$
$$\text{locally bounded below at every point of } D\}.$$

Later, in 10.6, we will show that the classes $\mathfrak{D}(D)$ actually depend on D.

By Theorem 6.2.2 every open set $G \subset D$ belongs to $\mathfrak{D}(D)$. Now we will prove the following

Lemma 9.7.1. *Let $D \subset \mathbb{R}^N$ be a convex and open set, and let $f : D \to \mathbb{R}$ be a convex function. Let $T \subset D$ be a set fulfilling one of the following conditions:*

(i) $T \in \mathcal{L}$ and $m(T) > 0$;

(ii) T is of the second category and with the Baire property.

Then[3]

$$m_f(x) < K \leqslant f(x) \qquad \text{for } x \in T \tag{9.7.1}$$

cannot hold for any real constant K.

Proof. Suppose that (9.7.1) holds with a certain constant $K \in \mathbb{R}$. If (*i*) holds, then take s to be an arbitrary density point of T. If (*ii*) holds, then

$$T = (G \cup P) \setminus R, \tag{9.7.2}$$

where $G \neq \varnothing$ is open, and the sets P, R are of the first category. Note that then for every set $U \subset G$ the set $U \setminus T$ is of the first category. Let s be an arbitrary point of the set $G \setminus R \subset T$.

Fix an $\varepsilon > 0$, and, for any $x \in \mathbb{R}^N$ and $r > 0$, let $K(x, r)$ denote the open ball centered at x and with the radius r. Put

$$S = K(s, 2\varepsilon).$$

If condition (*i*) is fulfilled, then we may choose ε so small that for every open ball $U \subset S$ centered at s we have

$$m(U \cap T) > \frac{1}{2} m(U), \tag{9.7.3}$$

and, moreover, $S \subset D$. If condition (*ii*) is fulfilled, then we assume that ε is so small that $S \subset G$ (whence also $S \subset D$; cf. (9.7.2)) so that for every set $U \subset S$ the set $U \setminus T$ is of the first category.

---

[3] Concerning the definition of m_f cf. 6.3.

Put
$$S_n = \frac{1}{2^n} S + \left(1 - \frac{1}{2^n}\right) s = K\left(s, \frac{2\varepsilon}{2^n}\right) \subset S, \qquad n \in \mathbb{N}, \qquad (9.7.4)$$

and
$$T_n = T \cap S_n, \qquad n \in \mathbb{N}. \qquad (9.7.5)$$

There exists an $L \in \mathbb{R}$ such that
$$m_f(s) < L < K,$$

and, by the definition of m_f, for every $n \in \mathbb{N}$ there exists a point $x_n \in D$ such that
$$|x_n - s| < \frac{\varepsilon}{2(2^n - 1)} \qquad \text{and} \qquad f(x_n) < L. \qquad (9.7.6)$$

Further put for $n \in \mathbb{N}$
$$\begin{aligned}
U_n &= 2^n S_{n+1} + (1 - 2^n)x_n = \frac{1}{2} S + 2^n \left(1 - \frac{1}{2^{n+1}}\right) s + (1 - 2^n)x_n \\
&= \frac{1}{2} S + 2^n(s - x_n) - \frac{1}{2}s + x_n = \frac{1}{2} S + \frac{1}{2} s + (2^n - 1)(s - x_n) \\
&= K(s, \varepsilon) + (2^n - 1)(s - x_n) \subset S,
\end{aligned}$$

since by (9.7.6) $|(2^n - 1)(s - x_n)| < \dfrac{\varepsilon}{2}$, and
$$B_n = 2^n T_{n+1} + (1 - 2^n)x_n \subset U_n \subset S.$$

We have by (9.7.5) $U_n \setminus B_n = 2^n(S_{n+1} \setminus T_{n+1}) + (1 - 2^n)x_n = 2^n(S_{n+1} \setminus T) + (1 - 2^n)x_n$. If, in particular, condition (ii) is fulfilled, then $S_{n+1} \setminus T$ is of the first category by (9.7.4), and hence also $U_n \setminus B_n$ is of the first category ($U_n \setminus B_n = g(S_{n+1} \setminus T)$), where $g(x) = 2^n x + (1 - 2^n)x_n$ is a homeomorphism) for every $n \in \mathbb{N}$. Note also that since U_n are open balls with a fixed radius ε, whose centers lie not further apart from s than $\dfrac{\varepsilon}{2}$, there is an open set $U_0 \subset U_n$ for all $n \in \mathbb{N}$. We may take, e.g., $U_0 = K\left(s, \dfrac{\varepsilon}{2}\right)$.

Put also
$$Q_n = \bigcup_{i=n}^{\infty} B_i, \qquad V_n = \bigcup_{i=n}^{\infty} U_i, \qquad (9.7.7)$$

$$Q = \bigcap_{n=1}^{\infty} Q_n, \qquad V = \bigcap_{n=1}^{\infty} V_n \qquad (9.7.8)$$

so that we have
$$Q_{n+1} \subset Q_n, \qquad V_{n+1} \subset V_n \qquad \text{for } n \in \mathbb{N}. \qquad (9.7.9)$$

Hence it follows that $U_0 \subset V_n$ for all $n \in \mathbb{N}$, and hence $U_0 \subset V$. Since U_0 is open (and non-empty), it is of the second category[4], and also V is of the second category.

---

[4] Cf. the proof of Theorem 2.9.1.

Now, if (i) is fulfilled, then we argue as follows. All U_n are open balls with a fixed radius ε, so there is a positive number c such that $m(U_n) = c$ for all $n \in \mathbb{N}$. Hence $m(V_n) \geqslant c$ for all $n \in \mathbb{N}$, and by (9.7.8) and (9.7.9)

$$m(V) = \lim_{n \to \infty} m(V_n) \geqslant c. \qquad (9.7.10)$$

Moreover, by (9.7.3),

$$m(T_n) = m(S_n \cap T) > \frac{1}{2} m(S_n), \qquad (9.7.11)$$

and by (9.7.11)

$$m(B_n) = 2^{nN} m(T_{n+1}) > 2^{nN-1} m(S_{n+1}) = \frac{1}{2} m(2^n S_{n+1}) = \frac{1}{2} c, \qquad n \in \mathbb{N},$$

whence also

$$m(Q_n) > \frac{1}{2} c, \qquad n \in \mathbb{N},$$

and by (9.7.8) and (9.7.9)

$$m(Q) = \lim_{n \to \infty} m(Q_n) \geqslant \frac{1}{2} c.$$

Hence

$$Q \neq \varnothing. \qquad (9.7.12)$$

If condition (ii) is fulfilled, then we argue as follows. For every $n \in \mathbb{N}$ the set $U_n \setminus B_n$ is of the first category, as pointed out above. Since $B_n \subset U_n$, this means that

$$U_n = B_n \cup (U_n \setminus B_n), \qquad n \in \mathbb{N},$$

and

$$V_n = Q_n \cup \bigcup_{i=n}^{\infty} (U_i \setminus B_i), \qquad n \in \mathbb{N}. \qquad (9.7.13)$$

Thus $V_n \setminus Q_n \subset \bigcup_{i=n}^{\infty} (U_i \setminus B_i)$ is a set of the first category for every $n \in \mathbb{N}$. Similarly, we obtain from (9.7.13) and (9.7.8)

$$V \subset Q \cup \bigcup_{i=1}^{\infty} (U_i \setminus B_i). \qquad (9.7.14)$$

Since V is of the second category, (9.7.14) implies that also Q is of the second category, whence (9.7.12) results.

Thus (9.7.12) is valid in either case. Now take an arbitrary $x \in Q \subset Q_m$, $m \in \mathbb{N}$. There exists an index $n \geqslant m$ such that $x \in B_n$. Therefore

$$x = 2^n t + (1 - 2^n) x_n$$

with a $t \in T$. Hence

$$t = \frac{1}{2^n}\, x + \left(1 - \frac{1}{2^n}\right) x_n,$$

and

$$f(t) \leqslant \frac{1}{2^n}\, f(x) + \left(1 - \frac{1}{2^n}\right) f(x_n).$$

Hence it follows (note that $1 - 2^n < 0$) that

$$f(x) \geqslant 2^n f(t) + (1 - 2^n)f(x_n) \geqslant 2^n K + (1 - 2^n)L = L + 2^n(K - L).$$

Letting $m \to \infty$ (whence also $n \to \infty$) we obtain hence, since $K - L > 0$,

$$f(x) = \infty,$$

which contradicts the assumption $f : D \to \mathbb{R}$. The contradiction obtained shows that (9.7.1) is impossible. $\square$

Part (i) of the above Lemma, similarly as Theorem 9.7.1 below, is due to A. Császár [50].

Theorem 9.7.1. *Let $D \subset \mathbb{R}^N$ be a convex and open set, and let $f : D \to \mathbb{R}$ be a convex function. Let $T \subset D$ be a measurable set such that $m(T) > 0$. Then for no measurable function $g : T \to \mathbb{R}$ we may have*

$$m_f(x) < g(x) \leqslant f(x) \qquad \text{for } x \in T. \tag{9.7.15}$$

Proof. Suppose that (9.7.15) holds for a certain measurable set $T \subset D$ of positive measure, and for a certain measurable function $g : T \to \mathbb{R}$. Put for $\lambda \in \mathbb{Q}$

$$T_\lambda = \{\, x \in T \mid m_f(x) < \lambda \leqslant g(x) \,\}.$$

$T_\lambda, \lambda \in \mathbb{Q}$, are measurable sets, and

$$T = \bigcup_{\lambda \in \mathbb{Q}} T_\lambda. \tag{9.7.16}$$

If all T_λ had measure zero, we would get from (9.7.16) $m(T) = 0$. Consequently there exists a $\lambda_0 \in \mathbb{Q}$ such that $m(T_{\lambda_0}) > 0$. For $x \in T_{\lambda_0}$ we have by (9.7.15)

$$m_f(x) < \lambda_0 \leqslant f(x),$$

contrary to Lemma 9.7.1. $\square$

We also prove the following Theorem of Hukuhara[5].

Theorem 9.7.2. *Let $D \subset \mathbb{R}^N$ be a convex and open set, and let $T \subset D$ be a set fulfilling one of conditions (i), (ii) of Lemma 9.7.1. Then $T \in \mathfrak{D}(D)$.*

---

[5] M. Hukuhara [148] proved part (i) of Theorem 9.7.2 in the case $N = 1$. This result has then been extended to higher dimensions by S. Marcus [218] and A. Császár [50].

Proof. Suppose the contrary. Let $f : D \to \mathbb{R}$ be a convex function bounded below on T:

$$f(x) \geqslant K \qquad \text{for } x \in T, \tag{9.7.17}$$

but locally unbounded below at a point of D. By Theorem 6.2.2 f is locally unbounded below at every point of D so that

$$m_f(x) = -\infty \qquad \text{for } x \in D. \tag{9.7.18}$$

Relations (9.7.17) and (9.7.18) imply (9.7.1), contrary to Lemma 9.7.1. $\qquad \square$

Of course, in Theorem 9.7.2 and Lemma 9.7.1 condition (i) may be replaced by the condition

$$m_i(T) > 0, \tag{i'}$$

since by the definition of the inner measure (3.1) condition (i') implies that there exists a measurable set $T_0 \subset T$ fulfilling (i), and (9.7.17) (resp. (9.7.1)) holds for $x \in T_0$ as well.

A more thorough study of the classes $\mathfrak{D}(D)$ is deferred to 10.6.

9.8 Restrictions of convex functions and additive functions

Theorem 6.1.1 says that for every convex function $f : D \to \mathbb{R}$ and for every $a, b \in D$ the restriction $f \mid Q(a, b)$ is continuous. We may ask what can be inferred from the fact that for a certain set $T \subset D$ the restriction $f \mid T$ is continuous. This gives rise to the following definitions (cf. Kominek-Kominek [171]).

$N \in \mathbb{N}$ being fixed, let $\mathfrak{A}_C$ denote the class of sets $T \subset \mathbb{R}^N$ such that if D is a convex and open set such that $T \subset D$, and $f : D \to \mathbb{R}$ is a convex function such that the restriction $f \mid T$ is continuous, then f is continuous in D:

$$\mathfrak{A}_C = \{ \, T \subset \mathbb{R}^N \mid \text{ every convex function } f : D \to \mathbb{R}, \text{ where } D \text{ is a convex and} \\ \text{open and } T \subset D \subset \mathbb{R}^N, \text{ such that the restriction } f \mid T \text{ is} \\ \text{continuous, is continuous}\}.$$

Similarly, $\mathfrak{B}_C$ is the class of sets $T \subset \mathbb{R}^N$ such that if $f : \mathbb{R}^N \to \mathbb{R}$ is an additive function such that the restriction $f \mid T$ is continuous, then f is continuous in $\mathbb{R}^N$:

$$\mathfrak{B}_C = \{ \, T \subset \mathbb{R}^N \mid \text{ every additive function } f : \mathbb{R}^N \to \mathbb{R} \text{ such that} \\ \text{the restriction } f \mid T \text{ is continuous, is continuous}\}.$$

Let us note that the definition of the class $\mathfrak{A}_C$ makes the latter independent of D.

The classes $\mathfrak{A}_C$ and $\mathfrak{B}_C$ will be studied also in the next chapter. In particular, the relation $\mathfrak{A}_C = \mathfrak{B}_C$ does not hold. But directly from the definition it follows, since every additive function is convex, that

$$\mathfrak{A}_C \subset \mathfrak{B}_C.$$

Before proving analogues of Theorems 9.3.1 and 9.3.2 we will prove some other criteria for a set $T \subset \mathbb{R}^N$ to belong to $\mathfrak{A}_C$ or $\mathfrak{B}_C$.

We start with a lemma which will be proved in Chapter 11 (cf. Theorem 11.8.1). The operation R has been defined in 9.2. For any operation $F : \mathcal{P}(\mathbb{R}^N) \to \mathcal{P}(\mathbb{R}^N)$ we define the iterates F^n, $n \in \mathbb{N}$, as follows

$$F^1 = F, \qquad F^{n+1}(A) = F\big(F^n(A)\big), \qquad n \in \mathbb{N}.$$

Lemma 9.8.1. *Let $H \subset \mathbb{R}^N$ be a Hamel basis. Then there exists an $n \in \mathbb{N}$ such that $m_e\big(R^n(H)\big) > 0$.*

Theorem 9.8.1 below is due to F. B. Jones [158].

Theorem 9.8.1. *Let $T \subset \mathbb{R}^N$ be an analytic set containing a Hamel basis. Then $T \in \mathfrak{B}_C$.*

Proof. Let $f : \mathbb{R}^N \to \mathbb{R}$ be an additive function such that $f \mid T$ is continuous. Put

$$T_k = \{ x \in T \mid |f(x)| < k \} = T \cap f^{-1}\big((-k,\, k)\big) = (f \mid T)^{-1}\big((-k,\, k)\big), \quad k \in \mathbb{N}.$$

Since $f \mid T$ is continuous, $(f \mid T)^{-1}\big((-k,\, k)\big)$ is open (in T). Hence there exists an open set $G_k \subset \mathbb{R}^N$ such that $T_k = T \cap G_k$. It follows that T_k is analytic (Theorems 2.5.5 and 2.5.6) for every $k \in \mathbb{N}$.

We have $T_k \subset T_{k+1}$, $k \in \mathbb{N}$, and $T = \bigcup\limits_{k=1}^{\infty} T_k$. We want to show that

$$R(T) = \bigcup_{k=1}^{\infty} R(T_k). \tag{9.8.1}$$

If $x \in R(T)$, then $x = u - v$, where $u, v \in T$. There exists a $k \in \mathbb{N}$ such that $u, v \in T_k$, whence $x \in R(T_k) \subset \bigcup\limits_{k=1}^{\infty} R(T_k)$. Thus $R(T) \subset \bigcup\limits_{k=1}^{\infty} R(T_k)$. The converse inclusion is obvious.

Obviously, we have also for arbitrary $A, B \subset \mathbb{R}^N$

$$A \subset B \Rightarrow R(A) \subset R(B). \tag{9.8.2}$$

Repeating the above argument and making use of (9.8.1), we prove by induction that

$$R^n(T) = \bigcup_{k=1}^{\infty} R^n(T_k), \qquad n \in \mathbb{N}. \tag{9.8.3}$$

If $A \subset \mathbb{R}^N$ is an analytic set, then by Theorem 2.5.3 $A \times A$ also is analytic, and since the function $g : A \times A \to \mathbb{R}^N$ defined as $g(x, y) = x - y$ is continuous, also the set $R(A) = g(A \times A)$ is analytic (Theorem 2.5.2). Hence it follows that all the sets $R^n(T_k), n, k \in \mathbb{N}$, are analytic.

Let $H \subset T$ be a Hamel basis. By Lemma 9.8.1 there exists an $n \in \mathbb{N}$ such that $m_e\big(R^n(H)\big) > 0$. By (9.8.2) $R^n(H) \subset R^n(T)$, whence also $m_e\big(R^n(T)\big) > 0$. Relation (9.8.3) implies (cf. Exercise 3.3) that

$$m_e\big(R^n(T)\big) = \lim_{k \to \infty} m_e\big(R^n(T_k)\big),$$

whence it follows that there exists a $k \in \mathbb{N}$ such that $m_e\big(R^n(T_k)\big) > 0$. But $R^n(T_k)$ is an analytic set, and therefore measurable (Theorem 3.1.8). Thus

$$m\big(R^n(T_k)\big) > 0. \tag{9.8.4}$$

We have $|f(x)| < k$ for $x \in T_k$. Now (9.8.4) implies in virtue of Theorem 9.3.1 that $R^n(T_k) \in \mathfrak{C}$ and by Theorems 9.2.3, 9.2.6 and 9.2.2 $T_k \in \mathfrak{C}$. f is bounded on T_k, and hence it is continuous. $\qquad \square$

Before stating the next theorem we need some more definitions. If $a, b \in \mathbb{R}^N$, $a = (a_1, \ldots, a_N)$, $b = (b_1, \ldots, b_N)$, then we say that $a < b$ resp. $a \leqslant b$ iff $a_i < b_i$ resp. $a_i \leqslant b_i$ for $i = 1, \ldots, N$. If $a < b$, then the open interval (a, b) and the closed interval $[a, b]$ are defined as in the one-dimensional case:

$$(a, b) = \{\, x \in \mathbb{R}^N \mid a < x < b \,\}, \qquad [a, b] = \{\, x \in \mathbb{R}^N \mid a \leqslant x \leqslant b \,\}.$$

If $N = 2$, then the interval (open or closed) is a rectangle (open or closed), its lower left and upper right vertices being the endpoints of the interval (the sides of the rectangle are parallel to the coordinate axes).

The following theorem is due to Z. Kominek [173].

Theorem 9.8.2. *Let $T \subset \mathbb{R}^N$ be a set such that for certain $a, b \in \mathbb{R}^N$, $a < b$, and for a certain $n \in \mathbb{N}$, either*

$$\{a\} \cup (a, b) \subset \frac{1}{n}\, S_n(T) \subset [a, b], \tag{9.8.5}$$

or

$$(a, b) \cup \{b\} \subset \frac{1}{n}\, S_n(T) \subset [a, b]. \tag{9.8.6}$$

Then $T \in \mathfrak{A}_C$.

Proof. We assume that relation (9.8.5) holds. If (9.8.6) is fulfilled, the proof is analogous.

Since $T \subset \dfrac{1}{n}\, S_n(T)$, we have $T \subset [a, b]$. Relation (9.8.5) implies that $a \in \dfrac{1}{n}\, S_n(T)$, i.e.,

$$a = \frac{1}{n}\, (t_1 + \cdots + t_n), \qquad t_i \in T, \qquad i = 1, \ldots, n. \tag{9.8.7}$$

Thus $t_i \in [a, b]$, in particular, $t_i \geqslant a$, $i = 1, \ldots, n$. If, for some i, we had $t_i \neq a$, we would get from (9.8.7) $\dfrac{1}{n} \sum_{i=1}^{n} t_i \neq a$. Consequently all $t_i = a$, which shows that $a \in T$.

Now let $D \subset \mathbb{R}^N$ be a convex and open set such that $T \subset D$, and let $f : D \to \mathbb{R}$ be a convex function such that $f \mid T$ is continuous. Let, for any $\varepsilon > 0$, $K(a, \varepsilon)$ be the open ball centered at a and with the radius ε. We will show that there exists an $\varepsilon > 0$ such that f is bounded above on $(a, b) \cap K(a, \varepsilon)$. Hence it will follow by Lemma 9.3.1 that f is continuous in D, i.e., $T \in \mathfrak{A}_C$.

Supposing the contrary, for every $p \in \mathbb{N}$ we can find a point $x_p \in (a, b) \cap K\left(a, \dfrac{1}{p}\right)$ such that $f(x_p) > p$. It follows that

$$\lim_{p \to \infty} x_p = a, \qquad \lim_{p \to \infty} f(x_p) = \infty. \tag{9.8.8}$$

By (9.8.5) $x_p \in \dfrac{1}{n} S_n(T)$, $p \in \mathbb{N}$, i.e.,

$$x_p = \frac{1}{n}(t_{p1} + \cdots + t_{pn}), \qquad t_{pi} \in T, \qquad i = 1, \ldots, n; \qquad p \in \mathbb{N}. \tag{9.8.9}$$

We have $t_{pi} \in [a, b]$, $p \in \mathbb{N}$, $i = 1, \ldots, n$. It is easily seen that (9.8.8) implies

$$\lim_{p \to \infty} t_{pi} = a \qquad \text{for } i = 1, \ldots, n. \tag{9.8.10}$$

Since $f \mid T$ is continuous, $a \in T$, and $t_{pi} \in T$ for $p \in \mathbb{N}$, $i = 1, \ldots, n$, (9.8.10) implies that

$$\lim_{p \to \infty} f(t_{pi}) = f(a), \qquad i = 1, \ldots, n. \tag{9.8.11}$$

By Lemma 5.3.1 and by (9.8.9)

$$f(x_p) \leqslant \frac{1}{n}[f(t_{p1}) + \cdots + f(t_{pn})], \qquad p \in \mathbb{N}, \qquad i = 1, \ldots, n. \tag{9.8.12}$$

Letting $p \to \infty$, we obtain from (9.8.8), (9.8.11) and (9.8.12)

$$\infty \leqslant f(a),$$

which is a contradiction and completes the proof. □

Since for the Cantor set $\mathbf{C}$ we have $\dfrac{1}{2} S_2(\mathbf{C}) = [0, 1]$ (cf. Exercise 2.13), we obtain from Theorem 9.8.2

Corollary 9.8.1. $\mathbf{C} \in \mathfrak{A}_C$.

The following two theorems are due to B. Kominek-Z. Kominek [171] and Z. Kominek [173].

Theorem 9.8.3. *Let $T \subset \mathbb{R}^N$ be a set such that $m_i(T) > 0$. Then $T \in \mathfrak{A}_C$.*

Proof. The condition $m_i(T) > 0$ implies that there exists a compact set $C \subset T$ such that $m(C) > 0$. Let $D \subset \mathbb{R}^N$ be a convex and open set such that $T \subset D$, and let $f : D \to \mathbb{R}$ be a convex function such that the restriction $f \mid T$ is continuous. Hence also the restriction $f \mid C = f \mid T \mid C$ is continuous, and hence bounded. Thus f is bounded above on C, and by Theorem 9.3.1 f is continuous. □

Theorem 9.8.4. *Let $T \subset \mathbb{R}^N$ be a set of the second category and with the Baire property. Then $T \in \mathfrak{A}_C$.*

Proof. The conditions on T imply that

$$T = (G \cup P) \setminus R,$$

where G is a non-empty open set, and P, R are of the first category. We may find $a, b \in G \setminus R$ such that $a < b$ and $[a, b] \subset G$. Put

$$T_0 = [a, b] \setminus R \subset [a, b].$$

Then also

$$\frac{T_0 + T_0}{2} \subset [a, b]. \tag{9.8.13}$$

Since $a \in G \setminus R$, we have $a \in T_0$, and hence

$$a \in \frac{T_0 + T_0}{2}. \tag{9.8.14}$$

Now take an $x \in (a, b)$. There exists a $y \in \mathbb{R}^N$ such that $x + y \in (a, b)$ and $x - y \in (a, b)$ (since (a, b) is open), and hence $y \in \left[(a, b) - x \right] \cap \left[x - (a, b) \right]$. Consequently the last set is non-empty, and being the intersection of two open sets, it is itself open. It follows that it is of the second category (cf. the proof of Theorem 2.9.1). Since the set $(R - x) \cup (x - R)$ is of the first category, we have

$$\left\{ \left[(a, b) - x \right] \cap \left[x - (a, b) \right] \right\} \setminus \left\{ (R - x) \cup (x - R) \right\} \neq \varnothing. \tag{9.8.15}$$

Further,

$$\left\{ \left[(a, b) - x \right] \cap \left[x - (a, b) \right] \right\} \setminus \left\{ (R - x) \cup (x - R) \right\}$$
$$= \left\{ \left[(a, b) - x \right] \setminus (R - x) \right\} \cap \left\{ \left[x - (a, b) \right] \setminus (x - R) \right\}$$
$$= \left\{ \left[(a, b) \setminus R \right] - x \right\} \cap \left\{ x - \left[(a, b) \setminus R \right] \right\}. \tag{9.8.16}$$

Write shortly $W = (a, b) \setminus R$. (9.8.15) and (9.8.16) yield

$$(W - x) \cap (x - W) \neq \varnothing.$$

Hence also the translation

$$W \cap (2x - W) = \left[(W - x) \cap (x - W) \right] + x \neq \varnothing.$$

Thus there exists a $u \in W \cap (2x - W)$. In other words, $u \in W$ and $u = 2x - v$, where $v \in W$. Hence $x = \frac{1}{2}(u + v)$, where $u, v \in W$, i.e., $x \in \frac{1}{2}(W + W) \subset \frac{1}{2}(T_0 + T_0)$. Since x has been arbitrary in (a, b), we obtain hence

$$(a, b) \subset \frac{T_0 + T_0}{2}. \tag{9.8.17}$$

Relations (9.8.13), (9.8.14) and (9.8.17) imply in virtue of Theorem 9.8.2 that $T_0 \in \mathfrak{A}_C$. Since $T_0 \subset T$, we infer hence that $T \in \mathfrak{A}_C$. $\qquad\square$

Theorems 9.8.3 and 9.8.4 will be obtained again on another way in 11.5.

Finally let us note that it is not known if the conditions of Theorem 9.8.1 imply that $T \in \mathfrak{A}_C$.

Exercises

1. Show that Theorem 9.2.7 fails to hold if α is not assumed to be rational (Kuczma [184]).
   [Hint: Let H be an arbitrary Hamel basis of $\mathbb{R}$ containing 1 and $\sqrt{2}$. Let $g : H \to \mathbb{R}$ be an arbitrary function such that $g(1) = 0$, $g(\sqrt{2}) = 1$. Let $f : \mathbb{R} \to \mathbb{R}$ be the additive extension of g. Take $T = \mathbb{Q}$ and $\alpha = \sqrt{2}$.]

2. Show by suitable examples that none of the conditions occurring in Theorem 9.3.5 is necessary for $T \in \mathfrak{C}$.

3. Show that conditions (iii) and (iv) occurring in Theorem 9.3.3 are equivalent; similarly, conditions (v) and (vi) are equivalent.

4. Let $D \subset \mathbb{R}^N$ be a convex and open set, let $f : D \to \mathbb{R}$ be a convex function, and let $W : \mathbb{R} \to \mathbb{R}$ be a polynomial, $W \neq$ const. Show that if the function $W(f)$ is continuous, then also f is continuous.

5. Let $A \subset \mathbb{R}^{N-1}$ be an arbitrary set, and let $\varphi : A \to \mathbb{R}$ be a continuous function. Show that $\text{Gr}(\varphi)$ is a set of N-dimensional measure zero.
   [Hint: It is enough to consider the case, where A is bounded. Let $G \subset \mathbb{R}^{N-1}$ be a bounded open set such that $A \subset G$. Fix an $\varepsilon > 0$. Let $\mathcal{Q}$ be the family of closed cubes $Q \subset G$ such that $\sup_{A \cap Q} f - \inf_{A \cap Q} f < \varepsilon$. Use Vitali theorem.]

6. Show that every subset of a circle with positive linear measure belongs to $\mathfrak{A}_2$.

7. A function $f : X \to Y$ (where X, Y are separable metric spaces) is said to satisfy the *Baire condition* iff for every Borel subset B of Y the set $f^{-1}(B)$ satisfies the Baire condition. This is equivalent to the condition that there exists a residual subset T of X such that $f \mid T$ is continuous (cf. Kuratowski [196]). Show that every convex function which satisfies the Baire condition is continuous.

Chapter 10

The Classes $\mathfrak{A}$, $\mathfrak{B}$, $\mathfrak{C}$

10.1 A Hahn-Banach theorem

The main objective of the present chapter is to prove equality (9.1.2), and to give some characterizations of the classes $\mathfrak{A}$, $\mathfrak{C}$ introduced in the preceding chapter. Thus we do not assume equality (9.1.2) as valid, it will be proved in the sequel[1]. To this aim we will need a rational version of the Hahn-Banach theorem, which we will presently prove. We start with a definition.

Let $A \subset \mathbb{R}^N$ be a set. We say that A is *Q-radial* at a point $a \in A$ iff for every $y \in \mathbb{R}^N$, $y \neq 0$, there exists an $\varepsilon_y > 0$ such that $a + \lambda y \in A$ for every $\lambda \in \mathbb{Q} \cap (0, \varepsilon_y)$.

Lemma 10.1.1. *Let $D \subset \mathbb{R}^N$ be a set Q-convex and Q-radial at a point $x_0 \in D$, and let $L \subset \mathbb{R}^N$ be a linear space (over $\mathbb{Q}$), $x_0 \in L$. Let $f : D \to \mathbb{R}$ be a function fulfilling the inequality*

$$f\big(\lambda x + (1 - \lambda)y\big) \leqslant \lambda f(x) + (1 - \lambda)f(y) \tag{10.1.1}$$

for all $x, y \in D$ and all $\lambda \in \mathbb{Q} \cap [0, 1]$, and let $g : L \to \mathbb{R}$ be a homomorphism[2]. Let $z \notin L$, and put $L_1 = E(L \cup \{z\})$. If

$$g(x) \leqslant f(x) \qquad for\ x \in D \cap L, \tag{10.1.2}$$

then there exists a homomorphism $G : L_1 \to \mathbb{R}$ such that

$$G(x) \leqslant f(x) \qquad for\ x \in D \cap L_1 \tag{10.1.3}$$

and $G \mid L = g$.

Proof. For every $x, y \in L$ and every λ, $\mu \in \mathbb{Q} \cap (0, \infty)$ such that $x + \mu z \in D$ and $y - \lambda z \subset D$ we have by (10.1.2) and (10.1.1)

$$\frac{\lambda}{\lambda + \mu}\, g(x) + \frac{\mu}{\lambda + \mu}\, g(y) = g\left(\frac{\lambda}{\lambda + \mu}\, x + \frac{\mu}{\lambda + \mu}\, y\right)$$

[1] Theorem 10.2.2. The results of Chapter 9 (where equality (9.1.2) was postulated) will not be used in 10.1–10.2 of the present chapter, except for Theorem 9.2.8, in whose proof relation (9.1.2) was not used.
[2] From the space $(L\,;\ \mathbb{Q}\,;\ +\,;\ \cdot)$ into $(\mathbb{R}\,;\ \mathbb{Q}\,;\ +\,;\ \cdot)$.

$$\leqslant f\left(\frac{\lambda}{\lambda+\mu}\,x+\frac{\mu}{\lambda+\mu}\,y\right)$$

$$= f\left(\frac{\lambda}{\lambda+\mu}\,(x+\mu z)+\frac{\mu}{\lambda+\mu}\,(y-\lambda z)\right)$$

$$\leqslant \frac{\lambda}{\lambda+\mu}\,f(x+\mu z)+\frac{\mu}{\lambda+\mu}\,f(y-\lambda z).$$

Hence

$$\frac{g(x)-f(x+\mu z)}{\mu}\leqslant\frac{f(y-\lambda z)-g(y)}{\lambda},$$

and

$$\alpha=\sup_{U}\frac{g(x)-f(x+\mu z)}{\mu}\leqslant\inf_{V}\frac{f(y-\lambda z)-g(y)}{\lambda}=\beta,\qquad(10.1.4)$$

where

$$U=\{(x,\mu)\in L\times\mathbb{Q}\mid\mu>0,\ x+\mu z\in D\},$$

$$V=\{(y,\lambda)\in L\times\mathbb{Q}\mid\lambda>0,\ y-\lambda z\in D\}.$$

Since D is Q-radial at x_0, we have $(x_0,\ \mu)\in U$ for $\mu\in\mathbb{Q}\cap(0,\ \varepsilon_z)$ and $(x_0,\ \lambda)\in V$ for $\lambda\in\mathbb{Q}\cap(0,\ \varepsilon_{-z})$. Consequently $U\neq\varnothing$ and $V\neq\varnothing$. Observe also that (10.1.4) implies that $-\infty<\alpha\leqslant\beta<+\infty$. In particular, $[\alpha,\ \beta]\neq\varnothing$.

Choose any $c\in[\alpha,\ \beta]$. Every $t\in L_1$ may be uniquely written as

$$t=x+\lambda z,\qquad(10.1.5)$$

where $x\in L$ and $\lambda\in\mathbb{Q}$. For such a t define

$$G(t)=g(x)-c\lambda.$$

It is easily seen that $G:L_1\to\mathbb{R}$ is a homomorphism and that $G\mid L=g$. Now take an arbitrary $t\in D\cap L_1$. t may be represented in form (10.1.5). Consider three cases:

I. $\lambda=0$. Then $t=x$, and by (10.1.2)

$$G(t)=g(x)\leqslant f(x)=f(t).$$

Consequently (10.1.3) holds.

II. $\lambda>0$. Since $t\in D$, we have $(x,\ \lambda)\in U$. Since $c\geqslant\alpha$, this implies

$$\frac{g(x)-f(x+\lambda z)}{\lambda}\leqslant c,$$

or

$$g(x)-f(x+\lambda z)\leqslant c\lambda,$$

i.e.,

$$G(t)=g(x)-c\lambda\leqslant f(x+\lambda z)=f(t).$$

Again we obtain (10.1.3).

III. $\lambda < 0$. Since $t \in D$, we have $(x, -\lambda) \in V$. Since $c \leqslant \beta$, this implies

$$c \leqslant \frac{f(x - (-\lambda)z) - g(x)}{-\lambda},$$

or

$$-c\lambda \leqslant f(x + \lambda z) - g(x),$$

i.e.,

$$G(t) = g(x) - c\lambda \leqslant f(x + \lambda z) = f(t).$$

Thus (10.1.3) holds in this case, too. $\qquad\square$

Hence we derive the rational version of the Hahn-Banach theorem (cf. M. E. Kuczma [191], Berz [31]).

Theorem 10.1.1. *Let $D \subset \mathbb{R}^N$ be a set Q-convex and Q-radial at a point $x_0 \in D$, and let $L \subset \mathbb{R}^N$ be a linear space (over $\mathbb{Q}$), $x_0 \in L$. Let $f : D \to \mathbb{R}$ be a function fulfilling inequality (10.1.1) for all $x, y \in D$ and all $\lambda \in \mathbb{Q} \cap [0, 1]$, and let $g : L \to \mathbb{R}$ be a homomorphism fulfilling (10.1.2). Then there exists a homomorphism (i.e., and additive function) $G : \mathbb{R}^N \to \mathbb{R}$ such that $G \mid L = g$ and (10.1.3) holds in D.*

Proof. Let $\mathcal{R}$ be the family of all couples (X, A), where X is a linear space (over $\mathbb{Q}$), $L \subset X \subset \mathbb{R}^N$, and $A : X \to \mathbb{R}$ is a homomorphism such that $A \mid L = g$ and

$$A(x) \leqslant f(x) \qquad \text{for } x \in D \cap X. \qquad (10.1.6)$$

$(L, g) \in \mathcal{R}$, so $\mathcal{R} \neq \varnothing$. We introduce the order in $\mathcal{R}$ in the usual manner: for (X_1, A_1), $(X_2, A_2) \in \mathcal{R}$ we agree that $(X_1, A_1) \leqslant (X_2, A_2)$ iff $X_1 \subset X_2$ and $A_2 \mid X_1 = A_1$. If $\mathcal{I} \subset \mathcal{R}$ is any chain, then put

$$Y = \bigcup_{(X,A)\in\mathcal{I}} X,$$

and define $B : Y \to \mathbb{R}$ putting $B(y) = A(y)$ if $y \in X$ and $(X, A) \in \mathcal{I}$. The couple (Y, B) is an upper bound of $\mathcal{I}$ in $\mathcal{R}$. In fact, if $x, y \in Y$ and $\alpha \in \mathbb{Q}$, then there exists an $(X, A) \in \mathcal{I}$ such that $x, y \in X$. Then also $x + y \in X \subset Y$ and $\alpha x \in X \subset Y$, which shows that Y is a linear space (over $\mathbb{Q}$). Since $L \subset X \subset \mathbb{R}^N$ for all X such that $(X, A) \in \mathcal{I}$, also $L \subset Y = \bigcup_{(X,A)\in\mathcal{I}} X \subset \mathbb{R}^N$. Similarly it is shown that B is a homomorphism fulfilling (10.1.6) and such that $B \mid L = g$.

By Theorem 1.8.1 (Lemma of Kuratowski-Zorn) in $\mathcal{R}$ there exists a maximal element (Z, G). The only thing we need to show is that $Z = \mathbb{R}^N$. Supposing the contrary, let $z \in \mathbb{R}^N \setminus Z$. By Lemma 10.1.1 there exists a homomorphism $G^* : E(Z \cup \{z\}) \to \mathbb{R}$ such that $G^* \mid Z = G$, whence $G^* \mid L = G \mid L = g$, and

$$G^*(x) \leqslant f(x) \qquad \text{for } x \in D \cap E(Z \cup \{z\}).$$

Consequently $(E(Z \cup \{z\}), G^*) \in \mathcal{R}$, and clearly $(Z, G) < (E(Z \cup \{z\}), G^*)$, which contradicts the maximality of (Z, G). Consequently we must have $Z = \mathbb{R}^N$. $\qquad\square$

Theorem 10.1.2. *Let $D \subset \mathbb{R}^N$ be a set which is symmetric with respect to 0 (i.e., $-D = D$), and let $f : D \to \mathbb{R}$ be an even function:*

$$f(-x) = f(x) \qquad \text{for } x \in D. \tag{10.1.7}$$

Further, let $L \subset \mathbb{R}^N$ be a linear space (over $\mathbb{Q}$), and let $g : L \to \mathbb{R}$ be a homomorphism fulfilling (10.1.2). Then

$$|g(x)| \leqslant f(x) \qquad \text{for } x \in D \cap L. \tag{10.1.8}$$

Proof. Take an arbitrary $x \in D \cap L$. We have by (10.1.2) and (10.1.7) $g(-x) \leqslant f(-x) = f(x)$, whence

$$g(x) = -g(-x) \geqslant -f(x).$$

This together with (10.1.2) yields

$$-f(x) \leqslant g(x) \leqslant f(x),$$

which is equivalent to (10.1.8). $\qquad\qquad\qquad\qquad\qquad\qquad\qquad\qquad\qquad\qquad\square$

10.2 The class $\mathfrak{B}$

First we prove the following lemma.

Lemma 10.2.1. *Let $T \subset \mathbb{R}$ be a set Q-convex and Q-radial at a point $x_0 \in T$. Then $T \in \mathfrak{B}$ if and only if $\operatorname{int} T \neq \varnothing$.*

Proof. If $\operatorname{int} T \neq \varnothing$, then every additive function bounded above on T is continuous, as results from Corollary 6.4.1. Consequently $T \in \mathfrak{B}$. In order to prove the necessity of the condition $\operatorname{int} T \neq \varnothing$ we will show that if $\operatorname{int} T = \varnothing$, then $T \notin \mathfrak{B}$.

At first assume that $x_0 = 0$. Since $\operatorname{int} T = \varnothing$ and T is Q-radial at zero, there exist $x, y \in \mathbb{R}$ such that $0 < x < y$, $x \notin T$, $y \in T$. Put

$$S = \{(\xi, \eta) \in \mathbb{Q} \times \mathbb{Q} \mid (1 + \xi)x + \eta y \in T, \; \eta > 0\}.$$

Observe that $(-1, 1) \in S$, so $S \neq \varnothing$. Write

$$s = \sup_{(\xi, \eta) \in S} \frac{\xi}{\eta}.$$

Since T is Q-radial at 0, there exists an $\varepsilon > 0$ ($\varepsilon = \varepsilon_{-y}$) such that $-\alpha y \in T$ for $\alpha \in \mathbb{Q} \cap (0, \varepsilon)$. We will show that

$$s \leqslant \frac{1}{\varepsilon}. \tag{10.2.1}$$

Supposing the contrary, we would be able to find $(\xi, \eta) \in S$ such that $\dfrac{\xi}{\eta} > \dfrac{1}{\varepsilon}$. Then necessarily $\xi > 0$, and we have

$$0 < \frac{\eta}{\xi} < \varepsilon.$$

Hence $-\dfrac{\eta}{\xi}\, y \in T$, and, by the definition of S, $(1+\xi)x + \eta y \in T$. Since T is Q-convex, this implies that

$$x = \frac{1}{1+\xi}\,\big[(1+\xi)x + \eta y\big] + \frac{\xi}{1+\xi}\left(-\frac{\eta}{\xi}\,y\right) \in T,$$

contrary to the choice of $x \notin T$. The contradiction obtained proves (10.2.1).

Since $(-1,1) \in S$, we have

$$s \geqslant -1. \tag{10.2.2}$$

Put $L = E(\{x,y\})$. Observe that x and y are linearly independent over $\mathbb{Q}$. Supposing the contrary, let $\alpha x + \beta y = 0$ for some $\alpha, \beta \in \mathbb{Q}$, $\alpha^2 + \beta^2 > 0$. Since $x, y \neq 0$, we must have $\alpha, \beta \neq 0$, $\alpha\beta < 0$. And since $0 < x < y$, we must have $|\beta| < |\alpha|$. Since $0, y \in T$ and T is Q-convex,

$$x = -\frac{\beta}{\alpha}\,y + \left(1 + \frac{\beta}{\alpha}\right) 0 \in T.$$

This contradiction shows that x and y are linearly independent over $\mathbb{Q}$, and consequently form a base for L. By Theorem 4.3.1 there exists a unique homomorphism $g : L \to \mathbb{R}$ such that

$$g(x) = 1, \qquad g(y) = -s.$$

We propose to show that

$$g(z) \leqslant 1 \qquad \text{for } z \in T \cap L. \tag{10.2.3}$$

Suppose that there exists a $z \in T \cap L$ such that $g(z) > 1$. We have

$$z = \lambda x + \mu y$$

with some $\lambda, \mu \in \mathbb{Q}$. Hence

$$g(z) = \lambda g(x) + \mu g(y) = \lambda - \mu s,$$

and $g(z) > 1$ means

$$\mu s < \lambda - 1. \tag{10.2.4}$$

We must distinguish three cases.

I. $\mu > 0$. Then $(\lambda - 1,\ \mu) \in S$, and the inequality (10.2.4) yields

$$s < \frac{\lambda - 1}{\mu},$$

which contradicts the definition of s.

II. $\mu = 0$. Then (10.2.4) turns into $\lambda > 1$. Moreover, $\lambda x = \lambda x + \mu y = z \in T$. Hence also

$$x = \frac{1}{\lambda}\,(\lambda x) + \left(1 - \frac{1}{\lambda}\right) 0 \in T,$$

a contradiction.

III. $\mu < 0$. Then, by (10.2.4),

$$s > \frac{\lambda - 1}{\mu} = \frac{1 - \lambda}{-\mu}.$$

Consequently there exists a pair $(\xi, \eta) \in S$ such that $\frac{\xi}{\eta} > \frac{1 - \lambda}{-\mu}$. This means that

$$\lambda \eta > \eta + \mu \xi. \qquad (10.2.5)$$

Put

$$\kappa = \frac{\lambda \eta - \mu(1 + \xi)}{\eta - \mu}.$$

By (10.2.5) $\lambda \eta - \mu(1 + \xi) > \eta - \mu$, whence $\kappa > 1$.

Since $(\xi, \eta) \in S$, we have

$$(1 + \xi)x + \eta y \in T.$$

Since $z \in T$, we have

$$\lambda x + \mu y \in T.$$

Hence also

$$u = \frac{1}{\kappa}\left(\lambda x + \mu y\right) = \frac{1}{\kappa}\left(\lambda x + \mu y\right) + \left(1 - \frac{1}{\kappa}\right)0 \in T,$$

and

$$v = \frac{1}{\kappa}\left[(1 + \xi)x + \eta y\right] = \frac{1}{\kappa}\left[(1 + \xi)x + \eta y\right] + \left(1 - \frac{1}{\kappa}\right)0 \in T.$$

Finally,

$$x = \frac{\eta}{\eta - \mu}\,u + \frac{-\mu}{\eta - \mu}\,v \in T,$$

a contradiction.

Thus (10.2.3) has been proved. By Theorem 10.1.1 (where $N = 1$, $f(x) = 1$) there exists an additive function $G : \mathbb{R} \to \mathbb{R}$ such that $G \mid L = g$ and

$$G(z) \leqslant 1 \qquad \text{for } z \in T. \qquad (10.2.6)$$

The function G is discontinuous. If G were continuous, then we would have $G(z) = cz$ for $z \in \mathbb{R}$ with a certain real c, whence in particular $G(x)/x = G(y)/y$. But then

$$\frac{1}{x} = \frac{G(x)}{x} = \frac{G(y)}{y} = -\frac{s}{y},$$

whence $-s = \frac{y}{x} > 1$, which is a contradiction with (10.2.2).

Thus there exists a discontinuous additive function bounded above on T (cf. (10.2.6)). Consequently $T \notin \mathfrak{B}$.

Now assume that $x_0 \neq 0$. Then the set $T - x_0$ is Q-convex and Q-radial at 0 and, similarly as T, has empty interior. By what has already been proved, $T - x_0 \notin \mathfrak{B}$. Consequently there exists a discontinuous additive function $G : \mathbb{R} \to \mathbb{R}$ bounded above on $T - x_0$. By Theorem 9.2.8 G is also bounded above on T. This means that $T \notin \mathfrak{B}$, which completes the proof of the lemma. $\qquad\square$

The above Lemma 10.2.1, similarly as all the theorems in the present section, is due to M. E. Kuczma [191].

Theorem 10.2.1. *Let $T \subset \mathbb{R}^N$ be a set Q-convex and Q-radial at a point x_0. Then $T \in \mathfrak{B}$ if and only if $\mathrm{int}\, T \neq \varnothing$.*

Proof. The sufficiency results from Corollary 6.4.1 similarly as in the proof of Lemma 10.2.1. Thus we need only prove that if $\mathrm{int}\, T = \varnothing$, then $T \notin \mathfrak{B}$.

First assume that $x_0 = 0$. Let $e_1, \dots, e_N$ be a base of $\mathbb{R}^N$ over $\mathbb{R}$. For $i = 1, \dots, N$, let

$$T_i = \{x \in \mathbb{R} \mid xe_i \in T\} \subset \mathbb{R}.$$

First we show that if $\mathrm{int}\, T_i \neq \varnothing$ for $i = 1, \dots, N$, then $\mathrm{int}\, T \neq \varnothing$.

So assume that $\mathrm{int}\, T_i \neq \varnothing$ for $i = 1, \dots, N$. Then there exist open intervals $J_i \subset T_i \setminus \{0\}$, $i = 1, \dots, N$. Put

$$G = \{\, x \in \mathbb{R}^N \mid x = \alpha_1 x_1 e_1 + \cdots + \alpha_N x_N e_N,$$
$$\alpha_i \in \mathbb{Q} \cap (0, 1),\ x_i \in J_i,\ i = 1, \dots, N,\ \sum_{i=1}^{N} \alpha_i = 1 \,\}.$$

The set $G \neq \varnothing$ is open. For let $x \in G$, $y \in \mathbb{R}^N$. x may be written as

$$x = \alpha_1 x_1 e_1 + \cdots + \alpha_N x_N e_N,\ \alpha_i \in \mathbb{Q} \cap (0, 1),\ x_i \in J_i,\ i = 1, \dots, N,\ \sum_{i=1}^{N} \alpha_i = 1.$$
$$(10.2.7)$$

The points $x_1 e_1, \dots, x_N e_N$ are linearly independent over $\mathbb{R}$ (just like $e_1, \dots, e_N$), and since $\dim \mathbb{R}^N = N$ (over $\mathbb{R}$), $\{x_1 e_1, \dots, x_N e_N\}$ is a base of $\mathbb{R}^N$ over $\mathbb{R}$. Hence y may be written as

$$y = \beta_1 x_1 e_1 + \cdots + \beta_N x_N e_N,\qquad \beta_i \in \mathbb{R},\qquad i = 1, \dots, N.$$

Hence (cf. (10.2.7))

$$y = x + (\beta_1 - \alpha_1) x_1 e_1 + \cdots + (\beta_N - \alpha_N) x_N e_N,$$

or

$$y = \alpha_1 \left(x_1 + \frac{\beta_1 - \alpha_1}{\alpha_1} x_1 \right) e_1 + \cdots + \alpha_N \left(x_N + \frac{\beta_N - \alpha_N}{\alpha_N} x_N \right) e_N.$$

In other words

$$y = \alpha_1 y_1 e_1 + \cdots + \alpha_N y_N e_N,$$

where

$$y_i = x_i + \frac{\beta_i - \alpha_i}{\alpha_i}\, x_i, \qquad i = 1, \ldots, N.$$

If y is sufficiently close to x, then $|\beta_i - \alpha_i|$ are arbitrarily small for $i = 1, \ldots, N$, and $y_i \in J_i$ for $i = 1, \ldots, N$. Hence $y \in G$. Consequently G is open.

Since T is Q-convex, we have by Theorem 5.1.3 $G \subset Q(T) = T$. This shows that $\operatorname{int} T \neq \varnothing$.

Thus if $\operatorname{int} T = \varnothing$, there exists an i, $1 \leqslant i \leqslant N$, such that $\operatorname{int} T_i = \varnothing$. Clearly T_i is Q-convex and Q-radial at zero (in $\mathbb{R}$). By Lemma 10.2.1 there exists a discontinuous additive function $g : \mathbb{R} \to \mathbb{R}$ bounded above on T_i. Let $L = \mathbb{R}e_i$, and define $G : L \to \mathbb{R}$ by

$$G(xe_i) = g(x).$$

It is easily checked that G is a homomorphism bounded above on $T \cap L$. By Theorem 10.1.1 G may be extended onto $\mathbb{R}^N$ to an additive function $A : \mathbb{R}^N \to \mathbb{R}$ bounded above on T. Since $A \mid L = G$ is discontinuous, also A is discontinuous. Consequently $T \notin \mathfrak{B}$.

The case where $x_0 \neq 0$ may be reduced to that where $x_0 = 0$ like in the proof of Lemma 10.2.1. $\square$

The following example (cf. M. E. Kuczma [191]) shows that the assumption that T is Q-radial at a point x_0 is essential.

Example 10.2.1. Let $H \subset \mathbb{R}^N$ be a Hamel basis and let V_0 be the cone defined in the proof of Theorem 9.3.4. We have $E^+(V_0) = V_0$, whence also $Q(V_0) = V_0$, i.e., V_0 is Q-convex. Let $f : \mathbb{R}^N \to \mathbb{R}$ be an additive function bounded above on V_0:

$$f(x) < M \qquad \text{for } x \in V_0. \tag{10.2.8}$$

Take an arbitrary $h \in H$. There is an $\alpha < \gamma$ such that $h = h_\alpha$. Then $\lambda h + h_{\alpha+1} \in V_0$ for every $\lambda \in \mathbb{Q}$. Hence by (10.2.8)

$$\lambda f(h) + f(h_{\alpha+1}) = f(\lambda h + h_{\alpha+1}) \leqslant M$$

and letting λ range over whole $\mathbb{Q}$ we obtain $f(h) = 0$. Consequently $f \mid H = 0$, and since both, 0 and f, are additive extensions of $f \mid H$ onto $\mathbb{R}^N$, we get by Theorem 5.2.2 $f = 0$. Consequently f is continuous, and thus $V_0 \in \mathfrak{B}$. On the other hand, $\operatorname{int} V_0 = \varnothing$, since arbitrarily close to any $x \in \mathbb{R}^N$ there are points which do not belong to V_0. In fact, if x has form

$$x = \lambda_1 h_{\alpha_1} + \cdots + \lambda_n h_{\alpha_n},$$

where $n \in \mathbb{N}$, $\lambda_i \in \mathbb{Q}$, $h_{\alpha_i} \in H$, $i = 1, \ldots, n$, $\alpha_1 < \cdots < \alpha_n$, then the points

$$y = x - \lambda h_{\alpha_n+1}, \qquad \lambda \in \mathbb{Q} \cap (0, \infty),$$

do not belong to V_0 and approach x when λ tends to zero.

The same argument shows also that V_0 is not Q-radial at any point.

Lemma 10.2.2. *Let $D \subset \mathbb{R}^N$ be a convex and open set, let $f : D \to \mathbb{R}$ be a convex function, and let $M \in \mathbb{R}$ be a constant such that $M > \inf\limits_{D} f$. Then the set*

$$T = \{x \in D \mid f(x) < M\}$$

is Q-convex and Q-radial at every point $x \in T$.

Proof. If $x, y \in T \subset D$ and $\lambda \in \mathbb{Q} \cap [0, 1]$, then $\lambda x + (1 - \lambda)y \in D$ and by Theorem 5.3.5

$$f\big(\lambda x + (1 - \lambda)y\big) \leqslant \lambda f(x) + (1 - \lambda)f(y) < \lambda M + (1 - \lambda)M = M.$$

Consequently $\lambda x + (1 - \lambda)y \in T$. This means that T is Q-convex.

If $x \in T$ and $y \in \mathbb{R}^N$, then by Theorem 6.1.1

$$\lim_{\substack{\lambda \to 0 \\ \lambda \in \mathbb{Q}}} f(x + \lambda y) = f(x),$$

since for $\lambda \in \mathbb{Q} \cap [0, 1]$ $x + \lambda y = (1 - \lambda)x + \lambda(x + y) \in Q(x, \, x + y)$. Thus there exists an $\varepsilon > 0$ such that for $\lambda \in \mathbb{Q} \cap (0, \, \varepsilon)$ we have $f(x + \lambda y) < M$, and $x + \lambda y \in T$. This means that T is Q-radial at x.   □

Now we prove the theorem which has been our main aim.

Theorem 10.2.2. $\mathfrak{A} = \mathfrak{B}$.

Proof. The inclusion $\mathfrak{A} \subset \mathfrak{B}$ is obvious, so we need only prove that $\mathfrak{B} \subset \mathfrak{A}$. Take a set $T \subset \mathbb{R}^N$ such that $T \notin \mathfrak{A}$. We must show that $T \notin \mathfrak{B}$.

Since $T \notin \mathfrak{A}$, there exist a convex and open set $D \subset \mathbb{R}^N$ and a discontinuous convex function $f : D \to \mathbb{R}$ such that $T \subset D$ and f is bounded above on T:

$$f(x) < M \qquad \text{for } x \in T. \tag{10.2.9}$$

Put

$$T_0 = \{x \in D \mid f(x) < M\}. \tag{10.2.10}$$

Thus $T \subset T_0$. By Lemma 10.2.2 T_0 is Q-convex and Q-radial at every its point.

If we had $\operatorname{int} T_0 \neq \varnothing$, the function f would be continuous by Corollary 6.4.1. Thus $\operatorname{int} T_0 = \varnothing$. By Theorem 10.2.1 $T_0 \notin \mathfrak{B}$, and hence also $T \notin \mathfrak{B}$.   □

Incidentally we can observe that at the same time we have obtained a new proof of the fact that Q is an $\mathfrak{A}$-conservative operation. In fact, if $f : \mathbb{R}^N \to \mathbb{R}$ is an additive function bounded above on a set T (i.e., (10.2.9) holds), then the set T_0 defined by (10.2.10) is Q-convex and contains T, and hence also contains the smallest set with these properties, i.e., $Q(T)$:

$$Q(T) \subset T_0.$$

Moreover, f is bounded above on T_0 (cf. (10.2.10)), and hence also on $Q(T)$.

10.3   The class $\mathfrak{C}$

The results in this section are due to J. Smítal [287] and J. Mościcki [231]. We start with some lemmas.

Lemma 10.3.1. *Let $T \subset \mathbb{R}^N$ be a set Q-radial at zero. Then T contains a Hamel basis.*

Proof. Put $L = E(T)$. Suppose that $L \neq \mathbb{R}^N$. Then there exists a $y \in \mathbb{R}^N \setminus L$. Since T is Q-radial at 0, there exists a positive $\lambda \in \mathbb{Q}$ such that $\lambda y \in T \subset E(T) = L$. Hence also $y = \dfrac{1}{\lambda}(\lambda y) \in L$, a contradiction with $y \in \mathbb{R}^N \setminus L$. By Corollary 4.2.2 there exists a Hamel basis contained in T. $\square$

Lemma 10.3.2. *Let $T \subset \mathbb{R}^N$ be a set which is Q-convex and symmetric with respect to zero. Then T contains a Hamel basis if and only if T is Q-radial at zero.*

Proof. If T is Q-radial at zero, then it contains a Hamel basis in virtue of Lemma 10.3.1. Conversely, let $H \subset T$ be a Hamel basis. Take an arbitrary $y \in \mathbb{R}^N$. y may be written as

$$y = \alpha_1 h_1 + \cdots + \alpha_n h_n,$$

where $\alpha_i \in \mathbb{Q}$, $\alpha_i > 0$, $h_i \in H \cup (-H) \subset T \cup (-T) = T$, since by the symmetry $-T = T$. Put $\alpha = \alpha_1 + \cdots + \alpha_n > 0$. Then

$$\frac{y}{\alpha} = \lambda_1 h_1 + \cdots + \lambda_n h_n,$$

where $\lambda_i \in \mathbb{Q}$, $\lambda_i > 0$, $\lambda_1 + \cdots + \lambda_n = 1$. Hence (cf. Lemma 5.1.3) $\dfrac{y}{\alpha} \in T$, since T is Q-convex. By the symmetry also $-\dfrac{y}{\alpha} \in T$, whence also

$$0 = \frac{1}{2}\left(\frac{y}{\alpha} + \left(-\frac{y}{\alpha}\right)\right) \in T,$$

and

$$\lambda y = \alpha\lambda\frac{y}{\alpha} + (1 - \alpha\lambda)\,0 \in T$$

for $\lambda \in \mathbb{Q} \cap \left(0, \dfrac{1}{\alpha}\right)$. This means that T is Q-radial at zero. $\square$

Theorem 10.3.1. *Let $T \subset \mathbb{R}^N$ be a set which is Q-convex and symmetric with respect to zero. Then $T \in \mathfrak{C}$ if and only if $\operatorname{int} T \neq \varnothing$.*

Proof. The sufficiency results from Lemma 9.3.1 and from (9.1.1). To prove the necessity, assume that $\operatorname{int} T = \varnothing$. We must distinguish two cases.

I. T is Q-radial at zero. Then by Theorem 10.2.1 $T \notin \mathfrak{B}$, i.e., there exists a discontinuous additive function $g : \mathbb{R}^N \to \mathbb{R}$ bounded above on T:

$$g(x) \leqslant M \qquad \text{for } x \in T.$$

Hence it follows by Theorem 10.1.2 that $\big($we take $f(x) = M\big)$

$$|g(x)| \leqslant M \qquad \text{for } x \in T,$$

i.e., g is bounded on T. Consequently $T \notin \mathfrak{C}$.

II. T is not Q-radial at zero. By Lemma 10.3.2 T does not contain any Hamel basis, and by Theorem 9.3.6 $T \notin \mathfrak{C}$. $\qquad\square$

Theorem 10.3.2. *Let $T \subset \mathbb{R}^N$ be an arbitrary set. The following conditions are equivalent:*

(i) $T \in \mathfrak{C}$,

(ii) $\operatorname{int} Q(R(T)) \neq \varnothing$,

(iii) $\operatorname{int} R(Q(T)) \neq \varnothing$,

(iv) $\operatorname{int} Q(T \cup (-T)) \neq \varnothing$.

Proof. Assume (i). The sets $Q(R(T))$ and $Q(T \cup (-T))$ obviously are Q-convex and, by Corollary 5.1.1, symmetric with respect to zero. Since $T \subset T \cup (-T) \subset Q(T \cup (-T))$, (i) implies that $Q(T \cup (-T)) \in \mathfrak{C}$. Take a $t \in T$. Then $T = (T - t) + t$, whence by Theorems 9.2.8 and 9.2.2 $T - t \in \mathfrak{C}$. Since $T - t \subset T - T = R(T) \subset Q(R(T))$, we get hence $Q(R(T)) \in \mathfrak{C}$. By Theorem 10.3.1 conditions (ii) and (iv) are fulfilled.

Now assume (ii). By Lemmas 5.1.6 and 5.1.5

$$Q(R(T)) = Q(T - T) = Q(T + (-T)) = Q(T) + Q(-T) = Q(T) - Q(T) = R(Q(T)).$$

Thus (ii) implies (iii).

Now assume (iii). By Theorems 9.2.5, 9.2.6, 9.2.3, 9.2.2 and by Lemma 9.3.1 $T \in \mathfrak{C}$, i.e., (i) is fulfilled.

Finally assume (iv). If $f : \mathbb{R}^N \to \mathbb{R}$ is an additive function bounded on T:

$$|f(x)| \leqslant M \quad \text{for } x \in T, \tag{10.3.1}$$

then for $x \in -T$ we have $|f(x)| = |-f(x)| = |f(-x)| \leqslant M$ by (10.3.1). Consequently $F : \mathcal{P}(\mathbb{R}^N) \to \mathcal{P}(\mathbb{R}^N)$ given by $F(T) = T \cup (-T)$ is a $\mathfrak{C}$-conservative operation. Hence, like above, it follows that $T \in \mathfrak{C}$, i.e., (i) is fulfilled. $\qquad\square$

10.4 The class $\mathfrak{A}$

Theorem 10.3.2 gives a characterization of the class $\mathfrak{C}$. In this section we prove a similar characterization for the class $\mathfrak{A} = \mathfrak{B}$. The results are due to J. Smítal [288] and J. Mościcki [231].

Theorem 10.4.1. *Let $T \subset \mathbb{R}^N$ be an arbitrary set. The following conditions are equivalent:*

(i) $T \in \mathfrak{A}$,

(ii) *For every set $A \subset \mathbb{R}^N$, Q-radial at a point, $\operatorname{int} Q(T + A) \neq \varnothing$,*

(iii) *For every set $A \subset \mathbb{R}^N$, Q-radial at a point, $\operatorname{int} [Q(T) + Q(A)] \neq \varnothing$,*

(iv) *For every set $A \subset \mathbb{R}^N$, Q-radial at a point, $\operatorname{int} Q(T \cup A) \neq \varnothing$.*

Proof. Assume (i). The sets $Q(T + A)$ and $Q(T \cup A)$ obviously are Q-convex and Q-radial at a point. We have $T \subset Q(T \cup A)$, whence $Q(T \cup A) \in \mathfrak{A}$. Similarly, with an $a \in A$, we have by Theorems 9.2.8 and 9.2.2 $T + a \in \mathfrak{A}$, and since $T + a \subset T + A \subset Q(T + A)$, we get hence $Q(T + A) \in \mathfrak{A}$. By Theorems 10.2.1 and 10.2.2 conditions (ii) and (iv) are fulfilled.

Now assume (ii). By Lemma 5.1.6 $Q(T + A) = Q(T) + Q(A)$, whence (iii) follows.

Finally assume that either (iii) or (iv) is fulfilled. Let $f : \mathbb{R}^N \to \mathbb{R}$ be an arbitrary additive function bounded above on T:

$$f(x) < M \qquad \text{for } x \in T.$$

Put

$$A = \{x \in \mathbb{R}^N \mid f(x) < M\}.$$

By Lemma 10.2.2 the set A is Q-convex and Q-radial at every point. Moreover $T \subset A$, whence $T \cup A = A$ and $Q(T) \subset Q(A) = A$ so that $Q(T) + Q(A) \subset A + A = S_2(A)$. Thus condition (iii) resp. (iv) implies that $\operatorname{int} S_2(A) \neq \varnothing$ resp. $\operatorname{int} A \neq \varnothing$. By Theorem 9.3.3 resp. Lemma 9.3.1 $A \in \mathfrak{A}$. Since the function f is bounded above on A, f is continuous. Consequently $T \in \mathfrak{B}$. By Theorem 10.2.2 $T \in \mathfrak{A}$. $\qquad\square$

We may observe that in conditions (ii), (iii), (iv) of Theorem 10.4.1 we may restrict ourselves to sets A which are Q-radial at a point and such that $T \subset A$. Hence we get the following

Theorem 10.4.2. *Let $T \subset \mathbb{R}^N$ be an arbitrary set. Then $T \in \mathfrak{A}$ if and only if for every set A which is Q-radial at a point and such that $T \subset A$ we have $\operatorname{int} Q(A) \neq \varnothing$.*

The proof of Theorem 10.4.2 does not differ from that of Theorem 10.4.1, since the set A constructed there has the property that $T \subset A$.

Similarly, in the "if" part of Theorem 10.4.2 it would be enough to assume that A is Q-radial at every point.

In Theorem 10.4.1 we can replace A by $-A$, since A is Q-radial at a point if and only if $-A$ is Q-radial at a point. It is less obvious that we can replace T by $-T$.

Theorem 10.4.3. *Let $T \subset \mathbb{R}^N$ be an arbitrary set. The following conditions are equivalent:*

(i) *$T \in \mathfrak{A}$,*

(ii) *For every set A, Q-radial at a point, $\operatorname{int} Q(A - T) \neq \varnothing$,*

(iii) *For every set A, Q-radial at a point, $\operatorname{int}\big[Q(A) - Q(T)\big] \neq \varnothing$.*

Proof. Let condition (i) be fulfilled. If A is an arbitrary set Q-radial at a point, then the set $-A$ also is Q-radial at a point. By Theorem 10.4.1 $\operatorname{int} Q(-A + T) \neq \varnothing$. But $Q(-A + T) = Q\big(-(A - T)\big) = -Q(A - T)$ (cf. Lemma 5.1.5). Hence also $\operatorname{int} Q(A - T) \neq \varnothing$, i.e., condition (ii) is fulfilled. By Lemmas 5.1.5 and 5.1.6

$$Q(A - T) = Q(A) - Q(T),$$

so (ii) implies (iii).

Now suppose that condition (iii) is fulfilled, and let $A \subset \mathbb{R}^N$ be an arbitrary set Q-radial at a point. Then also $-A$ is Q-radial at a point, and by (iii) int $\left[Q(-A) - Q(T)\right] \neq \varnothing$. But

$$Q(-A) - Q(T) = -\left[Q(A) + Q(T)\right],$$

so int $\left[Q(A) + Q(T)\right] \neq \varnothing$. By Theorem 10.4.1 $T \in \mathfrak{A}$. □

10.5 Set-theoretic operations

The theorems of 9.3 say that every reasonably large set belongs to the class $\mathfrak{A}$. Consequently the sets which do not belong to $\mathfrak{A}$ may be described as "small". Consequently it is rather surprising that the union of two "small" sets may be "large".

Theorem 10.5.1. *There exist sets* T_1, $T_2 \subset \mathbb{R}^N$ *such that* $T_i \notin \mathfrak{C}, i = 1, 2, T_1$ *is countable and* $T_1 \cup T_2 \in \mathfrak{A}$.

Proof. Let $H \subset \mathbb{R}^N$ be an arbitrary Hamel basis, and write

$$H = H_1 \cup H_2,$$

where H_1 is countable, $H_1 \cap H_2 = \varnothing$. Put

$$T_1 = E(H_1), \qquad T_2 = E(H_2).$$

Since H is linearly independent (over $\mathbb{Q}$), we have for arbitrary $h_1 \in H_1, h_2 \in H_2$

$$h_1 \notin T_2, \qquad h_2 \notin T_1.$$

Consequently $T_1 \neq \mathbb{R}^N$, $T_2 \neq \mathbb{R}^N$. Suppose that T_1 contains a Hamel basis H_0. Then $\mathbb{R}^N = E(H_0) \subset E(T_1) = T_1 \neq \mathbb{R}^N$, a contradiction. Similarly we show that also T_2 does not contain a Hamel basis. By Theorem 9.3.6 $T_1 \notin \mathfrak{C}$ and $T_2 \notin \mathfrak{C}$. Moreover, since H_1 is countable, so is also T_1, in virtue of Lemma 4.1.3.

Write $T = T_1 \cup T_2$, and let $f : \mathbb{R}^N \to \mathbb{R}$ be an arbitrary additive function bounded above on T. For every $h \in H$ we have $\lambda h \in T$ for every $\lambda \in \mathbb{Q}$. Hence $\lambda f(h) = f(\lambda h)$ remains bounded when λ varies over the whole $\mathbb{Q}$, which is impossible unless $f(h) = 0$. Consequently $f \mid H = 0$. Zero is an additive extension of $f \mid H$ onto $\mathbb{R}^N$, and so is also f, whence by Theorem 5.2.2 $f = 0$ in $\mathbb{R}^N$. Consequently f is continuous. This means that $T \in \mathfrak{B}$. By Theorem 10.2.2 $T \in \mathfrak{A}$. □

Now we are going to deal with cartesian products. Write $N = p + q$, p, $q \in \mathbb{N}$, and consider $\mathbb{R}^N$ as the product $\mathbb{R}^N = \mathbb{R}^p \times \mathbb{R}^q$.

Theorem 10.5.2. *Let* $T \subset \mathbb{R}^p$ *and* $S \subset \mathbb{R}^q$ *be arbitrary sets, and put* $W = T \times S \subset \mathbb{R}^N$. *Then* $W \in \mathfrak{A}_N$ *$[W \in \mathfrak{C}_N]$ if and only if* $T \in \mathfrak{A}_p$, $S \in \mathfrak{A}_q$ *$[T \in \mathfrak{C}_p, S \in \mathfrak{C}_q]$.*

Proof. Assume that, e.g., $T \notin \mathfrak{A}_p$. Then there exists a discontinuous additive function $g : \mathbb{R}^p \to \mathbb{R}$ bounded above on T. For every $x \in \mathbb{R}^N$ write $x = (x_p, x_q)$, where $x_p \in \mathbb{R}^p$, $x_q \in \mathbb{R}^q$. Put

$$f(x) = g(x_p).$$

The function f is additive, discontinuous, and bounded above on W. This shows that $W \notin \mathfrak{B}_N = \mathfrak{A}_N$ (cf. Theorem 10.2.2).

Conversely, let $W \notin \mathfrak{A}_N = \mathfrak{B}_N$. Then there exists a discontinuous additive function $f : \mathbb{R}^N \to \mathbb{R}$ bounded above on W. By Theorem 5.5.1

$$f(x) = f_p(x_p) + f_q(x_q), \tag{10.5.1}$$

where $f_p : \mathbb{R}^p \to \mathbb{R}$ and $f_q : \mathbb{R}^q \to \mathbb{R}$ are additive functions. Since f is discontinuous, at least one of the functions f_p and f_q, say f_p, must be discontinuous. Fix an $x_q \in S$. Then, as x_p varies over T,

$$f_p(x_p) = f(x) - f_q(x_q), \qquad x = (x_p, \, x_q),$$

remains bounded above. Consequently $T \notin \mathfrak{B}_p = \mathfrak{A}_p$.

The proof for the class $\mathfrak{C}$ is analogous.   □

Now consider an arbitrary set $W \subset \mathbb{R}^N$. For every $x_p \in \mathbb{R}^p$ and $x_q \in \mathbb{R}^q$ write

$$W_p[x_q] = \{x_p \in \mathbb{R}^p \mid (x_p, \, x_q) \in W\},$$
$$W_q[x_p] = \{x_q \in \mathbb{R}^q \mid (x_p, \, x_q) \in W\},$$

(cf. 2.1), and

$$W_p = \{x_p \in \mathbb{R}^p \mid \text{There exists an } x_q \in \mathbb{R}^q \text{ such that } (x_p, \, x_q) \in W\},$$
$$W_q = \{x_q \in \mathbb{R}^q \mid \text{There exists an } x_p \in \mathbb{R}^p \text{ such that } (x_p, \, x_q) \in W\}.$$

The sets W_p and W_q are the projections of W onto $\mathbb{R}^p$ and $\mathbb{R}^q$, respectively, the sets $W_p[x_q]$ and $W_q[x_p]$ are sections of W.

Theorem 10.5.3. *Let $W \subset \mathbb{R}^N$ be a set such that $W \notin \mathfrak{C}_N$ and the set W_q is bounded. If there exists an $\bar{x}_p \in \mathbb{R}^p$ such that $W_q[\bar{x}_p] \in \mathfrak{C}_q$, then $W_p \notin \mathfrak{C}_p$.*

Proof. There exists a discontinuous additive function $f : \mathbb{R}^N \to \mathbb{R}$ bounded on W. f can be written in form (10.5.1). Thus

$$f_q(x_q) = f(x) - f_p(\bar{x}_p), \qquad x = (\bar{x}_p, \, x_q).$$

When x_q varies over $W_q[\bar{x}_p]$, $f_q(x_q)$ remains bounded. Since $W_q[\bar{x}_p] \in \mathfrak{C}_q$, f_q is continuous. By Theorem 5.5.2 $f_q(x_q) = c x_q$ with a certain $c \in \mathbb{R}^q$. Hence f_q is bounded on bounded sets, in particular on W_q. Varying $x = (x_p, \, x_q)$ in such a manner that x_p runs over the whole W_p and $x \in W$, we see from (10.5.1) that $f_p(x_p)$ remains bounded. Moreover, f_p is discontinuous, since f is discontinuous and f_q is continuous. Consequently $W_p \notin \mathfrak{C}_p$.   □

The assumption that W_q is bounded cannot be dropped, as may be seen from the following example.

Example 10.5.1. Let $p = q = 1$, $N = 2$, and let $f_1 : \mathbb{R} \to \mathbb{R}$ be a discontinuous additive function. Put

$$W = \{x = (x_1, \, x_2) \in \mathbb{R}^2 \mid f_1(x_1) < x_2 < f_1(x_1) + 1\}.$$

For every $x_1 \in \mathbb{R}$ the set $W_2[x_1]$ is the open interval $\big(f_1(x_1), f_1(x_1) + 1\big)$ and consequently (Lemma 9.3.1) belongs to $\mathfrak{C}_1$. Also the set $W \notin \mathfrak{C}_2$, since the additive function

$$f(x) = x_2 - f_1(x_1)$$

is discontinuous and bounded on W:

$$0 < f(x) < 1 \qquad \text{for } x \in W.$$

But the set $W_1 = \mathbb{R} \in \mathfrak{C}_1$.

Theorem 10.5.3 says that if the set W_q is bounded and $W \notin \mathfrak{C}_N$, then either all sections $W_p[x_q]$ are not in $\mathfrak{C}_p$, or all sections $W_q[x_p]$ are not in $\mathfrak{C}_q$. (Of course, here the assumption of the boundedness of W_q could be replaced by that of the boundedness of W_p). One could expect that if $W \in \mathfrak{C}_N$, then all (or, in some sense, almost all) sections $W_p[x_q]$ and $W_q[x_p]$ are in $\mathfrak{C}_p$ resp. $\mathfrak{C}_q$. In particular, the following statement would be an analogue of the Fubini theorem: If $W \in \mathfrak{C}_N$, then

$$\{x_p \in \mathbb{R}^p \mid W_q[x_p] \in \mathfrak{C}_q\} \in \mathfrak{C}_p.$$

However, this is not the case, as may be seen from the following example.

Example 10.5.2. Let $p = q = 1$, $N = 2$, and let $I \subset \mathbb{R}$ be a non-trivial finite closed interval. Further, let $\varphi : I \to \mathbb{R}$ be non-linear and one-to-one continuous function [e.g., $\varphi(x) = x^3$ for $x \in I$]. Put $W = \mathrm{Gr}(\varphi)$. By Theorem 9.5.2 $W \in \mathfrak{A}_2 \subset \mathfrak{C}_2$. Also the sets $W_1 = I$ and $W_2 = \varphi(I)$ are bounded. But for all $x \in \mathbb{R}$ the sets $W_1[x]$ and $W_2[x]$ contain each at most one point, and thus are not in $\mathfrak{C}_1$.

10.6 The classes $\mathfrak{D}$

The classes $\mathfrak{D}(D)$ have been defined in 9.7. The results in this section are due to J. Smítal [289].

Our knowledge of the classes $\mathfrak{D}(D)$ is much narrower than that of the classes $\mathfrak{A}$, $\mathfrak{B}$, $\mathfrak{C}$. Here we give only some sufficient conditions for $T \in \mathfrak{D}(D)$. (Other such sufficient conditions are contained in Theorem 9.7.2). No characterization of the classes $\mathfrak{D}(D)$ is known.

Theorem 9.7.2 suggests a likeness between the classes $\mathfrak{D}(D)$ and $\mathfrak{A}$. In fact, we have the following

Theorem 10.6.1. *Let $D \subset \mathbb{R}^N$ be a convex and open set. We have*

$$\mathfrak{D}(D) \subset \mathfrak{A}. \tag{10.6.1}$$

Proof. Let $T \in \mathfrak{D}(D)$, and let $f : \mathbb{R}^N \to \mathbb{R}$ be an arbitrary additive function bounded above on T:

$$f(x) \leqslant M \qquad \text{for } x \in T.$$

Put $g = -f \mid D$. Then $g : D \to \mathbb{R}$ is a convex function bounded below on T:

$$g(x) \geqslant -M \qquad \text{for } x \in T.$$

Since $T \in \mathfrak{D}(D)$, this means that g is locally bounded below at every point of D, and hence f is locally bounded above at every point of D. By the Theorem of Bernstein-Doetsch (Theorem 6.4.2) f is continuous. Hence $T \in \mathfrak{B} = \mathfrak{A}$. $\qquad\square$

But actually the inclusion in (10.6.1) cannot be replaced by the equality, as may be seen from the following example.

Example 10.6.1. Let $N = 1$, and let $D \subset \mathbb{R}$ be an open interval such that $0 \in D$. There exists non-trivial compact interval $S \subset D$ symmetric with respect to zero $(-S = S)$. By Lemma 9.3.1 $S \in \mathfrak{C}$. By Theorem 9.3.6 S contains a Hamel basis H. Let B be a countable subset of H. Put

$$T = \{x \in D \mid \text{If } x = \alpha_1 h_1 + \cdots + \alpha_n h_n, \ \alpha_i \in \mathbb{Q}, \ h_i \in H, \ i = 1, \ldots, n,$$

$$\text{then } \alpha_i \notin (0,1) \text{ whenever } h_i \in B\}.$$

Take an arbitrary $x \in D$. Then

$$x = \alpha_1 h_1 + \cdots + \alpha_n h_n, \qquad \alpha_i \in \mathbb{Q}, \ h_i \in H, \ i = 1, \ldots, n.$$

Choose $h_{n+1}, h_{n+2} \in H \setminus B$ different from $h_1, \ldots, h_n$. Since $h_{n+1}, h_{n+2} \in H$, they are linearly independent over $\mathbb{Q}$, i.e., incommensurable. It follows from Lemma 3.8.1 that the set $E(\{h_{n+1}, h_{n+2}\})$ is dense in $\mathbb{R}$. Thus we can choose $\beta_1, \ldots, \beta_n, \gamma_1, \gamma_2 \in \mathbb{Q}$ in such a manner that

$$\beta_i > \max(1, \alpha_i), \qquad i = 1, \ldots, n,$$

and $\gamma_1 h_{n+1} + \gamma_2 h_{n+2}$ is close to $\sum_{i=1}^{n} (\beta_i - \alpha_i) h_i$. Put

$$y = \beta_1 h_1 + \cdots + \beta_n h_n - \gamma_1 h_{n+1} - \gamma_2 h_{n+2},$$

where γ_1, γ_2 have been chosen so that $y - x \in D$ and $y = x + (y - x) \in D$. By the choice of β_i we have also $y \in T$, $x - y \in T$. Hence $x = y + (x - y) \in T + T = S_2(T)$. Consequently $D \subset S_2(T)$, and hence by Theorem 9.3.3 $T \in \mathfrak{A}$.

On the other hand, $T \notin \mathfrak{D}(D)$. To show this, arrange B into a sequence $B = \{b_k\}_{k \in \mathbb{N}}$, and for every $k \in \mathbb{N}$, let $g_k : \mathbb{R} \to \mathbb{R}$ be the function

$$g_k(t) = \begin{cases} -2kt & \text{for } t < \dfrac{1}{2}, \\[2mm] 2k(t-1) & \text{for } t \geqslant \dfrac{1}{2}. \end{cases}$$

Thus g_k is a broken linear function, convex in $\mathbb{R}$, and with the properties

$$g_k(0) = g_k(1) = 0, \qquad g_k\left(\frac{1}{2}\right) = -k, \qquad g_k(t) \geqslant 0 \qquad \text{for } t \notin (0,1). \qquad (10.6.2)$$

Now we define a function $f : D \to \mathbb{R}$ as follows. If $x \in D$, then x may be written as

$$x = \alpha_1 \, b_{k_1} + \cdots + \alpha_n \, b_{k_n} + \beta_1 \, h_1 + \cdots + \beta_m \, h_m,$$

where $\alpha_i, \beta_j \in \mathbb{Q}$, $b_{k_i} \in B$, $h_j \in H \setminus B$, $i = 1, \ldots, n$; $j = 1, \ldots, m$. For such an x put

$$f(x) = \sum_{i=1}^{n} g_{k_i}(\alpha_i).$$

The function f is convex (cf. Example 5.3.2, where we must take $g_{b_k} = g_k$ for $b_k \in B$, $g_h = 0$ for $h \in H \setminus B$). For $x \in T$ we have for the corresponding coefficients $\alpha_i \notin (0, 1)$, whence $g_{k_i}(\alpha_i) \geqslant 0$, and consequently also $f(x) \geqslant 0$. Thus f is bounded below on T.

We have $b_k \in B \subset S$, $k \in \mathbb{N}$. Since $0 \in S$, also $\frac{1}{2} \, b_k \in S$ for $k \in \mathbb{N}$. Since S is compact, we can choose from the sequence $\left\{ \frac{1}{2} \, b_k \right\}_{k \in \mathbb{N}}$ a subsequence $\left\{ \frac{1}{2} \, b_{k_n} \right\}_{n \in \mathbb{N}}$ convergent to a point of S:

$$\lim_{n \to \infty} \frac{1}{2} \, b_{k_n} = \bar{x} \in S \subset D.$$

According to the definition of f we have by (10.6.2)

$$f\left(\frac{1}{2} \, b_{k_n} \right) = g_{k_n}\left(\frac{1}{2} \right) = -k_n$$

so that

$$\lim_{n \to \infty} f\left(\frac{1}{2} \, b_{k_n} \right) = -\infty.$$

This shows that f is not locally bounded below at $\bar{x}$. Consequently $T \notin \mathfrak{D}(D)$.

Example 10.6.1 shows also that the condition $\operatorname{int} S_2(T) \neq \varnothing$ is not sufficient for $T \in \mathfrak{D}(D)$.

Lemma 10.6.1. *Let $D \subset \mathbb{R}^N$ be a convex and open set such that $0 \in D$, and let $T \subset D$ be an arbitrary set. If there exists a point $a \in D$ such that for every set A which is Q-convex and Q-radial at zero $(0 \in A)$ $a \in \operatorname{int}(T - A)$, then $T \in \mathfrak{D}(D)$.*

Proof. Let $f : D \to \mathbb{R}$ be an arbitrary convex function bounded below on T:

$$f(x) > -M \qquad \text{for } x \in T, \tag{10.6.3}$$

and we may assume that $M > 0$ and $M > |f(0)|$. Put

$$B = \{ x \in D \mid f(x) < M \}. \tag{10.6.4}$$

By Lemma 10.2.2 the set B is Q-convex and Q-radial at zero. Take a $\lambda \in \mathbb{Q} \cap (1, 2)$ such that $\lambda a \in D$. Write

$$A = \left(1 - \frac{1}{\lambda} \right) B.$$

It is easily seen that also A is Q-convex and Q-radial at zero. By the conditions of the theorem $a \in \mathrm{int}(T - A)$, whence $\lambda a \in \mathrm{int}\,\lambda(T - A)$. Let

$$G = D \cap \mathrm{int}\,\lambda(T - A).$$

Then G is a non-empty ($\lambda a \in G$) open set.

Take an arbitrary $z \in G$. Then $z \in \lambda(T - A)$, which means that there exist a $t \in T$ and an $x \in A$ such that $z = \lambda t - \lambda x$. Hence $z = \lambda t - (\lambda - 1)b$, where $b \in B \subset D$. Consequently

$$t = \frac{1}{\lambda}\,z + \left(1 - \frac{1}{\lambda}\right) b.$$

By Theorem 5.3.5

$$f(t) \leqslant \frac{1}{\lambda}\,f(z) + \left(1 - \frac{1}{\lambda}\right) f(b),$$

whence by (10.6.3) and (10.6.4)

$$f(z) \geqslant \lambda f(t) - (\lambda - 1)f(b) \geqslant -\lambda M - (\lambda - 1)M = (-2\lambda + 1)M > -3M. \quad (10.6.5)$$

Relation (10.6.5) says that f is bounded below on G. By Theorem 6.2.2 f is locally bounded below in D, i.e., $D \in \mathfrak{D}(D)$. $\qquad\square$

It may be shown by a suitable example that the condition occurring in Lemma 10.6.1 is not necessary for $T \in \mathfrak{D}(D)$ (cf. Smítal [289]). Now, using Lemma 10.6.1, we prove a theorem which shows that the classes $\mathfrak{D}(D)$ actually depend on D.

Theorem 10.6.2. *Let $D_1 \subset D_2 \subset \mathbb{R}^N$ be convex and open sets such that $0 \in D_1$ and $D_1 \neq D_2$. Then there exists a set $T \subset D_1$ such that $T \in \mathfrak{D}(D_2) \setminus \mathfrak{D}(D_1)$.*

Proof. Let $H \subset \mathbb{R}^N$ be a Hamel basis. Since the linear spaces $E_{\mathbb{R}}(H)$ and $E_{\mathbb{Q}}(H)$ spanned by H over $\mathbb{R}$ and $\mathbb{Q}$, respectively, evidently fulfil $E_{\mathbb{Q}}(H) \subset E_{\mathbb{R}}(H) \subset \mathbb{R}^N$, and $E_{\mathbb{Q}}(H) = \mathbb{R}^N$ we have $E_{\mathbb{R}}(H) = \mathbb{R}^N$. Thus, by Corollary 4.2.2, H must contain a basis of $\mathbb{R}^N$ over $\mathbb{R}$. Let $\{b_1, \ldots, b_N\} \subset H$ be such a basis. Fix an $h_0 \in H \setminus \{b_1, \ldots, b_N\}$.

Let B denote the frontier of D_1, and let $f : D_1 \to \mathbb{R}$ be a bounded below convex function such that

$$\lim_{x \to B} f(x) = +\infty. \qquad (10.6.6)$$

(One can take, e.g.,

$$f(x) = \frac{1}{\varrho(x,\,B)},$$

where ϱ denotes the Euclidean distance in $\mathbb{R}^N$.) Define $g_0 : H \to \mathbb{R}$ putting

$$g_0(h_0) = 1, \qquad g_0(h) = 0 \qquad \text{for } h \in H \setminus \{h_0\}. \qquad (10.6.7)$$

By Theorem 5.2.2 the function g_0 can be extended to an additive function $g : \mathbb{R}^N \to \mathbb{R}$, and by Corollary 5.2.2 g is discontinuous, and hence locally unbounded below at any point of $\mathbb{R}^N$. (Otherwise $-g$ would be an additive function locally bounded above at a point of $\mathbb{R}^N$, and hence, by Theorem 6.4.2, continuous, and thus also g would be continuous.) We define the convex function $F : D_1 \to \mathbb{R}$ by

$$F = f + g.$$

Thus F is locally unbounded below at any point of $\mathbb{R}^N$. Consequently the set

$$T = \{x \in D_1 \mid F(x) > 0\} \tag{10.6.8}$$

does not belong to $\mathfrak{D}(D_1)$:

$$T \notin \mathfrak{D}(D_1). \tag{10.6.9}$$

We are going to show that

$$T \in \mathfrak{D}(D_2). \tag{10.6.10}$$

To this aim take an $a \in B \cap D_2$ and an arbitrary set A which is Q-convex and Q-radial at zero, $0 \in A$. Consequently there exists an $\varepsilon > 0$ such that for every $\beta \in \mathbb{Q}$ with $|\beta| \leqslant \varepsilon$ we have

$$\beta b_i \in A, \qquad i = 1, \ldots, N,$$

and clearly we can choose ε rational. Let

$$C = Q(\{\pm \varepsilon b_i\}_{i=1,\ldots,N}).$$

The set $\operatorname{cl} C$ is convex (cf. Exercise 5.2). Since for every $i = 1, \ldots, N$ the rational segment $Q(-\varepsilon b_i, \varepsilon b_i)$ is contained in C, the real segment $\overline{-\varepsilon b_i, \varepsilon b_i}$ is contained in $\operatorname{cl} C$. Hence it follows easily that $0 \in \operatorname{int} \operatorname{cl} C$. Consequently there exists an open ball S centered at the origin and contained in $\operatorname{cl} C$:

$$S \subset \operatorname{cl} C. \tag{10.6.11}$$

Since $a \in D_2$, and D_2 is open, there exists a small ball around a contained in D_2. In other words, there exists a $\lambda \in \mathbb{Q} \cap (0, 1)$ such that

$$a + \lambda S \subset D_2.$$

Take an arbitrary $u \in (a + \lambda S) \setminus D_1$. (Note that $(a + \lambda S) \setminus D_1 \neq \varnothing$, since a is a frontier point of D_1.) We have $(u + \lambda S) \cap D_1 \neq \varnothing$, since $a \in B$, and since by (10.6.11) $\lambda S \subset \operatorname{cl} C$, also $(u + \operatorname{cl} C) \cap D_1 \neq \varnothing$ and $(u + C) \cap D_1 \neq \varnothing$. Consequently there exists a $c \in C$ such that $u + c \in D_1$. Put

$$s = \inf \{ \xi \in \mathbb{Q} \cap [0, 1] \mid u + \xi c \in D_1 \} \tag{10.6.12}$$

Since $\xi = 1$ belongs to the set on the right-hand side of (10.6.12) and D_1 is open, we have

$$s < 1,$$

and since D_1 is convex, for every $\xi \in (s, 1] \cap \mathbb{Q}$ we have $u + \xi c \in D_1$, whence

$$F(u + \xi c) = f(u + \xi c) + g(u) + \xi g(c).$$

Now, $c \in C \subset E_{\mathbb{Q}}(\{b_1, \ldots, b_N\})$, so there exist $\alpha_1, \ldots, \alpha_N \in \mathbb{Q}$ such that $c = \alpha_1 b_1 + \cdots + \alpha_N b_N$. Since $b_i \neq h_0$ for $i = 1, \ldots, N$, we obtain hence by (10.6.7)

$$g(c) = \alpha_1 g_0(b_1) + \cdots + \alpha_N g_0(b_N) = 0,$$

and

$$F(u + \xi c) = f(u + \xi c) + g(u).$$

Since $u \notin D_1$ and $u + \xi c \in D_1$, and by (10.6.6) f becomes infinite at the frontier of D_1, there exists an $\eta \in \mathbb{Q} \cap (s, 1)$ such that $F(u + \eta c) > 0$. According to (10.6.8) this means that $u + \eta c \in T$.

Now, $\pm \varepsilon b_i \in A$, $i = 1, \ldots, N$, whence

$$C = Q(\{\pm \varepsilon b_i\}_{i=1,\ldots,N}) \subset Q(A) = A,$$

since A is Q-convex. Consequently $c \in A$, and $\eta c = \eta c + (1 - \eta)0 \in A$. Thus $u = (u + \eta c) - \eta c \in T - A$. Since u could have been arbitrary in $(a + \lambda S) \setminus D_1$, we infer that

$$(a + \lambda S) \setminus D_1 \subset T - A. \tag{10.6.13}$$

Similarly, for arbitrary $u \in (a + \lambda S) \cap D_1$ we have $(u + C) \setminus D_1 \neq \varnothing$, and so there exists a $c \in C$ such that $u + c \notin D_1$. Let

$$s = \sup\{\xi \in \mathbb{Q} \cap [0, 1] \mid u + \xi c \in D_1\}.$$

Since $u \in D_1$ and D_1 is open, we have $s > 0$. By the convexity of D_1 $u + \xi c \in D_1$ for every $\xi \in \mathbb{Q} \cap (0, s)$, and

$$F(u + \xi c) = f(u + \xi c) + g(u) + \xi g(c) = f(u + \xi c) + g(u).$$

Thus we can find an $\eta > 0$ such that $F(u + \eta c) > 0$. Similarly as above we show that $u \in T - A$, whence

$$(a + \lambda S) \cap D_1 \subset T - A. \tag{10.6.14}$$

Relations (10.6.13) and (10.6.14) prove that $a + \lambda S \subset T - A$, i.e.,

$$a \in \operatorname{int}(T - A).$$

By Lemma 10.6.1 we obtain hence (10.6.10). Relations (10.6.9) and (10.6.10) show that $T \in \mathfrak{D}(D_2) \setminus \mathfrak{D}(D_1)$. $\square$

10.7 The classes $\mathfrak{A}_C$ and $\mathfrak{B}_C$

The classes $\mathfrak{A}_C$ and $\mathfrak{B}_C$ have been defined in 9.8. We do not know any characterization of these classes, but we do know that they differ from the classes $\mathfrak{A}$ and $\mathfrak{B}$. The relations between the classes $\mathfrak{A}$, $\mathfrak{B}$, $\mathfrak{A}_C$ and $\mathfrak{B}_C$ have been established by B. Kominek-Z. Kominek [171] and Z. Kominek [173]. As pointed out in 9.8 it follows directly from the definition that

$$\mathfrak{A}_C \subset \mathfrak{B}_C. \tag{10.7.1}$$

Now we will prove

Theorem 10.7.1. $\mathfrak{A}_C \subset \mathfrak{A}$.

Proof. The inclusion asserted will be proved if we show that if a set $T \subset \mathbb{R}^N$ does not belong to $\mathfrak{A}$, then also $T \notin \mathfrak{A}_C$. Let $T \subset \mathbb{R}^N$ be an arbitrary set such that $T \notin \mathfrak{A}$. This means that there exists a convex and open set $D \subset \mathbb{R}^N$, $T \subset D$, and a discontinuous convex function $f : D \to \mathbb{R}$ such that f is bounded above on T:

$$f(x) \leqslant M \qquad \text{for } x \in T.$$

Put

$$T_0 = \{x \in D \mid f(x) \leqslant M\},$$

and define the function $F : D \to \mathbb{R}$:

$$F(x) = \begin{cases} M & \text{for } x \in T_0, \\ f(x) & \text{for } x \in D \setminus T_0. \end{cases}$$

Clearly $T \subset T_0 \subset D$ and $F \mid T$ is continuous (being constant). To prove that F is convex, take arbitrary $x, y \in D$. We must distinguish some cases.

1. $x, y \in D \setminus T_0$ and $\frac{1}{2}(x+y) \in D \setminus T_0$. Then

$$F\left(\frac{x+y}{2}\right) = f\left(\frac{x+y}{2}\right) \leqslant \frac{f(x) + f(y)}{2} = \frac{F(x) + F(y)}{2}.$$

2. $x, y \in D \setminus T_0$, but $\frac{1}{2}(x+y) \in T_0$. Then

$$F\left(\frac{x+y}{2}\right) = M = \frac{M+M}{2} < \frac{f(x)+f(y)}{2} = \frac{F(x)+F(y)}{2}.$$

3. $x, y \in T_0$. By Lemma 10.2.2 the set T_0 is Q-convex, and thus $\frac{1}{2}(x+y) \in T_0$. Hence

$$F\left(\frac{x+y}{2}\right) = M = \frac{M+M}{2} = \frac{F(x)+F(y)}{2}.$$

4. $x \in T_0$, $y \in D \setminus T_0$, $\frac{1}{2}(x+y) \in T_0$. Then

$$F\left(\frac{x+y}{2}\right) = M = \frac{M+M}{2} = \frac{F(x)+M}{2} < \frac{F(x)+f(y)}{2} = \frac{F(x)+F(y)}{2}.$$

5. $x \in T_0$, $y \in D \setminus T_0$, $\frac{1}{2}(x+y) \in D \setminus T_0$. Then

$$F\left(\frac{x+y}{2}\right) = f\left(\frac{x+y}{2}\right) \leqslant \frac{f(x)+f(y)}{2} \leqslant \frac{M+f(y)}{2} = \frac{F(x)+F(y)}{2}.$$

The cases where $x \in D \setminus T_0$, $y \in T_0$, are analogous to 4 and 5.

Thus F is convex. Suppose that F is continuous in D. Then the set T_0 is closed (in D), and hence Lebesgue measurable. Since f is a discontinuous convex function bounded above on T_0, we infer hence in virtue of Theorem 9.3.1 that $m(T_0) = 0$. Consequently $m(D \setminus T_0) > 0$. On $D \setminus T_0$ we have $f = F$, so $f \mid D \setminus T_0$ is continuous. By Theorem 9.8.3 f is continuous, a contradiction. Hence F is discontinuous. This shows that there exists a discontinuous convex function F such that the restriction $F \mid T$ is continuous. Consequently $T \notin \mathfrak{A}_C$. $\square$

Now we exhibit two examples which show that no other inclusion can be asserted. First we show that the inclusion in Theorem 10.7.1 is strict.

Example 10.7.1. Let $H \subset \mathbb{R}^N$ be a Hamel basis. Put $g(h) = 1$ for $h \in H$, and let $f : \mathbb{R}^N \to \mathbb{R}$ be the additive extension of g (cf. Theorem 5.2.2). By Corollary 5.2.2 f is discontinuous. Moreover, all the values of f are rational. For every $\lambda \in \mathbb{Q}$ put

$$T_\lambda = \{x \in \mathbb{R}^N \mid f(x) = \lambda\}.$$

Let $\{\lambda_n\}_{n \in \mathbb{N}}$ be a sequence of all rational numbers. We define a new sequence $\{\mu_n\}_{n \in \mathbb{N}}$ as follows

$$\mu_1 = \lambda_1, \quad \mu_2 = \lambda_2,$$
$$\mu_3 = \lambda_1, \quad \mu_4 = \lambda_2, \quad \mu_5 = \lambda_3,$$
$$\mu_6 = \lambda_1, \quad \mu_7 = \lambda_2, \quad \mu_8 = \lambda_3, \quad \mu_9 = \lambda_4,$$

$$\dots\dots\dots\dots\dots\dots\dots\dots\dots\dots\dots\dots$$

Every $\lambda \in \mathbb{Q}$ occurs in the sequence $\{\mu_n\}_{n \in \mathbb{N}}$ infinitely many times. Let

$$U_n = T_{\mu_n} \cap \{x \in \mathbb{R}^N \mid 2^{2n-1} < |x| < 2^{2n+1}\}, \qquad n \in \mathbb{N},$$

and

$$T = \{0\} \cup \bigcup_{n=1}^{\infty} U_n.$$

It is easily seen that the restriction $f \mid T$ is continuous (f is constant on every U_n, and the sets U_n are separated). Since f is discontinuous, this shows that

$$T \notin \mathfrak{B}_C. \tag{10.7.2}$$

Now take an arbitrary $x \in \mathbb{R}^N$, $x \neq 0$. Let $n \in \mathbb{N}$ be such that $f(x) = \lambda_n$. Then

$$f\left(\frac{1}{2}x\right) = \frac{1}{2}f(x) = \frac{1}{2}\lambda_n.$$

Let l be the straight line passing through the origin and x. Since $\operatorname{card} l = \mathfrak{c}$, whereas $f(l) \subset \mathbb{Q}$ is at most countable, $f \mid l$ cannot be one-to-one. Thus there exist $a, b \in l$, $a \neq b$, such that $f(a) = f(b)$, whence $f(a - b) = f(a) - f(b) = 0$. For every $\alpha \in \mathbb{Q}$ we have $\frac{1}{2}x + \alpha(a - b) \in l$ and $f\left(\frac{1}{2}x + \alpha(a - b)\right) = f\left(\frac{1}{2}x\right) + \alpha f(a - b) = f\left(\frac{1}{2}x\right) = \frac{1}{2}\lambda_n$. Consequently the set $T_{\frac{1}{2}\lambda_n}$ is dense on l.

It follows from the definition of $\{\mu_n\}$ and $\{U_n\}$ that there exists a $k \in \mathbb{N}$ such that

$$0 < |x| < 2^{2k-1} \quad \text{and} \quad f(t) = \frac{1}{2}\lambda_n \quad \text{for all } t \in U_k.$$

Take a $y \in U_k \cap l \cap \{x \in \mathbb{R}^N \mid 2^{2k} < |x| < 2^{2k+1}\} \subset U_k \subset T$ lying on the same semiline of l as x, and put $z = x - y$. Obviously $f(z) = f(x) - f(y) = \lambda_n - \frac{1}{2}\lambda_n = \frac{1}{2}\lambda_n$. Moreover, $z \in l$, whence $2^{2k-1} < |z| < 2^{2k+1}$, whence $z \in U_k \subset T$. Consequently $x = y + z \in T + T$. Evidently also $0 \in T + T$, whence $T + T = \mathbb{R}^N$. By Theorem 9.3.3 $T \in \mathfrak{A}$, and by Theorem 10.2.2 $T \in \mathfrak{B}$. Thus by (10.7.2) $T \in \mathfrak{B} \setminus \mathfrak{B}_C = \mathfrak{A} \setminus \mathfrak{B}_C$, whence in virtue of (10.7.1) $T \in \mathfrak{A} \setminus \mathfrak{A}_C$.

Example 10.7.1 shows also that the condition int $S_2(T) \neq \varnothing$ is not sufficient for $T \in \mathfrak{A}_C$ nor $T \in \mathfrak{B}_C$.

Example 10.7.2. For every $x \in \mathbb{R}^N$ we write $x = (\xi_1, \ldots, \xi_N)$, $\xi_i \in \mathbb{R}$, $i = 1, \ldots, N$, and let

$$e_1 = (1, 0, \ldots, 0), \ldots, e_N = (0, \ldots, 0, 1)$$

be the usual orthonormal base of $\mathbb{R}^N$ over $\mathbb{R}$. Put

$$V = \{x \in \mathbb{R}^N \mid \xi_1 \geqslant 0, \ldots, \xi_N \geqslant 0, \xi_1 + \cdots + \xi_N \leqslant 1\}$$

and

$$V_{\mathbb{Q}} = \{x \in \mathbb{R}^N \mid x \in V \text{ and } \xi_1 \in \mathbb{Q}, \ldots, \xi_N \in \mathbb{Q}\}.$$

We have $V_{\mathbb{Q}} \subset V$, moreover $V_{\mathbb{Q}}$ is dense in V. Also $e_1, \ldots, e_N \in V$ and int $V \neq \varnothing$, whence by Lemma 9.3.1 $V \in \mathfrak{A}$, and by Corollary 9.3.2 V contains a Hamel basis. Hence it follows that $E(V) = \mathbb{R}^N$. Since $e_1, \ldots, e_N$ are linearly independent over $\mathbb{R}$, they are also linearly independent over $\mathbb{Q}$. By Lemma 4.2.1 there exists a Hamel basis H of $\mathbb{R}^N$ such that

$$e_1, \ldots, e_N \in H \subset V. \tag{10.7.3}$$

Define $g_0 : H \to \mathbb{R}$ by $g_0(h) = 1$ for $h \in H$, and let $g : \mathbb{R}^N \to \mathbb{R}$ be the additive extension of g_0 (cf. Theorem 5.2.2). Let

$$T = \{x \in \mathbb{R}^N \mid 0 \leqslant g(x) \leqslant 1\}.$$

By Corollary 5.2.2 g is discontinuous, and since g is bounded on T, we obtain

$$T \notin \mathfrak{C}. \tag{10.7.4}$$

Moreover, $g \mid H = g_0 = 1$, whence $H \subset T$. Also, if $x \subset V_{\mathbb{Q}}$, then $x = \xi_1 e_1 + \cdots + \xi_N e_N$ with $\xi_i \in \mathbb{Q} \cap [0, \infty)$, $\xi_1 + \cdots + \xi_N \leqslant 1$. Hence by (10.7.3) $g(x) = \xi_1 g(e_1) + \cdots + \xi_N g(e_N) = \xi_1 g_0(e_1) + \cdots + \xi_N g_0(e_N) = \xi_1 + \cdots + \xi_N \in [0, 1]$. Thus $V_{\mathbb{Q}} \subset T$.

Let $f : \mathbb{R}^N \to \mathbb{R}$ be an arbitrary additive function such that the restriction $f \mid T$ is continuous. Put $c_i = f(e_i)$, $i = 1, \ldots, N$, and $c = (c_1, \ldots, c_N) \in \mathbb{R}^N$. For $x \in V_{\mathbb{Q}}$, $x = (\xi_1, \ldots, \xi_N)$, we have

$$f(x) = f(\xi_1 e_1 + \cdots + \xi_N e_N) = \xi_1 f(e_1) + \cdots + \xi_N f(e_N) = \xi_1 c_1 + \cdots + \xi_N c_N = cx.$$

Take an arbitrary $h \in H \subset V$. Since $V_{\mathbb{Q}}$ is dense in V, there exist $x_n \in V_{\mathbb{Q}}$, $n \in \mathbb{N}$, such that

$$\lim_{n \to \infty} x_n = h. \tag{10.7.5}$$

We have $V_{\mathbb{Q}} \cup H \subset T$ and $f \mid T$ is continuous, so (10.7.5) implies that

$$\lim_{n \to \infty} f(x_n) = f(h), \quad \text{i.e.,} \quad f(h) = \lim_{n \to \infty} cx_n = c \lim_{n \to \infty} x_n = ch.$$

Consequently for $h \in H$ we have $f(h) = ch$. By the uniqueness part of Theorem 5.2.2 $f(x) = cx$ in $\mathbb{R}^N$, i.e., f is continuous. Consequently $T \in \mathfrak{B}_C$ and by (10.7.4) $T \in \mathfrak{B}_C \setminus \mathfrak{C}$. By (9.1.1) $T \in \mathfrak{B}_C \setminus \mathfrak{B}$ and $T \in \mathfrak{B}_C \setminus \mathfrak{A}$, whence in virtue of Theorem 10.7.1 $T \in \mathfrak{B}_C \setminus \mathfrak{A}_C$.

Gathering together the conclusions following from the above examples, we have the following

Theorem 10.7.2. $\mathfrak{A} \setminus \mathfrak{A}_C \neq \varnothing$, $\mathfrak{B} \setminus \mathfrak{B}_C \neq \varnothing$, $\mathfrak{B}_C \setminus \mathfrak{B} \neq \varnothing$, $\mathfrak{B}_C \setminus \mathfrak{A}_C \neq \varnothing$ and $\mathfrak{B}_C \setminus \mathfrak{C} \neq \varnothing$.

We conclude this section with the following

Theorem 10.7.3. *Let $T \subset \mathbb{R}^N$ be an arbitrary set. If $T \in \mathfrak{A}_C$, then T contains a Hamel basis.*

Proof. This is a consequence of Theorem 10.7.1 and Corollary 9.3.2. $\qquad \square$

Investigations similar to those contained in 10.5, but for classes $\mathfrak{A}_C$ and $\mathfrak{B}_C$, were carried out by Z. Kominek [174].

Exercises

1. Let $T \subset \mathbb{R}^N$ be an arbitrary set. Show that if $T \in \mathfrak{A}$, then also $-T \in \mathfrak{A}$.
2. Let $T \subset \mathbb{R}^N$ be an arbitrary set. Show that if $T \in \mathfrak{A}$, then also $\alpha T + b \in \mathfrak{A}$ for arbitrary $\alpha \in \mathbb{Q} \setminus \{0\}$, $b \in \mathbb{R}^N$.
3. Let $T \subset \mathbb{R}^N$ be an arbitrary set. Show that if $T \in \mathfrak{C}$, then also $\alpha T + b \in \mathfrak{C}$ for arbitrary $\alpha \in \mathbb{R} \setminus \{0\}$, $b \in \mathbb{R}^N$.
4. Let $T \subset \mathbb{R}^N$ be an arbitrary set. Show that if $T \in \mathfrak{A}$, then also $\alpha T + b \in \mathfrak{A}$ for arbitrary $\alpha \in \mathbb{R} \setminus \{0\}$, $b \in \mathbb{R}^N$. (Generalization of Exercise 10.2).
5. Let $T \subset \mathbb{R}^N$ be an analytic set such that $E(T) = \mathbb{R}^N$. Show that, for every $\alpha \in \mathbb{R} \setminus \{0\}$ and $b \in \mathbb{R}^N$, we have $E(\alpha T + b) = \mathbb{R}^N$.
   [Hint: Use Exercise 10.3].
6. Let $\mathbf{C} \subset \mathbb{R}$ be the Cantor set, and let $\mathbf{C}^N$ be the N-fold cartesian product of $\mathbf{C}$. Show that $\mathbf{C}^N \in \mathfrak{A}_N$. (Note that $\mathbf{C}^N$ is nowhere dense in $\mathbb{R}^N$, and of N-dimensional measure zero.)
7. Let $D \subset \mathbb{R}^N$ be a convex and open set, and let $T \subset D$ be an arbitrary set. Show that if $T \in \mathfrak{D}(D)$, or $T \in \mathfrak{B}_C$, then T contains a Hamel basis.
8. Let $T \subset \mathbb{R}^N$ be the set constructed in Example 10.7.1. Show that T does not contain an analytic subset containing a Hamel basis.

Chapter 11

Properties of Hamel Bases

11.1 General properties

We recall that a Hamel basis is any base of the linear space $(\mathbb{R}^N; \mathbb{Q}; +; \cdot)$. We have constructed Hamel bases already many times in this book. Theorem 4.2.1 (cf., in particular, Corollary 4.2.1) asserts that there exist Hamel bases. More exactly (Lemma 4.2.1), for every set $A \subset C \subset \mathbb{R}^N$ such that A is linearly independent over $\mathbb{Q}$, and $E(C) = \mathbb{R}^N$, there exists a Hamel basis H of $\mathbb{R}^N$ such that $A \subset H \subset C$. In particular, every set belonging to any of the classes $\mathfrak{A} = \mathfrak{B}$, $\mathfrak{C}$, $\mathfrak{D}(D)$, $\mathfrak{A}_C$, $\mathfrak{B}_C$ contains a Hamel basis (Theorems 9.3.6 and 10.7.3 and Exercise 10.7). On the other hand, we have the following

Theorem 11.1.1. *No Hamel basis belongs to any of the classes* $\mathfrak{A} = \mathfrak{B}$, $\mathfrak{C}$, $\mathfrak{D}(D)$, $\mathfrak{A}_C$, $\mathfrak{B}_C$.

Proof. Let $H \subset \mathbb{R}^N$ be a Hamel basis. Let the function $g : H \to \mathbb{R}$ be defined by

$$g(h) = 1 \qquad \text{for } h \in H,$$

and let $f : \mathbb{R}^N \to \mathbb{R}$ be the additive extension of g (Theorem 5.2.2). By Corollary 5.2.2 f is discontinuous. $f \mid H = g = 1$ is bounded and continuous, which shows that H does not belong to $\mathfrak{C}$, nor to $\mathfrak{B}_C$, and hence $H \notin \mathfrak{A} = \mathfrak{B} \subset \mathfrak{C}$, $H \notin \mathfrak{D}(D) \subset \mathfrak{A}$ and $H \notin \mathfrak{A}_C \subset \mathfrak{B}_C$. $\qquad \square$

Theorem 4.2.3 says that every Hamel basis has the power of continuum. Consequently the linear space $\mathbb{R}^N$ over $\mathbb{Q}$ is infinitely (continuum) dimensional.

Every additive function $f : \mathbb{R}^N \to \mathbb{R}$ can be prescribed arbitrarily on a Hamel basis (Theorem 5.2.2), and then uniquely extended onto $\mathbb{R}^N$.

Now we prove the following (Kuczma [179])

Theorem 11.1.2. *Let* $C \subset \mathbb{R}^N$ *be a cone, and suppose that* C *contains a cone-base[1].* *Then* $C \notin \mathfrak{B}$.

---

[1] Cf. 4.2.

Proof. Let $B \subset C$ be a cone-base of C. Since B is linearly independent over $\mathbb{Q}$, there exists a Hamel basis H of $\mathbb{R}^N$ such that $B \subset H$. Define the function $g : H \to \mathbb{R}$ by

$$g(h) = -1 \qquad \text{for } h \in H,$$

and let $f : \mathbb{R}^N \to \mathbb{R}$ be the additive extension of g. By Corollary 5.2.2 f is discontinuous. $C = E^+(B)$, consequently every $x \in C$ can be written as

$$x = \sum_{i=1}^{n} \alpha_i \, b_i, \qquad \alpha_i \in \mathbb{Q} \cap [0, \infty), \qquad b_i \in B, \qquad i = 1, \ldots, n.$$

For such an x we have

$$f(x) = \sum_{i=1}^{n} \alpha_i f(b_i) = \sum_{i=1}^{n} \alpha_i g(b_i) = -\sum_{i=1}^{n} \alpha_i \leqslant 0.$$

Thus f is bounded above on C, which shows that $C \notin \mathfrak{B}$. $\square$

   Theorem 11.1.2 justifies the claim made in 4.2 that only very small cones may have a cone-base. In particular, we obtain from Theorem 11.1.2 and Lemma 9.3.1 the following result of J. Aczél and P. Erdős [9]

Corollary 11.1.1. *There does not exist a cone-base of* $[0, \infty) \subset \mathbb{R}$ *(over* $\mathbb{Q}$*).*

   Let $H \subset \mathbb{R}^N$ be a Hamel basis. Then H is a cone-base of $E^+(H)$. By Theorem 11.1.2 $E^+(H) \notin \mathfrak{B}$. But evidently $R(E^+(H)) = E^+(H) - E^+(H) = E(H) = \mathbb{R}^N$. Hence we obtain (Erdős [76]).

Corollary 11.1.2. *There exists a set* $T \subset \mathbb{R}^N$ *which does not belong to the class* $\mathfrak{B}$ *and such that* $R(T) = \mathbb{R}^N$.

Corollary 11.1.3. *The condition* $\operatorname{int} R(T) \neq \varnothing$ *is not sufficient for* $T \in \mathfrak{B}$.

This, together with Theorem 9.3.5, yields another proof of Theorem 9.1.1 ($\mathfrak{B} \neq \mathfrak{C}$).

11.2   Measure

Theorem 11.2.1 below is due to W. Sierpiński [276] (cf. also Marcus [216], Abian [1], Kuczma [189]).

Theorem 11.2.1. *Let* $H \subset \mathbb{R}^N$ *be a Hamel basis. Then* $m_i(H) = 0$.

Proof. This results from Theorems 11.1.1 and 9.3.1. $\square$

Corollary 11.2.1. *If* $H \subset \mathbb{R}^N$ *is a Hamel basis, and* $A \subset H$ *is a Lebesgue measurable set, then* $m(A) = 0$.

   In particular, every measurable Hamel basis has measure zero.

   It is worthwhile to observe that there exist measurable Hamel bases. In Chapter 9 we saw that in every space $\mathbb{R}^N$ there exist measurable sets $T \in \mathfrak{A}$ of measure zero. By Corollary 9.3.2 every such set contains a Hamel basis, which necessarily is Lebesgue measurable and of measure zero.

   Now let $H \subset \mathbb{R}^N$ be a Hamel basis, and let $H_0 \subset H$ be a finite or countable subset of H. We have (Sierpiński [276])

Theorem 11.2.2. *If $H \subset \mathbb{R}^N$ is a Hamel basis, and $H_0 \subset H$ is a finite or countable set, then the set $E(H \setminus H_0)$ is saturated non-measurable and has property $(*)$ from 3.3.*

Proof. First suppose that $E(H \setminus H_0)$ is Lebesgue measurable and of measure zero, or $E(H \setminus H_0)$ is of the first category. Take an arbitrary $x \in \mathbb{R}^N$. Then

$$x = \sum_{i=1}^{n} \alpha_i h_i, \qquad \alpha_i \in \mathbb{Q}, \qquad h_i \in H, \qquad i = 1, \ldots, n.$$

Let $h_i \in H \setminus H_0$ for $i = 1, \ldots, k$, and $h_i \in H_0$ for $i = k+1, \ldots, n$. Then

$$u = \sum_{i=1}^{k} \alpha_i h_i \in E(H \setminus H_0), \qquad v = \sum_{i=k+1}^{n} \alpha_i h_i \in E(H_0),$$

($u = 0$ if $k = 0$, $v = 0$ if $k = n$), and $x = u + v$. This means that $\mathbb{R}^N = E(H \setminus H_0) + E(H_0)$, or

$$\mathbb{R}^N = \bigcup_{v \in E(H_0)} E(H \setminus H_0) + v. \tag{11.2.1}$$

By Lemma 4.1.3 the union in (11.2.1) is countable. Thus our supposition leads to the conclusion that $\mathbb{R}^N$ itself is of measure zero, resp. of the first category, which is not the case. Consequently $m_e(E(H \setminus H_0)) > 0$, and $E(H \setminus H_0)$ is of the second category.

Next observe that every proper linear subspace of $\mathbb{R}^N$ over $\mathbb{R}$ (a k-dimensional hyperplane, where $k < N$) is of measure zero and of the first category (even nowhere dense). Since $E(H \setminus H_0) = E_{\mathbb{Q}}(H \setminus H_0) \subset E_{\mathbb{R}}(H \setminus H_0)$, we must have $E_{\mathbb{R}}(H \setminus H_0) = \mathbb{R}^N$. By Lemma 4.2.1 there exists a base $B = \{b_1, \ldots, b_N\} \subset H \setminus H_0$ of $\mathbb{R}^N$ over $\mathbb{R}$. The set $H_1 = H_0 \cup B$ is finite or countable, and therefore we may replace H_0 by H_1 in the argument above. Consequently $m_e(E(H \setminus H_1)) > 0$, and $E(H \setminus H_1)$ is of the second category.

In virtue of Lemma 4.2.3 the set $E(B)$ is dense in $\mathbb{R}^N$.

As is easily seen (cf. the argument above), we have

$$E(H \setminus H_0) = E(H \setminus H_1) + E(B).$$

By Theorems 3.6.1 and 3.6.2 $m_i(\mathbb{R}^N \setminus E(H \setminus H_0)) = 0$ and every set $T \subset \mathbb{R}^N \setminus E(H \setminus H_0)$ with the Baire property is of the first category. Suppose that $m_i(E(H \setminus H_0)) > 0$, or $E(H \setminus H_0)$ contains a set of the second category and with the Baire property. By Theorem 4.3.3 there exists an additive function $f : \mathbb{R}^N \to \mathbb{R}$ such that $\operatorname{Ker} f = E(H \setminus H_0)$. Since $B \subset H \setminus H_0$, we have $E(B) \subset E(H \setminus H_0)$, whence it follows that $\operatorname{Ker} f$ is dense in $\mathbb{R}^N$. If f were continuous, we would obtain hence $E(H \setminus H_0) = \operatorname{Ker} f = \mathbb{R}^N$, which is impossible. Consequently f must be discontinuous. But f is zero (and hence bounded) on $E(H \setminus H_0)$, and thus, by Theorem 9.3.1 resp. 9.3.2 it is continuous. This contradiction shows that $m_i(E(H \setminus H_0)) = 0$ and $E(H \setminus H_0)$ cannot contain a set of the second category and with the Baire property. Consequently $E(H \setminus H_0)$ is saturated non-measurable and has property $(*)$ from 3.3. $\qquad \square$

Theorem 11.2.2 shows that the existence of non-measurable sets can be derived directly (without a use of the Axiom of Choice or of an equivalent statement) from the existence of a Hamel basis. Of course, the Axiom of Choice is inherent in the existence of a Hamel basis, but once we know that a Hamel basis exists, then we are able to construct non-measurable sets by elementary devices, without making a further appeal to the Axiom of Choice (Sierpiński [276]).

The next theorem (Kurepa [199]) shows a certain irregularity of the operation "+".

Theorem 11.2.3. *There exist measurable sets A, $B \subset \mathbb{R}^N$ such that the set $A + B$ is non-measurable.*

Proof. Let $H \subset \mathbb{R}^N$ be a measurable Hamel basis. Then also all sets λH with $\lambda \in \mathbb{Q}$ are measurable.

We have

$$\mathbb{R}^N = \bigcup_{n=1}^{\infty} \bigcup_{\lambda_1,\ldots,\lambda_n \in \mathbb{Q}} \left(\lambda_1 H + \cdots + \lambda_n H\right) \tag{11.2.2}$$

(cf. the proof of Theorem 9.3.7, in particular formula (9.3.10)). Put $g(h) = 1$ for $h \in H$, and let $f : \mathbb{R}^N \to \mathbb{R}$ be the additive extension of g (Theorem 5.2.2). By Corollary 5.2.2 f is discontinuous.

Fix $\lambda_1, \ldots, \lambda_n \in \mathbb{Q}$, and let $\mu = \max(|\lambda_1|, \ldots, |\lambda_n|)$. Every $x \in \lambda_1 H + \cdots + \lambda_n H$ has a form

$$x = \lambda_1 h_1 + \cdots + \lambda_n h_n,$$

with $h_1, \ldots, h_n \in H$. For such an x we have

$$\begin{aligned} f(x) &= \lambda_1 f(h_1) + \cdots + \lambda_n f(h_n) = \lambda_1 g(h_1) + \cdots + \lambda_n g(h_n) \\ &= \lambda_1 + \cdots + \lambda_n \leqslant |\lambda_1| + \cdots + |\lambda_n| \leqslant n\mu. \end{aligned}$$

Thus f is bounded above (by $n\mu$) on $\lambda_1 H + \cdots + \lambda_n H$, whence $\lambda_1 H + \cdots + \lambda_n H \notin \mathfrak{B}$. It follows in virtue of Theorem 9.3.1 that

$$m_i(\lambda_1 H + \cdots + \lambda_n H) = 0. \tag{11.2.3}$$

Consequently all summands in (11.2.2) have inner measure zero. But since the union in (11.2.2) is countable, they cannot all have the outer measure zero.

Let $n_0 \in \mathbb{N}$ be the minimal positive integer such that there exist $\lambda_1, \ldots, \lambda_{n_0} \in \mathbb{Q}$ such that

$$m_e(\lambda_1 H + \cdots + \lambda_{n_0} H) > 0. \tag{11.2.4}$$

By what has been said at the beginning of the proof $n_0 > 1$. It follows from (11.2.3), (11.2.4) and Theorem 3.1.2 that the set $\lambda_1 H + \cdots + \lambda_{n_0} H$ is non-measurable, whereas by the minimal property of n_0 and by Theorem 3.1.3 the set $\lambda_1 H + \cdots + \lambda_{n_0-1} H$ is measurable (and of measure zero). Also the set $\lambda_{n_0} H$ is measurable.

Now, we have

$$\lambda_1 H + \cdots + \lambda_{n_0} H = (\lambda_1 H + \cdots + \lambda_{n_0-1} H) + (\lambda_{n_0} H). \tag{11.2.5}$$

Our theorem results now from (11.2.5) with $A = \lambda_1 H + \cdots + \lambda_{n_0-1} H$ and $B = \lambda_{n_0} H$. $\qquad\square$

11.3 Topological properties

Theorems 11.3.1, 11.3.2 and Corollary 11.3.1 (Sierpiński [282]; cf. also Marcus [216])
are topological analogues of Theorems 11.2.1 and 11.2.3 and of Corollary 11.2.1.
Theorem 11.3.1 results from Theorems 11.1.1 and 9.3.2, whereas the proof of Theorem
11.3.2 is identical as that of Theorem 11.2.3.

Theorem 11.3.1. *Every subset T of a Hamel basis such that T has the Baire property,
is of the first category.*

Corollary 11.3.1. *If a Hamel basis $H \subset \mathbb{R}^N$ has the Baire property, then H is of the
first category.*

Again we may observe that there exist Hamel bases $H \subset \mathbb{R}^N$ which are of the
first category, or even nowhere dense (and hence have the Baire property), since in
every space $\mathbb{R}^N$ there exist sets $T \in \mathfrak{A}$ (and consequently containing a Hamel basis;
cf. Corollary 9.3.2) which are nowhere dense in $\mathbb{R}^N$.

Theorem 11.3.2. *There exist sets A, $B \subset \mathbb{R}^N$ with the Baire property and such that
$A + B$ does not have the Baire property.*

As an immediate consequence of Theorems 11.1.1 and either Theorem 9.3.7 or
Theorem 9.8.1 we obtain the following (Sierpiński [276], Jones [158])

Theorem 11.3.3. *No Hamel basis is an analytic set.*

From Theorem 11.3.3 and Theorem 2.5.6 we immediately get (Sierpiński [276])

Corollary 11.3.2. *No Hamel basis is a Borel set.*

Also the topological part of Theorem 11.1.2 belongs in this section.

11.4 Burstin bases

In 11.2–11.3 we saw that there exist very small Hamel bases: of measure zero and
nowhere dense. Now we will try to answer the question: how large a Hamel basis can
be. Theorems 11.1.1, 11.2.1 and 11.3.1 suggest that not too large. However, as we will
presently see, there exist quite large Hamel bases.

Let $X \subset \mathbb{R}^N$. A Hamel basis $H \subset X$ is called a *Burstin basis relative to X* iff
H intersects every uncountable Borel subset of X. A Burstin basis relative to $\mathbb{R}^N$ is
simply called a *Burstin basis*. (After C. Burstin, who in Burstin [39] first considered
similar constructions. Cf. also Abian [1], Kuczma [189]).

Theorem 11.4.1. *If the set $X \subset \mathbb{R}^N$ is of the full measure (i.e., $m(\mathbb{R}^N \setminus X) = 0$),
then every Burstin basis relative to X is saturated non-measurable.*

Proof. Let $H \subset X$ be a Burstin basis relative to X, and suppose that H is not
saturated non-measurable. In view of Theorem 11.2.1 this means that $m_i(\mathbb{R}^N \setminus H) > 0$.
Thus there exists a closed set $F \subset \mathbb{R}^N \setminus H$ with $m(F) > 0$. Since $H \subset X$, we have
$\mathbb{R}^N \setminus H = (\mathbb{R}^N \setminus X) \cup (X \setminus H)$, whence $F = F \cap (\mathbb{R}^N \setminus H) = [F \cap (\mathbb{R}^N \setminus X)] \cup [F \cap (X \setminus H)]$,
and

$$F \cap (X \setminus H) = F \setminus [F \cap (\mathbb{R}^N \setminus X)]. \qquad (11.4.1)$$

Since $m(\mathbb{R}^N \setminus X) = 0$, also the set $F \cap (\mathbb{R}^N \setminus X)$ is measurable and of measure zero. F is measurable, being closed. Evidently $F \cap (\mathbb{R}^N \setminus X) \subset F$, thus (11.4.1) implies

$$m[F \cap (X \setminus H)] = m(F) - m[F \cap (\mathbb{R}^N \setminus X)] = m(F) > 0.$$

Consequently there exists a closed set $K \subset F \cap (X \setminus H)$ with $m(K) > 0$. Hence it follows that K is uncountable. (Otherwise we would have $m(K) = 0$). K, being closed, is a Borel set, and we have $K \subset X$. Thus by the definition of a Burstin basis we have $H \cap K \neq \varnothing$, which contradicts the relation $K \subset X \setminus H$.   □

A topological analogue of Theorem 11.4.1 is the following

Theorem 11.4.2. *If the set $X \subset \mathbb{R}^N$ is residual, then every Burstin basis relative to X fulfils condition $(*)$ from 3.3.*

Proof. Let $H \subset X$ be a Burstin basis relative to X, and suppose that H does not fulfil $(*)$. In view of Theorem 11.3.1 this means that the set $\mathbb{R}^N \setminus H$ contains a subset E of the second category and with the Baire property. Thus

$$E = (G \cup P) \setminus R,$$

where G is open and non-empty (since E is of the second category), and the sets P, R are of the first category. G, being open, is in particular of class $\mathcal{G}_\delta$ and

$$G \setminus (G \cap E) = G \setminus E = G \cap E' = G \cap [(G \cup P)' \cup R] \subset G \cap R \subset R,$$

whence $G \setminus (G \cap E)$ is of the first category. By Theorem 2.1.3 there exists a set $F \subset G \cap E \subset E$ such that $F \in \mathcal{G}_\delta$ and $G \setminus F$ is of the first category, whence it follows that F is of the second category. We have $F \subset E \subset \mathbb{R}^N \setminus H$, whence

$$F = F \cap (\mathbb{R}^N \setminus H) = [F \cap (X \setminus H)] \cup [F \cap (\mathbb{R}^N \setminus X)],$$

whence (11.4.1) follows. Since X is residual, $\mathbb{R}^N \setminus X$ is of the first category, and hence also $F \setminus [F \cap (X \setminus H)] = F \cap (\mathbb{R}^N \setminus X)$ is of the first category. Again by Theorem 2.1.3 there exists a set $K \subset F \cap (X \setminus H)$ such that $K \in \mathcal{G}_\delta$ and $F \setminus K$ is of the first category. Consequently K is of the second category, and thus uncountable. Since K belongs to $\mathcal{G}_\delta$, it is a Borel set, and we have $K \subset X$. Consequently $H \cap K \neq \varnothing$, which contradicts the relation $K \subset X \setminus H$.   □

Taking $X = \mathbb{R}^N$ we obtain from Theorems 11.4.1 and 11.4.2

Corollary 11.4.1. *Every Burstin basis is saturated non-measurable and has property $(*)$ from 3.3.*

We see that a Burstin basis yields an example of a comparatively "large" Hamel basis. The obvious question arises as to the existence of a Burstin basis. Presently we will answer this question.

Theorem 11.4.3. *Let $X \subset \mathbb{R}^N$ be a Borel set such that $E(X) = \mathbb{R}^N$. Then there exists a Burstin basis relative to X.*

Proof. Let $\mathcal{F}$ be the family of all uncountable Borel subsets of X. Note that if we had card $X \leqslant \aleph_0$, then by Lemma 4.1.3 we would have card $\mathbb{R}^N = \text{card } E(X) = \aleph_0$. Consequently X must be uncountable, and thus $X \in \mathcal{F}$. Consequently $\mathcal{F} \neq \varnothing$.

By Theorem 2.3.4 we get, since $\mathcal{F} \subset \mathcal{B}(\mathbb{R}^N)$,

$$\text{card } \mathcal{F} \leqslant \text{card } \mathcal{B}(\mathbb{R}^N) = \mathfrak{c}. \tag{11.4.2}$$

On the other hand, for every $x \in X$ the set $X \setminus \{x\}$ is an uncountable Borel subset of X, and hence belongs to $\mathcal{F}$. Thus $\mathcal{F}$ contains the family $\{X \setminus \{x\}\}_{x \in X}$. Consequently

$$\text{card } \mathcal{F} \geqslant \text{card } X = \mathfrak{c} \tag{11.4.3}$$

(cf. Theorem 2.8.3). Relations (11.4.2) and (11.4.3) show that card $\mathcal{F} = \mathfrak{c}$.

Let (cf. 1.5)

$$A = \{\alpha \in M(\mathfrak{c}) \mid \overline{\alpha} = \mathfrak{c}\},$$

and let γ be the smallest element of A. It follows from Theorem 1.7.1 that $A \neq \varnothing$, and thus γ exists in virtue of Theorem 1.4.3. Thus $\overline{\gamma} = \mathfrak{c}$, but for every ordinal number $\alpha < \gamma$ we have $\overline{\alpha} < \mathfrak{c}$.

Since card $\Gamma(\gamma) = \overline{\gamma} = \mathfrak{c} = \text{card } \mathcal{F}$, there exists a one-to-one mapping $f : \Gamma(\gamma) \to \mathcal{F}$ which is onto. For every $\alpha \in \Gamma(\gamma)$ write $F_\alpha = f(\alpha) \in \mathcal{F}$. Thus we have arranged $\mathcal{F}$ into a transfinite sequence $\mathcal{F} = \{F_\alpha\}_{\alpha < \gamma}$ of type γ.

Now we define a transfinite sequence $\{b_\alpha\}_{\alpha < \gamma}$ of elements of X, and for every $\alpha < \gamma$ we put $B_\alpha = \bigcup_{\beta < \alpha} \{b_\beta\}$. As b_0 we take an arbitrary element of F_0 with the only restriction $b_0 \neq 0$:

$$b_0 = w(F_0 \setminus \{0\}), \tag{11.4.4}$$

where $w : \mathcal{P}(X) \setminus \{\varnothing\} \to X$ is a choice function (cf. 1.1). Having already defined b_β for $\beta < \alpha < \gamma$, we take

$$b_\alpha = w(F_\alpha \setminus E(B_\alpha)). \tag{11.4.5}$$

Since card $B_\alpha = \overline{\alpha} < \mathfrak{c}$, we have by Lemmas 4.1.3 and 4.2.2 card $E(B_\alpha) < \mathfrak{c}$, whereas by Theorem 2.8.3 card $F_\alpha = \mathfrak{c}$. Consequently $F_\alpha \setminus E(B_\alpha) \neq \varnothing$, and b_α is well defined. By Theorem 1.6.2 there exists a transfinite sequence $\{b_\alpha\}_{\alpha < \gamma}$ fulfilling (11.4.4) and (11.4.5) for every $\alpha < \gamma$. Let $B = \bigcup_{\alpha < \gamma} \{b_\alpha\} = \bigcup_{\alpha < \gamma} B_\alpha$. We will prove by transfinite induction that B is linearly independent over $\mathbb{Q}$.

First we prove that for every $\alpha < \gamma$ the set $B_\alpha^* = B_\alpha \cup \{b_\alpha\} = \bigcup_{\beta \leqslant \alpha} \{b_\beta\}$ is linearly independent. For $\alpha = 0$ we have $B_0^* = \{b_0\}$, which is linearly independent, since $b_0 \neq 0$. Assuming that B_β^* are linearly independent for $\beta < \alpha$, we check that the set $\bigcup_{\beta < \alpha} B_\beta^*$ also is linearly independent. In fact, let $C \subset \bigcup_{\beta < \alpha} B_\beta^*$ be any finite set, $C = \{b_{\beta_1}, \ldots, b_{\beta_k}\}$, where $\beta_i < \alpha$ for $i = 1, \ldots, k$. Put $\beta = \max(\beta_1, \ldots, \beta_k) < \alpha$. We have $\beta_i \leqslant \beta$ for $i = 1, \ldots, k$, and since $B_{\beta'}^* \subset B_{\beta''}^*$ for $\beta' \leqslant \beta''$, we have $B_{\beta_i}^* \subset B_\beta^*$ for $i = 1, \ldots, k$, and so $b_{\beta_i} \in B_{\beta_i}^* \subset B_\beta^*$ for $i = 1, \ldots, k$. Hence $C \subset B_\beta^*$. Since $\beta < \alpha$, the set B_β^* is linearly independent, and consequently also C is linearly independent. This shows that the set $\bigcup_{\beta<\alpha} B_\beta^*$ is linearly independent.

Now, $\bigcup_{\beta<\alpha} B_\beta^* = B_\alpha$, and by (11.4.5) $b_\alpha \notin E(B_\alpha)$. By Lemma 4.1.2 the set $B_\alpha^* = B_\alpha \cup \{b_\alpha\}$ is linearly independent.

Thus by Theorem 1.6.1 the sets B_α^* are linearly independent for all $\alpha < \gamma$. By the same argument as above the set $B = \bigcup_{\alpha<\gamma} B_\alpha^*$ is linearly independent.

By Lemma 4.2.1 there exists a Hamel basis H such that $B \subset H \subset X$. Let $F \subset X$ be an uncountable Borel set. Then $F \in \mathcal{F}$, and hence there exists an $\alpha < \gamma$ such that $F = F_\alpha$. Then $b_\alpha \in F$ (cf. (11.4.5)), and $b_\alpha \in B \subset H$, whence $b_\alpha \in F \cap H$, and $F \cap H \neq \varnothing$. This proves that H is a Burstin basis relative to X. $\square$

Taking in Theorem 11.4.3 $X = \mathbb{R}^N$, we obtain

Corollary 11.4.2. *There exists a Burstin basis.*

Let $D \subset \mathbb{R}^N$ be a dense and countable set. Since D is countable, $m(D) = 0$, and hence for every $n \in \mathbb{N}$ there exists an open set G_n such that $D \subset G_n$ and $m(G_n) < n^{-1}$. Put $A = \bigcap_{n=1}^{\infty} G_n$, $B = \mathbb{R}^N \setminus A$. Then $A \in \mathcal{G}_\delta$, and hence $B \in \mathcal{F}_\sigma$. Since $D \subset A$, A is dense and int $B \neq \varnothing$. By Lemma 2.1.1 B is of the first category, i.e., A is residual. (In particular, A is of the second category and with the Baire property, whence by Theorem 9.3.2 $A \in \mathfrak{A}$). We have $m(A) \leqslant m(G_n) < n^{-1}$ for $n \in \mathbb{N}$, whence $m(A) = 0$. Hence B is of the full measure. (In particular, $m(B) = \infty > 0$, whence by Theorem 9.3.1 $B \in \mathfrak{A}$). Thus A and B are uncountable Borel sets, and by Corollary 9.3.2 $E(A) = E(B) = \mathbb{R}^N$. By Theorem 11.4.3 A contains a Burstin basis H_1 relative to A, and B contains a Burstin basis H_2 relative to B. By Theorem 11.4.1 H_2 is saturated non-measurable, whereas by Theorem 11.4.2 H_1 has property $(*)$ from 3.3. Since $m(A) = 0$, also H_1 has measure zero, and since B is of the first category, so is also H_2. Hence we obtain

Corollary 11.4.3. *There exists a Hamel basis which is of measure zero and has property $(*)$ from 3.3, and there exists a Hamel basis which is of the first category and is saturated non-measurable.*

11.5   Erdős sets

Let $H \subset \mathbb{R}^N$ be an arbitrary Hamel basis. With every such a basis we associate the set $Z(H)$ defined as follows (Erdős [76])

$$Z(H) = \{x \in \mathbb{R}^N \mid x = \kappa_1 h_1 + \cdots + \kappa_n h_n, \ \kappa_i \in \mathbb{Z}, \ h_i \in H, \ i = 1, \ldots, n; \ n \in \mathbb{N}\}.$$
$$(11.5.1)$$

The set $Z(H)$ will be referred to as the *Erdős set* associated with the basis H (Kuczma [188]).

Before proving more sophisticated properties of Erdős sets let us first observe some simple facts.

Lemma 11.5.1. *For every Hamel basis $H \subset \mathbb{R}^N$ we have*

$$Z(H) + Z(H) = Z(H). \tag{11.5.2}$$

Proof. If x and y are linear combinations with integral coefficients of elements of H, then so is also $x + y$, which shows that

$$Z(H) + Z(H) \subset Z(H). \tag{11.5.3}$$

On the other hand, always $0 \in Z(H)$ $\big($take in (11.5.1) all $\kappa_i = 0\big)$, whence

$$Z(H) = Z(H) + 0 \subset Z(H) + Z(H). \tag{11.5.4}$$

(11.5.3) and (11.5.4) yield (11.5.2). $\qquad\square$

Lemma 11.5.2. *For every Hamel basis $H \subset \mathbb{R}^N$ and for every $x \in \mathbb{R}^N$ there exists a $k \in \mathbb{N}$ such that $kx \in Z(H)$.*

Proof. There exist $\lambda_1, \ldots, \lambda_n \in \mathbb{Q}$ and $h_1, \ldots, h_n \in H$ such that

$$x = \lambda_1 h_1 + \cdots + \lambda_n h_n.$$

Let k be a common multiple of the denominators of $\lambda_1, \ldots, \lambda_n$. Then $k\lambda_1, \ldots, k\lambda_n \in \mathbb{Z}$, and so by (11.5.1)

$$kx = k\lambda_1 h_1 + \cdots + k\lambda_n h_n \in Z(H). \qquad\square$$

Corollary 11.5.1. *For every Hamel basis $H \subset \mathbb{R}^N$ we have*

$$\mathbb{R}^N = \bigcup_{n=1}^{\infty} \frac{1}{n} Z(H). \tag{11.5.5}$$

Proof. By Lemma 11.5.2 for every $x \in \mathbb{R}^N$ there exists a $k \in \mathbb{N}$ such that

$$x \in \frac{1}{k} Z(H) \subset \bigcup_{n=1}^{\infty} \frac{1}{n} Z(H).$$

This shows that $\mathbb{R}^N \subset \bigcup_{n=1}^{\infty} \frac{1}{n} Z(H)$. The converse inclusion is obvious. $\qquad\square$

Lemma 11.5.3. *For every Hamel basis $H \subset \mathbb{R}^N$ the set $Z(H)$ is dense in $\mathbb{R}^N$.*

Proof. Take an arbitrary $x \in \mathbb{R}^N$. By Lemma 11.5.2 there exist $k, l \in \mathbb{Z}$ such that $kx \in Z(H)$ and $\sqrt{2}\, lx \in Z(H)$. Hence also $pkx + ql\sqrt{2}\, x = k\left(p + q\dfrac{l}{k}\sqrt{2}\right) x \in Z(H)$ for every $p, q \in \mathbb{Z}$. But by Lemma 3.8.1 the set

$$\left\{ t \in \mathbb{R} \mid t = p + q\,\frac{l}{k}\,\sqrt{2},\ p, q \in \mathbb{Z} \right\}$$

is dense in $\mathbb{R}$, so there exist sequences $\{p_n\}_{n\in\mathbb{N}}$ and $\{q_n\}_{n\in\mathbb{N}}$, p_n, $q_n \in \mathbb{Z}$ for $n \in \mathbb{N}$, such that

$$\lim_{n\to\infty} \left(p_n + q_n \frac{l}{k} \sqrt{2} \right) = \frac{1}{k}.$$

Put $x_n = \left(p_n + q_n \dfrac{l}{k} \sqrt{2} \right) x$, $n \in \mathbb{N}$. Then $kx_n \in Z(H)$ for $n \in \mathbb{N}$ and $\lim\limits_{n\to\infty} kx_n = x$. Consequently $Z(H)$ is dense in $\mathbb{R}^N$. $\qquad\square$

Theorem 11.5.1. *For every Hamel basis $H \subset \mathbb{R}^N$ the set $Z(H)$ is saturated non-measurable and has property $(*)$ from 3.3.*

Proof. Take an $h_0 \in H$. Then, for every $x \in Z(H)$, we have $\dfrac{1}{2} h_0 + x \notin Z(H)$, i.e., $Z(H)$ is disjoint with $\dfrac{1}{2} h_0 + Z(H)$. By Lemma 11.5.3 the set $\dfrac{1}{2} h_0 + Z(H)$ is dense in $\mathbb{R}^N$, consequently

$$\text{int } Z(H) = \varnothing. \tag{11.5.6}$$

It follows from (11.5.2), (11.5.6), and Theorems 2.9.1 and 3.7.1 that $m_i\big(Z(H)\big) = 0$ and $Z(H)$ does not contain any subset of the second category and with the Baire property. If we had $m\big(Z(H)\big) = 0$, or if $Z(H)$ were of the first category, then according to (11.5.5) the same would be true about $\mathbb{R}^N$, which is not the case. Consequently $m_e\big(Z(H)\big) > 0$ and $Z(H)$ is of the second category.

Theorems 3.6.1 and 3.6.2 now imply in virtue of relation (11.5.2) and Lemma 11.5.3 that $m_i\big(\mathbb{R}^N \setminus Z(H)\big) = 0$ and $\mathbb{R}^N \setminus Z(H)$ does not contain any subset of the second category and with the Baire property. Hence the theorem follows. $\qquad\square$

Independently of how small Hamel basis H we start from, Theorem 11.5.1 asserts that the corresponding Erdős set is always large. This fact is also reflected in the relation

$$Z(H) \in \mathfrak{A}. \tag{11.5.7}$$

Relation (11.5.7) (Ger [98]) has already been mentioned in the proof of Lemma 9.3.1, but we will prove here a stronger result (Kuczma [188]; cf. also Kuczma [180]).

Lemma 11.5.4. *For every Hamel basis $H \subset \mathbb{R}^N$ and every non-empty open set $G \subset \mathbb{R}^N$ we have*

$$G \cap Z(H) \in \mathfrak{A}. \tag{11.5.8}$$

Proof. First consider the case where $0 \in G$. Let $f : \mathbb{R}^N \to \mathbb{R}$ be a discontinuous additive function.

Since $\mathbb{R}^N = E(H) = E_{\mathbb{Q}}(H) \subset E_{\mathbb{R}}(H) \subset \mathbb{R}^N$, we have $E_{\mathbb{R}}(H) = \mathbb{R}^N$, and thus, by Corollary 4.2.2, H contains a base $\{b_1, \ldots, b_N\}$ of $\mathbb{R}^N$ over $\mathbb{R}$. Every $x \in \mathbb{R}^N$ can be uniquely represented as

$$x = x_1 b_1 + \cdots + x_N b_N,$$

and $x_1, \ldots, x_N$ depend on x in a continuous manner. We have, since f is additive,

$$f(x) = f(x_1 b_1) + \cdots + f(x_N b_N) = \varphi_1(x_1) + \cdots + \varphi_N(x_N),$$

where $\varphi_i(t) = f(tb_i)$. If all functions $\varphi_1, \ldots, \varphi_N$ were continuous, so would also be f, so there exists an i_0, $1 \leqslant i_0 \leqslant N$, such that φ_{i_0} is discontinuous. In the sequel we suppress the index i_0 and write φ instead of φ_{i_0} and b instead of b_{i_0}. Note that $b \in H$.

It is easily seen that $\varphi : \mathbb{R} \to \mathbb{R}$ is an additive function. If we had $\varphi(t) = \varphi(1)t$, φ would be continuous, consequently there exists a $u \in \mathbb{R}$ such that $\varphi(u) \neq \varphi(1)u$. Write

$$\varphi(u) = \varphi(1)u + c \tag{11.5.9}$$

so that $c \neq 0$. It follows from Theorem 5.2.1 that u is irrational. By Lemma 11.5.2 there exists a $k \in \mathbb{N}$ such that $kub \in Z(H)$, and clearly ku is also irrational.

For every p, $q \in \mathbb{Z}$ we have $pb + qkub \in Z(H)$. Write

$$D = \{\alpha \in \mathbb{R} \mid \alpha = p + qku, \; p, \; q \in \mathbb{Z}\}.$$

By Lemma 3.8.1 the set D is dense in $\mathbb{R}$, so there exists a sequence $\{\alpha_n\}_{n \in \mathbb{N}}$ of elements of D such that

$$\alpha_n \neq 0 \quad \text{for } n \in \mathbb{N}, \tag{i}$$

$$\lim_{n \to \infty} \alpha_n = 0. \tag{ii}$$

We have also

$$\alpha_n b \in Z(H) \quad \text{for } n \in \mathbb{N}. \tag{iii}$$

Every α_n can be written as $p_n + q_n ku$, with p_n, $q_n \in \mathbb{Z}$, $n \in \mathbb{N}$. By (ii) the sequence $\{\alpha_n\}$ is bounded. If the sequence $\{q_n\}$ were bounded, also $\{p_n\}$ would have to be bounded, and thus the coefficients p_n and q_n would assume only a finite number of different values. Thus also the sequence $\{\alpha_n\}$ could only have a finite number of different terms, which is incompatible with (i) and (ii).

Thus the sequence $\{q_n\}$ must be unbounded. Note that properties (i)–(iii) of $\{\alpha_n\}$ remain unchanged if we replace the sequence $\{\alpha_n\}$ by $\{-\alpha_n\}$, or by a subsequence. Consequently we may assume that the sequence $\{\alpha_n\}$ has been chosen in such a manner that

$$\lim_{n \to \infty} c\, q_n = +\infty. \tag{11.5.10}$$

Now, we have by Theorem 5.2.1 and by (11.5.9)

$$f(\alpha_n b) = f(p_n b + q_n kub) = f(p_n b) + f(q_n kub) = p_n f(b) + q_n k f(ub)$$
$$= p_n \varphi(1) + q_n k \varphi(u) = (p_n + q_n ku)\varphi(1) + q_n kc = \alpha_n \varphi(1) + q_n kc,$$

whence by (ii) and (11.5.10)

$$\lim_{n \to \infty} f(\alpha_n b) = +\infty. \tag{11.5.11}$$

By (ii) $\alpha_n b \in G$ for n sufficiently large, and thus in view of (iii) $\alpha_n b \in G \cap Z(H)$ for n sufficiently large. Relation (11.5.11) shows now that f is unbounded above on $G \cap Z(H)$.

Consequently every discontinuous additive function $f : \mathbb{R}^N \to \mathbb{R}$ is unbounded above on $G \cap Z(H)$. Hence it follows that $G \cap Z(H) \in \mathfrak{B}$. By Theorem 10.2.2 we obtain hence (11.5.8).

Now let $0 \notin G$, and take an $h \in G \cap Z(H)$. (Such an h exists in virtue of Lemma 11.5.3). $G - h$ is a non-empty open set and $0 \in G - h$, so by the first part of the proof we have $(G - h) \cap Z(H) \in \mathfrak{A}$. Next

$$(G - h) \cap Z(H) + h = G \cap \big(Z(H) + h\big) = G \cap Z(H),$$

for[2] by Lemma 11.5.1 $Z(H) \subset Z(H) + h \subset Z(H) + Z(H) \subset Z(H)$, and so $Z(H) + h = Z(H)$. Consequently $(G - h) \cap Z(H) = G \cap Z(H) - h$. By Theorems 9.2.8 and 9.2.1 we obtain hence (11.5.8). $\qquad\square$

Note that we obtain (11.5.7) from (11.5.8) taking $G = \mathbb{R}^N$.

Lemma 11.5.5. *Let $H \subset \mathbb{R}^N$ be an arbitrary Hamel basis, and let $f : \mathbb{R}^N \to \mathbb{R}$ be a discontinuous additive function. Then, for every $M \in \mathbb{R}$, the set*

$$A_M = \{x \in Z(H) \mid f(x) > M\} \tag{11.5.12}$$

is saturated non-measurable and has property $(*)$ from 3.3.

Proof. It follows from (11.5.7) that there exists an $x_0 \in Z(H)$ such that $d = f(x_0) > 0$. Write

$$B_k = \{\, x \in Z(H) \mid kd < f(x) \leqslant (k+1)d \,\}, \quad k \in \mathbb{Z}.$$

First we prove that

$$B_{k+1} = B_k + x_0, \quad k \in \mathbb{Z}. \tag{11.5.13}$$

In fact, if $x \in B_k + x_0$, then $x = t + x_0$, where $t \in B_k$. Thus $t \in Z(H)$ and $kd < f(t) \leqslant (k+1)d$. Then by (11.5.2) $x = t + x_0 \in Z(H) + Z(H) = Z(H)$, and $f(x) = f(t) + f(x_0) = f(t) + d$ so that $(k+1)d < f(x) \leqslant (k+2)d$. Consequently $x \in B_{k+1}$, and $B_k + x_0 \subset B_{k+1}$. Conversely, if $x \in B_{k+1}$, then $x \in Z(H)$ and $(k+1)d < f(x) \leqslant (k+2)d$. Write $t = x - x_0$. Then, since evidently $-Z(H) = Z(H)$, we have $t \in Z(H) - Z(H) = Z(H) + Z(H) = Z(H)$, and $f(t) = f(x) - f(x_0) = f(x) - d$ so that $kd < f(t) \leqslant (k+1)d$. Consequently $t \in B_k$ and $x = t + x_0 \in B_k + x_0$, and $B_{k+1} \subset B_k + x_0$. This proves (11.5.13).

Relation (11.5.13) shows that all the sets B_k are congruent to each other under translation, and hence they all have the same measure and category. Clearly

$$\bigcup_{k=-\infty}^{+\infty} B_k = Z(H). \tag{11.5.14}$$

If the sets B_k had measure zero, or if they all were of the first category, then by (11.5.14) the same would be true about $Z(H)$, which contradicts Theorem 11.5.1. Consequently $m_e(B_k) > 0$ for $k \in \mathbb{Z}$, and all B_k are of the second category, $k \in \mathbb{Z}$.

---

[2] If $h \in Z(H)$, then also $-h \in Z(H)$, whence $Z(H) - h \subset Z(H) + Z(H) = Z(H)$, and $Z(H) \subset Z(H) + h$.

For $k \geqslant M/d$ we have $B_k \subset A_M$. It follows that, for every $M \in \mathbb{R}$, $m_e(A_M) > 0$ and A_M is of the second category.

Suppose that for a non-empty open set $G \subset \mathbb{R}^N$ we have $G \cap A_M = \varnothing$. This means that $f(x) \leqslant M$ for $x \in G \cap Z(H)$. Lemma 11.5.4 would then imply that f is continuous, which is not the case. This shows that, for every $M \in \mathbb{R}$, the set A_M is dense in $\mathbb{R}^N$.

Take arbitrary $x, y \in A_{\frac{1}{2}M}$. Then $x \in Z(H)$ and $y \in Z(H)$, whence $x + y \in Z(H) + Z(H) = Z(H)$ (cf. Lemma 11.5.1), and $f(x) > \frac{1}{2}M$ and $f(y) > \frac{1}{2}M$, whence $f(x + y) = f(x) + f(y) > \frac{1}{2}M + \frac{1}{2}M = M$. Consequently $x + y \in A_M$, and

$$A_{\frac{1}{2}M} + A_{\frac{1}{2}M} \subset A_M. \tag{11.5.15}$$

By Theorems 3.6.1 and 3.6.2 $m_i\big(\mathbb{R}^N \setminus (A_{\frac{1}{2}M} + A_{\frac{1}{2}M})\big) = 0$ and the set $\mathbb{R}^N \setminus (A_{\frac{1}{2}M} + A_{\frac{1}{2}M})$ does not contain any subset of the second category and with the Baire property. It follows from (11.5.15) that also $m_i(\mathbb{R}^N \setminus A_M) = 0$ and the set $\mathbb{R}^N \setminus A_M$ does not contain any subset of the second category with the Baire property.

According to definition (11.5.12) we have $A_M \subset Z(H)$. In virtue of Theorem 11.5.1 $m_i(A_M) = 0$ and A_M does not contain any subset of the second category with the Baire property. Consequently A_M is saturated non-measurable and has property $(*)$ from 3.3. $\qquad\square$

Now we can improve on Lemma 11.5.4.

Theorem 11.5.2. *Let $H \subset \mathbb{R}^N$ be an arbitrary Hamel basis, and let $A \subset \mathbb{R}^N$ be an arbitrary measurable set of positive measure. Then*

$$A \cap Z(H) \in \mathfrak{A}. \tag{11.5.16}$$

Proof. Let $f : \mathbb{R}^N \to \mathbb{R}$ be a discontinuous additive function, and let the sets A_M be defined by (11.5.12). By Lemma 11.5.5 and Theorem 3.3.1 we have $A \cap A_M \neq \varnothing$ for every $M \in \mathbb{R}$, which means that f is unbounded above on $A \cap Z(H)$. Thus every discontinuous additive function $f : \mathbb{R}^N \to \mathbb{R}$ is unbounded above on $A \cap Z(H)$. This means that $A \cap Z(H) \in \mathfrak{B}$. By Theorem 10.2.2 we obtain (11.5.16). $\qquad\square$

Theorem 11.5.3. *Let $H \subset \mathbb{R}^N$ be an arbitrary Hamel basis, and let $A \subset \mathbb{R}^N$ be an arbitrary set of the second category and with the Baire property. Then we have* (11.5.16).

Proof. Let $f : \mathbb{R}^N \to \mathbb{R}$ be a discontinuous additive function, and let the sets A_M be defined by (11.5.12). By Lemma 11.5.5 and Corollary 3.3.1 we have $A \cap A_M \neq \varnothing$ for every $M \in \mathbb{R}$, and we argue as in the proof of Theorem 11.5.2. $\qquad\square$

Theorem 11.5.4. *Let $H \subset \mathbb{R}^N$ be an arbitrary Hamel basis, and let $A \subset \mathbb{R}^N$ be an arbitrary measurable set of positive measure. Then*

$$A \cap Z(H) \in \mathfrak{A}_C. \tag{11.5.17}$$

Proof. Let $D \subset \mathbb{R}^N$ be a convex and open set such that $A \cap Z(H) \subset D$, and let $f : D \to \mathbb{R}$ be an arbitrary convex function such that the restriction $f \mid A \cap Z(H)$ is continuous. By the Lebesgue density theorem[3] (cf. also Theorem 3.5.1) $m(A \cap A^*) = m(A) > 0$. By Theorems 11.5.1 and 3.3.1 $A \cap A^* \cap Z(H) \neq \varnothing$. Thus there exists a point $a \in A \cap Z(H)$ which is a density point of A. Further, because of the continuity of $f \mid A \cap Z(H)$, there exists a neighbourhood U of a in $\mathbb{R}^N$ such that the function f is bounded on $U \cap A \cap Z(H)$. Since a is a density point of A and U is a neighbourhood of a, $m(U \cap A) > 0$. By Theorem 11.5.2 $U \cap A \cap Z(H) \in \mathfrak{A}$, whence it follows that f is continuous. This proves (11.5.17). $\qquad\square$

Theorem 11.5.5. *Let $H \subset \mathbb{R}^N$ be an arbitrary Hamel basis, and let $A \subset \mathbb{R}^N$ be an arbitrary set of the second category and with the Baire property. Then we have* (11.5.17).

Proof. Let $D \subset \mathbb{R}^N$ be a convex and open set such that $A \cap Z(H) \subset D$, and let $f : D \to \mathbb{R}$ be an arbitrary convex function such that the restriction $f \mid A \cap Z(H)$ is continuous. By Theorems 2.1.5, 2.3.1 and 2.2.1 the set[4] $A \cap D(A)$ has the Baire property. Moreover, by Theorem 2.1.6 the set $A \setminus D(A)$ is of the first category, whence $A \cap D(A)$ is of the second category, in view of the relation $A = [A \cap D(A)] \cup [A \setminus D(A)]$. By Theorem 11.5.1 and Corollary 3.3.1 we have $A \cap D(A) \cap Z(H) \neq \varnothing$. Thus there exists a point $a \in A \cap Z(H)$ at which the set A is locally of the second category. Further, there exists a neighbourhood U of a in $\mathbb{R}^N$ such that the function f is bounded on $U \cap A \cap Z(H)$. Since A is of the second category at a and U is a neighbourhood of a, the set $U \cap A$ is of the second category, and since U is open, $U \cap A$ has the Baire property (cf. Theorems 2.3.1 and 2.2.1). By Theorem 11.5.3 $U \cap A \cap Z(H) \in \mathfrak{A}$, whence it follows that f is continuous. This proves (11.5.17). $\quad\square$

Taking in (11.5.17) $A = \mathbb{R}^N$ we obtain from Theorem 11.5.4 or 11.5.5, for an arbitrary Hamel basis $H \subset \mathbb{R}^N$,

$$Z(H) \in \mathfrak{A}_C.$$

Let us observe that Theorems 9.3.1, 9.3.2, 9.8.3 and 9.8.4 are immediate consequences of Theorems 11.5.2, 11.5.3, 11.5.4 and 11.5.5, respectively.

Another example of a set $T \subset \mathbb{R}$ such that $A \cap T \in \mathfrak{A}$ for every measurable set $A \subset \mathbb{R}$ of positive measure can be found in Kuczma-Smítal [190].

11.6 Lusin sets

In 11.4 we saw that there exist "large" Hamel bases. In 11.5 we encountered the Erdős set $Z(H)$ associated with the Hamel basis H, which is "large" independently of what Hamel basis H we start with. The fact that in the definition of $Z(H)$ there occur positive as well as negative coefficients is essential, as we will presently see. In this section we meet very small Hamel bases.

---

[3] A^* is the set of the density points of A; cf. 3.5.
[4] $D(A)$ is the set of those points at which A is locally of the second category; cf. 2.1.

Lusin sets were defined in 3.4. W. Sierpiński [282] proved, assuming the continuum hypothesis (cf. 1.5) that there exists a Hamel basis which is a Lusin set. This is a simple corollary from the following theorem of P. Erdős [76].

Theorem 11.6.1. *Under assumption of the continuum hypothesis there exists a Hamel basis $H \subset \mathbb{R}^N$ such that $E^+(H)$ is a Lusin set.*

Proof. Let $\mathcal{F}$ be the collection of all nowhere dense perfect sets in $\mathbb{R}^N$. By Theorem 2.3.4 $\operatorname{card}\mathcal{F} \leqslant \mathfrak{c}$. If $N > 1$, then every closed segment in $\mathbb{R}^N$ is a nowhere dense perfect set, and there are continuum many closed segments in $\mathbb{R}^N$, whence $\operatorname{card}\mathcal{F} \geqslant \mathfrak{c}$. If $N = 1$, then every translation $\mathbf{C} + x$, $x \in \mathbb{R}$, of the Cantor set $\mathbf{C}$ is a nowhere dense perfect set, and there are continuum many such translations. We argue further as in the preceding case and arrive at the inequality $\operatorname{card}\mathcal{F} \geqslant \mathfrak{c}$. Thus finally we get $\operatorname{card}\mathcal{F} = \mathfrak{c}$.

Under assumption of the continuum hypothesis we have (cf. 1.5)

$$\operatorname{card}\Gamma(\Omega) = \overline{\Omega} = \aleph_1 = \mathfrak{c},$$

so there exist functions, one-to-one and onto, $f : \Gamma(\Omega) \to \mathcal{F}$ and $g : \Gamma(\Omega) \to \mathbb{R}^N$. For $\alpha \in \Gamma(\Omega)$ we put $F_\alpha = f(\alpha)$, $x_\alpha = g(\alpha)$ so that $\mathcal{F} = \{F_\alpha\}_{\alpha<\Omega}$, $\mathbb{R}^N = \{x_\alpha\}_{\alpha<\Omega}$. Put for $\alpha < \Omega$

$$F^{(\alpha)} = \bigcup_{\beta\leqslant\alpha} F_\beta.$$

Since the sets F_β are nowhere dense, and there are only countably many $\beta \leqslant \alpha$, the sets $F^{(\alpha)}$ are of the first category.

We will define a transfinite sequence $\{h_\alpha\}_{\alpha<\Omega}$ of linearly independent elements of $\mathbb{R}^N$, and we write $H_\alpha = \bigcup_{\beta<\alpha} \{h_\beta\}$ ($H_0 = \varnothing$), and $E^\alpha = E(H_\alpha)$, $\alpha < \Omega$. Since there are at most countably many $\beta < \alpha \ (< \Omega)$, the sets H_α are at most countable, and hence, by Lemma 4.1.3, the sets E^α are countable ($\alpha < \Omega$). Suppose that h_β have already been defined for $\beta < \alpha < \Omega$. (If $\alpha = 0$, no h_β have yet been defined so that $H_0 = \varnothing$, and we assume $E^0 = \{0\}$). Since E^α is countable, $\mathbb{R}^N \setminus E^\alpha \neq \varnothing$, so there exists the smallest $\delta < \Omega$ such that $x_\delta \notin E^\alpha$. We seek elements $u, v \in \mathbb{R}^N$ such that

(i) $u - v = x_\delta$;

(ii) For every $h \in E^\alpha \setminus \{0\}$ the set $\{u\} \cup \{v\} \cup \{h\}$ is linearly independent (over $\mathbb{Q}$);

(iii) For every $h \in E^\alpha$ and $\varrho', \varrho'' \in \mathbb{Q}$

$$\varrho'u + \varrho''v + h \in F^{(\alpha)} \text{ implies } \varrho' + \varrho'' = 0.$$

Inserting from (i) $u = v + x_\delta$, we may express (ii) and (iii) as follows:

(iv) For every $h \in E^\alpha$ and $\sigma', \sigma'' \in \mathbb{Q}$ such that $|\sigma'| + |\sigma''| > 0$

$$(\sigma' + \sigma'') v + \sigma' x_\delta + h \neq 0;$$

(v) For every $h \in E^\alpha$ and $\varrho', \varrho'' \in \mathbb{Q}$ such that $\varrho' + \varrho'' \neq 0$

$$(\varrho' + \varrho'') v + \varrho' x_\delta + h \notin F^{(\alpha)}.$$

If for some σ', $\sigma'' \in \mathbb{Q}$, $v \in \mathbb{R}^N$, and $h \in E^\alpha$ we have $(\sigma' + \sigma'') v + \sigma' x_\delta + h = 0$, then $\sigma' + \sigma'' \neq 0$, since otherwise we would have $\sigma' x_\delta + h = 0$, whence $x_\delta \in E^\alpha$, which is untrue. Thus (iv) may be expressed as

$$v \neq -\frac{\sigma' x_\delta + h}{\sigma' + \sigma''},$$

or

$$v \notin \bigcup_{\substack{\sigma', \sigma'' \in \mathbb{Q} \\ \sigma' + \sigma'' \neq 0}} \left(-\frac{\sigma' x_\delta + E^\alpha}{\sigma' + \sigma''} \right). \tag{11.6.1}$$

Similarly, (v) may be expressed as

$$v \notin \bigcup_{\substack{\varrho', \varrho'' \in \mathbb{Q} \\ \varrho' + \varrho'' \neq 0}} \bigcup_{h \in E^\alpha} \frac{F^{(\alpha)} - \varrho' x_\delta - h}{\varrho' + \varrho''}. \tag{11.6.2}$$

Every set under the union sign in (11.6.1) is countable, since E^α is countable, and hence the countable union of such sets is countable, and so of the first category. Similarly, every set under the union sign in (11.6.2) is of the first category, just like $F^{(\alpha)}$, and hence countable union of such sets is of the first category. Therefore the set

$$B_\alpha = \bigcup_{\substack{\sigma', \sigma'' \in \mathbb{Q} \\ \sigma' + \sigma'' \neq 0}} \left(-\frac{\sigma' x_\delta + E^\alpha}{\sigma' + \sigma''} \right) \cup \bigcup_{\substack{\varrho', \varrho'' \in \mathbb{Q} \\ \varrho' + \varrho'' \neq 0}} \bigcup_{h \in E^\alpha} \frac{F^\alpha - \varrho' x_\delta - h}{\varrho' + \varrho''}$$

is of the first category, and therefore $\mathbb{R}^N \setminus B_\alpha \neq \varnothing$. Let $w : \mathcal{P}(\mathbb{R}^N) \setminus \{\varnothing\} \to \mathbb{R}^N$ be a choice function. If we put

$$v = w(\mathbb{R}^N \setminus B_\alpha), \quad u = v + x_\delta,$$

then u, v so defined will satisfy conditions (i)–(iii). Next we define

$$h_\alpha = u, \quad h_{\alpha+1} = v. \tag{11.6.3}$$

In this way, by transfinite induction (cf. Theorem 1.6.2), we have defined a sequence $\{h_\alpha\}_{\alpha < \Omega}$ of linearly independent (compare condition (ii)) elements of $\mathbb{R}^N$. Let $H = \bigcup_{\alpha < \Omega} \{h_\alpha\} = \bigcup_{\alpha < \Omega} H_\alpha$. Then the set H is linearly independent[5]. We will show that

$$E(H) = \mathbb{R}^N. \tag{11.6.4}$$

Supposing (11.6.4) untrue, there exists an $x \in \mathbb{R}^N \setminus E(H)$. Then $x = x_\gamma$ for a certain $\gamma < \Omega$. Since $H_\alpha \subset H$ for every $\alpha < \Omega$, we have $x_\gamma \notin E(H_\alpha) = E^\alpha$ for every

---

[5] If H were linearly dependent, then there would exist an $\alpha < \Omega$ such that H_α, and hence also $H_{\alpha+2}$ (note that $H_\alpha \subset H_{\alpha+2}$) is linearly dependent. But $H_{\alpha+2} = H_\alpha \cup \{h_\alpha\} \cup \{h_{\alpha+1}\} = \{u\} \cup \{v\} \cup H_\alpha \subset \{u\} \cup \{v\} \cup E^\alpha$ (cf. (11.6.3)). Thus $\{u\} \cup \{v\} \cup E^\alpha$ would be linearly dependent, which contradicts (ii).

$\alpha < \Omega$. Since in the induction step described above the index δ was minimal such that $x_\delta \notin E^\alpha$, we must have $\delta \leqslant \gamma$ every time. But there are at most countably many $\delta \leqslant \gamma < \Omega$, whereas the induction involves uncountably many steps (at every step we define only two elements of the sequence $\{h_\alpha\}_{\alpha<\Omega}$, and the whole sequence has $\overline{\Omega} = \mathfrak{c}$ elements, and at every step we must take different δ, since by (i) $x_\delta = h_\alpha - h_{\alpha+1} \in E(H_{\alpha+2}) = E^{\alpha+2}$). The contradiction obtained shows that (11.6.4) actually holds.

Thus the set H is linearly independent and fulfils (11.6.4). Consequently H is a Hamel basis. It remains to show that $E^+(H)$ is a Lusin set. Note that $H \subset E^+(H)$, and since, by Theorem 4.2.3 H is uncountable, also $E^+(H)$ is uncountable.

Let $F \subset \mathbb{R}^N$ be a nowhere dense perfect set. Then there exists an $\alpha < \Omega$ such that $F = F_\alpha \subset F^{(\alpha)}$. Let

$$x \in E^+(H) \cap F^{(\alpha)}.$$

Then

$$x = \varrho_1 h_{\alpha_1} + \cdots + \varrho_n h_{\alpha_n}, \qquad (11.6.5)$$

where ϱ_i are rational and positive, and $h_{\alpha_i} \in H$, $i = 1, \ldots, n$. We may arrange h_{α_i} in such a manner that $\alpha_1 < \cdots < \alpha_n$. Suppose that $\alpha_n > \alpha$. h_{α_n} is either u or v in our construction, and possibly adding in (11.6.5) the suitable term with the coefficient zero, we may assume that $h_{\alpha_{n-1}} = u$, $h_{\alpha_n} = v$. But then $h_{\alpha_1}, \ldots, h_{\alpha_{n-2}} \in H_{\alpha_n}$, and $h = \varrho_1 h_{\alpha_1} + \cdots + \varrho_{n-2} h_{\alpha_{n-2}} \in E(H_{\alpha_n}) = E^{\alpha_n}$. We have $x = h + \varrho_{n-1}u + \varrho_n v \in F^{(\alpha)} \subset F^{(\alpha_n)}$, whence by (iii) $\varrho_{n-1} + \varrho_n = 0$. But ϱ_{n-1}, ϱ_n are non-negative, and at least one of them is strictly positive. Hence $\varrho_{n-1} + \varrho_n > 0$. This contradiction shows that $\alpha_n \leqslant \alpha$.

But there are at most countably many $\varrho_1, \ldots, \varrho_n \in \mathbb{Q}$ and $\alpha_1 < \cdots < \alpha_n \leqslant \alpha < \Omega$. Consequently the set $E^+(H) \cap F^{(\alpha)}$ is at most countable, and hence also the set $E^+(H) \cap F_\alpha = E^+(H) \cap F$ is at most countable. Consequently $E^+(H)$ is a Lusin set. $\qquad\square$

Corollary 11.6.1. *Under assumption of the continuum hypothesis there exists a Hamel basis $H \subset \mathbb{R}^N$ which is a Lusin set.*

Proof. Let H be the basis whose existence is ensured by Theorem 11.6.1. Then $E^+(H)$ is a Lusin set, $H \subset E^+(H)$, and by Theorem 4.2.3 H is uncountable. Consequently (cf. 3.4) H also is a Lusin set. $\qquad\square$

By a slight modification of the above construction we obtain the following (Smítal [286])

Lemma 11.6.1. *Under assumption of the continuum hypothesis, there exists a linearly independent set $H^{(1)} \subset \mathbb{R}^N$ such that $E(H^{(1)})$ is a Lusin set.*

Proof. Let $\mathcal{F} = \{F_\alpha\}_{\alpha<\Omega}$, $\{x_\alpha\}_{\alpha<\Omega}$, and $F^{(\alpha)}$ have the same meaning as in the proof of Theorem 11.6.1. By the transfinite induction we define a transfinite sequence $\{h_\alpha\}_{\alpha<\Omega}$ of linearly independent elements of $\mathbb{R}^N$. Write $H_\alpha^{(1)} = \bigcup_{\beta<\alpha} \{h_\beta\}$ for $\alpha < \Omega$.

Suppose that h_β have already been defined for $\beta < \alpha < \Omega$. We seek a $u \in \mathbb{R}^N$ with the properties

(vi) $\{u\} \cup H_\alpha^{(1)}$ is linearly independent;

(vii) For every $h \in E(H_\alpha^{(1)})$ and $\varrho \in \mathbb{Q}$

$$\varrho u + h \in F^{(\alpha)} \text{ implies } \varrho = 0.$$

By Lemma 4.1.2 condition (vi) will be fulfilled if $u \notin E(H_\alpha^{(1)})$, and by Lemma 4.1.3 $E(H_\alpha^{(1)})$ is at most countable (we assume $E(H_0^{(1)}) = \{0\}$), and hence of the first category. Similarly, condition (vii) means

$$u \notin \bigcup_{\substack{\varrho \in \mathbb{Q} \\ \varrho \neq 0}} \bigcup_{h \in E(H_\alpha^{(1)})} \frac{F^{(\alpha)} - h}{\varrho}. \tag{11.6.6}$$

Again, the set occurring in (11.6.6) is of the first category. Thus we may take

$$u = w \left(\mathbb{R}^N \setminus \left[E(H_\alpha^{(1)}) \cup \bigcup_{\substack{\varrho \in \mathbb{Q} \\ \varrho \neq 0}} \bigcup_{h \in E(H_\alpha^{(1)})} \frac{F^{(\alpha)} - h}{\varrho} \right] \right),$$

where $w : \mathcal{P}(\mathbb{R}^N) \setminus \{\varnothing\} \to \mathbb{R}^N$ is a choice function. Then we put $h_\alpha = u$.

Thus we have defined by transfinite induction a sequence $\{h_\alpha\}_{\alpha < \Omega}$, and the set $H^{(1)} = \bigcup_{\alpha < \Omega} \{h_\alpha\} = \bigcup_{\alpha < \Omega} H_\alpha^{(1)}$ is linearly independent. But this time it is not true that $E(H^{(1)}) = \mathbb{R}^N$, so $H^{(1)}$ is not a Hamel basis. Actually $E(H^{(1)})$ is a Lusin set.

To see this, first observe that $\operatorname{card} H^{(1)} = \overline{\Omega} = \mathfrak{c}$, so $H^{(1)}$, and hence also $E(H^{(1)})$, is uncountable. Next, suppose that for a certain $\alpha < \Omega$

$$x \in E(H^{(1)}) \cap F_\alpha \subset E(H^{(1)}) \cap F^{(\alpha)}. \tag{11.6.7}$$

Then we have (11.6.5), where $\varrho_i \in \mathbb{Q}$, $\varrho_i \neq 0$, $h_{\alpha_i} \in H^{(1)}$, $i = 1, \ldots, n$; $\alpha_1 < \cdots < \alpha_n$. Suppose that $\alpha_n > \alpha$, h_{α_n} is a u from our construction, and $h_{\alpha_1}, \ldots, h_{\alpha_{n-1}} \in H_{\alpha_n}^{(1)}$, whence $h = \varrho_1 h_{\alpha_1} + \cdots + \varrho_{n-1} h_{\alpha_{n-1}} \in E(H_{\alpha_n}^{(1)})$. Hence $x = h + \varrho_n u$, $\varrho_n \neq 0$, whence by (vii) $x \notin F^{(\alpha_n)}$, and also $x \notin F^{(\alpha)} \subset F^{(\alpha_n)}$, a contradiction with (11.6.7). By the same argument as in the proof of Theorem 11.6.1 we infer that the set $E(H^{(1)}) \cap F_\alpha$ is at most countable for every $\alpha < \Omega$. $\qquad\square$

Also the next theorem is due to J. Smítal [286].

Theorem 11.6.2. *Under assumption of the continuum hypothesis, there exists a Hamel basis $H \subset \mathbb{R}^N$ which admits a decomposition $H = H^{(1)} \cup H^{(2)}$, $H^{(1)} \cap H^{(2)} = \varnothing$, where $H^{(1)}$ and $H^{(2)}$ have the power of continuum, and $E(H^{(1)})$ is a Lusin set.*

Proof. Let $H^{(1)}$ be the set occurring in Lemma 11.6.1. By Theorem 4.2.1 there exists a Hamel basis H in $\mathbb{R}^N$ such that $H^{(1)} \subset H$. Put $H^{(2)} = H \setminus H^{(1)}$.

By Lemma 11.6.1 $E(H^{(1)})$ is a Lusin set, and $\operatorname{card} H^{(1)} = \mathfrak{c}$, as has been pointed out in the proof of Lemma 11.6.1. Suppose that $\operatorname{card} H^{(2)} < \mathfrak{c}$. Then, by the continuum hypothesis, $\operatorname{card} H^{(2)} \leqslant \aleph_0$, and by Lemma 4.1.3 $\operatorname{card} E(H^{(2)}) \leqslant \aleph_0$. Every $x \in \mathbb{R}^N =$

$E(H) = E(H^{(1)}) + E(H^{(2)})$ can be written as $x = u + v$, where $u \in E(H^{(1)})$, $v \in E(H^{(2)})$. Consequently

$$\mathbb{R}^N = \bigcup_{v \in E(H^{(2)})} \left(E(H^{(1)}) + v \right). \tag{11.6.8}$$

By Corollary 3.4.1 $E(H^{(1)})$ is of measure zero. Consequently also $\left(E(H^{(1)}) + v \right)$ is of measure zero for every $v \in E(H^{(2)})$. Since the union in (11.6.8) is at most countable, it follows that the set on the right-hand side of (11.6.8) is of measure zero. The contradiction obtained shows that card $H^{(2)} = \mathfrak{c}$. $\square$

11.7 Perfect sets

Corollary 11.3.2 says that no Hamel basis is a Borel set. The question arises as to whether a Hamel basis may contain an uncountable Borel set[6]. In the case of a Burstin basis the answer is "no" (Jones [158]).

Theorem 11.7.1. Let $H \subset \mathbb{R}^N$ be a Burstin basis. Then H contains no uncountable Borel subset.

Proof. Suppose that B is an uncountable Borel set and $B \subset H$. Then $2B$ also is an uncountable Borel set, and consequently $H \cap 2B \neq \varnothing$. Let $y \in H \cap 2B$. In particular $y \in 2B$, i.e., $y = 2x$, where $x \in B \subset H$. Thus $x \in H$ and $2x = y \in H$, which is impossible, since x and $2x$ are linearly dependent (over $\mathbb{Q}$). $\square$

But the situation as described in Theorem 11.7.1 is not general. Following F. B. Jones [158] we will prove

Theorem 11.7.2. There exists a Hamel basis $H \subset \mathbb{R}^N$ which contains a perfect set.

Proof. Let $\{\lambda_n\}_{n \in \mathbb{N}}$ be a sequence of all rational numbers such that $\lambda_1 = 0$. Choose arbitrarily $x_0, y_0 \in \mathbb{R}^N$ $(x_0 \neq y_0)$. Then the continuous functions

$$f_{ij}(x, y) = \lambda_i x + \lambda_j y, \quad i, j = 1, 2,$$

have for $x = x_0$ and $y = y_0$ values different from x_0 and y_0 except when $f_{ij}(x_0, y_0) = x_0$ or y_0. There exist small closed balls K_{11} and K_{12}, of diameter less than 1 and such that $x_0 \in K_{11}$, $y_0 \in K_{12}$, $K_{11} \cap K_{12} = \varnothing$, and

$$f_{ij}(x, y) \notin K_{11} \cup K_{12}$$

for $x \in K_{11}$, $y \in K_{12}$, except when $f_{ij}(x, y) = x$ or y.

By a similar argument we can find closed balls $K_{21}, K_{22}, K_{23}, K_{24}$, pairwise disjoint, of diameter less than $\frac{1}{2}$, and such that $K_{21} \cup K_{22} \subset K_{11}$, $K_{23} \cup K_{24} \subset K_{12}$, and

$$\lambda_i x + \lambda_j y + \lambda_k z + \lambda_m u \notin \bigcup_{\nu=1}^{4} K_{2\nu}, \quad i, j, k, m = 1, 2, 3, 4,$$

---

[6] Or, what ammounts to the same, whether a Hamel basis may contain an uncountable analytic subset. In fact, every Borel set is analytic, and every uncountable analytic set contains an uncountable Borel subset (a homeomorphic image of the Cantor set $\mathbf{C}$; cf. the proof of Theorem 2.8.3 and Exercise 2.12).

for all $x \in K_{21}$, $y \in K_{22}$, $z \in K_{23}$, $u \in K_{24}$, except when the combination is x, y, z, or u.

In the n-th step we find closed balls $K_{n1}, \ldots, K_{n,2^n}$, pairwise disjoint, of diameter less than $\frac{1}{n}$, and such that $K_{n1} \cup K_{n2} \subset K_{n-1,1}, \ldots, \ldots, K_{n,2^n-1} \cup K_{n,2^n-1} \subset K_{n-1,2^{n-1}}$, and

$$\lambda_{i_1} x_1 + \cdots + \lambda_{i_{2^n}} x_{2^n} \notin \bigcup_{\nu=1}^{2^n} K_{n\nu}, \quad i_1, \ldots, i_{2^n} = 1, \ldots, 2^n, \qquad (11.7.1)$$

for all $x_1 \in K_{n1}, \ldots, x_{2^n} \in K_{n,2^n}$, except when the combination is one of x_ν, $\nu = 1, \ldots, 2^n$.

In this way we have defined a sequence $\{K_{n\nu}\}$, $n \in \mathbb{N}$, $\nu = 1, \ldots, 2^n$, of closed balls with the properties described above. Put

$$F = \bigcap_{n=1}^{\infty} \bigcup_{\nu=1}^{2^n} K_{n\nu}. \qquad (11.7.2)$$

It is easily seen that F is a perfect set. We will show that F is linearly independent over $\mathbb{Q}$.

Supposing the contrary, let $\{x_1, \ldots, x_p\} \subset F$ be a finite linearly dependent system ($x_i \neq x_j$ for $i \neq j$). Then one of x's, say x_p, is a linear combination of the remaining ones:

$$x_p = \lambda_{i_1} x_1 + \cdots + \lambda_{i_{p-1}} x_{p-1}. \qquad (11.7.3)$$

Choose an $n \in \mathbb{N}$ so that $2^n \geqslant p$, $2^n \geqslant \max_{\kappa=1,\ldots,p-1} i_\kappa$, and $\frac{1}{n} < \min_{\substack{\mu,\nu=1,\ldots,p \\ \mu \neq \nu}} |x_\mu - x_\nu|$. By

(11.7.2) we have $x_1, \ldots, x_{p-1} \in \bigcup_{\nu=1}^{2^n} K_{n\nu}$, and since the diameters of the balls $K_{n\nu}$ are less than $\frac{1}{n}$, every x_k lies in a different ball. Choose arbitrarily points $\overline{x}_k$ in the remaining balls (one in each ball). Then (11.7.3) may be written as

$$x_p = \lambda_{i_1} x_1 + \cdots + \lambda_{i_{p-1}} x_{p-1} + \lambda_1 \overline{x}_p + \cdots + \lambda_1 \overline{x}_{2^n}. \qquad (11.7.4)$$

(We may choose the points so that none of them coincides with x_p.) All the coefficients λ_i in (11.7.4) fulfil the condition $1 \leqslant i \leqslant 2^n$. By (11.7.1)

$$x_p \notin \bigcup_{\nu=1}^{2^n} K_{n\nu}, \qquad \text{whereas by (11.7.2)} \qquad x_p \in F \subset \bigcup_{\nu=1}^{2^n} K_{n\nu}.$$

This contradiction shows that F is linearly independent (over $\mathbb{Q}$).

By Theorem 4.2.1 there exists a Hamel basis $H \subset \mathbb{R}^N$ such that $F \subset H$. H is the required basis. $\square$

It follows easily from Corollary 11.2.1 and Theorem 11.3.1 that if a Hamel basis contains a perfect set F, then F is nowhere dense and of measure zero.

11.8   The operations R and U

The operations R and U were introduced in 9.2:

$$R(T) = T - T, \quad U(T) = T + R(T).$$

The iterates of the operations R and U are defined by the recurrence:

$$R^1(T) = R(T), \qquad U^1(T) = U(T),$$
$$R^{n+1}(T) = R(R^n(T)), \quad U^{n+1}(T) = U(U^n(T)), \quad n \in \mathbb{N}.$$

Since $J(T), Q(T)$ and $S_n(T)$ for all $n \in \mathbb{N}$ are contained in $E^+(T)$, it follows from Theorem 11.6.1 that (under assumption of the continuum hypothesis) there exists a Hamel basis H such that $J(H), Q(H)$ and $S_n(H)$ for all $n \in \mathbb{N}$ are measurable sets of measure zero. For the iterates of the operations R and U we have, of course, the following

Lemma 11.8.1. *For every Hamel basis $H \subset \mathbb{R}^N$ and every $n \in \mathbb{N}$ we have*

$$m_i(R^n(H)) = m_i(U^n(H)) = 0.$$

Proof. By Theorems 9.2.3 and 9.2.6 the operations R^n and U^n are $\mathfrak{C}$-conservative for every $n \in \mathbb{N}$. If we had $m_i(R^n(H)) > 0$ or $m_i(U^n(H)) > 0$, then by Theorem 9.3.1 we would get $R^n(H) \in \mathfrak{C}$ or $U^n(H) \in \mathfrak{C}$, whence, by Theorem 9.2.2, $H \in \mathfrak{C}$, which contradicts Theorem 11.1.1. $\qquad\square$

But, contrary to the case of the operations J, Q, S_n, it cannot happen that $R^n(H)$ and $U^n(H)$ are of measure zero for all $n \in \mathbb{N}$. The following two theorems are due to F. B. Jones [158].

Theorem 11.8.1. *For every Hamel basis $H \subset \mathbb{R}^N$ there exists an $n \in \mathbb{N}$ such that $m_e(R^n(H)) > 0$.*

Proof. For every non-empty set $T \subset \mathbb{R}^N$ we have $0 \in R(T)$, whence $0 \in R^n(H)$ for all $n \in \mathbb{N}$. Hence $R^n(H) = R^n(H) - 0 \subset R^n(H) - R^n(H) = R^{n+1}(H)$, i.e.,

$$R^n(H) \subset R^{n+1}(H), \quad n \in \mathbb{N}. \tag{11.8.1}$$

Take an $x \in Z(H)$ (cf. 11.5) and an $a \in H$. x may be written as

$$x - \sum_{i=1}^{k} \varepsilon_i h_i = \sum_{i=1}^{k} \varepsilon_i (h_i - a) + la, \tag{11.8.2}$$

where $\varepsilon_i = \pm 1$, $h_i \in H$ $(i = 1, \ldots, k)$, $l = \sum_{i=1}^{k} \varepsilon_i \in \mathbb{Z}$. (Not all h_i's need be distinct). We will show that

$$\sum_{i=1}^{k} \varepsilon_i (h_i - a) \in R^k(H). \tag{11.8.3}$$

The proof runs by induction on k. For $k = 1$ we have $h_1 - a \in H - H = R(H)$, and $-(h_1 - a) = a - h_1 \in H - H = R(H)$. Assuming (11.8.3) true for a $k \in \mathbb{N}$, we have

$$\sum_{i=1}^{k+1} \varepsilon_i(h_i - a) = \sum_{i=1}^{k} \varepsilon_i(h_i - a) - \varepsilon_{k+1}(a - h_{k+1}).$$

By the induction hypothesis we have (11.8.3), whereas by what we have just shown and by (11.8.1) we have

$$\varepsilon_{k+1}(a - h_1) \in R(H) \subset R^k(H).$$

Hence

$$\sum_{i=1}^{k+1} \varepsilon_i(h_i - a) \in R^k(H) - R^k(H) = R(R^k(H)) = R^{k+1}(H).$$

Thus the validity of (11.8.3) has been established for every $k \in \mathbb{N}$.

Put

$$M = \bigcup_{k=1}^{\infty} R^k(H). \tag{11.8.4}$$

By (11.8.2) and (11.8.3)

$$x \in M + la \subset \bigcup_{l \in \mathbb{Z}} (M + la).$$

Since x has been arbitrary in $Z(H)$, we get hence

$$Z(H) \subset \bigcup_{l \in \mathbb{Z}} (M + la). \tag{11.8.5}$$

If M were of measure zero, also all sets $M + la$ would be of measure zero, whence by (11.8.5) we would get $m(Z(H)) = 0$, which is not the case (Theorem 11.5.1). Consequently $m_e(M) > 0$, whence by (11.8.4) $m_e(R^n(H)) > 0$ for a certain $n \in \mathbb{N}$. $\square$

Theorem 11.8.2. *For every Hamel basis $H \subset \mathbb{R}^N$ there exists an $n \in \mathbb{N}$ such that $m_e(U^n(H)) > 0$.*

Proof. We have, since $0 \in R^n(T)$ for every $n \in \mathbb{N}$ and every non-empty set $T \subset \mathbb{R}^N$,

$$U^n(H) = U^n(H) + 0 \subset U^n(H) + R(U^n(H)) = U(U^n(H)) = U^{n+1}(H), \tag{11.8.6}$$

and

$$H = H + 0 \subset H + R(H) = U(H) \subset U^n(H). \tag{11.8.7}$$

Put

$$M = \bigcup_{k=1}^{\infty} U^k(H).$$

If x, y, $z \in M$, then by (11.8.6) there exists a $k \in \mathbb{N}$ such that x, y, $z \in U^k(H)$, whence

$$x + (y - z) \in U^k(H) + R(U^k(H)) = U(U^k(H)) = U^{k+1}(H) \subset M. \qquad (11.8.8)$$

Take an $x \in Z(H)$ and an $a \in M$. Then

$$x = \sum_{i=1}^{k} \varepsilon_i h_i = \sum_{i=1}^{k} \varepsilon_i(h_i - a) + la,$$

where $\varepsilon_i = \pm 1$, $h_i \in H (i = 1, \ldots, k)$, $l = \sum_{i=1}^{k} \varepsilon_i \in \mathbb{Z}$. We have for every $k \in \mathbb{N}$

$$a + \sum_{i=1}^{k} \varepsilon_i(h_i - a) \in M. \qquad (11.8.9)$$

The proof runs by induction. For $k = 1$ we have

$$a + \varepsilon_1(h_1 - a) \in M$$

by (11.8.8), since $a \in M$ and $h_1 \in H \subset M$ (cf. (11.8.7)). Assuming (11.8.9) true for a $k \in \mathbb{N}$, we have

$$a + \sum_{i=1}^{k+1} \varepsilon_i(h_i - a) = a + \sum_{i=1}^{k} \varepsilon_i(h_i - a) + \varepsilon_{k+1}(h_{k+1} - a) \in M$$

by (11.8.9) and (11.8.8). Thus (11.8.9) is true for all $k \in \mathbb{N}$. Hence

$$a + x = a + \sum_{i=1}^{k} \varepsilon_i(h_i - a) + la \in M + la,$$

whence

$$x \in M + (l - 1)a,$$

and

$$Z(H) \subset \bigcup_{l \in \mathbb{Z}} (M + la).$$

Hence $m_e(M) > 0$ follows as in the proof of Theorem 11.8.1, whence $m_e(U^n(H)) > 0$ for a certain $n \in \mathbb{N}$. $\qquad \square$

The following results can be proved quite similarly as Lemma 11.8.1 and Theorems 11.8.1 and 11.8.2:

Lemma 11.8.2. *For every Hamel basis $H \subset \mathbb{R}^N$, every $n \in \mathbb{N}$, and every set $A \subset \mathbb{R}^N$ with the Baire property, if $A \subset R^n(H)$ or $A \subset U^n(H)$, then A is of the first category.*

Theorem 11.8.3. *For every Hamel basis $H \subset \mathbb{R}^N$ there exists an $n \in \mathbb{N}$ such that $R^n(H)$ is of the second category.*

Theorem 11.8.4. *For every Hamel basis $H \subset \mathbb{R}^N$ there exists an $n \in \mathbb{N}$ such that $U^n(H)$ is of the second category.*

We conclude this chapter with an open problem[7] (Aczél-Erdős [10]) concerning Hamel bases:

Problem.

– Does there exist a Hamel basis $H \subset \mathbb{R}$ such that $h^{-1} \in H$ for every $h \in H$?
– Does there exist a Hamel basis $H \subset \mathbb{R}$ such that $h^n \in H$ for every $h \in H$ and $n \in \mathbb{N}$?

On the other hand it is known that there exists no Hamel basis $H \subset \mathbb{R}$ such that $ab \in H$ whenever $a, b \in H$. (Cf. Exercise 14.7).

Exercises

1. Let $H \subset \mathbb{R}^N$ be a Hamel basis. Show that aH is a Hamel basis for every $a \in \mathbb{R} \setminus \{0\}$.
2. Let $H \subset \mathbb{R}^N$ be a Hamel basis, and let $h \in H$. Show that $H \cap (H + h) = \varnothing$.
3. Let $H \subset \mathbb{R}^N$ be a Hamel basis. Show that there is a $b \in \mathbb{R}^N$ such that the set $H + b$ contains no Hamel basis.
   [Hint: Let $b \in -H$. Show that $b \notin E(H + b)$].
4. Show that the conclusion of Exercise 10.5 fails to hold if we drop the analiticity assumption on T.
5. Let $H \subset \mathbb{R}^N$ be an arbitrary Hamel basis. Show that $D(\mathbb{R}^N \setminus H) = \mathbb{R}^N$ and $(\mathbb{R}^N \setminus H)^* = \mathbb{R}^N$ (cf. 2.1 and 3.5).
6. Show that under assumption of the continuum hypothesis there exists a Hamel basis $H \subset \mathbb{R}^N$ such that $\mathrm{card}\,(E^+(H) \cap G) \leqslant \aleph_0$ for every set $G \subset \mathbb{R}^N$ such that $G \in \mathcal{G}_\delta$ and $m(G) = 0$.
7. Show that under assumption of the continuum hypothesis there exists a Hamel basis $H \subset \mathbb{R}^N$ such that the set $E^+(H)$ is of the first category.
   [Hint: Use Exercise 11.6 and the decomposition $\mathbb{R}^N = A \cup B$, where $A \in \mathcal{G}_\delta$ and $m(A) = 0$, $B \in \mathcal{F}_\sigma$ and B is of the first category; cf. the end of 11.4.]
8. Let $H \subset \mathbb{R}^N$ be an arbitrary Hamel basis. Show that the set $Z(H)$ does not contain an analytic subset containing a Hamel basis. (Consequently Theorems 9.3.7 and 9.8.1 cannot be reversed).
   [Hint: Show that for any $A \subset Z(H)$ and $n \in \mathbb{N}$ we have $R^n(A) \subset R^n\bigl(Z(H)\bigr) = Z(H)$.]

---

[7] Added in proof. Recently solved in the positive (under assumption of the continuum hypothesis) by J. Smítal [290].

Chapter 12

Further Properties of Additive Functions and Convex Functions

12.1 Graphs

Let $D \subset \mathbb{R}^N$ be a convex and open set, and let $f : D \to \mathbb{R}$ be a convex function. Let m_f be the lower hull of f (cf. 6.3). By Theorem 6.3.1 either $m_f(x) = -\infty$ for all $x \in D$, or $m_f : D \to \mathbb{R}$ is a continuous and convex function.

Let

$$\mathrm{Gr}(f) = \{(x, y) \in \mathbb{R}^{N+1} \mid x \in D, \, y = f(x)\}$$

be the graph of f (cf. 9.5). We have

Theorem 12.1.1. *If $D \subset \mathbb{R}^N$ is a convex and open set and $f : D \to \mathbb{R}$ is a discontinuous convex function, then the set $\mathrm{Gr}(f)$ is dense in the set*

$$A_f = \{(x, y) \in \mathbb{R}^{N+1} \mid x \in D, \, y > m_f(x)\}.$$

Proof. Let $K \subset D$ be an arbitrary closed ball, and $J = [c, d] \subset \mathbb{R}$ an arbitrary closed interval such that

$$m_f(x) < c \quad \text{for } x \in K. \tag{12.1.1}$$

(Both, K and J, assumed non-degenerated.) Note that if $m_f(x) \equiv -\infty$, then (12.1.1) certainly is fulfilled, and if m_f is a continuous function, then it is bounded above on the compact set K. Every point of the set A_f can be included in a set $K \times J$, where K and J are as described above. Thus it is enough to show that

$$(K \times J) \cap \mathrm{Gr}(f) \neq \emptyset \tag{12.1.2}$$

Since f is discontinuous, it is not bounded above on K. Thus there exists an $x_1 \in K$ such that

$$f(x_1) > d. \tag{12.1.3}$$

By (12.1.1) there exists an $x_2 \in K$ such that

$$f(x_2) < c. \tag{12.1.4}$$

Let $L = Q(x_1, x_2)$ be the rational segment from x_1 to x_2, and $l = \mathrm{cl}\, L = \overline{x_1 x_2} = \{z \in \mathbb{R}^N \mid z = \lambda x_1 + (1 - \lambda)x_2, \ \lambda \in [0,1]\}$ be the closed segment joining x_1 and x_2. We have

$$L \subset l \subset K. \tag{12.1.5}$$

By Theorem 6.1.2 there exists a continuous function $g : l \to \mathbb{R}$ such that $g \mid L = f$. Since x_1, $x_2 \in L$, we have by (12.1.3) and (12.1.4)

$$g(x_1) > d, \quad g(x_2) < c.$$

Since g is continuous, there exists a non-degenerated segment $S \subset l$ such that $g(S) \subset [c, d]$. Clearly $S \cap L \neq \varnothing$. Hence there exists an $x_0 \in L$ such that

$$f(x_0) = g(x_0) \in J.$$

By (12.1.5) $x_0 \in K$, and thus the point $(x_0, f(x_0))$ belongs to the set on the left-hand side of (12.1.2). □

If $f : \mathbb{R}^N \to \mathbb{R}$ is a discontinuous additive function, then f is not locally bounded below at any point of $\mathbb{R}^N$. (Otherwise the additive function $-f$ would be locally bounded above at a point, and hence continuous.) Therefore $m_f = -\infty$. Since in this case $D = \mathbb{R}^N$, the set A_f becomes the whole space $\mathbb{R}^{N+1}$. Thus, as a particular case of Theorem 12.1.1, we obtain

Theorem 12.1.2. *If $f : \mathbb{R}^N \to \mathbb{R}$ is a discontinuous additive function, then the set* $\mathrm{Gr}(f)$ *is dense in* $\mathbb{R}^{N+1}$.

Theorem 12.1.2 can also be reformulated as follows.

Theorem 12.1.3. *If $f : \mathbb{R}^N \to \mathbb{R}$ is a discontinuous additive function, then for every (non-degenerated) interval $J \subset \mathbb{R}$ the set $f^{-1}(J)$ is dense in $\mathbb{R}^N$.*

In connection with Theorems 10.3.2 and 10.4.1 we will exhibit here two examples (Mościcki [231]) showing that the conditions $\mathrm{int}\, Q(T \cup (-T)) \neq \varnothing$ in Theorem 10.3.2 and $\mathrm{int}\, Q(T \cup A) \neq \varnothing$ in Theorem 10.4.1 cannot be replaced by $\mathrm{int}[Q(T) \cup Q(-T)] \neq \varnothing$ and $\mathrm{int}[Q(T) \cup Q(A)] \neq \varnothing$, respectively.

Example 12.1.1. Let $g : \mathbb{R}^N \to \mathbb{R}$ be an arbitrary discontinuous additive function, and put

$$T = \{x \in \mathbb{R}^N \mid g(x) < -1\}.$$

By Lemma 10.2.2 the set T is Q-convex:

$$Q(T) = T, \tag{12.1.6}$$

and, moreover, we have $T \in \mathfrak{C}$. To see this, let $f : \mathbb{R}^N \to \mathbb{R}$ be an arbitrary additive function bounded on T:

$$|f(x)| \leqslant M \quad \text{for } x \in T.$$

Take an arbitrary $x \in \mathbb{R}^N$. We must distinguish two cases.

I. $x \in T$. Then, for $k \in \mathbb{N}$, we have $g(kx) = kg(x) < -k \leqslant -1$, whence $kx \in T$, and

$$k\,|f(x)| = |f(kx)| \leqslant M \quad \text{for } k \in \mathbb{N}.$$

But this implies $f(x) = 0$. Thus $f \mid T = 0$.

II. $x \notin T$. Then $g(x) \geqslant -1$. Take an arbitrary $y \in T$. Then $g(y - x) = g(y) - g(x) < -1 + 1 = 0$. Hence, for $k \in \mathbb{N}$ sufficiently large $k(y - x) \in T$. Therefore (by I.) $f\big(k(y - x)\big) = 0$, i.e., $k\big[f(y) - f(x)\big] = 0$. But $f(y) = 0$, since $y \in T$, thus $-kf(x) = 0$, whence $f(x) = 0$. Consequently $f(x) = 0$ for all $x \in \mathbb{R}^N$, i.e., f is continuous.

Now, by Corollary 5.1.1 and by (12.1.6)

$$Q(T) \cup Q(-T) = Q(T) \cup \big[-Q(T)\big] = T \cup (-T),$$

and since

$$-T = \{x \in \mathbb{R}^N \mid g(x) > 1\},$$

we get

$$Q(T) \cup Q(-T) = \{x \in \mathbb{R}^N \mid |g(x)| > 1\}. \tag{12.1.7}$$

Let $G = \mathrm{int}\big[Q(T) \cup Q(-T)\big]$, and suppose that $G \neq \varnothing$. By Theorem 12.1.3 $g^{-1}([-1, 1]) \cap G \neq \varnothing$, i.e., there exists a point $x_0 \in G \subset Q(T) \cup Q(-T)$ such that $|g(x_0)| \leqslant 1$. But this contradicts (12.1.7).

Example 12.1.1 shows that the condition $T \in \mathfrak{C}$ is not equivalent to $\mathrm{int}\big[Q(T) \cup Q(-T)\big] \neq \varnothing$.

Example 12.1.2. Let $H \subset \mathbb{R}^N$ be a Burstin basis (cf. 11.4). H may be arranged into a transfinite sequence $H = \{h_\alpha\}_{\alpha < \gamma}$, where γ is the smallest ordinal number such that $\overline{\gamma} = \mathfrak{c}$. Every $x \in \mathbb{R}^N$ may be written as

$$x = \lambda_1 h_{\alpha_1} + \cdots + \lambda_n h_{\alpha_n}, \tag{12.1.8}$$

where $\lambda_i \in \mathbb{Q}$, $i = 1, \ldots, n$, and $\alpha_1 < \cdots < \alpha_n < \gamma$. We put

$$T = V_0 = \{x \in \mathbb{R}^N \mid \lambda_n > 0\} \cup \{0\},$$

where we assume that x has expansion (12.1.8) (cf. 10.2). In 10.2 it was shown that T is Q-convex (i.e., fulfils (12.1.6)) and that $T \in \mathfrak{B} = \mathfrak{A}$.

Let $g : H \to \mathbb{R}$ be given by $g(h) = 1$ for $h \in H$, and let $f : \mathbb{R}^N \to \mathbb{R}$ be the additive extension of g (Theorem 5.2.2). By Corollary 5.2.2 f is discontinuous. Put

$$A = \{x \in \mathbb{R}^N \mid f(x) < 0\}.$$

By Lemma 10.2.2 the set A is Q-convex and Q-radial at every point.

Suppose that there exists a non-empty open ball $K \subset Q(T) \cup Q(A) = T \cup A$. Since H is a Burstin basis, H is dense in $\mathbb{R}^N$. Hence

$$\bigcup_{0 < \alpha < \gamma} (K + h_\alpha) = \mathbb{R}^N.$$

Thus there exists an $\alpha > 0$ such that $h_0 \in K + h_\alpha$, i.e., $h_0 - h_\alpha \in K \subset T \cup A$. But $f(h_0 - h_\alpha) = f(h_0) - f(h_\alpha) = g(h_0) - g(h_\alpha) = 1 - 1 = 0$, so $h_0 - h_\alpha \notin A$. Also, the point $h_0 - h_\alpha$ has expansion (12.1.8) with $n = 2$, $\alpha_1 = 0$, $\alpha_2 = \alpha$, $\lambda_1 = 1$, $\lambda_2 = -1$, so $h_0 - h_\alpha \notin T$. This contradiction shows that $\mathrm{int}\big[Q(T) \cup Q(A)\big] = \varnothing$, i.e., the condition $T \in \mathfrak{B} = \mathfrak{A}$ is not equivalent to the condition that $\mathrm{int}\big[Q(T) \cup Q(A)\big] \neq \varnothing$ for every set A which is Q-convex and Q-radial at a point.

12.2 Additive functions

The contents of Theorem 12.1.2 can be expressed as follows: If $f : \mathbb{R}^N \to \mathbb{R}$ is a discontinuous additive function, then for every non-empty and open set $G \subset \mathbb{R}^N$ and every (non-degenerated) interval $J \subset \mathbb{R}$

$$(G \times J) \cap \mathrm{Gr}(f) \neq \varnothing. \tag{12.2.1}$$

The question arises whether in (12.2.1) G and J may be replaced by more general sets. In this section we will show that in (12.2.1) G may be an arbitrary Lebesgue measurable set of positive measure, or an arbitrary set of the second category and with the Baire property (Ostrowski [248], Kuczma [185], [189]). As expressed in Theorem 12.1.3, condition (12.2.1) with arbitrary non-empty and open set G is equivalent to the statement that $f^{-1}(J)$ is dense in $\mathbb{R}^N$. We shall prove that (12.2.1) holds for an arbitrary set G of positive measure or of the second category with the Baire property when we shall have proved that $f^{-1}(J)$ is saturated non-measurable and has property $(*)$ from 3.3. In other words this will mean that on an arbitrary set G of positive measure or of the second category with the Baire property f assumes values from every interval $J \subset \mathbb{R}$, i.e., that for every such set G the set $f(G)$ is dense in $\mathbb{R}$.

Theorem 12.2.1. *If $f : \mathbb{R}^N \to \mathbb{R}$ is a discontinuous additive function, then for every (non-degenerated) interval $J \subset \mathbb{R}, J \neq \mathbb{R}$, the set $f^{-1}(J)$ is saturated non-measurable and has property $(*)$ from 3.3.*

Proof. If [1] $J = \langle c, d \rangle,\ -\infty \leqslant c < d < +\infty$, then f is bounded above on $f^{-1}(J)$ by d. If $J = \langle c, d \rangle,\ -\infty < c < d \leqslant +\infty$, then f is bounded below on $f^{-1}(J)$ by c, and then the discontinuous additive function $-f$ is bounded above on $f^{-1}(J)$ by $-c$. At any case $f^{-1}(J) \notin \mathfrak{B} = \mathfrak{A}$, whence by Theorems 9.3.1 and 9.3.2 $m_i\big(f^{-1}(J)\big) = 0$ and $f^{-1}(J)$ does not contain any subset of the second category with the Baire property. It remains to show that also the complement $\mathbb{R}^N \setminus f^{-1}(J)$ of $f^{-1}(J)$ has the same properties, and this will be shown when we prove it for an arbitrary subinterval of J. Thus in the sequel we may assume that $J = [c, d]$, where $d/c > 0$ and $q = d/c \in \mathbb{Q}$.

By Theorem 5.2.1 we have $f(qx) = qf(x)$ for every $x \in \mathbb{R}^N$, whence

$$f^{-1}(q^n J) = q^n f^{-1}(J) \quad \text{for } n \in \mathbb{N}.$$

But $\bigcup_{n=1}^{\infty} q^n J$ is an infinite interval $I = [c, \infty)$ (if $0 < c < d$) or $I = (-\infty, d]$ (if $c < d < 0$). Hence

$$\bigcup_{n=1}^{\infty} q^n f^{-1}(J) = f^{-1}\Big(\bigcup_{n=1}^{\infty} q^n J \Big) = f^{-1}(I). \tag{12.2.2}$$

If $f^{-1}(J)$ were of measure zero or of the first category, then by (12.2.2) so would be also $f^{-1}(I)$, which means that $\mathbb{R}^N \setminus f^{-1}(I)$ would be of positive (infinite) measure or of the second category with the Baire property. But on $\mathbb{R}^N \setminus f^{-1}(I)$ the function f

---

[1] $\langle c, d \rangle$ denotes any of the intervals (c, d) (open), $[c, d]$ (closed), $(c, d]$ and $[c, d)$ (semiclosed).

takes values outside of I, i.e., it is bounded above by c (when $I = [c, \infty)$), or bounded below by d (when $I = (-\infty, d]$), which contradicts the fact that f is discontinuous.

Thus $m_e(f^{-1}(J)) > 0$ and $f^{-1}(J)$ is of the second category. This is valid for arbitrary non-degenerated interval $J \subset \mathbb{R}$.

Suppose that $0 < c < d$. Put $J_1 = \left[c, \dfrac{c+d}{2}\right]$, $J_2 = \left[0, \dfrac{d-c}{2}\right]$, $J_3 = \left[\dfrac{c+d}{2}, d\right]$. We will show that

$$f^{-1}(J) = f^{-1}(J_1) + f^{-1}(J_2). \tag{12.2.3}$$

Let $x \in f^{-1}(J)$. Then there exists a $u \in J$ such that $f(x) = u$. Either $u \in J_1$ or $u \in J_3$. If $u \in J_1$, then note that $f(0) = 0 \in J_2$, whence $0 \in f^{-1}(J_2)$, whereas $x \in f^{-1}(J_1)$. Then $x = x + 0 \in f^{-1}(J_1) + f^{-1}(J_2)$. Now suppose that $u \in J_3$. If $u < d$, then by Theorem 12.1.3 $f^{-1}\left(\left[u - \dfrac{c+d}{2}, d - \dfrac{c+d}{2}\right]\right) \neq \varnothing$.

If $u = d$, then for $w = \dfrac{1 - q^{-1}}{2} x$ we have according to Theorem 5.2.1, since $\dfrac{1 - q^{-1}}{2} \in \mathbb{Q}$ and $q^{-1}d = c$, $f(w) = \dfrac{1 - q^{-1}}{2} f(x) = \dfrac{1 - q^{-1}}{2} d = \dfrac{d - c}{2}$ so that $f^{-1}\left(\left[u - \dfrac{c+d}{2}, d - \dfrac{c+d}{2}\right]\right) \neq \varnothing$, too. Take a $y \in f^{-1}\left(\left[u - \dfrac{c+d}{2}, d - \dfrac{c+d}{2}\right]\right)$ and put $v = f(y) \in \left[u - \dfrac{c+d}{2}, d - \dfrac{c+d}{2}\right] \subset \left[0, \dfrac{d-c}{2}\right] = J_2$. Thus $y \in f^{-1}(J_2)$. Let $z = x - y$. Then $f(z) = f(x) - f(y) = u - v \in J_1$ so that $z \in f^{-1}(J_1)$. Now, $x = z + y \in f^{-1}(J_1) + f^{-1}(J_2)$. Thus we have obtained

$$f^{-1}(J) \subset f^{-1}(J_1) + f^{-1}(J_2). \tag{12.2.4}$$

Conversely, let $x \in f^{-1}(J_1) + f^{-1}(J_2)$. Then $x = z + y$, where $z \in f^{-1}(J_1)$, $y \in f^{-1}(J_2)$ so that $f(z) \in J_1$, $f(y) \in J_2$. Hence $f(x) = f(z) + f(y) \in J$, i.e., $x \in f^{-1}(J)$, and

$$f^{-1}(J_1) + f^{-1}(J_2) \subset f^{-1}(J). \tag{12.2.5}$$

Relations (12.2.4) and (12.2.5) prove (12.2.3).

Now, as we have already shown, $f^{-1}(J_1)$ is of the second category and of positive outer measure, whereas by Theorem 12.1.3 $f^{-1}(J_2)$ is dense in $\mathbb{R}^N$. By Theorems 3.6.1 and 3.6.2 the set $\mathbb{R}^N \setminus f^{-1}(J)$ has zero inner measure and does not contain any second category subset with the Baire property, whence the theorem follows. $\sqcup$

The next natural question is whether in relation (12.2.1) also J may be an arbitrary set of positive measure or of the second category with the Baire property. In other words, whether a discontinuous additive function on every set of positive measure assumes values from every set of positive measure (resp. the corresponding question for the category). The answer to this question clearly is in the negative. It was pointed out in 5.2 that there exist discontinuous additive functions $f : \mathbb{R}^N \to \mathbb{R}$

which admit only rational values. For such an f relation (12.2.1) with J being the set of irrational numbers is invalid even with $G = \mathbb{R}^N$.

But a result of A. Ostrowski [248] gives us a certain insight into the situation. We recall that if $f : \mathbb{R}^N \to \mathbb{R}$ is an additive function, then the *kernel of* f (cf. 4.3) is the set

$$\operatorname{Ker} f = f^{-1}(0) = \{x \in \mathbb{R}^N \mid f(x) = 0\}.$$

Lemma 12.2.1. *Let $N = 1$, and let $f : \mathbb{R} \to \mathbb{R}$ be an additive function. The set $\operatorname{Ker} f$ is dense in $\mathbb{R}$ if and only if f is non-invertible.*

Proof. If $\operatorname{Ker} f$ is dense in $\mathbb{R}$, then it contains infinitely many points, i.e., $f(x) = 0$ for several x, and consequently f cannot be invertible. If f is non-invertible, then there exist x_1, $x_2 \in \mathbb{R}$ such that $x_1 \neq x_2$ and $f(x_1) = f(x_2)$. Put $x_0 = x_1 - x_2 \neq 0$. We have $f(x_0) = f(x_1) - f(x_2) = 0$, whence, for every $\lambda \in \mathbb{Q}$, we obtain by Theorem 5.2.1 $f(\lambda x_0) = \lambda f(x_0) = 0$. Consequently $\mathbb{Q}x_0 \subset \operatorname{Ker} f$, and $\mathbb{Q}x_0$ is dense in $\mathbb{R}$, since $x_0 \neq 0$, and consequently so is also $\operatorname{Ker} f$. $\qquad\square$

For $N > 1$ Lemma 12.2.1 is no longer valid. If $x_0 \in \mathbb{R}^N$, $x_0 \neq 0$, then the set $\mathbb{Q}x_0$ is not dense in $\mathbb{R}^N$. $\mathbb{Q}x_0$ is a linear subspace (over $\mathbb{Q}$) of $\mathbb{R}^N$, and $\{x_0\}$ is its base. By Theorem 4.2.1 there exists a Hamel basis H of $\mathbb{R}^N$ such that $x_0 \in H$. We have by Theorem 4.2.3 $\operatorname{card}(H \setminus \{x_0\}) = \operatorname{card} H = \mathfrak{c} = \dim \mathbb{R}$. By Theorem 4.3.3 there exists an additive function $f : \mathbb{R}^N \to \mathbb{R}$ such that $\operatorname{Ker} f = \mathbb{Q}x_0$. This f clearly is non-invertible[2] (since $\operatorname{Ker} f$ is infinite), but $\operatorname{Ker} f$ is not dense in $\mathbb{R}^N$.

Some sufficient conditions for the set $\operatorname{Ker} f$ to be dense in $\mathbb{R}^N$ can be inferred from Lemma 4.3.1 and Lemmas 4.2.3 and 4.2.4.

Now we pass to the announced theorem of A. Ostrowski[3].

Theorem 12.2.2. *Let $f : \mathbb{R}^N \to \mathbb{R}$ be an additive function[4] such that $\operatorname{Ker} f$ is dense in $\mathbb{R}^N$. Then either on every set of positive measure f assumes values from every set of positive measure, or there exists a decomposition*

$$\mathbb{R}^N = A_1 \cup A_2, \tag{12.2.6}$$

such that $m(A_1) = 0$ and $m\big(f(A_2)\big) = 0$.

Proof. $\mathbb{R}^N$ is separable, so there exists a countable set $D \subset \operatorname{Ker} f$ dense in $\operatorname{Ker} f$, and hence in $\mathbb{R}^N$. Since $D \subset \operatorname{Ker} f$, we have $f(x) = 0$ for every $x \in D$.

Now suppose that there exist sets $A \subset \mathbb{R}^N$ of positive (N-dimensional) measure, and $B \subset \mathbb{R}$ of positive (one-dimensional) measure such that

$$f(A) \cap B = \varnothing. \tag{12.2.7}$$

---

[2] f is also discontinuous, since otherwise $\operatorname{Ker} f$ would have to be closed, which is not the case with $\mathbb{Q}x_0$. An example of continuous non-invertible $f : \mathbb{R}^N \to \mathbb{R}$ with the kernel non-dense in $\mathbb{R}^N$ would be easier. E.g., if $N = 2$, and $x = (x_1, x_2)$, then the function $f(x) = x_1 + x_2$ yields such an example.

[3] A. Ostrowski [248] proved this theorem in the one-dimensional case ($N = 1$) under the condition that f is non-invertible. But Lemma 12.2.1 ensures that in this case the latter condition is equivalent to the condition that $\operatorname{Ker} f$ is dense in $\mathbb{R}$.

[4] There is no explicit assumption that f is discontinuous. If f is continuous, then $\operatorname{Ker} f$ is closed, and, being dense in $\mathbb{R}^N$, equals the whole $\mathbb{R}^N$. Hence $f = 0$. For such an f the theorem is trivially true, viz. (12.2.6) holds with $A_1 = \varnothing$, $A_2 = \mathbb{R}^N$.

Let $C = A + D = \bigcup_{d \in D} (A + d)$, and $C' = \mathbb{R}^N \setminus C$. Every set $A + d$ is measurable and the union is countable, so C is measurable. By Theorem 3.6.1 $m_i(C') = 0$, and thus, since C is measurable,

$$m(C') = 0. \tag{12.2.8}$$

Let $t \in f(A)$. Then $t = f(x)$ for a certain $x \in A$. Take a $d \in D$. Then $x + d \in C$, and $t = f(x) = f(x) + f(d) = f(x + d) \in f(C)$. Thus $f(A) \subset f(C)$. Conversely, let $t \in f(C)$. Then $t = f(x)$ for a certain $x \in C$, i.e., $x = a + d$, where $a \in A$, $d \in D$. Hence $t = f(x) = f(a + d) = f(a) + f(d) = f(a) \in f(A)$. Thus $f(C) \subset f(A)$, and finally we obtain $f(A) = f(C)$, whence by (12.2.7)

$$f(C) \cap B = \varnothing. \tag{12.2.9}$$

Put

$$P = \bigcup_{\lambda \in \mathbb{Q}} \lambda C', \quad M = \bigcup_{\lambda \in \mathbb{Q}} \lambda B, \quad P' = \mathbb{R}^N \setminus P, \quad M' = \mathbb{R} \setminus M.$$

By (12.2.8) $m(\lambda C') = 0$ for every $\lambda \in \mathbb{Q}$, whence

$$m(P) = 0. \tag{12.2.10}$$

Further, with $B^+ = B \cap (0, \infty)$, $B^- = B \cap (-\infty, 0)$ and $B^0 = B \cap \{0\}$, we have $B = B^+ \cup B^- \cup B^0$. The sets B^+, B^- and B^0 are measurable, $m(B^0) = 0$, whence

$$0 < m(B) = m(B^+) + m(B^-).$$

Hence at least one of the sets B^+ and B^- must have positive measure. Let, e.g., $m(B^+) > 0$. Let $L : (0, \infty) \to \mathbb{R}$ be the logarithmic function:

$$L(x) = \log x.$$

The functions $-L$ and L^{-1} (the exponential function) are continuous and convex, and hence, by Theorem 7.4.6 are absolutely continuous. Therefore also L is absolutely continuous. Thus if one of the sets E, $L(E)$ is measurable, so is also the other, and if one of E, $L(E)$ has measure zero, so has also the other. It follows that $m(L(B^+)) > 0$. Now, with $\mathbb{Q}^+ = \mathbb{Q} \cap (0, \infty)$, $\mathbb{Q}^- = \mathbb{Q} \cap (-\infty, 0)$, we have

$$L\Big(\bigcup_{\lambda \in \mathbb{Q}^+} \lambda B^+ \Big) = \bigcup_{\lambda \in \mathbb{Q}^+} L(\lambda B^+) = \bigcup_{\lambda \in \mathbb{Q}^+} [L(B^+) + L(\lambda)] = L(B^+) + L(\mathbb{Q}^+).$$

The above set is measurable (as a countable union of measurable sets) and by Theorem 3.6.1 its complement has measure zero (since $L(\mathbb{Q}^+)$ is dense in $\mathbb{R}$). But

$$\mathbb{R} \setminus L\Big(\bigcup_{\lambda \in \mathbb{Q}^+} \lambda B^+ \Big) = L\Big((0, \infty) \setminus \bigcup_{\lambda \in \mathbb{Q}^+} \lambda B^+ \Big), \tag{12.2.11}$$

since L is one-to-one. The set occurring in (12.2.11) has measure zero, whence, since L^{-1} is absolutely continuous,

$$m\left((0, \infty) \setminus \bigcup_{\lambda \in \mathbb{Q}^+} \lambda B^+\right) = 0. \tag{12.2.12}$$

Further, we have $\bigcup_{\lambda \in \mathbb{Q}^-} \lambda B^+ = - \bigcup_{\lambda \in \mathbb{Q}^+} \lambda B^+$, whence by (12.2.12)

$$m\left((-\infty, 0) \setminus \bigcup_{\lambda \in \mathbb{Q}^-} \lambda B^+\right) = 0. \tag{12.2.13}$$

Now,

$$\bigcup_{\lambda \in \mathbb{Q}^-} \lambda B^+ \cup \bigcup_{\lambda \in \mathbb{Q}^+} \lambda B^+ \subset \bigcup_{\lambda \in \mathbb{Q}} \lambda B^+ \subset \bigcup_{\lambda \in \mathbb{Q}} \lambda B = M,$$

whence by (12.2.12) and (12.2.13)

$$m(M') = 0. \tag{12.2.14}$$

Suppose that there exists a $y \neq 0$ such that $y \in f(P') \cap M$. Since $y \in M$, $y \neq 0$, there exists a $\lambda \in \mathbb{Q}$, $\lambda \neq 0$, such that $y \in \lambda B$, i.e., $y/\lambda \in B$. Since $y \in f(P')$, there exists an x,

$$x \in P', \tag{12.2.15}$$

such that $y = f(x)$. Therefore, by Theorem 5.2.1,

$$\frac{y}{\lambda} = \frac{1}{\lambda} f(x) = f\left(\frac{x}{\lambda}\right),$$

and (12.2.9) implies that $y/\lambda \notin f(C)$, i.e., $x/\lambda \notin C$. Thus $x/\lambda \in C'$ and $x \in \lambda C' \subset P$, which contradicts (12.2.15).

Thus the set $f(P') \cap M$ may contain at most one element, viz. 0. Consequently

$$f(P') \subset M' \cup \{0\},$$

which together with (12.2.14) shows that

$$m\big(f(P')\big) = 0. \tag{12.2.16}$$

Now, in order to obtain decomposition (12.2.6), we take $A_1 = P$, $A_2 = P'$. Relations (12.2.10) and (12.2.16) show that this decomposition has the required properties. $\qquad\square$

By the same argument one can also prove the topological analogue of Theorem 12.2.2.

Theorem 12.2.3. *Let $f : \mathbb{R}^N \to \mathbb{R}$ be an additive function such that* Ker f *is dense in $\mathbb{R}^N$. Then either on every set of the second category and with the Baire property f assumes values from every set of the second category with the Baire property, or there exists a decomposition (12.2.6) such that A_1 is of the first category (in the topology of $\mathbb{R}^N$), and $f(A_2)$ is of the first category (in the topology of $\mathbb{R}$).*

The invariance of the Baire property and of the category under L will now result from the fact that L is a homeomorphism. Theorem 3.6.1 must everywhere be replaced by Theorem 3.6.2.

12.3 Convex functions

The next arising question is how far the results of 12.2 generalize to the case of convex functions. Unfortunately, we are unable to answer this question. Below we prove a result (Theorem 12.3.1) due to A. Ostrowski [249], which is a very weak analogue of Theorem 12.2.1.

Lemma 12.3.1. *Let $D \subset \mathbb{R}^N$ be a convex and open set, and let $f : D \to \mathbb{R}$ be a discontinuous convex function. Further, let $J \subset \mathbb{R}$ be an interval with a finite upper bound. Then $m_i\big(f^{-1}(J)\big) = 0$.*

Proof. Let $d \in \mathbb{R}$ be such that $t \leqslant d$ for $t \in J$. Then f is bounded above (by d) on $f^{-1}(J)$, whence the lemma results in view of Theorem 9.3.1 and of the fact that f is discontinuous. $\qquad\square$

Theorem 12.3.1. *Let $D \subset \mathbb{R}^N$ be a convex and open set, and let $f : D \to \mathbb{R}$ be a discontinuous convex function. Let the interval $J = [c, d] \subset \mathbb{R}$ be such that*

$$\inf_{x \in D} f(x) = \inf_{x \in D} m_f(x) < c < d < \infty. \tag{12.3.1}$$

Then $m_e\big(f^{-1}(J)\big) > 0$.

Proof. By (12.3.1) there exists an $x_0 \in D$ such that

$$f(x_0) < c. \tag{12.3.2}$$

For every $x \in D$, $x \neq x_0$, let $l(x)$ denote the closed segment joining x_0 and x:

$$l(x) - \{z \in D \mid z = x_0 + t(x - x_0), \ t \in [0, 1]\},$$

and let $L(x) = Q(x_0, x)$ denote the rational segment between x_0 and x:

$$L(x) = \{z \in D \mid z = x_0 + t(x - x_0), \ t \in \mathbb{Q} \cap [0, 1]\}.$$

By Theorem 6.1.2 for every $x \in D$, $x \neq x_0$, there exists a continuous function $g_x : l(x) \to \mathbb{R}$ such that $g_x \mid L(x) = f \mid L(x)$.

Put

$$C = \{x \in D \mid f(x) > d\} = f^{-1}\big((d, \infty)\big).$$

The complement $C' = D \setminus C$ of C is the set $C' = f^{-1}\big((-\infty, d]\big)$. If C had measure zero, then C' would have a positive measure, which, according to Lemma 12.3.1, is impossible. Consequently

$$m_e(C) > 0. \tag{12.3.3}$$

For every $x \in C$ we have $g_x(x) = f(x) > d$, whereas by (12.3.2) $g_x(x_0) = f(x_0) < c$. Consequently we have $g_x(z) \in [c, d]$ for z from a certain non-degenerated segment $S_x \subset l(x)$. In other words, there exists a (non-degenerated) segment $S_x \subset l(x)$ such that

$$S_x \subset g_x^{-1}\big([c, d]\big).$$

Put $I(x) = \dfrac{1}{|x - x_0|} (S_x - x_0)$. Thus $I(x)$ also is a non-degenerated segment con-

tained in $\dfrac{1}{|x - x_0|} (l(x) - x_0)$. Let $|I(x)|$ denote the length of $I(x)$. Consequently

$0 < |I(x)| \leqslant 1$.

Let

$$C_k = \left\{ x \in C \;\middle|\; \frac{1}{k+1} < |I(x)| \leqslant \frac{1}{k} \right\}, \quad k \in \mathbb{N}.$$

Clearly

$$\bigcup_{k=1}^{\infty} C_k = C, \tag{12.3.4}$$

and it follows by (12.3.3) and (12.3.4) that there exists a $k_0 \in \mathbb{N}$ such that

$$m_e(C_{k_0}) > 0. \tag{12.3.5}$$

Write $n = k_0 + 1$, and consider the points $x_i = x_i(x) = \dfrac{i}{n+1} \dfrac{x - x_0}{|x - x_0|}, i = 0, \ldots, n+1$.

We have $|x_{i+1} - x_i| = \dfrac{1}{n+1}$, and $|x - x_0| x_i = \left(x_0 + \dfrac{i}{n+1} (x - x_0) \right) - x_0 \in l(x) - x_0$,

whence $x_i \in \dfrac{1}{|x - x_0|} (l(x) - x_0)$. Since for $x \in C_{k_0}$ we have $|I(x)| > \dfrac{1}{n}$, for every such

x there must exist an i, $0 \leqslant i \leqslant n+1$, such that $x_i \in I(x)$. Let

$$E_i = \{ x \in C_{k_0} \mid x_i(x) \in I(x) \}, \quad i = 0, \ldots, n+1.$$

Again

$$\bigcup_{i=0}^{n+1} E_i = C_{k_0}, \tag{12.3.6}$$

and it follows by (12.3.5) and (12.3.6) that there exists an $i_0 \in \{ 0, \ldots, n+1 \}$ such
that

$$m_e(E_{i_0}) > 0. \tag{12.3.7}$$

Let $x \in E_{i_0}$. Then $x_{i_0}(x) \in I(x)$, i.e.,

$$\frac{i_0}{n+1} \frac{x - x_0}{|x - x_0|} \in \frac{1}{|x - x_0|} (S_x - x_0),$$

or

$$x_0 + \frac{i_0}{n+1} (x - x_0) \in S_x \subset g_x^{-1}([c, d]),$$

and

$$g_x \left(x_0 + \frac{i_0}{n+1} (x - x_0) \right) \in [c, d].$$

But $x_0 + \dfrac{i_0}{n+1} (x - x_0) \in L(x)$, whence

$$f \left(x_0 + \frac{i_0}{n+1} (x - x_0) \right) = g_x \left(x_0 + \frac{i_0}{n+1} (x - x_0) \right) \in [c, d].$$

In other words, for $x \in E_{i_0}$,

$$x_0 + \frac{i_0}{n+1}(x - x_0) \in f^{-1}([c, d]). \tag{12.3.8}$$

Write $E^* = x_0 + \frac{i_0}{n+1}(E_{i_0} - x_0)$. By (12.3.7) (cf. Corollary 3.2.2)

$$m_e(E^*) > 0, \tag{12.3.9}$$

and (12.3.8) shows that

$$E^* \subset f^{-1}([c, d]) = f^{-1}(J). \tag{12.3.10}$$

Now, the theorem results from (12.3.9) and (12.3.10). $\qquad\square$

From Lemma 12.3.1 and Theorem 12.3.1 we obtain

Corollary 12.3.1. *Let $D \subset \mathbb{R}^N$ be a convex and open set, and let $f : D \to \mathbb{R}$ be a discontinuous convex function. Then, for every finite and non-degenerated interval $J \subset \mathbb{R}$ such that*

$$\inf J > \inf_{x \in D} m_f(x) = \inf_{x \in D} f(x) \tag{12.3.11}$$

the set $f^{-1}(J)$ is non-measurable.

Proof. Since J is non-degenerated, it contains a non-degenerated interval $[c, d]$, and by (12.3.11) condition (12.3.1) is fulfilled. Hence, by Theorem 12.3.1,

$$m_e(f^{-1}(J)) \geqslant m_e(f^{-1}([c, d])) > 0. \tag{12.3.12}$$

On the other hand, since J is finite, it has a finite upper bound. Thus by Lemma 12.3.1

$$m_i(f^{-1}(J)) = 0. \tag{12.3.13}$$

Relations (12.3.12) and (12.3.13) imply, in virtue of Theorem 3.1.2, that $f^{-1}(J)$ is non-measurable. $\qquad\square$

Theorem 12.3.2. *Let $D \subset \mathbb{R}^N$ be a convex and open set, and let $f : D \to \mathbb{R}$ be a discontinuous convex function. Then, for every non-degenerated interval $J \subset \mathbb{R}$ fulfilling (12.3.11) the set $f^{-1}(J)$ is non-measurable.*

Proof. If $\sup J < \infty$, then J is finite, and the theorem results from Corollary 12.3.1. So we may restrict ourselves to the case where $J = (c, \infty)$, or $J = [c, \infty)$ with a certain $c \in \mathbb{R}$. Put $J_0 = \mathbb{R} \setminus J$, and

$$c_0 = \inf_{x \in D} m_f(x).$$

Then $c_0 < c < \infty$. Choose a $\gamma \in (c_0, c)$. By Corollary 12.3.1 the set $f^{-1}((\gamma, c))$ is non-measurable, whence

$$m_e(f^{-1}(J_0)) \geqslant m_e(f^{-1}((\gamma, c))) > 0.$$

On the other hand, by Lemma 12.3.1,

$$m_i\big(f^{-1}(J_0)\big) = 0.$$

Consequently $f^{-1}(J_0)$ is non-measurable. We have $f^{-1}(J_0) = f^{-1}(\mathbb{R}) \setminus f^{-1}(J) = D \setminus f^{-1}(J)$. Since D is measurable, and $f^{-1}(J_0)$ is not, the set $f^{-1}(J)$ also must be non-measurable.   □

The attribute "non-measurable" in Theorem 12.3.2 cannot be replaced by "saturated non-measurable". Clearly, $f^{-1}(J) \subset D$, and there exist convex and open $D \subset \mathbb{R}^N$ such that $m(\mathbb{R}^N \setminus D) > 0$. For such D we have $0 < m\big(\mathbb{R}^N \setminus D\big) \leqslant m_i\big(\mathbb{R}^N \setminus f^{-1}(J)\big)$. But if $J \subset \mathbb{R}$ is an interval fulfilling (12.3.11), write $d = \sup J$ and

$$D_0 = \{x \in D \mid m_f(x) \leqslant d\}.$$

If $x \in f^{-1}(J)$, then $m_f(x) \leqslant f(x) \leqslant d$, whence $x \in D_0$. Thus $f^{-1}(J) \subset D_0$. It is an open problem whether

$$m_i\big(D_0 \setminus f^{-1}(J)\big) = 0$$

for every non-degenerated interval $J \subset \mathbb{R}$ fulfilling (12.3.11).

By exactly the same argument as that employed in the respective proofs, we obtain also topological analogues of Lemma 12.3.1 and Theorems 12.3.1 and 12.3.2.

Lemma 12.3.2. *Let $D \subset \mathbb{R}^N$ be a convex and open set, and let $f : D \to \mathbb{R}$ be a discontinuous convex function. Further, let $J \subset \mathbb{R}$ be an interval with a finite upper bound. Then the set $f^{-1}(J)$ does not contain any set of the second category and with the Baire property.*

Theorem 12.3.3. *Let $D \subset \mathbb{R}^N$ be a convex and open set, and let $f : D \to \mathbb{R}$ be a discontinuous convex function. Let $J = [c, d] \subset \mathbb{R}$ be an interval fulfilling (12.3.1). Then the set $f^{-1}(J)$ is of the second category.*

Theorem 12.3.4. *Let $D \subset \mathbb{R}^N$ be a convex and open set, and let $f : D \to \mathbb{R}$ be a discontinuous convex function. Then, for every non-degenerated interval $J \subset \mathbb{R}$ fulfilling (12.3.11) the set $f^{-1}(J)$ does not have the Baire property.*

12.4   Big graph

Let $f : \mathbb{R}^N \to \mathbb{R}$ be a discontinuous additive function. In 12.1 we have shown that the set $\mathrm{Gr}(f)$ is dense in $\mathbb{R}^N$.

Nothing more can be said in general about $\mathrm{Gr}(f)$. In 5.2 we have seen that there exist (necessarily discontinuous) additive functions $f : \mathbb{R}^N \to \mathbb{R}$, $f \neq 0$, which assume only rational values: $f(\mathbb{R}^N) \subset \mathbb{Q}$. Take arbitrary $\lambda \in \mathbb{Q}$ and an arbitrary $x \in \mathbb{R}^N$ such that $f(x) \neq 0$. Then $f(x) \in \mathbb{Q}$, whence also $q = \lambda/f(x) \in \mathbb{Q}$. By Theorem 5.2.1 $f(qx) = qf(x) = \lambda$. Consequently $\mathbb{Q} \subset f(\mathbb{R}^N)$ and we get $f(\mathbb{R}^N) = \mathbb{Q}$. Thus there exist (necessarily discontinuous) additive functions $f : \mathbb{R}^N \to \mathbb{R}$ such that $f(\mathbb{R}^N)$ is countable. For such an f we have

$$\mathrm{Gr}(f) \subset \{\, (x, y) \in \mathbb{R}^{N+1} \mid x \in \mathbb{R}^N, \; y \in f(\mathbb{R}^N) \,\} = \bigcup_{y \in f(\mathbb{R}^N)} (\mathbb{R}^N \times \{y\}).$$

Every set $\mathbb{R}^N \times \{y\}$ is of ($N+1$-dimensional) measure zero, and nowhere dense (in the topology of $\mathbb{R}^{N+1}$). Hence $\mathrm{Gr}(f)$ also is of measure zero, and of the first category. Moreover, $\mathrm{Gr}(f)$ is disconnected, since for arbitrary $y_0 \notin f(\mathbb{R}^N)$ we have

$$\mathrm{Gr}(f) = \left[\mathrm{Gr}(f) \cap \left(\mathbb{R}^N \times (-\infty, y_0]\right)\right] \cup \left[\mathrm{Gr}(f) \cap \left(\mathbb{R}^N \times [y_0, \infty)\right)\right],$$

and the sets $\mathrm{Gr}(f) \cap \left(\mathbb{R}^N \times (-\infty, y_0]\right)$ and $\mathrm{Gr}(f) \cap \left(\mathbb{R}^N \times [y_0, \infty)\right)$ are non-empty[5], disjoint, and closed in the space $\mathrm{Gr}(f)$.

Every additive function $f : \mathbb{R}^N \to \mathbb{R}$ such that $f(\mathbb{R}^N)$ is countable $\big(\mathrm{card}\, f(\mathbb{R}^N) = \aleph_0\big)$ will henceforth be referred to as a *function with small graph*. We may summarize what we have already established in the following theorems.

Theorem 12.4.1. *There exist additive functions $f : \mathbb{R}^N \to \mathbb{R}$ with small graphs. Every such function is discontinuous.*

Theorem 12.4.2. *If $f : \mathbb{R}^N \to \mathbb{R}$ is an additive function with small graph, then the set $\mathrm{Gr}(f)$ is of measure zero and of the first category (in $\mathbb{R}^{N+1}$), and is not connected.*

Now, we want to show that there exist additive functions $f : \mathbb{R}^N \to \mathbb{R}$ which have particularly big graphs. Let $\pi : \mathbb{R}^{N+1} \to \mathbb{R}^N$ be the projection: if $p \in \mathbb{R}^{N+1}$ and $p = (x, y)$, $x \in \mathbb{R}^N$, $y \in \mathbb{R}$, then $\pi(p) = x$. An additive function $f : \mathbb{R}^N \to \mathbb{R}$ is called a *function with big graph* iff for every Borel set $F \subset \mathbb{R}^{N+1}$ such that[6] $\mathrm{card}\,\pi(F) = \mathfrak{c}$ we have

$$F \cap \mathrm{Gr}(f) \neq \varnothing. \tag{12.4.1}$$

We have[7]

Theorem 12.4.3. *There exist additive functions $f : \mathbb{R}^N \to \mathbb{R}$ with big graphs. Every such function is discontinuous.*

Proof. Let $\mathcal{F}$ be the family of all Borel sets $F \subset \mathbb{R}^{N+1}$ such that $\mathrm{card}\,\pi(F) = \mathfrak{c}$. Since $\mathcal{F} \subset \mathcal{B}(\mathbb{R}^{N+1})$, we get by Theorem 2.3.4 $\mathrm{card}\,\mathcal{F} \leqslant \mathfrak{c}$. On the other hand, $\mathcal{F}$ contains all open balls in $\mathbb{R}^{N+1}$, and the family of all such balls has the power of continuum. Hence $\mathrm{card}\,\mathcal{F} \geqslant \mathfrak{c}$, whence

$$\mathrm{card}\,\mathcal{F} = \mathfrak{c}. \tag{12.4.2}$$

Let γ be the smallest ordinal such that $\overline{\gamma} = \mathfrak{c}$ (cf. the proof of Theorem 11.4.3). Hence also $\mathrm{card}\,\Gamma(\gamma) = \mathfrak{c}$, and by (12.4.2) there exists a one-to-one mapping $F : \Gamma(\gamma) \to \mathcal{F}$ (onto). Instead of $F(\alpha)$ we write F_α. Thus $\mathcal{F} = \{F_\alpha\}_{\alpha < \gamma}$.

Now we define a transfinite sequence $\{(x_\alpha, y_\alpha)\}$ of points of $\mathbb{R}^{N+1}$. For every $\alpha < \gamma$ we write $X_\alpha = \bigcup_{\beta < \gamma} \{x_\beta\}$ and $E_\alpha = E(X_\alpha)$, except for $\alpha = 0$, where we assume

---

[5] If $f(\mathbb{R}^N)$ is countable, then there exists a $y_0 \in f(\mathbb{R}^N)$ such that $y_0 \neq 0$. Thus $y_0 = f(x_0)$ for an $x_0 \in \mathbb{R}^N$. For every $\lambda \in \mathbb{Q}$ we have $\lambda y_0 = \lambda f(x_0) = f(\lambda x_0) \in f(\mathbb{R}^N)$, whence $\mathbb{Q}y_0 \subset f(\mathbb{R}^N)$. Hence $f(\mathbb{R}^N)$ is dense in $\mathbb{R}$.
[6] It ammounts to the same to require that $\mathrm{card}\,\pi(F) > \aleph_0$. In fact, every Borel set is analytic, and the function π is continuous, hence the set $\pi(F)$ also is analytic (Theorems 2.5.6 and 2.5.2). Thus, by Theorem 2.8.3, the conditions $\mathrm{card}\,\pi(F) = \mathfrak{c}$ and $\mathrm{card}\,\pi(F) > \aleph_0$ are equivalent.
[7] Cf. Jones [157].

$E_0 = \{0\}$. We want to construct the sequence $\{(x_\alpha, y_\alpha)\}_{\alpha<\gamma}$ so that for every $\alpha < \gamma$ the set $\{x_\alpha\} \cup X_\alpha$ is linearly independent (over $\mathbb{Q}$). Let $w : \mathcal{P}(\mathbb{R}^{N+1}) \setminus \{\varnothing\} \to \mathbb{R}^{N+1}$ be a choice function. Suppose that we have already defined (x_β, y_β) for $\beta < \alpha < \gamma$, (if $\alpha = 0$, nothing has been defined yet) so that the sets $\{x_\beta\} \cup X_\beta$ are linearly independent. We put

$$(x_\alpha,\ y_\alpha) = w\big(F_\alpha \setminus [E_\alpha \times \mathbb{R}]\big). \tag{12.4.3}$$

Note that $\operatorname{card} X_\alpha = \overline{\alpha} < \mathfrak{c}$, since $\alpha < \gamma$, and by Lemmas 4.1.3 and 4.2.2 we have $\operatorname{card} E_\alpha < \mathfrak{c}$, whereas $\operatorname{card} \pi(F_\alpha) = \mathfrak{c}$. Consequently $F_\alpha \setminus [E_\alpha \times \mathbb{R}] \neq \varnothing$.

Suppose that the set X_α is linearly dependent. Then there exist $x_{\beta_1}, \ldots, x_{\beta_n} \in X_\alpha$ such that the system $\{x_{\beta_1}, \ldots, x_{\beta_n}\}$ is linearly dependent.

Let $\beta = \max(\beta_1, \ldots, \beta_n) < \alpha$. Then $\{x_{\beta_1}, \ldots, x_{\beta_n}\} \subset X_\beta \cup \{x_\beta\}$, which contradicts the fact that $\{x_\beta\} \cup X_\beta$ is linearly independent ($\beta < \alpha$). According to (12.4.3) we have $x_\alpha \notin E_\alpha$, whence it follows by Lemma 4.1.2 that the set $\{x_\alpha\} \cup X_\alpha$ is linearly independent. (If $\alpha = 0$, then we get from (12.4.3) $x_0 \neq 0$, and thus $\{x_0\}$ is linearly independent).

By Theorem 1.6.2 there exists a transfinite sequence $\{(x_\alpha,\ y_\alpha)\}_{\alpha<\gamma}$ fulfilling (12.4.3) for every $\alpha < \gamma$ and such that $\{x_\alpha\} \cup X_\alpha$ (and hence also X_α) is linearly independent for every $\alpha < \gamma$. Put

$$X = \bigcup_{\alpha<\gamma} X_\alpha = \bigcup_{\alpha<\gamma} \{x_\alpha\}.$$

The same argument as above shows that the set X is linearly independent. By Theorem 4.2.1 there exists a Hamel basis $H \subset \mathbb{R}^N$ such that $X \subset H$. Define the function $g : H \to \mathbb{R}$ putting

$$g(x) = \begin{cases} y_\alpha & \text{if } x = x_\alpha,\ \alpha < \gamma, \\ 0 & \text{if } x \in H \setminus X. \end{cases}$$

Let $f : \mathbb{R}^N \to \mathbb{R}$ be the additive extension of g (Theorem 5.2.2). Thus

$$(x_\alpha,\ y_\alpha) \in \operatorname{Gr}(f) \quad \text{for every } \alpha < \gamma. \tag{12.4.4}$$

Let $F \subset \mathbb{R}^{N+1}$ be a Borel set such that $\operatorname{card} \pi(F) = \mathfrak{c}$. Thus $F \in \mathcal{F}$, and there exists an $\alpha < \gamma$ such that $F = F_\alpha$. By (12.4.3) $(x_\alpha, y_\alpha) \in F_\alpha = F$, whence by (12.4.4)

$$(x_\alpha,\ y_\alpha) \in F \cap \operatorname{Gr}(f).$$

Hence we obtain (12.4.1). Consequently f is a function with big graph.

If f were continuous, then by Theorem 5.5.2 f would have the form $f(x) = cx$ with a $c \in \mathbb{R}^N$. Hence

$$\operatorname{Gr}(f) = \{(x, y) \in \mathbb{R}^{N+1} \mid x \in \mathbb{R}^N,\ y = cx\}. \tag{12.4.5}$$

But set (12.4.5) does not have property (12.4.1). Consequently f is discontinuous. $\square$

Let us also note the following

Lemma 12.4.1. *Let* $f : \mathbb{R}^N \to \mathbb{R}$ *be an additive function with big graph. Then* $f(\mathbb{R}^N) = \mathbb{R}$.

Proof. Let $y \in \mathbb{R}$ be arbitrary. The set $F = \mathbb{R}^N \times \{y\} \subset \mathbb{R}^{N+1}$ is closed, and hence Borel, and $\pi(F) = \mathbb{R}^N$, so card $\pi(F) = \mathfrak{c}$. Thus we must have (12.4.1). In other words, there exists a point

$$(x_0, y_0) \in F \cap \mathrm{Gr}(f).$$

Since $(x_0, y_0) \in F$, we must have $y_0 = y$. Since $(x_0, y_0) \in \mathrm{Gr}(f)$, we must have $f(x_0) = y_0 = y$. Consequently $y \in f(\mathbb{R}^N)$, whence the lemma follows. $\qquad \square$

Lemma 12.4.1 shows, in particular, that a function with big graph cannot be at the same time a function with small graph. This will follow also from further theorems.

A big graph has none of the properties specified in Theorem 12.4.2 for small graphs. Namely, we have (cf. also Marczewski [220], Marcus [215])

Theorem 12.4.4. *If* $f : \mathbb{R}^N \to \mathbb{R}$ *is an additive function with big graph, then the set* $\mathrm{Gr}(f)$ *is saturated non-measurable, and has property* (*) *from 3.3.*

Proof. Let $(x_1, y_1), (x_2, y_2) \in \mathrm{Gr}(f)$. This means that $y_1 = f(x_1)$, $y_2 = f(x_2)$. Hence

$$y_1 + y_2 = f(x_1) + f(x_2) = f(x_1 + x_2),$$

i.e., $(x_1, y_1) + (x_2, y_2) = (x_1 + x_2, y_1 + y_2) \in \mathrm{Gr}(f)$. Consequently[8]

$$\mathrm{Gr}(f) + \mathrm{Gr}(f) \subset \mathrm{Gr}(f). \tag{12.4.6}$$

On the other hand, $\mathrm{int}\, \mathrm{Gr}(f) = \varnothing$, since on every vertical line $l = \{x_0\} \times \mathbb{R}$, $x_0 \in \mathbb{R}^N$, there is only one point of $\mathrm{Gr}(f)$, viz. $(x_0, f(x_0))$. By Theorem 3.7.1

$$m_i(\mathrm{Gr}(f)) = 0. \tag{12.4.7}$$

Similarly, by Theorem 2.9.1 we infer that $\mathrm{Gr}(f)$ does not contain any second category set with the Baire property.

Now suppose that $m_i(\mathbb{R}^{N+1} \setminus \mathrm{Gr}(f)) > 0$. By the definition of the inner measure there must exist a closed (and hence a Borel) set

$$F \subset \mathbb{R}^{N+1} \setminus \mathrm{Gr}(f) \tag{12.4.8}$$

such that

$$m(F) > 0. \tag{12.4.9}$$

If the set $\pi(F)$ were countable, then F would be contained in a countable union of vertical lines:

$$F \subset \bigcup_{n=1}^{\infty} l_n. \tag{12.4.10}$$

In $\mathbb{R}^{N+1}$ every line l_n is of measure zero, and hence (12.4.10) would imply that $m(F) = 0$, contrary to (12.4.9). Therefore card $\pi(F) > \aleph_0$, whence by Theorem 2.8.3

---

[8] Actually, in (12.4.6) we have the equality. In fact, since $f(0) = 0$, we have $(0, 0) \in \mathrm{Gr}(f)$. Now, for an arbitrary $(x, y) \in \mathrm{Gr}(f)$ we have $(x, y) = (x, y) + (0, 0) \in \mathrm{Gr}(f) + \mathrm{Gr}(f)$, whence $\mathrm{Gr}(f) \subset \mathrm{Gr}(f) + \mathrm{Gr}(f)$. This together with (12.4.6) yields $\mathrm{Gr}(f) = \mathrm{Gr}(f) + \mathrm{Gr}(f)$.

$\operatorname{card} \pi(F) = \mathfrak{c}$. Thus we must have (12.4.1), which is incompatible with (12.4.8). Therefore we must have

$$m_i(\mathbb{R}^{N+1} \setminus \operatorname{Gr}(f)) = 0. \tag{12.4.11}$$

Relations (12.4.7) and (12.4.11) show that $\operatorname{Gr}(f)$ is saturated non-measurable.

Similarly, suppose that there exists a second category set B with the Baire property such that

$$B \subset \mathbb{R}^{N+1} \setminus \operatorname{Gr}(f). \tag{12.4.12}$$

Then

$$B = (G \cup P) \setminus R,$$

where G is open and non-empty, and P, R are of the first category. Put $C = G \setminus R \subset B$. Hence $G \setminus C \subset R$ is of the first category, and by Theorem 2.1.3 there exists a set $F \in \mathcal{G}_\delta$ such that

$$F \subset C \subset B, \tag{12.4.13}$$

and the set $G \setminus F$ is of the first category. Consequently F must be of the second category. If we had $\operatorname{card} \pi(F) \leqslant \aleph_0$, then F would be contained in a countable union of vertical lines, i.e., we would have (12.4.10). In $\mathbb{R}^{N+1}$ every line is nowhere dense, and hence the sets in (12.4.10) would be of the first category, which is impossible. Consequently $\operatorname{card} \pi(F) = \mathfrak{c}$, and we obtain (12.4.1), which contradicts (12.4.12) and (12.4.13).

Thus neither $\operatorname{Gr}(f)$, nor $\mathbb{R}^{N+1} \setminus \operatorname{Gr}(f)$, contains a second category set with the Baire property, which means that $\operatorname{Gr}(f)$ has property $(*)$. $\square$

Theorem 12.4.4 shows, in particular, that the graph of f is non-measurable and does not have the Baire property, whence $m_e(\operatorname{Gr}(f)) > 0$ and $\operatorname{Gr}(f)$ is of the second category. This is a very surprising property of a graph of a function. The existence of functions $f : \mathbb{R} \to \mathbb{R}$ such that $m_e(\operatorname{Gr}(f)) > 0$ was first shown (on another way) by W. Sierpiński [278].

Let us also note that every vertical line $\{x\} \times \mathbb{R}$, $x \in \mathbb{R}^N$, intersects $\operatorname{Gr}(f)$ at exactly one point $\big(\text{viz. } (x, f(x))\big)$, and consequently this intersection has measure zero. Nevertheless, $m_e(\operatorname{Gr}(f)) > 0$. This shows that in the Fubini theorem to the effect that if $A \in \mathcal{L}$, and for almost every vertical line l we have $m(A \cap l) = 0$, then $m(A) = 0$, the assumption of the measurability of A is essential.

Theorem 12.4.5. *If $f : \mathbb{R}^N \to \mathbb{R}$ is an additive function with big graph, then the set $\operatorname{Gr}(f)$ is connected.*

Proof. For an indirect proof suppose that $\operatorname{Gr}(f)$ is not connected. Then

$$\operatorname{Gr}(f) = A_0 \cup B_0, \tag{12.4.14}$$

where the sets A_0 and B_0 are separated:

$$(A_0 \cap \operatorname{cl} B_0) \cup (B_0 \cap \operatorname{cl} A_0) = \varnothing. \tag{12.4.15}$$

Write

$$A = \operatorname{cl} A_0, \quad B = \operatorname{cl} B_0, \quad F = A \cap B. \tag{12.4.16}$$

Suppose that there exists a point $p \in F \cap \mathrm{Gr}(f)$. By (12.4.14) p belongs either to A_0, or to B_0, and by (12.4.16) p belongs to A and to B. Hence $p \in (A_0 \cap B) \cup (B_0 \cap A)$, which is impossible in view of (12.4.15). Consequently

$$F \cap \mathrm{Gr}(f) = \varnothing, \tag{12.4.17}$$

which implies, since F is closed (and hence Borel) that (cf. Theorem 2.8.3)

$$\mathrm{card}\,\pi(F) \leqslant \aleph_0. \tag{12.4.18}$$

Note also that $A \cup B = \mathrm{cl}\,A_0 \cup \mathrm{cl}\,B_0 = \mathrm{cl}(A_0 \cup B_0) = \mathrm{cl}\,\mathrm{Gr}(f)$, whence by Theorem 12.1.2

$$A \cup B = \mathbb{R}^{N+1}. \tag{12.4.19}$$

Relation (12.4.19) implies that $A' = \mathbb{R}^{N+1} \setminus A \subset B$ and $B' = \mathbb{R}^{N+1} \setminus B \subset A$, whence $A' \cap B' = \varnothing$. Hence

$$A \setminus F = A \setminus B = A \cap B' = B',$$

i.e., $A \setminus F$ is open, and $A \setminus F \subset \mathrm{int}\,A$. On the other hand, if $p \in F \cap \mathrm{int}\,A$, then there exists a neighbourhood U of p such that $U \subset A$. But since $p \in B = \mathrm{cl}\,B_0$, we have $U \cap B_0 \neq \varnothing$, i.e., $A \cap B_0 \neq \varnothing$, which is incompatible with (12.4.15). Thus $F \cap \mathrm{int}\,A = \varnothing$ and $\mathrm{int}\,A \subset A \setminus F$. Hence $\mathrm{int}\,A = A \setminus F$. A similar argument shows also that $\mathrm{int}\,B = B \setminus F$. So

$$\mathrm{int}\,A = A \setminus F, \qquad \mathrm{int}\,B = B \setminus F. \tag{12.4.20}$$

Let $l \subset \mathbb{R}^{N+1}$ be any vertical line such that $l \cap F = \varnothing$. Then, by (12.4.19) and (12.4.20),

$$l = (l \cap \mathrm{int}\,A) \cup (l \cap \mathrm{int}\,B).$$

Sets (12.4.20) are clearly disjoint, and hence also the sets $l \cap \mathrm{int}\,A$ and $l \cap \mathrm{int}\,B$ are disjoint, and are open in l. Since l is connected, we must have either $l \cap \mathrm{int}\,A = \varnothing$, or $l \cap \mathrm{int}\,B = \varnothing$.

Now let $l \subset \mathbb{R}^{N+1}$ be an arbitrary vertical line. We will show that one of the sets $(l \setminus F) \cap \mathrm{int}\,A$ and $(l \setminus F) \cap \mathrm{int}\,B$ is empty. Supposing the contrary, let $p \in (l \setminus F) \cap \mathrm{int}\,A \subset l \cap \mathrm{int}\,A$, and $q \in (l \setminus F) \cap \mathrm{int}\,B \subset l \cap \mathrm{int}\,B$. There exist neighbourhoods U and V of p and q, respectively, such that $U \subset \mathrm{int}\,A$ and $V \subset \mathrm{int}\,B$. By (12.4.18)

$$\left[\pi(U) \cap \pi(V)\right] \setminus \pi(F) \neq \varnothing.$$

(Observe that p and q lie on the same vertical line l, whence $\pi(p) = \pi(q)$ and $\pi(U) \cap \pi(V) \neq \varnothing$. Since the function π is open, the sets $\pi(U)$ and $\pi(V)$ are open, and hence $\pi(U) \cap \pi(V)$ is a non-empty open set, and thus of the power of continuum.) Let $x_0 \in \left[\pi(U) \cap \pi(V)\right] \setminus \pi(F)$, and let $l_0 = \{x_0\} \times \mathbb{R}$ be the vertical line through x_0. Then

$$l_0 \cap U \neq \varnothing, \quad l_0 \cap V \neq \varnothing, \quad l_0 \cap F = \varnothing.$$

Hence

$$l_0 \cap \mathrm{int}\,A \neq \varnothing, \quad l_0 \cap \mathrm{int}\,B \neq \varnothing, \quad l_0 \cap F = \varnothing,$$

which, as we have already shown, is impossible.

Consequently, by (12.4.20), for every vertical line l either $l \cap \operatorname{int} A$ or $l \cap \operatorname{int} B$ is empty. This means that the sets $\pi(\operatorname{int} A)$ and $\pi(\operatorname{int} B)$ are disjoint. These sets are also open (since the function π is open), and since $\mathbb{R}^N$ is connected, there must exist an $x \in \mathbb{R}^N \setminus [\pi(\operatorname{int} A) \cup \pi(\operatorname{int} B)]$. Let $l = \{x\} \times \mathbb{R}$ be the vertical line through x. Then (cf. (12.4.20)) $l \cap [\operatorname{int} A \cup \operatorname{int} B] = l \cap [(A \cup B) \setminus F] = \varnothing$, i.e., $l \subset F$. We have $(x, f(x)) \in l \cap \operatorname{Gr}(f) \subset F \cap \operatorname{Gr}(f)$, which contradicts (12.4.17).

The contradiction obtained shows that $\operatorname{Gr}(f)$ is connected.   □

The connectedness of the graph of an additive function was also studied by F. B. Jones [157], F. Obreanu [245], S. Marcus [215] and W. Kulpa [195].

Since every function $f : \mathbb{R} \to \mathbb{R}$ without the Darboux property must have a disconnected graph, we get from Theorem 12.4.5

Corollary 12.4.1. *If $f : \mathbb{R} \to \mathbb{R}$ is an additive function with big graph, then f has the Darboux property.*

Thus an additive function $f : \mathbb{R} \to \mathbb{R}$ with big graph yields also an example of a discontinuous function with the Darboux property. The Darboux property of additive functions was also considered by J. Smítal [286], S. Marcus [215], A. M. Bruckner–J. G. Ceder–Max Weiss [38]. Cf. also Exercise 5.7.

12.5   Invertible additive functions

As is generally known, and as we have repeatedly seen before, discontinuous additive functions have many pathological properties. Therefore it is often believed that such functions cannot have any nice property. In particular, one could believe that a discontinuous additive function cannot be invertible. But this is not the case (Marczewski [220], Smítal [286], Makai [214]), as we are presently going to show.

We start with a rather well-known lemma, which is a particular case of a much more general fact, and which we prove here for the sake of completeness.

Lemma 12.5.1. *Let $f : \mathbb{R}^N \to \mathbb{R}$ be an additive function. Then f is invertible if and only if*

$$\operatorname{Ker} f = \{0\}. \tag{12.5.1}$$

Proof. If f is invertible, then there may exist only one $x \in \mathbb{R}^N$ such that $f(x) = 0$, which is of course $x = 0$. Thus (12.5.1) holds. Conversely, assume (12.5.1), and let $x_1, x_2 \in \mathbb{R}^N$ be such that $f(x_1) = f(x_2)$. Then

$$f(x_1 - x_2) = f(x_1) - f(x_2) = 0,$$

i.e., $x_1 - x_2 \in \operatorname{Ker} f$, and by (12.5.1), $x_1 - x_2 = 0$. Thus $x_1 = x_2$. This shows that f is one-to-one, and hence invertible.   □

Theorem 12.5.1. *There exist discontinuous, invertible additive functions $f : \mathbb{R}^N \to \mathbb{R}$.*

Proof. We shall distinguish two cases: $N = 1$ or $N > 1$.

At first consider the case $N = 1$. Let H be a Hamel basis of the space $(\mathbb{R}; \mathbb{Q}; +; \cdot)$. Let h_1 and h_2 be two fixed but different points of H. Define the function $g : H \to H$ putting

$$g(h_1) = h_2, \ g(h_2) = h_1, \ g(h) = h \text{ whenever } h \neq h_1 \text{ and } h \neq h_2.$$

Let $f : \mathbb{R} \to \mathbb{R}$ be the additive extension of g (cf. Theorem 5.2.2). It follows from Corollary 5.2.1 that f is discontinuous, since otherwise we would have $h_2/h_1 = h_1/h_2 = 1$.

Observe that f is invertible. In fact, since g is one-to-one and takes the values in H, we have (12.5.1). By Lemma 12.5.1 the function f is invertible.

Now consider the case $N > 1$. Let $H \subset \mathbb{R}^N$ and $B \subset \mathbb{R}$ be Hamel bases of $\mathbb{R}^N$ and $\mathbb{R}$, respectively. Since $\operatorname{card} H = \operatorname{card} B$ (cf. Theorem 4.2.3), there exists a bijection $g : H \to B$. Let $f : \mathbb{R}^N \to \mathbb{R}$ be the additive extension of g (cf. Theorem 5.2.2). Of course, f is invertible, because g is one-to-one and takes the values in B. Moreover, f is discontinuous, since otherwise we would have (cf. Theorem 5.2.2)

$$f(x) = cx$$

for every $x \in \mathbb{R}^N$ with a constant $c \in \mathbb{R}^N$, but such a function is not invertible. $\square$

Note that if $N > 1$, then there do not exist invertible continuous additive functions $f : \mathbb{R}^N \to \mathbb{R}$. (So the existence of discontinuous invertible additive functions can again be considered as a pathological property.) But if $N = 1$, functions $f(x) = cx$ with $c \neq 0$ are all continuous and invertible.

Actually, in the case $N = 1$ we can prove more (Makai [214]). Let X be an arbitrary set. A function $f : X \to X$ fulfilling for every $x \in X$ the relation

$$f[f(x)] = x \qquad\qquad (12.5.2)$$

is called *involutory* or an *involution* (cf. Kuczma [177]).

Lemma 12.5.2. *Let X be an arbitrary set, and let $f : X \to X$ be an involution. Then f is invertible and onto: $f(X) = X$.*

Proof. If $f(x_1) = f(x_2)$ for some $x_1, x_2 \in X$, then by (12.5.2) $x_1 = f[f(x_1)] = f[f(x_2)] = x_2$. Thus f is one-to-one, and hence invertible.

Further, we have

$$X = f[f(X)] \subset f(X) \subset X,$$

whence $f(X) = X$. $\square$

Condition (12.5.2) can be expressed as $f^{-1} = f$.

Theorem 12.5.2. *Let $B \subset \mathbb{R}$ be a non-empty finite or countable set. Then there exists a discontinuous additive function $f : \mathbb{R} \to \mathbb{R}$ fulfilling (12.5.2) for all $x \in \mathbb{R}$ and such that*

$$f(x) = x \quad \text{ for } x \in B.$$

Proof. Let $H \subset \mathbb{R}$ be a Hamel basis. Every $x \in B$ can be written in the form

$$x = \lambda_1^{(x)} h_1^{(x)} + \cdots + \lambda_{n_x}^{(x)} h_{n_x}^{(x)}, \tag{12.5.3}$$

where $\lambda_1^{(x)}, \ldots, \lambda_{n_x}^{(x)} \in \mathbb{Q}$, $h_1^{(x)}, \ldots, h_{n_x}^{(x)} \in H$. Put

$$H_0 = \bigcup_{x \in B} \bigcup_{i=1}^{n_x} \{h_i^{(x)}\}.$$

Since B is finite or countable, so is also H_0. Hence $H \setminus H_0$ is infinite (of the power of continuum). There exists a decomposition

$$H \setminus H_0 = H_1 \cup H_2,$$

where $H_1 \cap H_2 = \varnothing$ and $\operatorname{card} H_1 = \operatorname{card} H_2 \ (= \mathfrak{c})$. Hence there exists a one-to-one mapping $g_0 : H_1 \to H_2$ (onto). Thus the function g_0^{-1} is defined on H_2 and maps H_2 onto H_1.

Now we define the function $g : H \to H$ as follows:

$$g(h) = \begin{cases} g_0(h) & \text{for } h \in H_1, \\ g_0^{-1}(h) & \text{for } h \in H_2, \\ h & \text{for } h \in H_0. \end{cases} \tag{12.5.4}$$

Let $f : \mathbb{R} \to \mathbb{R}$ be the additive extension of g. For $h \in H_0$ we have $g(h)/h = 1$, and for $h \in H_1$ we have $g(h) = g_0(h) \in H_2$ so that $g(h) \neq h$ and $g(h)/h \neq 1$. By Corollary 5.2.1 f is discontinuous.

For arbitrary $h \in H_1$ we have $g(h) = g_0(h) \in H_2$, whence $g\big[g(h)\big] = g_0^{-1}\big[g_0(h)\big] = h$. Similarly, for $h \in H_2$ we have $g(h) = g_0^{-1}(h) \in H_1$ and $g\big[g(h)\big] = g_0\big[g_0^{-1}(h)\big] = h$. Obviously we have $g\big[g(h)\big] = h$ for $h \in H_0$. Thus $g\big[g(h)\big] = h$ for all $h \in H$.

Now take an arbitrary $x \in \mathbb{R}$. Then

$$x = \lambda_1 h_1 + \cdots + \lambda_n h_n, \quad \lambda_i \in \mathbb{Q}, \quad h_i \in H, \quad i = 1, \ldots, n,$$

whence

$$f(x) = \lambda_1 f(h_1) + \cdots + \lambda_n f(h_n) = \lambda_1 g(h_1) + \cdots + \lambda_n g(h_n),$$

and, since g maps H onto H,

$$\begin{aligned} f\big[f(x)\big] &= \lambda_1 f\big[g(h_1)\big] + \cdots + \lambda_n f\big[g(h_n)\big] \\ &= \lambda_1 g\big[g(h_1)\big] + \cdots + \lambda_n g\big[g(h_n)\big] \\ &= \lambda_1 h_1 + \cdots + \lambda_n h_n = x. \end{aligned}$$

Thus f fulfils (12.5.2).

Now take an arbitrary $x \in B$. Then x has a representation (12.5.3) with $\lambda_i^{(x)} \in \mathbb{Q}$, $h_i^{(x)} \in H_0$, $i = 1, \ldots, n_x$. Hence by (12.5.4)

$$f(x) = \lambda_1^{(x)} f\big(h_1^{(x)}\big) + \cdots + \lambda_{n_x}^{(x)} f\big(h_{n_x}^{(x)}\big)$$
$$= \lambda_1^{(x)} g\big(h_1^{(x)}\big) + \cdots + \lambda_{n_x}^{(x)} g\big(h_{n_x}^{(x)}\big) = \lambda_1^{(x)} h_1^{(x)} + \cdots + \lambda_{n_x}^{(x)} h_{n_x}^{(x)} = x.$$

Thus $f(x) = x$ for $x \in B$. $\qquad\square$

It can be shown that for the function f constructed above the set $\{x \in \mathbb{R} \mid f(x) = x\}$ is always larger than B. Generally, if $f : \mathbb{R} \to \mathbb{R}$ is an additive function, then so is also the function $\varphi : \mathbb{R} \to \mathbb{R}$ given by $\varphi(x) = f(x) - x$. Consequently, in virtue of Lemma 4.3.1, the set

$$\{x \in \mathbb{R} \mid f(x) = x\} = \operatorname{Ker} \varphi$$

is a linear subspace of $\mathbb{R}$.

In the proof of Theorem 12.5.2 the function f has been constructed in such a way that

$$f(H) = H \qquad\qquad (12.5.5)$$

for a certain Hamel basis $H \subset \mathbb{R}$. But there exist involutory additive functions such that (12.5.5) does not hold for any Hamel basis $H \subset \mathbb{R}$. Such is, e.g., the function[9] $f(x) = -x$. If $H \subset \mathbb{R}$ is an arbitrary Hamel basis and $h \in H$, then $f(h) = -h \notin H$ (note that h and $-h$ are linearly dependent). Consequently for this f we even have $H \cap f(H) = \varnothing$ for every Hamel basis $H \subset \mathbb{R}$. On the other hand, if $f : \mathbb{R} \to \mathbb{R}$ is an involutory additive function, and $H \subset \mathbb{R}$ is a Hamel basis, then $f(H)$ also is a Hamel basis. In order to show this we prove two lemmas.

Lemma 12.5.3. *Let $f : \mathbb{R}^N \to \mathbb{R}$ be an additive function. Then, for every linearly independent set $A \subset \mathbb{R}^N$ the set $f(A)$ also is linearly independent if and only if f is invertible.*

Proof. Let f be invertible. Take an arbitrary finite system $\{y_1, \ldots, y_n\} \subset f(A)$, an arbitrary linear combination

$$y = \lambda_1 y_1 + \cdots + \lambda_n y_n,$$

$\lambda_1, \ldots, \lambda_n \in \mathbb{Q}$, of $y_1, \ldots, y_n$, and suppose that $y = 0$. There exist $a_1, \ldots, a_n \in A$ such that $y_i = f(a_i)$, $i = 1, \ldots, n$. Hence

$$0 = y = \lambda_1 f(a_1) + \cdots + \lambda_n f(a_n) = f(\lambda_1 a_1 + \cdots + \lambda_n a_n),$$

and by Lemma 12.5.1 $\lambda_1 a_1 + \cdots + \lambda_n a_n = 0$. Since A is linearly independent, and $\{a_1, \ldots, a_n\} \subset A$, we must have $\lambda_1 = \cdots = \lambda_n = 0$, which proves that $y_1, \ldots, y_n$ are linearly independent. Hence also $f(A)$ is linearly independent.

Now suppose that $f(A)$ is linearly independent for every linearly independent set $A \subset \mathbb{R}^N$. Then, in particular, $f(x) \neq 0$ for $x \neq 0$. It follows that $\operatorname{Ker} f = \{0\}$. By Lemma 12.5.1 f is invertible. $\qquad\square$

---

[9] This function is continuous. We have been unable to find an example of a discontinuous involutory additive function $f : \mathbb{R} \to \mathbb{R}$ such that $f(H) \setminus H \neq \varnothing$ for every Hamel basis $H \subset \mathbb{R}$.

Lemma 12.5.4. *Let $f : \mathbb{R}^N \to \mathbb{R}$ be an additive function. Then $E\big(f(A)\big) = \mathbb{R}$ for every set $A \subset \mathbb{R}^N$ such that $E(A) = \mathbb{R}^N$ if and only if*

$$f(\mathbb{R}^N) = \mathbb{R}. \tag{12.5.6}$$

Proof. Assume that we have (12.5.6). Take an arbitrary $y \in \mathbb{R}$ and an arbitrary set $A \subset \mathbb{R}^N$ such that $E(A) = \mathbb{R}^N$. By (12.5.6) there exists an $x \in \mathbb{R}^N = E(A)$ such that $y = f(x)$. Further, there exist an $n \in \mathbb{N}$, $\lambda_1, \ldots, \lambda_n \in \mathbb{Q}$, and $a_1, \ldots, a_n \in A$ such that

$$x = \lambda_1 a_1 + \cdots + \lambda_n a_n.$$

Hence
$$y = \lambda_1 f(a_1) + \cdots + \lambda_n f(a_n) \in E\big(f(A)\big).$$

Thus $E\big(f(A)\big) = \mathbb{R}$.

Now suppose that $E\big(f(A)\big) = \mathbb{R}$ for every set $A \subset \mathbb{R}^N$ such that $E(A) = \mathbb{R}^N$. Let $H \subset \mathbb{R}^N$ be a Hamel basis, and take an arbitrary $y \in \mathbb{R}$. We have $E(H) = \mathbb{R}^N$, whence $E\big(f(H)\big) = \mathbb{R}$, and $y \in E\big(f(H)\big)$. Consequently there exist an $n \in \mathbb{N}$, $\lambda_1, \ldots, \lambda_n \in \mathbb{Q}$, and $y_1, \ldots, y_n \in f(H)$ such that

$$y = \lambda_1 y_1 + \cdots + \lambda_n y_n.$$

Further, there exist $h_1, \ldots, h_n \in H$ such that $y_i = f(h_i)$, $i = 1, \ldots, n$. Hence

$$y = \lambda_1 f(h_1) + \cdots + \lambda_n f(h_n) = f(\lambda_1 h_1 + \cdots + \lambda_n h_n) \in f(\mathbb{R}^N).$$

Thus $\mathbb{R} \subset f(\mathbb{R}^N)$. The converse inclusion is trivial, whence (12.5.6) follows. $\square$

Remark 12.5.1. This proof can be simplified by using the relation $E\big(f(A)\big) = f\big(E(A)\big)$ (cf. Exercise 4.4). Assuming (12.5.6) we have for every set $A \subset \mathbb{R}^N$ such that $E(A) = \mathbb{R}^N$

$$E\big(f(A)\big) = f\big(E(A)\big) = f(\mathbb{R}^N) = \mathbb{R}.$$

In the other direction observe that $E(\mathbb{R}^N) = \mathbb{R}^N$, whence $E\big(f(\mathbb{R}^N)\big) = \mathbb{R}$ and

$$f(\mathbb{R}^N) = f\big(E(\mathbb{R}^N)\big) = E\big(f(\mathbb{R}^N)\big) = \mathbb{R}.$$

Theorem 12.5.3. *Let $f : \mathbb{R}^N \to \mathbb{R}$ be an invertible additive function fulfilling condition (12.5.6). Then, for every Hamel basis $H \subset \mathbb{R}^N$, the set $f(H) \subset \mathbb{R}$ is a Hamel basis.*

Proof. Results from Lemmas 12.5.3 and 12.5.4. $\square$

Corollary 12.5.1. *Let $f : \mathbb{R} \to \mathbb{R}$ be an additive function fulfilling condition (12.5.2) for all $x \in \mathbb{R}$. Then, for every Hamel basis $H \subset \mathbb{R}$, the set $f(H) \subset \mathbb{R}$ is a Hamel basis.*

Proof. Results from Lemma 12.5.2 and Theorem 12.5.3. $\square$

12.6 Level sets

In Theorem 12.2.1 the restriction on the interval J to be non-degenerated is essential. If J is degenerated (i.e., reduces to a point), the set $f^{-1}(J)$ need not be large. E.g., if $f : \mathbb{R}^N \to \mathbb{R}$ is an invertible additive function, then for every $t \in \mathbb{R}$ the set $f^{-1}(t)$ contains at most one point. Of course, if $t \notin f(\mathbb{R}^N)$, then $f^{-1}(t) = \varnothing$. For the sake of convenience in the sequel we exclude such cases. Thus, if $f : \mathbb{R}^N \to \mathbb{R}$ is an arbitrary function, then under a *level set* of f we understand any set of the form $L = f^{-1}(t)$, where $t \in f(\mathbb{R}^N)$. If $f = 0$, then the only level set of f is $\operatorname{Ker} f = f^{-1}(0) = \mathbb{R}^N$. We will pay no attention to this trivial case, and so we will usually assume that the function in question is not identically zero. In the present section we are going to study the level sets of additive functions. We start with the following theorem of S. Ruziewicz [271].

Theorem 12.6.1. *Let $f : \mathbb{R}^N \to \mathbb{R}$ be an arbitrary additive function. Then any two level sets of f are congruent under translation.*

Proof. Let $L = f^{-1}(t)$ and $M = f^{-1}(s)$, $t, s \in f(\mathbb{R}^N)$, be arbitrary two level sets of f. Take arbitrary $u \in L$ and $v \in M$. Then $f(u) = t$, $f(v) = s$, and

$$f(u - v) = f(u) - f(v) = t - s.$$

Now let $x \in M$ be arbitrary. Then

$$f\big(x + (u - v)\big) = f(x) + f(u - v) = s + (t - s) = t,$$

whence $x + (u - v) \in L$. Thus $M + (u - v) \subset L$. Now take an arbitrary $x \in L$, and put $y = x - (u - v)$. Then

$$f(y) = f(x) - f(u - v) = t - (t - s) = s,$$

i.e., $y \in M$. On the other hand, $x = y + (u - v) \in M + (u - v)$, whence $L \subset M + (u - v)$. Consequently

$$L = M + (u - v),$$

which shows that L is a translation of M. $\qquad\square$

Theorem 12.6.1 shows that all level sets of an additive function are congruent to each other, and thus have analogical properties. Thus in the sequel, in order to establish a property of all level sets of an additive function f, we will establish this property for a particular level set, e.g., for $L = \operatorname{Ker} f = f^{-1}(0)$. (Note that, since $f(0) = 0$, always $0 \in f(\mathbb{R}^N)$ so that $f^{-1}(0)$ is a level set of f.)

The next two theorems are due to J. Smítal [286].

Theorem 12.6.2. *Let $f : \mathbb{R}^N \to \mathbb{R}$ be an arbitrary additive function. Then, for every level set L of f, we have either $\operatorname{card} L = 1$, or $\operatorname{card} L \geqslant \aleph_0$.*

Proof. Let $L = \operatorname{Ker} f$, and suppose that $\operatorname{card} L > 1$. Then there exists an $x \in L$ such that $x \neq 0$. By Lemma 4.3.1 L is a linear subspace of $\mathbb{R}^N$, whence $\lambda x \in L$ for every $\lambda \in \mathbb{Q}$. In other words $\mathbb{Q}x \subset L$. But, since $x \neq 0$, $\operatorname{card} \mathbb{Q}x = \aleph_0$. Hence $\operatorname{card} L \geqslant \aleph_0$. $\qquad\square$

Of course, if $f : \mathbb{R}^N \to \mathbb{R}$ is an invertible additive function (cf. Theorem 12.5.1), then we have card $L = 1$ for every level set L of f. On the other hand, we have the following complement to Theorem 12.6.2. Note that, for every level set L of an additive function $f : \mathbb{R}^N \to \mathbb{R}$, we have $L \subset \mathbb{R}^N$, whence card $L \leqslant \mathfrak{c}$.

Theorem 12.6.3. *For every cardinal number* $\mathfrak{m}$ *such that* $\aleph_0 \leqslant \mathfrak{m} \leqslant \mathfrak{c}$ *there exists an additive function* $f : \mathbb{R}^N \to \mathbb{R}$ *such that* $f(\mathbb{R}^N) = \mathbb{R}$ *and* card $L = \mathfrak{m}$ *for every level set* L *of* f.

Proof. Let $H \subset \mathbb{R}^N$ be a Hamel basis. According to Theorem 4.2.3 card $H = \mathfrak{c}$, so there exists a set $H_1 \subset H$ such that card $H_1 = \mathfrak{m}$, card$(H \setminus H_1) = \mathfrak{c}$. By Lemmas 4.1.3 and 4.2.2 card $E(H_1) = \mathfrak{m}$. Now, H_1 as a subset of H, is linearly independent, so H_1 is a base of $E(H_1)$. We have card$(H \setminus H_1) = \mathfrak{c} = \dim \mathbb{R}$. By Theorem 4.3.4 there exists an additive function $f : \mathbb{R}^N \to \mathbb{R}$ such that $f(\mathbb{R}^N) = \mathbb{R}$ and Ker $f = E(H_1)$. This function has the desired properties. $\qquad\square$

Now we turn to measure-theoretical and topological properties of the level sets of additive functions (Smítal [286], Erdős-Marcus [77], Halperin [132], Marcus [217]).

Theorem 12.6.4. *Let* $f : \mathbb{R}^N \to \mathbb{R}$ *be an additive function,* $f \neq 0$. *Then, for every level set* L *of* f, *we have* $m_i(L) = 0$ *and* L *does not contain any second category set with the Baire property.*

Proof. Let $L = \text{Ker } f$, and suppose that $m_i(L) > 0$ or L contains a second category set with the Baire property. By Theorems 9.3.1, 9.3.2 and Corollary 9.2.1 we have $L \in \mathfrak{A}$. By Corollary 9.3.2 there exists a Hamel basis $H \subset L$. By Lemma 4.3.1 we have $E(L) = L$. Thus

$$\mathbb{R}^N = E(H) \subset E(L) = L \subset \mathbb{R}^N$$

i.e., Ker $f = L = \mathbb{R}^N$, which means that $f = 0$, contrary to the assumption. $\qquad\square$

The following theorem yields more information.

Theorem 12.6.5. *Let* $f : \mathbb{R}^N \to \mathbb{R}$ *be an additive function,* $f \neq 0$. *Then, for every level set* L *of* f, *we have* $m(L) = 0$, *or* L *is saturated non-measurable. Similarly, either* L *is of the first category, or has property* $(*)$ *from* 3.3.

Proof. Let $L = \text{Ker } f$. By Lemma 4.3.1

$$L + L = L. \tag{12.6.1}$$

If $m_e(L) > 0$, then by Lemma 4.2.4 L is dense in $\mathbb{R}^N$, whence by Theorem 3.6.1 $m_i(\mathbb{R}^N \setminus L) = 0$. Since by Theorem 12.6.4 also $m_i(L) = 0$, L is saturated non-measurable.

Similarly, if L is of the second category, then L is dense in $\mathbb{R}^N$ according to Lemma 4.2.4. It follows from (12.6.1) and from Theorem 3.6.2 that the set $\mathbb{R}^N \setminus L$ does not contain any second category set with the Baire property. Theorem 12.6.4 asserts that L has the same property. This means that L has property $(*)$. $\qquad\square$

Again, if f is an invertible additive function, then the level sets of f consist of a single point each, and consequently are of measure zero and of the first category. The following two theorems assert that the other possibility in Theorem 12.6.5 also can happen.

Theorem 12.6.6. *Let $f : \mathbb{R}^N \to \mathbb{R}$ be an additive function with small graph. Then every level set of f is saturated non-measurable and has property $(*)$ from 3.3.*

Proof. Two different level sets of f are disjoint, and we have

$$\mathbb{R}^N = \bigcup_{t \in f(\mathbb{R}^N)} f^{-1}(t). \tag{12.6.2}$$

If the sets $f^{-1}(t)$ were of measure zero, or of the first category, then so would be also $\mathbb{R}^N$, since the union in (12.6.2) is countable. The theorem now results from Theorem 12.6.5. $\qquad\square$

Theorem 12.6.7. *Let $f : \mathbb{R}^N \to \mathbb{R}$ be an additive function with big graph. Then every level set of f is saturated non-measurable and has property $(*)$ from 3.3.*

Proof. Let $L = \operatorname{Ker} f$, and suppose that $m(L) = 0$. Then $m(\mathbb{R}^N \setminus L) = \infty$, and there exists a closed set $F_0 \subset \mathbb{R}^N \setminus L$ such that

$$m(F_0) > 0. \tag{12.6.3}$$

Similarly, suppose that L is of the first category. Then the set $\mathbb{R}^N \setminus L$ is residual, and by Theorem 2.1.2 there exists a set $F_0 \in \mathcal{G}_\delta$, $F_0 \subset \mathbb{R}^N \setminus L$, and such that

$$F_0 \text{ is residual.} \tag{12.6.4}$$

In both cases, by (12.6.3) or (12.6.4), we have $\operatorname{card} F_0 = \mathfrak{c}$. Now let $F = F_0 \times \{0\} \subset \mathbb{R}^{N+1}$. The set F is a Borel set (since F_0 is a Borel set), and $\pi(F) = F_0$, whence $\operatorname{card} \pi(F) = \mathfrak{c}$. Thus we must have (12.4.1). This means that there exists a point $(x, y) \in F \cap \operatorname{Gr}(f)$. Hence $x \in F_0 \subset \mathbb{R}^N \setminus L$, and $f(x) = 0$ so that $x \in \operatorname{Ker} f = L$. Consequently $x \in L \cap (\mathbb{R}^N \setminus L)$, which is absurd. The theorem results now from Theorem 12.6.5. $\qquad\square$

It follows from Theorem 12.6.7 that there exists an additive function $f : \mathbb{R}^N \to \mathbb{R}$ such that $f(\mathbb{R}^N) = \mathbb{R}$ (cf. Lemma 12.4.1) so that $\operatorname{card} f(\mathbb{R}^N) = \mathfrak{c}$, and the level sets of f are saturated non-measurable and have property $(*)$. Similarly, it follows from Theorem 12.6.6 that there exists an additive function $f : \mathbb{R}^N \to \mathbb{R}$ such that $\operatorname{card} f(\mathbb{R}^N) = \aleph_0$ and the level sets of f are saturated non-measurable and have property $(*)$. The following theorem (Halperin [132], Smítal [286], Erdős-Marcus [77]) yields a joint generalization of these statements.

Theorem 12.6.8. *For every cardinal number $\mathfrak{m}$ such that $\aleph_0 \leqslant \mathfrak{m} \leqslant \mathfrak{c}$ there exists an additive function $f : \mathbb{R}^N \to \mathbb{R}$ such that $\operatorname{card} f(\mathbb{R}^N) = \mathfrak{m}$ and the level sets of f are saturated non-measurable and have property $(*)$ from 3.3.*

Proof. Let $H \subset \mathbb{R}$ be a Hamel basis of $\mathbb{R}$. By Theorem 4.2.3 $\operatorname{card} H = \mathfrak{c}$, so there exists a set $H_1 \subset H$ such that $\operatorname{card} H_1 = \mathfrak{m}$, $\operatorname{card}(H \setminus H_1) = \mathfrak{c}$. By Lemma 4.2.2 $\operatorname{card} E(H_1) = \mathfrak{m}$. $H \setminus H_1$ is linearly independent set, so it is a base of $E(H \setminus H_1)$. We have $H \setminus (H \setminus H_1) = H_1$, whence $\operatorname{card}[H \setminus (H \setminus H_1)] = \operatorname{card} H_1 = \mathfrak{m} = \dim E(H_1)$. By Theorem 4.3.4 there exists an additive function $f_1 : \mathbb{R} \to \mathbb{R}$ such that $f_1(\mathbb{R}) = E(H_1)$, $\operatorname{Ker} f_1 = E(H \setminus H_1)$.

Let $f_2 : \mathbb{R}^N \to \mathbb{R}$ be an additive function with big graph, and put $f(x) = f_1[f_2(x)]$, $x \in \mathbb{R}^N$. Then $f : \mathbb{R}^N \to \mathbb{R}$ is an additive function and (cf. Lemma 12.4.1)

$$f(\mathbb{R}^N) = f_1[f_2(\mathbb{R}^N)] = f_1(\mathbb{R}) = E(H_1).$$

Consequently $\operatorname{card} f(\mathbb{R}^N) = \mathfrak{m}$. Now, $\operatorname{Ker} f = f^{-1}(0) = f_2^{-1}(f_1^{-1}(0)) = f_2^{-1}(E(H \setminus H_1))$. Take an arbitrary $t \in E(H \setminus H_1)$. Then $f_2^{-1}(t) \subset f_2^{-1}(E(H \setminus H_1)) = \operatorname{Ker} f$. By Theorem 12.6.7 the set $f_2^{-1}(t)$ is saturated non-measurable and has property $(*)$, so, in particular, $m_e(f_2^{-1}(t)) > 0$ and $f_2^{-1}(t)$ is of the second category. Consequently also $\operatorname{Ker} f$ has the same properties. By Theorem 12.6.5 $\operatorname{Ker} f$ is saturated non-measurable and has property $(*)$. $\quad\square$

As we have pointed out before, every invertible additive function yields an example of an additive function with level sets of measure zero. The question arises whether there exist additive functions with uncountable level sets of measure zero. Following J. Smítal [286] we will answer this question in the affirmative, assuming the continuum hypothesis.

Theorem 12.6.9. *Under assumption of the continuum hypothesis there exists an additive function $f : \mathbb{R}^N \to \mathbb{R}$ such that $f(\mathbb{R}^N) = \mathbb{R}$ and the level sets of f have the power of continuum and measure zero.*

Proof. By Theorem 11.6.2 there exists a Hamel basis $H \subset \mathbb{R}^N$ which admits a decomposition $H = H_1 \cup H_2$, $H_1 \cap H_2 = \varnothing$, such that $\operatorname{card} H_1 = \operatorname{card} H_2 = \mathfrak{c}$ and $E(H_1)$ is a Lusin set. H_1 is a base of $E(H_1)$, and $\operatorname{card}(H \setminus H_1) = \operatorname{card} H_2 = \mathfrak{c} = \dim \mathbb{R}$. By Theorem 4.3.4 there exists an additive function $f : \mathbb{R}^N \to \mathbb{R}$ such that $f(\mathbb{R}^N) = \mathbb{R}$, $\operatorname{Ker} f = E(H_1)$.

Since $E(H_1)$ is a Lusin set, it is uncountable, and by the continuum hypothesis $\operatorname{card} \operatorname{Ker} f = \operatorname{card} E(H_1) = \mathfrak{c}$. By Corollary 3.4.1 $m(\operatorname{Ker} f) = m(E(H_1)) = 0$. $\quad\square$

If $N \geqslant 2$, then for every continuous additive function $f : \mathbb{R}^N \to \mathbb{R}$ the level sets have the power of continuum and measure zero. But the function f constructed in the proof of Theorem 12.6.9 is discontinuous. In fact, if f were continuous, then the set $\operatorname{Ker} f = f^{-1}(0)$ would be closed. But actually, according to Theorem 3.4.2, $\operatorname{Ker} f$, being a Lusin set, does not have the Baire property.

12.7   Partitions

Let $f : \mathbb{R}^N \to \mathbb{R}$ be an additive function. Then the level sets of f are pairwise disjoint, fulfil (12.6.2), and, according to Theorem 12.6.1, they are congruent under translation. Thus every additive function $f : \mathbb{R}^N \to \mathbb{R}$ furnishes a partition (12.6.2) of $\mathbb{R}^N$ into disjoint congruent sets, and the number of constituents of the partition is equal to card $f(\mathbb{R}^N)$. Partitions of $\mathbb{R}^N$ were studied by Burstin [39], Ruziewicz [271], Hahn-Rosenthal ([128], §8.3.3), Sierpiński [283], Halperin [132], Erdős-Marcus [77], Smítal [286]. Directly from Theorem 12.6.8 we obtain the following (Halperin [132], Smítal [286], Ruziewicz [271])

Theorem 12.7.1. *For every cardinal number* $\mathfrak{m}$ *such that* $\aleph_0 \leqslant \mathfrak{m} \leqslant \mathfrak{c}$ *there exists a partition of* $\mathbb{R}^N$ *into* $\mathfrak{m}$ *pairwise disjoint saturated non-measurable sets with property* $(*)$ *from 3.3, congruent under translation.*

In this context the following question arises: if $\mathcal{A}$ is a family of pairwise disjoint sets congruent under translation and such that

$$\bigcup \mathcal{A} = \mathbb{R}^N, \tag{12.7.1}$$

does there exist an additive function $f : \mathbb{R}^N \to \mathbb{R}$ such that the sets $A \in \mathcal{A}$ are the level sets of f? The following theorem (Kuczma [180]) answers this question.

Observe that, since the sets from $\mathcal{A}$ are pairwise disjoint and fulfil (12.7.1), there exists a unique set $A_0 \in \mathcal{A}$ such that $0 \in A_0$.

Theorem 12.7.2. *Let* $\mathcal{A}$ *be a family of pairwise disjoint sets which are congruent under translation and fulfil* (12.7.1). *Let* $A_0 \in \mathcal{A}$ *be such that* $0 \in A_0$. *Then there exists an additive function* $f : \mathbb{R}^N \to \mathbb{R}$ *such that the sets* $A \in \mathcal{A}$ *are the level sets of* f *if and only if the set* A_0 *is a linear subspace (over* $\mathbb{Q}$) *of* $\mathbb{R}^N$.

Proof. Suppose that such an f exists. We have $f(0) = 0$, consequently $A_0 = \operatorname{Ker} f$, and by Lemma 4.3.1 A_0 is a linear subspace of $\mathbb{R}^N$.

Now suppose that the sets from $\mathcal{A}$ are pairwise disjoint, congruent under translation, fulfil (12.7.1), and A_0 is a linear subspace of $\mathbb{R}^N$. Since the sets $A \in \mathcal{A}$ are congruent under translation, for every $A \in \mathcal{A}$ there exists a $d_A \in \mathbb{R}^N$ such that

$$A = A_0 + d_A. \tag{12.7.2}$$

Let $B \subset A_0$ be a base of A_0. By Corollary 4.2.3 there exists a Hamel basis H of $\mathbb{R}^N$ such that $B \subset H$. We have $\operatorname{card}(H \setminus B) \leqslant \operatorname{card} H = \mathfrak{c} = \dim \mathbb{R}$. By Theorem 4.3.3 there exists an additive function $f : \mathbb{R}^N \to \mathbb{R}$ such that $\operatorname{Ker} f = A_0$. Put $t_A = f(d_A)$ for $A \in \mathcal{A}$.

Now take an $A \in \mathcal{A}$. If $x \in A$, then by (12.7.2) there exists an $x_0 \in A_0$ such that $x = x_0 + d_A$, whence $f(x) = f(x_0) + f(d_A) = f(d_A) = t_A$. Thus $x \in f^{-1}(t_A)$ and $A \subset f^{-1}(t_A)$.

Take an $x \in f^{-1}(t_A)$. Thus $f(x) = t_A = f(d_A)$ and

$$f(x - d_A) = f(x) - f(d_A) = 0,$$

i.e., $x - d_A \in \operatorname{Ker} f = A_0$ and $x \in A_0 + d_A = A$. Consequently $f^{-1}(t_A) \subset A$, and finally we get $A = f^{-1}(t_A)$. Thus every set $A \in \mathcal{A}$ is a level set of f. $\square$

Example 12.7.1. Let $H \subset \mathbb{R}^N$ be an arbitrary Hamel basis, and let $Z(H)$ be the associated Erdős set (cf. 11.5). Define the relation $\sim$ in $\mathbb{R}^N \times \mathbb{R}^N$ as follows:

$$x \sim y \Leftrightarrow x - y \in Z(H). \tag{12.7.3}$$

Since $Z(H)$ fulfils (11.5.2), $\sim$ is an equivalence relation. Consequently $\mathbb{R}^N$ can be partitioned into disjoint classes of equivalent elements. Let $\mathcal{A}$ be the collection of these classes. (Note that we have $\mathcal{A} = \mathbb{R}^N/Z(H)$.) Then the sets from $\mathcal{A}$ are pairwise disjoint and fulfil (12.7.1). Take arbitrary $A, B \in \mathcal{A}$, $u \in A$, $v \in B$, and put $d = v - u$. Take an $x \in A$. Then $x \sim u$, i.e., by (12.7.3), $x - u \in Z(H)$. Now let $y = x + d = x + v - u = v + (x - u)$, whence $y - v = x - u \in Z(H)$, and by (12.7.3) $y \sim v$. Hence $y \in B$, i.e., $x + d \in B$. This shows that $A + d \subset B$.

Now take an arbitrary $y \in B$. Thus $y \sim v$, i.e., $y - v \in Z(H)$. Let $x = y - d$. Then $x = y - (v - u) = u + (y - v)$, and $x - u = y - v \in Z(H)$. Consequently $x \sim u$, whence $x \in A$, and $y = x + d \in A + d$. Hence $B \subset A + d$, and finally $B = A + d$.

We have shown that any two sets from $\mathcal{A}$ can be obtained from each other by translation, i.e., the sets from $\mathcal{A}$ are congruent under translation. Observe that $0 \in Z(H)$, whence $A_0 = Z(H)$. The set $Z(H)$ is not a linear subspace of $\mathbb{R}^N$, consequently the sets from $\mathcal{A}$ *are not* the level sets of an additive function.

Now we are going to deal with the one-dimensional case. Let $I \subset \mathbb{R}$ be an interval. Then every partition (12.7.1) of $\mathbb{R}$ ($N = 1$) generates a partition of I:

$$I = \bigcup_{A \in \mathcal{A}} (A \cap I).$$

But now the sets $A \cap I$ need not be congruent under translation. We will say that two sets $A, B \subset I$ are *congruent by decomposition* iff there exist non-negative real numbers t, s such that $t + s = |I|$ and

$$[(A + t) \cap I] \cup [(A - s) \cap I] = B,$$

where $|I|$ denotes the length of I.

Lemma 12.7.1. *Let $f : \mathbb{R}^N \to \mathbb{R}$ be an additive function. If $x_0 \in \operatorname{Ker} f$ and $x_0 \neq 0$, then x_0 is a period of f.*

Proof. We have, for arbitrary $x \in \mathbb{R}^N$,

$$f(x + x_0) = f(x) + f(x_0) = f(x),$$

i.e., x_0 is a period of f. $\qquad\square$

Lemma 12.7.2. *Let $f : \mathbb{R} \to \mathbb{R}$ be a non-invertible additive function. Then f is microperiodic[10].*

---

[10] I.e., f has arbitrarily small positive periods.

Proof. Since f is non-invertible, the set $\operatorname{Ker} f$ is dense in $\mathbb{R}$ according to Lemma 12.2.1. Consequently there exists an arbitrarily small positive $x_0 \in \operatorname{Ker} f$. By Lemma 12.7.1 every such x_0 is a period of f. $\qquad \square$

Now we will prove a theorem (cf. Lusin-Sierpiński [212], Halperin [132]) on a partition of the unit interval.

Theorem 12.7.3. *Let $I = [0,1]$ be the unit interval. For every cardinal number $\mathfrak{m}$ such that $\aleph_0 \leqslant \mathfrak{m} \leqslant \mathfrak{c}$ there exists a partition*

$$I = \bigcup \mathcal{A} \qquad (12.7.4)$$

of I into $\mathfrak{m}$ sets which are congruent by decomposition and such that $m_i(A) = 0$, $m_e(A) = 1$ for every $A \in \mathcal{A}$. Also, if $A \in \mathcal{A}$, then neither A, nor $I \setminus A$, contains a second category subset with the Baire property.

Proof. There exists an additive function $\tilde{f} : \mathbb{R} \to \mathbb{R}$ such that $\operatorname{card} \tilde{f}(\mathbb{R}) = \mathfrak{m}$ and the level sets of $\tilde{f}$ are saturated non-measurable and have property $(*)$ from 3.3 (Theorem 12.6.8). Consequently there exists a $z \in \operatorname{Ker} \tilde{f}$ such that $z \neq 0$. Define the function $f : \mathbb{R} \to \mathbb{R}$ by $f(x) = \tilde{f}(zx)$. Then f is an additive function, $f(\mathbb{R}) = \tilde{f}(z\mathbb{R}) = \tilde{f}(\mathbb{R})$ so that $\operatorname{card} f(\mathbb{R}) = \mathfrak{m}$, and $\operatorname{Ker} f = z^{-1} \operatorname{Ker} \tilde{f}$ is saturated non-measurable and has property $(*)$ so that all level sets of f are saturated non-measurable and have property $(*)$. Moreover, $f(1) = 0$, whence, by Lemma 12.7.1, 1 is a period of f:

$$f(x+1) = f(x) \quad \text{for } x \in \mathbb{R}. \qquad (12.7.5)$$

For every $t \in f(\mathbb{R})$ put $A_t = f^{-1}(t)$, and let $\mathcal{A} = \{I \cap A_t\}_{t \in f(\mathbb{R})}$. Thus $\operatorname{card} \mathcal{A} = \mathfrak{m}$, and since $\mathbb{R} = \bigcup_{t \in f(\mathbb{R})} A_t$, we have

$$I = \bigcup_{t \in f(\mathbb{R})} (I \cap A_t). \qquad (12.7.6)$$

Formula (12.7.6) yields a partition of I into $\mathfrak{m}$ pairwise disjoint sets. Since the sets A_t are saturated non-measurable, we have by Theorem 3.3.1 $m_e(I \cap A_t) = |I| = 1$ for every $t \in f(\mathbb{R})$, whereas evidently $m_i(I \cap A_t) \leqslant m_i(A_t) = 0$, whence $m_i(I \cap A_t) = 0$ for $t \in f(\mathbb{R})$. Also, it follows from $(*)$ that neither $I \cap A_t \subset A_t$, nor $I \setminus A_t \subset \mathbb{R} \setminus A_t$ can contain a second category subset with the Baire property, $t \in f(\mathbb{R})$. It remains to show that the sets from $\mathcal{A}$ are congruent by decomposition.

Take an arbitrary $t \in f(\mathbb{R})$ and $x \in A_t = f^{-1}(t)$. By (12.7.5) $x + 1 \in A_t$, whence $A_t + 1 \subset A_t$. Similarly, if $x \in A_t + 1$, then $x = y + 1$, where $y \in A_t$. Thus $f(x) = f(y+1) = f(y) = t$, i.e., $x \in f^{-1}(t) = A_t$ and $A_t + 1 \subset A_t$. Hence

$$A_t + 1 = A_t \quad \text{for } t \in f(\mathbb{R}). \qquad (12.7.7)$$

Now take arbitrary $t, s \in f(\mathbb{R})$, $t \neq s$. The sets A_t and A_s are congruent under translation (Theorem 12.6.1), i.e., there exists a $d \in \mathbb{R}$ such that

$$A_t + d = A_s. \qquad (12.7.8)$$

Write $d = [d] + r$, where $[d]$ is the integral part of d, and $0 \leqslant r < 1$. By (12.7.8) and (12.7.7)

$$A_t + r = A_s, \tag{12.7.9}$$

and since $t \neq s$, whence $A_t \neq A_s$, we must have actually $0 < r < 1$. Subtracting 1 from both sides of (12.7.9) we obtain by (12.7.7)

$$A_t - (1 - r) = A_s. \tag{12.7.10}$$

By (12.7.9)

$$(A_t \cap I) + r = (A_t + r) \cap (I + r) = (A_t + r) \cap [r, r + 1] = A_s \cap [r, r + 1],$$

and

$$[(A_t \cap I) + r] \cap I = A_s \cap [r, \, r + 1] \cap I = A_s \cap [r, 1]. \tag{12.7.11}$$

Similarly, by (12.7.10)

$$(A_t \cap I) - (1 - r) = (A_t - (1 - r)) \cap (I - (1 - r)) = (A_t - (1 - r)) \cap [r - 1, r] = A_s \cap [r - 1, r],$$

whence

$$[(A_t \cap I) - (1 - r)] \cap I = A_s \cap [r - 1, r] \cap I = A_s \cap [0, r]. \tag{12.7.12}$$

From (12.7.11) and (12.7.12) we obtain

$$([[(A_t \cap I) + r] \cap I) \cup ([(A_t \cap I) - (1 - r)] \cap I) = A_s \cap I.$$

Consequently the sets $A_t \cap I$ and $A_s \cap I$ are congruent by decomposition.   □

It is clear that in the above theorem the closed interval $[0, 1]$ may be replaced by the open interval $(0, 1)$, or by the half-open interval $(0, 1]$ or $[0, 1)$, or by any other interval $I \subset \mathbb{R}$ (with the obvious modification in the formulation of the theorem).

As pointed out in 10.5 the sets from class $\mathfrak{A}$ may be considered as "large". Therefore also the following theorem (Kuczma [184]) may be of interest.

Theorem 12.7.4. *There exists a partition of* $\mathbb{R}^N$ *into* $\mathfrak{c}$ *pairwise disjoint, congruent (under translation) sets belonging to the class* $\mathfrak{A}$.

Proof. Consider the partition described in Example 12.7.1. By Lemma 11.5.4 (cf. formula (11.5.7)) $Z(H) \in \mathfrak{A}$, whence, by Theorem 9.2.8, every translation of $Z(H)$ belongs to $\mathfrak{A}$. It remains to show that $\operatorname{card} \mathcal{A} = \mathfrak{c}$.

Let D be the set of those $x \in \mathbb{R}^N$ which have expansions

$$x = \alpha_1 h_1 + \cdots + \alpha_n h_n, \quad n \in \mathbb{N}, \, h_i \in H, \, i = 1, \ldots, n, \tag{12.7.13}$$

with $\alpha_i \in \mathbb{Q} \cap (0, 1)$. Thus $\frac{1}{2} H \subset D$, whence (cf. Theorem 4.2.3) $\mathfrak{c} = \operatorname{card} H = \operatorname{card}\left(\frac{1}{2} H\right) \leqslant \operatorname{card} D \leqslant \mathfrak{c}$, and $\operatorname{card} D = \mathfrak{c}$. If y, $z \in Z(H) + d$ for a $d \in D$, then

$y = u + d$, $z = v + d$, where u, $v \in Z(H)$, whence $y - z = u - v \in Z(H)$. Thus $y \sim z$. Conversely, if $y \in Z(H) + d$ and $z \sim y$, then $z - y \in Z(H)$, $y = u + d$, $u \in Z(H)$. Hence $z = (z - y) + y = (z - y) + (u + d) = [(z - y) + u] + d$. By (11.5.2) we have $(z - y) + u \in Z(H) + Z(H) = Z(H)$, and thus $z \in Z(H) + d$. Thus $Z(H) + d$ is a class of equivalent elements of $\mathbb{R}^N$, whence $Z(H) + d \in \mathcal{A}$ for every $d \in D$. We must show that for $d_1 \neq d_2$, d_1, $d_2 \in D$, also $Z(H) + d_1 \neq Z(H) + d_2$. This will imply that $\operatorname{card} \mathcal{A} \geqslant \operatorname{card} D = \mathfrak{c}$, and since evidently $\operatorname{card} \mathcal{A} \leqslant \mathfrak{c}$, we will get hence $\operatorname{card} \mathcal{A} = \mathfrak{c}$.

First observe that $-Z(H) = Z(H)$, whence by (11.5.2) $Z(H) - Z(H) = Z(H)$. Now suppose that for certain $d_1, d_2 \in D$, $d_1 \neq d_2$, we have $Z(H) + d_1 = Z(H) + d_2$. Then

$$d_1 - d_2 \in Z(H) - Z(H) = Z(H). \qquad (12.7.14)$$

The point $x = d_1 - d_2$ has an expansion (12.7.13), where $\alpha_i \in \mathbb{Q} \cap (-1, 1)$, and, since $d_1 \neq d_2$, not all α_i are zero. Thus $x \notin Z(H)$. This contradicts (12.7.14). $\qquad \square$

12.8 Monotonicity

Now we are going to deal with quite a different problem. If $D \subset \mathbb{R}$ is an open interval and $f : D \to \mathbb{R}$ is a convex function, and if f is monotonic in D, then f is bounded on every closed interval $I \subset D$, and hence it is continuous (cf. also Corollary 12.8.1 below). This fact can be extended to higher dimensions (Bereanu [28]). Here we are going to present one of such extensions (Kuczma [180]).

Let $I \subset \mathbb{R}$ be an interval, and let $\varphi : I \to \mathbb{R}$ be a function. Let ω be a positive real number. The function φ is called ω-*increasing* (Anastassiadis [14]) iff

$$\varphi(x + \omega) \geqslant \varphi(x)$$

holds for all $x \in I$ such that $x + \omega \in I$. Similarly, φ is called ω-*decreasing* iff

$$\varphi(x + \omega) \leqslant \varphi(x)$$

holds for all $x \in I$ such that $x + \omega \in I$. Let $\Omega \subset (0, \infty)$ be a set of positive numbers. The function φ is called Ω-*increasing* [Ω-*decreasing*] iff it is ω-increasing [ω-decreasing] for every $\omega \in \Omega$. A function which is either Ω-increasing, or Ω-decreasing, is called Ω-*monotonic*.

There exist sets $\Omega \subset (0, \infty)$ and functions $\varphi : I \to \mathbb{R}$ which are Ω-monotonic, but not monotonic in the customary sense. To see this consider Example 12.7.1 ($N = 1$). Since, as was shown in the proof of Theorem 12.7.4, $\operatorname{card} \mathcal{A} = \mathfrak{c}$, there exists a one-to-one mapping $\psi : \mathcal{A} \to \mathbb{R}$ (onto). Define the function $\varphi : \mathbb{R} \to \mathbb{R}$ putting

$$\varphi(x) = \psi(A) \quad \text{if } x \in A \in \mathcal{A}.$$

Let

$$\Omega = Z(H) \cap (0, \infty). \qquad (12.8.1)$$

For every $\omega \in \Omega \subset Z(H)$ and every $x \in \mathbb{R}$ we have: if $x \in A \in \mathcal{A}$, then $x + \omega \in A$, whence $\varphi(x + \omega) = \varphi(x)$. Thus φ is ω-monotonic, and hence Ω-monotonic. But φ is not monotonic, since for every $t \in \varphi(\mathbb{R})$ we have $\varphi^{-1}(t) = A$ for a certain $A \in \mathcal{A}$, and these sets, being translations of $Z(H)$, are non-measurable (cf. Theorem 11.5.1). Consequently also the function φ is non-measurable.

Lemma 12.8.1. *Let $I \subset \mathbb{R}$ be an interval, let Ω_1, $\Omega_2 \subset (0, \infty)$ be arbitrary sets, and let $\varphi : I \to \mathbb{R}$ be a function which is Ω_1-increasing and Ω_2-increasing [Ω_1-decreasing and Ω_2-decreasing]. Then φ is $(\Omega_1 + \Omega_2)$-increasing [$(\Omega_1 + \Omega_2)$-decreasing].*

Proof. Take an $x \in I$ and an $\omega \in \Omega_1 + \Omega_2$ such that $x + \omega \in I$. Then $\omega = \omega_1 + \omega_2$, $\omega_1 \in \Omega_1$, $\omega_2 \in \Omega_2$, and $x + \omega_1 \in I$. Hence, if φ is Ω_1-increasing and Ω_2-increasing,

$$\varphi(x + \omega) = \varphi\big((x + \omega_1) + \omega_2\big) \geqslant \varphi(x + \omega_1) \geqslant \varphi(x),$$

i.e., φ is ω-increasing. If φ is Ω_1-decreasing and Ω_2-decreasing, the inequalities are reversed. $\square$

Lemma 12.8.2. *Let $I \subset \mathbb{R}$ be an interval, let $\Omega \subset (0, \infty)$ be a set, and let $\varphi : I \to \mathbb{R}$ be an Ω-monotonic function. For every $\delta \in (0, \infty)$ put $\Omega_\delta = \Omega \cap (0, \delta)$. If*

$$\operatorname{int}(\Omega_\delta + \Omega_\delta) \neq \varnothing \quad \text{for } \delta \in (0, \delta_0) \tag{12.8.2}$$

and for a $\delta_0 \in (0, \infty)$, then φ is monotonic.

Proof. Suppose that φ is Ω-increasing, and take arbitrary $x, y \in I$, $x < y$. Choose a $\delta < \min\left(\delta_0, \frac{1}{2}(y - x)\right)$. By (12.8.2) there exists an interval $(\alpha, \beta) \subset (\Omega + \Omega) \cap (0, 2\delta)$. By Lemma 12.8.1 φ is ω-increasing for every $\omega \in (\alpha, \beta)$. We have $\beta < 2\delta < y - x$. Note that by (12.8.2) $\Omega_{\delta'} \neq \varnothing$ for every $\delta' > 0$. Hence there exists an $\omega \in \Omega$ such that $0 < \omega < \beta - \alpha < y - x - \alpha$. Consider the interval $J = (y - x - \beta, y - x - \alpha) \subset (0, \infty)$. Since the length of J is $\beta - \alpha$, there exists a $k \in \mathbb{N}$ such that $k\omega \in J$, i.e., $y - x - k\omega \in (\alpha, \beta)$. Hence

$$\varphi(y) = \varphi\big((y - k\omega) + k\omega\big) \geqslant \varphi(y - k\omega) = \varphi\big(x + (y - x - k\omega)\big) \geqslant \varphi(x).$$

(Note that $y - k\omega \in (x + \alpha, x + \beta)$ and $k\omega > 0$, whence $x < y - k\omega < y$ and $y - k\omega \in I$.) Consequently φ is increasing.

If φ is Ω-decreasing, the proof is the same, except that the inequalities must be reversed. $\square$

Condition (12.8.2) is certainly fulfilled if $m_i(\Omega_\delta) > 0$ for $\delta \in (0, \delta_0)$ (Theorem 3.7.1) or if, for $\delta \in (0, \delta_0)$, Ω_δ contains a second category set with the Baire property (Theorem 2.9.1). But, e.g., for the set Ω given by (12.8.1) condition (12.8.2) is not fulfilled. In fact, $\Omega_\delta \subset \Omega \subset Z(H)$, whence by (11.5.2) $\Omega_\delta + \Omega_\delta \subset Z(H) + Z(H) = Z(H)$, and $\operatorname{int}(\Omega_\delta + \Omega_\delta) \subset \operatorname{int} Z(H) = \varnothing$. But we will prove the following

Theorem 12.8.1. *Let $D \subset \mathbb{R}^N$ be a convex and open set, and let $f : D \to \mathbb{R}$ be a convex function. Let $x_0 \in D$, and let $a_1, \ldots, a_N \in \mathbb{R}^N$ be linearly independent over $\mathbb{R}$. For every $i = 1, \ldots, N$ put $d_i = \sup\{t > 0 \mid x_0 + ta_i \in D\}$, and let the functions $\varphi_i : [0, d_i) \to \mathbb{R}$ be given by*

$$\varphi_i(t) = f(x_0 + ta_i), \quad i = 1, \ldots, N.$$

Suppose that $\varepsilon_i \in (0, d_i)$, $i = 1, \ldots, N$, and that we are given sets $\Omega_i \subset (0, \varepsilon_i)$ such that $\Omega_i \in \mathfrak{A}_1$, $i = 1, \ldots, N$. If, for every $i = 1, \ldots, N$, the function φ_i is Ω_i-monotonic in the interval $[0, \varepsilon_i] \subset [0, d_i)$, then f is continuous.

Proof. Write $b_i = x_0 + \varepsilon_i a_i \in D$, $i = 1, \ldots, N$. Let

$$T_i = \{x \in D \mid x = x_0 + ta_i, \ t \in \Omega_i\} \tag{12.8.3}$$

if φ_i is Ω_i-decreasing, and

$$T_i = \{x \in D \mid x = x_0 + (\varepsilon_i - t)a_i = b_i - ta_i, \ t \in \Omega_i\} \tag{12.8.4}$$

if φ_i is Ω_i-increasing, $i = 1, \ldots, N$. Put

$$T = \bigcup_{i=1}^{N} T_i.$$

We will show that f is bounded above on T by

$$M = \max\{f(x_0), \ f(b_1), \ldots, f(b_N)\}.$$

Take an arbitrary $x \in T$. There exists an i, $1 \leqslant i \leqslant N$, such that $x \in T_i$. Suppose that T_i is given by (12.8.3) so that φ_i is Ω_i-decreasing. Then $x = x_0 + ta_i$ with a $t \in \Omega_i$, so $\varphi_i(t) \leqslant \varphi_i(0)$. In other words,

$$f(x) = f(x_0 + ta_i) = \varphi_i(t) \leqslant \varphi_i(0) = f(x_0) \leqslant M.$$

Now suppose that T_i is given by (12.8.4) so that φ_i is Ω_i-increasing. Then $x = b_i - ta_i$ with a $t \in \Omega_i$. Since $\Omega_i \subset (0, \varepsilon_i)$, we have $\varepsilon_i - t \in (0, \varepsilon_i) \subset [0, d_i)$, and $\varphi_i(\varepsilon_i - t) \leqslant \varphi_i(\varepsilon_i)$. In other words,

$$f(x) = f(x_0 + (\varepsilon_i - t)a_i) = \varphi_i(\varepsilon_i - t) \leqslant \varphi_i(\varepsilon_i) = f(x_0 + \varepsilon_i a_i) = f(b_i) \leqslant M.$$

Thus f is bounded above on T. It remains to show that $T \in \mathfrak{A}_N$.

Let $g : \mathbb{R}^N \to \mathbb{R}$ be an arbitrary additive function bounded above on T:

$$g(x) \leqslant K \quad \text{for } x \in T.$$

Put

$$\gamma_i(t) = g(x_0 + ta_i) - g(x_0) = g(ta_i), \ i = 1, \ldots, N.$$

Thus for every $i = 1, \ldots, N$, $\gamma_i : \mathbb{R} \to \mathbb{R}$ is an additive function. Suppose that T_i is given by (12.8.3) and $t \in \Omega_i$. Then $x_0 + ta_i \in T_i$ and $\gamma_i(t) \leqslant K - g(x_0)$, i.e., γ_i is bounded above on Ω_i, and since $\Omega_i \in \mathfrak{A}_1 = \mathfrak{B}_1$ (cf. Theorem 10.2.2), γ_i is continuous.

Now suppose that T_i is given by (12.8.4) and $t \in \varepsilon_i - \Omega_i$. Then $t = \varepsilon_i - t'$, where $t' \in \Omega_i$ and $x_0 + ta_i = x_0 + (\varepsilon_i - t')a_i \in T_i$, whence $\gamma_i(t) \leqslant K - g(x_0)$, i.e., γ_i is bounded above on $\varepsilon_i - \Omega_i$. By Theorems 10.2.2, 9.2.8 and 9.2.1 we have $\varepsilon_i - \Omega_i \in \mathfrak{B}_1$. Hence γ_i is continuous.

Thus all the functions γ_i, $i = 1, \ldots, N$, are continuous. By Theorem 5.5.2 there exist constants $c_1, \ldots, c_N \in \mathbb{R}$ such that $\gamma_i(t) = c_i t$ for $i = 1, \ldots, N$. An arbitrary $x \in \mathbb{R}^N$ can be written as

$$x = x_1 a_1 + \cdots + x_N a_N, \quad x_i \in \mathbb{R}, \ i = 1, \ldots, N,$$

since $a_1, \ldots, a_N$ are linearly independent over $\mathbb{R}$. Then

$$g(x) = g(x_1 a_1) + \cdots + g(x_N a_N) = \gamma_1(x_1) + \cdots + \gamma_N(x_N) = \sum_{i=1}^{N} c_i x_i,$$

whence it follows that g is continuous. Thus $T \in \mathfrak{B}_N = \mathfrak{A}_N$ (Theorem 10.2.2).   □

Since every function monotonic in an interval $(0, \varepsilon)$ is $(0, \varepsilon)$-monotonic, and by Lemma 9.3.1 open intervals belong to $\mathfrak{A}_1$, it follows that (with the notation as in Theorem 12.8.1), if every function φ_i is monotonic in a right vicinity of zero, then f is continuous. In the particular case $N = 1$ we obtain hence

Corollary 12.8.1. *Let $D \subset \mathbb{R}$ be an open interval, and let $f : D \to \mathbb{R}$ be a convex function. If f is monotonic on an interval $I \subset D$, then f is continuous.*

The following theorem is in a sense converse to Theorem 12.8.1.

Theorem 12.8.2. *Let $D \subset \mathbb{R}^N$ be a convex and open set, and let $f : D \to \mathbb{R}$ be a continuous convex function. Let $x_0 \in D$ and $a \in \mathbb{R}^N$, $a \neq 0$. Let $d = \sup\{t > 0 \mid x_0 + ta \in D\}$, and let $\varphi : [0, d) \to \mathbb{R}$ be given by*

$$\varphi(t) = f(x_0 + at).$$

Then the function φ is monotonic in an interval $(0, \varepsilon) \subset (0, d)$.

Proof. Let $\psi = \varphi \mid (0, d)$. Then $\psi : (0, d) \to \mathbb{R}$ is a continuous convex function of a single real variable. By Theorem 7.3.5 either ψ is monotonic in $(0, d)$, or there exists an $\varepsilon \in (0, d)$ such that ψ is decreasing in $(0, \varepsilon)$. The theorem follows.   □

Exercises

1. Let $f : \mathbb{R}^N \to \mathbb{R}$ be a discontinuous additive function, and let $A \subset \mathbb{R}^N$ be a set which is either measurable of positive measure, or of the second category and with the Baire property. Show that $\operatorname{cl} f(A) = \mathbb{R}$.

2. Let $f : \mathbb{R}^N \to \mathbb{R}$ be a discontinuous additive function, and let $w : \mathbb{R} \to \mathbb{R}$ be an arbitrary function. Show that for every interval $I \subset \mathbb{R}$ we have

$$m_i\big(f^{-1}\big(w^{-1}\big(w(\mathbb{R}) \setminus w(I)\big)\big)\big) = 0.$$

3. Let $f : \mathbb{R} \to \mathbb{R}$ be an additive function, let $A \subset \mathbb{R} \setminus \{0\}$ be a set which is either measurable of positive measure, or of the second category and with the Baire property. Suppose that f satisfies the condition

$$f(x)f\left(\frac{1}{x}\right) > c \quad \text{for } x \in A$$

with a certain $c > 0$. Show that f is continuous[11].

---

[11] Aczél [7]; cf. also Kannappan [161]. The problem whether this result remains valid also when $c = 0$ remains still open (Rothberger [270]; cf. also Benz [26]). Added in proof: This has recently been solved by G. M. Bergman [29].

[Hint: Let $g : A \to \mathbb{R}$ be given by $g(x) = x + x^{-1}$. On $g(A)$ we have $f(t) = f(x) + f(x^{-1})$ with an $x \in A$, and $|f(t)| = |f(x)| + |f(x^{-1})| \geqslant 2|f(x)f(x^{-1})|^{1/2} > 2\sqrt{c}$ (cf. 8.3).]

4. Show that in Lemma 12.3.1 the assumption that J has a finite upper bound is essential.
   [Hint: Let $g : \mathbb{R}^N \to \mathbb{R}$ be a discontinuous additive function. Consider the function $f : \mathbb{R}^N \to \mathbb{R}$ given by $f(x) = |x| + |g(x)|$.]

5. Let $D \subset \mathbb{R}^N$ be a convex and open set, and let $f : D \to \mathbb{R}$ be a discontinuous convex function. Show that for every $c \in \mathbb{R}$ we have $m_i \left[f^{-1}((c, \infty)) \cap m_f^{-1}((-\infty, c)) \right] = 0$.

6. Let $f : \mathbb{R}^N \to \mathbb{R}$ be a function with big graph, and let $A \subset \mathbb{R}^N$ be a set which is either measurable of positive measure or of the second category with the Baire property. Show that $f(A) = \mathbb{R}$.

7. Let $f : \mathbb{R}^N \to \mathbb{R}$ be an additive function, and let $H \subset \mathbb{R}^N$ be a Hamel basis. Show that if $f(H) \neq \{0\}$ and $\operatorname{card} f(H) \leqslant \aleph_0$, then f is a function with small graph.
   [Hint: Use Exercise 4.4 and Lemma 4.1.3.]

8. Show that there exists a non-constant periodic convex function $f : \mathbb{R} \to \mathbb{R}$ which is not additive.
   [Hint: Let $g : \mathbb{R} \to \mathbb{R}$, $g \neq 0$, be a non-invertible additive function, and take $f = |g|$.]

9. Show by a suitable counter-example that the following extension of the Egorov theorem is invalid (Kuczma [189]):
   Let $A \subset \mathbb{R}$ be a measurable set with $m(A) < \infty$, and let $f : A \times (0, 1) \to \mathbb{R}$ be a function such that for every fixed $t \subset (0, 1)$ the function $f(x, t)$ as a function of x is measurable. Suppose that there exists a function $\hat{f} : A \to \mathbb{R}$ such that $\lim_{t \to 0} f(x, t) = \hat{f}(x)$ for almost every $x \in A$. Then, for every $\varepsilon > 0$ and $\eta > 0$, there exists a closed set $F \subset A$ and a $\delta > 0$ such that $m(A \setminus F) < \eta$ and $|f(x, t) - \hat{f}(x)| < \varepsilon$ uniformly on $F \times (0, \delta)$.
   [Hint: Let $g : \mathbb{R} \to \mathbb{R}$ be an invertible discontinuous additive function. Take $A = [0, 1]$, and $f(x, t) = 1$ when $(x, t) \in \operatorname{Gr}(g)$, $f(x, t) = 0$ when $(x, t) \notin \operatorname{Gr}(g)$.]

10. Show that there exists an additive function $f : \mathbb{R}^N \to \mathbb{R}$ such that $f(\mathbb{R}^N) = \mathbb{R}$, and the level sets of f are saturated non-measurable, have property $(*)$ from 3.3, and have the power of continuum (Smítal [286]).
    [Hint: Let $f_1 : \mathbb{R} \to \mathbb{R}$ be an additive function such that $f_1(\mathbb{R}) = \mathbb{R}$ and $\operatorname{card} \operatorname{Ker} f_1 = \mathfrak{c}$, and let $f_2 : \mathbb{R}^N \to \mathbb{R}$ be an additive function with big graph. Put $f = f_1 \circ f_2$.]

Part III

Related Topics

Chapter 13

Related Equations

13.1 The remaining Cauchy equations

The following functional equations are also referred to as Cauchy's equations (Cauchy [41]; cf. also Aczél [5])

$$f(x + y) = f(x)f(y), \tag{13.1.1}$$
$$f(xy) = f(x) + f(y), \tag{13.1.2}$$
$$f(xy) = f(x)f(y). \tag{13.1.3}$$

Equation (13.1.1) will be considered for functions $f : \mathbb{R}^N \to \mathbb{R}$, whereas equations (13.1.2) and (13.1.3) will be considered only for functions of a real variable, because of the operation of multiplication occurring in the argument.

Lemma 13.1.1. *Let $f : \mathbb{R}^N \to \mathbb{R}$ be a solution of equation* (13.1.1). *Then either $f = 0$, or $f > 0$ in $\mathbb{R}^N$.*

Proof. Suppose that there exists an $x_0 \in \mathbb{R}^N$ such that $f(x_0) = 0$. Then, for arbitrary $x \in \mathbb{R}^N$,

$$f(x) = f\big((x - x_0) + x_0\big) = f(x - x_0)f(x_0) = 0,$$

i.e., $f = 0$. Now suppose that $f \neq 0$. By what has just been shown, f is never zero in $\mathbb{R}^N$. For an arbitrary $x \in \mathbb{R}^N$ we have

$$f(x) = f(\tfrac{1}{2}x + \tfrac{1}{2}x) = [f(\tfrac{1}{2}x)]^2 \geqslant 0,$$

and since $f(x) \neq 0$, we obtain $f(x) > 0$. Thus f is positive in $\mathbb{R}^N$. $\qquad\square$

Theorem 13.1.1. *Let $f : \mathbb{R}^N \to \mathbb{R}$ be a solution of equation* (13.1.1). *Then either $f = 0$, or there exists an additive function $g : \mathbb{R}^N \to \mathbb{R}$ such that*

$$f = \exp g. \tag{13.1.4}$$

Proof. The function $f = 0$ clearly is a solution of (13.1.1). So suppose that $f \neq 0$. By Lemma 13.1.1 f is positive in $\mathbb{R}^N$. Put $g = \log f$. Then, by (13.1.1), $g : \mathbb{R}^N \to \mathbb{R}$ is an additive function, and the theorem follows. $\qquad\square$

On the other hand, it is easily seen that every function of form (13.1.4), where $g : \mathbb{R}^N \to \mathbb{R}$ is additive, actually satisfies (13.1.1).

Lemma 13.1.2. *Let $D \subset \mathbb{R}$ be such that $xy \in D$ for every $x, y \in D$, and $0 \in D$. The only solution $f : D \to \mathbb{R}$ of equation (13.1.2) is $f = 0$.*

Proof. Setting in (13.1.2) $y = 0$ we obtain $f(0) = f(x) + f(0)$, whence $f(x) = 0$ for every $x \in D$. Of course, $f = 0$ is a solution of (13.1.2). $\square$

But we can obtain non-trivial solutions of (13.1.2) if we exclude 0 from the domain of definition of f. Thus we have

Theorem 13.1.2. *Let $D = (0, \infty)$, or $D = (-\infty, 0) \cup (0, \infty)$. If a function $f : D \to \mathbb{R}$ satisfies equation (13.1.2), then there exists an additive function $g : \mathbb{R} \to \mathbb{R}$ such that*

$$f(x) = g\big(\log |x|\big) \quad for \ x \in D. \tag{13.1.5}$$

Proof. For $t \in \mathbb{R}$ put $g(t) = f(e^t)$ so that $g : \mathbb{R} \to \mathbb{R}$. We have by (13.1.2) for arbitrary $u, v \in \mathbb{R}$

$$g(u + v) = f(e^{u+v}) = f(e^u e^v) = f(e^u) + f(e^v) = g(u) + g(v),$$

i.e., g is additive. Moreover, $f(x) = f(|x|) = g(\log |x|)$ for $x > 0$. For $x = 1$ we obtain hence $f(1) = g(0) = 0$.

Now, if $D = (-\infty, 0) \cup (0, \infty)$, set in (13.1.2) $x = y = -1$. Then

$$0 = f(1) = f(-1) + f(-1) = 2f(-1),$$

whence $f(-1) = 0$. For $x < 0$, $y = -1$, equation (13.1.2) yields

$$f(|x|) = f(-x) = f(x) + f(-1) = f(x).$$

Hence we obtain (13.1.5) for $x < 0$. So (13.1.5) is generally valid. $\square$

On the other hand, it is easily seen that every function of form (13.1.5), where $g : \mathbb{R} \to \mathbb{R}$ is additive, actually satisfies (13.1.2).

Equation (13.1.3) is most complicated of all the three equations. We will solve it for functions $f : D \to \mathbb{R}$, where D is one of the sets

$$(0, 1), [0, 1), (-1, 1), (-1, 0) \cup (0, 1), (1, \infty), (0, \infty), [0, \infty), (-\infty, 0) \cup (0, \infty), \mathbb{R}. \tag{13.1.6}$$

Lemma 13.1.3. *Let D be one of sets (13.1.6). If $f : D \to \mathbb{R}$, $f \neq 0$, is a solution of (13.1.3), then $f(1) = 1$ if $1 \in D$; $f(-1) = 1$ or -1 if $-1 \in D$; $f(0) = 0$ or 1 if $0 \in D$. If $0 \in D$ and $f(0) = 1$, then $f(x) = 1$ for all $x \in D$.*

Proof. Setting in (13.1.3) in turn $x = y = 1$; $x = y = -1$; $x = y = 0$, we obtain

$$f(1) = f(1)f(1) = \big[f(1)\big]^2,$$

whence $f(1) = 0$ or 1; if $f(1) = 0$, then we have for arbitrary $x \in D$

$$f(x) = f(1)f(x) = 0,$$

contrary to the assumption that $f \neq 0$. So $f(1) = 1$. Next,

$$1 = f(1) = f(-1)f(-1) = [f(-1)]^2,$$

whence $f(-1) = 1$ or -1. And

$$f(0) = f(0)f(0) = [f(0)]^2,$$

whence $f(0) = 0$ or 1. If $f(0) = 1$, then we have for every $x \in D$

$$f(x) = f(0)f(x) = f(0) = 1,$$

and the proof is complete. □

Lemma 13.1.4.[1] *If D is one of the sets*

$$(0, 1), \quad [0, 1), \quad (1, \infty),$$

and if $f_0 : D \to \mathbb{R}$, $f_0 \neq 0$, is a solution of (13.1.3), *then the function $f : [D \cup \{1\} \cup (D \setminus \{0\})^{-1}] \to \mathbb{R}$ given by*

$$f(x) = \begin{cases} f_0(x) & \text{if } x \in D, \\ 1 & \text{if } x = 1, \\ [f_0(x^{-1})]^{-1} & \text{if } x \in (D \setminus \{0\})^{-1} \end{cases} \tag{13.1.7}$$

satisfies equation (13.1.3) *in $D \cup \{1\} \cup (D \setminus \{0\})^{-1}$ and $f_0 = f \mid D$.*

Proof. First note that $f_0(x) \neq 0$ for $x \in D$, $x \neq 0$. In fact, suppose that $f_0(x_0) = 0$ for an $x_0 \in D$, $x_0 \neq 0$. Suppose, for the argument's sake, that $x_0 \in (-1, 1) \cap D$. Take an arbitrary $x \in D$ such that $|x| < |x_0|$. Then $|x/x_0| < 1$ and $x/x_0 \in D$. Hence

$$f_0(x) = f_0\left(x_0 \frac{x}{x_0}\right) = f_0(x_0)f_0\left(\frac{x}{x_0}\right) = 0.$$

Now take an arbitrary $x \in D$. We have from (13.1.3) by induction

$$f_0(x^n) = [f_0(x)]^n \quad \text{for } n \in \mathbb{N}. \tag{13.1.8}$$

But for sufficiently large $n \in \mathbb{N}$ we have $|x^n| < |x_0|$. Thus $f_0(x^n) = 0$, and by (13.1.8) $f_0(x) = 0$. Consequently $f_0 = 0$ in D, contrary to the assumption.

Thus definition (13.1.7) is meaningful. The relation $f_0 = f|D$ results directly from (13.1.7). It remains to show that f given by (13.1.7) satisfies (13.1.3) for $x, y \in D \cup \{1\} \cup (D \setminus \{0\})^{-1}$.

---

[1] In the following part of this section, some corrections were made in the 2nd edition, based on M. Kuczma, *Note on multiplicative functions*, Ann. Math. Sil. **3(15)** (1990), 45–50, by K. Baron.

We must distinguish several cases.

I. $x, y \in D$. Then also $xy \in D$, and by (13.1.7) and (13.1.3) for f_0

$$f(xy) = f_0(xy) = f_0(x)f_0(y) = f(x)f(y).$$

II. x arbitrary, $y = 1$. Then, by (13.1.7), $f(y) = 1$. Hence

$$f(xy) = f(x) = f(x)f(y).$$

Of course, if $x = 1$ and y is arbitrary, then we obtain hence (13.1.3) in virtue of the commutativity of the operation of multiplication.

III. $x, y \in (D \setminus \{0\})^{-1}$. Then also $xy \in (D \setminus \{0\})^{-1}$, and by (13.1.7) and (13.1.3) for f_0 $f(x)f(y) = [f_0(x^{-1})]^{-1}[f_0(y^{-1})]^{-1} = [f_0(x^{-1})f_0(y^{-1})]^{-1} = [f_0((xy)^{-1})]^{-1} = f(xy)$.

IV. $x \in D$, $y \in (D \setminus \{0\})^{-1}$, $xy \in D$. Then by (13.1.7) $f(y) = [f_0(y^{-1})]^{-1}$ and $y^{-1} \in D$. We have

$$f_0(xy)f_0(y^{-1}) = f_0(xyy^{-1}) = f_0(x),$$

whence

$$f(x)f(y) = f_0(x)[f_0(y^{-1})]^{-1} = f_0(xy) = f(xy).$$

V. $x \in D$, $y \in (D \setminus \{0\})^{-1}$, $xy = 1$. Then $y^{-1} = x$, whence $f(y) = [f_0(y^{-1})]^{-1} = [f_0(x)]^{-1}$ and

$$f(x)f(y) = f_0(x)[f_0(x)]n^{-1} = 1 = f(xy).$$

VI. $x \in D$, $y \in (D \setminus \{0\})^{-1}$, $xy \in (D \setminus \{0\})^{-1}$. Then $x^{-1}y^{-1} = (xy)^{-1} \in D$, and

$$[f_0(x^{-1}y^{-1})]^{-1}[f_0(x)]^{-1} = [f_0(x^{-1}y^{-1})f_0(x)]^{-1} = [f_0(x^{-1}y^{-1}x)]^{-1} = [f_0(y^{-1})]^{-1},$$

whence

$$f(x)f(y) = f_0(x)[f_0(y^{-1})]^{-1} = [f_0(x^{-1}y^{-1})]^{-1} = f(xy).$$

Of course, if $x \in (D \setminus \{0\})^{-1}$ and $y \in D$, then we obtain (13.1.3) in virtue of the commutativity of the operation of the multiplication.   □

The next lemma shows that we may restrict (13.1.6) to the following sets

$$(0, \infty), \quad [0, \infty), \quad (-\infty, 0) \cup (0, \infty), \quad \mathbb{R}. \tag{13.1.9}$$

Lemma 13.1.5. *If D_0 is one of sets (13.1.6), and $f_0 : D_0 \to \mathbb{R}$ satisfies equation (13.1.3), then there exist a set D of form (13.1.9) and a function $f : D \to \mathbb{R}$ satisfying (13.1.3) such that $D_0 \subset D$ and $f_0 = f \mid D_0$.*

Proof. Clearly, we may assume that $f_0 \neq 0$ and by Lemma 13.1.4 it is enough to consider sets D_0 of the form

$$(-1, 1), \quad (-1, 0) \cup (0, 1)$$

only. Like in the proof of Lemma 13.1.4, we show that $f(x) \neq 0$ for $x \in D_0$, $x \neq 0$.

For arbitrary $x, y \in D_0 \setminus \{0\}$ we have $-x, -y \in D_0 \setminus \{0\}$ and by (13.1.3)

$$f_0(-x)\, f_0(y) = f_0(-xy) = f_0(x)\, f_0(-y),$$

whence

$$\frac{f_0(-x)}{f_0(x)} = \frac{f_0(-y)}{f_0(y)}. \tag{13.1.10}$$

Relation (13.1.10) means that

$$\varepsilon = \frac{f_0(-x)}{f_0(x)}, \quad x \in D_0 \setminus \{0\}, \tag{13.1.11}$$

is constant. Setting $y = -x$ in (13.1.10), we obtain

$$\varepsilon^2 = 1. \tag{13.1.12}$$

Relation (13.1.11) may be written as

$$f_0(-x) = \varepsilon f_0(x) \tag{13.1.13}$$

for $x \in D_0 \setminus \{0\}$. If $0 \in D_0$, then according to Lemma 13.1.3 $f(0)$ is either 0 or 1; in the latter case $f_0(x) = 1$ for all $x \in D_0$. Thus, (13.1.13) is fulfilled also for $x = 0$. Consequently, (13.1.13) holds in the whole of D_0.

Now, we define the function $f : \left[D_0 \cup \{1\} \cup (D_0 \setminus \{0\})^{-1} \cup \{-1\}\right] \to \mathbb{R}$ by (13.1.7) with D replaced by D_0 and

$$f(-1) = \varepsilon, \tag{13.1.14}$$

where ε is defined by (13.1.11). It remains to verify that f satisfies equation (13.1.3) in $D_0 \cup (D_0 \setminus \{0\})^{-1} \cup \{1\} \cup \{-1\}$. Beside the cases dealt with in the proof of Lemma 13.1.4, we need to consider also the cases where one of x, y, xy is -1.

If $x = -1$, $y \in D_0$, then (13.1.3) results from (13.1.7), (13.1.13) and (13.1.14). If $x = -1$, $y = 1$, then we obtain (13.1.3) from (13.1.7) and (13.1.14). If $x = y = -1$, then (13.1.7), (13.1.14) and (13.1.12) imply (13.1.3). If $x = -1$, $y \in (D_0 \setminus \{0\})^{-1}$, then also $-y \in (D_0 \setminus \{0\})^{-1}$ and by (13.1.7), (13.1.13), (13.1.12) and (13.1.14)

$$f(xy) = f(-y) = \left[f_0(-y^{-1})\right]^{-1} = \left[\varepsilon f_0(y^{-1})\right]^{-1} = \varepsilon\left[f_0(y^{-1})\right]^{-1} = f(x)\, f(y).$$

If $x \in D_0$, $y \in (D_0 \setminus \{0\})^{-1}$, $xy = -1$, then $y^{-1} = -x$ and

$$f(x)\, f(y) = f_0(x)\left[f_0(y^{-1})\right]^{-1} = f_0(x)\left[f_0(-x)\right]^{-1} = f_0(x)\left[\varepsilon f_0(x)\right]^{-1} = \varepsilon = f(xy)$$

in virtue of (13.1.7), (13.1.13), (13.1.12) and (13.1.14).

The cases where $y = -1$, or $x \in (D_0 \setminus \{0\})^{-1}$, $y \in D_0$, $xy = -1$, follow from those already considered in view of the commutativity of the multiplication of real numbers. □

Theorem 13.1.3. *Let D be one of the sets (13.1.6). If a function $f : D \to \mathbb{R}$ satisfies equation (13.1.3), then there exists an additive function $g : \mathbb{R} \to \mathbb{R}$ such that f has one of the following forms:*

$$f = 0, \tag{13.1.15}$$

$$f = 1, \tag{13.1.16}$$

$$f(x) = \begin{cases} \exp g(\log|x|) & \text{for } x \neq 0,\ x \in D, \\ 0 & \text{for } x = 0\ (\text{if } 0 \in D), \end{cases} \tag{13.1.17}$$

$$f(x) = \begin{cases} \exp g(\log|x|) & \text{for } x > 0,\ x \in D, \\ 0 & \text{for } x = 0\ (\text{if } 0 \in D), \\ -\exp g(\log|x|) & \text{for } x < 0,\ x \in D. \end{cases} \tag{13.1.18}$$

Proof. By Lemma 13.1.5 it is enough to consider sets D of form (13.1.9). Let $x > 0$, $x \in D$. Then by (13.1.3)

$$f(x) = \left[f(\sqrt{x})\right]^2 \geqslant 0. \tag{13.1.19}$$

Suppose that there exists an $x_0 \in D$, $x_0 \neq 0$, such that $f(x_0) = 0$. Take an arbitrary $x \in D$. Then $x/x_0 \in D$ and

$$f(x) = f\left(x_0 \frac{x}{x_0}\right) = f(x_0)f\left(\frac{x}{x_0}\right) = 0.$$

Thus $f = 0$ in D, and we obtain (13.1.15). Now suppose that $f(x) \neq 0$ for $x \in D \setminus \{0\}$. By Lemma 13.1.3 $f(0) = 0$ or 1. If $f(0) = 1$, then $f = 1$ in D in virtue of Lemma 13.1.3, and we obtain (13.1.16). So let $f(0) = 0$. By (13.1.19) $f(x) > 0$ for $x > 0$. Let $f_1(x) = \log f(x)$ for $x > 0$. Then $f_1 : (0, \infty) \to \mathbb{R}$ satisfies equation (13.1.2), whence by Theorem 13.1.2 there exists an additive function $g : \mathbb{R} \to \mathbb{R}$ such that $f_1(x) = g(\log x)$. Again by Lemma 13.1.3 $f(-1) = 1$ or -1 (if $-1 \in D$). If $-1 \in D$ and $f(-1) = 1$, then we have for arbitrary $x \in D$

$$f(-x) = f(-1)f(x) = f(x),$$

i.e., f is even. In this case we obtain (13.1.17). If $-1 \in D$ and $f(-1) = -1$, then for arbitrary $x \in D$

$$f(-x) = f(-1)f(x) = -f(x),$$

i.e., f is odd. Hence we obtain (13.1.18). $\qquad\qquad\qquad\qquad\qquad\qquad\square$

On the other hand, it is easy to check that every function of form (13.1.15), (13.1.16), (13.1.17), or (13.1.18), where $g : \mathbb{R} \to \mathbb{R}$ is additive, actually satisfies equation (13.1.3).

Now we are going to determine the continuous solutions of (13.1.1), (13.1.2) and (13.1.3).

Theorem 13.1.4. *A function $f : \mathbb{R}^N \to \mathbb{R}$ is a continuous solution of equation* (13.1.1) *if and only if either $f = 0$, or*

$$f(x) = e^{cx}, \quad x \in \mathbb{R}^N, \tag{13.1.20}$$

with a certain constant $c \in \mathbb{R}^N$.

Proof. Let $f \neq 0$. Since f satisfies (13.1.1), according to Theorem 13.1.1 f has form (13.1.4), where $g : \mathbb{R}^N \to \mathbb{R}$ is an additive function. Since f is continuous, $g = \log f$ also is continuous. By Theorem 5.5.2 $g(x) = cx$ with a certain $c \in \mathbb{R}^N$. Formula (13.1.20) results now from (13.1.4).

It is obvious that $f = 0$ and every f of form (13.1.20) is a continuous solution of (13.1.1). $\square$

Theorem 13.1.5. *Let $D = (0, \infty)$ or $D = (-\infty, 0) \cup (0, \infty)$. A function $f : D \to \mathbb{R}$ is a continuous solution of equation* (13.1.2) *in D if and only if*

$$f(x) = c \log |x|, \quad x \in D, \tag{13.1.21}$$

with a certain constant $c \in \mathbb{R}$.

Proof. By Theorem 13.1.2 f has form (13.1.5), where $g(t) = f(e^t)$ is an additive function. Since f is continuous, g also is continuous, and hence, by Theorem 5.5.2, $g(t) = ct$ with a certain $c \in \mathbb{R}$. Formula (13.1.21) results now from (13.1.5).

It is obvious that every function of form (13.1.21) is a continuous solution of (13.1.2). $\square$

Theorem 13.1.6. *Let D be one of sets* (13.1.6). *A function $f : D \to \mathbb{R}$ is a continuous solution of equation* (13.1.3) *if and only if either $f = 0$, or $f = 1$, or f has one of the following forms*

$$f(x) = |x|^c, \quad x \in D, \tag{13.1.22}$$

$$f(x) = |x|^c \operatorname{sgn} x, \quad x \in D, \tag{13.1.23}$$

with a certain $c \in \mathbb{R}$. If $0 \in D$, then $c > 0$.

Proof. By Theorem 13.1.3 either $f = 0$, or $f = 1$, or f has form (13.1.17) or (13.1.18), where $g : \mathbb{R} \to \mathbb{R}$ is an additive function. We have $g(t) = \log f(e^t)$, and since f is continuous, so is also g, and consequently $g(t) = ct$ with a certain $c \in \mathbb{R}$. Formulas (13.1.22) and (13.1.23) result now from (13.1.17) and (13.1.18). Suppose that $0 \in D$. If we had $c = 0$, then (13.1.22) yields $f(x) = 1$ for $x \neq 0$, and by the continuity we must have $f(0) = 1$. Thus we obtain the solution $f = 1$, already listed. Formula (13.1.23) with $c = 0$ yields $f(x) = 1$ for $x > 0$ and $f(x) = -1$ for $x < 0$, and thus f cannot be continuous. (If $D \subset [0, \infty)$, then both formulas (13.1.22) and (13.1.23) coincide). Similarly, if $c < 0$ then f given by (13.1.22) or (13.1.23) satisfies $\lim_{x \to 0+} f(x) = \infty$, and thus f cannot be continuous at zero.

It is obvious that every one of the specified functions is a continuous solution of (13.1.3). $\square$

Remark 13.1.1. If $0 \in D$, and f is assumed continuous only in $D \setminus \{0\}$, then besides the solutions listed in Theorem 13.1.6 we obtain also

$$f(x) = \begin{cases} |x|^c & \text{for } x \neq 0, \, x \in D, \\ 0 & \text{for } x = 0, \end{cases} \tag{13.1.24}$$

and

$$f(x) = \begin{cases} |x|^c \operatorname{sgn} x & \text{for } x \neq 0, \, x \in D, \\ 0 & \text{for } x = 0, \end{cases} \tag{13.1.25}$$

where $c \leqslant 0$. This follows from the proof of Theorem 13.1.6 and from Lemma 13.1.3.

Simple conditions ensuring the continuity of solutions of equations (13.1.1)–(13.1.3) may be inferred from (13.1.4), (13.1.5), (13.1.17) and (13.1.18). Below we formulate some such conditions.

Theorem 13.1.7. *Let $f : \mathbb{R}^N \to \mathbb{R}$ be a solution of equation* (13.1.1). *If f is measurable, or bounded above on a set $T \in \mathfrak{A}$, or bounded below by a positive constant on a set $T \in \mathfrak{A}$, then f is continuous.*

Proof. If $f = 0$, then it is already continuous. If $f \neq 0$, then, according to Theorem 13.1.1, f is given by (13.1.4), where $g : \mathbb{R}^N \to \mathbb{R}$ is an additive function. If f is measurable, then so is also $g = \log f$, and hence, by Theorem 9.4.2, g is continuous. Then by (13.1.4), f is continuous, too. If f is bounded above, or f is bounded below by a positive constant, on a set $T \subset \mathbb{R}^N$, then g is bounded above resp. bounded below on T. If $T \in \mathfrak{A}$, then g is continuous, and hence also f is continuous. $\quad\square$

Theorem 13.1.8. *Let $D = (0, \infty)$, or $D = (-\infty, 0) \cup (0, \infty)$, and let $f : D \to \mathbb{R}$ be a solution of equation* (13.1.2). *If f is measurable, or bounded above or below on a set $T \subset D$ such that $m_i(T) > 0$ or T is of the second category and with the Baire property, then f is continuous.*

Proof. By Theorem 13.1.2 f has form (13.1.5), where $g : \mathbb{R} \to \mathbb{R}$ is an additive function. (13.1.5) implies that $g(t) = f(e^t)$, so if f is measurable, then so is also g. By Theorem 9.4.2 g is then continuous, so also f is continuous. Suppose that f is bounded above or below on a set $T \subset D$. It follows from (13.1.5) that f is even, so f is also bounded (above or below) on the set $T_0 = (0, \infty) \cap [T \cup (-T)]$, and is bounded (above or below) on the set $\log T_0 = \{t \in \mathbb{R} \mid e^t \in T_0\}$. If $m_i(T) > 0$, then also $m_i(T_0) > 0$, and since the function log is absolutely continuous (Theorem 7.4.6), also $m_i(\log T_0) > 0$. If T is of the second category with the Baire property, then so is also T_0 and $\log T_0$, since the function log is a homeomorphism. By Theorems 9.3.1 and 9.3.2 g is continuous, and hence also f is continuous in D. $\quad\square$

For solutions of equation (13.1.3) the situation is more complicated. For example, if $0 \in D$, then functions (13.1.24) and (13.1.25) ($c \leqslant 0$) are measurable, bounded on suitable open sets, satisfy equation (13.1.3), but are not continuous at zero. However, we have the following theorem, which may be proved similarly as Theorem 13.1.7 and 13.1.8, and therefore its proof is left to the reader.

Theorem 13.1.9. *Let D be one of sets* (13.1.6), *and let $f : D \to \mathbb{R}$ be a solution of equation* (13.1.3). *If f is measurable, then it is continuous in $D \setminus \{0\}$.*

13.2 Jensen equation

The equation resulting on replacing in the Jensen inequality (5.3.1) the sign of inequality by that of equality

$$f\left(\frac{x+y}{2}\right) = \frac{f(x) + f(y)}{2} \tag{13.2.1}$$

is known as the *Jensen equation* (cf. Aczél [5]). We will consider equation (13.2.1) for functions $f : D \to \mathbb{R}$, where $D \subset \mathbb{R}^N$ is a convex set. If D were also open, then f satisfying (13.2.1) would be convex, and thus all the results established in the previous chapters of this book for convex functions would apply. But most of these results become invalid when the set D is no longer open. For example, let $N = 1$, $D = [-1, 1]$, and let $f : D \to \mathbb{R}$ be defined as

$$f(x) = \begin{cases} x^2 & \text{if } x \in (-1, 1), \\ 2 & \text{if } |x| = 1. \end{cases}$$

Such an f satisfies inequality (5.3.1) in $D = [-1, 1]$, is measurable and bounded in D, but is discontinuous at $x = 1$ and $x = -1$. So we treat equation (13.2.1) in a more general setting.

Lemma 13.2.1. *Let $D \subset \mathbb{R}^N$ be a convex set such that $0 \in D$, and let $f : D \to \mathbb{R}$ be a solution of equation* (13.2.1) *such that*

$$f(0) = 0. \tag{13.2.2}$$

Then, for every $x \in D$ and $n \in \mathbb{N}$,

$$f\left(\frac{x}{2^n}\right) = \frac{1}{2^n} f(x). \tag{13.2.3}$$

Proof. Take an $x \in D$. Since D is convex, $\frac{1}{2}(x + 0) \in D$, and by (13.2.1) and (13.2.2)

$$f\left(\frac{x}{2}\right) = f\left(\frac{x+0}{2}\right)$$
$$= \frac{f(x) + f(0)}{2} = \frac{f(x)}{2}. \tag{13.2.4}$$

Thus (13.2.3) holds for $n = 1$. Assuming it true for an $n \in \mathbb{N}$, we have

$$\frac{x}{2^{n+1}} = \frac{1}{2^{n+1}} x + \left(1 - \frac{1}{2^{n+1}}\right) 0 \in D,$$

and by (13.2.3) (for n) and (13.2.4)

$$f\left(\frac{x}{2^{n+1}}\right) = \frac{1}{2^n} f\left(\frac{x}{2}\right) = \frac{1}{2^{n+1}} f(x).$$

Induction completes the proof. $\qquad\square$

Lemma 13.2.2. *Let $D \subset \mathbb{R}^N$ be a convex set such that $\operatorname{int} D \neq \varnothing$, and let $f : D \to \mathbb{R}$ be a solution of equation (13.2.1). Fix an $x_0 \in \operatorname{int} D$, and define the function $f_0 : (D - x_0) \to \mathbb{R}$ by*

$$f_0(x) = f(x_0 + x) - f(x_0). \tag{13.2.5}$$

Then there exists a unique function $f_1 : \mathbb{R}^N \to \mathbb{R}$ satisfying equation (13.2.1) in $\mathbb{R}^N$ and such that

$$f_1(x) = f_0(x) \text{ for } x \in D - x_0. \tag{13.2.6}$$

Proof. Function (13.2.5) is defined for $x \in D - x_0$. First we verify that f_0 satisfies equation (13.2.1) in $D - x_0$. For every $x, y \in D - x_0$, we have $x_0 + x, x_0 + y \in D$, and by (13.2.5) and (13.2.1)

$$f_0\left(\frac{x+y}{2}\right) = f\left(x_0 + \frac{x+y}{2}\right) - f(x_0) = f\left(\frac{x_0 + x + x_0 + y}{2}\right) - f(x_0)$$

$$= \frac{1}{2}f(x_0 + x) + \frac{1}{2}f(x_0 + y) - f(x_0)$$

$$= \frac{1}{2}\left[f(x_0 + x) - f(x_0)\right] + \frac{1}{2}\left[f(x_0 + y) - f(x_0)\right]$$

$$= \frac{1}{2}\left[f_0(x) + f_0(y)\right].$$

Also, it is easily seen that $0 \in D - x_0$ and by (13.2.5)

$$f_0(0) = 0. \tag{13.2.7}$$

Now put $D_0 = D - x_0$ and

$$D_n = 2^n D_0, \quad n \in \mathbb{N}.$$

If $x \in D_n$, then $\dfrac{x}{2^n} \in D_0$. D_0 is convex, just like D, and $0 \in D_0$, whence $\dfrac{x}{2^{n+1}} = \dfrac{1}{2}\left[\dfrac{x}{2^n} + 0\right] \in D_0$, and $x \in 2^{n+1}D_0 = D_{n+1}$. Thus

$$D_n \subset D_{n+1}, \quad n \in \mathbb{N} \cup \{0\}. \tag{13.2.8}$$

Also, $0 \in \operatorname{int} D_0$, since $x_0 \in \operatorname{int} D$. For every $x \in \mathbb{R}^N$ we have $\lim_{n\to\infty} x/2^n = 0$, whence it follows that there exists an $n \in \mathbb{N} \cup \{0\}$ such that $x/2^n \in D_0$, whence $x \in D_n$. Hence

$$\bigcup_{n=0}^{\infty} D_n = \mathbb{R}^N. \tag{13.2.9}$$

Define the function $f_1 : \mathbb{R}^N \to \mathbb{R}$ as follows:

$$f_1(x) = 2^n f_0\left(\frac{x}{2^n}\right) \text{ if } x \in D_n, n \in \mathbb{N} \cup \{0\}. \tag{13.2.10}$$

First we must check whether definition (13.2.10) is correct. If $x \in D_n = 2^n D_0$, then $x/2^n \in D_0 = D - x_0$, and the expression $f_0(x/2^n)$ is meaningful. If $x \in D_n$ and

$x \in D_m$ for $n \neq m$, say $n < m$, then by (13.2.7) and Lemma 13.2.1

$$2^m f_0 \left(\frac{x}{2^m}\right) = 2^m \frac{1}{2^{m-n}} f_0 \left(\frac{x}{2^n}\right) = 2^n f_0 \left(\frac{x}{2^n}\right),$$

and thus (13.2.10) is unambiguous. By (13.2.9) the function f_1 is defined in the whole of $\mathbb{R}^N$. We must verify that it satisfies equation (13.2.1) in $\mathbb{R}^N$.

Take arbitrary $x, y \in \mathbb{R}^N$. By (13.2.9) and (13.2.8) there exists an $n \in \mathbb{N} \cup \{0\}$ such that $x, y \in D_n$. Hence $x/2^n, y/2^n \in D_0 = D - x_0$, and $\dfrac{x+y}{2^{n+1}} = \dfrac{1}{2}\left(\dfrac{x}{2^n} + \dfrac{y}{2^n}\right) \in D_0$, whence $\dfrac{x+y}{2} \in D_n$. Now,

$$f_1 \left(\frac{x+y}{2}\right) = 2^n f_0 \left(\frac{x+y}{2^{n+1}}\right)$$

$$= 2^n \left[\frac{1}{2} f_0 \left(\frac{x}{2^n}\right) + \frac{1}{2} f_0 \left(\frac{y}{2^n}\right)\right]$$

$$= \frac{1}{2} \left[2^n f_0 \left(\frac{x}{2^n}\right) + 2^n f_0 \left(\frac{y}{2^n}\right)\right] = \frac{f_1(x) + f_1(y)}{2}.$$

Relation (13.2.6) results from (13.2.10) for $n = 0$.

To prove the uniqueness, suppose that a function $f_2 : \mathbb{R}^N \to \mathbb{R}$ satisfies equation (13.2.1) in $\mathbb{R}^N$ and fulfils the condition

$$f_2(x) = f_0(x) \quad \text{for} \quad x \in D - x_0. \tag{13.2.11}$$

By (13.2.11) and (13.2.7) $f_2(0) = f_0(0) = 0$, and hence, by Lemma 13.2.1, $f_2 \left(\dfrac{x}{2^n}\right) = \dfrac{1}{2^n} f_2(x)$, $x \in \mathbb{R}^N$, $n \in \mathbb{N} \cup \{0\}$. Take an arbitrary $x \in \mathbb{R}^N$. By (13.2.9) there exists an $n \in \mathbb{N} \cup \{0\}$ such that $x \in D_n = 2^n D_0$, whence $x/2^n \in D_0 = D - x_0$. Thus we have by (13.2.11) and (13.2.10)

$$f_2(x) = 2^n f_2 \left(\frac{x}{2^n}\right) = 2^n f_0 \left(\frac{x}{2^n}\right) = f_1(x).$$

Consequently $f_2 = f_1$ in $\mathbb{R}^N$. □

Lemma 13.2.3. *Let a function* $f : \mathbb{R}^N \to \mathbb{R}$ *satisfy equation* (13.2.1) *and relation* (13.2.2). *Then* f *is additive.*

Proof. We have by Lemma 13.2.1 for arbitrary $x, y \in \mathbb{R}^N$

$$f(x + y) = 2f \left(\frac{x+y}{2}\right) = 2 \frac{f(x) + f(y)}{2} = f(x) + f(y),$$

i.e., f is additive. □

Theorem 13.2.1. *Let* $D \subset \mathbb{R}^N$ *be a convex set such that* $\operatorname{int} D \neq \varnothing$, *and let* $f : D \to \mathbb{R}$ *be a solution of equation* (13.2.1). *Then there exist an additive function* $g : \mathbb{R}^N \to \mathbb{R}$ *and a constant* $a \in \mathbb{R}$ *such that*

$$f(x) = g(x) + a \quad \text{for} \quad x \in D. \tag{13.2.12}$$

Proof. Fix an $x_0 \in \operatorname{int} D$ and define the function $f_0 : (D - x_0) \to \mathbb{R}$ by (13.2.5). By Lemma 13.2.2 there exists a function $f_1 : \mathbb{R}^N \to \mathbb{R}$ satisfying equation (13.2.1) and condition (13.2.6). Hence by (13.2.7) $f_1(0) = f_0(0) = 0$. By Lemma 13.2.3 f_1 is additive.

For arbitrary $x \in D$ we have $x - x_0 \in D - x_0$, whence by (13.2.5) and (13.2.6)

$$f(x) = f\big(x_0 + (x - x_0)\big) = f_0(x - x_0) + f(x_0) = f_1(x - x_0) + f(x_0).$$

Since f_1 is additive, we get hence

$$f(x) = f_1(x) - f_1(x_0) + f(x_0). \tag{13.2.13}$$

Put $g = f_1$, $a = f(x_0) - f_1(x_0)$. Relation (13.2.12) results now from (13.2.13). $\qquad \square$

It is easily seen that every function of form (13.2.12), where $g : \mathbb{R}^N \to \mathbb{R}$ is additive, actually satisfies equation (13.2.1).

Theorem 13.2.2. *Let $D \subset \mathbb{R}^N$ be a convex set such that $\operatorname{int} D \neq \varnothing$. A function $f : D \to \mathbb{R}$ is a continuous solution of equation (13.2.1) if and only if*

$$f(x) = cx + a, \quad x \in D, \tag{13.2.14}$$

with certain constants $c \in \mathbb{R}^N$, $a \in \mathbb{R}$.

Proof. By Theorem 13.2.1 we have (13.2.12) with an additive $g : \mathbb{R}^N \to \mathbb{R}$. We have $g = f - a$ in D, and since f is continuous, also g is continuous in $\operatorname{int} D$ and hence (Theorem 6.4.3) in $\mathbb{R}^N$. By Theorem 5.5.2 $g(x) = cx$ with a certain constant $c \in \mathbb{R}^N$. Thus (13.2.14) results from (13.2.12).

It is obvious that every function of form (13.2.14) is a continuous solution of (13.2.1). $\qquad \square$

Even if D is convex and open, then, using the fact that f fulfilling (13.2.1) is convex, we can infer on its continuity only from the boundedness above on a suitable set. If D is not open, then we cannot do even this, as it has been pointed out at the beginning of this section. But from Theorem 13.2.1 we obtain the following result.

Theorem 13.2.3. *Let $D \subset \mathbb{R}^N$ be a convex set such that $\operatorname{int} D \neq \varnothing$, and let $f : D \to \mathbb{R}$ be a solution of equation (13.2.1). If f is measurable, or is bounded above, or below, on a set $T \in \mathfrak{A}$, then f is continuous.*

Proof. By Theorem 13.2.1 we have (13.2.12), and if f is measurable, then so is also $g \mid \operatorname{int} D$. The function $g \mid \operatorname{int} D$ is convex (Theorem 5.1.5), and hence, by Theorem 9.4.1 is continuous. By Theorem 6.4.3 g is continuous, and thus, by (13.2.12), f is continuous.

If f is bounded above, or below, on a set $T \subset D$, then so is also g. Since $T \in \mathfrak{A}$, this implies that g is continuous (if g is bounded below on T, then $-g$ is bounded above on T), and hence, by (13.2.12), f is continuous. $\qquad \square$

13.3 Pexider equations

These are the equations (Pexider [254], Aczél [5])

$$f(x+y) = g(x) + h(y), \qquad (13.3.1)$$
$$f(x+y) = g(x)h(y), \qquad (13.3.2)$$
$$f(xy) = g(x) + h(y), \qquad (13.3.3)$$
$$f(xy) = g(x)h(y), \qquad (13.3.4)$$

with three unknown functions f, g, h. Equation (13.3.1) has a simple interpretation: the value of f at $x + y$ is the sum of two values, of which one depends only on x, and the other only on y (separation of variables). Similar interpretations can also be given for equations (13.3.2)–(13.3.4).

Equations (13.3.1)–(13.3.4) can easily be reduced to the Cauchy equations (5.2.1) and (13.1.1)–(13.1.3). Below we give theorems to this effect.

Theorem 13.3.1. *Let* $f : \mathbb{R}^N \to \mathbb{R}$, $g : \mathbb{R}^N \to \mathbb{R}$, *and* $h : \mathbb{R}^N \to \mathbb{R}$ *satisfy equation* (13.3.1). *Then there exist an additive function* $f_0 : \mathbb{R}^N \to \mathbb{R}$ *and constants* $a, b \in \mathbb{R}$ *such that*

$$f = f_0 + a + b, \quad g = f_0 + a, \quad h = f_0 + b. \qquad (13.3.5)$$

Proof. Put $a = g(0)$, $b = h(0)$, $f_0 = f - a - b$. Setting in (13.3.1) $y = 0$ we obtain $f(x) = g(x) + b$, $x \in \mathbb{R}^N$, or $f_0 + a + b = g + b$, i.e., $g = f_0 + a$. Similarly, setting in (13.3.1) $x = 0$ and writing x instead of y, we obtain $f(x) = a + h(x)$, $x \in \mathbb{R}^N$, or $f_0 + a + b = a + h$, i.e., $h = f_0 + b$. The formula $f = f_0 + a + b$ results from the definition of f_0.

It remains to check that $f_0 : \mathbb{R}^N \to \mathbb{R}$ is additive. We have by (13.3.5) and (13.3.1)

$$f_0(x+y) = f(x+y) - a - b = g(x) - a + h(y) - b = f_0(x) + f_0(y).$$

Consequently f_0 is additive. $\qquad \square$

On the other hand, it is easily seen that every triple (f, g, h) of form (13.3.5), where $f_0 : \mathbb{R}^N \to \mathbb{R}$ is additive, and $a, b \in \mathbb{R}$ are arbitrary constants, actually satisfies equation (13.3.1).

Before we give the general solution of equation (13.3.2), note that every triple as follows:

$$\begin{cases} f = 0, \\ g \text{ arbitrary}, \\ h = 0, \end{cases} \qquad \begin{cases} f = 0, \\ g = 0, \\ h \text{ arbitrary}, \end{cases} \qquad (13.3.6)$$

is a solution of (13.3.2).

Theorem 13.3.2. *Let* $f : \mathbb{R}^N \to \mathbb{R}$, $g : \mathbb{R}^N \to \mathbb{R}$, *and* $h : \mathbb{R}^N \to \mathbb{R}$ *satisfy equation* (13.3.2). *Then either* f, g, h *have one of forms* (13.3.6), *or there exist a function* $f_0 : \mathbb{R}^N \to \mathbb{R}$ *satisfying equation* (13.1.1) *and constants* $a, b \in \mathbb{R} \setminus \{0\}$ *such that*

$$f = abf_0, \quad g = af_0, \quad h = bf_0. \qquad (13.3.7)$$

Proof. If $f(0) = 0$, then setting in (13.3.2) $x = y = 0$ we obtain that either $g(0) = 0$, or $h(0) = 0$. If $g(0) = 0$, then setting in (13.3.2) $x = 0$ we obtain $f = 0$. Similarly, if $h(0) = 0$, then setting in (13.3.2) $y = 0$ we obtain that $f = 0$. Thus $f(0) = 0$ implies that $f = 0$ in $\mathbb{R}^N$. If there existed $x, y \in \mathbb{R}^N$ such that $g(x) \neq 0$ and $h(y) \neq 0$, then we would have by (13.3.2) $f(x + y) = g(x)h(y) \neq 0$, which is impossible. Consequently at least one of the functions g, h must be identically zero. Thus f, g, h must have one of forms (13.3.6).

So let $f(0) \neq 0$. Then also $g(0) \neq 0$ and $h(0) \neq 0$ $\big($as may be seen from (13.3.2) for $x = y = 0\big)$. Put $a = g(0)$, $b = h(0)$, $f_0 = a^{-1}b^{-1}f$. Then, setting in (13.3.2) $y = 0$ we obtain $f = bg$, or $abf_0 = bg$, i.e., $g = af_0$. Similarly setting in (13.3.2) $x = 0$ we obtain $f = ah$, or $abf_0 = ah$, i.e., $h = bf_0$. The formula $f = abf_0$ results from the definition of f_0. Now, for arbitrary $x, y \in \mathbb{R}^N$,

$$f_0(x + y) = a^{-1}b^{-1}f(x + y) = a^{-1}b^{-1}g(x)h(y) = f_0(x)f_0(y),$$

i.e., f_0 satisfies (13.1.1). $\qquad\qquad\qquad\qquad\qquad\qquad\qquad\qquad\qquad\qquad\qquad\quad$ $\square$

On the other hand, it is easily seen that every triple of form (13.3.7), where $f_0 : \mathbb{R}^N \to \mathbb{R}$ satisfies (13.1.1), and $a, b \in \mathbb{R} \setminus \{0\}$ are arbitrary constants, actually satisfies equation (13.3.2).

Theorem 13.3.3. *Let $D = (0, \infty)$, or $D = [0, \infty)$, or $D = (-\infty, 0) \cup (0, \infty)$, or $D = \mathbb{R}$, and let $f : D \to \mathbb{R}$, $g : D \to \mathbb{R}$, and $h : D \to \mathbb{R}$ satisfy equation (13.3.3). Then there exist a function $f_0 : D \to \mathbb{R}$ satisfying equation (13.1.2) and constants $a, b \in \mathbb{R}$ such that (13.3.5) holds*[2].

Proof. Put $a = g(1)$, $b = h(1)$, $f_0 = f - a - b$. Setting in (13.3.3) in turn $y = 1$ and $x = 1$ we obtain (13.3.5). Then

$$f_0(xy) = f(xy) - a - b = g(x) - a + h(y) - b = f_0(x) + f_0(y),$$

i.e., f_0 satisfies (13.1.2). $\qquad\qquad\qquad\qquad\qquad\qquad\qquad\qquad\qquad\qquad\qquad\quad$ $\square$

On the other hand, it is easily seen that every triple of form (13.3.5), where $f_0 : D \to \mathbb{R}$ satisfies (13.1.2), and $a, b \in \mathbb{R}$ are arbitrary constants, actually satisfies equation (13.3.3).

Lemma 13.3.1. *Let D be one of sets (13.1.6), and let $f : D \to \mathbb{R}$, $g : D \to \mathbb{R}$ and $h : D \to \mathbb{R}$ satisfy equation (13.3.4). If there exists a point $x_0 \in D$, $x_0 \neq 0$, at which one of the functions f, g, h is zero, then f, g, h have one of forms (13.3.6) in D.*

Proof. We will give the proof for $D = \mathbb{R}$. If $D \neq \mathbb{R}$, then simply some parts of the proof can be omitted.

If $f(x_0) = 0$, then $g\left(\sqrt{|x_0|}\right) h\left(\sqrt{|x_0|}\operatorname{sgn} x_0\right) = f(x_0) = 0$, whence either $g\left(\sqrt{|x_0|}\right) = 0$ or $h\left(\sqrt{|x_0|}\operatorname{sgn} x_0\right) = 0$. So we may assume that $g(x_0) = 0$ or $h(x_0) = 0$. If $x_0 < 0$, then $g(-x_0)h(-x_0) = f(x_0^2) = g(x_0)h(x_0) = 0$, whence either $g(-x_0) =$

---

[2] If $0 \in D$, then, according to Lemma 13.1.2, necessarily $f_0 = 0$. Then $f = a + b = \text{const}$, $g = a = \text{const}$, and $h = b = \text{const}$.

0, or $h(-x_0) = 0$. So we may assume that $x_0 > 0$. For the argument's sake we assume that $x_0 \in (0, 1)$. If $x_0 \in [1, \infty)$ the proof is similar.

For $x \in (0, x_0)$ we have $f(x) = f\left(x_0 \dfrac{x}{x_0}\right) = g(x_0)h\left(\dfrac{x}{x_0}\right) = g\left(\dfrac{x}{x_0}\right)h(x_0) = 0$.

So f is zero in $(0, x_0)$.

For every $x \in (0, x_0)$ also $x^2 \in (0, x_0)$, whence $0 = f\left(x^2\right) = g(x)h(x)$ and either $g(x) = 0$, or $h(x) = 0$. We will show that

$$f(x) = 0 \text{ and either } g(x) = 0 \text{ or } h(x) = 0 \tag{13.3.8}$$

for every $x \in \left(0, x_0^{2-n}\right)$, $n \in \mathbb{N} \cup \{0\}$. We have already proved this for $n = 0$. Suppose it true for an $n \in \mathbb{N} \cup \{0\}$, and take an arbitrary $x \in \left(0, x_0^{2-(n+1)}\right)$. Then $x^2 \in \left(0, x_0^{2-n}\right)$, and

$$0 = f\left(x^2\right) = g(x)h(x),$$

whence either $g(x) = 0$ or $h(x) = 0$. Now take a $y \in \left(x, x_0^{2-(n+1)}\right)$. Then

$$f(x) = f\left(y\dfrac{x}{y}\right) = g(y)h\left(\dfrac{x}{y}\right) = g\left(\dfrac{x}{y}\right)h(y) = 0.$$

Consequently (13.3.8) holds for $x \in \left(0, x_0^{2-(n+1)}\right)$. Induction shows that (13.3.8) holds in every interval $\left(0, x_0^{2-n}\right)$, $n \in \mathbb{N} \cup \{0\}$. Since every $x \in (0, 1)$ belongs to a certain interval $\left(0, x_0^{2-n}\right)$ for an $n \in \mathbb{N} \cup \{0\}$, (13.3.8) holds for all $x \in (0, 1)$.

Now take an $x \in (1, \infty)$ and a $y \in (0, 1)$. Then $x/y > 1$ and $f(x) = f\left(y\dfrac{x}{y}\right) = g(y)h\left(\dfrac{x}{y}\right) = g\left(\dfrac{x}{y}\right)h(y) = 0$. Thus f is zero in $(1, \infty)$. For arbitrary $x \in (1, \infty)$ also $x^2 \in (1, \infty)$, and $0 = f(x^2) = g(x)h(x)$, whence either $g(x) = 0$, or $h(x) = 0$. Consequently we have (13.3.8) for every $x \in (0, 1) \cup (1, \infty)$. Now take an arbitrary $x \in (0, 1) \cup (1, \infty)$. Then

$$f(1) = f\left(xx^{-1}\right) = g(x)h\left(x^{-1}\right) = 0.$$

Hence $0 = f(1) = g(1)h(1)$, and either $g(1) = 0$, or $h(1) = 0$. Thus (13.3.8) holds for $x \in (0, \infty)$. Now, for any $x \in (0, \infty)$, we have

$$f(0) = f(0x) = g(0)h(x) = g(x)h(0) = 0,$$

whence $0 = f(0) = g(0)h(0)$, and either $g(0) = 0$, or $h(0) = 0$. Thus (13.3.8) holds for $x \in [0, \infty)$.

Take an $x < 0$. Then $x^2 \in (0, \infty)$, and $0 = f(x^2) = g(x)h(x)$, whence either $g(x) = 0$, or $h(x) = 0$. Further, $\sqrt{|x|} \in (0, \infty)$, whence $f(x) = g\left(\sqrt{|x|}\right)h\left(-\sqrt{|x|}\right) = g\left(-\sqrt{|x|}\right)h\left(\sqrt{|x|}\right) = 0$. Thus (13.3.8) holds for every $x \in \mathbb{R}$.

If there existed $x, y \in \mathbb{R}$ such that $g(x) \neq 0$ and $h(y) \neq 0$, then we would have $f(xy) = g(x)h(y) \neq 0$, contrary to (13.3.8). Consequently one of the functions g, h must be identically zero, and we obtain (13.3.6). $\square$

Theorem 13.3.4. *Let D be one of sets (13.1.6), and let $f : D \to \mathbb{R}$, $g : D \to \mathbb{R}$, $h : D \to \mathbb{R}$ satisfy equation (13.3.4). Then either f, g, h have one of forms (13.3.6), or there exist a function $f_0 : D \to \mathbb{R}$ satisfying equation (13.1.3) and constants $a, b \in \mathbb{R} \setminus \{0\}$ such that (13.3.7) holds in D.*

Proof. Suppose that f, g, h have not one of forms (13.3.6). By Lemma 13.3.1 $f(x) \neq 0$, $g(x) \neq 0$, $h(x) \neq 0$ for all $x \in D \setminus \{0\}$. If $1 \in D$, then we put $a = g(1)$, $b = h(1)$, $f_0 = a^{-1}b^{-1}f$, and the proof runs as that of Theorem 13.3.3. The case where $1 \notin D$, is more difficult. We will give here a proof for $D = (-1, 1)$. For other D of form (13.1.6) the proof is similar.

By (13.3.4) we have $f(xy) = g(x)h(y) = g(y)h(x)$, whence for $x \in (-1, 0) \cup (0, 1) = D \setminus \{0\}$

$$\frac{h(x)}{g(x)} = \frac{h(y)}{g(y)}.$$

Thus $h(x)/g(x) = c = \text{const}$, and

$$h(x) = cg(x), \quad x \in (-1, 0) \cup (0, 1). \tag{13.3.9}$$

Equation (13.3.4) takes now the form

$$f(xy) = cg(x)g(y), \quad x, y \in (-1, 0) \cup (0, 1). \tag{13.3.10}$$

Now we show that each of the functions f, g has a constant sign in $(0, 1)$. For $x \in (0, 1)$ we have by (13.3.10)

$$f(x) = c\left[g\left(\sqrt{x}\right)\right]^2,$$

and so $\text{sgn } f(x) = \text{sgn } c = \text{const}$. If there existed $x, y \in (0, 1)$ such that $g(x)g(y) < 0$, then we would have

$$f(xy) = -c|g(x)g(y)|, \tag{13.3.11}$$

whence $\text{sgn } c = \text{sgn } f(xy) = -\text{sgn } c$, and $\text{sgn } c = 0$, which is impossible. Consequently also g has a constant sign in $(0, 1)$. Similarly we prove that g has a constant sign in $(-1, 0)$ (possibly different from the sign of g in $(0, 1)$). If there existed $x, y \in (-1, 0)$ such that $g(x)g(y) < 0$, then $xy \in (0, 1)$, $\text{sgn } f(xy) = \text{sgn } c$, and we obtain (13.3.11), which again leads us to a contradiction.

Put $\varphi_0(t) = \log|f(e^t)|$, $\gamma_0(t) = \log|g(e^t)|$, $\lambda = \log|c|$, $t \in (-\infty, 0)$. Then, for arbitrary $u, v \in (-\infty, 0)$ $\varphi_0(u + v) = \log|f(e^{u+v})| = \log|f(e^u e^v)| = \log|cg(e^u)g(e^v)| = \log|g(e^u)| + \log|g(e^v)| + \log|c| = \gamma_0(u) + \gamma_0(v) + \lambda$, or, with $\varphi = \varphi_0 + \lambda$, $\gamma = \gamma_0 + \lambda$,

$$\varphi(u + v) = \gamma(u) + \gamma(v). \tag{13.3.12}$$

Hence

$$\gamma(u) + \gamma(v) = \varphi(u + v) = \gamma\left(\frac{u+v}{2}\right) + \gamma\left(\frac{u+v}{2}\right) = 2\gamma\left(\frac{u+v}{2}\right),$$

i.e.,

$$\gamma\left(\frac{u+v}{2}\right) = \frac{\gamma(u) + \gamma(v)}{2}.$$

By Theorem 13.2.1 there exist an additive function $\alpha : \mathbb{R} \to \mathbb{R}$ and a constant $\mu \in \mathbb{R}$ such that

$$\gamma(t) = \alpha(t) + \mu \quad \text{for } t \in (-\infty, 0).$$

Hence, by (13.3.12), for $t \in (-\infty, 0)$,

$$\varphi(t) = \gamma\left(\frac{t}{2}\right) + \gamma\left(\frac{t}{2}\right) = 2\gamma\left(\frac{t}{2}\right) = \alpha(t) + 2\mu.$$

Thus, with $\lambda_1 = \mu - \lambda$, $\lambda_2 = 2\mu - \lambda$,

$$\varphi_0(t) = \alpha(t) + \lambda_2, \quad \gamma_0(t) = \alpha(t) + \lambda_1, \quad t \in (-\infty, 0),$$

and

$$|f(e^t)| = \exp\left(\alpha(t) + \lambda_2\right), \quad |g(e^t)| = \exp\left(\alpha(t) + \lambda_1\right), \quad t \in (-\infty, 0),$$

or with $c_1' = \exp\lambda_1$, $c_2' = \exp\lambda_2$,

$$|f(e^t)| = c_2' \exp\alpha(t), \quad |g(e^t)| = c_1' \exp\alpha(t), \quad t \in (-\infty, 0).$$

Writing $c_1 = c_1' \operatorname{sgn} g$, $c_2 = c_2' \operatorname{sgn} f$ (the sign in $(0,1)$), we have

$$f(x) = c_2 \exp\alpha(\log x), \quad g(x) = c_1 \exp\alpha(\log x), \quad x \in (0,1). \tag{13.3.13}$$

The function $f_1(x) = \exp\alpha(\log x)$, $x \in (0,1)$, satisfies equation (13.1.3):

$$f_1(xy) = \exp\alpha(\log xy) = \exp\alpha(\log x + \log y)$$
$$= \exp\left[\alpha(\log x) + \alpha(\log y)\right] = \left[\exp\alpha(\log x)\right]\left[\exp\alpha(\log y)\right]$$
$$= f_1(x)f_1(y).$$

Hence, by (13.3.13) and (13.3.9), for $x \in (0,1)$,

$$f(x) = c_2 f_1(x), \quad g(x) = c_1 f_1(x), \quad h(x) = cc_1 f_1(x). \tag{13.3.14}$$

If we write $a = c_1$, $b = cc_1$, and insert (13.3.14) into (13.3.4), we obtain

$$c_2 f_1(xy) = f(xy) = g(x)h(y) = af_1(x)bf_1(y) = abf_1(xy),$$

whence $c_2 = ab$. So we obtain from (13.3.14) for $x \in (0,1)$

$$f(x) = abf_1(x), \quad g(x) = af_1(x), \quad h(x) = bf_1(x). \tag{13.3.15}$$

Now, for $x \in (-1, 0)$ we have by (13.3.10) and (13.3.15)

$$c[g(x)]^2 = f(x^2) = c[g(|x|)]^2 = ca^2[f_1(|x|)]^2,$$

whence

$$g(x) = \varepsilon af_1(|x|), \quad x \in (-1, 0), \tag{13.3.16}$$

where $\varepsilon = +1$ or -1 (remember that g has a constant sign in $(-1,0)$!). With a $y \in \big(|x|, 1\big)$ we get hence according to (13.3.10), (13.3.15) and the fact that f_1 satisfies (13.1.3)

$$f(x) = f\left(y\frac{x}{y}\right) = cg(y)g\left(\frac{x}{y}\right) = c\varepsilon a^2 f_1(y)f_1\left(\frac{|x|}{y}\right)$$
$$= c\varepsilon a^2 f_1\big(|x|\big) = \varepsilon a b f_1\big(|x|\big). \tag{13.3.17}$$

Similarly, for $x \in (-1, 0)$, we have by (13.3.9), and (13.3.16)

$$h(x) = cg(x) = c\varepsilon a f_1\big(|x|\big) = \varepsilon b f_1\big(|x|\big). \tag{13.3.18}$$

If

$$g(0) = 0, \tag{13.3.19}$$

then by (13.3.4)

$$f(0) = 0, \tag{13.3.20}$$

and, taking an $x \in (0, 1)$, we obtain from (13.3.4) $0 = f(0) = f(x0) = g(x)h(0)$, whence

$$h(0) = 0, \tag{13.3.21}$$

since $g(x) \neq 0$ for $x \neq 0$. If $g(0) \neq 0$, then by (13.3.4), with an $x \in (0, 1)$, $g(x)h(0) = f(0) = g(0)h(x) \neq 0$, whence also $f(0) \neq 0$ and $h(0) \neq 0$. Let

$$a' = g(0) \neq 0, \quad b' = h(0) \neq 0. \tag{13.3.22}$$

By (13.3.4)

$$f(0) = g(0)h(0) = a'b'. \tag{13.3.23}$$

Now take an arbitrary $x \in (0, 1)$. Then by (13.3.22), (13.3.23) and (13.3.15)

$$a'b' = f(0) = g(0)h(x) = a'bf_1(x),$$

whence

$$f_1(x) = \frac{b'}{b} = \text{const.} \tag{13.3.24}$$

Since f_1 satisfies (13.1.3), we have for arbitrary $x, y \in (0, 1)$

$$\frac{b'}{b} = f_1(xy) = f_1(x)f_1(y) = \left(\frac{b'}{b}\right)^2,$$

whence $b'/b = 1$ (since $b'/b \neq 0$), and

$$b' = b. \tag{13.3.25}$$

Next, by (13.3.25), (13.3.23), (13.3.4), (13.3.15), (13.3.22), again (13.3.25), and (13.3.24)

$$a'b = a'b' = f(0) = g(x)h(0) = abf_1(x) = ab' = ab,$$

whence

$$a' = a, \tag{13.3.26}$$

and

$$a'b' = ab. \tag{13.3.27}$$

Now put

$$f_0(x) = \begin{cases} f_1(x) & \text{for } x \in (0,1), \\ \eta & \text{for } x = 0, \\ \varepsilon f_1(|x|) & \text{for } x \in (-1,0), \end{cases} \tag{13.3.28}$$

where $\eta = 0$ if $g(0) = 0$, and $\eta = 1$ if $g(0) \neq 0$ (note that in the latter case $f_1 = 1$). We will show that function (13.3.28) satisfies equation (13.1.3) in $(-1,1)$.

If $x > 0$, $y > 0$, then $xy > 0$, and $f_0(xy) = f_1(xy) = f_1(x)f_1(y) = f_0(x)f_0(y)$. If $x < 0$, $y < 0$, then $xy > 0$, and $f_0(xy) = f_1(xy) = f_1(|xy|) = f_1(|x|) f_1(|y|) = \varepsilon^2 f_1(|x|) f_1(|y|) = [\varepsilon f_1(|x|)] [\varepsilon f_1(|y|)] = f_0(x)f_0(y)$. If $x > 0$, $y < 0$, then $xy < 0$, and $f_0(xy) = \varepsilon f_1(|xy|) = \varepsilon f_1(x|y|) = f_1(x)\varepsilon f_1(|y|) = f_0(x)f_0(y)$. If $x < 0$, $y > 0$, the argument is analogous.

If $x = y = 0$, then $xy = 0$, and $f_0(xy) = f_0(0) = \eta = \eta^2 = f_0(0)f_0(0) = f_0(x)f_0(y)$.

In order to consider the cases, where one of x, y is zero, we will distinguish two cases.

I. $g(0) = 0$. Then $\eta = 0$. If $x = 0$, then we have for arbitrary $y \in (-1, 1)$ $f_0(xy) = f_0(0) = 0 = 0 f_0(y) = f_0(x)f_0(y)$. If $y = 0$, the argument is analogous.

II. $g(0) \neq 0$. Then $\eta = 1$, $g = a$ and $\varepsilon = 1$. If $x = 0$, $y \neq 0$, then $xy = 0$, and $f_0(xy) = 1 = f_0(x)f_0(y)$. If $x \neq 0$, $y = 0$, the argument is analogous.

Now, by (13.3.15)–(13.3.23) and (13.3.25)–(13.3.27) we have (13.3.7) for all $x \in (-1, 1) = D$. The proof is completed. $\qquad\square$

On the other hand, it is easily seen that every triple of form (13.3.7), where $f_0 : D \to \mathbb{R}$ satisfies (13.1.3), and $a, b \in \mathbb{R} \setminus \{0\}$ are arbitrary constants, actually satisfies equation (13.3.4).

Now we will determine the continuous solutions of equations (13.3.1)–(13.3.4).

Theorem 13.3.5. *Functions $f : \mathbb{R}^N \to \mathbb{R}$, $g : \mathbb{R}^N \to \mathbb{R}$ and $h : \mathbb{R}^N \to \mathbb{R}$ are a continuous solution of equation (13.3.1) if and only if there exist constants $a, b \in \mathbb{R}$ and $c \in \mathbb{R}^N$ such that*

$$f(x) = cx + a + b, \quad g(x) = cx + a, \quad h(x) = cx + b, \quad x \in \mathbb{R}^N. \tag{13.3.29}$$

Proof. The functions f, g, h have form (13.3.5). Hence if any one of f, g, h is continuous, then so is also f_0, and by Theorem 5.5.2 $f_0(x) = cx$ with a certain $c \in \mathbb{R}^N$. (13.3.29) results now from (13.3.5).

It is obvious that every triple of form (13.3.29), where $a, b \in \mathbb{R}$ and $c \in \mathbb{R}^N$ are arbitrary constants, represents a continuous solution of equation (13.3.1). $\qquad\square$

An analogous theorem is not true for equation (13.3.2). The functions

$$\begin{cases} f = 0, \\ g \text{ arbitrary continuous,} \\ h = 0, \end{cases} \qquad \begin{cases} f = 0, \\ g = 0, \\ h \text{ arbitrary continuous,} \end{cases} \qquad (13.3.30)$$

are continuous solutions of (13.3.2), and nothing more can be said about g resp. h in (13.3.30). But we have the following

Theorem 13.3.6. *Functions $f : \mathbb{R}^N \to \mathbb{R}$, $g : \mathbb{R}^N \to \mathbb{R}$ and $h : \mathbb{R}^N \to \mathbb{R}$ are a continuous solution of equation (13.3.2) if either they have one of the forms (13.3.30), or there exist constants $a, b \in \mathbb{R} \setminus \{0\}$ and $c \in \mathbb{R}^N$ such that*

$$f(x) = abe^{cx}, \quad g(x) = ae^{cx}, \quad h(x) = be^{cx}, \quad x \in \mathbb{R}^N.$$

Proof. This results from Theorems 13.3.2 and 13.1.4. (Note that the case, where $f_0 = 0$, is contained in (13.3.30).) □

Theorem 13.3.7. *Let $D = (0, \infty)$, or $D = [0, \infty)$, or $D = (-\infty, 0) \cup (0, \infty)$, or $D = \mathbb{R}$. Functions $f : D \to \mathbb{R}$, $g : D \to \mathbb{R}$ and $h : D \to \mathbb{R}$ are a continuous solution of equation (13.3.3) if and only if there exist constants $a, b, c \in \mathbb{R}$ such that*

$$f(x) = a + b, \quad g(x) = a, \quad h(x) = b, \quad x \in D,$$

whenever $0 \in D$, or

$$f(x) = c \log |x| + a + b, \quad g(x) = c \log |x| + a, \quad h(x) = c \log |x| + b, \quad x \in D,$$

whenever $0 \notin D$.

Proof. This results from Theorems 13.3.3 and 13.1.5, and Lemma 13.1.2. □

Theorem 13.3.8. *Let D be one of sets (13.1.6). Functions $f : D \to \mathbb{R}$, $g : D \to \mathbb{R}$ and $h : D \to \mathbb{R}$ are a continuous solution of equation (13.3.4) if and only if either they have one of forms (13.3.30), or there exist constants $a, b \in \mathbb{R} \setminus \{0\}$ and $c \in \mathbb{R}$ (moreover, $c > 0$ if $0 \in D$) such that for $x \in D$*

$$f(x) = ab, \quad g(x) = a, \quad h(x) = b,$$

or

$$f(x) = ab|x|^c, \quad g(x) = a|x|^c, \quad h(x) = b|x|^c,$$

or

$$f(x) = ab|x|^c \operatorname{sgn} x, \quad g(x) = a|x|^c \operatorname{sgn} x, \quad h(x) = b|x|^c \operatorname{sgn} x.$$

Proof. This results from Theorems 13.3.4 and 13.1.6. (The case, where $f_0 = 0$, is contained in (13.3.30).) □

Remark 13.3.1. If f, g, h are assumed to be continuous only in $D \setminus \{0\}$, then the condition $c > 0$ disappears, and the value of f, g, h at zero (unless f, g, h are constant) is zero.

Simple proofs of the following theorems are left to the reader.

Theorem 13.3.9. *Let* $f : \mathbb{R}^N \to \mathbb{R}$, $g : \mathbb{R}^N \to \mathbb{R}$ *and* $h : \mathbb{R}^N \to \mathbb{R}$ *satisfy equation* (13.3.1). *If one of the functions* f, g, h *is measurable, or bounded above, or below, on a set* $T \in \mathfrak{A}$, *then* f, g, h *are continuous.*

Theorem 13.3.10. *Let* $f : \mathbb{R}^N \to \mathbb{R}$, $g : \mathbb{R}^N \to \mathbb{R}$ *and* $h : \mathbb{R}^N \to \mathbb{R}$ *satisfy equation* (13.3.2). *If* $f \neq 0$, *and one of the functions* f, g, h *is measurable, or bounded on a set* $T \in \mathfrak{C}$, *then* f, g, h *are continuous.*

Theorem 13.3.11. *Let* $D = (0, \infty)$ *or* $D = (-\infty, 0) \cup (0, \infty)$, *and let* $f : D \to \mathbb{R}$, $g : D \to \mathbb{R}$ *and* $h : D \to \mathbb{R}$ *satisfy equation* (13.3.3). *If one of the functions* f, g, h *is measurable, or bounded above, or below, on a set* $T \subset D$ *such that* $m_i(T) > 0$, *or* T *is of the second category and with the Baire property, then* f, g, h *are continuous.*

Theorem 13.3.12. *Let* D *be one of sets* (13.1.6), *and let* $f : D \to \mathbb{R}$, $g : D \to \mathbb{R}$ *and* $h : D \to \mathbb{R}$ *satisfy equation* (13.3.4). *If* $f \neq 0$, *and one of the functions* f, g, h *is measurable, then* f, g, h *are continuous in* $D \setminus \{0\}$.

13.4 Multiadditive functions

Let $p \in \mathbb{N}$. A function $f : \mathbb{R}^{pN} \to \mathbb{R}$ is called *p-additive*, iff, for every i, $1 \leqslant i \leqslant p$, and for every $x_1, \ldots, x_p, y_i \in \mathbb{R}^N$

$$f(x_1, \ldots, x_{i-1}, x_i + y_i, x_{i+1}, \ldots, x_p)$$
$$= f(x_1, \ldots, x_p) + f(x_1, \ldots, x_{i-1}, y_i, x_{i+1}, \ldots, x_p), \quad (13.4.1)$$

i.e., f is additive in each of its variables $x_i \in \mathbb{R}^N$, $i = 1, \ldots, p$. A 2-additive function is called *biadditive*.

An example of a p-additive function is furnished by the product

$$f_1(x_1) \cdots f_p(x_p),$$

where $f_i : \mathbb{R}^N \to \mathbb{R}$, $i = 1, \ldots, p$, are additive.

At first let us note the following obvious

Lemma 13.4.1. *Let* $f_1, f_2 : \mathbb{R}^{pN} \to \mathbb{R}$ *be p-additive functions, and let* $a, b \in \mathbb{R}$ *be arbitrary constants. Then the function* $f = af_1 + bf_2$ *also is p-additive.*

Lemma 13.4.2. *Let* $f : \mathbb{R}^{pN} \to \mathbb{R}$ *be a p-additive function, and let* $(i_1, \ldots, i_p)$ *be an arbitrary permutation of* $(1, \ldots, p)$. *Define the function* $g : \mathbb{R}^{pN} \to \mathbb{R}$ *by* $g(x_1, \ldots, x_p) = f(x_{i_1}, \ldots, x_{i_p})$. *Then* g *is* p *additive.*

Keeping the $p - 1$ variables $x_1, \ldots, x_{i-1}, x_{i+1}, \ldots, x_p$ fixed, we obtain from Theorem 5.2.1

Theorem 13.4.1. *If* $f : \mathbb{R}^{pN} \to \mathbb{R}$ *is p-additive, then for every* $x_1, \ldots, x_p \in \mathbb{R}^N$, *every* $\lambda \in \mathbb{Q}$, *and every* i, $1 \leqslant i \leqslant p$,

$$f(x_1, \ldots, x_{i-1}, \lambda x_i, x_{i+1}, \ldots, x_p) = \lambda f(x_1, \ldots, x_p). \quad (13.4.2)$$

Hence we obtain

Lemma 13.4.3. *If $f : \mathbb{R}^{pN} \to \mathbb{R}$ is a p-additive function, and f is bounded in $\mathbb{R}^{pN}$, then $f = 0$.*

Proof. Suppose that there exists a point $(\overline{x}_1, \ldots, \overline{x}_p) \in \mathbb{R}^{pN}$ such that $f(\overline{x}_1, \ldots, \overline{x}_p) \neq 0$. Then we have by Theorem 13.4.1

$$f(k\overline{x}_1, \overline{x}_2, \ldots, \overline{x}_p) = kf(\overline{x}_1, \ldots, \overline{x}_p)$$

for all $k \in \mathbb{Z}$, and thus f cannot be bounded in $\mathbb{R}^{pN}$. $\square$

A multiadditive function cannot, in general, be expressed by additive functions, but its general construction can be found with the aid of Hamel basis (Aczél [4]).

Theorem 13.4.2. *Let $H \subset \mathbb{R}^N$ be a Hamel basis, and let $g : H^p \to \mathbb{R}$ be an arbitrary function. Then there exists a unique p-additive function $f : \mathbb{R}^{pN} \to \mathbb{R}$ such that $f \mid H^p = g$.*

Proof. Suppose that $f : \mathbb{R}^{pN} \to \mathbb{R}$ is a p-additive function such that $f \mid H^p = g$. Take arbitrary $x_1, \ldots, x_p \in \mathbb{R}^N$. Every x_i has a representation

$$x_i = \sum_{j=1}^{n_i} \lambda_{ij} h_{ij}, \quad i = 1, \ldots, p, \tag{13.4.3}$$

where $\lambda_{ij} \in \mathbb{Q}$, $h_{ij} \in H$, $i = 1, \ldots, p; j = 1, \ldots, n_i$. Then by (13.4.1) and (13.4.2)

$$f(x_1, \ldots, x_p) = \sum_{j_1=1}^{n_1} \sum_{j_2=1}^{n_2} \cdots \sum_{j_p=1}^{n_p} \left(\prod_{i=1}^{p} \lambda_{ij_i} \right) f\left(h_{1j_1}, \ldots, h_{pj_p} \right) \tag{13.4.4}$$

$$= \sum_{j_1=1}^{n_1} \sum_{j_2=1}^{n_2} \cdots \sum_{j_p=1}^{n_p} \left(\prod_{i=1}^{p} \lambda_{ij_i} \right) g\left(h_{1j_1}, \ldots, h_{pj_p} \right).$$

Consequently, if such an f exists, it must have form (13.4.4), and hence it is unique. On the other hand, it is easily checked that a function $f : \mathbb{R}^{pN} \to \mathbb{R}$ given by (13.4.4) is p-additive. If $x_1 = h_1, \ldots, x_p = h_p$ are elements of H, then in (13.4.3) every $n_i = 1$, and $\lambda_{ij} = \lambda_{i1} = 1$, $i = 1, \ldots, p$. Hence by (13.4.4)

$$f(h_1, \ldots, h_p) = \prod_{i=1}^{p} \lambda_{i1} g(h_1, \ldots, h_p) = g(h_1, \ldots, h_p).$$

Thus $f \mid H^p = g$. $\square$

It is obvious that the above theorem gives the general construction of p-additive functions, since every p-additive function $f : \mathbb{R}^{pN} \to \mathbb{R}$ is the unique p-additive extension of $g = f \mid H^p$.

Corollary 13.4.1. *Let $H \subset \mathbb{R}^N$ be a Hamel basis, and let $g : H^p \to \mathbb{R}$ be a function. Further, let $f : \mathbb{R}^{pN} \to \mathbb{R}$ be the p-additive extension of g. Then f is symmetric [antisymmetric] if and only if g is symmetric [antisymmetric].*

Proof. If f is symmetric [antisymmetric] then so is also $g = f \mid H^p$. Now suppose that g is symmetric [antisymmetric]. Then $g\big(h_{i_1 j_{i_1}}, \ldots, h_{i_p j_{i_p}}\big) = \varepsilon g\big(h_{1 j_1}, \ldots, h_{p j_p}\big)$, where $(i_1, \ldots, i_p)$ is an arbitrary permutation of $(1, \ldots, p)$, and $\varepsilon = +1$ if g is symmetric or g is antisymmetric and $(i_1, \ldots, i_p)$ is an even permutation, and $\varepsilon = -1$ if g is antisymmetric and the permutation $(i_1, \ldots, i_p)$ is odd. But then, according to (13.4.4),

$$
f(x_{i_1}, \ldots, x_{i_p}) = \sum_{j_{i_1}=1}^{n_{i_1}} \cdots \sum_{j_{i_p}=1}^{n_{i_p}} \left(\prod_{\kappa=1}^{p} \lambda_{i_\kappa j_{i_\kappa}} \right) g\big(h_{i_1 j_{i_1}}, \ldots, h_{i_p j_{i_p}}\big)
$$

$$
= \sum_{j_1=1}^{n_1} \cdots \sum_{j_p=1}^{n_p} \left(\prod_{i=1}^{p} \lambda_{iji} \right) \varepsilon g\big(h_{1 j_1}, \ldots, h_{p j_p}\big)
$$

$$
= \varepsilon f(x_1, \ldots, x_p),
$$

which means that f is also symmetric [antisymmetric]. $\qquad \square$

Theorem 13.4.3. *Let $f : \mathbb{R}^{pN} \to \mathbb{R}$ be a continuous p-additive function. Then there exist constants $c_{j_1 \ldots j_p} \in \mathbb{R}$, $j_1, \ldots, j_p = 1, \ldots, N$, such that*

$$
f(x_1, \ldots, x_p) = \sum_{j_1=1}^{N} \cdots \sum_{j_p=1}^{N} c_{j_1 \ldots j_p} x_{1 j_1} \cdots x_{p j_p}, \tag{13.4.5}
$$

where $x_i = (x_{i1}, \ldots, x_{iN})$, $i = 1, \ldots, p$.

Proof. We will prove (13.4.5) by induction on p. For $p = 1$ (13.4.5) holds in virtue of Theorem 5.5.2. Suppose (13.4.5) true for a $p \in \mathbb{N}$, and let $f : \mathbb{R}^{(p+1)N} \to \mathbb{R}$ be a continuous $(p + 1)$-additive function. Keep the p variables $x_1, \ldots, x_p$ fixed. Then the function $f(x_1, \ldots, x_p, y)$ as a function of y is additive and continuous, whence by Theorem 5.5.2 there exists a constant $c = (c_1, \ldots, c_N) \in \mathbb{R}^N$ such that

$$
f(x_1, \ldots, x_p, y) = cy = \sum_{i=1}^{N} c_i y_i \tag{13.4.6}
$$

where $y = (y_1, \ldots, y_N)$. Of course c, and hence every c_i, depends on $x_1, \ldots, x_p$:

$$
c_i = c_i(x_1, \ldots, x_p), \quad i = 1, \ldots, N. \tag{13.4.7}
$$

Take $\overline{y} = (0, \ldots, 0, 1, 0, \ldots, 0)$, where 1 stands at i-th place. Then we get by (13.4.6) and (13.4.7)

$$
c_i(x_1, \ldots, x_p) = f(x_1, \ldots, x_p, \overline{y}). \tag{13.4.8}
$$

Relation (13.4.8) shows, since f is $(p+1)$-additive, that c_i is p-additive. By the induction hypothesis

$$c_i(x_1, \ldots, x_p) = \sum_{j_1=1}^{N} \cdots \sum_{j_p=1}^{N} \bar{c}_{ij_1 \ldots j_p} x_{1j_1} \cdots x_{pj_p},$$

with certain constants $\bar{c}_{ij_1 \ldots j_p}$, whence by (13.4.6)

$$f(x_1, \ldots, x_p, x_{p+1}) = \sum_{j_{p+1}=1}^{N} \sum_{j_1=1}^{N} \cdots \sum_{j_p=1}^{N} \bar{c}_{j_{p+1}j_1 \ldots j_p} x_{1j_1} \cdots x_{pj_p} x_{p+1,j_{p+1}}.$$

Putting $c_{j_1 \ldots j_p j_{p+1}} = \bar{c}_{j_{p+1}j_1 \ldots j_p}$, we obtain hence (13.4.5) for $p+1$. $\qquad \square$

In the proof of Theorem 13.4.3 only the continuity of f with respect to each variable separately was used. Therefore the theorem remains valid if f is supposed separately continuous with respect to each variable.

Corollary 13.4.2. *Let $f : \mathbb{R}^{pN} \to \mathbb{R}$ be a p-additive function. If f is separately continuous with respect to each variable, then f is continuous.*

Proof. If f is separately continuous with respect to each variable, then the proof of Theorem 13.4.3 shows that f has form (13.4.5), and hence it is continuous. $\qquad \square$

Theorem 13.4.4. *Let $f : \mathbb{R}^{pN} \to \mathbb{R}$ be a measurable p-additive function. Then f is continuous.*

Proof. We will prove, by induction on p, that f has form (13.4.5). For $p = 1$ this follows from Theorems 9.4.3 and 5.5.2. Assume it true for a $p \in \mathbb{N}$, and let $f : \mathbb{R}^{(p+1)N} \to \mathbb{R}$ be a measurable $(p+1)$-additive function. Then, for almost all $(x_1, \ldots, x_p) \in \mathbb{R}^{pN}$, say for $(x_1, \ldots, x_p) \in A \subset \mathbb{R}^{pN}$, where $m(\mathbb{R}^{pN} \setminus A) = 0$, the function $f(x_1, \ldots, x_p, y)$, as a function of y, is a measurable additive function, and therefore (13.4.6) holds with a certain $c = (c_1, \ldots, c_N) \in \mathbb{R}^N$ depending on $x_1, \ldots, x_p$. For almost all $(\bar{x}_1, \ldots, \bar{x}_{p-1}) \in \mathbb{R}^{(p-1)N}$ the set

$$B[\bar{x}_1, \ldots, \bar{x}_{p-1}] = \{x_p \in \mathbb{R}^N \mid (\bar{x}_1, \ldots, \bar{x}_{p-1}, x_p) \in A\}$$

has full measure (i.e., $m(\mathbb{R}^N \setminus B) = 0$). Take $\bar{x}_1, \ldots, \bar{x}_{p-1}$ such that the set $B[\bar{x}_1, \ldots, \bar{x}_{p-1}]$ has full measure, and take an arbitrary $x_p \in \mathbb{R}^N$. The set $x_p - B[\bar{x}_1, \ldots, \bar{x}_{p-1}]$ also has full measure, and consequently also the set

$$B[\bar{x}_1, \ldots, \bar{x}_{p-1}] \cap (x_p - B[\bar{x}_1, \ldots, \bar{x}_{p-1}]) \tag{13.4.9}$$

has full measure. Thus there exists an x' in set (13.4.9), which means that x' and $x_p - x'$ both belong to $B[\bar{x}_1, \ldots, \bar{x}_{p-1}]$, or

$$(\bar{x}_1, \ldots, \bar{x}_{p-1}, x') \in A \quad \text{and} \quad (\bar{x}_1, \ldots, \bar{x}_{p-1}, x_p - x') \in A.$$

Therefore, for arbitrary $y = (y_1, \ldots, y_N) \in \mathbb{R}^N$,

$$f(\overline{x}_1, \ldots, \overline{x}_{p-1}, x', y) = \sum_{i=1}^{N} c_i(\overline{x}_1, \ldots, \overline{x}_{p-1}, x')y_i,$$

and

$$f(\overline{x}_1, \ldots, \overline{x}_{p-1}, x_p - x', y) = \sum_{i=1}^{N} c_i(\overline{x}_1, \ldots, \overline{x}_{p-1}, x_p - x')y_i.$$

If we put

$$c_i(\overline{x}_1, \ldots, \overline{x}_{p-1}, x_p) = c_i(\overline{x}_1, \ldots, \overline{x}_{p-1}, x') + c_i(\overline{x}_1, \ldots, \overline{x}_{p-1}, x_p - x'), \ i = 1, \ldots, N,$$

we will have by (13.4.1)

$$f(\overline{x}_1, \ldots, \overline{x}_{p-1}, x_p, y) = f(\overline{x}_1, \ldots, \overline{x}_{p-1}, x', y) + f(\overline{x}_1, \ldots, \overline{x}_{p-1}, x_p - x', y)$$
$$= \sum_{i=1}^{N} c_i(\overline{x}_1, \ldots, \overline{x}_{p-1}, x_p)y_i$$

for all $y \in \mathbb{R}^N$, all $x_p \in \mathbb{R}^N$, and almost all $(\overline{x}_1, \ldots, \overline{x}_{p-1}) \in \mathbb{R}^{(p-1)N}$. Repeating this procedure p times we obtain formula (13.4.6) with (13.4.7) valid for all $y \in \mathbb{R}^N$ and all $(x_1, \ldots, x_p) \in \mathbb{R}^{pN}$. Further the proof runs as that of Theorem 13.4.3. $\square$

Similarly one can prove

Theorem 13.4.5. *Let $f : \mathbb{R}^{pN} \to \mathbb{R}$ be a p-additive function, bounded above, or below, on a set $T \subset \mathbb{R}^{pN}$, where either $m_i(T) > 0$, or T is of the second category and with the Baire property. Then f is continuous.*

13.5 Cauchy equation on an interval

Now let $J \subset I \subset \mathbb{R}$ be (non-degenerated) intervals, and suppose that $J + J \subset I$. Suppose that we are given a function $f : I \to \mathbb{R}$ such that

$$f(x + y) = f(x) + f(y) \text{ for } x, y \in J. \tag{13.5.1}$$

Following J. Aczél [3] we will prove the following

Theorem 13.5.1. *Let $J \subset I \subset \mathbb{R}$ be intervals such that $J + J \subset I$, and let $f : I \to \mathbb{R}$ be a function satisfying equation (13.5.1). Then there exist an additive function $g : \mathbb{R} \to \mathbb{R}$ and a constant $a \in \mathbb{R}$ such that*

$$f(x) = \begin{cases} g(x) + a & \text{for } x \in J, \\ g(x) + 2a & \text{for } x \in J + J. \end{cases} \tag{13.5.2}$$

Proof. Take arbitrary $x, y \in J$. Then also $\frac{1}{2}(x+y) \in J$. By (13.5.1)

$$f(x) + f(y) = f(x+y) = f\left(\frac{x+y}{2}\right) + f\left(\frac{x+y}{2}\right) = 2f\left(\frac{x+y}{2}\right),$$

whence (13.2.1) follows. Consequently $f \mid J$ satisfies the Jensen equation (13.2.1). By Theorem 13.2.1 there exist an additive function $g : \mathbb{R} \to \mathbb{R}$ and a constant $a \in \mathbb{R}$ such that $f(x) = g(x) + a$ for $x \in J$. Now, if $x \in J + J$, then there exist $u, v \in J$ such that $x = u + v$. Hence we get by what has already been shown

$$f(x) = f(u+v) = f(u) + f(v) = g(u) + a + g(v) + a = g(u+v) + 2a = g(x) + 2a.$$

Thus formula (13.5.2) has been completely shown. $\square$

Theorem 13.5.2. *Let $J \subset I \subset \mathbb{R}$ be intervals such that $J + J \subset I$, and let $f : I \to \mathbb{R}$ be a function satisfying equation (13.5.1). If $J \cap 2J \neq \varnothing$, then there exists an additive function $g : \mathbb{R} \to \mathbb{R}$ such that*

$$f(x) = g(x) \quad for \ x \in J \cup (J+J). \tag{13.5.3}$$

If $J \cap 2J = \varnothing$, then there exist an additive function $g : \mathbb{R} \to \mathbb{R}$ and a constant $a \in \mathbb{R}$ such that (13.5.2) holds.

Proof. First let us note that

$$2J = J + J. \tag{13.5.4}$$

In fact, the inclusion $2J \subset J + J$ is evident. On the other hand, if $x \in J + J$, then there exist $u, v \in J$ such that $x = u + v$. Then also $w = \frac{1}{2}(u+v) \in J$, since J, being an interval, is convex. Moreover, $x = 2w \in 2J$. Thus $J + J \subset 2J$, and we obtain (13.5.4).

Now, if $J \cap 2J = \varnothing$, then (13.5.2) results from Theorem 13.5.1. (Note that in this case every function of form (13.5.2) actually satisfies equation (13.5.1).) If there exists an $x_0 \in J \cap 2J$, then by Theorem 13.5.1 and (13.5.4) $f(x_0) = g(x_0) + a$ and $f(x_0) = g(x_0) + 2a$. Hence $g(x_0) + a = g(x_0) + 2a$, and $a = 0$. Formula (13.5.3) results now from (13.5.2). (Again, every function of form (13.5.3) actually satisfies eqution (13.5.1)). $\square$

Theorem 13.5.3. *Let $J \subset \mathbb{R}$ be an interval such that $0 \in \mathrm{cl}\, J$, and let $I = J + J = 2J$. If a function $f : I \to \mathbb{R}$ satisfies equation (13.5.1), then f can be uniquely extended onto $\mathbb{R}$ to an additive function.*

Proof. Since $0 \in \mathrm{cl}\, J$ and J is convex, with every $x \in J$ also $\frac{1}{2}x \in J$, whence $x = 2\left(\frac{1}{2}x\right) \in 2J$. Thus $J \subset 2J$ so that $J \cap 2J = J \neq \varnothing$, and $J \cup (J+J) = J \cup 2J = 2J = I$. The function g occurring in Theorem 13.5.2 is the desired extension. The uniqueness of the extension results from the fact that, since J is non-degenerated, int $J \neq \varnothing$, whence by Lemma 9.3.1 $J \in \mathfrak{A}$, and by Corollary 9.3.2 J contains a Hamel basis H. Thus any additive extension of f must also be an additive extension of $f \mid H$., and by Theorem 5.2.2 such an extension is unique. $\qquad \square$

The particular case $J = I = (0, \infty)$ was dealt with by J. Aczél–P. Erdős [9].

The form $\big($in $J \cup (J + J)\big)$ of the continuous solutions of equation (13.5.1), and conditions ensuring the continuity $\big($on $J \cup (J + J)\big)$ of an f satisfying (13.5.1) will be easily found by the reader.

The results of the present section for the case, where the interval J is open, will be extended to higher dimensions in the next section.

13.6 The restricted Cauchy equation

Now we are going to consider the case (cf. Daróczy- Losonczi [59], and also Gołąb-Losonczi [119], Daróczy-Győry [58], Székelyhidi [309], Lajkó [204]), where f is a function satisfying the equation

$$f(x + y) = f(x) + f(y) \qquad (13.6.1)$$

only for $(x, y) \in G$, where $G \subset \mathbb{R}^{2N}$ is a certain set. As usually $K(x, r) \subset \mathbb{R}^N$ will denote the open ball centered at x and with the radius r.

Lemma 13.6.1. *Let $g_1 : \mathbb{R}^N \to \mathbb{R}$ and $g_2 : \mathbb{R}^N \to \mathbb{R}$ be additive functions, and let $a, b \in \mathbb{R}$ be arbitrary constants. If $U \subset \mathbb{R}^N$ is a non-empty open set and*

$$g_1(x) + a = g_2(x) + b \text{ for } x \in U,$$

then $g_1 = g_2$ and $a = b$.

Proof. For $x \in U$ we have

$$g_1(x) - g_2(x) = b - a = \text{const.}$$

The function $g_1 - g_2$ is additive, and is constant on the open set U. Thus U is contained in a level set of $g_1 - g_2$. By Theorem 12.6.4 $g_1 - g_2 = 0$, whence $g_1 = g_2$ and $a = b$. $\quad \square$

Lemma 13.6.2. *Let $U = K(0, r) \subset \mathbb{R}^N$ with a certain $r > 0$, and let $D \subset \mathbb{R}^N$ be a set such that $U + U = K(0, 2r) \subset D$. If a function $f : D \to \mathbb{R}$ satisfies equation (13.6.1) for all $x, y \in U$, then there exists a unique additive function $g : \mathbb{R}^N \to \mathbb{R}$ such that $f(x) = g(x)$ for $x \in U + U$.*

Proof. First observe that we have $U = K(0, r) \subset K(0, 2r) = U + U$.

Take arbitrary $x, y \in U$. Then $\dfrac{x + y}{2} \in U$, and by (13.6.1)

$$f(x) + f(y) = f(x + y) = f\left(\frac{x + y}{2}\right) + f\left(\frac{x + y}{2}\right) = 2f\left(\frac{x + y}{2}\right),$$

i.e., f satisfies in U the Jensen equation (13.2.1). By Theorem 13.2.1 there exist an additive function $g : \mathbb{R}^N \to \mathbb{R}$ and a constant $a \in \mathbb{R}$ such that

$$f(x) = g(x) + a \quad \text{for } x \in U. \tag{13.6.2}$$

Take an $x \in U + U$. Then $x = u + v$, where $u, v \in U$, and by (13.6.1) and (13.6.2)

$$f(x) = f(u+v) = f(u)+f(v) = g(u)+a+g(v)+a = g(u+v)+2a = g(x)+2a. \tag{13.6.3}$$

By (13.6.1) $f(0) = 0$, and similarly we have $g(0) = 0$, whence by (13.6.3)

$$0 = f(0) = g(0) + 2a = 2a,$$

and $a = 0$. Thus we get from (13.6.3) $f(x) = g(x)$ for $x \in U + U$. The uniqueness results from Lemma 13.6.1. $\qquad \square$

Now let $G \subset \mathbb{R}^{2N}$ be an arbitrary set. Every point $p \in \mathbb{R}^{2N}$ can be written as $p = (x, y)$, where $x, y \in \mathbb{R}^N$. For every $G \subset \mathbb{R}^{2N}$ we define sets $G_1, G_2, G_3 \subset \mathbb{R}^N$ as follows:

$$G_1 = \{x \in \mathbb{R}^N \mid \text{There exists a } y \text{ such that } (x, y) \in G\},$$
$$G_2 = \{y \in \mathbb{R}^N \mid \text{There exists an } x \text{ such that } (x, y) \in G\},$$
$$G_3 = \{z \in \mathbb{R}^N \mid z = x + y, (x, y) \in G\}.$$

Thus the sets G_1 and G_2 are the projections of G onto the suitable spaces.

For every $p = (x, y) \in \mathbb{R}^{2N}$ and $r > 0$ let $M(p, r) = K(x, r) \times K(y, r)$. Every $M(p, r)$ is an open subset of $\mathbb{R}^{2N}$.

Let $G \subset \mathbb{R}^{2N}$, and let $D \subset \mathbb{R}^N$ be a set such that

$$G_1 \cup G_2 \cup G_3 \subset D. \tag{13.6.4}$$

We say that a function $f : D \to \mathbb{R}$ *satisfies equation* (13.6.1) *on G* iff (13.6.1) holds for every x, y such that $(x, y) \in G$.

Lemma 13.6.3. *Let $G = M(p, r)$ with a certain $p \in \mathbb{R}^{2N}$ and a certain $r > 0$, and let $D \subset \mathbb{R}^N$ be a set fulfilling* (13.6.4). *If a function $f : D \to \mathbb{R}$ satisfies equation* (13.6.1) *on G, then there exist a unique additive function $g : \mathbb{R}^N \to \mathbb{R}$ and unique constants $a, b \in \mathbb{R}$ such that*

$$\begin{cases} f(x) = g(x) + a & \text{for } x \in G_1, \\ f(x) = g(x) + b & \text{for } x \in G_2, \\ f(x) = g(x) + a + b & \text{for } x \in G_3. \end{cases} \tag{13.6.5}$$

Proof. Let $p = (u, v)$. We have $G_1 = K(u, r), G_2 = K(v, r)$ and

$$G_3 = G_1 + G_2 = K(u + v, 2r). \tag{13.6.6}$$

In fact, if $z \in G_3$, then $z = x + y$, where $(x, y) \in G$. Then $x \in G_1$, $y \in G_2$, and $z = x + y \in G_1 + G_2 = K(u + v, 2r)$. If $z \in K(u + v, 2r)$, then $z = u + v + x$, where $|x| < 2r$. Then $\left|\frac{1}{2}x\right| < r$ so that $u + \frac{1}{2}x \in K(u, r) = G_1$ and $v + \frac{1}{2}x \in K(v, r) = G_2$. Consequently $z = \left(u + \frac{1}{2}x\right) + \left(v + \frac{1}{2}x\right) \in K(u, r) + K(v, r) = G_1 + G_2$. Hence we obtain (13.6.6).

Now take arbitrary $x, y \in K(0, r)$ and write $x' = u + x \in K(u, r)$, $y' = v + y \in K(v, r)$. Thus $(x', y') \in K(u, r) \times K(v, r) = G$. So we get by (13.6.1)

$$f(x' + y') = f(x') + f(y') = f(u + x) + f(v + y). \tag{13.6.7}$$

We have also $(u, y') \in K(u, r) \times K(v, r) = G$, whence by (13.6.1)

$$f(u + y') = f(u) + f(y') = f(u) + f(v + y). \tag{13.6.8}$$

Similarly, $(x', v) \in K(u, r) \times K(v, r) = G$, whence by (13.6.1)

$$f(x' + v) = f(x') + f(v) = f(u + x) + f(v). \tag{13.6.9}$$

Relations (13.6.7), (13.6.8), (13.6.9) yield

$$f(u + v + x + y) = f(x' + y') = f(x' + v) - f(v) + f(u + y') - f(u) \tag{13.6.10}$$
$$= f(u + v + x) - f(v) + f(u + v + y) - f(u).$$

Obviously, since $(u, v) \in G$,

$$f(u + v) = f(u) + f(v). \tag{13.6.11}$$

Subtracting (13.6.11) from (13.6.10) we obtain

$$f(u+v+x+y) - f(u+v) = f(u+v+x) - f(u+v) + f(u+v+y) - f(u+v). \tag{13.6.12}$$

Relations (13.6.6) and (13.6.4) show that $K(0,r) \subset K(0,r) + K(0,r) = K(0, 2r) = G_3 - (u+v) \subset D - (u+v)$. Define the function $f_0 : (D - (u + v)) \to \mathbb{R}$ by

$$f_0(x) = f(u + v + x) - f(u + v). \tag{13.6.13}$$

Relation (13.6.12) shows that f_0 satisfies equation (13.6.1) for $x, y \in K(0, r)$. By Lemma 13.6.2 there exists an additive function $g : \mathbb{R}^N \to \mathbb{R}$ such that $f_0(x) = g(x)$ for $x \in K(0, r)$.

Now, if $x \in G_1 = K(u, r)$, then $x - u \in K(0, r)$. We have, since $x \in K(u, r)$ and $v \in K(v, r)$, and thus $(x, v) \in K(u, r) \times K(v, r) = G$, $f(x + v) = f(x) + f(v)$, i.e., $f(u + v + (x - u)) = f(x) + f(v)$, and by (13.6.13) and (13.6.11)

$$f(x) = f(x + v) - f(v) = f(u + v + (x - u)) - f(v)$$
$$= f_0(x - u) + f(u + v) - f(v) = g(x - u) + f(u)$$
$$= g(x) - g(u) + f(u).$$

With $a = f(u) - g(u)$ we obtain hence the first relation in (13.6.5). The second is established similarly. The third results from the first and the second: if $x \in G_3$, then $x = t + s$, where $(t, s) \in G$. Hence $t \in G_1$ and $s \in G_2$, and $f(t) = g(t) + a$, $f(s) = g(s) + b$. Since $(t, s) \in G$, we have by (13.6.1)

$$f(x) = f(t + s) = f(t) + f(s) = g(t) + a + g(s) + b = g(t + s) + a + b = g(x) + a + b.$$

The uniqueness results from Lemma 13.6.1. $\square$

From Lemma 13.6.3 we derive the following result (Daróczy-Losonczi [59]).

Theorem 13.6.1. *Let $G \subset \mathbb{R}^{2N}$ be an open and connected set, and let $D \subset \mathbb{R}^N$ be a set fulfilling* (13.6.4). *If a function $f : D \to \mathbb{R}$ satisfies equation* (13.6.1) *on G, then there exist a unique additive function $g : \mathbb{R}^N \to \mathbb{R}$ and constants $a, b \in \mathbb{R}$ such that* (13.6.5) *holds.*

Proof. For every $x \in G$ there exists a set $M(x, r_x) \subset G$. The family $\{M(x, r_x)\}_{x \in G}$ forms an open cover of G. Since $\mathbb{R}^{2N}$ is separable, it follows from Lindelöf's theorem that there exists a countable family $\mathcal{M} = \{M(x_n, r_n)\}_{n \in \mathbb{N}} \subset \{M(x, r_x)\}_{x \in G}$ such that $G \subset \bigcup \mathcal{M}$. Since the converse inclusion is trivial, we must have

$$G = \bigcup \mathcal{M}.$$

Now we re-arrange $\mathcal{M}$. Let $M_0 = M(x_1, r_1)$. Suppose that M_β have already been defined for $\beta < \alpha < \Omega$ in such a manner that $\left(\bigcup_{\beta < \xi} M_\beta\right) \cap M_\xi \neq \varnothing$ for all $\xi < \alpha$. If $\bigcup_{\beta < \alpha} M_\beta = G$, we end the induction here. If $\bigcup_{\beta < \alpha} M_\beta \neq G$, and if we had $\left(\bigcup_{\beta < \alpha} M_\beta\right) \cap \left(\bigcup_{M \in \mathcal{M} \setminus \{M_\beta\}_{\beta < \alpha}} M\right) = \varnothing$, then G would be a union of two disjoint open sets

$$G = \left(\bigcup_{\beta < \alpha} M_\beta\right) \cup \left(\bigcup_{M \in \mathcal{M} \setminus \{M_\beta\}_{\beta < \alpha}} M\right),$$

which contradicts the connectedness of G. Thus there exists a set M_α such that

$$\left(\bigcup_{\beta < \alpha} M_\beta\right) \cap M_\alpha \neq \varnothing. \tag{13.6.14}$$

We cannot have $\bigcup_{\beta < \alpha} M_\beta \neq G$ for all $\alpha < \Omega$, since then we would have card $\mathcal{M} \geqslant \aleph_1 > \aleph_0$. So there exists a $\gamma < \Omega$ such that

$$\bigcup_{\beta < \gamma} M_\beta = G,$$

and we stop there. By Theorem 1.6.2 there exists a transfinite sequence $\{M_\beta\}_{\beta < \gamma}$, where $\gamma < \Omega$ (so that $\overline{\gamma} = \aleph_0$) fulfilling (13.6.14) for all $\alpha < \gamma$.

By Lemma 13.6.3 for every $\beta < \gamma$ there exists an additive function $g : \mathbb{R}^N \to \mathbb{R}$ and constants $a_\beta, b_\beta \in \mathbb{R}$ such that

$$f(x) = g_\beta(x) + a_\beta \text{ for } x \in [M_\beta]_1, \; f(x) = g_\beta(x) + b_\beta \text{ for } x \in [M_\beta]_2, \quad (13.6.15)$$

$$f(x) = g_\beta(x) + a_\beta + b_\beta \text{ for } x \in [M_\beta]_3.$$

We will prove that

$$g_\beta = g_0, \quad a_\beta = a_0, \quad b_\beta = b_0 \quad (13.6.16)$$

for all $\beta < \gamma$. For $\beta = 0$ this is certainly true. Assume (13.6.16) for $\beta < \alpha < \gamma$. It follows by (13.6.14) that there exists a $\beta < \alpha$ such that $M_\beta \cap M_\alpha \neq \varnothing$. Thus $M_\beta \cap M_\alpha$ is a non-empty open set, and so is also the set $[M_\beta]_1 \cap [M_\alpha]_1$. By the induction hypothesis we have (13.6.16), whence by (13.6.15), for $x \in [M_\beta]_1 \cap [M_\alpha]_1$ $g_0(x) + a_0 = g_\beta(x) + a_\beta = g_\alpha(x) + a_\alpha$. By Lemma 13.6.1 $g_\alpha = g_0$ and $a_\alpha = a_0$. The relation $b_\alpha = b_0$ is shown similarly. By Theorem 1.6.1 (13.6.16) holds for all $\beta < \gamma$.

Put $g = g_0$, $a = a_0$, $b = b_0$. For $x \in G_1 = \left[\bigcup_{\beta < \gamma} M_\beta \right]_1 = \bigcup_{\beta < \gamma} [M_\beta]_1$ we have by (13.6.15) and (13.6.16) $f(x) = g(x) + a$. Similarly $f(x) = g(x) + b$ for $x \in G_2 = \left[\bigcup_{\beta < \gamma} M_\beta \right]_2 = \bigcup_{\beta < \gamma} [M_\beta]_2$. If $x \in G_3$, then $x = t + s$, where $(t, s) \in G$. Then $t \in G_1$ and $s \in G_2$. Consequently $f(t) = g(t) + a$, $f(s) = g(s) + b$, whence

$$f(x) = f(t+s) = f(t) + f(s) = g(t) + a + g(s) + b = g(t+s) + a + b = g(x) + a + b.$$

The uniqueness of g, a, b results from Lemma 13.6.1. $\qquad \square$

Corollary 13.6.1. *Let $G \subset \mathbb{R}^{2N}$ be an open and connected set such that $G_1 \cap G_2 \neq \varnothing$, and let $D \subset \mathbb{R}^N$ be a set fulfilling (13.6.4). If a function $f : D \to \mathbb{R}$ satisfies equation (13.6.1) on G, then there exist a unique additive function $g : \mathbb{R}^N \to \mathbb{R}$ and a constant $a \in \mathbb{R}$ such that*

$$f(x) = g(x) + a \qquad \text{for } x \in G_1 \cup G_2,$$
$$f(x) = g(x) + 2a \qquad \text{for } x \in G_3.$$

Proof. This results from Theorem 13.6.1 and Lemma 13.6.1. $\qquad \square$

Corollary 13.6.2. *Let $G \subset \mathbb{R}^{2N}$ be an open and connected set such that $G_1 \cap G_2 \cap G_3 \neq \varnothing$, and let $D \subset \mathbb{R}^N$ be a set fulfilling (13.6.4). If a function $f : D \to \mathbb{R}$ satisfies equation (13.6.1) on G, then there exist a unique additive function $g : \mathbb{R}^N \to \mathbb{R}$ such that $f(x) = g(x)$ for $x \in G_1 \cup G_2 \cup G_3$.*

Proof. This results from Theorem 13.6.1 and Lemma 13.6.1. $\qquad \square$

Note that if $(0, 0) \in G$, then $0 \in G_1 \cap G_2 \cap G_3$. Thus we obtain from Corollary 13.6.2

Corollary 13.6.3. *Let $D = K(0, 2r) \subset \mathbb{R}^N$, with a certain $r > 0$. If a function $f : D \to \mathbb{R}$ satisfies equation (13.6.1) for all $x, y \in K(0, r)$, then f can be uniquely extended onto $\mathbb{R}^N$ to an additive function.*

The next theorem(Daróczy-Losonczi [59]) refers to the case $N = 1$. Fix an $r > 0$. Let $I = [0, r]$, and let $H \subset \mathbb{R}^2$ be the set

$$H = \{p = (x, y) \in \mathbb{R}^2 \mid x \in I, \, y \in I, \text{ and } x + y \in I\}.$$

Theorem 13.6.2. *Let $f : I \to \mathbb{R}$ be a function satisfying equation (13.6.1) on H. Then there exists a unique additive function $g : \mathbb{R} \to \mathbb{R}$ such that $f = g \mid I$.*

Proof. Let $J = \left[0, \dfrac{1}{2}r\right]$ so that $I = 2J = J + J$. Take arbitrary $x, y \in J$. Then $x, y \in I$ and $x + y \in I$, whence $(x, y) \in H$, and (13.6.1) holds. Thus f satisfies (13.6.1) for $x, y \in J$. By Theorem 13.5.3 f can be uniquely extended onto $\mathbb{R}$ to an additive function $g : \mathbb{R} \to \mathbb{R}$. The function g has all the properties specified in the theorem. $\square$

Theorem 13.6.2 with $r = 1$ is particularly important in the probability theory. The variables x, y are then to be interpreted as probabilities, and the conditions of the theorem say that in the case where the sum of probabilities is again a probability, then the value of f for such a sum is the sum of the values of f for the corresponding probabilities.

It is clear that in Theorem 13.6.2 the closed interval $I = [0, r]$ may be replaced by any one of the intervals $(0, r)$, $[0, r)$ and $(0, r]$.

Equation (13.6.1) postulated not for all x, y from the space considered, but only for x, y satisfying a certain additional condition (in the case of Theorem 13.6.1 the condition was $(x, y) \in G$) is called the *Cauchy equation on a restricted domain*, or *a restricted Cauchy equation* or *a conditional Cauchy equation*. A review of results and references pertinent to restricted Cauchy equations may be found in Kuczma [187], Dhombres-Ger [69], [70], Ger [108]. In the present book we only touch upon such problems. A complete treatment of these questions is found in the book Dhombres [68].

13.7 Hosszú equation

The functional equation[3] (Blanuša [32], Daróczy [56], [57], Davison [60],[61],[62], Davison-Redlin [63], Fenyő [78], Głowacki-Kuczma [117], Lajkó [204], Świa̧tak [300], [301], [304])

$$f(x + y - xy) + f(xy) = f(x) + f(y) \tag{13.7.1}$$

is referred to as the *Hosszú equation*. Its connection with the theory of additive functions is expressed by the following (Daróczy [57]; cf. also Blanuša [32], Davison [60], Świa̧tak [304])

Theorem 13.7.1. *Let a function $f : \mathbb{R} \to \mathbb{R}$ satisfy equation (13.7.1). Then there exist an additive function $g : \mathbb{R} \to \mathbb{R}$ and a constant $a \in \mathbb{R}$ such that*

$$f(x) = g(x) + a. \tag{13.7.2}$$

---

[3] This equation was mentioned for the first time by M. Hosszú at the International Symposium on Functional Equations held in Zakopane (Poland) in October 1967; cf. Hosszú [147].

Proof. Define the function $G : \mathbb{R}^2 \to \mathbb{R}$ by

$$G(x, y) = f(x) + f(y) - f(xy). \tag{13.7.3}$$

It is easily seen from (13.7.3) that we have for arbitrary $u, v, w \in \mathbb{R}$

$$G(uv, w) + G(u, v) = G(u, vw) + G(v, w). \tag{13.7.4}$$

By (13.7.1) $G(x, y) = f(x + y - xy)$. Inserting this into (13.7.4) yields

$$f(uv + w - uvw) + f(u + v - uv) = f(u + vw - uvw) + f(v + w - vw). \tag{13.7.5}$$

Setting in (13.7.5) $w = v^{-1}$, we obtain the relation

$$f\left(uv + v^{-1} - u\right) + f(u + v - uv) = f(1) + f\left(v + v^{-1} - 1\right) \tag{13.7.6}$$

valid for all $u \in \mathbb{R}$ and $v \in \mathbb{R} \setminus \{0\}$.

Now take arbitrary $x, y \in \mathbb{R}$ such that $x + y > 0$. We are looking for $u, v \in \mathbb{R}$, $v \neq 0$, fulfilling the equations

$$uv + v^{-1} - u = x + 1, \quad u + v - uv = y + 1. \tag{13.7.7}$$

Adding equations (13.7.7) we obtain $v + v^{-1} = x + y + 2$, i.e.,

$$v^2 - (x + y + 2)v + 1 = 0. \tag{13.7.8}$$

Now, $(x + y + 2)^2 - 4 = x^2 + y^2 + 2xy + 4x + 4y = (x + y)(x + y + 4) > 0$. Thus there exist two distinct real values of v fulfilling (13.7.8), and at least one of them must be different from zero[4]. Let v_0 be this value. We cannot have $v_0 = 1$, since $v_0 + v_0^{-1} = x + y + 2 > 2$.

Now, if $u_0 = \dfrac{y + 1 - v_0}{1 - v_0}$, then u_0, v_0 satisfy the system of equations (13.7.7).

Define the function $g : \mathbb{R} \to \mathbb{R}$ as

$$g(x) = f(x + 1) - f(1). \tag{13.7.9}$$

We will show that g is additive.

Take arbitrary $x, y \in \mathbb{R}$ such that $x + y > 0$. Let u, v satisfy (13.7.7), $v \neq 0$. Then we get from (13.7.6)

$$f(x + 1) + f(y + 1) = f(1) + f(x + y + 1),$$

i.e., after subtracting from both sides $2f(1)$

$$f(x + y + 1) - f(1) = f(x + 1) - f(1) + f(y + 1) - f(1),$$

whence by (13.7.9)

$$g(x + y) = g(x) + g(y). \tag{13.7.10}$$

---

[4] Actually both roots of (13.7.8) are different from zero, since their product is 1.

If $x + y \leqslant 0$, then we can find a $t \in \mathbb{R}$ such that $t + x > 0$ and $t + x + y > 0$. Then

$$g(t + x) = g(t) + g(x),$$
$$g(t + x + y) = g(t + x) + g(y) = g(t) + g(x) + g(y),$$
$$g(t + x + y) = g(t) + g(x + y).$$

Thus $g(t) + g(x + y) = g(t) + g(x) + g(y)$, whence again we obtain (13.7.10). Consequently g fulfills (13.7.10) for all $x, y \in \mathbb{R}$, which means that g is additive.

By (13.7.9)

$$f(x) = g(x - 1) + f(1) = g(x) - g(1) + f(1),$$

whence we obtain (13.7.2) with $a = f(1) - g(1)$.   □

On the other hand, it is easily seen that every function of form (13.7.2), where $g : \mathbb{R} \to \mathbb{R}$ is additive and $a \in \mathbb{R}$ is an arbitrary constant, actually satisfies equation (13.7.1).

Theorem 13.7.1 remains true also for functions $f : F \to \Gamma$, where $(F; +; \cdot)$ is an arbitrary field with card $F \geqslant 5$, and $(\Gamma, +)$ is a commutative group. (An additive function is then understood as a function $g : F \to \Gamma$ satisfying equation (13.7.10) for all $x, y \in F$; the constant a in (13.7.2) must be taken from Γ). The proof is then more difficult (cf. Davison [61]).

Equation (13.7.1) was also studied on other structures (Davison [60],[62], Głowacki-Kuczma [117], Świątak [304]). In particular, on $\mathbb{Z}$ equation (13.7.1) has solutions which are not of form (13.7.2). For example, the function

$$f(n) = \begin{cases} 1 & \text{if } n \text{ is even,} \\ 0 & \text{if } n \text{ is odd} \end{cases}$$

is such a solution.

By Theorem 13.7.1 equation (13.7.1) has exactly the same solutions as the Jensen equation (13.2.1) (cf. Theorem 13.2.1). Thus Theorem 13.7.1 is often formulated as follows.

Corollary 13.7.1. *For functions $f : \mathbb{R} \to \mathbb{R}$ the Hosszú equation and the Jensen equation are equivalent.*

13.8 Mikusiński equation

At the Symposium on Functional and Differential Equations held in Zawoja (Poland) in October 1971, J. Mikusiński mentioned the equation

$$f(x + y)[f(x + y) - f(y) - f(x)] = 0, \tag{13.8.1}$$

which since has been named after him (cf. Dubikajtis-Ferens-Ger-Kuczma [72], Baron-Ger [20], Ger [105], [107]; cf. also Kannappan-Kuczma [162], Ger [109] concerning more general equations of this type). Our primary aim is to show that every function $f : \mathbb{R}^N \to \mathbb{R}$ fulfilling (13.8.1) is additive. But we will carry out our considerations for functions $f : X \to Y$, where $(X, +)$ and $(Y, +)$ are arbitrary (not necessarily commutative) groups (cf. 4.5).

If the range of f is a ring without divisors of zero (e.g., a field or an integral domain), then equation (13.8.1) can be written as

$$\text{if } f(x + y) \neq 0, \quad \text{then } f(x + y) = f(x) + f(y). \tag{13.8.2}$$

Equation (13.8.2) is meaningful for functions f with the range in a group.

Lemma 13.8.1. Let $(X, +)$ and $(Y, +)$ be arbitrary groups, and let $f : X \to Y$ satisfy (13.8.2). Write $K = f^{-1}(0) = \{x \in X \mid f(x) = 0\}$. Then $(K, +)$ is a subgroup of $(X, +)$.

Proof. If we had $f(0) \neq 0$, then we would get by (13.8.2) $f(0) = f(0) + f(0)$, whence $f(0) = 0$, a contradiction. Consequently $0 \in K$ so that $K \neq \varnothing$.

Suppose that for an $x \in K$ we have $-x \notin K$, i.e., $f(-x) \neq 0$. Hence by (13.8.2)

$$f(-x) = f(-2x + x) = f(-2x) + f(x) = f(-2x), \tag{13.8.3}$$

since $x \in K$ means $f(x) = 0$. By (13.8.3) $f(-2x) = f(-x) \neq 0$, whence by (13.8.2)

$$f(-2x) = f(-x) + f(-x). \tag{13.8.4}$$

By (13.8.3) and (13.8.4) $f(-x) = f(-x) + f(-x)$, whence $f(-x) = 0$, contrary to the supposition.

Now take arbitrary $x, y \in K$, and suppose that $x + y \notin K$. Then $f(x + y) \neq 0$, and by (13.8.2)

$$f(x + y) = f(x) + f(y). \tag{13.8.5}$$

But $f(x) = f(y) = 0$, whence $f(x + y) = 0$, contrary to the supposition.

For arbitrary $x, y \in K$ we have $-y \in K$, and $x - y = x + (-y) \in K$. By Lemma 4.5.1 $(K, +)$ is a subgroup of $(X, +)$. $\qquad\square$

Theorem 13.8.1. Let $(X, +)$ and $(Y, +)$ be arbitrary groups, and let a function $f : X \to Y$ satisfy equation (13.8.2). Then either f satisfies (13.8.5) for all $x, y \in X$ (i.e., f is a homomorphism), or there exist a subgroup $(K, +)$ of $(X, +)$, of index 2, and a constant $c \in Y \setminus \{0\}$ such that

$$f(x) = \begin{cases} 0 & \text{for } x \in K, \\ c & \text{for } x \notin K. \end{cases} \tag{13.8.6}$$

Proof. [5] Suppose that $f : X \to Y$ satisfies (13.8.2), but does not satisfy (13.8.5). Then there exist $u, v \in X$ such that

$$f(u + v) \neq f(u) + f(v). \tag{13.8.7}$$

---

[5] Ger [107].

We will prove that for every $x \in X$

$$\text{either } f(u + v + x) = 0, \text{ or } f(v + x) = 0. \tag{13.8.8}$$

Suppose that (13.8.8) does not hold, i.e., there exists a $z \in X$ such that

$$f(u + v + z) \neq 0 \text{ and } f(v + z) \neq 0. \tag{13.8.9}$$

Hence, by (13.8.2),

$$f(u + v + z) = f(u) + f(v + z) = f(u) + f(v) + f(z). \tag{13.8.10}$$

On the other hand, we have by (13.8.9) and (13.8.2)

$$f(u + v + z) = f(u + v) + f(z). \tag{13.8.11}$$

Now, (13.8.10) and (13.8.11) yield $f(u + v) = f(u) + f(v)$, contrary to (13.8.7).

Now put $K = f^{-1}(0)$ and take an arbitrary $x \in X$. By (13.8.8) either $u + v + x \in K$, or $v + x \in K$, i.e., either $x \in -v - u + K$, or $x \in -v + K$. In other words, $X \subset (-v - u + K) \cup (-v + K) = -v - u + [K \cup (u + K)]$. This implies that $X = X + u + v \subset K \cup (u + K)$, and since the converse inclusion is obvious, we have

$$K \cup (u + K) = X. \tag{13.8.12}$$

Put $c = f(u) \in Y$, and take an arbitrary $x \in X \setminus K$. By (13.8.12) $x \in u + K$, i.e., there exists a $y \in K$ such that $x = u + y$. The condition $x \in X \setminus K$ means $f(u + y) = f(x) \neq 0$, whereas the condition $y \in K$ means $f(y) = 0$. Thus by (13.8.2)

$$f(x) = f(u + y) = f(u) + f(y) = f(u) = c.$$

Hence we obtain (13.8.6) in view of the definition of K. We also have $c = f(x) \neq 0$, whence $c \in Y \setminus \{0\}$.

By Lemma 13.8.1 $(K, +)$ is a subgroup of $(X, +)$. Take arbitrary $x, y \in K' = X \setminus K$. By (13.8.6) $f(x) = f(y) = c$. If we had $f(x + y) \neq 0$, then we would get by (13.8.6) and (13.8.2)

$$c = f(x + y) = f(x) + f(y) = 2c,$$

whence $c = 0$, which is not the case. Consequently $f(x + y) = 0$, i.e., $x + y \in K$. This shows that $K' + K' \subset K$. By Lemma 4.5.4 the index of $(K, +)$ is 2. $\qquad \square$

Corollary 13.8.1. *Let $(X, +)$ and $(Y, +)$ be arbitrary groups, and let $f : X \to Y$ satisfy equation (13.8.2) but not (13.8.5). Put $K = f^{-1}(0)$. Then there exists a $u \in X$ such that (13.8.12) holds.*

Clearly, if f satisfies (13.8.5), then it satisfies also (13.8.2). If f has form (13.8.6), where $(K, +)$ is a subgroup of $(X, +)$ of index 2, then f satisfies (13.8.2), too. In fact, let $x, y \in X$ be arbitrary. If $x, y \in K$, then $x + y \in K$, since $(K, +)$ is group (a subgroup of $(X, +)$), and

$$f(x + y) = 0 = 0 + 0 = f(x) + f(y),$$

and we have (13.8.2). If $x \in K$, $y \notin K$, then by Lemma 4.5.3 $x + y \notin K$, and by (13.8.6) $f(x + y) = c$, $f(y) = c$. Thus

$$f(x + y) = c = 0 + c = f(x) + f(y),$$

and (13.8.2) is fulfilled. Similarly, if $x \notin K$, $y \in K$, then by Lemma 4.5.3 $x + y \notin K$, and

$$f(x + y) = c = c + 0 = f(x) + f(y).$$

Finally, if $x, y \notin K$, then by Lemma 4.5.4 $x + y \in K$, whence $f(x+y) = 0$, and again (13.8.2) is fulfilled.

Note that if $2c \neq 0$, then function (13.8.6) does not satisfy (13.8.5) for $x, y \notin K$, since otherwise we would have, since by Lemma 4.5.4 $x + y \in K$,

$$0 = f(x + y) = f(x) + f(y) = 2c,$$

a contradiction.

Since by Corollary 4.5.3 the group $(\mathbb{R}^N, +)$ has no subgroup of index 2, we obtain from Theory 13.8.1

Theorem 13.8.2. *If a function $f : \mathbb{R}^N \to \mathbb{R}$ satisfies equation (13.8.1), then it is additive.*

We will prove also the following

Theorem 13.8.3. *Let $(X, +)$ be a second category topological group*[6] *such that for every neighbourhood V of zero*

$$\bigcup_{n \in \mathbb{N}} nV = X. \tag{13.8.13}$$

Let $(Y, +)$ be an arbitrary group. If a function $f : X \to Y$ satisfies (13.8.2) and the set $K = f^{-1}(0)$ has the Baire property, then f satisfies (13.8.5) for all $x, y \in X$.

Proof. If K is of the first category, then so are the sets $x + K$. Suppose that f does not satisfy (13.8.5). Then, by Corollary 13.8.1, there exists a $u \in X$ such that (13.8.12) holds. (13.8.12) implies that $K' = X \setminus K \subset u + K$. If K were of the first category, then so would be also K', and hence also $X = K \cup K'$ would be of the first category, contrary to the assumption.

Thus K is a set of the second category and with the Baire property. By Theorem 2.9.1 $\mathrm{int}(K+K) \neq \varnothing$. But, since by Lemma 13.8.1 $(K, +)$ is a group, we have $K+K \subset K$, whence $\mathrm{int}\, K \neq \varnothing$. Let $x \in \mathrm{int}\, K$. Then the set $V = \mathrm{int}\, K - x$ is a neighbourhood of zero, and, by Lemmas 13.8.1 and 4.5.3, $\mathrm{int}\, K - x \subset K - K = K + K = K$. Thus $V \subset K$, and hence also $nV \subset K$ for every $n \in \mathbb{N}$. By (13.8.13)

$$X = \bigcup_{n \in \mathbb{N}} nV \subset K,$$

and thus $K = X$, which is incompatible with the fact that $(K, +)$ is a subgroup of $(X, +)$ of index 2. $\qquad\square$

Finally let us observe that equation (13.8.1) (especially in form (13.8.2)) can be regarded as a conditional Cauchy equation, the condition being $f(x + y) \neq 0$.

---

[6]This means that $(X, +)$ is a topological group (cf. 2.9) and the set X in the topological space X is of the second category (i.e., cannot be represented as a countable union of nowhere dense sets).

13.9 An alternative equation

If the function $f : \mathbb{R}^N \to \mathbb{R}$ is additive, then it satisfies also

$$[f(x+y)]^2 = [f(x) + f(y)]^2. \tag{13.9.1}$$

The converse implication is less trivial. (13.9.1) yields immediately only

$$\text{either } f(x+y) = f(x) + f(y), \text{ or } f(x+y) = -[f(x) + f(y)]. \tag{13.9.2}$$

Since equation (13.9.2) has the form of an alternative, it is often referred to (as are also similar functional equations) as an *alternative functional equation* (cf. Hosszú [145], Vincze [318], [319], Świątak [303], Świątak–Hosszú [307], Fischer [81], Fischer–Muszély [83, 84], Kuczma [186]; more general equations are dealt with in Świątak [299], [302], [305], Świątak–Hosszú [306], Vincze [320], [321]. Cf. also Kuczma [187] concerning a review of results and further references). In this sense also the Mikusiński functional equation (13.8.1) and related equations (Kannappan–Kuczma [162], Ger [109]) are alternative functional equations. Alternative functional equations may be regarded as functional equations on restricted domains (conditional equations).

Similarly as in the case of Mikusiński's functional equation (13.8.1) we will carry out our considerations in more general structures. If $(X, +)$ is a semigroup, and $(Y; +, \cdot)$ is an integral domain (cf. Chapter 4), then equation (13.9.1) for functions $f : X \to Y$ can be written as (13.9.2). It is in this form that we are going to deal with equation (13.9.1).

Lemma 13.9.1. *Let $(X, +)$ be a semigroup, let $(Y; +, \cdot)$ be an integral domain, and let a function $f : X \to Y$ satisfy equation (13.9.1) and the condition*

$$f(2x) = 2f(x) \tag{13.9.3}$$

for all $x \in X$. Then f satisfies equation (13.8.5) for all $x, y \in X$.

Proof. Suppose that for a certain $x, y \in X$ (13.8.5) does not hold. By (13.9.1) (or, equivalently, (13.9.2))

$$f(x+y) = -f(x) - f(y). \tag{13.9.4}$$

We have

$$f(x + x + y) = f[x + (x+y)] = e_1[f(x) + f(x+y)] = -e_1 f(y),$$

and by (13.9.3)

$$f(x + x + y) = f(2x + y) = e_2[f(2x) + f(y)] = e_2[2f(x) + f(y)],$$

where $e_1, e_2 = \pm 1$. Hence

$$2f(x) + f(y) = e_3 f(y), \tag{13.9.5}$$

where $e_3 = \pm 1$. Similarly

$$2f(y) + f(x) = e_4 f(x), \tag{13.9.6}$$

where $e_4 = \pm 1$. If $e_3 = e_4 = 1$, then we get by (13.9.5) and (13.9.6) $2f(x) = 2f(y) = 0$, i.e.,

$$2[f(x) + f(y)] = 0. \tag{13.9.7}$$

If $e_3 = -1$, then (13.9.7) results from (13.9.5), and if $e_4 = -1$, then (13.9.7) results from (13.9.6). Thus (13.9.7) holds in every case. (13.9.7) means that

$$f(x) + f(y) = -[f(x) + f(y)] = -f(x) - f(y),$$

and by (13.9.4) we get

$$f(x+y) = f(x) + f(y),$$

contrary to the supposition. □

Lemma 13.9.2. *Let $(X, +)$ be a semigroup, let $(Y; +, \cdot)$ be an integral domain, and let $f : X \to Y$ satisfy equation (13.9.1). If $f(2x_0) \neq 2f(x_0)$ for a certain $x_0 \in X$, then $d_0 = f(x_0)$ fulfils the conditions*

$$2d_0 \neq 0, \quad 4d_0 \neq 0, \tag{13.9.8}$$

and

$$6d_0 = 0. \tag{13.9.9}$$

Proof. We have by (13.9.1) (or, equivalently, (13.9.2))

$$f(2x_0) = -2f(x_0), \tag{13.9.10}$$

whence

$$f(3x_0) = f(2x_0 + x_0) = e_1[f(2x_0) + f(x_0)] = e_1[-2f(x_0) + f(x_0)] = -e_1 f(x_0) = -e_1 d_0,$$

where $e_1 = \pm 1$. Hence

$$f(4x_0) = f(3x_0 + x_0) = e_2[f(3x_0) + f(x_0)] = e_2(-e_1 + 1)d_0,$$

and, on the other hand

$$f(4x_0) = f(2x_0 + 2x_0) = e_3[f(2x_0) + f(2x_0)] = -4e_3 d_0,$$

where $e_2, e_3 = \pm 1$. Hence

$$(k_1 - k_2)d_0 = 0,$$

where $k_1 = e_2(-e_1 + 1)$, $k_2 = -4e_3$. Thus the only possible values of $k_1 - k_2$ are $\pm 2, \pm 4, \pm 6$.

If we had $4d_0 = 0$, then we would have $2d_0 = -2d_0$, and by (13.9.10)

$$f(2x_0) = -2f(x_0) = -2d_0 = 2d_0 = 2f(x_0),$$

contrary to the assumption. Thus $2d_0 = 0$ also is impossible, for otherwise we would have $4d_0 = 2(2d_0) = 0$. Hence we get (13.9.8). Consequently we must have (13.9.9), or $-6d_0 = 0$, which also implies (13.9.9). □

Theorem 13.9.1. *Let $(X, +)$ be a semigroup, and let $(Y; +, \cdot)$ be an integral domain of characteristic different from 3. If a function $f : X \to Y$ satisfies equation (13.9.1), then it satisfies also equation (13.8.5).*

Proof. Suppose that f does not satisfy (13.8.5). By Lemma 13.9.1 there exists an $x_0 \in X$ such that $f(2x_0) \neq 2f(x_0)$. By Lemma 13.9.2 $d_0 = f(x_0)$ fulfils (13.9.8) and (13.9.9). Thus $y_0 = 2d_0 \in Y$ is an element of order 3, and hence the characteristic of Y must be 3, contrary to the assumption. $\square$

Note that if the characteristic of Y is 3, then every constant function $f : X \to Y$, $f \neq 0$, satisfies equation (13.9.1), but not (13.8.5). In fact, then for every $d \in Y$ we have $3d = 0$, whence $2d = -d$, and if $f(x) = d = \text{const}$ in X, then $f(x) + f(y) = d + d = 2d = -d$, and

$$\left[f(x) + f(y)\right]^2 = (-d)^2 = d^2 = \left[f(x + y)\right]^2.$$

On the other hand $(d \neq 0)$, $f(x + y) = d \neq -d = f(x) + f(y)$, since otherwise we would have $2d = 0$, which together with $3d = 0$ implies $d = 0$.

Since $(\mathbb{R}; +, \cdot)$ is a field, and hence an integral domain, and the characteristic of $\mathbb{R}$ is zero, we get from Theorem 13.9.1

Theorem 13.9.2. *If a function $f : \mathbb{R}^N \to \mathbb{R}$ satisfies equation (13.9.1), then it is additive.*

Equation (13.9.1) for functions $f : X \to Y$, where $(X, +)$ is a semigroup, and $(Y; +, \cdot)$ is a commutative ring, as well as equation (13.9.2) for functions $f : X \to Y$, where $(X, +)$ is a semigroup, and $(Y, +)$ is a group, are dealt with in Kuczma [186].

13.10   The general linear equation

The general linear equation[7]

$$f(ax + by + c) = Af(x) + Bf(y) + C, \quad abAB \neq 0, \tag{13.10.1}$$

has been studied by J. Aczél [2], Z. Daróczy [54], [55], and L. Losonczi [209]; cf. also Aczél [5]. The Cauchy equation (5.2.1) is the particular case $a = b = A = B = 1$, $c = 0$, $C = 0$, of (13.10.1). The Jensen equation (13.2.1) corresponds to $a = b = A = B = \frac{1}{2}$, $c = 0$, $C = 0$. The question arises what are the solutions $f : \mathbb{R}^N \to \mathbb{R}$ of (13.10.1) for other values of a, b, c, A, B, C.

First we determine the constant solutions of (13.10.1). If a function $f(x) = \beta = \text{const}$ satisfies (13.10.1), then $\beta = (A + B)\beta + C$, i.e.,

$$(A + B - 1)\beta + C = 0. \tag{13.10.2}$$

---

[7] In equation (13.10.1) it is understood that $a, b, A, B, C \in \mathbb{R}$ and $c \in \mathbb{R}^N$ are fixed constants. So the expression "$f : \mathbb{R}^N \to \mathbb{R}$ satisfies (13.10.1)" means that (13.10.1) holds for all $x, y \in \mathbb{R}^N$ with fixed $a, b, A, B, C \in \mathbb{R}$ and $c \in \mathbb{R}^N$ such that none of a, b, A, B is zero.

If $A + B \neq 1$, then equation (13.10.2) has the unique solution $\beta = -C(A + B - 1)^{-1}$, and the function

$$f(x) = -\frac{C}{A + B - 1}, \quad x \in \mathbb{R}^N, \tag{13.10.3}$$

is the only constant solution of (13.10.1). If $A + B = 1$ and $C = 0$, then every $\beta \in \mathbb{R}$ satisfies (13.10.2), and consequently every constant function $f : \mathbb{R}^N \to \mathbb{R}$ is a solution of (13.10.1). Finally, if $A + B = 1$ and $C \neq 0$, then (13.10.2) has no solutions, whence also (13.10.1) has no constant solutions.

We summarize this discussion in the following

Theorem 13.10.1. *If $A + B \neq 1$, then function (13.10.3) is the only constant solution of (13.10.1). If $A + B = 1$ and $C = 0$, then every constant function $f : \mathbb{R}^N \to \mathbb{R}$ satisfies (13.10.1). If $A + B = 1$ and $C \neq 0$, then equation (13.10.1) has no constant solution.*

Now we turn to non-constant solutions of (13.10.1). We start with the following

Lemma 13.10.1. *If a function $f : \mathbb{R}^N \to \mathbb{R}$ satisfies equation (13.10.1), then there exist an additive function $g : \mathbb{R}^N \to \mathbb{R}$ and a constant $\beta \in \mathbb{R}$ such that*

$$f(x) = g(x) + \beta, \quad x \in \mathbb{R}^N. \tag{13.10.4}$$

Proof. Take arbitrary $u, v \in \mathbb{R}^N$. Equation (13.10.1) with $x = u/a$, $y = (v - c)/b$ goes into

$$f(u + v) = Af\left(\frac{u}{a}\right) + Bf\left(\frac{v - c}{b}\right) + C. \tag{13.10.5}$$

Setting in (13.10.1) $x = u/a$, $y = -c/b$, we obtain

$$f(u) = Af\left(\frac{u}{a}\right) + Bf\left(-\frac{c}{b}\right) + C, \tag{13.10.6}$$

and setting in (13.10.1) $x = 0$, $y = (v - c)/b$, we get

$$f(v) = Af(0) + Bf\left(\frac{v - c}{b}\right) + C. \tag{13.10.7}$$

Finally, with $x = 0$, $y = -c/b$, equation (13.10.1) becomes

$$f(0) = Af(0) + Bf\left(-\frac{c}{b}\right) + C. \tag{13.10.8}$$

Relations (13.10.5), (13.10.6), (13.10.7), (13.10.8) yield

$$f(u + v) - f(u) - f(v) + f(0) = 0. \tag{13.10.9}$$

Define the function $g : \mathbb{R}^N \to \mathbb{R}$ by

$$g(x) = f(x) - f(0), \quad x \in \mathbb{R}^N. \tag{13.10.10}$$

Then we have by (13.10.9) and (13.10.10)

$$g(u + v) = g(u) + g(v),$$

i.e., g is additive, and (13.10.4) follows with $\beta = f(0)$. $\qquad\square$

But now not every function of form (13.10.4) satisfies equation (13.10.1).

Theorem 13.10.2. *If a non-constant function $f : \mathbb{R}^N \to \mathbb{R}$ satisfies equation (13.10.1), and f is measurable, or bounded above, or below, on a set $T \in \mathfrak{A}$, then necessarily $a = A$, $b = B$, and there exists a constant $\alpha \in \mathbb{R}^N \setminus \{0\}$ such that*

$$f(x) = \alpha x + \beta, \quad x \in \mathbb{R}^N, \tag{13.10.11}$$

where, in the case where $A + B \neq 1$, $\beta = (\alpha c - C)(A + B - 1)^{-1}$. If $A + B = 1$, then α additionally fulfils the relation $\alpha c = C$, and then $\beta \in \mathbb{R}$ may be arbitrary. In particular, f is continuous.

Proof. By Lemma 13.10.1 f has form (13.10.4), where $g : \mathbb{R}^N \to \mathbb{R}$ is additive. The conditions on f enforce the analogous conditions on g, therefore g must be continuous (cf., in particular, Theorem 9.4.2), and it follows from Theorem 5.5.2 that there exists an $\alpha \in \mathbb{R}^N$ such that $g(x) = \alpha x$, and so f has form (13.10.11). We must have $\alpha \neq 0$, since otherwise f would be constant. Inserting (13.10.11) into (13.10.1) yields

$$a\alpha x + b\alpha y + \alpha c + \beta = A\alpha x + A\beta + B\alpha y + B\beta + C$$

for all $x, y \in \mathbb{R}^N$, which is possible only if $a = A$ and $b = B$. Hence we get $(A + B - 1)\beta = \alpha c - C$. Thus, if $A + B \neq 1$, β must have the form asserted. If $A + B = 1$, then α must fulfil the condition $\alpha c = C$, and there is no constraint on β. $\qquad\square$

The argument presented shows also that every function $f : \mathbb{R}^N \to \mathbb{R}$ of the form described actually satisfies equation (13.10.1).

In order to determine non-measurable solutions $f : \mathbb{R}^N \to \mathbb{R}$ of (13.10.1) we will need the following

Lemma 13.10.2. *Let $f : \mathbb{R}^N \to \mathbb{R}$ be a non-constant solution of (13.10.1), and let $r \in \mathbb{Q}(x, y)$ be a rational function in x, y, with the coefficients from $\mathbb{Q}$. If one of the expressions $r(a, b)$, $r(A, B)$ has sense, then the other has sense, too, and function (13.10.10) satisfies*

$$g(r(a, b)x) = r(A, B)g(x) \tag{13.10.12}$$

for every x in $\mathbb{R}^N$.

Proof. First we prove that for every $x \in \mathbb{R}^N$ and $k \in \mathbb{N} \cup \{0\}$

$$g(a^k x) = A^k g(x), \quad g(b^k x) = B^k g(x). \tag{13.10.13}$$

For $k = 0$ (13.10.13) is trivial. Assume it valid for a $k \in \mathbb{N} \cup \{0\}$. Replacing in (13.10.1) x by $a^k x$, and setting $y = -c/b$, we obtain

$$f(a^{k+1}x) = Af(a^k x) + Bf\left(-\frac{c}{b}\right) + C.$$

Setting in (13.10.1) $x = 0$, $y = -c/b$, we get

$$f(0) = Af(0) + Bf\left(-\frac{c}{b}\right) + C.$$

Subtracting the two expressions, and making use of the induction hypothesis (13.10.13), we obtain

$$g(a^{k+1}x) = f(a^{k+1}x) - f(0) = A[f(a^k x) - f(0)] = Ag(a^k x) = A^{k+1}g(x).$$

Induction shows that the first formula in (13.10.13) is generally valid. The second formula in (13.10.13) is established similarly.

Now let $p \in \mathbb{Q}[x, y]$ be an arbitrary polynomial in x, y with rational coefficients:

$$p(x, y) = \sum_{k=0}^{n} \sum_{i=0}^{k} \alpha_{ki} x^i y^{k-i}, \quad \alpha_{ki} \in \mathbb{Q}, \quad k = 1, \ldots, n; \quad i = 1, \ldots, k.$$

It follows from Lemma 13.10.1 that g is additive. In fact, if f has form (13.10.4), then $f(0) = \beta$, and $g(x) = f(x) - \beta = f(x) - f(0)$ coincides with function (13.10.10). Hence, by Theorem 5.2.1, $g(\alpha_{ki}u) = \alpha_{ki}g(u)$ for every $u \in \mathbb{R}^N$ and $k = 1, \ldots, n$; $i = 1, \ldots, k$, whence by (13.10.13), for every $x \in \mathbb{R}^N$,

$$g(p(a, b)x) = g\left(\left(\sum_{k=0}^{n} \sum_{i=0}^{k} \alpha_{ki} a^i b^{k-i}\right)x\right)$$

$$= \left(\sum_{k=0}^{n} \sum_{i=0}^{k} \alpha_{ki} A^i B^{k-i}\right)g(x) = p(A, B)g(x). \tag{13.10.14}$$

Since f is non-constant, there exists a $u \in \mathbb{R}^N$ such that $g(u) \neq 0$. Suppose that $p(a, b) \neq 0$. Setting in (13.10.14) $x = u/p(a, b)$ we obtain $g(u) = p(A, B)g(x)$, whence $p(A, B) \neq 0$. If $p(A, B) \neq 0$, then $g(p(a, b)u) = p(A, B)g(u) \neq 0$, whence $p(a, b) \neq 0$, since $g(0) = 0$.

Now take an arbitrary $r \in \mathbb{Q}(x, y)$:

$$r(x, y) = \frac{p(x, y)}{q(x, y)},$$

where $p, q \in \mathbb{Q}[x, y]$. If one of $q(a, b), q(A, B)$ is different from zero, so is the other, as has been just shown, and consequently if one of $r(a, b), r(A, B)$ has sense, then the other has sense, too. Now for arbitrary $x \in \mathbb{R}^N$, we have by (13.10.14), if $r(a, b)$ and $r(A, B)$ have sense,

$$p(A, B)g(x) = g(p(a, b)x) = g(q(a, b)r(a, b)x) = q(A, B)g(r(a, b)x),$$

whence we obtain (13.10.12). $\qquad\square$

Now consider the extensions $\mathbb{Q}(a, b)$ and $\mathbb{Q}(A, B)$ of $\mathbb{Q}$ by a, b and A, B, respectively (cf. 4.7).

Lemma 13.10.3. *Let $\varphi : \mathbb{Q}(a, b) \to \mathbb{Q}(A, B)$ be an isomorphism such that*

$$\varphi(a) = A, \quad \varphi(b) = B. \tag{13.10.15}$$

Then for every $x \in \mathbb{Q}$ we have $\varphi(x) = x$, and for every rational function $r \in \mathbb{Q}(x, y)$ such that $r(a, b), r(A, B)$ have sense

$$\varphi(r(a, b)) = r(A, B). \tag{13.10.16}$$

Proof. Since φ is an isomorphism, φ is one-to-one. Now, $\varphi(0) = \varphi(0) + \varphi(0)$, whence $\varphi(0) = 0$, and so $\varphi(x) \neq 0$ for $x \neq 0$. We have $\varphi(1) = \varphi(1)\varphi(1)$, whence $\varphi(1) = 1$. By induction $\varphi(k) = k$ for $k \in \mathbb{N}$. Further $\varphi(1) = \varphi(-1)\varphi(-1)$, whence $\varphi(-1) = -1$, since φ is one-to-one and $1 = \varphi(1)$. So $\varphi(-k) = \varphi(-1)\varphi(k) = -\varphi(k) = -k$ for $k \in \mathbb{N}$, and since $\varphi(0) = 0$, we have $\varphi(k) = k$ for $k \in \mathbb{Z}$. Now, for arbitrary $k \in \mathbb{Z}$ and $m \in \mathbb{N}$ we have $m\varphi\left(\dfrac{k}{m}\right) = \varphi(k) = k$, whence $\varphi(x) = x$ for $x \in \mathbb{Q}$.

Since φ is an isomorphism, we have by (13.10.15) $\varphi\big(r(a,b)\big) = r\big(\varphi(a), \varphi(b)\big) = r(A, B)$. $\qquad\square$

$\mathbb{R}^N$ may be regarded as the linear space $\big(\mathbb{R}^N; \mathbb{Q}(a,b); +; \cdot\big)$ over $\mathbb{Q}(a,b)$, and as such has a base H in virtue of Corollary 4.2.1. If $c \neq 0$, then $\{c\}$ is linearly independent over $Q(a,b)$, and by Theorem 4.2.1 we may require that $c \in H$.

Theorem 13.10.3. *Suppose that there exists an isomorphism* $\varphi : \mathbb{Q}(a,b) \to \mathbb{Q}(A,B)$ *fulfilling* (13.10.15). *Let* $H \subset \mathbb{R}^N$ *be a base of* $\mathbb{R}^N$ *over* $\mathbb{Q}(a,b)$ *such that* $c \in H$ *whenever* $c \neq 0$. *Put* $H_0 = H \cup \{0\}$. *Further assume that if* $A + B = 1$ *and* $c = 0$, *then* $C = 0$. *For every function* $f_0 : H_0 \to \mathbb{R}$ *such that*

$$f_0(c) = (A + B)f_0(0) + C \tag{13.10.17}$$

there exists a unique function $f : \mathbb{R}^N \to \mathbb{R}$ *satisfying equation* (13.10.1) *and such that* $f(x) = f_0(x)$ *for* $x \in H_0$.

Proof. Every $x \in \mathbb{R}^N$ has a representation

$$x = \alpha_1 h_1 + \cdots + \alpha_n h_n, \tag{13.10.18}$$

where $\alpha_i \in \mathbb{Q}(a,b)$, $h_i \in H$, $i = 1, \ldots, n$, and representation (13.10.18) is unique up to terms with coefficients zero. For such an x put

$$f(x) = \sum_{i=1}^{n} \varphi(\alpha_i)\big[f_0(h_i) - f_0(0)\big] + f_0(0). \tag{13.10.19}$$

Thus the function $f : \mathbb{R}^N \to \mathbb{R}$ is unambiguously defined in the whole $\mathbb{R}^N$; note that, since $\varphi(0) = 0$ (Lemma 13.10.3), adding to (13.10.18) terms with coefficients zero does not affect (13.10.19).

Let $c \neq 0$ so that $c \in H$. Take arbitrary $x, y \in \mathbb{R}^N$, and let x have a representation (13.10.18). y has a representation

$$y = \beta_1 h_1 + \cdots + \beta_n h_n. \tag{13.10.20}$$

We may assume that the same h_i occur in (13.10.18) and (13.10.20), adding, if necessary, suitable terms with coefficients zero. For the same reason we may assume that one of $h_1, \ldots, h_n$ is c, and thus c can be written as

$$c = \gamma_1 h_1 + \cdots + \gamma_n h_n, \tag{13.10.21}$$

where all γ_i are zero, except one, which is one. Now, by (13.10.18)–(13.10.21),

$$f(ax + by + c) = f\left(\sum_{i=1}^{n}(\alpha_i a + \beta_i b + \gamma_i)h_i\right)$$

$$= \sum_{i=1}^{n}\varphi(\alpha_i a + \beta_i b + \gamma_i)\big[f_0(h_i) - f_0(0)\big] + f_0(0)$$

$$= \sum_{i=1}^{n}\big[\varphi(\alpha_i)A + \varphi(\beta_i)B + \varphi(\gamma_i)\big]\big[f_0(h_i) - f_0(0)\big] + f_0(0)$$

$$= A\sum_{i=1}^{n}\varphi(\alpha_i)\big[f_0(h_i) - f_0(0)\big] + B\sum_{i=1}^{n}\varphi(\beta_i)\big[f_0(h_i) - f_0(0)\big]$$

$$+ \sum_{i=1}^{n}\varphi(\gamma_i)\big[f_0(h_i) - f_0(0)\big] + f_0(0)$$

$$= A\big[f(x) - f_0(0)\big] + B\big[f(y) - f_0(0)\big] + f(c)$$

$$= Af(x) + Bf(y) + C - \big[(A + B)f_0(0) - f(c) + C\big]$$

$$= Af(x) + Bf(y) + C,$$

because the expression in brackets is zero in virtue of (13.10.17) and of the fact that, since $c \in H$, we have by (13.10.19) $f(c) = \varphi(1)\big[f_0(c) - f_0(0)\big] + f_0(0) = f_0(c) - f_0(0) + f_0(0) = f_0(c)$.

If $c = 0$, the argument is the same, only we do not assume that c is one of $h_1, \ldots, h_n$, and in representation (13.10.21) all γ_i are zeros.

Thus f satisfies equation (13.10.1). For $x \in H$ we have by (13.10.19)

$$f(x) = \varphi(1)\big[f_0(x) - f_0(0)\big] + f_0(0) = f_0(x).$$

Similarly, for $x = 0$ we have $x = 0h$ with an arbitrary $h \in H$, whence by (13.10.19)

$$f(0) = \varphi(0)\big[f_0(h) - f_0(0)\big] + f_0(0) = f_0(0).$$

Thus $f(x) = f_0(x)$ for $x \in H_0$.

In order to prove uniqueness, suppose that a function $f : \mathbb{R}^N \to \mathbb{R}$ satisfies equation (13.10.1) and $f \mid H_0 = f_0$. By Lemma 13.10.1 f has form (13.10.4), where $g : \mathbb{R}^N \to \mathbb{R}$ is additive, and $\beta \in \mathbb{R}$ is a constant. Since $g(0) = 0$, we have $\beta = f(0)$, and thus g fulfils (13.10.10). Now take an arbitrary $x \in \mathbb{R}^N$, and let x have representation (13.10.18). Since $\alpha_1, \ldots, \alpha_n \in \mathbb{Q}(a, b)$, it follows by Corollary 4.8.1 that for every $i - 1, \ldots, n$ there exists a rational function $r_i \in \mathbb{Q}(x, y)$ such that $\alpha_i = r_i(a, b)$. By Lemmas 13.10.2 and 13.10.3

$$g(x) = \sum_{i=1}^{n}g(\alpha_i h_i) = \sum_{i=1}^{n}g\big(r_i(a, b)h_i\big) = \sum_{i=1}^{n}r_i(A, B)g(h_i)$$

$$= \sum_{i=1}^{n}\varphi\big(r_i(a, b)\big)g(h_i) = \sum_{i=1}^{n}\varphi(\alpha_i)g(h_i),$$

whence

$$f(x) = g(x) + f(0) = \sum_{i=1}^{n} \varphi(\alpha_i) g(h_i) + f(0) = \sum_{i=1}^{n} \varphi(\alpha_i) \big[f(h_i) - f(0) \big] + f(0),$$

i.e., since $h_i \in H_0$, $0 \in H_0$, and $f \mid H_0 = f_0$,

$$f(x) = \sum_{i=1}^{n} \varphi(\alpha_i) \big[f_0(h_i) - f_0(0) \big] + f_0(0).$$

Thus f must be given by formula (13.10.19), which proves the uniqueness of f. $\qquad \square$

Theorem 13.10.4. *Suppose that equation* (13.10.1) *has a non-constant solution* f : $\mathbb{R}^N \to \mathbb{R}$. *Then there exists an isomorphism* $\varphi : \mathbb{Q}(a,b) \to \mathbb{Q}(A,B)$ *fulfilling* (13.10.15). *Moreover, if* $A + B = 1$ *and* $c = 0$, *then also* $C = 0$.

Proof. Define the function $\varphi : \mathbb{Q}(a,b) \to \mathbb{Q}(A,B)$ by (13.10.16). We shall show that this definition is correct. Suppose that $r_1(a,b) = r_2(a,b)$ for some $r_1, r_2 \in \mathbb{Q}(x,y)$. Function (13.10.10) is not constant, since otherwise f would be constant. So there exists a $u \in \mathbb{R}^N$ such that $g(u) \neq 0$. We have by Lemma 13.10.2

$$r_1(A,B) g(u) = g\big(r_1(a,b) u \big) = g\big(r_2(a,b) u \big) = r_2(A,B) g(u),$$

whence also $r_1(A,B) = r_2(A,B)$. Also, if $\alpha = r(a,b) \in \mathbb{Q}(a,b)$, then $r(A,B)$ has sense, and so $\varphi(\alpha) = \varphi(r(a,b))$ is by (13.10.16) well defined.

It follows by Corollary 4.8.1 that φ is defined on the whole $\mathbb{Q}(a,b)$, and maps it onto $\mathbb{Q}(A,B)$. Also, the rational functions $r_x(x,y) = x/1$, and $r_y(x,y) = y/1$ belong to $\mathbb{Q}(x,y)$, whence $\varphi(a) = \varphi(r_x(a,b)) = r_x(A,B) = A$, and $\varphi(b) = \varphi(r_y(a,b)) = r_y(A,B) = B$. Thus φ satisfies (13.10.15).

Now we show that φ is a homomorphism. Let $\alpha, \beta \in \mathbb{Q}(a,b)$. Then $\alpha = r_1(a,b)$, $\beta = r_2(a,b)$ for certain $r_1, r_2 \in \mathbb{Q}(x,y)$. Then $\alpha + \beta = r_1(a,b) + r_2(a,b) = (r_1 + r_2)(a,b)$ and $\alpha\beta = r_1(a,b) r_2(a,b) = (r_1 r_2)(a,b)$. Of course, $r_1 + r_2$, $r_1 r_2 \in \mathbb{Q}(x,y)$. Hence

$$\varphi(\alpha + \beta) = \varphi\big((r_1 + r_2)(a,b) \big) = (r_1 + r_2)(A,B) = r_1(A,B) + r_2(A,B) = \varphi(\alpha) + \varphi(\beta),$$

$$\varphi(\alpha\beta) = \varphi\big((r_1 r_2)(a,b) \big) = (r_1 r_2)(A,B) = r_1(A,B) r_2(A,B) = \varphi(\alpha) \varphi(\beta).$$

Suppose that for a certain $\alpha = r(a,b) \in \mathbb{Q}(a,b)$ we have $\varphi(\alpha) = r(A,B) = 0$. Let $u \in \mathbb{R}^N$ be such that $g(u) \neq 0$. Suppose that $\alpha \neq 0$. Then by Lemma 13.10.2

$$0 = r(A,B) g \left(\frac{u}{r(a,b)} \right) = g(u) \neq 0.$$

Consequently $\varphi(\alpha) = 0$ implies $\alpha = 0$, i.e., if $\varphi(\alpha) = \varphi(\beta)$ for some $\alpha, \beta \in \mathbb{Q}(a,b)$, then $\varphi(\alpha - \beta) = \varphi(\alpha) - \varphi(\beta) = 0$, and $\alpha = \beta$. Thus φ is one-to-one, i.e., it is a monomorphism. We have pointed out above that φ is an epimorphism, and so φ is an isomorphism.

Setting in equation (13.10.1) $x = y = 0$, we obtain

$$f(c) = (A + B)f(0) + C.$$

Hence, if $A + B = 1$ and $c = 0$, we obtain $C = 0$.   □

Theorems 13.10.3 and 13.10.4 imply the following

Theorem 13.10.5. *Equation* (13.10.1) *has a non-constant solution $f : \mathbb{R}^N \to \mathbb{R}$ if and only if there exists[8] an isomorphism $\varphi : \mathbb{Q}(a, b) \to \mathbb{Q}(A, B)$ fulfilling* (13.10.15) *and $C = 0$ whenever $A + B = 1$ and $c = 0$.*

Proof. Since $\mathbb{R} \neq \mathbb{Q}(a, b)$ for any $a, b \in \mathbb{R}$, we have $\dim (\mathbb{R}; \mathbb{Q}(a, b); +; \cdot) > 1$, and hence, in Theorem 13.10.3, $\operatorname{card} H > 1$. Thus f_0 fulfilling (13.10.17) can be chosen non-constant, and then also f is non-constant. The theorem results now from Theorems 13.10.3 and 13.10.4.   □

Exercises

1. Find solutions $f : (0, 1] \to \mathbb{R}$ of equation (13.1.3).
2. Determine all functions $f : \mathbb{R}^N \to \mathbb{R}$ satisfying the condition: For every x, y, u, $v \in \mathbb{R}^N$, if $x + y = u + v$, then $f(x) + f(y) = f(u) + f(v)$.
3. Let $I = [0, 1]$, and let the operation $\dotplus$ be defined for $x, y \in I$ by

$$x \dotplus y = \begin{cases} x + y & \text{if } x + y \leqslant 1, \\ x + y - 1 & \text{if } x + y > 1. \end{cases}$$

   Find all monotonic solutions $f : I \to I$ of the equation

$$f(x \dotplus y) = f(x) \dotplus f(y). \tag{$*$}$$

   [Hint: Note that if $f : I \to I$ satisfies $(*)$, and $g(x) = 1 - f(x)$ for $x \in I$, then also g satisfies $(*)$.]
4. Find solutions $f : \mathbb{R}^N \to \mathbb{R}$ of the equation

$$f(xy)[f(xy) - f(x) - f(y)] = 0.$$

5. Let $f : \mathbb{R}^N \to \mathbb{R}$ and $g : \mathbb{R}^N \to \mathbb{R}$ be additive functions. Prove that if $[f(x)]^2 = [g(x)]^2$ for $x \in \mathbb{R}^N$, then either $f = g$, or $f = -g$ in $\mathbb{R}^N$. (Grząślewicz [123].)

---

[8]A condition for the existence of such an isomorphism is found in Theorem 4.12.2. (Note that by Lemma 13.10.3 we have $\varphi \mid \mathbb{Q} = $ identity.)

Chapter 14

Derivations and Automorphisms

14.1 Derivations

In this chapter we will deal with functions satisfying the Cauchy equation (5.2.1) and also, simultaneously, another equations of a similar type.

Let $(Q; +, \cdot)$ be a commutative ring, and let $(P; +, \cdot)$ be a subring of $(Q; +, \cdot)$. A function $f : P \to Q$ is called a *derivation*[1] (or a *darivation of P*) iff it satisfies both the equations

$$f(x + y) = f(x) + f(y), \tag{14.1.1}$$
$$f(xy) = xf(y) + yf(x) \tag{14.1.2}$$

for all $x, y \in P$. Before we give some examples, let us note the following simple fact.

Lemma 14.1.1. *Let $P \subset Q$ be commutative rings, let $f_1, f_2 : P \to Q$ be derivations (of P), and let $a, b \in Q$ be arbitrary constants. Then $f = af_1 + bf_2$ also is a derivation of P.*

Proof. Obvious. □

Example 14.1.1. Let $(F; +, \cdot)$ be a field, and let $P = Q = F[x]$ be the ring of polynomials with coefficients from F. Let the function $f : F[x] \to F[x]$ be defined as $f(p) = p'$, where the derivative p' is defined in 4.7. We have clearly

$$f(p + q) = (p + q)' = p' + q' = f(p) + f(q),$$
$$f(pq) = (pq)' = pq' + qp' = pf(q) + qf(p).$$

Consequently f is a derivation.

Example 14.1.2. Let $(F; +, \cdot)$ be a field, and suppose that we are given a derivation $f : F \to F$. We define a function $f_0 : F[x] \to F[x]$ as follows. If $p \in F[x]$, $p(x) = \sum_{k=0}^{n} a_k x^k$, then $f_0(p)$ is the polynomial (in sequel denoted by p^f):

$$f_0(p) = p^f(x) = \sum_{k=0}^{n} f(a_k) x^k.$$

---

[1] The present exposition of the theory of derivations is based on Zariski-Samuel [325] (cf. also Horinouchi-Kannappan [144] concerning a generalization).

If $q \in F[x]$, $q(x) = \sum_{k=0}^{n} b_k x^k$ is another polynomial (we can write p and q with the same n, adding to that of the smaller degree terms with coefficients zero), then by (14.1.1)

$$f_0(p+q) = \sum_{k=0}^{n} f(a_k + b_k) x^k = \sum_{k=0}^{n} f(a_k) x^k + \sum_{k=0}^{n} f(b_k) x^k = f_0(p) + f_0(q).$$

We have $pq = \sum_{k=0}^{2n} \left(\sum_{i=0}^{k} a_i b_{k-i} \right) x^k$ (where $a_j = b_j = 0$ for $j > n$), whence by (14.1.1) and (14.1.2)

$$f_0(pq) = \sum_{k=0}^{2n} f\left(\sum_{i=0}^{k} a_i b_{k-i} \right) x^k = \sum_{k=0}^{2n} \left(\sum_{i=0}^{k} f(a_i b_{k-i}) \right) x^k$$

$$= \sum_{k=0}^{2n} \left(\sum_{i=0}^{k} \left(a_i f(b_{k-i}) + b_{k-i} f(a_i) \right) \right) x^k$$

$$= \sum_{k=0}^{2n} \left(\sum_{i=1}^{k} a_i f(b_{k-i}) \right) x^k + \sum_{k=0}^{2n} \left(\sum_{i=0}^{k} f(a_i) b_{k-i} \right) x^k$$

$$= p f_0(q) + q f_0(p).$$

(Note that by (14.1.1) $f(0) = 0$ so that $f(a_j) = f(b_j)$ for $j > n$.) Consequently f_0 is a derivation.

The derivations described in the above two examples have rather a fundamental importance for we have the following

Lemma 14.1.2. *Let $(K; +, \cdot)$ be a field, and let $(F; +, \cdot)$ be a subfield of $(K; +, \cdot)$, and let $f : F \to K$ be a derivation. Then we have, for every $a \in F$ and every polynomial $p \in F[x]$,*

$$f(p(a)) = p^f(a) + f(a)p'(a).$$

Proof. For $k \in \mathbb{N}$ we have

$$f(a^k) = ka^{k-1}f(a). \tag{14.1.3}$$

(14.1.3) is evidently true for $k = 1$. Assuming it valid for a $k \in \mathbb{N}$, we have by (14.1.2)

$$f(a^{k+1}) = f(aa^k) = af(a^k) + a^k f(a) = aka^{k-1}f(a) + a^k f(a) = (k+1)a^k f(a),$$

i.e., we obtain (14.1.3) for $k + 1$. Thus (14.1.3) is generally true.

Moreover, setting in (14.1.2) $x = y = 1$ we obtain $f(1) = 2f(1)$, whence

$$f(1) = 0. \tag{14.1.4}$$

Now let $p(x) = \sum\limits_{k=0}^{n} a_k x^k$. Then by (14.1.3) and (14.1.4)

$$f\big(p(a)\big) = f\Big(\sum_{k=0}^{n} a_k a^k\Big) = \sum_{k=0}^{n} f\big(a_k a^k\big) = \sum_{k=0}^{n} a^k f\big(a_k\big) + \sum_{k=1}^{n} a_k f\big(a^k\big)$$

$$= p^f(a) + \sum_{k=1}^{n} a_k k a^{k-1} f(a) = p^f(a) + f(a)p'(a),$$

which proves the lemma. $\qquad\square$

In $\mathbb{R}$ the function $f = 0$ evidently is a (trivial) derivation (of $\mathbb{R}$). It is difficult to find another example. We have, in particular,

Lemma 14.1.3. *If $f : \mathbb{R} \to \mathbb{R}$ is a derivation, then $f(x) = 0$ for every $x \in \mathbb{Q}$.*

Proof. By (14.1.1) f is an additive function. Hence we have by (14.1.4) and Theorem 5.2.1 $f(x) = xf(1) = 0$ for every $x \in \mathbb{Q}$. $\qquad\square$

This result can be strengthened as follows.

Lemma 14.1.4. *If $f : \mathbb{R} \to \mathbb{R}$ is a derivation, then $f(x) = 0$ for every $x \in \mathrm{alg\,cl}\,\mathbb{Q}$.*

Proof. Let $a \in \mathrm{alg\,cl}\,\mathbb{Q}$, and let $p \in \mathbb{Q}\,[x]$ be its minimal polynomial. Write $p(x) = \sum\limits_{k=0}^{n} a_k x^k$, $a_k \in \mathbb{Q}$, $k = 0, \ldots, n$. By Lemma 14.1.3 $f(a_k) = 0$ for $k = 0, \ldots, n$, whence

$$p^f(x) = \sum_{k=0}^{n} f\big(a_k\big) x^k = 0.$$

So by Lemma 14.1.2 $f\big(p(a)\big) = f(a)p'(a)$. Now, $p(a) = 0$, whence $f\big(p(a)\big) = f(0) = 0$. On the other hand $p'(a) \neq 0$, because $\mathrm{degree}\,p'(a) < \mathrm{degree}\,p(a)$, and p is the minimal polynomial of a. Hence $f(a) = 0$. $\qquad\square$

From Lemma 14.1.3 we get also

Theorem 14.1.1. *If $f : \mathbb{R} \to \mathbb{R}$ is a derivation, and f is measurable, or bounded above, or below, on a set $T \in \mathfrak{A}$, then $f = 0$.*

Proof. Since, by (14.1.1), f is additive, the conditions on f imply its continuity. Since $\mathbb{Q}$ is dense in $\mathbb{R}$, we obtain from Lemma 14.1.3 that $f = 0$ identically in $\mathbb{R}$. $\qquad\square$

It is not difficult to solve equation (14.1.2) alone for $f : \mathbb{R} \to \mathbb{R}$. Taking $x, y \neq 0$ and dividing (14.1.2) by xy, we obtain

$$\frac{f(xy)}{xy} = \frac{f(x)}{x} + \frac{f(y)}{y},$$

which means that the function $f_0(x) = f(x)/x$ satisfies equation (13.1.2) in $(-\infty, 0) \cup (0, \infty)$. By Theorem 13.1.2 there exists an additive function $g : \mathbb{R} \to \mathbb{R}$ such that $f_0(x) = g\big(\log|x|\big)$, whence

$$f(x) = xg\big(\log|x|\big). \qquad (14.1.5)$$

Formula (14.1.5), established so far only for $x \neq 0$, remains true also for $x = 0$ if we agree that $0g(\log 0) = 0$.

But, in general, function (14.1.5) will not be additive, i.e., it will not satisfy (14.1.1). Is it possible to choose g in such a manner that f given by (14.1.5) should be additive? In the light of our experience up to now it seems doubtful whether there may exist non-trivial derivations $f : \mathbb{R} \to \mathbb{R}$. In the next section we will answer this question.

14.2   Extensions of derivations

In the present section we study the possibility of extending a derivation from its domain of definition onto a larger algebraic structure.

Lemma 14.2.1. *Let $(P; +, \cdot)$ be an integral domain, let $(F; +, \cdot)$ be its field of fractions, and let $(K; +, \cdot)$ be a field, $P \subset F \subset K$. If $f : P \to K$ is a derivation, then there exists a unique derivation $g : F \to K$ such that $g \mid P = f$.*

Proof. Every $x \in F$ can be written as $x = u/v$, where $u, v \in P$, $v \neq 0$. For such an x we put

$$g(x) = g\left(\frac{u}{v}\right) = \frac{vf(u) - uf(v)}{v^2}. \tag{14.2.1}$$

We must check that definition (14.1.1) is unambiguous, i.e., if $u/v = z/w$, $v, w \neq 0$, then $g(u/v) = g(z/w)$. Now, $u/v = z/w$ means

$$uw = vz, \tag{14.2.2}$$

whence by (14.1.2) $uf(w) + wf(u) = vf(z) + zf(v)$, i.e.,

$$vf(z) - uf(w) = wf(u) - zf(v).$$

Multiplying this by uw we get

$$v^2wf(z) - uvwf(w) = vw^2f(u) - zvwf(v),$$

or, by (14.2.2)

$$v^2wf(z) - v^2zf(w) = vw^2f(u) - uw^2f(v).$$

Dividing this by v^2w^2 we obtain

$$\frac{wf(z) - zf(w)}{w^2} = \frac{vf(u) - uf(v)}{v^2},$$

i.e., $g(z/w) = g(u/v)$. Thus expression (14.2.1) does not depend on the representation of x as a fraction u/v.

Now take arbitrary $x, y \in F$, $x = u/v$, $y = z/w$, $u, v, z, w \in P$, $v, w \neq 0$. Then by (14.1.1) and (14.1.2)

$$g(x+y) = g\left(\frac{u}{v} + \frac{z}{w}\right) = g\left(\frac{uw+zv}{vw}\right) = \frac{vwf(uw+zv) - (uw+zv)f(uw)}{v^2w^2}$$

$$= \frac{vw\big(uf(w) + wf(u) + zf(v) + vf(z)\big) - (uw+zv)\big(vf(w) + wf(v)\big)}{v^2w^2}$$

$$= \frac{vw^2 f(u) + v^2 w f(z) - uw^2 f(v) - zv^2 f(w)}{v^2w^2}$$

$$= \frac{vf(u) - uf(v)}{v^2} + \frac{wf(z) - zf(w)}{w^2} = g\left(\frac{u}{v}\right) + g\left(\frac{z}{w}\right)$$

$$= g(x) + g(y),$$

and

$$g(xy) = g\left(\frac{uz}{vw}\right) = \frac{vwf(uz) - uzf(vw)}{v^2w^2}$$

$$= \frac{vw\big(uf(z) + zf(u)\big) - uz\big(vf(w) + wf(v)\big)}{v^2w^2}$$

$$= \frac{uv\big(wf(z) - zf(w)\big) + zw\big(vf(u) - uf(v)\big)}{v^2w^2}$$

$$= \frac{u}{v}\frac{wf(z) - zf(w)}{w^2} + \frac{z}{w}\frac{vf(u) - uf(v)}{v^2}$$

$$= \frac{u}{v}g\left(\frac{z}{w}\right) + \frac{z}{w}g\left(\frac{u}{v}\right) = xg(y) + yg(x).$$

Consequently g is a derivation of F. If $x \in P$, then $x = x/1$ and by (14.1.4)

$$g(x) = g\left(\frac{x}{1}\right) = \frac{f(x) - xf(1)}{1} = f(x),$$

i.e., $g \mid P = f$.

Now let $g : F \to K$ be an arbitrary derivation such that $g \mid P = f$. Take an arbitrary $x \in F$, $x = u/v$, $u, v \in P$, $v \neq 0$. We have $f(u) = g(u) = g(vx) = vg(x) + xg(v) = vg(x) + xf(v)$, whence

$$g(x) = \frac{f(u) - xf(v)}{v} = \frac{vf(u) - uf(v)}{v^2}.$$

This means that g has form (14.2.1), which proves the uniqueness of the extension. $\square$

Lemma 14.2.2. *Let $(K; +, \cdot)$ be a field, let $(F; +, \cdot)$ be a subfield of $(K; +, \cdot)$, and let $f : F \to K$ be a derivation. Further, let $a, u \in K$. There exists a derivation $g : F(a) \to K$ such that $g \mid F = f$ and $g(a) = u$ if and only if*

$$r^f(a) + ur'(a) = 0 \tag{14.2.3}$$

for every $r \in F[x]$ such that $r(a) = 0$. If it exists, the extension g is unique.

Proof. First we define a derivation $f_0 : F[a] \to K$ such that $f_0 \mid F = f$ and $f_0(a) = u$.

If $x \in F[a]$, then there exists a polynomial $p \in F[t]$ such that $p(a) = x$. For such an x we put

$$f_0(x) = f_0(p(a)) = p^f(a) + up'(a). \qquad (14.2.4)$$

We must check that definition (14.2.4) is unambiguous, i.e., if $p(a) = q(a)$, $q \in F[t]$, then

$$p^f(a) + up'(a) = q^f(a) + uq'(a). \qquad (14.2.5)$$

Put $r = p - q$. Then $r(a) = 0$, whence it follows that (14.2.3) is fulfilled. But $r^f = p^f - q^f$ and $r' = p' - q'$, so (14.2.3) means (14.2.5).

It is easily seen that f_0 is a derivation, and if $x \in F$ so that $x = p(a)$ for $p(t) = x = \text{const}$, then by (14.2.4)

$$f_0(x) = f_0(p(a)) = p^f(a) + up'(a) = f(x),$$

since $p^f(t) = f(x)$ for every $t \in F$, and $p' = 0$. Thus $f_0 \mid F = f$. Similarly, $a = p(a)$ for $p(t) = t$. By (14.2.4)

$$f_0(a) = f_0(p(a)) = p^f(a) + up'(a) = u,$$

because $p^f(t) = f(1)t = 0$ by (14.1.4), and $p'(t) = 1$. Thus $f_0(a) = u$.

The extension f_0 is unique, for if $f_0 : F[a] \to K$ is an arbitrary derivation such that $f_0 \mid F = f$ and $f_0(a) = u$, then, by Lemma 14.1.2, for arbitrary $x \in F[a]$, $x = p(a)$, $p \in F[t]$,

$$f_0(x) = f_0(p(a)) = p^{f_0}(a) + f_0(a)p'(a) = p^f(a) + up'(a).$$

So f_0 has to have form (14.2.4).

Before proceeding further, let us make sure that $F[a]$ is an integral domain. $F[a]$ clearly is commutative, and the constant polynomial $p(t) = 1$ fulfils $p(a) = 1$, so $1 \in F[a]$. Now let $w, v \in F[a]$ be such that $wv = 0$. Since $F[a] \subset K$, we have $w, v \in K$. But a field has no divisors of zero, so $wv = 0$ implies that either $w = 0$, or $v = 0$. Consequently $F[a]$ has no divisors of zero, and so it is an integral domain.

By Lemma 4.8.1 $F(a)$ is the field of fractions of $F[a]$. By Lemma 14.2.1 f_0 can be uniquely extended onto $F(a)$ to a derivation $g : F(a) \to K$ such that $g \mid F[a] = f_0$. Hence $g \mid F = g \mid F[a] \mid F = f_0 \mid F = f$, and, since $a \in F[a]$, $g(a) = f_0(a) = u$.

In the opposite direction, if g is the required extension, then, in particular, g satisfies equation (14.1.1) for $x, y \in F(a)$, whence $g(0) = 0$. Hence, by Lemma 14.1.2, if $r \in F[t]$ is such that $r(a) = 0$, then

$$r^f(a) + ur'(a) = r^g(a) + g(a)r'(a) = g(r(a)) = g(0) = 0.$$

Consequently condition (14.2.3) is fulfilled.   □

Lemma 14.2.3. *Let $(K; +, \cdot)$ be a field, let $(F; +, \cdot)$ be a subfield of $(K; +, \cdot)$ and let $S \subset K$ algebraically independent over F. Let $f : F \to K$ be a derivation, and let $u : S \to K$ be an arbitrary function. Then there exists a unique derivation $g : F(S) \to K$ such that $g \mid F = f$ and $g \mid S = u$.*

Proof. Let $\mathcal{R}$ be the collection of all couples (S_α, g_α) such that $S_\alpha \subset S$, $g_\alpha : F(S_\alpha) \to K$ is a derivation, and $g_\alpha \mid F = f$, $g_\alpha \mid S_\alpha = u$. The couple $(\varnothing, f) \in \mathcal{R}$ so that $\mathcal{R} \neq \varnothing$. We order $\mathcal{R}$ as follows: $(S_\alpha, g_\alpha) \leqslant (S_\beta, g_\beta)$ iff $S_\alpha \subset S_\beta$ and $g_\beta \mid F(S_\alpha) = g_\alpha$. Thus $(\mathcal{R}, \leqslant)$ in an ordered set, and if $\mathcal{L} \subset \mathcal{R}$ is a chain, then the couple (S_0, g_0) such that $S_0 = \bigcup_{(S_\alpha, g_\alpha) \in \mathcal{L}} S_\alpha$, $g_0 \mid F(S_\alpha) = g_\alpha$ for $(S_\alpha, g_\alpha) \in \mathcal{L}$, is an upper bound of $\mathcal{L}$ in $\mathcal{R}$. By Theorem 1.8.1 there exists a maximal element $(S_{\max}, g_{\max})$ in $\mathcal{R}$. Thus, in particular, $S_{\max} \subset S$. Suppose that there exists an $a \in S \setminus S_{\max}$. Since S is algebraically independent over F and $a \in S \setminus S_{\max}$, we have (cf. Exercise 4.19) $r(a) \neq 0$ for every $r \in F(S_{\max})[x]$. So the condition in Lemma 14.2.2 is trivially fulfilled. By Lemma 14.2.2 the derivation $g_{\max}$ can be extended onto $F(S_{\max})(a)$ to a derivation $g^* : F(S_{\max})(a) \to K$ such that $g^* \mid F(S_{\max}) = g_{\max}$, $g^*(a) = u(a)$. Hence g^* satisfies $g^* \mid F = g^* \mid F(S_{\max}) \mid F = g_{\max} \mid F = f$, $g^* \mid S_{\max} = g_{\max} \mid S_{\max} = u$, whence $g^* \mid (S_{\max} \cup \{a\}) = u$. Writing $S^* = S_{\max} \cup \{a\}$ we obtain hence that $(S^*, g^*) \in \mathcal{R}$ and $(S_{\max}, g_{\max}) < (S^*, g^*)$, which contradicts the maximality of $(S_{\max}, g_{\max})$.

Consequently $S_{\max} = S$, and $g = g_{\max}$ is the required extension.

It remains to prove the uniqueness. Let $x \in F(S)$. By Lemma 4.8.2 there exists a finite set $S_1 = \{a_1, \ldots, a_n\} \subset S$ such that $x \in F(S_1)$. But

$$F(S_1) = F(a_1)(a_2) \cdots (a_n).$$

Since, by Lemma 14.2.2, the extension of a derivation onto a simple extension of its field of definition (with the prescribed value at the element by which we extend the basic field) is unique, g is uniquely determined on $F(a_1) \ldots (a_k)$ for every $k = 1, \ldots, n$, and so, in particular, $g(x)$ is uniquely determined. Thus g is uniquely determined at every point $x \in F(S)$, and so g is unique. $\qquad \square$

Lemma 14.2.4. *Let $(K; +, \cdot)$ be a field of characteristic zero, let $(K_1; +, \cdot)$ be a subfield of $(K; +, \cdot)$, and let $(F; +, \cdot)$ be a subfield of $(K_1; +, \cdot)$ such that $K_1 \subset \mathrm{alg\,cl}\,F$. Let $f : F \to K$ be a derivation. Then there exists a unique derivation $g : K_1 \to K$ such that $g \mid F = f$.*

Proof. Let $\mathcal{R}$ be the collection of all couples (K_α, g_α) such that $(K_\alpha; +, \cdot)$ is a subfield of $(K_1; +, \cdot)$, $F \subset K_\alpha$, $g_\alpha : F_\alpha \to K$ is a derivation and $g_\alpha \mid F = f$. $(F, f) \in \mathcal{R}$, so $\mathcal{R} \neq \varnothing$. We order $\mathcal{R}$ similarly as in the proof of Lemma 14.2.3: $(K_\alpha, g_\alpha) \leqslant (K_\beta, g_\beta)$ iff $(K_\alpha; +, \cdot)$ is a subfield of $(K_\beta; +, \cdot)$ and $g_\beta \mid K_\alpha = g_\alpha$. Then $(\mathcal{R}, \leqslant)$ is an ordered set, and as previously we verify that every chain in $\mathcal{R}$ has an upper bound $(K_0, g_0) \in \mathcal{R}$. By Theorem 1.8.1 there exists in $\mathcal{R}$ a maximal element $(K_{\max}, g_{\max})$. In particular, $(K_{\max}; +, \cdot)$ is a subfield of $(K_1; +, \cdot)$. Suppose that there exists an $a \in K_1 \setminus K_{\max}$. Thus a is algebraic over F, and let $p \in F[x]$ be its minimal polynomial. We have $p' \neq 0$, since the characteristic of K is zero (whence also the characteristic of F is zero), whence $p'(a) \neq 0$, since degree $p' <$ degree p, and p is the minimal polynomial of a. Put $u = -p^{g_{\max}}(a)/p'(a)$, and let $r \in K_{\max}[x]$ be a polynomial such that $r(a) = 0$. There exist polynomials $q, s \in K_{\max}[x]$ such that $r = qp + s$ and degree $s <$ degree p. Hence $0 = r(a) = q(a)p(a) + s(a) = s(a)$, whence it follows that $s = 0$, and $r = qp$.

Now

$$r^{g_{\max}}(a) + ur'(a) = q(a)p^{g_{\max}}(a) + p(a)q^{g_{\max}}(a) + up'(a)q(a) + up(a)q'(a)$$
$$= q(a)\big(p^{g_{\max}}(a) + up'(a)\big),$$

since $p(a) = 0$. By the choice of u we have $p^{g_{\max}}(a) + up'(a) = 0$, whence also $r^{g_{\max}}(a) + ur'(a) = 0$.

By Lemma 14.2.2 there exists a derivation $g^* : K_{\max}(a) \to K$ such that $g^* \mid K_{\max} = g_{\max}$. Write $K^* = K_{\max}(a)$. Since $K_{\max} \subset K_1$ and $a \in K_1$, the field $(K^*; +, \cdot)$ is a subfield of $(K_1; +, \cdot)$, and, of course, $F \subset K_{\max} \subset K^*$. Moreover, $g^* \mid F = g_{\max} \mid F = f$. Thus $(K^*, g^*) \in \mathcal{R}$, and $(K_{\max}, g_{\max}) < (K^*, g^*)$, which contradicts the maximality of $(K_{\max}, g_{\max})$. Hence $K_{\max} = K_1$, and $g = g_{\max}$ is the desired extension.

To prove uniqueness, suppose that $g : K_1 \to K$ is a derivation such that $g \mid F = f$. Take an $a \in K_1$. Let $p \in F[x]$ be the minimal polynomial of a. As we have seen above, we have $p'(a) \neq 0$. By Lemma 14.1.2

$$p^f(a) + g(a)p'(a) = p^g(a) + g(a)p'(a) = g\big(p(a)\big) = g(0) = 0.$$

Hence $g(a) = -p^f(a)/p'(a)$ is uniquely determined. Thus all the values of g on K_1 are uniquely determined, whence g is unique. $\qquad\square$

Theorem 14.2.1. *Let $(K; +, \cdot)$ be a field of characteristic zero, let $(F; +, \cdot)$ be a subfield of $(K; +, \cdot)$, let S be an algebraic base of K over F, if it exists, and let $S = \varnothing$ otherwise. Let $f : F \to K$ be a derivation. Then, for every function $u : S \to K$, there exists a unique derivation $g : K \to K$ such that $g \mid F = f$ and $g \mid S = u$.*

Proof. If $K = \operatorname{alg\,cl} F$, this results from Lemma 14.2.4. (Then $S = \varnothing$, and there is no u involved). If $K \neq \operatorname{alg\,cl} F$, then, by Theorem 4.10.1, there exists an algebraic base S of K over F so that $\operatorname{alg\,cl} F(S) = K$. S is algebraically independent over F, so by Lemma 14.2.3 there exists a unique derivation $g_0 : F(S) \to K$ such that $g_0 \mid F = f$ and $g_0 \mid S = u$. By Lemma 14.2.4 the derivation g_0 can be uniquely extended onto $\operatorname{alg\,cl} F(S) = K$ to a derivation $g : K \to K$ such that $g \mid F(S) = g_0$, whence $g \mid F = g \mid F(S) \mid F = g_0 \mid F = f$, and $g \mid S = g \mid F(S) \mid S = g_0 \mid S = u$. $\qquad\square$

Theorem 14.2.2. *There exist non-trivial derivations of $\mathbb{R}$.*

Proof. We have $\mathbb{R} \neq \operatorname{alg\,cl} \mathbb{Q}$, so there exists, by Theorem 4.10.1, an algebraic base of $\mathbb{R}$ over $\mathbb{Q}$. Let $u : S \to \mathbb{R}$ be an arbitrary function, $u \neq 0$. The characteristic of $\mathbb{R}$ is zero. Now take in Theorem 14.2.1 $F = \mathbb{Q}$, $K = \mathbb{R}$. The trivial derivation $f_0 : \mathbb{Q} \to \mathbb{R}$, $f_0 = 0$, can, by Theorem 14.2.1, be uniquely extended onto $\mathbb{R}$ to a derivation $f : \mathbb{R} \to \mathbb{R}$ such that $f \mid S = u$, whence $f \neq 0$. $\qquad\square$

Incidentally, in this way we have obtained a description of all the derivations of $\mathbb{R}$. Every such derivation can be arbitrary prescribed on an algebraic base S of $\mathbb{R}$ over $\mathbb{Q}$, and then it is already uniquely determined. Unless identically equal to zero, every such derivation is discontinuous (and hence non-measurable, unbounded above and below on every set $T \in \mathfrak{A}$) in virtue of Theorem 14.1.1 .

14.3 Relations between additive functions

Let $f : \mathbb{R} \to \mathbb{R}$ be a derivation, and let $x \in \mathbb{R} \setminus \{0\}$. We have by Lemma 14.1.3

$$0 = f(1) = f\left(x\frac{1}{x}\right) = xf\left(\frac{1}{x}\right) + \frac{1}{x}f(x),$$

whence

$$f(x) = -x^2 f\left(\frac{1}{x}\right). \tag{14.3.1}$$

We will show that relation (14.3.1) characterizes derivations among additive functions[2] $f : \mathbb{R} \to \mathbb{R}$.

Theorem 14.3.1. *Let $f : \mathbb{R} \to \mathbb{R}$ be an additive function satisfying condition (14.3.1) for all real $x \neq 0$. Then f is a derivation.*

Proof. Take an arbitrary $x \in \mathbb{R}$, $x \neq 0, 1, -1$. Then we have by (14.3.1) and (14.1.1)

$$f(x) + \frac{1}{x^2}f(x) = f(x) - f\left(\frac{1}{x}\right) = f\left(x - \frac{1}{x}\right) = f\left(\frac{x^2 - 1}{x}\right)$$

$$= -\left(\frac{x^2 - 1}{x}\right)^2 f\left(\frac{x}{x^2 - 1}\right) = -\left(\frac{x^2 - 1}{x}\right)^2 f\left(\frac{1}{x - 1} - \frac{1}{x^2 - 1}\right)$$

$$= -\left(\frac{x^2 - 1}{x}\right)^2 f\left(\frac{1}{x - 1}\right) + \left(\frac{x^2 - 1}{x}\right)^2 f\left(\frac{1}{x^2 - 1}\right)$$

$$= \left(\frac{x^2 - 1}{x}\right)^2 \left(\frac{1}{x - 1}\right)^2 f(x - 1) - \left(\frac{x^2 - 1}{x}\right)^2 \left(\frac{1}{x^2 - 1}\right)^2 f(x^2 - 1)$$

$$= \left(\frac{x + 1}{x}\right)^2 f(x - 1) - \frac{1}{x^2}f(x^2 - 1).$$

Setting in (14.3.1) $x = 1$ or -1, we obtain

$$f(1) = f(-1) = 0. \tag{14.3.2}$$

Hence $f(u - 1) = f(u) - f(1) = f(u)$ for every $u \in \mathbb{R}$, and we get

$$f(x) + \frac{1}{x^2}f(x) = \left(\frac{x + 1}{x}\right)^2 f(x) - \frac{1}{x^2}f(x^2),$$

$$x^2 f(x) + f(x) = (x + 1)^2 f(x) - f(x^2),$$

i.e.,

$$f(x^2) = 2xf(x). \tag{14.3.3}$$

---

[2]Kurepa [201]; cf. also Jurkat [159], Kurepa [202], Nishiyama-Horinouchi [243], Kannappan-Kurepa [163], Grząślewicz [122], [123]. Also the remaining results in this section are due to S. Kurepa [201], [202].

By (14.1.1) $f(0) = 0$. Thus, in view of (14.3.2), relation (14.3.3), so far obtained for $x \neq 0, 1, -1$, is valid for all $x \in \mathbb{R}$. We have by (14.3.3) for every $x, y \in \mathbb{R}$

$$f\left((x+y)^2\right) = 2(x+y) f(x+y),$$

i.e.,

$$f\left(x^2\right) + 2f(xy) + f\left(y^2\right) = 2xf(x) + 2xf(y) + 2yf(x) + 2yf(y),$$

and by (14.3.3)

$$2xf(x) + 2f(xy) + 2yf(y) = 2xf(x) + 2xf(y) + 2yf(x) + 2yf(y). \qquad (14.3.4)$$

Relation (14.3.4) yields (14.1.2), i.e., f is a derivation. $\qquad\square$

Lemma 14.3.1. *Let $f : \mathbb{R} \to \mathbb{R}$ and $g : \mathbb{R} \to \mathbb{R}$ be additive functions, $f \neq 0$, and let $P : \mathbb{R} \to \mathbb{R}$ be a continuous function. If*

$$g(x) = P(x)f\left(\frac{1}{x}\right) \qquad (14.3.5)$$

for all $x \in \mathbb{R}$, $x \neq 0$, then

$$P(x) = P(1)x^2 \qquad (14.3.6)$$

for all $x \in \mathbb{R}$.

Proof. Since $f(0) = 0$, but $f \neq 0$, there exists an $x_0 \in \mathbb{R} \setminus \{0\}$ such that $f(x_0) \neq 0$. Put $u = x_0^{-1}$, and take an arbitrary $\alpha \in \mathbb{Q}$, $\alpha \neq 0$. We have by (14.3.5) and Theorem 5.2.1

$$\alpha g(u) = g(\alpha u) = P(\alpha u) f\left(\frac{1}{\alpha u}\right) = \frac{1}{\alpha} P(\alpha u) f(x_0).$$

On the other hand, also by (14.3.5)

$$\alpha g(u) = \alpha P(u) f(x_0).$$

Hence

$$\left[\alpha P(u) - \frac{1}{\alpha} P(\alpha u)\right] f(x_0) = 0,$$

and, since $f(x_0) \neq 0$,

$$P(\alpha u) = \alpha^2 P(u)$$

for all $\alpha \in \mathbb{Q} \setminus \{0\}$. Setting $\alpha u = y$ we obtain hence

$$P(y) = y^2 \frac{P(u)}{u^2} \qquad (14.3.7)$$

for all $y \in \mathbb{Q}u \setminus \{0\}$. Since P is continuous and the set $\mathbb{Q}u \setminus \{0\}$ is dense in $\mathbb{R}$, it follows that (14.3.7) is valid for all $y \in \mathbb{R}$. Setting $y = 1$, we obtain from (14.3.7) $P(1) = P(u)/u^2$, and so (14.3.7) implies (14.3.6). $\qquad\square$

Theorem 14.3.2. *Let $f : \mathbb{R} \to \mathbb{R}$ and $g : \mathbb{R} \to \mathbb{R}$ be additive functions satisfying the relation*

$$g(x) = x^2 f\left(\frac{1}{x}\right) \qquad (14.3.8)$$

for all $x \in \mathbb{R}$, $x \neq 0$. Then the function $f + g$ is continuous, and the functions $F, G : \mathbb{R} \to \mathbb{R}$:

$$F(x) = f(x) - x f(1), \quad G(x) = g(x) - x g(1) \qquad (14.3.9)$$

are derivations.

Proof. We have by (14.3.8) $g(1) = f(1)$, whence by (14.3.8) and (14.3.9), for $x \neq 0$,

$$x^2 F\left(\frac{1}{x}\right) = x^2 f\left(\frac{1}{x}\right) - x f(1) = g(x) - x g(1) = G(x)$$

so that

$$G(x) = x^2 F\left(\frac{1}{x}\right) \qquad (14.3.10)$$

for every $x \neq 0$. Now, $F(1) = G(1) = 0$ by (14.3.9), whence, since F and G clearly are additive, $F(x+1) = F(x) + F(1) = F(x)$ and $G(x+1) = G(x) + G(1) = G(x)$ for every $x \in \mathbb{R}$. Hence and by (14.3.10) we have for every $x \neq -1$

$$G(x) = G(x+1) = (x+1)^2 F\left(\frac{1}{x+1}\right) = (x+1)^2 F\left(1 - \frac{x}{x+1}\right)$$

$$= -(x+1)^2 F\left(\frac{x}{x+1}\right) = -(x+1)^2 \left(\frac{x}{x+1}\right)^2 G\left(\frac{x+1}{x}\right)$$

$$= -x^2 G\left(1 + \frac{1}{x}\right) = -x^2 G\left(\frac{1}{x}\right) = -F(x),$$

i.e.,

$$G(x) = -F(x). \qquad (14.3.11)$$

For $x = -1$ we have $G(-1) = -G(1) = 0 = F(1) = -F(-1)$, so (14.3.11) holds for $x = -1$, too, and thus (14.3.11) is valid for all $x \in \mathbb{R}$. (14.3.11) yields $g(x) - x g(1) = -f(x) + x f(1)$, whence

$$f(x) + g(x) = x(f(1) + g(1)),$$

and consequently $f + g$ is a continuous function. Further, we get by (14.3.10) and (14.3.11)

$$F(x) = -x^2 F\left(\frac{1}{x}\right)$$

for $x \neq 0$. By Theorem 14.3.1 F is a derivation, and by Lemma 14.1.1 $G = -F$ also is a derivation. $\qquad \square$

Theorem 14.3.3. *Let $f : \mathbb{R} \to \mathbb{R}$ be an additive function, and let $P : \mathbb{R} \to \mathbb{R}$ be a continuous function. If*

$$f(x) = P(x) f\left(\frac{1}{x}\right) \tag{14.3.12}$$

for all $x \in \mathbb{R}$, $x \neq 0$, then if $P(1) = -1$, f is a derivation, and if $P(1) \neq -1$, then f is continuous.

Proof. By Lemma 14.3.1

$$P(x) = cx^2, \quad x \in \mathbb{R}, \tag{14.3.13}$$

where $c = P(1)$. If $c = 0$, then $f = 0$ by (14.3.12), and f is continuous. If $c \neq 0$, write $g(x) = c^{-1} f(x)$, $x \in \mathbb{R}$. Thus $g : \mathbb{R} \to \mathbb{R}$ is an additive function, and (14.3.12) with (14.3.13) imply (14.3.8). By Theorem 14.3.2 $f + g = \left(1 + c^{-1}\right) f$ is continuous, and hence f is continuous if $c \neq -1$. If $c = -1$, then $P(x) = -x^2$, and by Theorem 14.3.1 f is a derivation. $\qquad\square$

14.4   Automorphisms of $\mathbb{R}$

Let $(F; +, \cdot)$ and $(K; +, \cdot)$ be fields. As introduced in 14.2 a function $f : F \to K$ is called a *homomorphism* iff the equations

$$f(x + y) = f(x) + f(y) \tag{14.4.1}$$

and

$$f(xy) = f(x)f(y) \tag{14.4.2}$$

are fulfilled for every $x, y \in F$. If $K = F$, then f fulfilling (14.4.1) and (14.4.2) is called an *endomorphism*. If $K = F$ and f is one-to-one and onto, then f is called an *automorphism* of F.

In this and the next section we will be concerned with the automorphisms of $\mathbb{R}$ and $\mathbb{C}$. In $\mathbb{R}$ the situation is very simple (Darboux [53]).

Theorem 14.4.1. *The only function $f : \mathbb{R} \to \mathbb{R}$ satisfying both equations (14.4.1) and (14.1.2) are $f_1 = 0$ and $f_2 = $ identity.*

Proof. Let a function $f : \mathbb{R} \to \mathbb{R}$ satisfy (14.4.1) and (14.4.2). Then f is additive, and we have by (14.4.2) for every $x > 0$

$$f(x) = \left[f\left(\sqrt{x}\right)\right]^2 \geqslant 0.$$

Thus f is bounded below on $(0, \infty)$. By Lemma 9.3.1 f is continuous, and hence (Theorem 5.5.2) $f(x) = cx$ with a certain $c \in \mathbb{R}$. Inserting this into (14.4.2) we obtain, $cxy = c^2 xy$ for every $x, y \in \mathbb{R}$, whence (on setting $x = y = 1$) $c = c^2$, and either $c = 0$, or $c = 1$. Thus f is either zero, or the identity. $\qquad\square$

The function $f(x) = x$ clearly is an automorphism (of every field). We call it a *trivial automorphism*. On the other hand, the function $f = 0$ is neither invertible, nor onto, and thus it is not an automorphism. So we obtain from Theorem 14.4.1 the following

Theorem 14.4.2. *The only automorphism of $\mathbb{R}$ is the trivial automorphism $f(x) = x$.*

By Theorem 14.4.1 the only endomorphisms of $\mathbb{R}$ are $f = 0$ and $f(x) = x$ (trivial endomorphisms).

14.5 Automorphisms of $\mathbb{C}$

As we shall presently see, the situation in $\mathbb{C}$ is completely different (Kestelman [170]).

Lemma 14.5.1. *The only continuous endomorphisms $f : \mathbb{C} \to \mathbb{C}$ are $f = 0$, $f(x) = x$ and $f(x) = \overline{x}$ [3].*

Proof. By Theorem 5.6.1, if $f : \mathbb{C} \to \mathbb{C}$ is a continuous solution of equation (14.4.1), there exist constants $a, b \in \mathbb{C}$ such that

$$f(x) = ax + b\overline{x}. \tag{14.5.1}$$

Inserting (14.5.1) into (14.4.2) we obtain

$$axy + b\overline{x}\,\overline{y} = a^2 xy + b^2 \overline{x}\,\overline{y} + ab\,(x\overline{y} + \overline{x}y) \tag{14.5.2}$$

for every $x, y \in \mathbb{C}$. Setting in (14.5.2) $x = y = 1$ we get

$$a + b = a^2 + b^2 + 2ab, \tag{14.5.3}$$

whereas setting in (14.5.2) $x = y = i$ we obtain

$$-a - b = -a^2 - b^2 + 2ab. \tag{14.5.4}$$

Adding (14.5.3) and (14.5.4) yields $4ab = 0$, whence either $a = 0$, or $b = 0$. If $a = b = 0$, then we get by (14.5.1) $f = 0$. If $a \neq 0$, then $b = 0$, and (14.5.3) reduces to $a = a^2$, whence $a = 1$, and by (14.5.1) $f(x) = x$ for $x \in \mathbb{C}$. If $b \neq 0$, then $a = 0$, and (14.5.3) reduces to $b = b^2$, whence $b = 1$, and by (14.5.1) $f(x) = \overline{x}$. It is readily seen that all the three functions actually satisfy both equations (14.4.1) and (14.4.2). $\square$

The endomorphisms furnished by Lemma 14.5.1 are referred to as *trivial*. Of these $f = 0$ is only an endomorphism, whereas $f(x) = x$ and $f(x) = \overline{x}$ are automorphisms. So far the problem of the existence of non-trivial endomorphisms and automorphisms is unsettled. By Lemma 14.5.1, such endomorphisms and automorphisms, if they exist, must necessarily be discontinuous.

In the sequel we use the terminology and notation from 4.12.

Lemma 14.5.2. *Let $(F; +, \cdot)$ be a subfield of $(\mathbb{C}; +, \cdot)$, and let $a \in \mathbb{C}$ be transcendental over F. Further, let $f^* : F \to F$ be an automorphism of F. Then there exists an automorphism $f : F(a) \to F(a)$ such that $f \mid F = f^*$.*

Proof. Since a is transcendental over F, $a \neq 0$ and a is f^*-conjugate with itself. By Theorem 4.12.1 there exists an isomorphism $f : F(a) \to F(a)$ (i.e., an automorphism of $F(a)$) such that $f(a) = a$ and $f \mid F = f^*$. f is the desired extension. $\square$

---

[3] The complex conjugate of x.

Lemma 14.5.3. *Let $(F; +, \cdot)$ be a subfield of $(\mathbb{C}; +, \cdot)$, and let $a \in \mathbb{C}$ be algebraic over F. Further, let $f^* : F \to \mathbb{C}$ be a homomorphism, $f \neq 0$. Then there exists a homomorphism $f : F(a) \to \mathbb{C}$ such that $f \mid F = f^*$.*

Proof. It is easily checked that $(f^*(F); +; \cdot)$ is a field (subfield of $(\mathbb{C}; +, \cdot)$). By Lemma 4.12.2 f^* is a monomorphism, and hence an isomorphism of $(F; +, \cdot)$ onto $(f^*(F); +; \cdot)$. Since f^* satisfies (14.4.2), we have (on setting $x = y = 1$) $f^*(1) = 1$ (for otherwise we would have $f^*(1) = 0$, whence $f^* = 0$ results on setting $y = 1$ in (14.4.2)).

Let $q \in F[x]$ be the minimal polynomial of a, and put $Q = I_{f^*}(q)$. If we had $Q = Q_1 Q_2$, $Q_1, Q_2 \in f^*(F)[x]$, then since by Lemma 4.12.1 I_{f^*} is an isomorphism, so is also $I_{f^*}^{-1}$, and we would have $q = I_{f^*}^{-1}(Q) = I_{f^*}^{-1}(Q_1)I_{f^*}^{-1}(Q_2)$ and q would be reducible, which is not the case. So also Q is irreducible. The coefficient of x in the highest power in Q is $f^*(1) = 1$. Consequently Q is the minimal polynomial of every of its roots. By the fundamental theorem of algebra Q has a root $A \in \mathbb{C}$. Thus Q is the minimal polynomial of A.

Consequently a and A are f^*-conjugate. By Theorem 4.12.1 there exists an isomorphism $f : F(a) \to f^*(F)(A) \subset \mathbb{C}$ such that $f(a) = A$ and $f \mid F = f^*$. f is a homomorphism of $(F; +, \cdot)$ into $(\mathbb{C}; +, \cdot)$ fulfilling $f \mid F = f^*$, and hence it is the desired extension. $\qquad\square$

Let $(F_0; +, \cdot)$ be a subfield of $(\mathbb{C}; +, \cdot)$, and let $\mathcal{F}$ be the collection of all couples (F, f) such that $F_0 \subset F$, $(F; +, \cdot)$ is a subfield of $(\mathbb{C}; +, \cdot)$ and $f : F \to \mathbb{C}$ is a homomorphism. The order $\leqslant$ is defined in $\mathcal{F}$ as usually: for $(F_1, f_1), (F_2, f_2) \in \mathcal{F}$ we write $(F_1, f_1) \leqslant (F_2, f_2)$ iff $F_1 \subset F_2$ and $f_2 \mid F_1 = f_1$.

Lemma 14.5.4. *Let $\mathcal{F}$ be as described above, and let $\mathcal{L} \subset \mathcal{F}$ be a chain. Define $F^* = \bigcup\limits_{(F,f)\in\mathcal{L}} F$, and let $f^* : F^* \to \mathbb{C}$ be defined by the condition $f^* \mid F = f$ for every $(F, f) \in \mathcal{L}$. Then $(F^*, f^*) \in \mathcal{F}$, and $(F, f) \leqslant (F^*, f^*)$ for every $(F, f) \in \mathcal{L}$.*

Proof. Take arbitrary $x, y \in F^*$. Since $\mathcal{L}$ is a chain, there exists an $(\tilde{F}, \tilde{f}) \in \mathcal{L}$ such that $x, y \in \tilde{F}$. By Lemma 4.7.1 $x - y \in \tilde{F} \subset F^*$ and, if $y \neq 0$, $xy^{-1} \in \tilde{F} \subset F^*$. Thus again by Lemma 4.7.1, $(F^*; +, \cdot)$ is a field (a subfield of $(\mathbb{C}; +, \cdot)$), and clearly $F_0 \subset F^*$. Consequently we have also $x + y, xy \in \tilde{F}$. Hence

$$f^*(x + y) = \tilde{f}(x + y) = \tilde{f}(x) + \tilde{f}(y) = f^*(x) + f^*(y),$$

$$f^*(xy) = \tilde{f}(xy) = \tilde{f}(x)\tilde{f}(y) = f^*(x)f^*(y),$$

since $\tilde{f}$ is a homomorphism. Consequently f^* also is a homomorphism. Thus $(F^*, f^*) \in \mathcal{F}$, and evidently $(F, f) \leqslant (F^*, f^*)$ for every $(F, f) \in \mathcal{L}$. $\qquad\square$

$u = \sqrt{2}$ is algebraic over $\mathbb{Q}$ of degree 2, $q(x) = x^2 - 2$ being its minimal polynomial. By Lemma 4.11.3 the system $\{1, u\}$ forms a base of the linear space $(\mathbb{Q}(u); \mathbb{Q}; +, \cdot)$. Consequently every $x \in \mathbb{Q}(u)$ can be uniquely written as

$$x = \alpha + \beta u, \quad \alpha, \beta \in \mathbb{Q}. \tag{14.5.5}$$

Define the function $f_0 : \mathbb{Q}(u) \rightarrow \mathbb{Q}(u)$ by

$$f_0(x) = f_0(\alpha + \beta u) = \alpha - \beta u. \qquad (14.5.6)$$

Thus, unless $\beta = 0$, $f_0(x) \neq x = \bar{x}$ and $f_0(x) \neq 0$, and so f_0 is not a restriction to $\mathbb{Q}(u)$ of a trivial automorphism of $\mathbb{C}$.

Lemma 14.5.5. *Let $u = \sqrt{2}$, and let the function $f_0 : \mathbb{Q}(u) \rightarrow \mathbb{Q}(u)$ be defined by* (14.5.6). *Then f_0 is an automorphism of $\mathbb{Q}(u)$.*

Proof. Take arbitrary $x, y \in \mathbb{Q}(u)$. Then x can be written in form (14.5.5), and similarly

$$y = \gamma + \delta u, \quad \gamma, \delta \in \mathbb{Q}.$$

Hence $(x + y) = (\alpha + \gamma) + (\beta + \delta)u$, $xy = (\alpha\gamma + 2\beta\delta) + (\alpha\delta + \beta\gamma)u$, and

$$f_0(x + y) = (\alpha + \gamma) - (\beta + \delta)u = (\alpha - \beta u) - (\gamma - \delta u) = f_0(x) + f_0(y),$$
$$f_0(xy) = (\alpha\gamma + 2\beta\delta) - (\alpha\delta + \beta\gamma)u = (\alpha - \beta u)(\gamma - \delta u) = f_0(x)f_0(y).$$

Consequently f_0 is a homomorphism. Moreover, we have

$$f_0\big[f_0(x)\big] = f_0(\alpha - \beta u) = \alpha + \beta u = x.$$

By Lemma 15.5.2 f_0 is one-to-one and onto. Consequently f_0 is an automorphism of $\mathbb{Q}(u)$. $\qquad\square$

Theorem 14.5.1. *There exist non-trivial automorphisms of $\mathbb{C}$.*

Proof. Let $u = \sqrt{2}$. Let $\mathcal{R}$ be the collection of all couples (F, f) such that $(F; +, \cdot)$ is a subfield of $(\mathbb{C}; +, \cdot)$, $\mathbb{Q}(u) \subset F$, $f : F \rightarrow F$ is an automorphism of F, and $f \mid \mathbb{Q}(u) = f_0$, where f_0 is defined by (14.5.6). By Lemma 14.5.5 $(\mathbb{Q}(u), f_0) \in \mathcal{R}$, so $\mathcal{R} \neq \varnothing$. Let $\mathcal{L} \subset \mathcal{R}$ be an arbitrary chain, and put $F^* = \bigcup_{(F,f) \in \mathcal{L}} F$, and let $f^* : F^* \rightarrow F^*$ be defined by the condition that $f^* \mid F = f$ for every $(F, f) \in \mathcal{L}$. We will show that $(F^*, f^*) \in \mathcal{R}$. By Lemma 14.5.4 $(F^*; +, \cdot)$ is a subfield of $(\mathbb{C}; +, \cdot)$, and f^* is a homomorphism. Clearly $\mathbb{Q}(u) \subset F^*$ and $f^* \neq 0$, since $f^* \mid \mathbb{Q}(u) = f_0$. By Lemma 4.12.2 f^* is one-to-one. Take an arbitrary $y \in F^*$. Then there exists an $(F, f) \in \mathcal{L}$ such that $y \in F$. Since f is an automorphism of F, there exists an $x \in F \subset F^*$ such that $f(x) = y$. Hence also $f^*(x) = f(x) = y$. Thus $F^* \subset f^*(F^*)$. On the other hand, since for every $(F, f) \in \mathcal{L}$ we have $f(F) = F \subset F^*$, we have $f^*(F^*) \subset F^*$. Consequently f^* maps F^* onto F^*, and so is an automorphism. Consequently $(F^*, f^*) \in \mathcal{R}$.

By Lemma 14.5.4 (F^*, f^*) is an upper bound of $\mathcal{L}$. Thus in virtue of Theorem 1.8.1 there exists in $\mathcal{R}$ a maximal element $\big(\tilde{F}, \tilde{f}\big)$. Suppose that there exists an $a \in \mathbb{C}$ transcendental over $\tilde{F}$. Then, by Lemma 14.5.2, $\tilde{f}$ can be extended onto $\tilde{F}(a)$ to an automorphism f of $\tilde{F}(a)$, and obviously $\big(\tilde{F}(a), f\big) \in \mathcal{R}$ and $\big(\tilde{F}, \tilde{f}\big) < \big(\tilde{F}(a), f\big)$, which contradicts the maximality of $\big(\tilde{F}, \tilde{f}\big)$.

Consequently every element $x \in \mathbb{C}$ is algebraic over $\tilde{F}$. Let $\mathcal{F}$ be the collection of all pairs (F, f) such that $(F; +, \cdot)$ is a subfield of $(\mathbb{C}; +, \cdot)$, $\tilde{F} \subset F$, and $f : F \rightarrow \mathbb{C}$

is a homomorphism such that $f \mid \tilde{F} = \tilde{f}$. We have $(\tilde{F}, \tilde{f}) \in \mathcal{F}$, so $\mathcal{F} \neq \varnothing$. If we order $\mathcal{F}$ in the same way as $\mathcal{R}$ above, then it follows from Lemma 14.5.4 that every chain in $\mathcal{F}$ has an upper bound. By Theorem 1.8.1 there exists in $\mathcal{F}$ a maximal element $(F_{\max}, f_{\max})$. Suppose that $F_{\max} \neq \mathbb{C}$, and let $a \in \mathbb{C} \setminus F_{\max}$. Then a is algebraic over $\tilde{F}$, and hence also over $F_{\max}$. By Lemma 14.5.3 $f_{\max}$ can be extended onto $F_{\max}(a)$ to a homomorphism $f : F_{\max}(a) \to \mathbb{C}$. But then $(F_{\max}, f_{\max}) < (F_{\max}(a), f) \in \mathcal{F}$, which contradicts the maximality of $(F_{\max}, f_{\max})$. Consequently $F_{\max} = \mathbb{C}$.

Thus $f_{\max} : \mathbb{C} \to \mathbb{C}$ is a homomorphism, $f_{\max} \mid \tilde{F} = \tilde{f}$, and $\tilde{f} \mid \mathbb{Q}(u) = f_0 \neq 0$, whence $f_{\max} \neq 0$. By Lemma 4.12.2 $f_{\max}$ is one-to-one. We shall show that $f_{\max}(\mathbb{C}) = \mathbb{C}$. Since $\mathbb{C}$ is algebraically closed and $f_{\max}$ is an isomorphism of $\mathbb{C}$ onto $f_{\max}(\mathbb{C})$, also $f_{\max}(\mathbb{C})$ is algebraically closed (cf. Exercise 4.18). Since $\tilde{F} \subset \mathbb{C}$, we have $f_{\max}(\tilde{F}) \subset f_{\max}(\mathbb{C})$. But $f_{\max} \mid \tilde{F} = \tilde{f}$, so $f_{\max}(\tilde{F}) = \tilde{f}(\tilde{F}) = \tilde{F}$, since $\tilde{f}$ is an automorphism. Hence $\tilde{F} = f_{\max}(F) \subset f_{\max}(\mathbb{C})$.

Now, every $x \in \mathbb{C}$ is algebraic over $\tilde{F}$. This means that

$$\mathbb{C} = \mathrm{alg\,cl}\,\tilde{F} \subset \mathrm{alg\,cl}\,f_{\max}(\mathbb{C}) \subset f_{\max}(\mathbb{C})$$

(cf. Exercise 4.17). Consequently $f_{\max}(\mathbb{C}) = \mathbb{C}$. Thus $f_{\max}$ is onto $\mathbb{C}$, and consequently it is an automorphism of $\mathbb{C}$.

We have $f_{\max} \mid \mathbb{Q}(u) = f_{\max} \mid \tilde{F} \mid \mathbb{Q}(u) = \tilde{f} \mid \mathbb{Q}(u) = f_0$. If $f_{\max}$ were a trivial automorphism, then f_0 would be a restriction of a trivial automorphism, which, as pointed out before Lemma 14.5.5, is not the case. Consequently $f_{\max}$ is a non-trivial automorphism of $\mathbb{C}$. $\qquad\square$

14.6 Non-trivial endomorphisms of $\mathbb{C}$

Now we will prove a number of theorems indicating at certain irregularities of discontinuous functions $f : \mathbb{C} \to \mathbb{C}$ satisfying (14.4.1) and (14.4.2) (non-trivial endomorphisms of $\mathbb{C}$). If f is such a non-trivial endomorphism, then, in particular, $f \neq 0$, and hence by Lemma 4.12.2 f is one-to-one. Since, as results from (14.4.1), $f(0) = 0$, we have $f(x) \neq 0$ for every $x \in \mathbb{C}, x \neq 0$.

Theorem 14.6.1. *If $f : \mathbb{C} \to \mathbb{C}$ is a non-trivial endomorphism, then $f \mid \mathbb{R}$ is discontinuous.*

Proof. Write $f_0 = f \mid \mathbb{R}$. For every $x \in \mathbb{C}$ we have $x = \mathrm{Re}\,x + i\,\mathrm{Im}\,x$, whence by (14.4.1) and (14.4.2)

$$f(x) = f(\mathrm{Re}\,x + i\,\mathrm{Im}\,x) = f(\mathrm{Re}\,x) + f(i)f(\mathrm{Im}\,x) = f_0(\mathrm{Re}\,x) + f(i)f_0(\mathrm{Im}\,x).$$

If f_0 were continuous, then also f would be continuous, and f would be a trivial endomorphism. $\qquad\square$

Lemma 14.6.1. *If $f : \mathbb{C} \to \mathbb{C}$ satisfies (14.4.1), then we have for every $\alpha \in \mathbb{Q}$ and $x \in \mathbb{C}$*

$$f(\alpha x) = \alpha f(x).$$

Proof. This results from Theorems 5.6.1 and 5.2.1. $\qquad\square$

Theorem 14.6.2. *Let* $f : \mathbb{C} \to \mathbb{C}$ *be a non-trivial endomorphism, and let* $A \subset \mathbb{R}$, $m_i(A) > 0$, *or* A *of the second category with the Baire property. Then* f *is unbounded on* A, *and* $f(A)$ *is not contained in* $\mathbb{R}$.

Proof. Suppose that f is bounded on A. Write $f_0 = f \mid \mathbb{R}$ and $g = |f_0|$. It follows from Lemma 14.6.1 that $g : \mathbb{R} \to \mathbb{R}$ is a convex function, and our supposition implies that g is bounded on A. By Theorems 9.3.1 and 9.3.2 g is continuous. In particular

$$\lim_{\substack{t \to 0 \\ t \in \mathbb{R}}} |f_0(t)| = \lim_{\substack{t \to 0 \\ t \in \mathbb{R}}} g(t) = g(0) = |f_0(0)| = f(0) = 0.$$

Hence $\lim_{\substack{t \to 0 \\ t \in \mathbb{R}}} f_0(t) = 0$, and for every $x \in \mathbb{R}$

$$\lim_{\substack{t \to 0 \\ t \in \mathbb{R}}} f_0(x + t) = \lim_{\substack{t \to 0 \\ t \in \mathbb{R}}} \left[f_0(x) + f_0(t) \right] = f_0(x),$$

i.e., f_0 is continuous, which contradicts Theorem 14.6.1.

Suppose that $f(A) \subset \mathbb{R}$. Let $x \in A + A$. Then $x = u + v$, $u, v \in A$, and $f(x) = f(u) + f(v) \in \mathbb{R}$. Thus $f(A + A) \subset \mathbb{R}$. By Theorems 3.7.1 and 2.9.1 the set $A + A$ contains an interval I. Take an $x_0 \in \mathrm{int}\, I$. If $x \in I - x_0$, then $x = y - x_0$, where $y \in I \subset A + A$, and $f(x) = f(y) - f(x_0) \in \mathbb{R}$. Thus $f(I - x_0) \subset \mathbb{R}$. But $I - x_0$ is a neighbourhood of 0. Consequently for every $x \in \mathbb{R}$ there exists an $n \in \mathbb{N}$ such that $x/n \in I - x_0$. Hence $f(x/n) \in \mathbb{R}$, and by Lemma 14.6.1 $f(x) = nf(x/n) \in \mathbb{R}$. Consequently $f(\mathbb{R}) \subset \mathbb{R}$, and f_0 is an endomorphism of $\mathbb{R}$. By Theorem 14.4.1 f_0 is continuous, which is incompatible with Theorem 14.6.1. $\qquad\square$

Theorem 14.6.3. *Let* $f : \mathbb{C} \to \mathbb{C}$ *be a non-trivial endomorphism. Then* $\mathrm{cl}\, f(\mathbb{R}) = \mathbb{C}$.

Proof. By Theorem 14.6.2 there exists a $u \in \mathbb{R}$ such that $f(u) \notin \mathbb{R}$. The set

$$Z = \{ z \in \mathbb{C} \mid z = \alpha + \beta f(u),\ \alpha, \beta \in \mathbb{Q} \} \tag{14.6.1}$$

is dense in $\mathbb{C}$. For every $z \in Z$ we have by (14.4.1), (14.4.2) and Lemma 14.6.1

$$z = \alpha + \beta f(u) = f(\alpha + \beta u) \in f(\mathbb{R}).$$

Consequently $Z \subset f(\mathbb{R})$, and $f(\mathbb{R})$ is dense in $\mathbb{C}$. $\qquad\square$

Lemma 14.6.2. *Let* $f : \mathbb{C} \to \mathbb{C}$ *be a non-trivial endomorphism, and let* $U = \{ x \in \mathbb{C} \mid |x - x_0| < r \} \subset \mathbb{C}$ *be an open disc in* $\mathbb{C}$ *centered at an* $x_0 \in \mathbb{C}$ *and with a radius* $r > 0$. *Then the set* $A = \mathbb{R} \cap f^{-1}(U) \subset \mathbb{R}$ *fulfils* $m_i(A) = 0$, $m_e(A) > 0$. *Also,* A *is of the second category, but does not contain any subset of the second category with the Baire property.*

Proof. Let $B = \{ x \in \mathbb{R} \mid f(x) \in \mathbb{R} \}$. Thus for every $u \in \mathbb{R} \setminus B$ we have $f(u) \notin \mathbb{R}$ and set (14.6.1) is dense in $\mathbb{C}$. It will remain dense when we subject β to the additional condition $\beta \neq 0$. Consequently for every $u \in \mathbb{R} \setminus B$ there exist $\alpha, \beta \in \mathbb{Q}$, $\beta \neq 0$, such that $f(\alpha + \beta u) = \alpha + \beta f(u) \in U$, whence $\alpha + \beta u \in f^{-1}(U)$. Since evidently $\alpha + \beta u \in \mathbb{R}$, we have $\alpha + \beta u \in A$, whence $u \in \frac{1}{\beta}(A - \alpha)$. Consequently

$$\mathbb{R} \setminus B \subset \bigcup_{\substack{\alpha, \beta \in \mathbb{Q} \\ \beta \neq 0}} \frac{1}{\beta}(A - \alpha).$$

If we had $m(A) = 0$ or A were of the first category, then the same would also be true about $\mathbb{R} \setminus B$, whence B would have positive measure or would be of the second category with the Baire property. By the definition of B we have $f(B) \subset \mathbb{R}$, which is incompatible with Theorem 14.6.2. Consequently $m_e(A) > 0$ and A is of the second category.

On the other hand, for $x \in A$ we have $f(x) \in U$, which means that f is bounded on A. By Theorem 14.6.2 $m_i(A) = 0$ and A cannot contain any subset of the second category and with the Baire property. $\square$

Lemma 14.6.3. *Let $A \subset \mathbb{R}$ be a measurable set. Then $m(A) > 0$ if and only if $m\big((A \setminus \{0\})^{-1}\big) > 0$. Similarly, let $A \subset \mathbb{R}$ have the Baire property. Then A is of the second category if and only if $(A \setminus \{0\})^{-1}$ is of the second category with the Baire property.*

Proof. Let $\varphi : (-\infty, 0) \cup (0, \infty) \to \mathbb{R}$ be given by $\varphi(x) = x^{-1}$, and let $\varphi_1 : (-\infty, 0) \to (-\infty, 0)$ and $\varphi_2 : (0, \infty) \to (0, \infty)$ be defined as $\varphi_1 = \varphi \mid (-\infty, 0)$, $\varphi_2 = \varphi \mid (0, \infty)$. The function φ_2 is continuous and convex, and φ_1 is continuous and concave. Moreover, φ_1 and φ_2 are involutions (cf. 12.5), whence $\varphi_1 = \varphi_1^{-1}$ and $\varphi_2 = \varphi_2^{-1}$. By Theorem 7.4.6 φ_1 and φ_2 are absolutely continuous, whence also φ_1^{-1} and φ_2^{-1} are absolutely continuous. Also, evidently, φ_1 and φ_2 are homeomorphisms.

For any set $A \subset \mathbb{R}$ write $A^+ = A \cap (0, \infty)$, and $A^- = A \cap (-\infty, 0)$. Then, the following statement are equivalent:

(i) A is measurable and $m(A) > 0$;

(ii) A^+ and A^- are measurable and $m(A^+) > 0$ or $m(A^-) > 0$;

(iii) $\varphi_2(A^+)$ and $\varphi_1(A^-)$ are measurable, and $m\big(\varphi_2(A^+)\big) > 0$ or $m\big(\varphi_1(A^-)\big) > 0$;

(iv) $\varphi(A \setminus \{0\})$ is measurable, and $m\big(\varphi(A \setminus \{0\})\big) > 0$.

Similarly, since φ_1 and φ_2 are homeomorphisms, the following statements are equivalent:

(i') A is of the second category and with the Baire property;

(ii') A^+ and A^- have the Baire property, and A^+ or A^- is of the second category;

(iii') $\varphi_2(A^+)$ and $\varphi_1(A^-)$ have the Baire property, and $\varphi_2(A^+)$ or $\varphi_1(A^-)$ is of the second category;

(iv') $\varphi(A \setminus \{0\})$ is of the second category and with the Baire property.

$\square$

Lemma 14.6.4. *Let $f : \mathbb{C} \to \mathbb{C}$ be a non-trivial endomorphism, and let $U \subset \mathbb{C}$ be an open disc. Put $A = \mathbb{R} \cap f^{-1}(U)$. Then $m_i(\mathbb{R} \setminus A) = 0$ and $\mathbb{R} \setminus A$ does not contain a subset of the second category and with the Baire property.*

Proof. At first we consider the case, where $U = \{ x \in \mathbb{C} \mid |x| < r \}$ is centered at the origin. Let $B \subset \mathbb{R}$ be an arbitrary measurable set with $m(B) > 0$ or a second category set with the Baire property. Suppose that $|f(x)| \geqslant r$ for $x \in B$. Take a $y \in (B \setminus \{0\})^{-1}$. Then $y^{-1} \in B \setminus \{0\} \subset B$, and $1 = f(1) = |f(1)| = |f(yy^{-1})| = |f(y)| |f(y^{-1})| \geq r |f(y)|$, whence $|f(y)| \leqslant 1/r$. Thus f is bounded on $(B \setminus \{0\})^{-1}$. By Lemma 14.6.3 $m((B \setminus \{0\}))^{-1} > 0$ or $(B \setminus \{0\})^{-1}$ is of the second category and with the Baire property, and we get a contradiction with Theorem 14.6.2.

Consequently there must exist an $x \in B \subset \mathbb{R}$ such that $|f(x)| < r$, i.e., $f(x) \in U$. Thus $x \in \mathbb{R} \cap f^{-1}(U) = A$, i.e., $A \cap B \neq \varnothing$. So we have $A \cap B \neq \varnothing$ for every measurable set $B \subset \mathbb{R}$ with $m(B) > 0$ and for every set $B \subset \mathbb{R}$ of the second category with the Baire property. By Theorem 3.3.1 $m_i(\mathbb{R} \setminus A) = 0$. It is also evident that $\mathbb{R} \setminus A$ cannot contain a subset of the second category and with the Baire property, for otherwise, if $B \subset \mathbb{R} \setminus A$ were such a set, then we would have $A \cap B = \varnothing$, contrary to what has just been established.

Now let $U = V + f(d)$, where $d \in \mathbb{R}$ and $V \subset \mathbb{C}$ is an open disc centered at the origin. Take an arbitrary $x \in \mathbb{R} \cap (f^{-1}(V) + d)$. Then there exists a $t \in f^{-1}(V)$ such that $x = t + d$. Thus $t = x - d \in \mathbb{R}$, since $x \in \mathbb{R}$ and $d \in \mathbb{R}$. We have also $f(x) = f(t + d) = f(t) + f(d) \in V + f(d) = U$. Thus $x \in f^{-1}(U)$, and since $x \in \mathbb{R}$, $x \in A$. Hence $\mathbb{R} \cap (f^{-1}(V) + d) \subset A$. By what we have already shown $m(\mathbb{R} \setminus (\mathbb{R} \cap f^{-1}(V))) = 0$ and $\mathbb{R} \setminus (\mathbb{R} \cap f^{-1}(V))$ does not contain any subset of the second category and with the Baire property. Further,

$$\mathbb{R} \setminus [\mathbb{R} \cap (f^{-1}(V) + d)] = [\mathbb{R} \setminus (\mathbb{R} \cap f^{-1}(V))] + d. \qquad (14.6.2)$$

By Corollary 3.3.2 the set on the right-hand side of (14.6.2) has inner measure zero, and since translation is a homeomorphism, this set does not contain any subset of the second category and with the Baire property. Consequently also $\mathbb{R} \setminus A$ has the analogical properties.

Finally, let $U \subset \mathbb{C}$ be an arbitrary open disc. By Theorem 14.6.3 there exists a $d \in \mathbb{R}$ such that $f(d) \in U$. Further, there exists an open disc W centered at $f(d)$ and contained in U. Put $V = W - f(d)$. Then V is an open disc centered at the origin, and $W = V + f(d)$. By what we have already proved the set $\mathbb{R} \setminus (\mathbb{R} \cap f^{-1}(W))$ has inner measure zero and does not contain any subset of the second category and with the Baire property. But we have $W \subset U$, whence $f^{-1}(W) \subset f^{-1}(U)$ and $\mathbb{R} \cap f^{-1}(W) \subset \mathbb{R} \cap f^{-1}(U) = A$. Consequently $\mathbb{R} \setminus A \subset \mathbb{R} \setminus (\mathbb{R} \cap f^{-1}(W))$, and it follows that $\mathbb{R} \setminus A$ has inner measure zero and does not contain any subset of the second category and with the Baire property. $\qquad \square$

Corollary 14.6.1. *Let $f : \mathbb{C} \to \mathbb{C}$ be a non-trivial endomorphism, and let $U \subset \mathbb{C}$ be an open disc. Then the set $A = \mathbb{R} \cap f^{-1}(U)$ is a saturated non-measurable and has property $(*)$ from 3.3.*

Proof. This results from Lemmas 14.6.2 and 14.6.4.   □

Theorem 14.6.4. *Let $f : \mathbb{C} \to \mathbb{C}$ be a non-trivial endomorphism, and let $S \subset \mathbb{C}$ be a set such that $\operatorname{int} S \neq \varnothing$ and $\operatorname{int}(\mathbb{C} \setminus S) \neq \varnothing$. Then the set $A = \mathbb{R} \cap f^{-1}(S)$ is saturated non-measurable and has property $(*)$ from 3.3.*

Proof. There exist open discs $U \subset S$ and $V \subset \mathbb{C} \setminus S$. We have $S \subset \mathbb{C} \setminus V$ and $\mathbb{C}\setminus S \subset \mathbb{C}\setminus U$, whence $f^{-1}(S) \subset f^{-1}(\mathbb{C}\setminus V) \subset \mathbb{C}\setminus f^{-1}(V)$ and $\mathbb{C}\setminus f^{-1}(S) \subset \mathbb{C}\setminus f^{-1}(U)$. Hence $A = \mathbb{R} \cap f^{-1}(S) \subset \mathbb{R} \cap [\mathbb{C}\setminus f^{-1}(V)] = \mathbb{R}\setminus (\mathbb{R}\cap f^{-1}(V))$. By Lemma 14.6.4 A has inner measure zero and does not contain any subset of the second category and with the Baire property. Similarly, $\mathbb{R}\setminus A = \mathbb{R}\setminus (\mathbb{R}\cap f^{-1}(S)) = \mathbb{R}\cap (\mathbb{C}\setminus f^{-1}(S)) \subset \mathbb{R}\cap(\mathbb{C}\setminus f^{-1}(U)) = \mathbb{R}\setminus(\mathbb{R}\cap f^{-1}(U))$. By Lemma 14.6.4 also $\mathbb{R}\setminus A$ has inner measure zero and does not contain any subset of the second category and with the Baire property.   □

Theorem 14.6.5. *Let $f : \mathbb{C} \to \mathbb{C}$ be a non-trivial endomorphism, let $l \subset \mathbb{C}$ be a straight line, and let $S \subset \mathbb{C}$ be a set such that $\operatorname{int} S \neq \varnothing$ and $\operatorname{int}(\mathbb{C} \setminus S) \neq \varnothing$. Then the set $l \cap f^{-1}(S)$ is saturated non-measurable and has property $(*)$ from 3.3.*

Proof. Let $x_0, x_1 \in l$, $x_0 \neq x_1$. The assertion of the theorem means that the set

$$A = \{t \in \mathbb{R} \mid x_0 + t(x_1 - x_0) \in f^{-1}(S)\}$$

is saturated non-measurable and has property $(*)$ from 3.3 in $\mathbb{R}$.

By Lemma 4.12.2 $f(x_0) \neq f(x_1)$. Put

$$B = \mathbb{R} \cap f^{-1}\left(\frac{S - f(x_0)}{f(x_1) - f(x_0)}\right).$$

If $t \in B$, then $f(t) \in [S - f(x_0)]/[f(x_1) - f(x_0)]$ and $f(x_0) + f(t)[f(x_1) - f(x_0)] \in S$. But by (14.4.1) and (14.4.2)

$$f(x_0) + f(t)[f(x_1) - f(x_0)] = f(x_0 + t(x_1 - x_0)). \tag{14.6.3}$$

Thus $x_0 + t(x_1 - x_0) \in f^{-1}(S)$, i.e., $t \in A$. Hence $B \subset A$. On the other hand, if $t \in A$, then $t \in \mathbb{R}$ and $x_0 + t(x_1 - x_0) \in f^{-1}(S)$. This means that, by (14.6.3), $f(x_0) + f(t)[f(x_1) - f(x_0)] \in S$, and $t \in B$, whence $A \subset B$. Consequently $A = B$.

Now, the set $S_1 = [S - f(x_0)]/[f(x_1) - f(x_0)]$ fulfils $\operatorname{int} S_1 \neq \varnothing$ and $\operatorname{int}(\mathbb{C}\setminus S_1) \neq \varnothing$. The theorem results now from Theorem 14.6.4.   □

Theorem 14.6.6. *Let $f : \mathbb{C} \to \mathbb{C}$ be a non-trivial endomorphism, and let $A \subset \mathbb{R}$ be a set of positive inner measure or of the second category and with the Baire property. Then $\operatorname{cl} f(A) = \mathbb{C}$.*

Proof. Let $U \subset \mathbb{C}$ be an arbitrary open disc. By Theorem 14.6.4 the set $B = \mathbb{R} \cap f^{-1}(U)$ is saturated non-measurable and has property $(*)$ from 3.3. Consequently (Theorem 3.3.1) B intersects every measurable set in $\mathbb{R}$ of positive measure.

If $m_i(A) > 0$, there exists a measurable set $C \subset A$ with $m(C) > 0$. Therefore $B \cap C \neq \emptyset$, whence also

$$B \cap A \neq \emptyset. \tag{14.6.4}$$

If A is of the second category and with the Baire property, then (14.6.4) results from Corollary 3.3.1.

Let $t \in B \cap A$. Then $f(t) \in U$ and $f(t) \in f(A)$, whence $f(t) \in U \cap f(A)$. Consequently $U \cap f(A) \neq \emptyset$. Consequently $f(A)$ intersects every open disc contained in $\mathbb{C}$, i.e., $f(A)$ is dense in $\mathbb{C}$. $\qquad \square$

In the sequel m^2 denotes the two-dimensional Lebesgue measure on the complex plane $\mathbb{C}$. Also, the expression "the set A is saturated non-measurable in $\mathbb{C}$" means that $m_i^2(A) = m_i^2(\mathbb{C} \setminus A) = 0$. Similarly, the expression "A has property $(*)$ in $\mathbb{C}$" means that neither A, nor $\mathbb{C} \setminus A$, contains a subset which is of the second category and has the Baire property in the topology of $\mathbb{C}$.

Theorem 14.6.7. *Let $f : \mathbb{C} \to \mathbb{C}$ be a non-trivial endomorphism, and let $S \subset \mathbb{C}$ be a set such that $m_i^2(S) > 0$ or S is of the second category and with the Baire property (in the topology of $\mathbb{C}$). Then $\operatorname{cl} f(S) = \mathbb{C}$.*

Proof. If $m_i^2(S) > 0$, then there exists a measurable (with respect to m^2) set $C \subset S$ with $m^2(C) > 0$. By the Fubini theorem there exists a straight line $l \subset \mathbb{C}$ such that the one-dimensional measure $m(l \cap C) > 0$. Hence $m_i(l \cap C) > 0$.

If S is of the second category and with the Baire property (in the topology of $\mathbb{C}$), then $S = (G \cup P) \setminus R$, where G is open (and non-empty), and P, R are of the first category (in the topology of $\mathbb{C}$). It follows from Theorem 2.1.7 that there exists a straight line $l \subset \mathbb{C}$ such that the sets $l \cap P$ and $l \cap R$ are of the first category (in the topology of l), whereas $l \cap G \neq \emptyset$. Thus $l \cap S = [(l \cap G) \cup (l \cap P)] \setminus (l \cap R)$ is of the second category and with the Baire property.

Suppose that $f(S)$ is not dense in $\mathbb{C}$. Then there exists an open disc $U \subset \mathbb{C}$ such that $f(S) \cap U = \emptyset$. Hence $f(S) \subset \mathbb{C} \setminus U$ and $S \subset f^{-1}(\mathbb{C} \setminus U)$. But $\mathbb{C} \setminus U$ fulfils $\operatorname{int}(\mathbb{C} \setminus U) \neq \emptyset$ and $\operatorname{int}[\mathbb{C} \setminus (\mathbb{C} \setminus U)] \neq \emptyset$. By Theorem 14.6.5 the set $l \cap f^{-1}(\mathbb{C} \setminus U)$ is saturated non-measurable and has property $(*)$ (in the topology of l). Since $l \cap S \subset l \cap f^{-1}(\mathbb{C} \setminus U)$, we obtain hence that $l \cap S$ has inner (one-dimensional) measure zero and does not contain a subset which is of the second category and with the Baire property, a contradiction. $\qquad \square$

Theorem 14.6.8. *Let $f : \mathbb{C} \to \mathbb{C}$ be a non-trivial endomorphism, and let $S \subset \mathbb{C}$ be a set such that $\operatorname{int} S \neq \emptyset$ and $\operatorname{int}(\mathbb{C} \setminus S) \neq \emptyset$. Then the set $f^{-1}(S)$ is saturated non-measurable in $\mathbb{C}$ and has property $(*)$ in $\mathbb{C}$.*

Proof. If we had $m_i^2(f^{-1}(S)) > 0$ or $f^{-1}(S)$ contained a second category subset S_1 with the Baire property, then $f(f^{-1}(S))$ resp. $f(f^{-1}(S_1)) \subset f(f^{-1}(S))$ would be dense in $\mathbb{C}$ according to Theorem 14.6.7, and since $f(f^{-1}(S)) = S$, the set S would be dense in $\mathbb{C}$, which is not the case, since $\operatorname{int}(\mathbb{C} \setminus S) \neq \emptyset$. Consequently $m_i^2(f^{-1}(S)) = 0$ and $f^{-1}(S)$ does not contain any subset of the second category and with the Baire property.

Because of the symmetry of the assumptions we may interchange S and $\mathbb{C} \setminus S$, and we get $m_i^2(f^{-1}(\mathbb{C} \setminus S)) = 0$ and $f^{-1}(\mathbb{C} \setminus S)$ does not contain any subset of the second category and with the Baire property. But $f^{-1}(\mathbb{C} \setminus S) = \mathbb{C} \setminus f^{-1}(S)$, whence also inner (two-dimensional) measure of $\mathbb{C} \setminus f^{-1}(S)$ is zero and $\mathbb{C} \setminus f^{-1}(S)$ does not contain any subset of the second category and with the Baire property.   □

Theorem 14.6.9. *Let $f : \mathbb{C} \to \mathbb{C}$ be a non-trivial endomorphism. Then $(f(\mathbb{R}); +, \cdot)$ is a proper subfield of $(\mathbb{C}; +, \cdot)$, card $f(\mathbb{R}) = \mathfrak{c}$, and either $m^2(f(\mathbb{R})) = 0$, or $f(\mathbb{R})$ is saturated non-measurable in $\mathbb{C}$. Similarly, $f(\mathbb{R})$ either is of the first category, or has property $(*)$ in $\mathbb{C}$.*

Proof. Take arbitrary $x, y \in f(\mathbb{R})$. Then there exist $u, v \in \mathbb{R}$ such that $f(u) = x, f(v) = y$. We have $u - v \in \mathbb{R}$, whence $x - y = f(u) - f(v) = f(u - v) \in f(\mathbb{R})$. Similarly, if $y \neq 0$, then by Lemma 4.12.2 also $v \neq 0$, since $f(0) = 0$. Thus $u/v \in \mathbb{R}$, and we have $u = v(u/v)$, whence by (14.4.2) $f(u) = f(v)f(u/v)$ and $f(u/v) = f(u)/f(v)$. Consequently $x/y = f(u/v) \in f(\mathbb{R})$. By Lemma 4.7.1 $(f(\mathbb{R}); +, \cdot)$ is a field, and thus a subfield of $(\mathbb{C}, +, \cdot)$. We have card $\mathbb{R} = \mathfrak{c}$, whence also card $f(\mathbb{R}) = \mathfrak{c}$, since by Lemma 4.12.2 f is one-to-one. Lemma 4.12.2 implies also that $f(\mathbb{R}) \neq f(\mathbb{C})$, whence it follows that necessarily $f(\mathbb{R}) \neq \mathbb{C}$, since $f(\mathbb{R}) \subset f(\mathbb{C}) \subset \mathbb{C}$, and thus $(f(\mathbb{R}); +, \cdot)$ is a proper subfield of $(\mathbb{C}; +, \cdot)$.

It follows easily from the fact that $(f(\mathbb{R}); +, \cdot)$ is a field, that

$$f(\mathbb{R}) + f(\mathbb{R}) = f(\mathbb{R}) - f(\mathbb{R}) = f(\mathbb{R}). \qquad (14.6.5)$$

Assume that we do not have $m^2(f(\mathbb{R})) = 0$. Then $m_e^2(f(\mathbb{R})) > 0$. By Theorem 14.6.3 $f(\mathbb{R})$ is dense in $\mathbb{C}$. By Theorem 3.6.1 ($\mathbb{C}$ is to be regarded as $\mathbb{R}^2$) $m_i^2(\mathbb{C} \setminus f(\mathbb{R})) = 0$. Suppose that $m_i^2(f(\mathbb{R})) > 0$. By (14.6.5) and Theorem 3.7.1 int $f(\mathbb{R}) \neq \varnothing$. Take an $x_0 \in \text{int } f(\mathbb{R}) \subset f(\mathbb{R})$. The set $V = \text{int } f(\mathbb{R}) - x_0$ is open, contains the origin, and by (14.6.5)

$$V \subset f(\mathbb{R}) - f(\mathbb{R}) = f(\mathbb{R}). \qquad (14.6.6)$$

Thus V is a neighbourhood of zero. For every $x \in \mathbb{C}$ there exists an $n \in \mathbb{N}$ such that $x/n \in V$. In other words, there exists a $y \in V$ such that $x = ny$. But by (14.6.6) and using repeatedly (14.6.5) we obtain $ny \in f(\mathbb{R})$. Hence $\mathbb{C} \subset f(\mathbb{R})$, i.e., $f(\mathbb{R}) = \mathbb{C}$, contrary to what we have already established. Thus $m_i^2(f(\mathbb{R})) = 0$, and $f(\mathbb{R})$ is saturated non-measurable in $\mathbb{C}$.

If we assume that $f(\mathbb{R})$ is of the second category, then from (14.6.5), Theorem 14.6.3, and Theorem 3.6.2 we infer that $\mathbb{C} \setminus f(\mathbb{R})$ does not contain any subset of the second category and with the Baire property. If the set $f(\mathbb{R})$ contained a subset of the second category and with the Baire property, then by (14.6.5) and Theorem 2.9.1 we would get int $f(\mathbb{R}) \neq \varnothing$, which, as we have just seen, leads to a contradiction. Consequently also $f(\mathbb{R})$ does not contain any subset of the second category and with the Baire property, and $f(\mathbb{R})$ has property $(*)$ in $\mathbb{C}$.   □

Exercises

1. Show that there exists a discontinuous additive function $f : \mathbb{R} \to \mathbb{R}$ such that the series $\sum_{n=0}^{\infty} a_n f(x^n)$ converges if the series $\sum_{n=0}^{\infty} a_n x^n$ converges (Baker-Segal [17]; cf. also Kemperman [166]).

2. Let $f_1 : \mathbb{R} \to \mathbb{R}$ and $f_2 : \mathbb{R} \to \mathbb{R}$ be derivations. Show that the composition $f_1 \circ f_2$ is a derivation if and only if either $f_1 = 0$ or $f_2 = 0$.

3. Let $f_1 : \mathbb{R} \to \mathbb{R}$ and $f_2 : \mathbb{R} \to \mathbb{R}$ be derivations. Show that the product $f_1 f_2$ is a derivation if and only if either $f_1 = 0$ or $f_2 = 0$.

4. Let R be a ring. Show that if $f_1, f_2 : R \to R$ are derivations, then so is also

$$[f_1, f_2] = f_2 \circ f_1 - f_1 \circ f_2.$$

   [The derivation $[f_1, f_2]$ is called the *bracket* of f_1 and f_2.]

5. Let F be a field and $K = F(x_1, \ldots, x_n)$ be the field of rational functions (in n variables) over F. Prove that, for each i ($1 \le i \le n$), there exists a unique derivation $D_i : K \to K$ such that

$$D_i \mid F = 0, \quad D_i(x_i) = 1, \quad D_i(x_j) = 0 \quad \text{for } j \ne i, \ 1 \le j \le n.$$

   [The derivation D_i is called the *partial derivation* (with respect to x_i) of K.]

6. Let F and K be as in Exercise 14.5, and let $\hat{K}$ be a field such that $K \subset \hat{K}$. Prove that the family $\mathcal{D}_0$ of all derivations $f : K \to \hat{K}$ such that $f \mid F = 0$ is a linear space over $\hat{K}$ and the partial derivations $D_1, \ldots, D_n$ form its base.

   [Hint: Show that for every $f \in \mathcal{D}_0$ we have $f = \sum_{i=1}^{n} f(x_i) D_i$.]

7. Show that there exists no Hamel basis H of $\mathbb{R}$ such that $ab \in H$, whenever $a, b \in H$.

   [Hint: Let $H \subset \mathbb{R}$ be a Hamel basis such that $ab \in H$ whenever $a, b \in H$, and put $f_0(h) = 1$ for $h \in H$. Let $f : \mathbb{R} \to \mathbb{R}$ be the additive extension of f_0. Show that f satisfies equation (14.4.2).]

Chapter 15

Convex Functions of Higher Orders

15.1 The difference operator

Let $D \subset \mathbb{R}^N$ be a convex set, let $f : D \to \mathbb{R}$ be an arbitrary function, and let $h \in \mathbb{R}^N$ be arbitrary. The difference operator Δ_h with the span h is defined by the equality

$$\Delta_h f(x) = f(x+h) - f(x). \tag{15.1.1}$$

Thus $\Delta_h f$ is a real-valued function defined for $x \in D$ such that $x + h \in D$.

If $g : \mathbb{R}^N \to \mathbb{R}$ is another function, then in general the expressions $\Delta_h \big[f\big(g(x)\big) \big] = f\big(g(x+h)\big) - f\big(g(x)\big)$ and $(\Delta_h f)\big(g(x)\big) = f\big(g(x)+h\big) - f\big(g(x)\big)$ do not coincide. In other words the order of effecting the operations of taking the difference and of substitution is essential. Only if g is a translation: $g(x) = x + a$, with a constant $a \in \mathbb{R}^N$, then

$$\Delta_h \big[f\big(g(x)\big) \big] = (\Delta_h f) g(x),$$

and in the sequel this common value will be denoted simply by $\Delta_h f\big(g(x)\big) = \Delta_h f(x+a)$.

The iterates Δ_h^p of Δ_h, $p = 0, 1, 2, \ldots$, are defined by the recurrence

$$\Delta_h^0 f = f, \quad \Delta_h^{p+1} = \Delta_h(\Delta_h^p f), \quad p = 0, 1, 2 \ldots \tag{15.1.2}$$

In particular , we have $\Delta_h^1 f = \Delta_h f$. More generally, the superposition of several difference operators will be denoted shortly

$$\Delta_{h_1 \ldots h_p} f = \Delta_{h_1} \Delta_{h_2} \ldots \Delta_{h_p} f, \quad p \in \mathbb{N}. \tag{15.1.3}$$

Of course, if $h_1 = \cdots = h_p = h$, the expression (15.1.3) reduces to

$$\Delta_{\underbrace{h \ldots h}_{p \text{ times}}} f = \Delta_h^p f, \quad p \in \mathbb{N}. \tag{15.1.4}$$

Expression (15.1.3) is a function defined for all $x \in D$ such that $x + \varepsilon_1 h_1 + \cdots + \varepsilon_p h_p \in D$ for every choice of $\varepsilon_i = 0$ or 1, $i = 1, \ldots, p$. Generally, in the sequel of this section we will write expressions containing difference operators without mentioning the domain of definition of the functions represented by these expressions, being understood that the identities derived are valid for x in the domain of definition of f such that all the occurring expressions are meaningful.

Lemma 15.1.1. *The operator Δ_h (and hence also all operators (15.1.3)) is linear, i.e., for arbitrary functions f_1, f_2 and for arbitrary constants $a, b \in \mathbb{R}$, we have*

$$\Delta_h(af_1 + bf_2) = a\Delta_h f_1 + b\Delta_h f_2.$$

Proof. This results immediately from (15.1.1).   □

Lemma 15.1.2. *For arbitrary $h_1, h_2 \in \mathbb{R}^N$ the operators Δ_{h_1} and Δ_{h_2} commute:*

$$\Delta_{h_1}\Delta_{h_2} f = \Delta_{h_2}\Delta_{h_1} f.$$

Proof. We have by (15.1.1)

$$\begin{aligned}
\Delta_{h_1}\Delta_{h_2} f &= \Delta_{h_1}\big(f(x+h_2) - f(x)\big) \\
&= f(x+h_1+h_2) - f(x+h_1) - f(x+h_2) + f(x) \\
&= \Delta_{h_2}\Delta_{h_1} f(x),
\end{aligned}$$

which was to be shown.   □

Corollary 15.1.1. *Operator (15.1.3) is symmetric under the permutation of $h_1, \ldots, h_p$.*

Lemma 15.1.3. *For arbitrary $h_1, h_2 \in \mathbb{R}^N$ we have*

$$\Delta_{h_1+h_2} f - \Delta_{h_1} f - \Delta_{h_2} f = \Delta_{h_1 h_2} f.$$

Proof. We have by (15.1.1)

$$\begin{aligned}
\Delta_{h_1+h_2} f(x) - \Delta_{h_1} f(x) - \Delta_{h_2} f(x) &= f(x+h_1+h_2) - f(x) - f(x+h_1) \\
&\quad + f(x) - f(x+h_2) + f(x) \\
&= f(x+h_1+h_2) - f(x+h_1) - f(x+h_2) + f(x) \\
&= \Delta_{h_1 h_2} f(x)
\end{aligned}$$

(cf. the proof of Lemma 15.1.2).   □

The following two lemmas are immediate consequences of (15.1.1).

Lemma 15.1.4. *Let $D \subset \mathbb{R}^N$ be a convex set, and let $H \subset \mathbb{R}^{2N}$ be the set of those $(x, h) \in D \times \mathbb{R}^N$ for which $x + h \in D$. If the function $f : D \to \mathbb{R}$ is continuous in D, then the function $g : H \to \mathbb{R}$ given by $g(x, h) = \Delta_h f(x)$ is continuous in H.*

Lemma 15.1.5. *If $\{f_n\}_{n \in \mathbb{N}}$ is a pointwise convergent sequence of functions, then*

$$\lim_{n \to \infty} \Delta_h f_n = \Delta_h \lim_{n \to \infty} f_n.$$

The next lemma, of a slightly similar character, refers to the one-dimensional case $N = 1$.

Lemma 15.1.6. *Let $J \subset \mathbb{R}$ be an interval, and let $f : J \to \mathbb{R}$ be a continuous function. Then, for arbitrary $x \in J$ and α, β, $h \in \mathbb{R}$ such that $x + h$, $x + \alpha$, $x + \beta$, $x + \alpha + h$, $x + \beta + h \in J$ we have*

$$\Delta_h \int\limits_{x+\alpha}^{x+\beta} f(t)dt = \int\limits_{x+\alpha}^{x+\beta} \Delta_h f(t)dt.$$

Proof. We have

$$\Delta_h \int\limits_{x+\alpha}^{x+\beta} f(t)dt = \int\limits_{x+\alpha+h}^{x+\beta+h} f(t)dt - \int\limits_{x+\alpha}^{x+\beta} f(t)dt = \int\limits_{x+\alpha}^{x+\beta} f(t+h)dt - \int\limits_{x+\alpha}^{x+\beta} f(t)dt$$

$$= \int\limits_{x+\alpha}^{x+\beta} \left[f(t+h) - f(t) \right] dt = \int\limits_{x+\alpha}^{x+\beta} \Delta_h f(t)dt,$$

which is the desired formula. $\qquad\square$

Theorem 15.1.1. *We have for $p \in \mathbb{N}$*

$$\Delta_{h_1 \ldots h_p} f(x) = \sum_{\varepsilon_1, \ldots, \varepsilon_p = 0}^{1} (-1)^{p - (\varepsilon_1 + \cdots + \varepsilon_p)} f(x + \varepsilon_1 h_1 + \cdots + \varepsilon_p h_p). \qquad (15.1.5)$$

Proof. For $p = 1$ formula (15.1.5) becomes

$$\Delta_{h_1} f(x) = \sum_{\varepsilon_1 = 0}^{1} (-1)^{1 - \varepsilon_1} f(x + \varepsilon_1 h_1) = -f(x) + f(x + h_1) = f(x + h_1) - f(x),$$

which is consistent with (15.1.1). Now assume (15.1.5) true for a $p \in \mathbb{N}$. Then we have by Corollary 15.1.1, induction hypothesis, and Lemma 15.1.1,

$$\Delta_{h_1 \ldots h_p h_{p+1}} f(x) = \Delta_{h_{p+1}} \left[\Delta_{h_1 \ldots h_p} f(x) \right]$$

$$= \Delta_{h_{p+1}} \sum_{\varepsilon_1, \ldots, \varepsilon_p = 0}^{1} (-1)^{p - (\varepsilon_1 + \cdots + \varepsilon_p)} f(x + \varepsilon_1 h_1 + \cdots + \varepsilon_p h_p)$$

$$= \sum_{\varepsilon_1, \ldots, \varepsilon_p = 0}^{1} (-1)^{p - (\varepsilon_1 + \cdots + \varepsilon_p)} \Delta_{h_{p+1}} f(x + \varepsilon_1 h_1 + \cdots + \varepsilon_p h_p)$$

$$= \sum_{\varepsilon_1, \ldots, \varepsilon_p = 0}^{1} (-1)^{p - (\varepsilon_1 + \cdots + \varepsilon_p)} \sum_{\varepsilon_{p+1} = 0}^{1} (-1)^{1 - \varepsilon_{p+1}} f(x + \varepsilon_1 h_1 + \cdots + \varepsilon_p h_p + \varepsilon_{p+1} h_{p+1})$$

$$= \sum_{\varepsilon_1, \ldots, \varepsilon_{p+1} = 0}^{1} (-1)^{p+1 - (\varepsilon_1 + \cdots + \varepsilon_{p+1})} f(x + \varepsilon_1 h_1 + \cdots + \varepsilon_{p+1} h_{p+1}),$$

and so we obtain formula (15.1.5) for $p + 1$. Induction completes the proof. $\qquad\square$

Corollary 15.1.2. *We have for $p \in \mathbb{N}$*

$$\Delta_h^p f(x) = \sum_{k=0}^{p} (-1)^{p-k} \binom{p}{k} f(x + kh). \tag{15.1.6}$$

Proof. If exactly k among $\varepsilon_1, \ldots, \varepsilon_p$ in (15.1.5) are ones, the expression under the sum sign (with $h_1 = \cdots = h_p = h$) reduces to $(-1)^{p-k} f(x + kh)$. But there are exactly $\binom{p}{k}$ choices of k ones from among $\varepsilon_1, \ldots, \varepsilon_p$, where k may run from 0 to p. Thus (15.1.6) results from (15.1.4) and (15.1.5). $\qquad\square$

Corollary 15.1.3. *We have for $p \in \mathbb{N}$*

$$\Delta_{-h}^p f(x) = (-1)^p \Delta_h^p f(x - ph).$$

Proof. We have by (15.1.6)

$$\Delta_{-h}^p f(x) = \sum_{k=0}^{p} (-1)^{p-k} \binom{p}{k} f(x - kh)$$

$$= \sum_{k=0}^{p} (-1)^{p-k} \binom{p}{k} f(x - ph + (p-k)h)$$

$$= \sum_{j=0}^{p} (-1)^{j} \binom{p}{p-j} f(x - ph + jh)$$

$$= (-1)^p \sum_{j=0}^{p} (-1)^{p-j} \binom{p}{j} f(x - ph + jh)$$

$$= (-1)^p \Delta_h^p f(x - ph),$$

due to the equalities $\binom{p}{j} = \binom{p}{p-j}$ and $(-1)^{2p-j} = (-1)^j$, $j = 0, \ldots, p$. $\qquad\square$

Theorem 15.1.2.[1] *Let $f : \mathbb{R}^N \to \mathbb{R}$ be an arbitrary function, and let $h_1, \ldots, h_p \in \mathbb{R}^N$ be arbitrary. For any $\varepsilon_1, \ldots, \varepsilon_p \in \mathbb{R}$ put*

$$h'_{\varepsilon_1 \ldots \varepsilon_p} = -\sum_{j=1}^{p} \varepsilon_j h_j / j, \quad h''_{\varepsilon_1 \ldots \varepsilon_p} = \sum_{j=1}^{p} \varepsilon_j h_j.$$

Then we have for every $x \in \mathbb{R}^N$

$$\Delta_{h_1 \ldots h_p} f(x) = \sum_{\varepsilon_1, \ldots, \varepsilon_p = 0}^{1} (-1)^{\varepsilon_1 + \cdots + \varepsilon_p} \Delta_{h'_{\varepsilon_1 \ldots \varepsilon_p}}^p f(x + h''_{\varepsilon_1 \cdots \varepsilon_p}). \tag{15.1.7}$$

---

[1] A private communication of J. H. B. Kemperman.

Proof. We will prove the formula

$$(-1)^p \Delta_{y_1\, 2y_2\ldots py_p} f(x) = \sum_{\varepsilon_1,\cdots,\varepsilon_p=0}^{1} (-1)^{p-(\varepsilon_1+\cdots+\varepsilon_p)} \Delta^p_{Y'_{\varepsilon_1\ldots\varepsilon_p}} f(x + Y''_{\varepsilon_1\ldots\varepsilon_p}), \quad (15.1.8)$$

where

$$Y'_{\varepsilon_1\ldots\varepsilon_p} = -\sum_{j=1}^{p} \varepsilon_j y_j, \qquad Y''_{\varepsilon_1\ldots\varepsilon_p} = \sum_{j=1}^{p} j\varepsilon_j y_j,$$

valid for all $x \in \mathbb{R}^N$ and all $y_1,\ldots,y_p \in \mathbb{R}^N$. (15.1.7) results from (15.1.8) on multiplying by $(-1)^p$ and putting $h_j = jy_j$, $j = 1,\ldots,p$.

We have by Corollary 15.1.2

$$\sum_{\varepsilon_1,\ldots,\varepsilon_p=0}^{1} (-1)^{p-(\varepsilon_1+\cdots+\varepsilon_p)} \Delta^p_{Y'_{\varepsilon_1\ldots\varepsilon_p}} f(x + Y''_{\varepsilon_1\ldots\varepsilon_p})$$

$$= \sum_{\varepsilon_1,\ldots,\varepsilon_p=0}^{1} (-1)^{p-(\varepsilon_1+\cdots+\varepsilon_p)} \sum_{k=0}^{p} (-1)^{p-k} \binom{p}{k} f\left(x + Y''_{\varepsilon_1\ldots\varepsilon_p} + kY'_{\varepsilon_1\ldots\varepsilon_p}\right)$$

$$= \sum_{\varepsilon_1,\ldots,\varepsilon_p=0}^{1} (-1)^{p-(\varepsilon_1+\cdots+\varepsilon_p)} \sum_{k=0}^{p} (-1)^{p-k} \binom{p}{k} f\left(x + \sum_{j=1}^{p}(j-k)\varepsilon_j y_j\right)$$

$$= \sum_{k=0}^{p} (-1)^{p-k} \binom{p}{k} \sum_{\varepsilon_1,\ldots,\varepsilon_p=0}^{1} (-1)^{p-(\varepsilon_1+\cdots+\varepsilon_p)} f\left(x + \sum_{j=1}^{p}(j-k)\varepsilon_j y_j\right)$$

$$= \sum_{k=0}^{p} (-1)^{p-k} \binom{p}{k} \Delta_{(1-k)\,y_1\ldots(p-k)\,y_k} f(x),$$

in virtue of Theorem 15.1.1. For every $k = 1,\ldots,p$ one of the increments $(1-k)y_1,\ldots,$ $(p-k)y_k$ is zero, whence by Corollary 15.1.1

$$\Delta_{(1-k)y_1\ldots(p-k)y_k} = \Delta_{(1-k)y_1\ldots-y_{k-1}y_{k+1}\ldots(p-k)y_k} \big[\Delta_0 f(x)\big] = 0,$$

since by (15.1.1) $\Delta_0 f(x) = 0$. Consequently in the last sum only the term corresponding to $k = 0$ remains, which yields (15.1.8). $\qquad\square$

Theorem 15.1.3. *Let $D \subset \mathbb{R}^N$ be a convex set, and let $f : D \to \mathbb{R}$ be an arbitrary function. Then for every $\alpha_1,\ldots,\alpha_p \in \mathbb{Q} \cap (0,\infty)$ there exists an $\alpha \in \mathbb{Q} \cap (0,\infty)$ and $m, k_0,\ldots,k_m \in \mathbb{N}$ such that for every $x \in D$ and $h \in \mathbb{R}^N$ such that $x + (\alpha_1 + \cdots + \alpha_p)h \in D$ we have $x + (m + p)\alpha h \in D$ and*

$$\Delta_{\alpha_1 h \ldots \alpha_p h} f(x) = \sum_{i=0}^{m} k_i \Delta^p_{\alpha h} f(x + i\alpha h). \qquad (15.1.9)$$

Proof. By (15.1.1)

$$\Delta_h f(x + ih) = f(x + (i+1)h) - f(x + ih). \tag{15.1.10}$$

Summing up (15.1.10) over i from 0 to an $n - 1 \in \mathbb{N} \cup \{0\}$ we obtain (supposing that x, $x + nh \in D$)

$$\Delta_{nh} f(x) = f(x + nh) - f(x) = \sum_{i=0}^{n-1} \Delta_h f(x + ih), \quad n \in \mathbb{N}. \tag{15.1.11}$$

Fix arbitrarily $x \in D$, $h \in \mathbb{R}^N$, and take arbitrary $\alpha_1, \ldots, \alpha_p \in \mathbb{Q} \cap (0, \infty)$ such that $x + (\alpha_1 + \cdots + \alpha_p)h \in D$. There exist numbers $M, m_1 \ldots, m_p \in \mathbb{N}$ such that $\alpha_i = m_i/M$, $i = 1, \ldots, p$. Put $\alpha = 1/M$ so that $\alpha \in \mathbb{Q} \cap (0, \infty)$, and $M_q = m_1 + \cdots + m_q \in \mathbb{N}$, $q = 1, \ldots, p$. We have $\alpha_i = m_i \alpha$, $i = 1, \ldots, p$, and $x + M_p \alpha h = x + (m_1 + \cdots + m_p)\alpha h = x + (\alpha_1 + \cdots + \alpha_p)h \in D$, whence by the convexity of D we have $x + M_q \alpha h \in D$ for $q = 1, \ldots, p$.

We will prove by induction on q that for every $q = 1, \ldots, p$ there exist numbers $k_{0q}, \ldots, k_{M_q - q, q} \in \mathbb{N}$ such that

$$\Delta_{\alpha_1 h \ldots \alpha_q h} f(x) = \sum_{i=0}^{M_q - q} k_{iq} \Delta_{\alpha h}^q f(x + i\alpha h). \tag{15.1.12}$$

For $q = 1$ we have by (15.1.11)

$$\Delta_{\alpha_1 h} f(x) = \Delta_{m_1 \alpha h} f(x) = \sum_{i=0}^{m_1 - 1} \Delta_{\alpha h} f(x + i\alpha h) = \sum_{i=0}^{M_1 - 1} \Delta_{\alpha h}^1 f(x + i\alpha h),$$

so (15.1.12) holds with $k_{01} = \cdots = k_{M_1 - 1, 1} = 1$. Assume (15.1.12) true for a $q \in \{1, \ldots, p-1\}$. We have by Corollary 15.1.1, induction hypothesis, and Lemma 15.1.1,

$$\Delta_{\alpha_1 h \ldots \alpha_q h \alpha_{q+1} h} f(x) = \Delta_{\alpha_{q+1} h} \left[\Delta_{\alpha_1 h \ldots \alpha_q h} f(x) \right] = \Delta_{\alpha_{q+1} h} \sum_{i=0}^{M_q - q} k_{iq} \Delta_{\alpha h}^q f(x + i\alpha h)$$

$$= \sum_{i=0}^{M_q - q} k_{iq} \Delta_{\alpha_{q+1} h} \Delta_{\alpha h}^q f(x + i\alpha h),$$

and again by Corollary 15.1.1, (15.1.11), and Lemma 15.1.1,

$$\Delta_{\alpha_1 h \ldots \alpha_{q+1} h} f(x) = \sum_{i=0}^{M_q - q} k_{iq} \Delta_{\alpha h}^q \Delta_{m_{q+1} \alpha h} f(x + i\alpha h)$$

$$= \sum_{i=0}^{M_q - q} k_{iq} \Delta_{\alpha h}^q \sum_{j=0}^{m_{q+1} - 1} \Delta_{\alpha h} f(x + i\alpha h + j\alpha h)$$

$$= \sum_{i=0}^{M_q-q} \sum_{j=0}^{m_{q+1}-1} k_{iq}\Delta_{\alpha h}^{q+1} f\left(x + (i+j)\alpha h\right)$$

$$= \sum_{i=0}^{M_q-q} \sum_{l=i}^{m_{q+1}+i-1} k_{iq}\Delta_{\alpha h}^{q+1} f(x + l\alpha h)$$

$$= \sum_{l=0}^{M_q+m_{q+1}-q-1} \sum_{i=\max(0,l-m_{q+1}+1)}^{\min(l,M_q-q)} k_{iq}\Delta_{\alpha h}^{q+1} f(x + l\alpha h)$$

$$= \sum_{l=0}^{M_{q+1}-q-1} k_{l,q+1}\Delta_{\alpha h}^{q+1} f(x + l\alpha h),$$

where $k_{l,q+1} = \displaystyle\sum_{i=\max(0,l-m_{q+1}+1)}^{\min(l,M_q-q)} k_{iq}$. Thus we have obtained (15.1.12) for $q+1$.
Induction shows that (15.1.12) is valid for $q = 1,\ldots,p$. For $q = p$ we obtain hence
(15.1.9) with $m = M_p - p$ and $k_i = k_{ip}$, $i = 0,\ldots,m$. Moreover, we have $x + (m + p)\alpha h = x + M_p\alpha h \in D$. $\qquad\square$

15.2 Divided differences

In the sequel of this chapter $\mathbb{R}_+^N$ denotes the set of those $x \in \mathbb{R}^N$ whose first non-zero coordinate is positive:

$$\mathbb{R}_+^N = \bigcup_{i=1}^{N} \left\{x = (\xi_1,\ldots,\xi_N) \in \mathbb{R}^N \mid \xi_1 = \cdots = \xi_{i-1} = 0,\, \xi_i > 0\right\}.$$

Then $\mathbb{R}^N = \mathbb{R}_+^N \cup (-\mathbb{R}_+^N) \cup \{0\}$. For $x, y \in \mathbb{R}^N$ we write $x > y$ iff $x - y \in \mathbb{R}_+^N$ and $x < y$ iff $y > x$. Then for any $x, y \in \mathbb{R}^N$ we have either $x = y$, or $x < y$, or $x > y$. If $x > y$ and $\alpha \in \mathbb{R}$ is positive, then $\alpha x > \alpha y$, and if $\alpha < 0$, then $\alpha x < \alpha y$. We write

$$\operatorname{sgn} x = \begin{cases} 1 & \text{if } x > 0, \\ 0 & \text{if } x = 0, \\ -1 & \text{if } x < 0. \end{cases}$$

Let $D \subset \mathbb{R}^N$ be a convex set, and let $f : D \to \mathbb{R}$ be a function. Let $x_1,\ldots,x_p \in D$ be distinct collinear points. Put

$$h = \frac{x_p - x_1}{|x_p - x_1|}\operatorname{sgn}(x_p - x_1) \tag{15.2.1}$$

so that always $h > 0$. Since $x_1,\ldots,x_p$ are collinear, they may be represented in the form

$$x_i = x_1 + \lambda_i h, \quad i = 1,\ldots,p. \tag{15.2.2}$$

The *divided difference* $[x_1, \ldots, x_p; f]$ of f at points $x_1, \ldots, x_p$ is defined by recurrence (Nörlund [244], Popoviciu [259], Ger [101]):

$$[x_1; f] = f(x_1),$$

$$[x_1, \ldots, x_p; f] = \frac{[x_2, \ldots, x_p; f] - [x_1, \ldots, x_{p-1}; f]}{\lambda_p - \lambda_1}, \quad p \geqslant 2. \tag{15.2.3}$$

It is seen from (15.2.3) that the divided difference $[x_1, \ldots, x_p; f]$ depends on the differences of λ's rather than on λ's themselves. Therefore if we represent $x_1, \ldots, x_p$ with the aid of another collinear point x_0 as $x_i = x_0 + \lambda_i' h$, where h is given by (15.2.1), and $i = 1, \ldots, p$, and form the divided difference (15.2.3) with the aid of λ_i' instead of λ_i, the result will be the same. This can also be shown by induction from (15.2.3). Also, if we build h according to (15.2.1) with other points $x_i' \neq x_p'$ lying on the same straight line, we obtain the same result.

For $p = 2$ we have always $\lambda_1 = 0$, $\lambda_2 = |x_2 - x_1| \operatorname{sgn}(x_2 - x_1)$, whence by (15.2.3)

$$[x_1, x_2; f] = \frac{f(x_2) - f(x_1)}{|x_2 - x_1| \operatorname{sgn}(x_2 - x_1)} = \frac{\Delta_{x_2 - x_1} f(x_1)}{|x_2 - x_1| \operatorname{sgn}(x_2 - x_1)}. \tag{15.2.4}$$

For higher values of p the situation is more complicated.

Let, for collinear $x_1, \ldots, x_p \in D$, $U(x_1, \ldots, x_p; f)$ denote the determinant

$$U(x_1, \ldots, x_p; f) = \begin{vmatrix} 1, & \lambda_1, & \lambda_1^2, & \ldots, & \lambda_1^{p-2}, & f(x_1) \\ 1, & \lambda_2, & \lambda_2^2, & \ldots, & \lambda_2^{p-2}, & f(x_2) \\ \cdots\cdots\cdots\cdots\cdots\cdots\cdots\cdots\cdots \\ 1, & \lambda_p, & \lambda_p^2, & \ldots, & \lambda_p^{p-2}, & f(x_p) \end{vmatrix}, \tag{15.2.5}$$

where $\lambda_1, \ldots, \lambda_p$ are defined by (15.2.2) and (15.2.1). Let $V(x_1,\ldots,x_p) = U(x_1,\ldots,x_p; |x - x_1|^{p-1})$ be the Van der Monde determinant of $\lambda_1, \ldots, \lambda_p$. We have, as is well known from the elementary theory of determinants,

$$V(x_1, \ldots, x_p) = \begin{vmatrix} 1, & \lambda_1, & \lambda_1^2, & \ldots, & \lambda_1^{p-2}, & \lambda_1^{p-1} \\ 1, & \lambda_2, & \lambda_2^2, & \ldots, & \lambda_2^{p-2}, & \lambda_2^{p-1} \\ \cdots\cdots\cdots\cdots\cdots\cdots\cdots\cdots\cdots \\ 1, & \lambda_p, & \lambda_p^2, & \ldots, & \lambda_p^{p-2}, & \lambda_p^{p-1} \end{vmatrix} = \prod_{\substack{i,j=1 \\ i>j}}^{p} (\lambda_i - \lambda_j), \tag{15.2.6}$$

and so $V(x_1, \ldots, x_p) \neq 0$ if and only if the points $x_1, \ldots, x_p$ are distinct; and if, moreover, $x_1 < \cdots < x_p$, then $V(x_1, \ldots, x_p) > 0$.

We have the following

Theorem 15.2.1. *Let $D \subset \mathbb{R}^N$ be a convex set, and let $f : D \to \mathbb{R}$ be an arbitrary function. Then, for every $p \geqslant 2$, $p \in \mathbb{N}$, and for every p distinct collinear points $x_1, \ldots, x_p \in D$, we have*

$$[x_1, \ldots, x_p; f] = \frac{U(x_1, \ldots, x_p; f)}{V(x_1, \ldots, x_p)}. \tag{15.2.7}$$

Proof. The proof runs by induction. For $p = 2$ we have by (15.2.5) and (15.2.6)

$$U(x_1, x_2; f) = \begin{vmatrix} 1 & f(x_1) \\ 1 & f(x_2) \end{vmatrix} = f(x_2) - f(x_1),$$

$$V(x_1, x_2) = \begin{vmatrix} 1 & \lambda_1 \\ 1 & \lambda_2 \end{vmatrix} = \lambda_2 - \lambda_1.$$

By (15.2.2) and (15.2.1) $\lambda_1 = 0$, $\lambda_2 = |x_2 - x_1| \operatorname{sgn}(x_2 - x_1)$, and thus (15.2.7) holds in virtue of (15.2.4).

Suppose (15.2.7) true for a $p \in \mathbb{N}$, $p \geqslant 2$. We have by (15.2.3)

$$[x_1, \ldots, x_{p+1}; f] = \frac{[x_2, \ldots, x_{p+1}; f] - [x_1, \ldots, x_p; f]}{\lambda_{p+1} - \lambda_1},$$

i.e., by the induction hypothesis,

$$[x_1, \ldots, x_{p+1}; f] = \frac{1}{\lambda_{p+1} - \lambda_1} \left[\frac{U(x_2, \ldots, x_{p+1}; f)}{V(x_2, \ldots, x_{p+1})} - \frac{U(x_1, \ldots, x_p; f)}{V(x_1, \ldots, x_p)} \right].$$

Developing the determinants U according to the last column we have[2]

$$U(x_2, \ldots, x_{p+1}; f) = \sum_{k=2}^{p+1} (-1)^{p+k+1} f(x_k) V(x_2, \ldots, x_{k-1}, x_{k+1}, \ldots, x_{p+1})$$

and

$$U(x_1, \ldots, x_p; f) = \sum_{k=1}^{p} (-1)^{p+k} f(x_k) V(x_1, \ldots, x_{k-1}, x_{k+1}, \ldots, x_p).$$

On the other hand, by (15.2.6),

$$V(x_1, \ldots, x_{p+1}) = V(x_2, \ldots, x_{p+1}) \prod_{i=2}^{p+1} (\lambda_i - \lambda_1)$$

$$= (\lambda_{p+1} - \lambda_1) V(x_2, \ldots, x_{p+1}) \prod_{i=2}^{p} (\lambda_i - \lambda_1),$$

---

[2] The remarks after formula (15.2.3) apply also to the determinants U and V. Therefore during the whole this proof we may calculate the occurring differences and determinants with the same λ's determined from (15.2.1) and (15.2.2) for $i = 1, \ldots, p + 1$.

$$V(x_1, \ldots, x_{p+1}) = V(x_1, \ldots, x_p) \prod_{i=1}^{p} (\lambda_{p+1} - \lambda_i)$$

$$= (\lambda_{p+1} - \lambda_1) V(x_1, \ldots, x_p) \prod_{i=2}^{p} (\lambda_{p+1} - \lambda_i).$$

Hence we get

$$[x_1, \ldots, x_{p+1}; f] = \frac{1}{V(x_1, \ldots, x_{p+1})} \tag{15.2.8}$$

$$\times \Bigg[\sum_{k=2}^{p+1} (-1)^{p+k+1} f(x_k) V(x_2, \ldots, x_{k-1}, x_{k+1}, \ldots, x_{p+1}) \prod_{i=2}^{p} (\lambda_i - \lambda_1)$$

$$+ \sum_{k=1}^{p} (-1)^{p+k+1} f(x_k) V(x_1, \ldots, x_{k-1}, x_{k+1}, \ldots, x_p) \prod_{i=2}^{p} (\lambda_{p+1} - \lambda_i) \Bigg].$$

Next again by (15.2.6), for $k = 2, \ldots, p$,

$$V(x_1, \ldots, x_{k-1}, x_{k+1}, \ldots, x_{p+1})(\lambda_{p+1} - \lambda_k) \tag{15.2.9}$$

$$= (\lambda_{p+1} - \lambda_1) V(x_1, \ldots, x_{k-1}, x_{k+1}, \ldots, x_p) \prod_{i=2}^{p} (\lambda_{p+1} - \lambda_i),$$

$$V(x_1, \ldots, x_{k-1}, x_{k+1}, \ldots, x_{p+1})(\lambda_k - \lambda_1) \tag{15.2.10}$$

$$= (\lambda_{p+1} - \lambda_1) V(x_2, \ldots, x_{k-1}, x_{k+1}, \ldots, x_{p+1}) \prod_{i=2}^{p} (\lambda_i - \lambda_1).$$

From (15.2.8) , (15.2.9) and (15.2.10) we get

$$[x_1, \ldots, x_{p+1}; f] = \frac{1}{V(x_1, \ldots, x_{p+1})} \Bigg\{ f(x_{p+1}) V(x_2, \ldots, x_p) \prod_{i=2}^{p} (\lambda_i - \lambda_1)$$

$$+ (-1)^p f(x_1) V(x_2, \ldots, x_p) \prod_{i=2}^{p} (\lambda_{p+1} - \lambda_i)$$

$$+ \sum_{k=2}^{p} (-1)^{p+k+1} f(x_k) \Bigg[\frac{\lambda_k - \lambda_1}{\lambda_{p+1} - \lambda_1} V(x_1, \ldots, x_{k-1}, x_{k+1}, \ldots, x_{p+1})$$

$$+ \frac{\lambda_{p+1} - \lambda_k}{\lambda_{p+1} - \lambda_1} V(x_1, \ldots, x_{k-1}, x_{k+1}, \ldots, x_{p+1}) \Bigg] \Bigg\}$$

$$= \frac{1}{V(x_1, \ldots, x_{p+1})} \sum_{k=1}^{p+1} (-1)^{p+k+1} f(x_k) V(x_1, \ldots, x_{k-1}, x_{k+1}, \ldots, x_{p+1})$$

$$= \frac{U(x_1, \ldots, x_{p+1}; f)}{V(x_1, \ldots, x_{p+1})}.$$

Thus we obtain (15.2.7) for $p + 1$. Induction completes the proof. $\qquad\square$

From Theorem 15.2.1 we get the following lemmas.

Lemma 15.2.1. *Let $D \subset \mathbb{R}^N$ be a convex set, and let $f_n : D \to \mathbb{R}$, $n \in \mathbb{N}$, be pointwise convergent sequence of functions. Then, for arbitrary p distinct collinear points $x_1, \ldots, x_p \in D$,*

$$\lim_{n \to \infty} [x_1, \ldots, x_p; f_n] = \left[x_1, \ldots, x_p; \lim_{n \to \infty} f_n \right].$$

Lemma 15.2.2. *Let $D \subset \mathbb{R}^N$ be a convex set, and let $f : D \to \mathbb{R}$ be a continuous function. Then the divided difference $[x_1, \ldots, x_p; f]$ is a continuous function of $x_1, \ldots, x_p \in D$, under the condition that these points vary in such a way that they remain collinear and distinct.*

Lemma 15.2.3. *Let $D \subset \mathbb{R}^N$ be a convex set, and let $f : D \to \mathbb{R}$ be an arbitrary function. Then, for every distinct collinear points $x_1, \ldots, x_p \in D$, the divided difference $[x_1, \ldots, x_p; f]$ is symmetric with respect to permutations of $x_1, \ldots, x_p$.*

Proof. If we permute $x_1, \ldots, x_p$ in (15.2.7), then $U(x_1, \ldots, x_p; f)$ and $V(x_1, \ldots, x_p)$ either remain unchanged (if the permutation is even), or both change the sign (if the permutation is odd). Consequently $[x_1, \ldots, x_p; f]$ remains unchanged in both cases. $\square$

Let $D \subset \mathbb{R}^N$ be a convex set, let $f : D \to \mathbb{R}$ be a function, and let $x_1, \ldots, x_p \in D$ be distinct and collinear. For $x \in \mathbb{R}^N$ of the form $x = x_1 + \lambda h$, where h is given by (15.2.1), the function P is defined by the formula

$$\begin{vmatrix} 1, & \lambda_1 & , \ldots, & \lambda_1^{p-2}, & f(x_1) \\ \cdots\cdots\cdots\cdots\cdots\cdots\cdots\cdots\cdots\cdots \\ 1, & \lambda_{p-1} & , \ldots, & \lambda_{p-1}^{p-2}, & f(x_{p-1}) \\ 1, & \lambda & , \ldots, & \lambda^{p-2}, & P(x) \end{vmatrix} = 0. \qquad (15.2.11)$$

It is seen that P is polynomial in λ of degree at most $p - 2$ fulfilling the conditions $P(x_i) = f(x_i)$ for $i = 1, \ldots, p - 1$. (The Lagrange interpolation polynomial). As is well known, and also easily seen, such a polynomial is unique. In the sequel P defined by (15.2.11) will be denoted by $P(x_1, \ldots, x_{p-1}; f \mid x)$.

Lemma 15.2.4. *Let $D \subset \mathbb{R}^N$ be a convex set, and let $f : D \to \mathbb{R}$ be an arbitrary function. Let $p \geqslant 2$, and let $x_1, \ldots, x_p \subset D$ be distinct and collinear. Then for every $x = x_1 + \lambda h$ lying on the straight line determined by x_1 and x_p ($x \neq x_1, \ldots, x_{p-1}$) we have*

$$f(x) - P(x_1, \ldots, x_{p-1}; f \mid x) = \frac{V(x_1, \ldots, x_{p-1}, x)}{V(x_1, \ldots, x_{p-1})} [x_1, \ldots, x_{p-1}, x; f]. \qquad (15.2.12)$$

Proof. We have, developing the determinant U according to the last row,

$$U(x_1, \ldots, x_{p-1}, x; f) = V(x_1, \ldots, x_{p-1}) f(x) + R, \qquad (15.2.13)$$

where the rest R depends only on λ and on the first $p-1$ rows of the determinant (15.2.5). Similarly, by (15.2.11),

$$0 = V(x_1,\ldots,x_{p-1})P(x_1,\ldots,x_{p-1};f \mid x) + R, \qquad (15.2.14)$$

where R is the same expression as in (15.2.13). From (15.2.13) and (15.2.14) we obtain

$$f(x) - P(x_1,\ldots,x_{p-1};f \mid x) = \frac{U(x_1,\ldots,x_{p-1},x;f)}{V(x_1,\ldots,x_{p-1})},$$

whence (15.2.12) follows in virtue of Theorem 15.2.1.   □

The next lemma concerns the case, where the points $x_1,\ldots,x_p$ are equidistributed on the segment $\overline{x_1 x_p}$.

Lemma 15.2.5. *Let $D \subset \mathbb{R}^N$ be a convex set, and let $f : D \to \mathbb{R}$ be an arbitrary function. Let $x_1 \in \mathbb{R}^N$, $d \in \mathbb{R}_+^N$, and let the points $x_2,\ldots,x_p$ be given by*

$$x_i = x_1 + (i-1)d, \quad i = 2,\ldots,p. \qquad (15.2.15)$$

Then

$$[x_1,\ldots,x_p;f] = \frac{\Delta_d^{p-1}f(x_1)}{(p-1)!\,|d|^{p-1}}. \qquad (15.2.16)$$

Proof. The proof runs by induction on p. For $p=2$ we have by (15.2.4)

$$[x_1,x_2;f] = \frac{\Delta_{x_2-x_1}f(x_1)}{|x_2-x_1|\operatorname{sgn}(x_2-x_1)} = \frac{\Delta_d f(x_1)}{|d|},$$

since by (15.2.15) $x_2 - x_1 = d > 0$. Thus (15.2.16) holds for $p=2$. Now we assume (15.2.16) true for a $p-1 \in \mathbb{N}$, $p-1 \geqslant 2$. By (15.2.1) we have $\lambda_1 = 0$, $\lambda_p h = x_p - x_1$, whereas by (15.2.15) $x_p - x_1 = (p-1)d$. Hence $\lambda_p h = (p-1)d$, and since $|h| = 1$, we have $|\lambda_p| = (p-1)|d|$. Finally $h = \big[(p-1)/\lambda_p\big]d$, and since $h > 0$, $d > 0$, $p-1 > 0$, we obtain $\lambda_p > 0$ and $\lambda_p = (p-1)|d|$. Thus we obtain from (15.2.3) by the induction hypothesis

$$[x_1,\ldots,x_p;f] = \frac{[x_2,\ldots,x_p;f] - [x_1,\ldots,x_{p-1};f]}{\lambda_p - \lambda_1}$$

$$= \frac{1}{(p-1)|d|}\left[\frac{\Delta_d^{p-2}f(x_2)}{(p-2)!\,|d|^{p-2}} - \frac{\Delta_d^{p-2}f(x_1)}{(p-2)!\,|d|^{p-2}}\right]$$

$$= \frac{1}{(p-1)!\,|d|^{p-1}}\left[\Delta_d^{p-2}f(x_2) - \Delta_d^{p-2}f(x_1)\right]$$

$$= \frac{1}{(p-1)!\,|d|^{p-1}}\left[\Delta_d^{p-2}f(x_1+d) - \Delta_d^{p-2}f(x_1)\right]$$

$$= \frac{1}{(p-1)!\,|d|^{p-1}}\Delta_d^{p-1}f(x_1).$$

So (15.2.16) holds true also for p. Induction ends the proof.   □

Now let $x_1, \ldots, x_p$, with (15.2.2) and (15.2.1), be arbitrary p distinct collinear points in D, $p \geqslant 2$. We prove the formula

$$(\lambda_p - \lambda_1) [x_1, \ldots, x_{i-1}, x_{i+1}, \ldots, x_p; f]$$
$$= (\lambda_i - \lambda_1) [x_1, \ldots, x_{p-1}; f] + (\lambda_p - \lambda_i) [x_2, \ldots, x_p; f] \quad (15.2.17)$$

for $i = 1, \ldots, p$. For $i = 1$ or p (15.2.17) is trivial, so we need only consider the case, where $1 < i < p$. By (15.2.3) and Lemma 15.2.3

$$[x_1, \ldots, x_p; f] = [x_1, \ldots, x_{i-1}, x_{i+1}, \ldots, x_p, x_i; f]$$
$$= \frac{[x_2, \ldots, x_{i-1}, x_{i+1}, \ldots, x_p, x_i; f] - [x_1, \ldots, x_{i-1}, x_{i+1}, \ldots, x_p; f]}{\lambda_i - \lambda_1}.$$

But, according to Lemma 15.2.3, $[x_2, \ldots, x_{i-1}, x_{i+1}, \ldots, x_p, x_i; f] = [x_2, \ldots, x_p; f]$. Hence and by (15.2.3) we obtain

$$\frac{[x_2, \ldots, x_p; f] - [x_1, \ldots, x_{p-1}; f]}{\lambda_p - \lambda_1} = \frac{[x_2, \ldots, x_p; f] - [x_1, \ldots, x_{i-1}, x_{i+1}, \ldots, x_p; f]}{\lambda_i - \lambda_1}.$$

After simple calculations we obtain hence (15.2.17).

Lemma 15.2.6. *Let $D \subset \mathbb{R}^N$ be a convex set, and let $f : D \to \mathbb{R}$ be an arbitrary function. Let $x_1, \ldots, x_p \in D$ be p distinct collinear points, and let $\{x'_1, \ldots, x'_m\}$ be a subsequence of $\{x_1, \ldots, x_p\}$. Then there exist constants $A_0, \ldots, A_{p-m} \in \mathbb{R}$ such that $\sum_{i=0}^{p-m} A_i = 1$ and*

$$[x'_1, \ldots, x'_m; f] = \sum_{i=0}^{p-m} A_i [x_{i+1}, \ldots, x_{i+m}; f]. \quad (15.2.18)$$

If, moreover, $x_1 < \cdots < x_p$, then $A_i \geqslant 0$ for $i = 0, \ldots, p - m$.

Proof. Write $k = p - m$. Introduce $\lambda_1, \ldots, \lambda_p$ by means of (15.2.2) and (15.2.1). Note that $x_1 < \cdots < x_p$ means $\lambda_1 < \cdots < \lambda_p$.

For $k = 0$, the lemma is trivial. If $k = 1$, the lemma results from (15.2.17). Observe that the coefficients $(\lambda_i - \lambda_1)/(\lambda_p - \lambda_1)$ and $(\lambda_p - \lambda_i)/(\lambda_p - \lambda_1)$ are non-negative whenever $\lambda_1 < \cdots < \lambda_p$.

Now assume the lemma true for a $k \in \mathbb{N}$. Take p distinct collinear points $x_1, \ldots, x_p \in D$, and let $\{x'_1, \ldots, x'_m\}$ be a subsequence of $\{x_1, \ldots, x_p\}$, where $p - m = k + 1$. Take another subsequence $\{x''_1, \ldots, x''_{m+1}\}$ of $\{x_1, \ldots, x_p\}$ containing all the points $x'_1, \ldots, x'_m$ and another point $x''_j \notin \{x'_1, \ldots, x'_m\}$. There exist constants B_0, B_1, where $B_0 + B_1 = 1$, such that

$$[x'_1, \ldots, x'_m; f] = B_0 [x''_1, \ldots, x''_m; f] + B_1 [x''_2, \ldots, x''_{m+1}; f] \quad (15.2.19)$$

(the lemma for $k = 1$). If $x_1 < \cdots < x_p$, then B_0 and B_1 are non-negative.

Now, $\{x''_1, \ldots, x''_m\}$ is a subsequence of $\{x_1, \ldots, x_{p-1}\}$, and $\{x''_2, \ldots, x''_{m+1}\}$ is a subsequence of $\{x_2, \ldots, x_p\}$, and in both cases we have $(p-1) - m = p - m - 1 = k$.

By the induction hypothesis there exist constants $A'_0, \ldots, A'_k$ and $A''_1, \ldots, A''_{k+1}$ (non-negative, whenever $x_1 < \cdots < x_p$) such that

$$\sum_{i=0}^{k} A'_i = \sum_{i=1}^{k+1} A''_i = 1 \tag{15.2.20}$$

and

$$[x''_1, \ldots, x''_m; f] = \sum_{i=0}^{k} A'_i [x_{i+1}, \ldots, x_{i+m}; f],$$

$$[x''_2, \ldots, x''_{m+1}; f] = \sum_{i=1}^{k+1} A''_i [x_{i+1}, \ldots, x_{i+m}; f].$$

Hence and by (15.2.19)

$$[x'_1, \ldots, x'_m; f] = \sum_{i=0}^{k} B_0 A'_i [x_{i+1}, \ldots, x_{i+m}; f] + \sum_{i=1}^{k+1} B_1 A''_i [x_{i+1}, \ldots, x_{i+m}; f]$$

$$= \sum_{i=0}^{k+1} A_i [x_{i+1}, \ldots, x_{i+m}; f],$$

where

$$A_0 = B_0 A'_0, \quad A_{k+1} = B_1 A''_{k+1}, \quad A_i = B_0 A'_i + B_1 A''_i \quad \text{for } i = 1, \ldots, k.$$

Hence and by (15.2.20)

$$\sum_{i=0}^{k+1} A_i = B_0 \sum_{i=0}^{k} A'_i + B_1 \sum_{i=1}^{k+1} A''_i = B_0 + B_1 = 1.$$

If $x_1 < \cdots < x_p$, then $B_0, B_1, A'_0, \ldots, A'_k, A''_1, \ldots, A''_{k+1}$ are non-negative, whence also $A_1, \ldots, A_{k+1}$ are non-negative.

Thus we have obtained (15.2.18) for $p - m = k + 1$. Induction ends the proof. $\square$

Lemma 15.2.7. *Let $D \subset \mathbb{R}^N$ be a convex set, and let $f : D \to \mathbb{R}$ be an arbitrary function. Let $x_1, \ldots, x_p \in D$ and $y_1, \ldots, y_p \in D$ be two systems of p distinct points such that $x_1, \ldots, x_p, y_1, \ldots, y_p$ are collinear. Suppose that for h defined by (15.2.1) and for a certain x_0 collinear with points x_i, y_i, we have*

$$x_i = x_0 + \lambda_i h, \quad y_i = x_0 + \mu_i h, \quad i = 1, \ldots, p.$$

If $x_i \neq y_j$ for $i \neq j$, then

$$[x_1, \ldots, x_p; f] - [y_1, \ldots, y_p; f] = \sum_{i=1}^{p} (\lambda_i - \mu_i) [x_1, \ldots, x_i, y_i, \ldots, y_p; f], \tag{15.2.21}$$

where

$$(\lambda_i - \mu_i)[x_1, \ldots, x_i, y_i, \ldots, y_p; f] = 0 \quad \text{whenever } x_i = y_i, \quad i = 1, \ldots, p. \tag{15.2.22}$$

Proof. As observed immediately after formula (15.2.3), in the definition of the divided differences we may use the linear coordinates $\lambda_1, \ldots, \lambda_p, \mu_1, \ldots, \mu_p$ related to an arbitrary point x_0 collinear with $x_1, \ldots, x_p, y_1, \ldots, y_p$. We have by (15.2.3) and Lemma 15.2.3, for arbitrary i such that $x_i \neq y_i$,

$$(\lambda_i - \mu_i)\,[x_1, \ldots, x_i, y_i, \ldots, y_p; f]$$
$$= (\lambda_i - \mu_i)\,[y_i, x_1, \ldots, x_{i-1}, y_{i+1}, \ldots, y_p, x_i; f]$$
$$= [x_1, \ldots, x_{i-1}, y_{i+1}, \ldots, y_p, x_i; f] - [y_i, x_1, \ldots, x_{i-1}, y_{i+1}, \ldots, y_p; f]$$
$$= [x_1, \ldots, x_{i-1}, x_i, y_{i+1}, \ldots, y_p; f] - [x_1, \ldots, x_{i-1}, y_i, y_{i+1}, \ldots, y_p; f],$$

and, due to convention (15.2.22), this formula remains valid also if $x_i = y_i$. Summing up over $i = 1, \ldots, p$ we obtain (15.2.21). $\square$

15.3 Convex functions of higher order

Let $D \subset \mathbb{R}^N$ be an open and convex set, and let $f : D \to \mathbb{R}$ be a function. The function f is called *convex of order p*, or shortly *p-convex* $(p \in \mathbb{N})$ iff

$$\Delta_h^{p+1} f(x) \geqslant 0 \tag{15.3.1}$$

for every $x \in D$ and every $h \in \mathbb{R}_+^N$ such that $x + (p+1)h \in D$. In virtue of Corollary 15.1.3 if p is odd and $f : D \to \mathbb{R}$ is p-convex, then $\Delta_h^{p+1} f(x) \geqslant 0$ for every $x \in D$ and every $h \in \mathbb{R}^N$ such that $x + (p+1)h \in D$.

Lemma 15.3.1. *Let $D \subset \mathbb{R}^N$ be a convex and open set, let $f_1, f_2 : D \to \mathbb{R}$ be p-convex functions, and let $a, b \in \mathbb{R}$ be non-negative constants. Then the function $f = af_1 + bf_2$ also is p-convex.*

Proof. This results from Lemma 15.1.1. $\square$

If

$$\Delta_h^{p+1} f(x) \leqslant 0 \tag{15.3.2}$$

for every $x \in D$ and every $h \in \mathbb{R}_+^N$ such that $x + (p+1)h \in D$, then f is called *concave of order p*, or shortly *p-concave* $(p \in \mathbb{N})$. A function f is p-concave if and only if $-f$ is p-convex (cf. Lemma 15.1.1), so in the sequel we will deal only with p-convex functions. p-concave functions will not be considered separately.

If a function $f : \mathbb{R}^N \to \mathbb{R}$ is at the same time p-convex and p-concave, i.e., if we have

$$\Delta_h^{p+1} f(x) = 0 \tag{15.3.3}$$

for every $x \in \mathbb{R}^N$ and (cf. Corollary 15.1.3) every $h \in \mathbb{R}^N$, then f is called *polynomial function of order p* $(p \in \mathbb{N})$. Conversely, every function $f : \mathbb{R}^N \to \mathbb{R}$ fulfilling (15.3.3) for all $x, h \in \mathbb{R}^N$ is at the same time p-convex and p-concave. Polynomial functions will be considered more in detail in the last section of this chapter.

The above notions are due to T. Popoviciu [259], who also proved a number of basic properties of p-convex functions. Concerning a generalization cf. Moldovan [228], [229], [230] and Kemperman [167].

Lemma 15.3.2. *Let* $f_1, f_2 : \mathbb{R}^N \to \mathbb{R}$ *be polynomial functions of order* p, *and let* $a, b \in \mathbb{R}$ *be arbitrary constants. Then the function* $f = af_1 + bf_2$ *also is a polynomial function of order* p.

Proof. This results from Lemma 15.1.1.   □

Let $D \subset \mathbb{R}^N$ be a convex and open set, and let $f : D \to \mathbb{R}$ be a convex function of order 1. Then, by Corollary 15.1.2

$$f(x + 2h) - 2f(x + h) + f(x) \geqslant 0, \tag{15.3.4}$$

and, as pointed out above, since $p = 1$ is odd, (15.3.4) holds for every $x \in D$ and $h \in \mathbb{R}^N$ such that $x + 2h \in D$. Take arbitrary $u, v \in D$, and put $x = u$, $h = \dfrac{1}{2}(v - u)$. Then $x + h = \dfrac{1}{2}(u + v)$, $x + 2h = v \in D$, and we get by (15.3.4)

$$f(v) - 2f\left(\frac{u + v}{2}\right) + f(u) \geqslant 0,$$

or

$$f\left(\frac{u + v}{2}\right) \leqslant \frac{f(u) + f(v)}{2}.$$

Thus f fulfils in D the Jensen inequality (5.3.1). In other words, convex functions of order 1 are ordinary convex functions as defined in 5.3. (It is evident that every convex function $f : D \to \mathbb{R}$ satisfies inequality (15.3.4) for every $x \in D$ and $h \in \mathbb{R}^N$ such that $x + 2h \in D$, i.e., is convex of order 1). The notion of p-convex functions is a generalization of the notion of ordinary convex functions.

Similarly, if $f : \mathbb{R}^N \to \mathbb{R}$ is a polynomial function of order 1, then f satisfies the Jensen equation (13.2.1). Thus the notion of polynomial functions generalizes the notion of functions fulfilling the Jensen equation, i.e., those functions $f : \mathbb{R}^N \to \mathbb{R}$ which are of the form (cf. Theorem 13.2.1)

$$f(x) = g(x) + a,$$

where $g : \mathbb{R}^N \to \mathbb{R}$ is additive, and $a \in \mathbb{R}$ is a constant.

Theorem 15.3.1. *Let* $D \subset \mathbb{R}^N$ *be a convex and open set, and let* $f : D \to \mathbb{R}$ *be a* p-*convex function. Further, let* $x_1, \ldots, x_{p+2} \in D$ *be* $p + 2$ *distinct collinear points. Let* $h \in \mathbb{R}^N_+$ *and* $\lambda_1, \ldots, \lambda_{p+2}$ *be such that*

$$x_i = x_1 + \lambda_i h, \quad i = 1, \ldots, p + 2.$$

If $\lambda_1, \ldots, \lambda_{p+2}$ *are rational, then*

$$[x_1, \ldots, x_{p+2}; f] \geqslant 0. \tag{15.3.5}$$

Proof. According to Lemma 15.2.3 we may assume that $x_1 < \cdots < x_{p+2}$. Let[3] $\lambda_i = k_i/m$, $k_i, m \in \mathbb{N}$, $i = 1, \ldots, p+2$, and put

$$y_i = x_1 + \frac{i-1}{m} h, \quad i = 1, \ldots, q = k_{p+2} + 1.$$

Then $x_i = y_{k_i+1}$, $i = 1, \ldots, p+2$, so that $\{x_1, \ldots, x_{p+2}\}$ is a subsequence of $\{y_1, \ldots, y_q\}$. By Lemma 15.2.6

$$[x_1, \ldots, x_{p+2}; f] = \sum_{i=0}^{q-p-2} A_i [y_{i+1}, \ldots, y_{i+p+2}; f],$$

where $A_i \geqslant 0$, $i = 0, \ldots, q - p - 2$. Put

$$d = \frac{h}{m} = \frac{y_{i+p+2} - y_i}{p+2}.$$

We have by Lemma 15.2.5, since $y_{i+j} = y_{i+1} + (j-1)d$, $j = 1, \ldots, p+2$ ($1 \leqslant i \leqslant q - p - 2$),

$$[y_{i+1}, \ldots, y_{i+p+2}; f] = \frac{\Delta_d^{p+1} f(y_{i+1})}{(p+1)!|d|^{p+1}}, \quad i = 1, \ldots, q - p + 2.$$

Hence

$$[x_1, \ldots, x_{p+2}; f] = \frac{1}{(p+1)!|d|^{p+1}} \sum_{i=0}^{q-p-2} A_i \Delta_d^{p+1} f(y_{i+1}) \geqslant 0,$$

i.e., we have (15.3.5). $\qquad\square$

Similarly we have

Theorem 15.3.2. *Let $D \subset \mathbb{R}^N$ be a convex and open set, and let $f : D \to \mathbb{R}$ be a p-convex function. Let $\lambda_1, \ldots, \lambda_{p+1} \in \mathbb{Q}$ be positive. Then for every $x \in D$ and $h \in \mathbb{R}^N_+$ such that $x + (\lambda_1 + \cdots + \lambda_{p+1})h \in D$, we have*

$$\Delta_{\lambda_1 h} \ldots \Delta_{\lambda_{p+1} h} f(x) \geqslant 0. \tag{15.3.6}$$

Proof. By Theorem 15.1.3 there exist $\alpha \in \mathbb{Q} \cap (0, \infty)$ and $m, k_1, \ldots, k_m \in \mathbb{N}$ such that

$$\Delta_{\lambda_1 h} \ldots \Delta_{\lambda_{p+1} h} f(x) = \sum_{i=1}^{m} k_i \Delta_{\alpha h}^{p+1} f(x + i\alpha h) \geqslant 0$$

so that (15.3.6) holds. $\qquad\square$

Theorem 15.3.3. *If $f : \mathbb{R}^N \to \mathbb{R}$ is a polynomial function of order p, then*

$$\Delta_{h_1 \ldots h_{p+1}} f(x) = 0$$

for every $x, h_1, \ldots, h_{p+1} \in \mathbb{R}^N$.

Proof. This results from Theorem 15.1.2. $\qquad\square$

---

[3] $x_1 < \cdots < x_{p+2}$ implies $0 = \lambda_1 < \cdots < \lambda_{p+2}$ and $0 = k_1 < \cdots < k_{p+2}$, whence $k_i \geqslant i - 1$ for $i = 1, \ldots, p+2$.

15.4   Local boundedness of p-convex functions

First we note that it is purposeless to consider boundedness above of p-convex functions for $p > 1$. A p-convex function $f : D \to \mathbb{R}$ can be bounded above in D, and nevertheless unbounded and discontinuous in D. E. g., let $g : \mathbb{R}^N \to \mathbb{R}$ be a discontinuous additive function, and put $f(x) = -\big[g(x)\big]^2$ for $x \in \mathbb{R}^N$. Then $f : \mathbb{R}^N \to \mathbb{R}$ is bounded above (by zero) in $\mathbb{R}^N$, but unbounded below at every point of $\mathbb{R}^N$, and is discontinuous at every point of $\mathbb{R}^N$. We also have by (15.1.6), for every $x, h \in \mathbb{R}^N$,

$$
\begin{aligned}
\Delta_h^3 f(x) &= -\big[g(x+3h)\big]^2 + 3\big[g(x+2h)\big]^2 - 3\big[g(x+h)\big]^2 + \big[g(x)\big]^2 \\
&= -\big[g(x)+3g(h)\big]^2 + 3\big[g(x)+2g(h)\big]^2 - 3\big[g(x)+g(h)\big]^2 + \big[g(x)\big]^2 \\
&= -\big[g(x)\big]^2 - 6g(x)g(h) - 9\big[g(h)\big]^2 + 3\big[g(x)\big]^2 + 12g(x)g(h) \\
&\quad + 12\big[g(h)\big]^2 - 3\big[g(x)\big]^2 - 6g(x)g(h) - 3\big[g(h)\big]^2 + \big[g(x)\big]^2 = 0 \, ,
\end{aligned}
$$

whence also

$$
\Delta_h^{p+1} f(x) = \Delta_h^{p-2} \Delta_h^3 f(x) = \Delta_h^{p-2} 0 = 0
$$

for $p > 2$. Thus f is a polynomial function of every order $p \geqslant 2$, and hence f is p-convex for every $p \geqslant 2$.

Theorem 15.4.1. *Let $D \subset \mathbb{R}^N$ be a convex and open set, and let $f : D \to \mathbb{R}$ be a p-convex function. If f is locally bounded at a point $x_0 \in D$, then f is locally bounded at every point $x \in D$.*

Proof. Suppose that f is locally bounded at a point $x_0 \in D$, and take an arbitrary $z \in D$, $z \neq x_0$. Since D is open, we can find a point $y \in D$, collinear with x_0 and z, and such that z lies between y and x_0.

There exists an open ball $K = K(x_0, r)$, centered at x_0 and with the radius $r > 0$, such that f is bounded on K, i.e., for a certain $M > 0$ we have

$$
|f(x)| \leqslant M \quad \text{for } x \in K \, . \tag{15.4.1}
$$

We may choose r so small that $r < |x_0 - z|$ so that $z \notin \operatorname{cl} K$.

Let C be the part of D formed by the rays issuing from y and passing through the points of the ball $K_1 = K\left(x_0, \frac{1}{2}r\right)$ centered at x_0 and with the radius $\frac{1}{2}r$. There exists an open ball B centered at z and contained in C, $B \cap K \neq \varnothing$. We will show that f is bounded on B.

Take an arbitrary $x \in B$, and assume that $y < x$. (If $y > x$, the proof is similar). Let l be the ray from y through x. The ray l intersects K_1, and hence also intersects K along a segment S, whose length is greater than $\sqrt{3}\, r$. Choose on S points $x_1 < \cdots < x_p$ (then we have also $x < x_1$) in such a manner that all the ratios $|x_i - y|/|x - y|$ are rational, and $|x_{i+1} - x_i| \geqslant \frac{1}{2p}\sqrt{3}\, r$, which is possible, because the length of S is greater that $\sqrt{3}\, r$, and the points $u \in S$ such that $|u - y|/|x - y| \in \mathbb{Q}$ from a dense subset of S. Let $h = (x - y)/|x - y|$, and

$$
x = y + \lambda_0 h \, , \quad x_i = y + \lambda_i h \, , \quad i = 1, \dots, p \, .
$$

Further put $\lambda_{-1} = 0$ so that $y = y + \lambda_{-1}h$. The conditions on x_i imply that $|\lambda_{i+1} - \lambda_i| \geqslant \dfrac{1}{2p}\sqrt{3}\,r$ for $i = 1, \ldots, p$, $0 = \lambda_{-1} < \lambda_0 < \lambda_1 < \cdots < \lambda_p$, and all the ratios $\alpha_i = \lambda_i/\lambda_0$, $i = -1, 0, 1, \ldots, p$ are rational. Put $h_0 = \lambda_0 h \in \mathbb{R}^N_+$. Then

$$y = y + \alpha_{-1}h_0\,, \quad x = y + \alpha_0 h_0\,, \quad x_i = y + \alpha_i h_0\,, \quad i = 1, \ldots, p.$$

We have by Theorems 15.2.1 and 15.3.1

$$\frac{U\,(y, x, x_1, \ldots, x_p; f)}{V\,(y, x, x_1, \ldots, x_p)} = [y, x, x_1, \ldots, x_p; f] \geqslant 0\,.$$

Also $V\,(y, x, x_1, \ldots, x_p) > 0$, since $y < x < x_1 < \cdots < x_p$. Consequently

$$U\,(y, x, x_1, \ldots, x_p; f) \geqslant 0\,. \tag{15.4.2}$$

Developing the determinant U in (15.4.2) according to the last column, we have

$$(-1)^{p+1}V\,(x, x_1, \ldots, x_p)\,f(y) + (-1)^p V\,(y, x_1, \ldots, x_p)\,f(x)$$

$$+ \sum_{i=1}^{p}(-1)^{p+i}\,V(y, x, x_1, \ldots, x_{i-1}, x_{i+1}, \ldots, x_p)f(x_i) \geqslant 0,$$

whence, if p is odd,

$$f(x) \leqslant \frac{V(x, x_1, \ldots, x_p)}{V(y, x_1, \ldots, x_p)} f(y) \tag{15.4.3}$$

$$+ \sum_{i=1}^{p}(-1)^{i-1}\frac{V(y, x, x_1, \ldots, x_{i-1}, x_{i+1}, \ldots, x_p)}{V(y, x_1, \ldots, x_p)} f(x_i).$$

If p is even, we obtain the opposite inequality. The Van der Monde determinants $V(y, x_1, \ldots, x_p)$, $V(x, x_1, \ldots, x_p)$, and $V(y, x, x_1, \ldots, x_{i-1}, x_{i+1}, \ldots, x_p)$ are bounded above and below by constants, which may be made independent of x, since all the differences $\lambda_i - \lambda_j$, $i, j = -1, 0, 1, \ldots, p$, $i > j$, may be estimated by r and the distances between y and B and between y and K, and the radius of B. Moreover, $f(x_i)$ fulfil (15.4.1), $i = 1, \ldots, p$. Consequently we obtain from (15.4.3) that f is bounded above (bounded below, if p is even) on B.

The boundedness of f on B in the opposite direction may be established similarly. One must choose points $x_1, \cdots, x_{p+1} \in S$ such that $x_1 < \cdots < x_{p+1}$, all the distances $|x_i - x|$ are rational, $i = 1, \ldots, p+1$, and $|x_{i+1} - x_i| > \dfrac{1}{2p}\sqrt{3}\,r$, $i = 1, \ldots, p$. Then it is proved quite similarly as above that $U(x, x_1, \ldots, x_{p+1}; f) \geqslant 0$, and further the argument is the same. $\qquad\square$

Theorem 15.4.2. *Let $D \subset \mathbb{R}^N$ be a convex and open set, and let $f : D \to \mathbb{R}$ be a p-convex function. If f is locally bounded at a point $x_0 \in D$, then f is continuous at x_0.*

Proof. It is enough to show that

$$\lim_{\substack{x \to x_0 \\ x < x_0}} f(x) = \lim_{\substack{x \to x_0 \\ x > x_0}} f(x) = f(x_0). \tag{15.4.4}$$

We will show that $\lim_{\substack{x \to x_0 \\ x < x_0}} f(x) = f(x_0)$. The other equality in (15.4.4) can be established similarly.

Let f be bounded on a ball $K = K(x_0, r)$ centered at x_0 and with the radius r. Take an arbitrary $x \in K$ such that $x < x_0$, and $|x_0 - x| < \frac{1}{2}r$. Let l be the straight line passing through x and x_0, and write $S = l \cap K$. We choose on S two sequences of points:

I. $x_1 < \cdots < x_p$ such that $x_p < x$. Put $h = (x_0 - x_1)/|x_0 - x_1| > 0$, and write

$$x = x_1 + \lambda_{p+1}h, \quad x_0 = x_1 + \lambda_{p+2}h, \quad x_i = x_1 + \lambda_i h.$$

We choose $x_1, \ldots, x_p \in S$ in such a manner that all the ratios λ_i/λ_{p+2}, $i = 1, \ldots, p+2$, are rational, and the distances $|\lambda_{i+1} - \lambda_i|$, $i = 1, \ldots, p$, are bounded below by a constant which may be made independent of x. Clearly $\lambda_{p+1} \to \lambda_{p+2}$ when $x \to x_0$.

II. $x_1' < \cdots < x_{p-1}' < y$ such that $x_{p-1}' < x < x_0 < y$. Write

$$x = x_1' + \lambda_p' h, \quad x_0 = x_1' + \lambda_{p+1}' h, \quad y = x_1' + \lambda_{p+2}' h, \quad x_i = x_1' + \lambda_i' h, \quad i = 1, \ldots, p-1.$$

We choose $x_1', \ldots, x_{p-1}', y \in S$ in such a manner that all the ratios $\lambda_i'/\lambda_{p+2}'$, $i = 1, \ldots, p+2$, are rational, and $|\lambda_{i+1}' - \lambda_i'|$, $i = 1, \ldots, p-1, p+1$, are bounded below by a constant which may be made independent of x. Now $x \to x_0$ implies $\lambda_p' \to \lambda_{p+1}'$.

Similarly as in the proof of Theorem 15.4.1 we have

$$U(x_1, \ldots, x_p, x, x_0; f) \geqslant 0 \ \text{ and } \ U(x_1', \ldots, x_{p-1}', x, x_0, y; f) \geqslant 0,$$

whence

$$f(x_0) - f(x) \geqslant f(x_0) \left[1 - \frac{V(x_1, \ldots, x_p, x)}{V(x_1, \ldots, x_p, x_0)} \right]$$

$$+ \sum_{i=1}^{p} (-1)^{p+i+1} \frac{V(x_1, \ldots, x_{i-1}, x_{i+1}, \ldots, x_p, x, x_0)}{V(x_1, \ldots, x_p, x_0)} f(x_i), \quad (15.4.5)$$

and

$$f(x_0) - f(x) \leqslant f(x_0) \left[1 - \frac{V(x_1', \ldots, x_{p-1}', x, y)}{V(x_1', \ldots, x_{p-1}', x_0, y)} \right]$$

$$+ \sum_{i=1}^{p-1} (-1)^{p+i} \frac{V(x_1', \ldots, x_{i-1}', x_{i+1}', \ldots, x_{p-1}', x, x_0, y)}{V(x_1', \ldots, x_{p-1}', x_0, y)}$$

$$+ \frac{V(x_1', \ldots, x_{p-1}', x, x_0)}{V(x_1', \ldots, x_{p-1}', x_0, y)} f(y). \tag{15.4.6}$$

Every determinant $V(x_1, \ldots, x_{i-1}, x_{i+1}, \ldots, x_p, x, x_0)$ contains the factor $(\lambda_{p+2} - \lambda_{p+1})$ (cf. (15.2.6)), and hence tends to zero when $x \to x_0$, whereas $f(x_i)$ remain bounded. The determinant $V(x_1, \ldots, x_p, x_0)$ can be estimated from below by a constant independent of x. The ratio $V(x_1, \ldots, x_p, x)/V(x_1, \ldots, x_p, x_0)$ tends to 1 when $x \to x_0$. So the right-hand side of (15.4.5) tends to zero when $x \to x_0$. Similarly, the right-hand side of (15.4.6) tends to zero when $x \to x_0$. Hence $f(x)$ approaches $f(x_0)$ when x approaches x_0. $\square$

Theorem 15.4.3. *Let $D \subset \mathbb{R}^N$ be a convex and open set, and let $f : D \to \mathbb{R}$ be a p-convex function. If f is bounded on a set $T \subset D$ such that $\operatorname{int} T \neq \varnothing$, then f is continuous in D.*

Proof. Results from Theorems 15.4.1 and 15.4.2. $\square$

15.5 Operation H

Fix a $p \in \mathbb{N}$. For every set $T \subset \mathbb{R}^N$ we define the set $H(T)$ as follows (Ger [98], [101], [104]):

$$H(T) = \{x \in \mathbb{R}^N \mid \text{there exists an } h \in \mathbb{R}^N \tag{15.5.1}$$
$$\text{such that } x + ih \in T, \, i = \pm 1, \ldots, \pm(p+1)\}.$$

Choosing, in particular, $h = 0$, we get from (15.5.1) $T \subset H(T)$.

Lemma 15.5.1. *If $D \subset \mathbb{R}^N$ is a convex set, and $T \subset D$, then $H(T) \subset D$.*

Proof. Suppose that $x \in H(T)$. Then there exists an $h \in \mathbb{R}^N$ such that $x + h \in T$ and $x - h \in T$. Hence also $x = \dfrac{1}{2}[(x+h) + (x-h)] \in D$. $\square$

The main property of the operation H is contained in the following (Ger [98], [101],[104])

Theorem 15.5.1. *Let $D \subset \mathbb{R}^N$ be a convex and open set, and let $f : D \to \mathbb{R}$ be a p-convex function. If f is bounded on a set $T \subset D$, then f is bounded also on $H(T)$.*

Proof. Let

$$|f(x)| \leqslant M \quad \text{for } x \in T. \tag{15.5.2}$$

Take an $x \in H(T)$. By (15.5.1) there exists an $h \in \mathbb{R}^N$ such that $x + ih \in T$ for $i = \pm 1, \ldots, \pm(p+1)$. Replacing, if necessary, h by $-h$, we may assume that $h > 0$. (If $h = 0$, then $x \in T$, and we get $|f(x)| \leqslant M$ by (15.5.2)).

We must distinguish two cases.

1. p is even. We have $\Delta_h^{p+1} f(x) \geqslant 0$, or, by Corollary 15.1.2,

$$\sum_{k=0}^{p+1} (-1)^{p+1-k} \binom{p+1}{k} f(x+kh) \geqslant 0. \tag{15.5.3}$$

Hence, by (15.5.2), since p is even,

$$f(x) \leqslant \sum_{k=1}^{p+1} (-1)^{p+1-k} \binom{p+1}{k} f(x+kh) \leqslant \sum_{k=1}^{p+1} \binom{p+1}{k} M = (2^{p+1} - 1)M. \quad (15.5.4)$$

Similarly, we have $\Delta_h^{p+1} f(x-h) \geqslant 0$, whence

$$\sum_{k=0}^{p+1} (-1)^{p+1-k} \binom{p+1}{k} f(x-h+kh) \geqslant 0, \qquad (15.5.5)$$

i.e.,

$$\sum_{k=-1}^{p} (-1)^{p-k} \binom{p+1}{k+1} f(x+kh) \geqslant 0,$$

and

$$(p+1)f(x) \geqslant (-1)^p f(x-h) + \sum_{k=1}^{p} (-1)^{p-k-1} \binom{p+1}{k+1} f(x+kh) \qquad (15.5.6)$$

$$\geqslant -M - \sum_{k=1}^{p} \binom{p+1}{k+1} M = -(2^{p+1} - p - 1)M.$$

We clearly have

$$\frac{2^{p+1} - p - 1}{p+1} \leqslant \frac{2^{p+1}}{p+1} \leqslant 2^{p+1} - 1,$$

whence by (15.5.6)

$$f(x) \geqslant -(2^{p+1} - 1)M.$$

This together with (15.5.4) yields

$$|f(x)| \leqslant (2^{p+1} - 1)M. \qquad (15.5.7)$$

2. p is odd. Again we have (15.5.3), whence

$$f(x) \geqslant \sum_{k=1}^{p+1} (-1)^{p-k} \binom{p+1}{k} f(x+kh) \geqslant -\sum_{k=1}^{p+1} \binom{p+1}{k} M = -(2^{p+1} - 1)M,$$

whereas by (15.5.5)

$$(p+1)f(x) \leqslant (-1)^{p+1} f(x-h) + \sum_{k=1}^{p} (-1)^{p-k} \binom{p+1}{k+1} f(x+kh)$$

$$\leqslant (2^{p+1} - p - 1)M \leqslant (2^{p+1} - 1)M,$$

and consequently we obtain again (15.5.7).

Since $2^{p+1} - 1 > 1$, if $h = 0$, then we obtain (15.5.7) from (15.5.2). Thus (15.5.7) holds for all $x \in H(T)$, i.e., f is bounded on H(T). $\qquad \square$

The following two lemmas (Ger [98], [101], [104]) show an important feature of the operation H.

Lemma 15.5.2. *Let $T \subset \mathbb{R}^N$ be such that $m_i(T) > 0$. Then* int $H(T) \neq \varnothing$.

Proof. There exists a measurable set $T_0 \subset T$ such that $m(T_0) > 0$. Let x_0 be a density point of T_0. Then 0 is a density point of $T_0 - x_0$ as well as of $\lambda(T_0 - x_0)$ for all $\lambda \in \mathbb{R}$, $\lambda \neq 0$. There exists a cube K centered at 0 such that, for every $i \in I = \{-(p+1), -p, \ldots, -1, +1, \ldots, p, (p+1)\}$,

$$m\left(K \setminus \frac{1}{i}(T_0 - x_0)\right) < \frac{1}{2^{N+1}(p+1)} m(K). \tag{15.5.8}$$

Let $U = \frac{1}{2}K$ so that $U + U \subset K$. We will show that

$$\bigcap_{i \in I} \frac{1}{i}(T_0 - x_0 - x) \neq \varnothing \tag{15.5.9}$$

for every $x \in U$. If $x \in U$, then $x/i \in U$ for $i \in I$, and hence $U + (x/i) \in U + U \subset K$, i.e., $U \subset K - (x/i)$ for $i \in I$, and hence $U \subset \bigcap_{i \in I}(K - (x/i))$. On the other hand,

$$K = \left(K \cap \frac{1}{i}(T_0 - x_0)\right) \cup \left(K \setminus \frac{1}{i}(T_0 - x_0)\right),$$

and

$$U \subset \bigcap_{i \in I}\left(K - \frac{1}{i}x\right) = \bigcap_{i \in I}\left\{\left[\left(K \cap \frac{1}{i}(T_0 - x_0)\right) - \frac{1}{i}x\right]\right.$$
$$\left. \cup \left[\left(K \setminus \frac{1}{i}(T_0 - x_0)\right) - \frac{1}{i}x\right]\right\}$$
$$\subset \bigcap_{i \in I}\left[\left(K \cap \frac{1}{i}(T_0 - x_0)\right) - \frac{1}{i}x\right] \cup \bigcup_{i \in I}\left[\left(K \setminus \frac{1}{i}(T_0 - x_0)\right) - \frac{1}{i}x\right]$$
$$= \bigcap_{i \in I}\left[\left(K \cap \frac{1}{i}(T_0 - x_0)\right) - \frac{1}{i}x\right] \cup Q, \tag{15.5.10}$$

where by (15.5.8) $m(Q) < m(K)/2^N$, whereas $m(U) = m(K)/2^N$. Hence

$$\bigcap_{i \in I}\left[\left(K \cap \frac{1}{i}(T_0 - x_0)\right) - \frac{1}{i}x\right] \neq \varnothing. \tag{15.5.11}$$

But

$$\bigcap_{i \in I}\left[\left(K \cap \frac{1}{i}(T_0 - x_0)\right) - \frac{1}{i}x\right] \subset \bigcap_{i \in I}\left[\frac{1}{i}(T_0 - x_0) - \frac{1}{i}x\right] = \bigcap_{i \in I}\frac{1}{i}(T_0 - x_0 - x)$$

Thus we have proved (15.5.9). Now take an arbitrary $x \in U$. By (15.5.9) there exists an

$$h \in \frac{1}{i}\left(T_0 - x_0 - x\right) \subset \frac{1}{i}\left(T - x_0 - x\right)$$

for every $i \in I$, which means that $x + ih \in T - x_0$ for $i \in I$. Thus for every $x \in U + x_0$ we have $x + ih \in T$ for $i \in I$. Consequently $x \in H(T)$ and $U + x_0 \subset H(T)$. Hence int $H(T) \neq \varnothing$. $\square$

Lemma 15.5.3. *Let $T \subset \mathbb{R}^N$ be of the second category and with the Baire property. Then* int $H(T) \neq \varnothing$.

Proof. The proof is similar to that of Lemma 15.5.2. Since T has the Baire property, we have

$$T = (G \cup P) \setminus R,$$

where G is a non-empty open set, and the sets P, R are of the first category. Put $T_0 = G \setminus R$, and let x_0 be an arbitrary point of T_0. Let K be an open ball centered at zero and contained in $\bigcap_{i \in I} \frac{1}{i}\left(G - x_0\right)$. (Note that 0 belongs to this intersection.) Then, for every $i \in I$, we have

$$K \setminus \frac{1}{i}\left(T_0 - x_0\right) \subset \frac{1}{i}\left(G - x_0\right) \setminus \frac{1}{i}\left(T_0 - x_0\right) = \frac{1}{i}\left[(G \setminus T_0) - x_0\right].$$

But $G \setminus T_0 = G \setminus (G \setminus R) \subset R$, and so $G \setminus T_0$ is of the first category, whence also $K \setminus \frac{1}{i}\left(T_0 - x_0\right)$ is of the first category for $i \in I$. This replaces relation (15.5.8). Define $U = \frac{1}{2}K$. Again we have (15.5.10) for every $x \in U$, where now Q is a set of the first category, and (15.5.11) follows, which implies (15.5.9). Further the proof runs exactly like that of Lemma 15.5.2. $\square$

From Lemmas 15.5.2 and 15.5.3 and from Theorems 15.5.1 and 15.4.3 we obtain the following theorem of Ciesielski (Ciesielski [46]; cf. also Kurepa [201]).

Theorem 15.5.2. *Let $D \subset \mathbb{R}^N$ be a convex and open set, and let $f : D \to \mathbb{R}$ be a p-convex function. If f is bounded on a set $T \subset D$ such that $m_i(T) > 0$, or T is of the second category and with the Baire property, then f is continuous in D.*

The iterates H^n of H are defined as usually:

$$H^1(T) = H(T), \quad H^{n+1}(T) = H\big(H^n(T)\big), \quad n = 1, 2, 3, \ldots$$

The following theorem, which may be regarded as a generalization of Theorem 15.5.2, is due to R. Ger [98], [101], [104].

Theorem 15.5.3. *Let $D \subset \mathbb{R}^N$ be a convex and open set, and let $f : D \to \mathbb{R}$ be a p-convex function. If f is bounded on a set $T \subset D$ such that for a certain $n \in \mathbb{N}$ we have $m_i\big(H^n(T)\big) > 0$ or $H^n(T)$ is of the second category and has the Baire property, then f is continuous in D.*

Proof. This is an immediate consequence of Theorems 15.5.1 and 15.5.2. $\square$

From Theorem 15.5.2 we obtain also the following (cf. also Gajda [95]).

Theorem 15.5.4. *Let $D \subset \mathbb{R}^N$ be a convex and open set, and let $f : D \to \mathbb{R}$ be a p-convex function. If f is measurable, then it is continuous in D.*

Proof. Put

$$E_n = \{x \in D \mid |f(x)| \leqslant n\}, \quad n \in \mathbb{N}. \tag{15.5.12}$$

The sets E_n are measurable, since f is measurable, and

$$\bigcup_{n \in \mathbb{N}} E_n = D.$$

Therefore for a certain $n_0 \in \mathbb{N}$ the set E_{n_0} must have positive measure: $m(E_{n_0}) > 0$. According to (15.5.12) the function f is bounded on E_{n_0}. By Theorem 15.5.2 f is continuous in D. $\qquad\square$

We conclude this section with an example (Ger [98]) showing that there exist sets T, of measure zero and nowhere dense, such that $\operatorname{int} H(T) \neq \varnothing$. Thus if f is a p-convex function bounded on such a set, its continuity can be inferred from Theorem 15.5.3, but not from Theorem 15.5.2.

Example 15.5.1. Let $N = 1$. Put $q = 2p + 4$, and let T be the set of those $x \in [0,1]$ which may be written as

$$x = \sum_{i=1}^{\infty} \frac{c_i}{q^i}, \tag{15.5.13}$$

where $c_i \in \{0, \ldots, p, p+2, \ldots, 2p+3\}$, $i \in \mathbb{N}$. The set T has measure zero (Theorem 3.6.4) and is nowhere dense.

Take an arbitrary $x \subset [0,1]$, and suppose that x has a representation (15.5.13), where $c_i \in \{0, \ldots, 2p+3\}$, $i \in \mathbb{N}$. Define the number h as

$$h = \sum_{i=1}^{\infty} \frac{d_i}{q^i},$$

where

$$d_i = \begin{cases} 0 & \text{if } c_i \neq p+1, \\ 1 & \text{if } c_i = p+1, \end{cases} \quad i \in \mathbb{N}.$$

Then $x + kh \in T$ for $k = \pm 1, \ldots, \pm(p+1)$, which means that $x \in H(T)$. Thus[4] $[0,1] \subset H(T)$, whence $\operatorname{int} H(T) \neq \varnothing$.

15.6 Continuous p convex functions

Let $D \subset \mathbb{R}^N$ be a convex and open set, and let $f : D \to \mathbb{R}$ be a p-convex function. If f is continuous in D, then we obtain by Theorem 15.3.1 and Lemma 15.2.2 that

$$[x_1, \ldots, x_{p+2}; f] \geqslant 0 \tag{15.6.1}$$

for every system of distinct collinear points $x_1, \ldots, x_{p+2} \in D$.

---

[4] By Lemma 15.5.1 $H(T) \subset [0,1]$ so that actually $H(T) = [0,1]$.

The converse is also true:

Theorem 15.6.1. *Let $D \subset \mathbb{R}^N$ be a convex and open set, and let $f : D \to \mathbb{R}$ be a function. The function f is p-convex and continuous in D if and only if (15.6.1) holds for every system of distinct collinear points $x_1, \ldots, x_{p+2} \in D$.*

Proof. As pointed out above, the necessity results from Theorem 15.3.1 and Lemma 15.2.2. Now assume that (15.6.1) holds for every system of distinct collinear points $x_1, \ldots, x_{p+2} \in D$. Take an arbitrary $x \in D$ and $h \in \mathbb{R}_+^N$ such that $x + (p+1)h \in D$. Put

$$x_i = x + (i - 1)h, \quad i = 1, \ldots, p + 2.$$

By Lemma 15.2.5 and by (15.6.1)

$$\Delta_h^{p+1} f(x) = p! |h|^p [x_1, \ldots, x_{p+2}; f] \geq 0,$$

whence $\Delta_h^{p+1} f(x) \geq 0$ for every $x \in D$, and $h \in \mathbb{R}_+^N$ such that $x + (p+1)h \in D$. Thus f is p-convex. Consequently in order to prove that f is continuous, it is sufficient to prove that f is bounded on a subset of D with a non-empty interior (cf. Theorem 15.4.3).

We will prove by induction on k that f is bounded on every closed k-dimensional rectangular parallelopiped P contained in D, $k = 1, \ldots, N$. For $k = N$ we have $\mathrm{int}\, P \neq \varnothing$, and the theorem follows.

For $k = 1$, P is a closed segment contained in D. Since D is open, there exists a closed segment $S \subset D$ such that $P \subset \mathrm{int}\, S$. Let $S = \overline{ab}$, $P = \overline{cd}$, and let, e.g., $a < c < d < b$. (The other possibility is $b < d < c < a$, which may be reduced to the former by re-naming the points a, b, c, d). Fix $p + 1$ distinct points $x_1, \ldots, x_{p+1}$ on the segment $\overline{ac}$ such that $a \leq x_1 < \cdots < x_{p+1} < c$. We have by (15.6.1) for arbitrary $x \in P$

$$[x_1, \ldots, x_{p+1}, x; f] \geq 0$$

whence also, according to Theorem 15.2.1,

$$U(x_1, \ldots, x_{p+1}, x; f) \geq 0, \tag{15.6.2}$$

since $x_1 < \cdots < x_{p+1} < x$ so that $V(x_1, \ldots, x_{p+1}, x) > 0$. Developing the determinant in (15.6.2) according to the last column we obtain

$$f(x) \geq \sum_{i=1}^{p+1} (-1)^{p+1+i} \frac{V(x_1, \ldots, x_{i-1}, x_{i+1}, \ldots x_{p+1}, x)}{V(x_1, \ldots, x_{p+1})} f(x_i). \tag{15.6.3}$$

The determinant $V(x_1, \ldots, x_{i-1}, x_{i+1}, \ldots x_{p+1}, x)$ (for every $i = 1, \ldots, p + 1$) is a continuous function of x, and thus remains bounded when x ranges over the compact set P. The remaining expressions on the right-hand side of (15.6.3) are independent of x (the points $x_1, \ldots, x_{p+1}$ are fixed). Consequently the whole expression on the right-hand side of (15.6.3) remains bounded when x ranges over P, and consequently f is bounded below on P.

Similarly, we have for arbitrary $x \in P$

$$[x_1, \ldots, x_p, x, b; f] \geqslant 0,$$

whence also

$$U(x_1, \ldots, x_p, x, b; f) \geqslant 0,$$

and

$$f(x) \leqslant \sum_{i=1}^{p} (-1)^{p+i} \frac{V(x_1, \ldots, x_{i-1}, x_{i+1}, \ldots x_p, x, b)}{V(x_1, \ldots, x_p, b)} f(x_i) + \frac{V(x_1, \ldots, x_p, x)}{V(x_1, \ldots, x_p, b)} f(b).$$

Consequently f is also bounded above on P, and hence f is bounded on P.

Now suppose that our statement is true for a certain $k < N$, and let $P \subset D$ be a closed $(k+1)$-dimensional rectangular parallelopiped. Let $P = S \times P_0$, where S is a closed segment and P_0 is a closed k-dimensional rectangular parallelopiped. We may find a closed segment S_0 such that $S \subset \operatorname{int} S_0$ and $S_0 \times P_0 \subset D$. Let $S_0 = \overline{ab}$, $S = \overline{cd}$. Again we may assume that $a < c < d < b$. Choose $p+1$ distinct points $x_1, \ldots, x_{p+1}$ on the segment $\overline{ac}$ such that $a \leqslant x_1 < \cdots < x_{p+1} < c$. Let

$$P_b = \{b\} \times P_0, \quad P_i = \{x_i\} \times P_0, \quad i = 1, \ldots, p+1.$$

$P_1, \ldots, P_{p+1}, P_b$ are closed k-dimensional rectangular parallelopipeds, consequently, by the induction hypothesis, the function f is bounded in the set

$$E = P_b \cup \bigcup_{i=1}^{p+1} P_i.$$

For every $x \in P$ let $l(x)$ be the straight line passing through x and parallel to S. Write

$$x_b' = l(x) \cap P_b, \quad x_i' = l(x) \cap P_i, \quad i = 1, \ldots, p+1.$$

Thus, for every $x \in P$, $x_1', \ldots, x_{p+1}', x, x_b'$ are distinct collinear points fulfilling $x_1' < \cdots < x_{p+1}' < x < x_b'$, and

$$x_1', \ldots, x_{p+1}', x_b' \in E. \tag{15.6.4}$$

The Van der Monde determinants $V(x_1', \ldots, x_{p+1}')$, $V(x_1', \ldots, x_{i-1}', x_{i+1}', \ldots x_{p+1}', x)$, $V(x_1', \ldots, x_{i-1}', x_{i+1}', \ldots x_p', x, x_b')$, $V(x_1', \ldots x_p', x_b')$ and $V(x_1', \ldots x_p', x)$ depend on $x_1', \ldots, x_{p+1}', x_b'$ only through their distances, which remain constant when x ranges over P, so as a function of x the above mentioned determinants are continuous, and hence bounded on P. Also, by (15.6.4), the values $f(x_1'), \ldots, f(x_{p+1}'), f(x_b')$ remain bounded when x ranges over P. From the inequalities

$$[x_1', \ldots, x_{p+1}', x; f] \geqslant 0, \quad [x_1', \ldots, x_p', x, x_b'; f] \geqslant 0,$$

we infer that f is bounded on P quite similarly as in the case $k = 1$. Induction ends the proof. $\qquad \square$

15.7   Continuous p-convex functions. Case $N = 1$

In this and in the next section, in order to avoid burdensome details, we restrict ourselves to the one-dimensional case $N = 1$. Moreover, many results are not longer valid in the general, N-dimensional case. At first let us note the following consequence of Theorem 15.3.2 and Lemma 15.1.4:

Theorem 15.7.1. *Let $D \subset \mathbb{R}$ be an open interval, and let $f : D \to \mathbb{R}$ be a continuous p-convex function. Then, for arbitrary $x \in D$ and arbitrary positive $h_1, \ldots, h_{p+1}$ such that $x + h_1 + \cdots + h_{p+1} \in D$, we have*

$$\Delta_{h_1 \ldots h_{p+1}} f(x) \geqslant 0. \tag{15.7.1}$$

But now condition (15.7.1) does not imply that f is continuous. E.g., let $f : \mathbb{R} \to \mathbb{R}$ be a discontinuous additive function. Then we have by (15.1.5) for arbitrary $x, h_1, h_2 \in \mathbb{R}$

$$\Delta_{h_1 h_2} f(x) = f(x + h_1 + h_2) - f(x + h_1) - f(x + h_2) + f(x) = 0,$$

but f is not continuous. On the other hand, condition (15.7.1) clearly implies that f is p-convex. (15.3.1) results from (15.7.1) with $h_1 = \cdots = h_{p+1} = h$.

Hence we get the following

Theorem 15.7.2. *Let $D \subset \mathbb{R}$ be an open interval, and let $f : D \to \mathbb{R}$ be a continuous p-convex function, $p > 1$. Let $h > 0$ be such that $D_h = D \cap (D - h) \neq \varnothing$, and define the function $g : D_h \to \mathbb{R}$ by*

$$g(x) = \Delta_h f(x).$$

Then g is $(p - 1)$-convex.

Proof. The set D_h evidently is open, and by Theorems 5.1.1 and 5.1.2 D_h is convex. By Theorem 15.7.1 f satisfies (15.7.1). Take $h_{p+1} = h$. Then (15.7.1) yields

$$\Delta_{h_1 \ldots h_p} g(x) \geqslant 0$$

for every $x \in D_h$ and every positive $h_1, \ldots, h_p$ such that $x + h_1 + \cdots + h_p \in D_h$. This implies that g is $(p - 1)$-convex. $\square$

Concerning the case $p = 1$, cf. Theorem 7.3.4.

According to the remarks after formula (15.2.3), in the one-dimensional case in the definition of the divided differences $[x_1, \ldots, x_p; f]$ we may take $\lambda_i = x_i$. Similarly, due to the uniqueness of $P(x_1, \ldots, x_{p+1}; f)$, in the definition (15.2.11) we may take $\lambda_i = x_i$. Now, making use of Lemma 15.2.4 we can give a geometrical interpretation of the p-convexity for continuous functions. Let the points $x_1, \ldots, x_{p+1}, x \in D$ satisfy $x_1 < \cdots < x_{p+1} < x$. Then $V(x_1, \ldots, x_{p+1}, x) > 0$ and $V(x_1, \ldots, x_{p+1}) > 0$, and consequently, by (15.2.12), $f(x) - P(x_1, \ldots, x_{p+1}; f \mid x) \geqslant 0$ (cf. Theorem 15.6.1). Thus if we take arbitrary $p + 1$ points $x_1, \ldots, x_{p+1} \in D$, $x_1 < \cdots < x_{p+1}$, and draw the Lagrange interpolation polynomial P through the points $\bigl(x_1, f(x_1)\bigr), \ldots, \bigl(x_{p+1}, f(x_{p+1})\bigr)$, then to the right of x_{p+1} the graph of f lies above (or on) that of P. When the point x moves leftward, the graph of f will lie in the intervals $(x_p, x_{p+1}), \ldots, (x_1, x_2)$ alternatively below and above (or on) that of P, since with every crossing of a point x_i the determinant $V(x_1, \ldots, x_{p+1}, x)$ changes the sign. Finally, to the left of x, the graph of f lies below resp. above (or on) that of P, depending on whether p is even or odd.

In the case $p = 1$, when f is continuous and convex in the usual sense, if we draw a straight line L (a polynomial of degree 1) through the points $(u, f(u))$ and $(v, f(v))$, $u, v \in D$, then the graph of f lies below (or on) L between u and v, and above (or on) L outside of the segment $\overline{uv}$, which is the usual geometrical interpretation of convexity for continuous functions (cf. 7.1).

Below we prove some lemmas concerning the boundedness of the divided differences of a function.

Lemma 15.7.1. *Let $D \subset \mathbb{R}$ be a finite interval, and let $f : D \to \mathbb{R}$ be a function. If, for a $p \in \mathbb{N}$, $p > 1$, the divided difference $[x_1, \ldots, x_p; f]$ remains bounded for all $x_1, \ldots, x_p \in D$, $x_1, \ldots, x_p$ distinct, then also the divided difference $[x_1, \ldots, x_{p-1}; f]$ remains bounded for all $x_1, \ldots, x_{p-1} \in D$, $x_1, \ldots, x_{p-1}$ distinct.*

Proof. There exist an $M > 0$ such that

$$| [t_1, \ldots, t_p; f] | \leqslant M \tag{15.7.2}$$

for all $t_1, \ldots, t_p \in D$, $t_1, \ldots, t_p$ distinct. Fix arbitrarily $p - 1$ distinct points $y_1, \ldots, y_{p-1} \in D$, and take arbitrary $p - 1$ distinct points $x_1, \ldots, x_{p-1} \in D$. By Lemma 15.2.3 we may assume that the point $x_1, \ldots, x_{p-1}$ have been numbered in such a manner that if some of the points $x_1, \ldots, x_{p-1}$ coincide with some of the points $y_1, \ldots, y_{p-1}$, then they have the same indices as the corresponding y's, i.e., if $x_j = y_i$, then $j = i$.

Put

$$A = | [y_1, \ldots, y_{p-1}; f] | . \tag{15.7.3}$$

A is a fix finite number.

By Lemma 15.2.7

$$[x_1, \ldots, x_{p-1}; f] - [y_1, \ldots, y_{p-1}; f] = \sum_{i=1}^{p-1} (x_i - y_i) [x_1, \ldots, x_i, y_i, \ldots, y_{p-1}; f],$$

whence by (15.7.2) and (15.7.3)

$$|[x_1, \ldots, x_{p-1}; f]| \leqslant A + M \sum_{i=1}^{p-1} |x_i - y_i| \leqslant A + (p-1) M |D|,$$

where $|D|$ is the length of D. $\qquad\square$

Now fix $m > p + 1$ points $x_1, \ldots, x_m \in D$ such that $x_1 < \cdots < x_m$, and put

$$d_i = [x_i, \ldots, x_{i+p}; f], \quad i = 1, \ldots, m - p. \tag{15.7.4}$$

Lemma 15.7.2. *Let $D \subset \mathbb{R}$ be an open interval, and let $f : D \to \mathbb{R}$ be a continuous p-convex function. Then, for every $x_1, \ldots, x_m \in D$, $x_1 < \cdots < x_m$, $m > p+1$, sequence (15.7.4) is increasing.*

Proof. We have by (15.2.3)

$$\frac{d_{i+1} - d_i}{x_{i+p+1} - x_i} = \frac{[x_{i+1}, \ldots, x_{i+p+1}; f] - [x_i, \ldots, x_{i+p}; f]}{x_{i+p+1} - x_i} = [x_i, \ldots, x_{i+p+1}; f] \geqslant 0$$

(cf. Theorem 15.6.1). Hence $d_i \leqslant d_{i+1}$, since $x_i < x_{i+p+1}$, $i = 1, \ldots, m - p$.   □

Lemma 15.7.3. *Let $D \subset \mathbb{R}$ be an open interval, and let $f : D \to \mathbb{R}$ be a continuous p-convex function. Then, for every compact interval $I \subset D$, the divided difference $[x_1, \ldots, x_{p+1}; f]$ remains bounded for all $x_1, \ldots, x_{p+1} \in I$, $x_1, \ldots, x_{p+1}$ distinct.*

Proof. Let $D = (a, b)$, $I = [c, d]$, $a < c < d < b$. Fix arbitrarily $p + 1$ distinct points $y_1, \ldots y_{p+1} \in (a, c)$ and $p + 1$ distinct points $z_1, \ldots, z_{p+1} \in (d, b)$ such that $y_1 < \cdots < y_{p+1} < z_1 < \cdots < z_{p+1}$. Take arbitrary $p + 1$ distinct points $x_1, \ldots, x_{p+1} \in I$. According to Lemma 15.2.3 we may assume that $x_1, \ldots, x_{p+1}$ have been numbered in such a manner that $x_1 < \cdots < x_{p+1}$. Then we have $y_1 < \cdots < y_{p+1} < x_1 < \cdots < x_{p+1} < z_1 < \cdots < z_{p+1}$. By Lemma 15.7.2 ($m = 3p + 3$) we have

$$[y_1, \ldots, y_{p+1}; f] \leqslant [x_1, \ldots, x_{p+1}; f] \leqslant [z_1, \ldots, z_{p+1}; f] ,$$

and the lemma follows.   □

From Lemmas 15.7.1 and 15.7.3 we obtain

Corollary 15.7.1. *Let $D \subset \mathbb{R}$ be an open interval, and let $f : D \to \mathbb{R}$ be a continuous p-convex function. Then the divided differences of f of all orders $\leqslant p + 1$ are bounded on every compact interval $I \subset D$.*

15.8   Differentiability of p-convex functions

Also in this section we restrict ourselves to the one-dimensional case $N = 1$.

Theorem 15.8.1. *Let $D \subset \mathbb{R}$ be an open interval, and let $f : D \to \mathbb{R}$ be a continuous p-convex function, $p > 1$. Then f is of class C^1 in D, and the derivative $f' : D \to \mathbb{R}$ is a $(p - 1)$-convex function.*

Proof. Take an arbitrary $x \in D$ and an $\eta > 0$ such that $I = [x - \eta, x + \eta] \subset D$. By Corollary 15.7.1 the divided difference $[t_1, t_2, t_3; f]$ is bounded in I, i.e., there exists a constant $M > 0$ such that

$$\left| [t_1, t_2, t_3; f] \right| \leqslant M$$

for every distinct points $t_1, t_2, t_3 \in I$. In other words, by (15.2.3)

$$\left| \frac{[t_2, t_3; f] - [t_1, t_2; f]}{t_3 - t_1} \right| \leqslant M$$

for every distinct points $t_1, t_2, t_3 \in I$. Hence

$$\left| [t_2, t_3; f] - [t_1, t_2; f] \right| \leqslant M \left| t_3 - t_1 \right| \tag{15.8.1}$$

for every distinct points $t_1, t_2, t_3 \in I$. Now take arbitrary distinct points x_1, x_2, x_1', $x_2' \in I$. We have by (15.8.1)

$$\left|[x_1', x_2'; f] - [x_1, x_2; f]\right| \leqslant \left|[x_2, x_1'; f] - [x_1, x_2; f]\right| +$$
$$+ \left|[x_1', x_2'; f] - [x_2, x_1'; f]\right|$$
$$\leqslant M\left|x_1' - x_1\right| + M\left|x_2' - x_2\right| \leqslant 4M\eta \qquad (15.8.2)$$

Inequality (15.8.2) shows that there exists the limit

$$\lim_{\substack{x_1, x_2 \to x \\ x_1 \neq x_2}} [x_1, x_2; f] = \lim_{\substack{x_1, x_2 \to x \\ x_1 \neq x_2}} \frac{f(x_2) - f(x_1)}{x_2 - x_1} = f'(x).$$

Consequently f is differentiable at every point $x \in D$. Moreover, it follows from (15.8.2), when $x_1, x_2 \to x \in I$, $x_1 \neq x_2$, and $x_1', x_2' \to y \in I$, $x_1' \neq x_2'$, that

$$\left|f'(y) - f'(x)\right| \leqslant 2M\left|y - x\right|,$$

which proves the continuity of f' in D. Consequently f is of class C^1 in D.

Since the derivative f' exists in D, we have for every $x \in D$

$$\lim_{h \to 0} \frac{\Delta_h f(x)}{h} = f'(x).$$

Take arbitrary $x \in D$ and arbitrary positive $h_1, \ldots, h_p \in D$ such that $x + h_1 + \cdots + h_p \in D$. For sufficiently small $h > 0$ we have $x + h_1 + \cdots + h_p + h \in D$, whence by Theorem 15.7.1

$$\frac{1}{h}\Delta_{h_1 \ldots h_p h} f(x) \geqslant 0.$$

Hence also (cf. Lemma 15.1.5)

$$\Delta_{h_1 \ldots h_p} f'(x) = \Delta_{h_1 \ldots h_p} \lim_{h \to 0} \frac{\Delta_h f(x)}{h} = \lim_{h \to 0} \frac{1}{h}\Delta_{h_1 \ldots h_p h} f(x) \geqslant 0.$$

Consequently f' is $(p-1)$-convex. □

Theorem 15.8.2. *Let $D \subset \mathbb{R}$ be an open interval, and let $f : D \to \mathbb{R}$ be a function of class C^1 such that f' is $(p-1)$-convex $(p > 1)$. Then f is p-convex.*

Proof. We have for arbitrary $x \in D$ and $h > 0$ such that $x + h \in D$

$$\Delta_h f(x) = f(x + h) - f(x) = \int\limits_x^{x+h} f'(t)\,dt,$$

whence by Lemma 15.1.6

$$\Delta_{h_1 \ldots h_p h} f(x) = \Delta_{h_1 \ldots h_p} \Delta_h f(x) = \Delta_{h_1 \ldots h_p} \int\limits_x^{x+h} f'(t)\,dt = \int\limits_x^{x+h} \Delta_{h_1 \ldots h_p} f'(t)\,dt \geqslant 0,$$

for arbitrary positive $h_1, \ldots, h_p, h$ such that $x + h_1 + \cdots + h_p + h \in D$. Consequently f is p-convex. □

Using Theorem 15.8.1 several times, we arrive at the following

Theorem 15.8.3. *Let $D \subset \mathbb{R}$ be an open interval, and let $f : D \to \mathbb{R}$ be a continuous p-convex function $(p > 1)$. Then f is of class C^{p-1} in D, and for every $r \in \mathbb{N}$, $1 \leqslant r \leqslant p - 1$, the derivative $f^{(r)}$ is $(p-r)$-convex.*

From Theorems 15.8.2 and 15.8.3 we obtain

Theorem 15.8.4. *Let $D \subset \mathbb{R}$ be an open interval, and let $f : D \to \mathbb{R}$ be a continuous function. Then f is p-convex $(p > 1)$ if and only if f is of class C^{p-1} in D and the derivative $f^{(p-1)}$ is convex.*

Hence and from Theorems 7.5.1 and 7.5.2 we obtain

Theorem 15.8.5. *Let $D \subset \mathbb{R}$ be an open interval, and let $f : D \to \mathbb{R}$ be a function of class C^p in D. Then f is p-convex if and only if the derivative $f^{(p)}$ is increasing in D.*

Theorem 15.8.6. *Let $D \subset \mathbb{R}$ be an open interval, and let $f : D \to \mathbb{R}$ be a function of class C^{p+1} in D. Then f is p-convex if and only if the derivative $f^{(p+1)}$ is non-negative in D.*

15.9   Polynomial functions

The main objective of this section is to prove the basic theorem about the representation of polynomial functions with the aid of multiadditive functions (cf. 13.4). The results are due to S. Mazur-W. Orlicz [222] (cf. also McKiernan [223], Hosszú [146], Djoković [71], Székelyhidi [310]).

Given a function $F : \mathbb{R}^{pN} \to \mathbb{R}$, by the *diagonalization* of F we understand the function $f : \mathbb{R}^N \to \mathbb{R}$ arising from F by putting all the variables (from $\mathbb{R}^N$) equal:

$$f(x) = F(\underbrace{x, \ldots, x}_{p \text{ times}}), \quad x \in \mathbb{R}^N.$$

We start with some preliminary lemmas.

Lemma 15.9.1. *Let $F : \mathbb{R}^{kN} \to \mathbb{R}$ be a symmetric k-additive function, and let $f : \mathbb{R}^N \to \mathbb{R}$ be the diagonalization of F. Then, for every $x, h \in \mathbb{R}^N$,*

$$\Delta_h f(x) = \sum_{i=0}^{k-1} \binom{k}{i} F(\underbrace{x, \ldots, x}_{i}, \underbrace{h, \ldots, h}_{k-i}). \tag{15.9.1}$$

Proof. It is enough to prove that

$$F(\underbrace{x+h, \ldots, x+h}_{k}) = \sum_{i=0}^{k} \binom{k}{i} F(\underbrace{x, \ldots, x}_{i}, \underbrace{h, \ldots, h}_{k-i}), \tag{15.9.2}$$

as (15.9.1) is an immediate consequence of (15.9.2). The proof will be by induction on k. For $k = 1$ relation (15.9.2) becomes

$$F(x+h) = F(x) + F(h),$$

which is true in virtue of the additivity (i.e., 1-additivity) of F. Now assume (15.9.2) true for a $k \in \mathbb{N}$, and consider a symmetric $(k+1)$-additive function $F : \mathbb{R}^{(k+1)N} \to \mathbb{R}$. We have for every fixed $u \in \mathbb{R}^N$, according to the induction hypothesis,

$$F(\underbrace{x+h,\ldots,x+h}_{k},u) = \sum_{i=0}^{k} \binom{k}{i} F(\underbrace{x,\ldots,x}_{i}, \underbrace{h,\ldots,h}_{k-i},u),$$

whence, on setting $u = x + h$ and using the $(k+1)$-additivity and symmetry of F,

$$F(\underbrace{x+h,\ldots,x+h}_{k+1}) = \sum_{i=0}^{k} \binom{k}{i} F(\underbrace{x,\ldots,x}_{i},\underbrace{h,\ldots,h}_{k-i},x+h)$$

$$= \sum_{i=0}^{k} \binom{k}{i} F(\underbrace{x,\ldots,x}_{i},\underbrace{h,\ldots,h}_{k-i},x) + \sum_{i=0}^{k} \binom{k}{i} F(\underbrace{x,\ldots,x}_{i},\underbrace{h,\ldots,h}_{k+1-i})$$

$$= \sum_{i=0}^{k} \binom{k}{i} F(\underbrace{x,\ldots,x}_{i+1},\underbrace{h,\ldots,h}_{k-i}) + \sum_{i=0}^{k} \binom{k}{i} F(\underbrace{x,\ldots,x}_{i},\underbrace{h,\ldots,h}_{k+1-i}).$$

Replacing in the first sum $i+1$ by j, and in the second i by j, and then making use of the well-known identity $\binom{k}{j-1} + \binom{k}{j} = \binom{k+1}{j}$ we obtain

$$F(\underbrace{x+h,\ldots,x+h}_{k+1}) = F(\underbrace{x,\ldots,x}_{k+1}) + \sum_{j=1}^{k} \binom{k}{j-1} F(\underbrace{x,\ldots,x}_{j}, \underbrace{h,\ldots,h}_{k+1-j})$$

$$+ \sum_{j=1}^{k} \binom{k}{j} F(\underbrace{x,\ldots,x}_{j},\underbrace{h,\ldots,h}_{k+1-j}) + F(\underbrace{h,\ldots,h}_{k+1})$$

$$= F(\underbrace{x,\ldots,x}_{k+1}) + \sum_{j=1}^{k} \left[\binom{k}{j-1} + \binom{k}{j} \right] F(\underbrace{x,\ldots,x}_{j}, \underbrace{h,\ldots,h}_{k+1-j}) + F(\underbrace{h,\ldots,h}_{k+1})$$

$$= \sum_{j=0}^{k+1} \binom{k+1}{j} F(\underbrace{x,\ldots,x}_{j}, \underbrace{h,\ldots,h}_{k+1-j}).$$

Thus we have obtained (15.9.2) for $k+1$. Induction completes the proof. □

Lemma 15.9.2. *Let $F : \mathbb{R}^{kN} \to \mathbb{R}$ be a symmetric k-additive function, and let $f : \mathbb{R}^N \to \mathbb{R}$ be the diagonalization of F. For every $n \in \mathbb{N}$, $n \geqslant k$, and for every $x, h_1 \ldots, h_n \in \mathbb{R}^N$ we have*

$$\Delta_{h_1 \ldots h_n} f(x) = \begin{cases} k! \, F(h_1,\ldots,h_k) & \text{if } n = k, \\ 0 & \text{if } n > k. \end{cases} \tag{15.9.3}$$

Proof. The proof is by induction on k. If $k = 1$, then $F = f$ is an additive function, and

$$\Delta_h f(x) = f(x + h) - f(x) = f(x) + f(h) - f(x) = F(h)$$

for every $x, h \in \mathbb{R}^N$. Hence also, for $n > 1$,

$$\Delta_{h_1 \ldots h_n} f(x) = \Delta_{h_1 \ldots h_{n-1}} \Delta_{h_n} f(x) = \Delta_{h_1 \ldots h_{n-1}} f(h_n) = 0,$$

since $f(h_n)$, as a function of x, is constant.

Now assume that (15.9.3) is true for $1, \ldots, k \in \mathbb{N}$, and consider a $(k+1)$-additive function $F : \mathbb{R}^{(k+1)N} \to \mathbb{R}$, and its diagonalization $f : \mathbb{R}^N \to \mathbb{R}$. We have by Lemmas 15.9.1 and 15.1.1, for arbitrary $x, h_1 \ldots, h_{k+1} \in \mathbb{R}^N$,

$$\Delta_{h_1 \ldots h_{k+1}} f(x) = \Delta_{h_1 \ldots h_k} \Delta_{h_{k+1}} f(x)$$

$$= \sum_{i=0}^{k} \binom{k+1}{i} \Delta_{h_1 \ldots h_k} F(\underbrace{x, \ldots, x}_{i}, \underbrace{h_{k+1}, \ldots, h_{k+1}}_{k+1-i}).$$

For every i, $1 \leqslant i \leqslant k$, the function $F(\underbrace{x, \ldots, x}_{i}, \underbrace{h_{k+1}, \ldots, h_{k+1}}_{k+1-i})$ is i-additive, whereas for $i = 0$ the expression $F(h_{k+1}, \ldots, h_{k+1})$ (as a function of x) is constant. By the induction hypothesis we have ($0 \leqslant i \leqslant k$)

$$\Delta_{h_1 \ldots h_k} F(\underbrace{x, \ldots, x}_{i}, \underbrace{h_{k+1}, \ldots, h_{k+1}}_{k+1-i}) = \begin{cases} k! \, F(h_1, \ldots, h_{k+1}) & \text{if } i = k, \\ 0 & \text{if } i < k. \end{cases}$$

Hence

$$\Delta_{h_1 \ldots h_{k+1}} f(x) = (k+1) \, k! \, F(h_1, \ldots, h_{k+1}) = (k+1)! \, F(h_1, \ldots, h_{k+1}).$$

For $n > k + 1$ we have hence, since $F(h_1 \ldots, h_{k+1})$ (as a function of x) is constant, for arbitrary $x, h_1, \ldots, h_n \in \mathbb{R}^N$, according to Corollary 15.1.1

$$\Delta_{h_1 \ldots h_n} f(x) = \Delta_{h_{k+2} \ldots h_n} \Delta_{h_1 \ldots h_{k+1}} f(x) = (k+1)! \, \Delta_{h_{k+2} \ldots h_n} F(h_1, \ldots, h_{k+1}) = 0.$$

Consequently (15.9.3) is true also for $k + 1$, and induction completes the proof. $\quad\square$

Corollary 15.9.1. *Let $F : \mathbb{R}^{kN} \to \mathbb{R}$ be a symmetric k-additive function, and let $f : \mathbb{R}^N \to \mathbb{R}$ be the diagonalization of F. If $f = 0$ in $\mathbb{R}^N$, then $F = 0$ in $\mathbb{R}^{kN}$.*

Proof. Since $f = 0$, we have also for arbitrary $x, x_1, \ldots, x_k \in \mathbb{R}^N$

$$\Delta_{x_1 \ldots x_k} f(x) = 0,$$

whence by Lemma 15.9.2 $F(x_1, \ldots, x_k) = 0$. $\quad\square$

In the sequel by a 0-additive function we understand a constant. A diagonalization of such a function is also the same constant. Observe that Lemma 15.9.2 remains valid for $k = 0$.

Lemma 15.9.3. *Let $F_k : \mathbb{R}^{kN} \to \mathbb{R}$, $k = 0, \ldots, p$, be symmetric k-additive functions, and let $f_0, \ldots, f_p$ be the diagonalizations of $F_0, \ldots, F_p$, respectively. Let the function $f : \mathbb{R}^N \to \mathbb{R}$ be given by*

$$f(x) = \sum_{k=0}^{p} f_k(x), \quad x \in \mathbb{R}^N. \tag{15.9.4}$$

If f is bounded in $\mathbb{R}^N$, then $F_k = 0$ for $k = 1, \ldots, p$.

Proof. By Theorem 13.4.1 we have for every $x \in \mathbb{R}^N$, $\lambda \in \mathbb{Q}$, and for $k = 0, \ldots, p$

$$f_k(\lambda x) = \lambda^k f_k(x).$$

Hence

$$f(\lambda x) = \sum_{k=0}^{p} \lambda^k f_k(x). \tag{15.9.5}$$

For every $x \in \mathbb{R}^N$ expression (15.9.5) is a polynomial in λ, which, according to the assumptions of the lemma, remains bounded for all rational values of λ. But this is possible only if the polynomial is constant, i.e., if

$$f_k(x) = 0 \quad \text{for } k = 1, \ldots, p. \tag{15.9.6}$$

(15.9.6) holds for every $x \in \mathbb{R}^N$. In virtue of Corollary 15.9.1 this implies that

$$F_k = 0 \quad \text{for } k = 1, \ldots, p,$$

which was to be shown. $\qquad\square$

Theorem 15.9.1. *Let $F_k : \mathbb{R}^{kN} \to \mathbb{R}$, $k = 0, \ldots, p$, be symmetric k-additive functions, and let $f_0, \ldots, f_p$ be the diagonalizations of $F_0, \ldots, F_p$, respectively. If the function $f : \mathbb{R}^N \to \mathbb{R}$ is given by (15.9.4), then f is a polynomial function of order p.*

Proof. By (15.1.4) and Lemma 15.9.2 we have for every $x, h \in \mathbb{R}^N$

$$\Delta_h^{p+1} f_k(x) = 0, \quad k = 0, \ldots, p,$$

whence by Lemma 15.1.1 $\Delta_h^{p+1} f(x) = 0$, which means that f is a polynomial function of order p. $\qquad\square$

The converse is expressed by the following

Theorem 15.9.2. *Let $f : \mathbb{R}^N \to \mathbb{R}$ be a polynomial function of order p. Then there exist unique k-additive functions $F_k : \mathbb{R}^{kN} \to \mathbb{R}$, $k = 0, \ldots, p$, such that (15.9.4) holds, where $f_0, \ldots, f_p$ are the diagonalizations of $F_0, \ldots, F_p$, respectively.*

Proof. [5] First we prove the existence of F_k, $k = 0, \ldots, p$. The proof will be by induction on p. If $p = 1$, then f satisfies $\Delta_h^2 f(x) = 0$ for every $x, h \in \mathbb{R}^N$, which means,

---

[5] Albert-Baker [11]

as was pointed out in 15.3, that f satisfies the Jensen equation (13.2.1). By Theorem 13.2.1

$$f(x) = f_1(x) + f_0,$$

where $f_1 : \mathbb{R}^N \to \mathbb{R}$ is an additive function, and $f_0 \in \mathbb{R}$ is a constant. Consequently f admits representation (15.9.4).

Now assume that for every polynomial function $g : \mathbb{R}^N \to \mathbb{R}$ of an order $p-1 \in \mathbb{N}$ there exist symmetric k-additive functions $F_k : \mathbb{R}^{kN} \to \mathbb{R}$, $k = 0, \dots, p-1$, such that

$$g(x) = \sum_{k=0}^{p-1} f_k(x), \tag{15.9.7}$$

where $f_0, \dots, f_{p-1}$ are the diagonalizations of $F_0, \dots, F_{p-1}$, respectively. Let $f : \mathbb{R}^N \to \mathbb{R}$ be a polynomial function of order p, and define the function $F_p : \mathbb{R}^{pN} \to \mathbb{R}$ by

$$F_p(x_1, \dots, x_p) = \frac{1}{p!} \Delta_{x_1 \dots x_p} f(0), \quad x_1, \dots, x_p \in \mathbb{R}^N. \tag{15.9.8}$$

By Corollary 15.1.1 F_p is symmetric, and we have by Corollary 15.1.1, Lemma 15.1.1, Lemma 15.1.3 and Theorem 15.3.3 for every i, $1 \leqslant i \leqslant p$, and for every $x_1, \dots, x_p$, $y_i \in \mathbb{R}^N$

$$F_p(x_1, \dots, x_{i-1}, x_i + y_i, x_{i+1}, \dots, x_p) - F_p(x_1, \dots, x_p) - F_p(x_1, \dots, x_{i-1}, y_i, x_{i+1}, \dots, x_p)$$
$$= \frac{1}{p!} \Delta_{x_1 \dots x_{i-1}, x_{i+1} \dots x_p} \big[\Delta_{x_i + y_i} f(0) - \Delta_{x_i} f(0) - \Delta_{y_i} f(0) \big] = \frac{1}{p!} \Delta_{x_1 \dots x_p y_i} f(0) = 0,$$

which means that F_p is p-additive. Also, for arbitrary $x, x_1, \dots, x_p \in \mathbb{R}^N$ we have by Lemma 15.1.1 and Theorem 15.3.3

$$\Delta_{x_1 \dots x_p} f(x) - \Delta_{x_1 \dots x_p} f(0) = \Delta_{x_1 \dots x_p} \big[f(x) - f(0) \big]$$
$$= \Delta_{x_1 \dots x_p} \Delta_x f(0)$$
$$= \Delta_{x_1 \dots x_p x} f(0) = 0,$$

whence by (15.9.8)

$$\Delta_{x_1 \dots x_p} f(x) = p! \, F_p(x_1, \dots, x_p). \tag{15.9.9}$$

Now let $f_p : \mathbb{R}^N \to \mathbb{R}$ be the diagonalization of F_p, and put $g = f - f_p$. We have by Lemmas 15.1.1 and 15.9.2 and by (15.9.9), for arbitrary $x, h_1, \dots, h_p \in \mathbb{R}^N$,

$$\Delta_{h_1 \dots h_p} g(x) = \Delta_{h_1 \dots h_p} f(x) - \Delta_{h_1 \dots h_p} f_p(x) = \Delta_{h_1 \dots h_p} f(x) - p! \, F_p(h_1, \dots, h_p) = 0.$$

This implies that g is a polynomial function of order $p - 1$. By the induction hypothesis there exist symmetric k-additive functions $F_k : \mathbb{R}^{kN} \to \mathbb{R}$, $k = 0, \dots, p-1$, such that (15.9.7) holds, where $f_0, \dots, f_{p-1}$ are the diagonalizations of $F_0, \dots, F_{p-1}$, respectively. Hence we obtain (15.9.4), which ends the induction.

To prove the uniqueness suppose that f has also a representation

$$f(x) = \sum_{k=0}^{p} g_k(x), \tag{15.9.10}$$

where each g_k is the diagonalization of a symmetric k-additive function $G_k : \mathbb{R}^{kN} \to \mathbb{R}$, $k = 0, \ldots, p$. Every function $F_k - G_k$ is symmetric and k-additive (cf. Lemma 13.4.1), and $f_k - g_k$ is the diagonalization of $F_k - G_k$, $k = 0, \ldots, p$. By (15.9.4) and (15.9.10)

$$\sum_{k=0}^{p} \big(f_k(x) - g_k(x) \big) = 0, \tag{15.9.11}$$

whence by Lemma 15.9.3 $F_k = G_k$ for $k = 1, \ldots, p$, and (15.9.11) reduces to $f_0 - g_0 = 0$. Hence $F_0 = f_0 = g_0 = G_0$ so that $F_k = G_k$ for $k = 0, \ldots, p$, which completes the proof of the uniqueness. $\qquad\square$

Corollary 15.9.2. *Let* $f : \mathbb{R}^N \to \mathbb{R}$ *be a polynomial function. If* f *is bounded in* $\mathbb{R}^N$, *then* $f = \mathrm{const}$ *in* $\mathbb{R}^N$.

Proof. This results from Theorem 15.9.2 and Lemma 15.9.3. $\qquad\square$

Since the general construction of symmetric multiadditive functions is known (cf. Theorem 13.4.2 and Corollary 13.4.1), Theorems 15.9.1 and 15.9.2 yield implicitly the general construction of polynomial functions. Let us note also that R. Ger [100] proved that for every $p \in \mathbb{N}$ there exists a set $H_p \subset \mathbb{R}^N$ such that every function $g : H_p \to \mathbb{R}$ can be uniquely extended onto $\mathbb{R}^N$ to a polynomial function of order p.

Every polynomial function of order p is, in particular, p-convex, therefore every condition implying the continuity of a p-convex function (cf. 15.5) implies also the continuity of a polynomial function of order p (cf. also Gajda [94] concerning analogues of Theorems 9.3.7 and 9.8.1). For continuous polynomial functions we have the following

Theorem 15.9.3. *A function* $f : \mathbb{R}^N \to \mathbb{R}$ *is a continuous polynomial function of order* p *if and only if*

$$[x_1, \ldots, x_{p+2}; f] = 0 \tag{15.9.12}$$

for every system of $p + 2$ *distinct collinear points* $x_1, \ldots, x_{p+2} \in \mathbb{R}^N$.

Proof. A function $f : \mathbb{R}^N \to \mathbb{R}$ is a polynomial function of order p if and only if it is p-convex and p-concave, or, equivalently, if and only if the functions f and $-f$ are p-convex. If f is moreover continuous, then, by Theorem 15.6.1, this is equivalent to the condition that

$$[x_1, \ldots, x_{p+2}; f] \geqslant 0 \quad \text{and} \quad [x_1, \ldots, x_{p+2}; -f] \geqslant 0 \tag{15.9.13}$$

for every system of $p + 2$ distinct collinear points $x_1, \ldots, x_{p+2} \in \mathbb{R}^N$. But, as is easily seen from Theorem 15.2.1, we have

$$[x_1, \ldots, x_{p+2}; -f] = -[x_1, \ldots, x_{p+2}; f],$$

and so condition (15.9.13) is equivalent to (15.9.12). Conversely, if $f : \mathbb{R}^N \to \mathbb{R}$ satisfies (15.9.12), then, by Theorem 15.6.1, it is continuous and p-convex, and also $-f$ is p-convex, whence f is p-concave. Consequently f is a continuous polynomial function of order p. $\qquad\square$

Actually we can say much more. First we prove the following

Lemma 15.9.4. *If a function* $f : \mathbb{R}^N \to \mathbb{R}$,

$$f(x) = f(\xi_1, \dots, \xi_N)$$

is a polynomial separately in each variable ξ_i, $i = 1, \dots, N$, *then* f *is a polynomial jointly in all variables.*

Proof. The proof runs by induction on N. If $N = 1$, the lemma is trivial. Assume it true for an $N - 1 \in \mathbb{N}$, and let $f : \mathbb{R}^N \to \mathbb{R}$ be a function which is a polynomial separately in each variable. Let p be the maximal degree of f with respect to particular variables. Thus we have

$$f(\xi_1, \dots, \xi_N) = \sum_{i=0}^{p} A_i^N(\xi_1, \dots, \xi_{N-1})\xi_N^i = \cdots = \sum_{i=0}^{p} A_i^1(\xi_2, \dots, \xi_N)\xi_1^i. \tag{15.9.14}$$

In particular,

$$\sum_{i=0}^{p} A_i^N(\xi_1, \dots, \xi_{N-1})\xi_N^i = \sum_{i=0}^{p} A_i^{N-1}(\xi_1, \dots, \xi_{N-2}, \xi_N)\xi_{N-1}^i. \tag{15.9.15}$$

Setting in (15.9.15) $\xi_1 = \bar{\xi}_1, \dots, \xi_{N-2} = \bar{\xi}_{N-2}$ and $\xi_N = 0$, we obtain that $A_0^N(\bar{\xi}_1, \dots, \bar{\xi}_{N-2}, \xi_{N-1})$ is a polynomial in ξ_{N-1}. Setting next in (15.9.15) $\xi_1 = \bar{\xi}_1, \dots, \xi_{N-2} = \bar{\xi}_{N-2}$, and ξ_N in turn $1, \dots, p$, we obtain a system of linear equations for $A_i^N(\bar{\xi}_1, \dots, \bar{\xi}_{N-2}, \xi_{N-1})$, $i = 1, \dots, p$, whose determinant is

$$\begin{vmatrix} 1 & \cdots & 1 \\ 2 & \cdots & 2^p \\ \vdots & & \vdots \\ p & \cdots & p^p \end{vmatrix} \neq 0,$$

and the right-hand sides are polynomials in ξ_{N-1}. Consequently also the solutions are polynomials in ξ_{N-1}. Using other equations in (15.9.14) we obtain that all A_i^N, $i = 0, \dots, p$, are polynomials in each variable separately. By the induction hypothesis A_i^N, $i = 0, \dots, p$, are polynomials. Thus we see from (15.9.14) that also f is a polynomial. $\qquad\square$

Theorem 15.9.4. *A function* $f : \mathbb{R}^N \to \mathbb{R}$ *is a continuous polynomial function of order p if and only if f is a polynomial of degree at most p.*

Proof. Consider a monomial $f : \mathbb{R}^N \to \mathbb{R}$ of a degree $k \in \mathbb{N}$:

$$f(x) = f(\xi_1, \dots, \xi_N) = a\xi_1^{\alpha_1} \cdots \xi_N^{\alpha_N}, \quad \alpha_1 + \cdots + \alpha_N = k.$$

Put $\alpha_0 = 0$, and define the function $F : \mathbb{R}^{kN} \to \mathbb{R}$ as

$$F(x_1, \dots, x_k) = F(\xi_{11}, \dots, \xi_{1N}, \dots, \xi_{k1}, \dots, \xi_{kN}) = \frac{a}{k!} \sum \prod_{j=1}^{N} \prod_{m=\alpha_{j-1}+1}^{\alpha_{j-1}+\alpha_j} \xi_{imj},$$

where the sum extends over all permutations $(i_1, \ldots, i_k)$ of numbers $(1, \ldots, k)$, and we assume the convention that $\prod_{r+1}^{r} = 1$ for every $r \in \mathbb{N} \cup \{0\}$. It is easily seen that the function F is symmetric and k-additive, and f is the diagonalization of F.

In virtue of Theorem 15.9.1 f is a polynomial function of every order $p \geqslant k$. A polynomial of degree at most p is a sum of monomials of degrees $\leqslant p$, and thus of polynomial functions of order p. In virtue of Lemma 15.3.2 such a polynomial itself is a polynomial function of order p. Evidently every polynomial is continuous.

Conversely, suppose that a continuous function $f : \mathbb{R}^N \to \mathbb{R}$,

$$f(x) = f(\xi_1, \ldots, \xi_N),$$

is a polynomial function of order p. Let $l \subset \mathbb{R}^N$ be an arbitrary straight line, and take arbitrary distinct points $x_1, \ldots, x_{p+1} \in l$. By Lemma 15.2.4 and Theorem 15.9.3 we have for every $x \in l$

$$f(x) = P(x_1, \ldots, x_{p+1}; f \mid x). \tag{15.9.16}$$

(By the continuity of f and P (15.9.16) holds also if x equals one of $x_1, \ldots, x_{p+1}$). Consequently on every straight line f is a polynomial of degree at most p in the linear coordinate of x. In particular, f is a polynomial separately in each variable $\xi_1, \ldots, \xi_N$. By Lemma 15.9.4 f is a polynomial. It remains to estimate the degree of f.

Let the degree of f be q, and let

$$f(x) = \sum_{k=0}^{q} g_k(x), \tag{15.9.17}$$

where, for $k = 0, \ldots, q$, g_k is the homogeneous part of f of degree k. Since the degree of f is q, we have $g_q \neq 0$, and consequently there exists an $x_0 \in \mathbb{R}^N$, $x_0 \neq 0$, such that $g_q(x_0) \neq 0$. Put $h = (x_0/|x_0|) \operatorname{sgn} x_0$ and $\lambda_0 = |x_0| \operatorname{sgn} x_0$. We have $g_q(x_0) = g_q(\lambda_0 h) = \lambda_0^q g_q(h)$, whence $g_q(h) \neq 0$. By (15.9.17)

$$f(\lambda h) = \sum_{k=0}^{q} \lambda^k g_k(h),$$

which is a polynomial in λ if degree q. On the other hand, by (15.9.16), on the straight line

$$l = \{x \in \mathbb{R}^N \mid x = \lambda h, \ \lambda \in \mathbb{R}\}$$

f is a polynomial of a degree not exceeding p. Consequently $q \leqslant p$. $\qquad\square$

Thus, as we see, the polynomial functions generalize ordinary polynomials, and reduce to the latter under mild regularity assumptions. Consequently relation (15.3.3) has often been used to define polynomials in abstract spaces, and functions fulfilling this and related conditions have been thoroughly investigated. In this connection cf., e.g., Fréchet [92], Mazur-Orlicz [222], Highberg [141], Hille-Phillips [142], Hyers [152].

Exercises

1. Let $D \subset \mathbb{R}$ be an open interval, and let $f : D \to \mathbb{R}$ be a continuous p-convex function, $p > 1$. Show that for every r, $1 \leqslant r \leqslant p - 1$, there exist at most $r + 1$ points $c_1, \ldots, c_{r+1} \in D$, $c_1 < \cdots < c_{r+1}$, such that, if $D = (a, b)$, $c_0 = a$, $c_{r+2} = b$, then the function f is either $(p - r)$-convex or $(p - r)$-concave in each interval (c_i, c_{i+1}), $i = 0, \ldots, r+1$. (The statement is true also for $r = p$, if under 0-convex [0-concave] function we understand increasing [decreasing] function.)

2. Let $D \subset \mathbb{R}$ be an open interval, and let $f_n : D \to \mathbb{R}$, $n \in \mathbb{N}$, be continuous p-convex functions. Show that if the sequence $\{f_n\}_{n \in \mathbb{N}}$ converges pointwise on a dense subset of D, then it converges uniformly on every compact subset of D, and its limit is a continuous p-convex function.

3. Let $D \subset \mathbb{R}$ be an open interval, and let $f : D \to \mathbb{R}$ be a continuous p-convex function. Without using the results of 15.8 show that f satisfies a Lipschitz condition on every compact subinterval of D.

4. Let $F_k : \mathbb{R}^{kN} \to \mathbb{R}$ be symmetric k-additive functions, $k = 0, \ldots, p$, and let $f_0, \ldots, f_p$ be the diagonalizations of $F_0, \ldots, F_p$, respectively. Show that if the function $\sum_{k=0}^{p} f_k(x)$ is continuous in $\mathbb{R}^N$, then every function F_k, $k = 0, \ldots, p$, is continuous in $\mathbb{R}^{kN}$.

Chapter 16

Subadditive Functions

16.1 General properties

The Jensen inequality (5.3.1) is not the natural counterpart of the Cauchy equation (5.2.1). The natural counterpart of the Cauchy equation would be the inequality

$$f(x + y) \leqslant f(x) + f(y). \tag{16.1.1}$$

It seems so the more interesting and astounding that the convex functions and the additive functions share so many properties that such is not the case of additive functions and functions satisfying inequality (16.1.1).

Functions[1] $f : \mathbb{R}^N \to [-\infty, +\infty]$ satisfying (16.1.1) for all $x, y \in \mathbb{R}^N$ are called *subadditive*. Functions satisfying the converse inequality

$$f(x + y) \geqslant f(x) + f(y) \tag{16.1.2}$$

for all $x, y \in \mathbb{R}^N$ are called *superadditive*.

Every additive function is, in particular, subadditive and superadditive[2]. Thus there exist non-measurable subadditive functions. Many less trivial examples of subadditive functions may be found in Rosenbaum [269].

Subadditive functions (and superadditive functions) may also be considered which are defined not necessarily on the whole $\mathbb{R}^N$, but on subsets C of $\mathbb{R}^N$ such that $x + y \in C$ whenever $x, y \in C$, i.e.,

$$C + C \subset C. \tag{16.1.3}$$

According to Lemma 4.5.2, if a non-empty set $C \subset \mathbb{R}^N$ fulfils (16.1.3), then $(C, +)$ is semigroup (a subsemigroup of $(\mathbb{R}^N, +)$). An example of a set $C \subset \mathbb{R}^N$ fulfilling

---

[1] So we admit also infinite values of f. Hereby it is understood that the expression $f(x) + f(y)$ on the right hand side of (16.1.1) is always meaningful. Therefore a subadditive function cannot have both $+\infty$ and $-\infty$ in its range. For if we had $f(x) = +\infty$ and $f(y) = -\infty$ for some x, y in the domain of f, then $f(x) + f(y) = \infty - \infty$ would be meaningless.
[2] The converse is also true: a function $f : \mathbb{R}^N \to \mathbb{R}$ which is both subadditive and superadditive is additive.

(16.1.3) is a cone[3] as defined in 4.1. If $C \subset \mathbb{R}^N$ is a set fulfilling (16.1.3) and $f : C \to [-\infty, \infty]$ is a function satisfying inequality (16.1.1) resp. (16.1.2) for all $x, y \in C$, then f also will be called subadditive resp. superadditive. Subadditive functions defined on sets C other than $\mathbb{R}^N$ will only briefly be mentioned in the present work.

A function $f : \mathbb{R}^N \to \mathbb{R}$ which is subadditive and satisfies

$$f(nx) = nf(x) \qquad \text{for } n \in \mathbb{N}, \, x \in \mathbb{R}^N, \tag{16.1.4}$$

will be called *sublinear* (Berz [31]). Sublinear functions are considered in 16.4–16.5.

Except in the last section we will be mainly concerned with subadditive functions $f : C \to \mathbb{R}$, i.e., with functions with only finite values. Subadditive functions admitting also infinite values will be referred to as *infinitary* (cf. Rosenbaum [269], Hille-Phillips [142]). Such functions will be treated in the last section.

First let us note the following two obvious lemmas.

Lemma 16.1.1. *If $C \subset \mathbb{R}^N$ is a set fulfilling (16.1.3), $f_1, f_2 : C \to [-\infty, \infty]$ are subadditive functions, and $a, b \in [0, \infty)$ are non-negative constants, then the function $f = af_1 + bf_2$, if well defined, is subadditive.*

Lemma 16.1.2. *If $C \subset \mathbb{R}^N$ is a set fulfilling (16.1.3), and $f : C \to [-\infty, \infty]$ is a subadditive function, then the function $-f$ is superadditive, and conversely, if f is superadditive, then $-f$ is subadditive.*

Because of Lemma 16.1.2 we will not deal with superadditive functions. In the present section we prove some elementary general properties of subadditive functions (Hille-Phillips [142], Cooper [49]).

Lemma 16.1.3. *Let $C \subset \mathbb{R}^N$ be a set fulfilling (16.1.3) such that $0 \in C$, and let $f : C \to \mathbb{R}$ be a subadditive function. Then $f(0) \geqslant 0$.*

Proof. We have from (16.1.1) for $x = y = 0$

$$f(0) \leqslant f(0) + f(0),$$

whence $f(0) \geqslant 0$. $\qquad\qquad\qquad\qquad\qquad\qquad\qquad\qquad\qquad\qquad\qquad\qquad$ □

Lemma 16.1.4. *Let $C \subset \mathbb{R}^N$ be a cone, and let $f : C \to [-\infty, \infty]$ be a subadditive function. Then*

$$f(nx) \leqslant nf(x), \tag{16.1.5}$$

$$f\left(\frac{1}{n}x\right) \geqslant \frac{1}{n}f(x) \tag{16.1.6}$$

for all $x \in C$ and $n \in \mathbb{N}$.

---

[3] All theorems in the present chapter where f is supposed to be defined on a cone $C \subset \mathbb{R}^N$ remain valid if C is replaced by $C_0 = C \setminus \{0\}$.

Proof. For $n = 1$ (16.1.5) is trivial. Assume (16.1.5) to hold for an $n \in \mathbb{N}$ and for all $x \in C$. Then by (16.1.1)

$$f\left((n+1)x\right) = f(nx + x) \leqslant f(nx) + f(x) \leqslant nf(x) + f(x) = (n+1)f(x).$$

It follows by induction that (16.1.6) holds for all $n \in \mathbb{N}$ and $x \in C$. Replacing in (16.1.5) x by $\frac{1}{n}x$ we obtain (16.1.6). $\qquad\square$

Lemma 16.1.5. *Let $f : \mathbb{R}^N \to \mathbb{R}$ be a subadditive function. Then $f(-x) \geqslant -f(x)$ for $x \in \mathbb{R}^N$.*

Proof. We have by (16.1.1) and Lemma 16.1.3 for arbitrary $x \in \mathbb{R}^N$

$$0 \leqslant f(0) = f(x - x) \leqslant f(x) + f(-x),$$

whence $f(-x) \geqslant -f(x)$. $\qquad\square$

Lemma 16.1.6. *Let $C \subset \mathbb{R}^N$ be a set fulfilling (16.1.3), and let $f_\alpha : C \to [-\infty, +\infty]$, $\alpha \in A$, be a family of subadditive functions. Let, for $x \in C$, $f(x) = \sup_\alpha f_\alpha(x)$. If $f(x) > -\infty$ for all $x \in C$, or $f(x) < \infty$ for all $x \in C$, then f is subadditive.*

Proof. Fix arbitrary $x, y \in C$. By the subadditivity of f_α and the definition of f we have

$$f_\alpha(x + y) \leqslant f_\alpha(x) + f_\alpha(y) \leqslant f(x) + f(y)$$

for all $\alpha \in A$, whence

$$f_\alpha(x + y) \leqslant f(x) + f(y)$$

for all $\alpha \in A$. That the supremum over α we obtain (16.1.1). $\qquad\square$

Lemma 16.1.7. *Let $C \subset \mathbb{R}^N$ be a set fulfilling (16.1.3), and let $f_n : C \to [-\infty, +\infty]$, $n \in \mathbb{N}$, be subadditve functions. Assume that for every $x \in C$ there exists a (finite or infinite) limit $f(x) = \lim_{n \to \infty} f_n(x)$. If $f(x) > -\infty$ for all $x \in C$, or $f(x) < \infty$ for all $x \in C$, then f is subadditive.*

Proof. We have for all $x, y \in C$

$$f_n(x + y) \leqslant f_n(x) + f_n(y), \qquad n \in \mathbb{N},$$

whence (16.1.1) follows on letting $n \to \infty$. $\qquad\square$

Lemma 16.1.8. *Let $f : \mathbb{R}^N \to \mathbb{R}$ be a subadditive function. If f is even, then $f(x) \geqslant 0$ for $x \in \mathbb{R}^N$.*

Proof. By Lemma 16.1.5 we have for every $x \in \mathbb{R}^N$

$$f(x) = f(-x) \geqslant -f(x),$$

i.e., $2f(x) \geqslant 0$, whence also $f(x) \geqslant 0$. $\qquad\square$

Lemma 16.1.9. *Let $f : \mathbb{R}^N \to \mathbb{R}$ be a subadditive function. If f is odd, then f is additive.*

Proof. By (16.1.1)

$$f(x) = f(x + y - y) \leqslant f(x + y) + f(-y) = f(x + y) - f(y)$$

for arbitrary $x, y \in \mathbb{R}^N$, since, due to the fact that f is odd, we have $f(-y) = -f(y)$. Hence

$$f(x) + f(y) \leqslant f(x + y).$$

This together with (16.1.1) yields (5.2.1).   □

Lemma 16.1.10. *Let $f : \mathbb{R}^N \to \mathbb{R}$ be a sublinear function. Then*

$$f(0) = 0 \tag{16.1.7}$$

and

$$f(\lambda x) = \lambda f(x) \tag{16.1.8}$$

for all $x \in \mathbb{R}^N$ and $\lambda \in \mathbb{Q} \cap [0, \infty)$.

Proof. Let $\lambda \in \mathbb{Q} \cap (0, \infty)$, $\lambda = p/q$, $p, q \in \mathbb{N}$. Then we have by (16.1.4), for arbitrary $x \in \mathbb{R}^N$,

$$pf(x) = f(px) = f(q\lambda x) = qf(\lambda x),$$

whence (16.1.8) follows. In particular, for $\lambda = 2$ and $x = 0$, we obtain

$$f(0) = 2f(0),$$

which implies (16.1.7). Thus (16.1.8) is valid also for $\lambda = 0$, which completes the proof.   □

Lemma 16.1.11. *Let $f : \mathbb{R}^N \to \mathbb{R}$ be a sublinear function. Then f is convex.*

Proof. We have by (16.1.1) and Lemma 16.1.10, for arbitrary $x, y \in \mathbb{R}^N$,

$$f\left(\frac{x + y}{2}\right) = \frac{1}{2}f(x + y) \leqslant \frac{1}{2}\left[f(x) + f(y)\right] = \frac{f(x) + f(y)}{2},$$

which means that f is convex.   □

16.2   Boundedness. Continuity

For a function $f : \mathbb{R}^N \to \mathbb{R}$ the functions $m_f : \mathbb{R}^N \to [-\infty\, \infty)$ and $M_f : \mathbb{R}^N \to (-\infty, +\infty]$ are defined by (6.3.1) and (6.3.2), respectively. We always have

$$m_f(x) \leqslant f(x) \leqslant M_f(x), \qquad x \in \mathbb{R}^N, \tag{16.2.1}$$

and f is continuous at a point x if and only if $m_f(x) = M_f(x)$ (cf. Exercise 6.8). In this section $K(t, r)$ denotes the open ball centered at t and with the radius r.

Lemma 16.2.1. *Let $f : \mathbb{R}^N \to \mathbb{R}$ be a subadditive function. Then the functions m_f and M_f also are subadditive.*

Proof. We will prove the subadditivity of m_f, the proof for M_f is similar. Take an $r > 0$, arbitrary $x, y \in \mathbb{R}^N$, and $a, b \in \mathbb{R}$ such that

$$a > \inf_{K(x,r)} f, \qquad b > \inf_{K(y,r)} f.$$

There exist $u \in K(x, r)$ and $v \in K(y, r)$ such that

$$a > f(u) \geqslant \inf_{K(x,r)} f, \qquad b > f(v) \geqslant \inf_{K(y,r)} f.$$

Then we have $u + v \in K(x + y, 2r)$, whence

$$\inf_{K(x+y,2r)} f \leqslant f(u + v) \leqslant f(u) + f(v) < a + b.$$

When $a \to \inf\limits_{K(x,r)} f$ and $b \to \inf\limits_{K(y,r)} f$, we obtain hence

$$\inf_{K(x+y,2r)} f \leqslant \inf_{K(x,r)} f + \inf_{K(y,r)} f.$$

Letting $r \to 0$, we get

$$m_f(x + y) \leqslant m_f(x) + m_f(y),$$

i.e., m_f is subadditive. $\qquad\qquad\square$

Lemma 16.2.2. *Let $f : \mathbb{R}^N \to \mathbb{R}$ be subadditive. Then, for every $x \in \mathbb{R}^N$,*

$$0 \leqslant M_f(x) - m_f(x) \leqslant M_f(0). \qquad (16.2.2)$$

Proof. The first inequality in (16.2.2) results from (16.2.1). Now take an arbitrary $r > 0$, an arbitrary $x \in \mathbb{R}^N$, and arbitrary $a, b \in \mathbb{R}$ such that

$$a > \inf_{K(x,r)} f, \qquad b < \sup_{K(x,r)} f.$$

Then there exist $u, v \in K(x, r)$ such that

$$a > f(u) \geqslant \inf_{K(x,r)} f, \qquad b < f(u) \leqslant \sup_{K(x,r)} f.$$

Hence

$$b - a < f(v) - f(u).$$

On the other hand, $f(v) = f(u+v-u) \leqslant f(u)+f(v-u)$, whence $f(v)-f(u) \leqslant f(v-u)$. Thus $b - a < f(v - u)$. But $v - u \in K(0, 2r)$, whence $f(v - u) \leqslant \sup\limits_{K(0,2r)} f$ and

$$b - a < \sup_{K(0,2r)} f.$$

Letting a tend to $\inf\limits_{K(x,r)} f$, and b tend to $\sup\limits_{K(x,r)} f$, we obtain hence

$$\sup_{K(x,r)} f - \inf_{K(x,r)} f \leqslant \sup_{K(0,2r)} f.$$

When $r \to 0$, we get hence the second inequality in (16.2.2). $\qquad\square$

Theorem 16.2.1. *Let* $f : \mathbb{R}^N \to \mathbb{R}$ *be subadditive. If* f *is continuous at zero and* $f(0) = 0$, *then* f *is continuous in* $\mathbb{R}^N$.

Proof. The continuity of f at zero implies in view of (16.2.1) that

$$m_f(0) = M_f(0) = f(0) = 0.$$

Condition (16.2.2) then yields $M_f(x) = m_f(x)$ for every $x \in \mathbb{R}^N$, i.e., f is continuous in $\mathbb{R}^N$. $\qquad\square$

Theorem 16.2.2. *Let* $f : \mathbb{R}^N \to \mathbb{R}$ *be subadditive. If* f *is locally bounded above at a point, then* f *is locally bounded at every point of* $\mathbb{R}^N$.

Proof. Let f be locally bounded above at a point $t_0 \in \mathbb{R}^N$. Suppose that f is not locally bounded above at 0. Then there exists a sequence $\{t_n\}_{n \in \mathbb{N}}$ such that $\lim\limits_{n \to \infty} t_n = 0$ and $\lim\limits_{n \to \infty} f(t_n) = \infty$. But then $\lim\limits_{n \to \infty} (t_0 + t_n) = t_0$, and by (16.1.1)

$$f(t_n) = f(t_0 + t_n - t_0) \leqslant f(t_0 + t_n) + f(-t_0),$$

i.e., $f(t_0 + t_n) \geqslant f(t_n) - f(-t_0)$, whence $\lim\limits_{n \to \infty} f(t_0 + t_n) = \infty$. But this means that f is locally unbounded above at t_0, contrary to the hypothesis. Consequently f is locally bounded above at 0 so that $M_f(0) < \infty$.

By Lemma 16.2.2 $M_f(x) - m_f(x) < \infty$, which implies that $M_f(x) < \infty$ and $m_f(x) > -\infty$, for every $x \in \mathbb{R}^N$. Consequently f is locally bounded (above and below) at every point $x \in \mathbb{R}^N$. $\qquad\square$

Theorem 16.2.2 is fully analogous to the situation for convex functions (cf. Theorem 6.2.3). But now the local boundedness of f does not imply its continuity, as may be seen from the following

Example 16.2.1. Let $\mathbb{Q}^N \subset \mathbb{R}^N$ be the set of those $x \in \mathbb{R}^N$ which have rational coordinates, and let the function $f : \mathbb{R}^N \to \mathbb{R}$ be given by

$$f(x) = \begin{cases} 0 & \text{if } x \in \mathbb{Q}^N, \\ b & \text{if } x \notin \mathbb{Q}^N, \end{cases}$$

where $b > 0$. Then f is subadditive and bounded on $\mathbb{R}^N$, but discontinuous at every point of $\mathbb{R}^N$.

The condition of local boundedness above in Theorem 16.2.2 may still be weakened.

Theorem 16.2.3. *Let* $f : \mathbb{R}^N \to \mathbb{R}$ *be subadditive. If* f *is bounded above on a set* $T \subset \mathbb{R}^N$ *such that* $m_i(T) > 0$, *or* T *is of the second category and with the Baire property, then* f *is locally bounded at every point of* $\mathbb{R}^N$.

Proof. If $t \in T + T$, then $t = x + y$, where $x, y \in T$. Hence

$$f(t) = f(x + y) \leqslant f(x) + f(y) \tag{16.2.3}$$

so that f is bounded above on $T + T$. By Theorems 3.7.1 and 2.9.1 $\operatorname{int}(T + T) \neq \varnothing$. The theorem results now from Theorem 16.2.2. $\qquad\square$

The operation S_n is defined in 9.2.

Theorem 16.2.4. *Let* $f : \mathbb{R}^N \to \mathbb{R}$ *be subadditive. If* f *is bounded above on a set* $T \subset \mathbb{R}^N$ *such that for a certain* $n \in \mathbb{N}$ *we have* $m_i(S_n(T)) > 0$ *or* $S_n(T)$ *is of the second category and with the Baire property, then* f *is locally bounded at every point of* $\mathbb{R}^N$.

Proof. By a repeated use of (16.2.3) we find that f is bounded above on $S_n(T)$, and the theorem follows from Theorem 16.2.3. $\qquad\square$

If $f : \mathbb{R}^N \to \mathbb{R}$ is a measurable subadditive function, then f is bounded above on a set $T \subset \mathbb{R}^N$ of positive measure (cf. the proof of Theorem 9.4.1), and hence is locally bounded at every point of $\mathbb{R}^N$. But actually we can prove this result for functions whose domain of definition is not necessarily the whole $\mathbb{R}^N$.

Lemma 16.2.3. *Let* $C \subset \mathbb{R}^N$ *be a cone such that* $\operatorname{int} C \neq \varnothing$, *and let* $f : C \to [-\infty, \infty)$ *be a measurable subadditive function. Then* f *is locally bounded above at every point of* $\operatorname{int} C$.

Proof. Suppose that f is not locally bounded above at a point $t_0 \in \operatorname{int} C$. We can find an $r > 0$ such that $K(t_0, 2r) \subset \operatorname{int} C$. Write

$$D = K(0, r) \cap \operatorname{int} C. \qquad (16.2.4)$$

C is a convex set, so we have by Theorem 5.1.6 $\operatorname{cl int} C = \operatorname{cl} C$, and $0 \in C \subset \operatorname{cl} C = \operatorname{cl int} C$. Consequently $D \neq \varnothing$. On the other hand evidently D is open, whence $m(D) > 0$.

There exists a sequence $\{t_n\}_{n \in \mathbb{N}}$ converging to t_0 and such that

$$f(t_n) \geqslant 2n, \qquad n \in \mathbb{N}. \qquad (16.2.5)$$

For every $n \in \mathbb{N}$ define the set E_n as

$$E_n = \{t \in K(t_0, 2r) \cup D \mid f(t) \geqslant n\}.$$

Since f is measurable, also E_n is measurable for every $n \in \mathbb{N}$. For n sufficiently large, say $n > n_0$, we have $t_n \in K(t_0, r)$. Take an $n > n_0$, an $x \in D \backslash E_n$, and put $y_n = t_n - x$. We have by (16.2.4)

$$|y_n - t_0| = |t_n - x - t_0| \leqslant |t_n - t_0| + |x| < 2r,$$

and so $y_n \in K(t_0, 2r)$. Moreover, by (16.1.1) and (16.2.5)

$$2n \leqslant f(t_n) = f(t_n - x + x) \leqslant f(t_n - x) + f(x) = f(y_n) + f(x). \qquad (16.2.6)$$

Since $x \notin E_n$, we have $f(x) < n$, whence $f(y_n) > n$. Consequently $y_n \in E_n$, i.e., $x = t_n - y_n \in t_n - E_n$. Thus, if $x \in D$, then for $n > n_0$ either $x \in E_n$, or $x \in t_n - E_n$, whence

$$D \subset E_n \cup (t_n - E_n), \qquad n > n_0,$$

and by Corollary 3.2.2

$$m(D) \leqslant m[E_n \cup (t_n - E_n)] \leqslant m(E_n) + m(t_n - E_n) = 2m(E_n), \qquad n > n_0.$$

Hence

$$m(E_n) \geqslant \frac{1}{2} m(D), \qquad n > n_0,$$

and (cf. Exercise 3.4), since $E_{n+1} \subset E_n \subset K(t_0, 2r) \cup K(0, r)$ so that $m(E_n) < \infty$, $n \in \mathbb{N}$,

$$m\left(\bigcap_{n=1}^{\infty} E_n\right) = \lim_{n \to \infty} m(E_n) \geqslant \frac{1}{2} m(D) > 0.$$

Thus $\bigcap_{n=1}^{\infty} E_n \neq \varnothing$, and if an $x_0 \in \bigcap_{n=1}^{\infty} E_n$, then $x_0 \in E_n$ for all $n \in \mathbb{N}$, and $f(x_0) \geqslant n$ for all $n \in \mathbb{N}$, which contradicts the fact that $f < \infty$ in C. $\qquad \square$

Theorem 16.2.5. *Let $C \subset \mathbb{R}^N$ be a cone such that $\operatorname{int} C \neq \varnothing$, and let $f : C \to [-\infty, \infty)$ be a measurable subadditive function. Then f is locally bounded at every point of $\operatorname{int} C$.*

Proof. By Lemma 16.2.3 f is locally bounded above at every point of $\operatorname{int} C$. Now suppose that f is not locally bounded below at a point $t_0 \in \operatorname{int} C$. Then there exists a sequence $\{t_n\}_{n \in \mathbb{N}}$ such that $\lim_{n \to \infty} t_n = t_0$ and $\lim_{n \to \infty} f(t_n) = -\infty$. Let $U \subset \operatorname{int} C$ be a neighbourhood of t_0 such that f is bounded above on U:

$$f(x) \leqslant M \qquad \text{for } x \in U.$$

There exists an $n_0 \in \mathbb{N}$ such that $2t_0 - t_n \in U$ for $n > n_0$. Thus we have by (16.1.1) for $n > n_0$

$$f(2t_0) = f(t_n + 2t_0 - t_n) \leqslant f(t_n) + f(2t_0 - t_n) \leqslant f(t_n) + M,$$

whence $-f(t_n) \leqslant M - f(2t_0)$, and

$$\infty = \lim_{n \to \infty} \left[-f(t_n) \right] \leqslant M - f(2t_0) < \infty.$$

The contradiction obtained completes the proof. $\qquad \square$

Theorem 16.2.6. *Let $f : \mathbb{R}^N \to \mathbb{R}$ be subadditive. If f is either measurable, or bounded above on a set $T \subset \mathbb{R}^N$ such that for a certain $n \in \mathbb{N}$ either $m_i\big(S_n(T)\big) > 0$, or $S_n(T)$ is of the second category and with the Baire property, then f is bounded on every bounded subset of $\mathbb{R}^N$.*

Proof. By Theorems 16.2.4 and 16.2.5 f is locally bounded at every point of $\mathbb{R}^N$. For every $x \in \mathbb{R}^N$ let K_x be an open ball centered at x such that f is bounded on K_x. Let $A \subset \mathbb{R}^N$ be a bounded set. Then the set $\operatorname{cl} A$ is compact. Clearly $\operatorname{cl} A \subset \bigcup_{x \in \operatorname{cl} A} K_x$, so there exists a finite sequence of points $x_1, \ldots, x_n \in \operatorname{cl} A$ such that $A \subset \operatorname{cl} A \subset \bigcup_{i=1}^{n} K_{x_i}$.

The function f, being bounded on every K_{x_i}, $i = 1, \ldots, n$, is also bounded on $\bigcup_{i=1}^{n} K_{x_i}$, and hence on A. $\qquad \square$

Theorem 16.2.7. *Let $f : \mathbb{R}^N \to [-\infty, \infty]$ be subadditive, and put $\lambda = \liminf_{x \to 0} f(x)$. Then $\lambda \geqslant 0$, or $\lambda = -\infty$. If $|\lambda| = \infty$, then f is either non-measurable, or infinitary.*

Proof. There exists a sequence $\{x_n\}_{n \in \mathbb{N}}$ of points $x_n \in \mathbb{R}^N$ such that

$$\lim_{n \to \infty} x_n = 0 \quad \text{and} \quad \lim_{n \to \infty} f(x_n) = \lambda.$$

By Lemma 16.1.4

$$\lambda \leqslant \liminf_{n \to \infty} f(2x_n) \leqslant \liminf_{n \to \infty} 2f(x_n) = 2 \liminf_{n \to \infty} f(x_n) = 2\lambda.$$

If λ is finite, this yields $\lambda \geqslant 0$. If f is measurable and finite, $f : \mathbb{R}^N \to \mathbb{R}$, then, by Theorem 16.2.5, f is locally bounded at 0, whence $\lambda = \liminf_{x \to 0} f(x) \neq \pm\infty$. Hence if f is measurable and $|\lambda| = \infty$, then f must be infinitary. $\qquad\square$

For subadditive functions assuming negative values we have the following

Theorem 16.2.8. *Let $f : \mathbb{R}^N \to \mathbb{R}$ be a measurable subadditive function, and suppose that there exists an $x_0 \in \mathbb{R}^N$ such that $f(x_0) < 0$. Then $f(tx_0) < 0$ for all $t \in \mathbb{R}$ sufficiently large, and $f(tx_0) \geqslant 0$ for all $t \leqslant 0$.*

Proof. By Theorem 16.2.6 there exists an $M > 0$ such that

$$|f(tx_0)| \leqslant M \qquad \text{for } t \in [1, 2]. \tag{16.2.7}$$

Now take an arbitrary $t > 2$, and let $n = [t]$ be the integral part of t so that $n \leqslant t < n + 1$. Hence $t - (n-1) \in [1, 2]$. Thus by (16.1.1), (16.2.7), and by Lemma 16.1.4

$$f(tx_0) = f\big((n-1)x_0 + (t-n+1)x_0\big) \leqslant f\big((n-1)x_0\big) + f\big((t-n+1)x_0\big)$$
$$\leqslant (n-1)f(x_0) + M = (n+1)f(x_0) + M - 2f(x_0) \leqslant tf(x_0) + M - 2f(x_0).$$

Since $f(x_0) < 0$, $f(tx_0)$ becomes negative for large values of $t > 0$. By Lemma 16.1.5 $f(-tx_0) \geqslant -f(tx_0)$, whence $f(tx_0) > 0$ for large negative values of t. Suppose that there exists a $t_0 < 0$ such that $f(t_0 x_0) \leqslant 0$. By Lemma 16.1.4

$$f(nt_0 x_0) \leqslant nf(t_0 x_0) \leqslant 0 \qquad \text{for } n \in \mathbb{N},$$

which contradicts the fact (already established) that $f(tx_0) > 0$ for large negative t. Thus $f(tx_0) > 0$ for $t < 0$. Together with Lemma 16.1.3 this yields $f(tx_0) \geqslant 0$ for $t \leqslant 0$. $\qquad\square$

This proof shows also that, for large values of t, the function $f(tx_0)$ is majorized by a linear function of t. A better information about the behavior of $f(tx_0)$ for $t \to \infty$ (without any assumption about the sign of $f(x_0)$) is obtained from the following

Theorem 16.2.9. *Let $f : \mathbb{R}^N \to \mathbb{R}$ be a measurable subadditive function. Then for every $x \in \mathbb{R}^N$ there exists the limit*

$$F(x) = \lim_{t \to \infty} \frac{f(tx)}{t}. \tag{16.2.8}$$

The function F is finite, continuous in $\mathbb{R}^N$, positively homogeneous, and subadditive[4].

---
[4] And hence sublinear.

Proof. If $x = 0$, then $F(0) = 0$ exists and is finite. Now fix an $x \in \mathbb{R}^N$, $x \neq 0$, and put

$$\beta = \inf_{t>0} \frac{f(tx)}{t}.$$

For every $t > 0$ the expression $f(tx)/t$ is finite, whence $-\infty \leqslant \beta < \infty$. Take an arbitrary $b > \beta$ and choose a $t_0 > 0$ such that $f(t_0 x)/t_0 < b$. Take an arbitrary $t > 3t_0$ and let $m = [t/t_0]$ be the integral part of t/t_0. Put $n = m - 2$. Clearly $n \in \mathbb{N}$, and by Theorem 16.2.6 f is bounded on the segment $\overline{2t_0 x, 3t_0 x}$:

$$|f(sx)| \leqslant M \qquad \text{for } s \in [2t_0, 3t_0].$$

In particular, $t - nt_0 \in [2t_0, 3t_0]$. Hence, by Lemma 16.1.4

$$f(tx) = f\big((nt_0 + t - nt_0)x\big) \leqslant f(nt_0 x) + f\big((t - nt_0)x\big) \leqslant nf(t_0 x) + M,$$

and

$$\beta \leqslant \frac{f(tx)}{t} \leqslant \frac{nt_0}{t} \frac{f(t_0 x)}{t_0} + \frac{M}{t} < \frac{nt_0}{t} b + \frac{M}{t}. \tag{16.2.9}$$

When $n \to \infty$, the expression nt_0/t approaches[5] 1, and we get from (16.2.9)

$$\beta \leqslant \liminf_{t\to\infty} \frac{f(tx)}{t} \leqslant \limsup_{t\to\infty} \frac{f(tx)}{t} \leqslant b.$$

Letting $b \to \beta$ we obtain hence in view of (16.2.8)

$$F(x) = \beta < \infty. \tag{16.2.10}$$

Now put

$$\alpha = \sup_{t<0} \frac{f(tx)}{t}.$$

We have $-\infty < \alpha \leqslant \infty$. But

$$\alpha = -\inf_{t>0} \frac{f\big(t(-x)\big)}{t} = -F(-x).$$

So

$$-F(-x) > -\infty. \tag{16.2.11}$$

By Lemma 16.1.5 we have $f(tx) + f(-tx) \geqslant 0$, whence

$$F(x) + F(-x) = \lim_{t\to\infty} \left[\frac{f(tx)}{t} + \frac{f(-tx)}{t} \right] \geqslant 0.$$

Consequently $F(x) \geqslant -F(-x) > -\infty$, by (16.2.11). Hence by (16.2.10) $|F(x)| < \infty$, i.e., F is finite.

---

[5] This follows from the inequalities $n + 2 \leqslant t/t_0 < n + 3$, whence $\dfrac{n}{n+3} < \dfrac{nt_0}{t} \leqslant \dfrac{n}{n+2}$.

Take arbitrary $x, y \in \mathbb{R}^N$. We have for all $t > 0$

$$f\big(t(x+y)\big) = f(tx + ty) \leqslant f(tx) + f(ty),$$

whence

$$F(x+y) = \lim_{t \to \infty} \frac{f\big(t(x+y)\big)}{t} \leqslant \lim_{t \to \infty} \left[\frac{f(tx)}{t} + \frac{f(ty)}{t} \right] = F(x) + F(y).$$

Thus F is subadditive.

Similarly, for arbitrary $x \in \mathbb{R}^N$ and $u > 0$ we have

$$F(ux) = \lim_{t \to \infty} \frac{f(tux)}{t} = u \lim_{t \to \infty} \frac{f(tux)}{tu} = uF(x).$$

So F is positively homogeneous, and, in particular, sublinear. By Lemma 16.1.11 F is convex. Since f is measurable, for every fixed $t > 0$ the function $f(tx)/t$ (as a function of x) is measurable. Consequently also function (16.2.8) is measurable. By Theorem 9.4.2 F is continuous in $\mathbb{R}^N$. □

Theorem 16.2.9 (Rosenbaum [269], Hille-Phillips [142]) shows that a subadditive function cannot increase too fast as $x \to \infty$. If f is defined only on a cone $C \subset \mathbb{R}^N$, the theorem, and the argument presented, remain valid, except that now we cannot prove that $F(x) > -\infty$. In such a cone it can actually happen that $F(x) = -\infty$. Let $N = 1$, $C = [0, \infty)$, let $g : C \to \mathbb{R}$ be an arbitrary increasing positive function, and let $f(x) = -xg(x)$ for $x \in C$. Then f is subadditive, and by a suitable choice of g we can make f to tend to $-\infty$, when $x \to \infty$, as fast as we wish. In particular, if $\lim_{x \to \infty} g(x) = \infty$, then for our f function (16.2.8) is identically $-\infty$.

16.3 Differentiability

Theorem 16.2.1 shows that in the theory of subadditive functions the origin plays a distinguished role. In particular, if $f(0) = 0$, then we can infer from the continuity of f at zero about its continuity at any other point. With the differentiability the situation is similar (cf., in particular, Wetzel [322]).

Theorem 16.3.1. *Let $f : \mathbb{R}^N \to \mathbb{R}$ be subadditive, and let $g : \mathbb{R}^N \to \mathbb{R}$ be a function such that $g(0) = 0$ and g has the Stolz differential at 0. If*

$$f(x) \leqslant g(x), \qquad x \in \mathbb{R}^N, \tag{16.3.1}$$

then there exists a $c \in \mathbb{R}^N$ such that

$$f(x) = cx, \qquad x \in \mathbb{R}^N. \tag{16.3.2}$$

In particular, f is of class C^1 in $\mathbb{R}^N$.

Proof. Put $c = \nabla g(0)$. Then

$$g(x) = cx + r(x), \qquad x \in \mathbb{R}^N, \tag{16.3.3}$$

where

$$\lim_{x \to 0} \frac{r(x)}{|x|} = 0 \tag{16.3.4}$$

(cf. 7.7). Now take an arbitrary $x \in \mathbb{R}^N$, $x \neq 0$. We have for $n \in \mathbb{N}$ by Lemma 16.1.4 and by (16.3.1) and (16.3.3)

$$f(x)/n \leqslant f(x/n) \leqslant g(x/n) = cx/n + r(x/n),$$

whence

$$f(x) - cx \leqslant nr(x/n). \tag{16.3.5}$$

By (16.3.4)

$$\lim_{n \to \infty} nr(x/n) = |x| \lim_{n \to \infty} \frac{r(x/n)}{|x/n|} = 0,$$

and so we obtain from (16.3.5), on letting $n \to \infty$,

$$f(x) \leqslant cx. \tag{16.3.6}$$

Hence and by Lemma 16.1.5

$$cx = -c(-x) \leqslant -f(-x) \leqslant f(x). \tag{16.3.7}$$

From (16.3.6) and (16.3.7) we obtain (16.3.2).   □

As an immediate consequence of Theorem 16.3.1 we obtain the following

Theorem 16.3.2. *Let $f : \mathbb{R}^N \to \mathbb{R}$ be subadditive. If $f(0) = 0$ and f has the Stolz differential at zero, then there exists a $c \in \mathbb{R}^N$ such that (16.3.2) holds. In particular, f is additive and of class C^1 in $\mathbb{R}^N$.*

Proof. Results from Theorem 16.3.1 on taking $g = f$.   □

The following example shows that the assumption $f(0) = 0$ in Theorem 16.3.2 is essential.

Example 16.3.1. Let $N = 1$, and let $f : \mathbb{R} \to \mathbb{R}$ be defined by

$$f(x) = \begin{cases} 1 & \text{for } x \leqslant 0, \\ e^{-x^2} & \text{for } x > 0. \end{cases}$$

If $x, y \leqslant 0$, then also $x + y \leqslant 0$ and $f(x + y) = 1 < 1 + 1 = f(x) + f(y)$. If $x \leqslant 0$, $y > 0$, and $x + y \leqslant 0$, then $f(x + y) = 1 < 1 + e^{-y^2} = f(x) + f(y)$. Similarly, if $x > 0$, $y \leqslant 0$, and $x + y \leqslant 0$, then $f(x + y) < f(x) + f(y)$. If $x > 0$, $y \leqslant 0$, and $x + y > 0$, then $f(x + y) = e^{-(x+y)^2} \leqslant 1 < e^{-x^2} + 1 = f(x) + f(y)$. Similarly if $x \leqslant 0$, $y > 0$, and $x + y > 0$, then $f(x + y) < f(x) + f(y)$. Finally, if $x > 0$ and $y > 0$, then also $x + y > 0$, and $f(x + y) = e^{-(x+y)^2} < e^{-x^2} < e^{-x^2} + e^{-y^2} = f(x) + f(y)$. So f is subadditive, and f is differentiable[6] at zero, but f has not form (16.3.2) for any $c \in \mathbb{R}$.

---

[6] If $N = 1$, then the differentiability of a function at a point, and the existence of the Stolz differential at this point, are equivalent.

In the sequel of this section we assume that $N = 1$.

Theorem 16.3.3. *Let $f : \mathbb{R} \to \mathbb{R}$ be a measurable subadditive function, and let*

$$A = \inf_{t<0} \frac{f(t)}{t}, \qquad B = \sup_{t>0} \frac{f(t)}{t}. \tag{16.3.8}$$

If A resp. B is finite, then

$$A = \lim_{h\to 0-} \frac{f(h)}{h} \quad resp. \qquad B = \lim_{h\to 0+} \frac{f(h)}{h}. \tag{16.3.9}$$

Formulas (16.3.9) remain valid for A and/or B infinite under the additional assumption that $\lim_{x\to 0} f(x) = 0$, or $\liminf_{x\to 0} f(x) > 0$. Moreover, in every case,

$$A \leqslant B \tag{16.3.10}$$

Proof. We prove only the first equality in (16.3.9), the proof of the other one is similar. Evidently we have $-\infty \leqslant A < \infty$. Assume that A is finite, or $A = -\infty$ and

$$\lim_{x\to 0-} f(x) = 0. \tag{16.3.11}$$

Suppose that A is finite and (16.3.11) does not hold. By Theorem 16.2.7

$$\limsup_{x\to 0-} f(x) \geqslant \liminf_{x\to 0-} f(x) \geqslant \liminf_{x\to 0} f(x) \geqslant 0.$$

If we had $\limsup_{x\to 0-} f(x) = 0$, then the above inequalities yield $\liminf_{x\to 0-} f(x) = 0$, whence (16.3.11) follows contrary to the supposition. Thus we must have $\limsup_{x\to 0-} f(x) > 0$, which means that there exists a sequence $\{x_n\}_{n\in\mathbb{N}}$ such that $x_n < 0$ for $n \in \mathbb{N}$, $\lim_{n\to\infty} x_n = 0$, and there exists an $\alpha > 0$ such that $f(x_n) > \alpha$ for $n \in \mathbb{N}$. Then

$$\frac{f(x_n)}{x_n} < \frac{\alpha}{x_n} \to -\infty,$$

which shows that $A = -\infty$. Consequently, if A is finite, relation (16.3.11) must hold. Thus we have (16.3.11) in either case.

Now take an arbitrary $a > A$. There exists a point $x_0 < 0$ such that

$$\frac{f(x_0)}{x_0} < a,$$

i.e., $f(x_0) > ax_0$. Choose an arbitrary h such that $x_0 < h < 0$, and let $n = [x_0/h]$ be the integral part of x_0/h so that $n \leqslant x_0/h < n+1$, whence $nh \geqslant x_0 > (n+1)h$. Put $\delta = x_0 - nh$ so that

$$0 \geqslant \delta > h. \tag{16.3.12}$$

Further, we have $x_0 = nh + \delta$. Hence by Lemma 16.1.4

$$ax_0 < f(x_0) = f(nh + \delta) \leqslant f(nh) + f(\delta) \leqslant nf(h) + f(\delta),$$

whence, since $x_0 < 0$,

$$a > \frac{f(x_0)}{x_0} \geqslant \frac{nf(h)}{x_0} + \frac{f(\delta)}{x_0} = \frac{nh}{x_0}\frac{f(h)}{h} + \frac{f(\delta)}{x_0}.$$

We have $f(h)/h \geqslant A$ and $nh/x_0 > 0$, whence

$$a > \frac{nh}{x_0}\frac{f(h)}{h} + \frac{f(\delta)}{x_0} \geqslant \frac{nh}{x_0}A + \frac{f(\delta)}{x_0}. \tag{16.3.13}$$

Now we let $h \to 0-$. Then, by (16.3.12), also $\delta \to 0-$, and by (16.3.11) $f(\delta) \to 0$. On the other hand, $nh/x_0 = 1 - \delta/x_0 \to 1$. Thus we get by (16.3.13)

$$a \geqslant \limsup_{h \to 0-} \frac{f(h)}{h} \geqslant \liminf_{h \to 0-} \frac{f(h)}{h} \geqslant A.$$

Letting a tend to A, we obtain the first equality in (16.3.9).

If $\liminf\limits_{x \to 0} f(x) > 0$ (which implies $A = -\infty$), then there exists an $\alpha > 0$ such that $f(x) > \alpha$ in a neighbourhood of zero. Then also

$$\lim_{h \to 0-} \frac{f(h)}{h} = -\infty = A.$$

The second equality in (16.3.9) is established similarly.

By Lemma 16.1.5 $-f(-x) \leqslant f(x)$, whence for $x > 0$

$$\frac{f(-x)}{-x} \leqslant \frac{f(x)}{x},$$

and

$$A = \inf_{x<0} \frac{f(x)}{x} = \inf_{x>0} \frac{f(-x)}{-x} \leqslant \sup_{x>0} \frac{f(-x)}{-x} \leqslant \sup_{x>0} \frac{f(x)}{x} = B.$$

This proves relation (16.3.10). $\square$

Theorem 16.3.3 may be formulated that for a measurable subadditive function $f : \mathbb{R} \to \mathbb{R}$ we have

$$\lim_{h \to 0+} \frac{f(h)}{h} = \sup_{h>0} \frac{f(h)}{h}, \qquad \lim_{h \to 0-} \frac{f(h)}{h} = \inf_{h<0} \frac{f(h)}{h}$$

provided that these quantities are finite (or, if they are infinite, provided that some further conditions concerning the behaviour of f near 0 are fulfilled). On the other hand, we have shown in the proof of Theorem 16.2.9 that

$$\lim_{h \to \infty} \frac{f(h)}{h} = \inf_{h>0} \frac{f(h)}{h}, \qquad \lim_{h \to -\infty} \frac{f(h)}{h} = \sup_{h<0} \frac{f(h)}{h},$$

without any further assumptions except the measurability and subadditivity of f.

In the next theorem there occur the Dini derivatives (cf. 7.5).

Theorem 16.3.4. *Let $f : \mathbb{R} \to \mathbb{R}$ be a measurable subadditive function. Then, with notation (16.3.8), we have for every $x \in \mathbb{R}$*

$$D^+ f(x) \leqslant B, \qquad D^- f(x) \leqslant B, \qquad d^+ f(x) \geqslant A, \qquad d^- f(x) \geqslant A. \qquad (16.3.14)$$

In particular, if $A = B$, then

$$f(x) = Ax \quad for \quad x \in \mathbb{R}. \qquad (16.3.15)$$

Proof. If $B = \infty$, then the first two inequalities in (16.3.14) are trivial. Similarly, if $A = -\infty$, then the last two inequalities in (16.3.14) are trivial.

Suppose that $B < \infty$, whence by (16.3.8) B is finite. We have for arbitrary $x \in \mathbb{R}$ and $h > 0$

$$f(x + h) \leqslant f(x) + f(h),$$

whence

$$\frac{f(x + h) - f(x)}{h} \leqslant \frac{f(h)}{h}, \qquad (16.3.16)$$

and by Theorem 16.3.3

$$D^+ f(x) = \limsup_{h \to 0+} \frac{f(x + h) - f(x)}{h} \leqslant \lim_{h \to 0+} \frac{f(h)}{h} = B.$$

Similarly, since

$$f(x) \leqslant f(h + x - h) \leqslant f(h) + f(x - h),$$

we get for $h > 0$

$$\frac{f(x) - f(x - h)}{h} \leqslant \frac{f(h)}{h}, \qquad (16.3.17)$$

and by Theorem 16.3.3

$$D^- f(x) = \limsup_{h \to 0-} \frac{f(x + h) - f(x)}{h} = \limsup_{h \to 0+} \frac{f(x) - f(x - h)}{h} \leqslant \lim_{h \to 0+} \frac{f(h)}{h} = B.$$

Now let $A > -\infty$ (and so, by (16.3.8), A is finite). For $h < 0$ the inequalities in (16.3.16) and (16.3.17) are reversed, whence by Theorem 16.3.3

$$d^+ f(x) = \liminf_{h \to 0+} \frac{f(x + h) - f(x)}{h} = \liminf_{h \to 0-} \frac{f(x) - f(x - h)}{h} \geqslant \lim_{h \to 0-} \frac{f(h)}{h} = A,$$

and

$$d^- f(x) = \liminf_{h \to 0-} \frac{f(x + h) - f(x)}{h} \geqslant \lim_{h \to 0-} \frac{f(h)}{h} = A.$$

If $A = B$, then both these quantities are finite, and $f'(x) = A$ for all $x \in \mathbb{R}$, whence $f(x) = Ax + b$, with a certain $b \in \mathbb{R}$. If we had $b \neq 0$, then

$$A = \lim_{h \to 0-} \frac{f(h)}{h} = \lim_{h \to 0-} \left(A + \frac{b}{h} \right)$$

would be infinite. Consequently $b = 0$, and we obtain formula (16.3.15). $\qquad \square$

Theorem 16.3.5. *Let* $f : \mathbb{R} \to \mathbb{R}$ *be a measurable subadditive function. If, with notation* (16.3.8), *A and B are finite, then* f *fulfils in* $\mathbb{R}$ *a Lipschitz condition.*

Proof. Take an $L > \max\left(|A|, |B|\right)$. It follows from Theorem 16.3.4 that for every $x \in \mathbb{R}$ there exists an $r_x > 0$ such that

$$\left| \frac{f(x) - f(y)}{x - y} \right| < L \tag{16.3.18}$$

for $y \in (x - r_x, x + r_x)$, $y \neq x$. Take an arbitrary $x, y \in \mathbb{R}$, and let, e.g., $x < y$. We have

$$[x, y] \subset \bigcup_{t \in [x,y]} (t - r_t, t + r_t).$$

Since the interval $[x, y]$ is compact, we can choose a finite cover

$$[x, y] \subset \bigcup_{i=1}^{n} (t_i - r_{t_i}, t_i + r_{t_i}).$$

We may assume that x, y occur in the sequence $\{t_1, \ldots, t_n\}$. Suppose that $x = t_1$. If we had

$$(x - r_x, x + r_x) \cap \bigcup_{i=2}^{n} (t_i - r_{t_i}, t_i + r_{t_i}) = \varnothing,$$

the interval $[x, y]$ would be disconnected. So in the sequence $\{t_2, \ldots, t_n\}$ there is $t_j = x_1$ such that (we write $r_{x_1} = r_1$)

$$(x - r_x, x + r_x) \cap (x_1 - r_1, x_1 + r_1) \neq \varnothing.$$

Proceeding in this way, and writing $x = x_0$, we may find a sequence

$$x = x_0 < x_1 < \cdots < x_n = y$$

such that (with the notation $r_{x_i} = r_i$, $i = 0, \ldots, n$)

$$(x_i - r_i, x_i + r_i) \cap (x_{i+1} - r_{i+1}, x_{i+1} + r_{i+1}) \neq \varnothing \qquad \text{for } i = 0, \ldots, n - 1.$$

Choose points $y_0, \ldots, y_{n-1}$ so that

$$y_i \in (x_i - r_i, x_i + r_i) \cap (x_{i+1} - r_{i+1}, x_{i+1} + r_{i+1}), \qquad i = 0, \ldots, n - 1,$$

and

$$x = x_0 < y_0 < x_1 < y_1 < \cdots < y_{n-1} < x_n = y.$$

Then we have by (16.3.18)

$$|f(y) - f(x)| \leqslant \sum_{i=0}^{n-1} |f(x_{i+1}) - f(y_i)| + \sum_{i=0}^{n-1} |f(y_i) - f(x_i)|$$

$$\leqslant L \sum_{i=0}^{n-1} |x_{i+1} - y_i| + L \sum_{i=0}^{n-1} |y_i - x_i| = L|y - x|.$$

This means that f fulfils in $\mathbb{R}$ the Lipschitz condition with the constant L. (For $x = y$ the inequality is trivial, and if $y < x$, we reverse the roles of x and y). $\qquad \square$

Theorem 16.3.6. *Let $f : \mathbb{R} \to \mathbb{R}$ be a measurable subadditive function. If, with notation (16.3.8), A and B are finite, then f is absolutely continuous in $\mathbb{R}$.*

Proof. Results from Theorem 16.3.5. □

16.4 Sublinear functions

These are functions $f : \mathbb{R}^N \to \mathbb{R}$ fulfilling (16.1.1) and (16.1.4). As we will show, there is a close connection between sublinear functions and additive functions (Berz [31]).

Lemma 16.4.1. *Let $f : \mathbb{R}^N \to \mathbb{R}$ be a sublinear function, and let $x_0 \in \mathbb{R}^N$. There exists an additive function $g : \mathbb{R}^N \to \mathbb{R}$ such that*

$$g(x_0) = f(x_0) \tag{16.4.1}$$

and

$$g(x) \leqslant f(x) \qquad for \ x \in \mathbb{R}^N. \tag{16.4.2}$$

Proof. Let $L = \mathbb{Q}x_0$. L is a linear subspace of $\mathbb{R}^N$ (over $\mathbb{Q}$). We define the function $g_0 : L \to \mathbb{R}$ by

$$g_0(x) = g_0(\lambda x_0) = \lambda f(x_0), \qquad \lambda \in \mathbb{Q}. \tag{16.4.3}$$

By Lemma 16.1.10, if $\lambda \geqslant 0$, we have $g_0(x) = g_0(\lambda x_0) = \lambda f(x_0) = f(\lambda x_0) = f(x)$. If $\lambda < 0$, then by Lemmas 16.1.5 and 16.1.10 $g_0(x) = g_0(\lambda x_0) = \lambda f(x_0) = -|\lambda| f(x_0) \leqslant |\lambda| f(-x_0) = f(-|\lambda| x_0) = f(x)$. Thus

$$g_0(x) \leqslant f(x) \qquad for \ x \in L.$$

Now, by Lemma 16.1.11 the function f is convex, and so it fulfils inequality (10.1.1). The set $D = \mathbb{R}^N$ evidently is Q-convex and Q-radial at every point. It follows from (16.4.3) that g_0 is a homomorphism from L into $\mathbb{R}$. By Theorem 10.1.1 there exists an additive function $g : \mathbb{R}^N \to \mathbb{R}$ fulfilling inequality (16.4.2) and the condition $g \mid L = g_0$. Hence by (16.4.3)

$$g(x_0) = g_0(x_0) = f(x_0),$$

i.e., g fulfils (16.4.1). □

Theorem 16.4.1. *Let $f : \mathbb{R}^N \to \mathbb{R}$ be a sublinear function. Then, for every $x \in \mathbb{R}^N$,*

$$f(x) = \sup \{g(x) \mid g : \mathbb{R}^N \to \mathbb{R} \text{ is additive and } g \leqslant f\}. \tag{16.4.4}$$

Proof. Let

$$h(x) = \sup \{g(x) \mid g : \mathbb{R}^N \to \mathbb{R} \text{ is additive and } g \leqslant f\}.$$

Then, of course,

$$h(x) \leqslant f(x) \qquad for \ x \in \mathbb{R}^N. \tag{16.4.5}$$

On the other hand, by Lemma 16.4.1, for every $x_0 \in \mathbb{R}^N$ there exists an additive function $g : \mathbb{R}^N \to \mathbb{R}$ fulfilling (16.4.1) and (16.4.2). Hence

$$h(x_0) \geqslant f(x_0) \qquad for \ x_0 \in \mathbb{R}^N. \tag{16.4.6}$$

Inequalities (16.4.5) and (16.4.6) yield (16.4.4). □

Let $\mathcal{F}$ be the family of all sublinear functions $f : \mathbb{R}^N \to \mathbb{R}$. $(\mathcal{F}, \leqslant)$ is an ordered space.

Theorem 16.4.2. *Additive functions $g : \mathbb{R}^N \to \mathbb{R}$ are the minimal elements of $(\mathcal{F}, \leqslant)$.*

Proof. Let $g : \mathbb{R}^N \to \mathbb{R}$ be additive, let $f : \mathbb{R}^N \to \mathbb{R}$ be sublinear, and suppose that $f \leqslant g$. Hence we have for every $x \in \mathbb{R}^N$

$$-f(-x) \geqslant -g(-x) = g(x),$$

whence by Lemma 16.1.5

$$f(x) \geqslant -f(-x) \geqslant g(x),$$

i.e., $f \geqslant g$. Thus $f = g$, which means that g is a minimal element of $(\mathcal{F}, \leqslant)$.

On the other hand, if f is a minimal element of $(\mathcal{F}, \leqslant)$, then, in particular, f is sublinear, and by Lemma 16.4.1 there exists an additive function $g : \mathbb{R}^N \to \mathbb{R}$ fulfilling (16.4.2). Since f is minimal, (16.4.2) implies that $f = g$, i.e., f is additive. □

In the one-dimensional case $N = 1$ we have the following

Theorem 16.4.3. *Let $f : \mathbb{R} \to \mathbb{R}$ be a measurable sublinear function. Then there exist real constants $a \geqslant b$ such that*

$$f(x) = \begin{cases} ax & \text{for } x \geqslant 0, \\ bx & \text{for } x < 0. \end{cases} \tag{16.4.7}$$

Conversely, every function $f : \mathbb{R} \to \mathbb{R}$ of form (16.4.7), where $a \geqslant b$, is measurable and sublinear.

Proof. Since f is measurable, by Theorem 9.4.1 every additive function $g : \mathbb{R} \to \mathbb{R}$ fulfilling (16.4.2) is continuous, and hence (Theorem 5.5.2) of the form $g(x) = c_g x$, $x \in \mathbb{R}$, with a certain constant $c_g \in \mathbb{R}$. Hence, by Theorem 16.4.1,

$$f(x) = \sup \{ c_g x \mid c_g t \leqslant f(t) \text{ for } t \in \mathbb{R} \}.$$

Put

$$a = \sup \{ c_g \mid c_g t \leqslant f(t) \text{ for } t \in \mathbb{R} \},$$
$$b = - \sup \{ -c_g \mid c_g t \leqslant f(t) \text{ for } t \in \mathbb{R} \}.$$

Then for $x \geqslant 0$

$$f(x) = \sup \{ c_g x \mid c_g t \leqslant f(t) \text{ for } t \in \mathbb{R} \} = x \sup \{ c_g \mid c_g t \leqslant f(t) \text{ for } t \in \mathbb{R} \} = ax,$$

whereas for $x < 0$

$$f(x) = \sup \{ -c_g(-x) \mid c_g t \leqslant f(t) \text{ for } t \in \mathbb{R} \}$$
$$= -x \sup \{ -c_g \mid c_g t \leqslant f(t) \text{ for } t \in \mathbb{R} \} = bx.$$

This proves relation (16.4.7). Moreover, we have $a = f(1)$, $b = -f(-1)$, so the inequality $a \geqslant b$ results from Lemma 16.1.5.

Conversely, let a function $f : \mathbb{R} \to \mathbb{R}$ be given by (16.4.7), where $a \geqslant b$. Then $f(x) = \max(ax, bx)$, and by Lemma 16.1.6 f is subadditive. Further, if $x \geqslant 0$ and $n \in \mathbb{N}$, then $nx \geqslant 0$, and $f(nx) = anx = nax = nf(x)$. If $x < 0$ and $n \in \mathbb{N}$, then $nx < 0$, and $f(nx) = bnx = nbx = nf(x)$. Consequently f satisfies (16.1.4). This means that f is sublinear. Clearly, a function of form (16.4.7) is continuous, and so the more measurable. $\qquad\square$

Theorems 16.4.1, 16.4.2, 16.4.3 are due to E. Berz [31].

16.5 Norm

Subadditive functions have many interesting applications in other branches of mathematics (cf., e.g., Hille-Phillips [142]). In particular, if E is any linear space (over $\mathbb{R}$), then every norm $f = \| \cdot \|$ in E (if it exists) is a subadditive function (and even sublinear). So is also the function $f = \|g\|$, where $g : \mathbb{R}^N \to E$ fulfils the Cauchy equation (5.2.1). P. Fischer [82] has expressed the following conjecture:

If $f : (0, \infty) \to \mathbb{R}$ is a non-negative subadditive function such that

$$f(\lambda x) = \lambda f(x) \qquad \text{for } x \in (0, \infty), \qquad \lambda \in \mathbb{Q} \cap (0, \infty), \tag{16.5.1}$$

then there exist a normed space E and a function $g : \mathbb{R} \to E$ satisfying equation (5.2.1) such that

$$f(x) = \|g(x)\| \qquad \text{for } x > 0. \tag{16.5.2}$$

Whereas this conjecture is not true in general (cf. Moszner [232]), we have the following (Berz [31])

Theorem 16.5.1. *If $f : \mathbb{R}^N \to \mathbb{R}$ is an even sublinear function, then there exist a normed space E (over $\mathbb{R}$) and a function $g : \mathbb{R}^N \to E$ satisfying equation (5.2.1) such that*

$$f(x) = \|g(x)\| \qquad \text{for } x \in \mathbb{R}^N. \tag{16.5.3}$$

Proof. Let $\{g_\alpha\}_{\alpha \in A}$ be the family of all additive functions $g_\alpha : \mathbb{R}^N \to \mathbb{R}$ such that $g_\alpha \leqslant f$, indexed by indices α from a certain set A. By Theorem 10.1.2 we have $|g_\alpha| \leqslant f$ for $\alpha \in A$. So $g_\alpha \leqslant |g_\alpha| \leqslant f$ for $\alpha \in A$, whence by Theorem 16.4.1

$$f(x) = \sup_{\alpha \in A} g_\alpha(x) \leqslant \sup_{\alpha \in A} |g_\alpha(x)| \leqslant f(x)$$

so that

$$f(x) = \sup_{\alpha \in A} |g_\alpha(x)|. \tag{16.5.4}$$

Let E be the set of all families $\{y_\alpha\}_{\alpha \in A}$, where $y_\alpha \in \mathbb{R}$ for $\alpha \in A$, such that $\sup_{\alpha \in A} |y_\alpha| < \infty$. With the operations defined in the usual manner:

$$\{y_\alpha\}_{\alpha \in A} + \{z_\alpha\}_{\alpha \in A} = \{y_\alpha + z_\alpha\}_{\alpha \in A},$$

$$\lambda\{y_\alpha\}_{\alpha \in A} = \{\lambda y_\alpha\}_{\alpha \in A} \qquad \text{for } \lambda \in \mathbb{R},$$

E becomes a linear space over $\mathbb{R}$. We introduce in E the norm:

$$\|\{y_\alpha\}_{\alpha \in A}\| = \sup_{\alpha \in A} |y_\alpha|. \qquad (16.5.5)$$

With this norm E becomes a normed linear space. Let the function $g : \mathbb{R}^N \to E$ be defined by

$$g(x) = \{g_\alpha(x)\}_{\alpha \in A} \qquad x \in \mathbb{R}^N.$$

Then, since g_α are additive

$$g(x + y) = \{g_\alpha(x + y)\}_{\alpha \in A} = \{g_\alpha(x) + g_\alpha(y)\}_{\alpha \in A}$$
$$= \{g_\alpha(x)\}_{\alpha \in A} + \{g_\alpha(y)\}_{\alpha \in A} = g(x) + g(y),$$

i.e., g satisfies equation (5.2.1). By (16.5.4) and (16.5.5) we get (16.5.3). $\square$

Theorem 16.5.2. *Let $f : (0, \infty) \to \mathbb{R}$ be an arbitrary function. Then the following conditions are equivalent:*

  (i) *f can be extended onto $\mathbb{R}$ to an even sublinear function $f^* : \mathbb{R} \to \mathbb{R}$;*
 (ii) *f fulfils condition (16.5.1) and (setting $f(0) = 0$)*

$$f(|x + y|) \leqslant f(|x|) + f(|y|) \qquad for\ x, y \in \mathbb{R}; \qquad (16.5.6)$$

(iii) *there exist a normed linear space E (over $\mathbb{R}$) and a function $g : \mathbb{R} \to E$ satisfying equation (5.2.1) such that (16.5.2) holds.*

Proof. Suppose that f can be extended onto $\mathbb{R}$ to an even sublinear function $f^* : \mathbb{R} \to \mathbb{R}$. Then, by Lemma 16.1.10, f^*, and hence also $f = f^* \mid (0, \infty)$, fulfils (16.5.1). Moreover, by Lemma 16.1.10, $f^*(0) = 0$, so if we put $f(0) = 0$, then we have $f = f^* \mid [0, \infty)$. Since f^* is even, we have $f^*(t) = f^*(|t|) = f(|t|)$ for every $t \in \mathbb{R}$. Hence

$$f(|x + y|) = f^*(|x + y|) = f^*(x + y) \leqslant f^*(x) + f^*(y)$$
$$= f^*(|x|) + f^*(|y|) = f(|x|) + f(|y|)$$

for arbitrary $x, y \in \mathbb{R}$, i.e., f satisfies (16.5.6). Consequently condition (i) implies (ii).

Conversely, suppose that f fulfils (ii) (we set $f(0) = 0$), and put

$$f^*(x) = f(|x|) \qquad for\ x \in \mathbb{R}. \qquad (16.5.7)$$

Formula (16.5.7) yields an extension of f onto $\mathbb{R}$, and it is obvious that f^* is even. By (16.5.1) f^* fulfils (16.1.4). Moreover, we have by (16.5.6)

$$f^*(x + y) = f(|x + y|) \leqslant f(|x|) + f(|y|) = f^*(x) + f^*(y)$$

for arbitrary $x, y \in \mathbb{R}$. Consequently f^* is subadditive, and hence sublinear. Thus (i) is true, and hence conditions (i), (ii) are equivalent.

If (iii) is satisfied, then the formula

$$f^*(x) = \|g(x)\| \qquad for\ x \in \mathbb{R}$$

defines an even sublinear function $f^* : \mathbb{R} \to \mathbb{R}$, and by (16.5.2) $f^* \mid (0, \infty) = f$. Consequently (*iii*) implies (*i*). The converse implication results from Theorem 16.5.1. Thus all the three conditions (*i*), (*ii*), (*iii*) are equivalent to each other. □

Theorem 16.5.2 (Berz [31]; cf. also Moszner [232]) gives precise conditions when Fischer's conjecture is true: a function $f : (0, \infty) \to \mathbb{R}$ fulfilling the hypotheses of Fischer's conjecture admits representation (16.5.2) if and only if it satisfies also condition (16.5.6).

16.6 Infinitary subadditive functions

The following theorem shows that infinitary subadditive functions only exceptionally can assume the value $-\infty$.

Theorem 16.6.1. *Let* $f : \mathbb{R}^N \to [-\infty, \infty]$ *be subadditive. If there exists an* $x_0 \in \mathbb{R}^N$ *such that* $f(x_0) = -\infty$*, then* $f(x) = -\infty$ *for all* $x \in \mathbb{R}^N$.

Proof. Since $f(x_0) = -\infty$, we have $f \neq +\infty$ in $\mathbb{R}^N$, whence for arbitrary $x \in \mathbb{R}^N$

$$f(x) = f(x - x_0 + x_0) \leqslant f(x - x_0) + f(x_0) = f(x - x_0) - \infty = -\infty.$$

Hence $f(x) = -\infty$. □

Hence it follows that if $f : \mathbb{R}^N \to [-\infty, \infty]$ satisfies equation (5.2.1) and assumes infinite values, then either $f(x) = -\infty$ for all $x \in \mathbb{R}^N$, or $f(x) = \infty$ for all $x \in \mathbb{R}^N$ (Halperin [131]). In fact, such a function is both subadditive and superadditive, and for superadditive functions Theorem 16.6.1 holds with $-\infty$ replaced by $+\infty$.

If the domain of f is only a subset of $\mathbb{R}^N$, we must assume more.

Theorem 16.6.2. *Let* $C \subset \mathbb{R}^N$ *be a set fulfilling (16.1.3) and let* $f : C \to [-\infty, \infty]$ *be a subadditive function. If there exists a sequence* $\{t_n\}_{n \in \mathbb{N}}$ *of points* $t_n \in C$ *such that* $\lim_{n \to \infty} t_n = 0$ *and* $f(t_n) = -\infty$ *for* $n \in \mathbb{N}$*, then* $f(x) = -\infty$ *for all* $x \in \text{int } C$.

Proof. Let $x \in \text{int } C$. Since f assumes the value $-\infty$, we have $f \neq +\infty$ in C. Further, there exists an $n \in \mathbb{N}$ such that $x - t_n \in C$. Hence

$$f(x) = f(t_n + x - t_n) \leqslant f(t_n) + f(x - t_n) = -\infty + f(x - t_n) = -\infty.$$

Thus $f(x) = -\infty$. □

With the value $+\infty$ the situation is more complicated. A subadditive function may be equal $+\infty$ on a proper subset of $\mathbb{R}^N$.

Example 16.6.1. Define the function $f : \mathbb{R}^N \to [-\infty, \infty]$ by (cf. Example 16.2.1)

$$f(x) = \begin{cases} +\infty & \text{for } x \notin \mathbb{Q}^N, \\ b & \text{for } x \in \mathbb{Q}^N, \end{cases}$$

where $b > 0$. Then f is subadditive and infinitary, but $f(x) \neq \infty$ on a dense subset of $\mathbb{R}^N$.

Theorem 16.6.3. *Let $f : \mathbb{R}^N \to [-\infty, \infty]$ be a subadditive function. If there exists an $r > 0$ such that $|f(x)| < \infty$ for $|x| < r$, then $|f(x)| < \infty$ for all $x \in \mathbb{R}^N$.*

Proof. Take an $x \in \mathbb{R}^N$. There exists an $n \in \mathbb{N}$ such that $|x/n| = |x|/n < r$. Hence $f(x/n) < \infty$, and by Lemma 16.1.4

$$f(x) \leqslant nf(x/n) < \infty. \tag{16.6.1}$$

If we had $f(x) = -\infty$, then we would get by Theorem 16.6.1 $f \equiv -\infty$, which is incompatible with the assumption that $|f(t)| < \infty$ for $|t| < r$. Consequently $f(x) > -\infty$, which together with (16.6.1) implies that $|f(x)| < \infty$. $\square$

If $f : \mathbb{R}^N \to [-\infty, \infty]$ is a subadditive function, we put

$$P_f = \{x \in \mathbb{R}^N \mid f(x) < \infty\}.$$

This definition allows that $f(x) = -\infty$ for some $x \in P_f$. But in view of Theorem 16.6.1, unless $f \equiv -\infty$, we have

$$P_f = \{x \in \mathbb{R}^N \mid |f(x)| < \infty\}.$$

The set P_f is characterized by the following property.

Theorem 16.6.4. *Let $C \subset \mathbb{R}^N$ be an arbitrary set. There exists a subadditive function $f : \mathbb{R}^N \to [-\infty, \infty]$ such that $P_f = C$ if and only if C fulfils (16.1.3).*

Proof. If $f : \mathbb{R}^N \to [-\infty, \infty]$ is subadditive, and if $u \in P_f + P_f$, then there exist $x, y \in P_f$ such that $u = x + y$, whence

$$f(u) = f(x + y) \leqslant f(x) + f(y) < \infty,$$

i.e., $u \in P_f$. Hence $P_f + P_f \subset P_f$.

On the other hand, if a set $C \subset \mathbb{R}^N$ fulfils (16.1.3), then the function

$$f(x) = \begin{cases} +\infty & \text{for } x \notin C, \\ 0 & \text{for } x \in C \end{cases}$$

is subadditive, and $P_f = C$. $\square$

Lemma 16.6.1. *Let $B \subset \mathbb{R}^N$ be an open ball centered at a point $x_0 \in \mathbb{R}^N$. Then there exists an $\alpha > 0$ such that $tx_0 \in \bigcup_{k=1}^{\infty} kB$ for $t > \alpha$.*

Proof. Let $r > 0$ denote the radius of B, and let $\alpha \in \mathbb{N}$ be such that $\alpha r > |x_0|$. Take an arbitrary $t > \alpha$, and let $n = [t]$ be the integral part of t. Then $n \geqslant \alpha$ and $|t - n| < 1$, whence

$$|tx_0 - nx_0| = |t - n||x_0| \leqslant |x_0| < \alpha r \leqslant nr.$$

But this means that tx_0 belongs to the open ball centered at nx_0 and with the radius nr, i.e., $tx_0 \in nB \subset \bigcup_{k=1}^{\infty} kB$. $\square$

Theorem 16.6.5. *Let $f : \mathbb{R}^N \to [-\infty, \infty]$ be a measurable subadditive function. If there exists a density point x_0 of P_f such that[7] $x_0 \in P_f$ and $sx_0 \in P_f$ for a certain $s < 0$, then $P_f = \mathbb{R}^N$.*

Proof. Since f is measurable, $P_f \in \mathcal{L}$. By Theorem 16.6.4 $P_f + P_f \subset P_f$.

By Theorem 3.7.2 $2x_0 = x_0 + x_0 \in \operatorname{int}(P_f + P_f) \subset \operatorname{int} P_f$. Let $B \subset \mathbb{R}^N$ be an open ball centered at $2x_0$ and contained in P_f. By Lemma 16.6.1 there exists an $\alpha > 0$ such that $tx_0 \in \bigcup_{k=1}^{\infty} kB$ for $t > \alpha$. Choose an $m \in \mathbb{N}$ such that $-ms > \alpha$. Then there exists an $n \in \mathbb{N}$ such that $-msx_0 \in nB$. The set $U = msx_0 + nB$ is a neighbourhood of zero. If $z \in U$, then $z = msx_0 + nx$, where $x \in B$. Hence, by Lemma 16.1.4

$$f(z) = f(msx_0 + nx) \leqslant f(msx_0) + f(nx) \leqslant mf(sx_0) + nf(x) < \infty,$$

i.e., $z \in P_f$. Consequently $U \subset P_f$ and there exists an $r > 0$ such that $x \in P_f$ whenever $|x| < r$. If $f(x) = -\infty$ for an $x \in \mathbb{R}^N$, then by Theorem 16.6.1 $f(x) = -\infty$ for every $x \in \mathbb{R}^N$, i.e., $P_f = \mathbb{R}^N$. If $f > -\infty$ in $\mathbb{R}^N$, then $|f(x)| < \infty$ for $|x| < r$, and by Theorem 16.6.3 $|f| < \infty$ in $\mathbb{R}^N$. Consequently $P_f = \mathbb{R}^N$. $\square$

The same argument may be used to prove the following

Theorem 16.6.6. *Let $f : \mathbb{R}^N \to [-\infty, \infty]$ be subadditive. If $m_i(P_f) > 0$, or P_f contains a set of the second category, and with the Baire property, and if there exists[8] an $x_0 \in \operatorname{int} P_f$ such that $sx_0 \in P_f$ for a certain $s < 0$, then $P_f = \mathbb{R}^N$.*

Theorem 16.6.7. *Let $f : \mathbb{R}^N \to [-\infty, \infty]$ be subadditive. If there exists an $x_0 \in \mathbb{R}^N$ such that $f(x_0) = \infty$ and $tx_0 \in P_f$ for all $t > 1$, then $f(tx_0) = \infty$ for all $t < 0$.*

Proof. Since $f(x_0) = \infty$, we have $f \neq -\infty$ in $\mathbb{R}^N$. Consequently $|f(tx_0)| < \infty$ for $t > 1$.

If $t < 0$, then $1 - t > 1$, and

$$\infty = f(x_0) = f\big(tx_0 + (1 - t)x_0\big) \leqslant f(tx_0) + f\big((1 - t)x_0\big).$$

But since $\big|f\big((1 - t)x_0\big)\big| < \infty$, this implies that $f(tx_0) = \infty$. $\square$

Theorem 16.2.9 says that if a measurable subadditive function $f : \mathbb{R}^N \to \mathbb{R}$ is finite-valued, then, for every $x_0 \in \mathbb{R}^N$,

$$\beta = \inf_{t > 0} \frac{f(tx_0)}{t} \tag{16.6.2}$$

is finite (cf., in particular, formula (16.2.10)). This is no longer true if f is infinitary. But in such a case we have the following

Theorem 16.6.8. *Let $f : \mathbb{R}^N \to [-\infty, \infty]$ be measurable and subadditive, and suppose that for a certain $x_0 \in \mathbb{R}^N$ we have $\beta = -\infty$, where β is given by (16.6.2). If $|f(tx_0)| < \infty$ for all $t > 0$, then $f(tx_0) = \infty$ for all $t < 0$.*

---

[7] The condition $x_0 \in P_f$ is not indispensable and can be omitted.
[8] From Theorems 3.7.1, 2.9.1 and 16.6.4 it follows that $\operatorname{int} P_f \neq \varnothing$.

Proof. Suppose that there exists a $t_0 < 0$ such that $f(t_0 x_0) < \infty$. For every $t \leqslant 0$ there exist a $t' \in [-t_0, -2t_0) \subset (0, \infty)$ and a $k \in \mathbb{N}$ such that $t = t' + k t_0$. Hence by Lemma 16.1.4

$$f(t x_0) = f(t' x_0 + k t_0 x_0) \leqslant f(t' x_0) + f(k t_0 x_0) \leqslant f(t' x_0) + k f(t_0 x_0) < \infty.$$

On the other hand, in view of Theorem 16.6.1, $f(t x_0) \neq -\infty$, since $f(\tau x_0) \neq -\infty$ for $\tau > 0$. Thus $|f(t x_0)| < \infty$ for all $t \in \mathbb{R}$. Consequently the function $\varphi : \mathbb{R} \to \mathbb{R}$ given by $\varphi(t) = f(t x_0)$, $t \in \mathbb{R}$, is finite, measurable, and subadditive, whence also

$$\beta = \inf_{t>0} \frac{\varphi(t)}{t}$$

is finite (cf. the proof of Theorem 16.2.9), contrary to the supposition.   □

We have also the following annex to Theorem 16.2.7.

Theorem 16.6.9. *Let $f : \mathbb{R}^N \to [-\infty, \infty)$ be a measurable subadditive function. If*

$$\lambda = \liminf_{x \to 0} f(x) = -\infty,$$

then $f = -\infty$ in $\mathbb{R}^N$.

Proof. There exists a sequence $\{t_n\}_{n \in \mathbb{N}}$ such that

$$\lim_{n \to \infty} t_n = 0 \quad \text{and} \quad \lim_{n \to \infty} f(t_n) = -\infty.$$

Take an arbitrary $x \in \mathbb{R}^N$. By Lemma 16.2.3 there exists a neighbourhood U of x and a constant $M > 0$ such that

$$f(t) < M \qquad \text{for } t \in U. \tag{16.6.3}$$

Further, there exist an $n_0 \in \mathbb{N}$ such that

$$x - t_n \in U \qquad \text{for } n > n_0. \tag{16.6.4}$$

Now, we have by (16.6.3) and (16.6.4), for $n > n_0$,

$$f(x) = f(t_n + x - t_n) \leqslant f(t_n) + f(x - t_n) < f(t_n) + M.$$

Letting $n \to \infty$, we obtain hence $f(x) = -\infty$.   □

In the one-dimensional case ($N = 1$) we can get further information about the set P_f. Theorems 16.6.10–16.6.12 say that if $f : \mathbb{R} \to (-\infty, \infty]$ is subadditive and infinitary, then the set of those $x \in \mathbb{R}$ at which $f(x) = \infty$ is rather large.

In the sequel I_1 denotes either $(0, \infty)$ or $(-\infty, 0)$, and $I_2 = (\mathbb{R} \setminus \{0\}) \setminus I_1$ is either $(-\infty, 0)$, or $(0, \infty)$, respectively.

Theorem 16.6.10. *Let $f : \mathbb{R} \to [-\infty, \infty]$ be subadditive. If $I_2 \cap P_f \neq \varnothing$, then either $P_f = \mathbb{R}$, or $f(x) = \infty$ on a dense subset of I_1.*

Proof. Suppose that the set

$$\{x \in I_1 \mid f(x) = \infty\} = I_1 \setminus P_f$$

is not dense in I_1. Then $\text{int}(I_1 \cap P_f) \neq \varnothing$, whence also $\text{int} P_f \neq \varnothing$, and thus $m_i(P_f) > 0$. Take an arbitrary $x_0 \in \text{int}(I_1 \cap P_f) \subset \text{int} P_f$. By hypothesis there exists a $y \in I_2 \cap P_f$. So $s = y/x_0 < 0$, and $y = sx_0$. By Theorem 16.6.6 $P_f = \mathbb{R}$. $\square$

Theorem 16.6.11. *Let* $f : \mathbb{R} \to [-\infty, \infty]$ *be a measurable subadditive function. If* $a \in I_1$ *and* $f(a) = \infty$, *then*

$$m(\{x \in I_1 \mid |x| < |a| \text{ and } f(x) = \infty\}) \geqslant \frac{1}{2}|a|. \tag{16.6.5}$$

Proof. Let

$$A = \{x \in I_1 \mid |x| < |a| \text{ and } f(x) = \infty\} = (I_1 \cap (-|a|, |a|)) \setminus P_f.$$

Since f is measurable, also the set A is measurable.

Let $x \in I_1 \cap (-|a|, |a|)$, and suppose that $x \notin A$ so that $x \in P_f$. Put $y = a - x$. Then $y \in I_1 \cap (-|a|, |a|)$ and $x + y = a$, whence

$$f(a) = f(x + y) \leqslant f(x) + f(y). \tag{16.6.6}$$

Since $f(x) < \infty$ and $f(a) = \infty$, (16.6.6) implies that $f(y) = \infty$. Consequently $y \in A$ and $x = a - y \in a - A$. So if $x \in I_1 \cap (-|a|, |a|)$, then either $x \in A$, or $x \in a - A$, whence $I_1 \cap (-|a|, |a|) = A \cup (a - A)$, and

$$|a| = m(I_1 \cap (-|a|, |a|)) \leqslant m(A) + m(a - A) = 2m(A).$$

Hence we obtain (16.6.5) $\square$

Theorem 16.6.12. *Let* $f : \mathbb{R} \to [-\infty, \infty]$ *be a measurable subadditive function. Then either* $I_1 \cap P_f = \varnothing$, *or* $I_2 \cap P_f = \varnothing$, *or* $m(P_f) = 0$, *or* $P_f = \mathbb{R}$.

Proof. Suppose that $I_1 \cap P_f \neq \varnothing$, $I_2 \cap P_f \neq \varnothing$, and $m(P_f) > 0$ (P_f is measurable since f is measurable). Then either $m(I_1 \cap P_f) > 0$, or $m(I_2 \cap P_f) > 0$. Let I denote that of sets I_1, I_2 for which $m(I \cap P_f) > 0$, and let J denote the other. By the Lebesgue density theorem (cf. also 3.5) the set $I \cap P_f$ contains a density point x_0 of P_f, whereas $J \cap P_f \neq \varnothing$, and hence there exists a $y \in J \cap P_f$. Then $s = y/x_0 < 0$, and $y = sx_0$. By Theorem 16.6.5 $P_f = \mathbb{R}$. $\square$

Theorem 16.6.12 admits an extension to the N-dimensional case (Rosenbaum [269]), which we state here without proof.

Theorem 16.6.13. *Let* $f : \mathbb{R}^N \to [-\infty, \infty]$ *be a measurable subadditive function. Then either there exists an* $(N-1)$-*dimensional hyperplane* π *through the origin such that* P_f *lies entirely on one side of* π, *or* $m(P_f) = 0$, *or* $P_f = \mathbb{R}^N$.

Exercises

1. Let for $x = (x_1, \ldots, x_N) \in \mathbb{R}^N$ such that $x_i \geqslant 0$, $i = 1, \ldots, N$, the function f be defined by

$$f(x) = \left(\sum_{i=1}^N x_i^r \right)^{1/r}$$

with a certain $r > 1$. Show that f is subadditive.

2. Suppose that $\{A_n\}_{n \in \mathbb{N}}$ is a sequence of arbitrary sets $A_n \subset \mathbb{R}^N$, and $\{c_n\}_{n \in \mathbb{N}}$ is a sequence of arbitrary positive constants. Further, let $a, b \in \mathbb{R}$ be such that $0 \leqslant a \leqslant 2b$ and $0 \leqslant b \leqslant 2a$. Let the functions $f_n : \mathbb{R}^N \to \mathbb{R}$ be defined by

$$f_n(x) = \begin{cases} a & \text{for } x \in A_n, \\ b & \text{for } x \notin A_n, \end{cases}$$

and put

$$f(x) = \sum_{n=1}^{\infty} c_n f_n(x).$$

Show that the function $f : \mathbb{R}^N \to [0, +\infty]$ is subadditive. The condition $b \leqslant 2a$ may be dropped if all the sets A_n, $n \in \mathbb{N}$, fulfil the condition (16.1.3).

3. Let $f : (0, \infty) \to \mathbb{R}$ be a measurable subadditive function. Show that if f is convex, then $f(x)/x$ is decreasing in $(0, \infty)$ (Hille-Philips [142]).

4. Let $C \subset \mathbb{R}^N$ be a set such that for every $x \in C$ and $t > 1$ also $tx \in C$. A function $f : C \to \mathbb{R}$ is called *quasi-homogeneous* iff $f(tx) \leqslant tf(x)$ for all $x \in C$ and $t > 1$. The function f is called *weakly-quasi-homogeneous* iff for every $x \in C$ and for every $n \in \mathbb{N}$ there exists a $t_n > n$ such that

$$f(t_n x) \leqslant t_n f(x). \qquad (*)$$

Let $f : (0, \infty) \to \mathbb{R}$ be a measurable function. Show that (Rosenbaum [269]):

   (i) If f is concave and $\lim\limits_{x \to 0+} f(x) \geqslant 0$, then f is quasi-homogeneous;

   (ii) f is quasi-homogeneous if and only if $f(x)/x$ is decreasing in $(0, \infty)$;

   (iii) If f is quasi-homogeneous, then f is subadditive;

   (iv) If f is subadditive, then f is weakly-quasi-homogeneous.

5. Let $f : \mathbb{R}^N \to \mathbb{R}$ be a continuous convex function. Show that f is subadditive if and only if f is quasi-homogeneous (Rosenbaum [269]).
[Hint: To show that f is quasi-homogeneous, represent $t > 1$ as $t = \lambda n + (1 - \lambda)(n + 1)$, where $n = [t]$ is the integral part of t.]

6. Let $f : \mathbb{R}^N \to \mathbb{R}$ be a weakly-quasi-homogeneous continuous convex function. Show that f is subadditive (Rosenbaum [269]).
[Hint: Use the relation $f(x + y) = \lim\limits_{n \to \infty} f\big(t_n^{-1}(t_n x) + (1 - t_n^{-1})y\big)$, where t_n are chosen according to $(*)$.]

7. Let $f : \mathbb{R}^N \to \mathbb{R}$ be a subadditive function such that $\lim\limits_{x \to 0} f(x) = 0$. Show that f is continuous in $\mathbb{R}^N$.

8. Show that in Theorem 16.2.1 the condition $f(0) = 0$ is essential (Hille-Philips [142]).

   [Hint: Take in Exercise 16.2 $N = 1$, $b = \dfrac{1}{2}$, $a = 1$, $c_n = 2^{-n}$, $A_n = \dfrac{1}{n}(\mathbb{N} \cup -(\mathbb{N})) = \dfrac{1}{n}(\mathbb{Z} \setminus \{0\})$, $n \in \mathbb{N}$. Then f is discontinuous at the points $x \in \mathbb{Q} \setminus \{0\}$, and continuous elsewhere. Actually we have $\lim\limits_{y \to x} f(y) = \dfrac{1}{2}$ for every $x \in \mathbb{R}$.]

9. Let $f : \mathbb{R} \to \mathbb{R}$ be a measurable subadditive function, and put

$$I_f(x) = \limsup_{h \to 0} \frac{1}{2h} \int_{-h}^{h} f(x+u)du,$$

$$i_f(x) = \liminf_{h \to 0} \frac{1}{2h} \int_{-h}^{h} f(x+u)du.$$

   Show that the function I_f, i_f are subadditive, and

$$0 \leqslant I_f(x) - i_f(x) \leqslant I_f(0)$$

   for every $x \in \mathbb{R}$ (Hille-Philips [142]).
   [Hint: Integrate the inequality $f(x+y+(\alpha+\beta)u) \leqslant f(x+\alpha u) + f(y+\beta u)$ with α and β suitably chosen.]

10. Let $f : (0, \infty) \to \mathbb{R}$ be a measurable and subadditive function, and set $M = \int_0^\infty \exp f(t)dt$. Show that if $M < \infty$, then $f(x) \leqslant 2 \log \dfrac{M}{x}$ for $x > 0$ (Phillips [255]).

    [Hint: Put $g = \exp f$ and show that $4g(x) \leqslant [g(x-y) + g(y)]^2$. Then use the relation $x[g(x)]^{1/2} = 2 \int_0^{x/2} [g(x)]^{1/2}dy$.]

11. Let $f : \mathbb{R}^N \to \mathbb{R}$ be an arbitrary function. Show that if the limit $\lim\limits_{t \to \infty} f(tx)/t$ exists for every $x \in \mathbb{R}^N$ and is a convex function of x, then $f = \varphi + \psi$, where φ is subadditive and continuous, and $\lim\limits_{t \to \infty} \Psi(tx)/t = 0$ for every $x \in \mathbb{R}^N$ (Rosenbaum [269]).

12. Let $f : \mathbb{R} \to [-\infty, \infty]$ be a measurable subadditive function. Show that if $\lim\limits_{x \to 0} f(x) = \infty$, then $f(x) = \infty$ almost everywhere in $(-\infty, 0)$ or almost everywhere in $(0, \infty)$ (Hille-Phillips [142]).

Chapter 17

Nearly Additive Functions and Nearly Convex Functions

17.1 Approximately additive functions

Under nearly additive and nearly convex functions we understand two modifications of additive functions and of convex functions: approximtely additive functions and approximtely convex functions, where the defining equation (5.2.1) resp. inequality (7.1.1) is satisfied only with some degree of accuracy, and almost additive functions and almost convex functions, where (5.2.1) resp. (5.3.1) is postulated only almost everywhere in the domain in question. (An interesting combination of both these conditions was considered by R. Ger [114].) Similar considerations will be carried out also for some other related classes of functions.

We start our considerations with approximtely additive functions.

A function $f : \mathbb{R}^N \to \mathbb{R}$ is called ε-*additive* iff the inequality

$$|f(x + y) - f(x) - f(y)| \leqslant \varepsilon \tag{17.1.1}$$

holds for all $x, y \in \mathbb{R}^N$. A function $f : \mathbb{R}^N \to \mathbb{R}$ is called *approximately additive* iff it is ε-additive with a certain $\varepsilon > 0$.

The main result concerning approximtely additive functions is due to D. H. Hyers [150] (however, cf. Pólya-Szegő [257], Chapter 3, 3.1, Problem 99; cf. also Albert-Baker [11], Rätz [264]; and also Baker [15], Baker-Lawrence-Zorzitto [16], Cholewa [42], [43], [44], Forti [88], Moszner [233], Tabor [313], and further sections of this chapter concerning related results).

Theorem 17.1.1. *Let* $f : \mathbb{R}^N \to \mathbb{R}$ *be an ε-additive function. Then there exists a unique additive function* $g : \mathbb{R}^N \to \mathbb{R}$ *such that*

$$|f(x) - g(x)| \leqslant \varepsilon \tag{17.1.2}$$

for $x \in \mathbb{R}^N$.

Proof. Setting in (17.1.1) $y = x$ we obtain

$$|f(2x) - 2f(x)| \leqslant \varepsilon,$$

whence

$$\left|\frac{1}{2}f(2x) - f(x)\right| \leqslant \frac{1}{2}\varepsilon. \tag{17.1.3}$$

By induction we show that for every $n \in \mathbb{N}$ we have

$$\left|2^{-n}f(2^n x) - f(x)\right| \leqslant (1 - 2^{-n})\varepsilon. \tag{17.1.4}$$

For $n = 1$ (17.1.4) reduces to (17.1.3). Assuming (17.1.4) true for an $n \in \mathbb{N}$, we have by (17.1.3)

$$\left|2^{-(n+1)}f(2^{n+1}x) - f(x)\right| \leqslant 2^{-n}\left|\frac{1}{2}f(2 \cdot 2^n x) - f(2^n x)\right| + \left|2^{-n}f(2^n x) - f(x)\right|$$

$$\leqslant 2^{-(n+1)}\varepsilon + (1 - 2^{-n})\varepsilon = (1 - 2^{-(n+1)})\varepsilon,$$

and thus we obtain (17.1.4) for $n + 1$, which completes the induction.

Write

$$g_n(x) = 2^{-n}f(2^n x), \quad x \in \mathbb{R}^N, \quad n \in \mathbb{N}.$$

It follows by (17.1.4) that for arbitrary $n, m \in \mathbb{N}$ and $x \in \mathbb{R}^N$ we have

$$|g_{n+m}(x) - g_n(x)| = 2^{-n}\left|2^{-m}f(2^m \cdot 2^n x) - f(2^n x)\right| \leqslant 2^{-n}(1 - 2^{-m})\varepsilon < 2^{-n}\varepsilon,$$

which means that for every $x \in \mathbb{R}^N$ the sequence $\{g_n(x)\}_{n \in \mathbb{N}}$ is a Cauchy sequence, and consequently converges. Let $g(x)$ be its limit. We have by (17.1.1)

$$2^{-n}\left|f(2^n(x + y)) - f(2^n x) - f(2^n y)\right| \leqslant 2^{-n}\varepsilon,$$

i.e.,

$$|g_n(x + y) - g_n(x) - g_n(y)| \leqslant 2^{-n}\varepsilon,$$

whence, on letting $n \to \infty$, we obtain

$$g(x + y) - g(x) - g(y) = 0,$$

which means that g is additive. (17.1.2) results from (17.1.4) on letting $n \to \infty$.

Now suppose that (17.1.2) holds with an additive function $g : \mathbb{R}^N \to \mathbb{R}$. By Theorem 5.2.1 $g(nx) = ng(x)$ for $n \in \mathbb{N}$ and $x \in \mathbb{R}^N$, whence by (17.1.2)

$$|f(nx) - ng(x)| \leqslant \varepsilon.$$

Dividing by n and passing to the limit as $n \to \infty$ we obtain

$$g(x) = \lim_{n \to \infty} \frac{f(nx)}{n}. \tag{17.1.5}$$

This proves the uniqueness of g. $\qquad \square$

Corollary 17.1.1. *Let $f : \mathbb{R}^N \to \mathbb{R}$ be an ε-additive function, and let $g : \mathbb{R}^N \to \mathbb{R}$ be the additive function fulfilling (17.1.2). Then g is given by formula (17.1.5).*

Let us observe that every function $f : \mathbb{R}^N \to \mathbb{R}$ fulfilling inequality (17.1.2), where $g : \mathbb{R}^N \to \mathbb{R}$ is an additive function, is approximtely additive. In fact, we have for arbitrary $x, y \in \mathbb{R}^N$, since g is additive,

$$f(x+y) - f(x) - f(y) = f(x+y) - f(x) - f(y) - g(x+y) + g(x) + g(y),$$

whence

$$|f(x+y) - f(x) - f(y)| \leqslant |f(x+y) - g(x+y)| + |f(x) - g(x)| + |f(y) - g(y)| \leqslant 3\varepsilon.$$

Consequently f is 3ε-additive.

Theorem 17.1.2. *Let $f : \mathbb{R}^N \to \mathbb{R}$ be an ε-additive function, and let $g : \mathbb{R}^N \to \mathbb{R}$ be the additive function fulfilling (17.1.2). If f is measurable, or bounded above, or below, on a set $T \in \mathfrak{A}$, then g is continuous.*

Proof. Inequlity (17.1.2) can be written as

$$f(x) - \varepsilon \leqslant g(x) \leqslant f(x) + \varepsilon.$$

If f is bounded above, or below, on a set $T \subset \mathbb{R}^N$, then so is also g. If $T \in \mathfrak{A}$, then g is continuous in virtue of Theorems 9.1.2 and 10.2.2 and of the definition of the class $\mathfrak{B}$. If f is measurable, then g is continuous in virtue of Theorem 9.4.1. $\square$

17.2 Approximately multiadditive functions

The result in this section are due to M. Albert and J. A. Baker [11].

Let $p \in \mathbb{N}$ be fixed.

Theorem 17.2.1. *If a function $f : \mathbb{R}^{pN} \to \mathbb{R}$ satisfies for every $x_1, \ldots, x_p, y_1, \ldots, y_p \in \mathbb{R}^N$ the system of inequalities*

$$|f(x_1, \ldots, x_{i-1}, x_i + y_i, x_{i+1}, \ldots, x_p) - f(x_1, \ldots, x_p)$$
$$- f(x_1, \ldots, x_{i-1}, y_i, x_{i+1}, \ldots, x_p)| \leqslant \varepsilon_i, \quad i = 1, \ldots, p, \quad (17.2.1)$$

where $\varepsilon_1, \ldots, \varepsilon_p$ are positive numbers, then there exists a unique p-additive function $g : \mathbb{R}^{pN} \to \mathbb{R}$ such that

$$|f(x_1, \ldots, x_p) - g(x_1, \ldots, x_p)| \leqslant \varepsilon = \min(\varepsilon_1, \ldots, \varepsilon_p) \qquad (17.2.2)$$

for all $x_1, \ldots, x_p \in \mathbb{R}^N$. Moreover, if f is symmetric, then g also is symmetric.

Proof. By a suitable renumeration of the variables we may achieve, without violating (17.2.1), that

$$\min(\varepsilon_1, \ldots, \varepsilon_p) = \varepsilon_1. \qquad (17.2.3)$$

By Theorem 17.1.1 and Corollary 17.1.1, for every $x_1, \ldots, x_p \in \mathbb{R}^N$ the limit

$$g(x_1, \ldots, x_p) = \lim_{n \to \infty} \frac{f(nx_1, \ldots, x_p)}{n}$$

exists, and with fixed $x_2, \ldots, x_p$ is an additive function of x_1 and

$$|f(x_1, \ldots, x_p) - g(x_1, \ldots, x_p)| \leqslant \varepsilon_1$$

for every $x_1, \ldots, x_p \in \mathbb{R}^N$, which together with (17.2.3) yields (17.2.2). For every $i = 2, \ldots, p$, and every $x_1, \ldots, x_p, y_i \in \mathbb{R}^N$ we have by (17.2.1)

$$|f(nx_1, x_2, \ldots, x_{i-1}, x_i + y_i, x_{i+1}, \ldots, x_p) - f(nx_1, x_2, \ldots, x_p) - \cdots$$
$$\cdots - f(nx_1, x_2, \ldots, x_{i-1}, y_i, x_{i+1}, \ldots, x_p)| \leqslant \varepsilon_i,$$

whence, on dividing by n and passing to the limit as $n \to \infty$, we obtain that g is p-additive. After permutating variables so as to return to the original numeration, g remains p-additive, and (17.2.2) still holds.

To prove the uniqueness, assume that (17.2.2) holds with two p-additive functions g_1 and g_2, and put $g = g_1 - g_2$. Then $g : \mathbb{R}^{pN} \to \mathbb{R}$ also is p-additive (Lemma 13.4.1), and

$$\begin{aligned}|g(x_1, \ldots, x_p)| &= |g_1(x_1, \ldots, x_p) - g_2(x_1, \ldots, x_p)| \\ &\leqslant |g_1(x_1, \ldots, x_p) - f(x_1, \ldots, x_p)| + |f(x_1, \ldots, x_p) - g_2(x_1, \ldots, x_p)| \\ &\leqslant 2\varepsilon\end{aligned}$$

so that g is bounded in $\mathbb{R}^N$. By Lemma 13.4.3 $g = 0$, i.e., $g_1 = g_2$.

Now assume that f is symmetric, and let $(i_1, \ldots, i_p)$ be an arbitrary permutation of the numbers $(1, \ldots, p)$. Define the function $h : \mathbb{R}^{pN} \to \mathbb{R}$ by

$$h(x_1, \ldots, x_p) = g(x_{i_1}, \ldots, x_{i_p}). \tag{17.2.4}$$

By Lemma 13.4.2 h is p-additive, and we have due to the symmetry of f

$$|f(x_1, \ldots, x_p) - h(x_1, \ldots, x_p)| = |f(x_{i_1}, \ldots, x_{i_p}) - g(x_{i_1}, \ldots, x_{i_p})| \leqslant \varepsilon,$$

whence by the uniqueness part of the theorem

$$h(x_1, \ldots, x_p) = g(x_1, \ldots, x_p).$$

This together with (17.2.4) yields

$$g(x_{i_1}, \ldots, x_{i_p}) = g(x_1, \ldots, x_p),$$

i.e., g is symmetric. $\qquad\square$

17.3   Functions with bounded differences

Now we will derive analogous results for polynomial functions (Hyers [151][1], [152], Albert-Baker [11]; cf. also Whitney [323], [324]).

---

[1] Added in the 2nd edition by K. Baron.

Theorem 17.3.1. *Let $f : \mathbb{R}^N \to \mathbb{R}$ satisfy (with a certain $\varepsilon > 0$)*

$$\left| \Delta_{h_1 \ldots h_{p+1}} f(x) \right| \leqslant \varepsilon \tag{17.3.1}$$

for all $x, h_1, \ldots, h_{p+1} \in \mathbb{R}^N$, where $p \in \mathbb{N} \cup \{0\}$ is fixed. Then there exist symmetric k-additive functions $F_k : \mathbb{R}^{kN} \to \mathbb{R}$, $k = 0, \ldots, p$ unique except for F_0, such that

$$\left| f(x) - g(x) \right| \leqslant \varepsilon \tag{17.3.2}$$

for all $x \in \mathbb{R}^N$, where

$$g(x) = \sum_{k=0}^{p} f_k(x), \tag{17.3.3}$$

and $f_0, \ldots, f_p$ are the diagonalizations of $F_0, \ldots, F_p$, respectively.

Proof. The proof (Albert-Baker [11]) is similar to that of Theorem 15.9.2, and the proof of the existence of g will be by induction on p.

For $p = 0$ f satisfies $|\Delta_x f(0)| \leqslant \varepsilon$ for every $x \in \mathbb{R}^N$, i.e.,

$$\left| f(x) - f(0) \right| \leqslant \varepsilon.$$

With $F_0 = f_0 = f(0)$ we obtain hence (17.3.2) with (17.3.3).

Now assume that the theorem is true for a $p - 1 \in \mathbb{N} \cup \{0\}$, and let $f : \mathbb{R}^N \to \mathbb{R}$ be a function fulfilling (17.3.1) for all $x, h_1, \ldots, h_{p+1} \in \mathbb{R}^N$. For every $x \in \mathbb{R}^N$ define the function $F_x : \mathbb{R}^{pN} \to \mathbb{R}$ by

$$F_x(x_1, \ldots, x_p) = \Delta_{x_1, \ldots, x_p} f(x). \tag{17.3.4}$$

We have by Corollary 15.1.3, Lemma 15.1.3, and by (17.3.1), for every fixed $x \in \mathbb{R}^N$ and arbitrary i, $1 \leqslant i \leqslant p$, and arbitrary $x_1, \ldots, x_p, y_i \in \mathbb{R}^N$,

$$\begin{aligned}
&\left| F_x(x_1, \ldots, x_{i-1}, x_i + y_i, x_{i+1}, \ldots, x_p) - F_x(x_1, \ldots, x_p) \right. \\
&\left. - F_x(x_1, \ldots, x_{i-1}, y_i, x_{i+1}, \ldots, x_p) \right| \\
&= \left| \Delta_{x_1 \ldots x_{i-1} x_{i+1} \ldots x_p} \left[\Delta_{x_i + y_i} f(x) - \Delta_{x_i} f(x) - \Delta_{y_i} f(x) \right] \right| \\
&= \left| \Delta_{x_1 \ldots x_{i-1} x_{i+1} \ldots, x_p} \Delta_{x_i y_i} f(x) \right| = \left| \Delta_{x_1 \ldots x_p y_i} f(x) \right| \leqslant \varepsilon.
\end{aligned}$$

By Theorem 17.2.1, for every $x \in \mathbb{R}^N$, there exists a unique p-additive function $G_x : \mathbb{R}^{pN} \to \mathbb{R}$ such that

$$\left| F_x(x_1, \ldots, x_p) - G_x(x_1, \ldots, x_p) \right| \leqslant \varepsilon \tag{17.3.5}$$

for all $x_1, \ldots, x_p \in \mathbb{R}^N$. Moreover, since by Corollary 15.1.3 function (17.3.4) is symmetric, also the function G_x is symmetric (Theorem 17.2.1).

For arbitrary $x, y \in \mathbb{R}^N$ and arbitrary $x_1, \ldots, x_p \in \mathbb{R}^N$ we have by (17.3.4), (17.3.1), and Lemma 15.1.1

$$\begin{aligned}
\left| F_y(x_1, \ldots, x_p) - F_x(x_1, \ldots, x_p) \right| &= \left| \Delta_{x_1 \ldots x_p} \left[f(y) - f(x) \right] \right| \tag{17.3.6} \\
&= \left| \Delta_{x_1 \ldots x_p} \left[f(x + (y - x)) - f(x) \right] \right| \\
&= \left| \Delta_{x_1 \ldots x_p} \Delta_{y-x} f(x) \right| \\
&= \left| \Delta_{x_1 \ldots x_p (y-x)} f(x) \right| \leqslant \varepsilon,
\end{aligned}$$

whereas by (17.3.5)

$$|F_y(x_1, \ldots, x_p) - G_y(x_1, \ldots, x_p)| \leqslant \varepsilon. \tag{17.3.7}$$

Relations (17.3.5), (17.3.6) and (17.3.7) yield

$$|G_y - G_x| \leqslant |G_y - F_y| + |F_y - F_x| + |F_x - G_x| \leqslant 3\varepsilon.$$

Consequently the function $G_y - G_x$ is p-additive (Lemma 13.4.1) and bounded in $\mathbb{R}^{pN}$. By Lemma 13.4.3 $G_x = G_y$. This shows that G_x is independent of x, and we may define a symmetric p-additive function $F_p : \mathbb{R}^{pN} \to \mathbb{R}$ by

$$F_p(x_1, \ldots, x_p) = \frac{1}{p!} G_x(x_1, \ldots, x_p). \tag{17.3.8}$$

Let $f_p : \mathbb{R}^N \to \mathbb{R}$ be the diagonalization of F_p, and define $g_p : \mathbb{R}^N \to \mathbb{R}$ as $g_p = f - f_p$. We have by (17.3.4), (17.3.8), (17.3.5), and by Lemmas 15.1.1 and 15.9.2, for arbitrary $h_1, \ldots, h_p \in \mathbb{R}^N$,

$$\begin{aligned}
\left|\Delta_{h_1 \ldots h_p} g_p(x)\right| &= \left|\Delta_{h_1 \ldots h_p} f(x) - \Delta_{h_1 \ldots h_p} f_p(x)\right| \\
&= \left|F_x(h_1, \ldots, h_p) - p! F_p(h_1, \ldots, h_p)\right| \\
&= \left|F_x(h_1, \ldots, h_p) - G_x(h_1, \ldots, h_p)\right| \leqslant \varepsilon.
\end{aligned}$$

By the induction hypothesis there exist symmetric k-additive functions $F_k : \mathbb{R}^{kN} \to \mathbb{R}$, $k = 0, \ldots, p - 1$, such that

$$\left|g_p(x) - \sum_{k=0}^{p-1} f_k(x)\right| \leqslant \varepsilon, \tag{17.3.9}$$

where $f_0, \ldots, f_{p-1}$ are the diagonalizations of $F_0, \ldots, F_{p-1}$, respectively. Define $g : \mathbb{R}^N \to \mathbb{R}$ by (17.3.3). Then (17.3.9) goes into (17.3.2) in view of the definition of g_p.

To prove the uniqueness, suppose that besides $F_0, \ldots, F_p$, there exist also symmetric k-additive functions $\widehat{F}_k : \mathbb{R}^{kN} \to \mathbb{R}$, $k = 0, \ldots, p$, such that

$$|f(x) - \hat{g}(x)| \leqslant \varepsilon \tag{17.3.10}$$

in $\mathbb{R}^N$, where

$$\hat{g}(x) = \sum_{k=0}^{p} \hat{f}_k(x),$$

and $\hat{f}_0, \ldots, \hat{f}_p$ are the diagonalizations of $\widehat{F}_0, \ldots, \widehat{F}_p$, respectively. By (17.3.2) and (17.3.10)

$$|g(x) - \hat{g}(x)| \leqslant 2\varepsilon. \tag{17.3.11}$$

We have

$$g(x) - \hat{g}(x) = \sum_{k=0}^{p} \left(f_k(x) - \hat{f}_k(x)\right), \tag{17.3.12}$$

and every function $f_k - \hat{f}_k$ is the diagonalization of the symmetric k-additive function $F_k - \widehat{F}_k$, $k = 0, \ldots, p$. Relation (17.3.11) says that function (17.3.12) is bounded in $\mathbb{R}^N$. By Lemma 15.9.3 $F_k = \widehat{F}_k$ for $k = 0, \ldots, p$. □

Theorem 17.3.2. *Let $f : \mathbb{R}^N \to \mathbb{R}$ satisfy (with a certain $\varepsilon > 0$)*

$$\left| \Delta_h^{p+1} f(x) \right| \leqslant \varepsilon \qquad (17.3.13)$$

for all $x, h \in \mathbb{R}^N$, where $p \in \mathbb{N}$ is fixed. Then there exists a polynomial function $g : \mathbb{R}^N \to \mathbb{R}$ of order p, unique up to an additive constant, such that

$$|f(x) - g(x)| \leqslant 2^{p+1} \varepsilon \qquad (17.3.14)$$

for all $x \in \mathbb{R}^N$.

Proof. In view of Theorem 15.1.2 relation (17.3.13) implies that

$$\left| \Delta_{h_1 \ldots h_{p+1}} f(x) \right| \leqslant 2^{p+1} \varepsilon$$

for all $x, h_1, \ldots, h_{p+1} \in \mathbb{R}^N$. By Theorem 17.3.1 there exist symmetric k-additive functions $F_k : \mathbb{R}^{kN} \to \mathbb{R}$, $k = 0, \ldots, p$, such that (17.3.14) holds in $\mathbb{R}^N$, where g is given by (17.3.3), and $f_0, \ldots, f_p$ are the diagonalizations of $F_0, \ldots, F_p$, respectively. By Theorem 15.9.1 g is a polynomial function of order p.

Now suppose that $g_1, g_2 : \mathbb{R}^N \to \mathbb{R}$ are polynomial functions of order p fulfilling

$$|f(x) - g_1(x)| \leqslant 2^{p+1} \varepsilon \quad \text{and} \quad |f(x) - g_2(x)| \leqslant 2^{p+1} \varepsilon$$

for all $x \in \mathbb{R}^N$. Then

$$|g_1(x) - g_2(x)| \leqslant 2^{p+2} \varepsilon. \qquad (17.3.15)$$

By Lemma 15.3.2 $g_1 - g_2$ is a polynomial function, and by (17.3.15) $g_1 - g_2$ is bounded in $\mathbb{R}^N$. In virtue of Corollary 15.9.2 $g_1 - g_2 = \text{const}$. □

The estimation in (17.3.14) can be improved. Actually, it can be shown (Whitney [323]) that, under the conditions of Theorem 17.3.2, the function g fulfils the inequality

$$|f(x) - g(x)| \leqslant 2\varepsilon / \max_k \binom{p+1}{k}$$

for all $x \in \mathbb{R}^N$.

Theorem 17.3.3. *Let $f : \mathbb{R}^N \to \mathbb{R}$ satisfy (17.3.13) (with certain $\varepsilon > 0$ and $p \in \mathbb{N}$), and suppose that f is measurable or bounded on a set $T \subset \mathbb{R}^N$, where $m_i(T) > 0$ or T is of the second category and with the Baire property. Then there exists a polynomial $g : \mathbb{R}^N \to \mathbb{R}$ of degree at most p, unique up to an additive constant, such that (17.3.14) holds for all $x \in \mathbb{R}^N$.*

Proof. If f is measurable, then the argument in the proof of Theorem 15.5.4 shows that f is bounded on a set of positive measure, so we need only consider the case where f is bounded on a suitble set $T \subset \mathbb{R}^N$. By Theorem 17.3.2 there exists a polynomial function $g : \mathbb{R}^N \to \mathbb{R}$ of order p, unique up to an additive constant and such that (17.3.14) holds in $\mathbb{R}^N$. Consequently g is also bounded on T, and by Theorem 15.5.2 g is continuous. By Theorem 15.9.4 g is a polynomial of degree at most p. □

17.4 Approximately convex functions

For convex functions we do not have a full analogue of Theorem 17.1.1 (cf. also Cholewa [43]). The corresponding result is weaker, and the proof (cf. Hyers-Ulam [153]) is much more complicated.

Let $D \subset \mathbb{R}^N$ be a convex and open set. A function $f : D \to \mathbb{R}$ is called ε-convex iff

$$f\big(\lambda x + (1 - \lambda)y\big) \leqslant \lambda f(x) + (1 - \lambda)f(y) + \varepsilon \qquad (17.4.1)$$

for every $x, y \in D$ and $\lambda \in [0, 1]$. A function $f : D \to \mathbb{R}$ is called *approximately convex* iff it is ε-convex with a certain $\varepsilon > 0$.

Lemma 17.4.1. *Let $D \subset \mathbb{R}^N$ be a convex and open set, let $n \in \mathbb{N}$, let $x_0, \ldots, x_n \in D$, and let $\alpha_0, \ldots, \alpha_n$ be non-negative numbers such that $\alpha_0 + \cdots + \alpha_n = 1$. Let $x = \alpha_0 x_0 + \cdots + \alpha_n x_n$. If $f : D \to \mathbb{R}$ is an ε-convex function, then*

$$f(x) \leqslant \sum_{i=0}^{n} \alpha_i f(x_i) + \frac{n^2 + 3n}{2n + 2}\varepsilon. \qquad (17.4.2)$$

Proof. First note that by Lemma 5.1.3 $x \in D$, so that $f(x)$ is meaningful. The proof of (17.4.2) will be by induction on n.

For $n = 1$ (17.4.2) reduces to (17.4.1). Assume (17.4.2) to hold for an $n \in \mathbb{N}$. Take arbitrary points $x_0, \ldots, x_{n+1} \in D$ and non-negative real numbers $\alpha_0, \ldots, \alpha_{n+1}$ such that

$$\sum_{i=0}^{n+1} \alpha_i = 1, \qquad (17.4.3)$$

and put $x = \alpha_0 x_0 + \cdots + \alpha_{n+1} x_{n+1}$. We may assume that the points x_i have been numbered in such a way that $\alpha_{n+1} \geqslant \alpha_i$ for $i = 0, \ldots, n$. If $\alpha_{n+1} = 1$, then by (17.4.3) $\alpha_i = 0$ for $i = 0, \ldots, n$, $x = x_{n+1}$, and the inequality to be proved

$$f(x) \leqslant f(x_{n+1}) + \frac{(n + 1)^2 + 3(n + 1)}{2(n + 1) + 2}\varepsilon$$

is trivial. If $\alpha_{n+1} < 1$, then put

$$\lambda = 1 - \alpha_{n+1} > 0, \qquad (17.4.4)$$

$\beta_i = \alpha_i/\lambda$, $i = 0, \ldots, n$, $z = \beta_0 x_0 + \cdots + \beta_n x_n$. All β's are non-negative, and by (17.4.3) and (17.4.4)

$$\sum_{i=0}^{n} \beta_i = \frac{1}{\lambda}\sum_{i=0}^{n} \alpha_i = \frac{1}{\lambda}(1 - \alpha_{n+1}) = 1.$$

By the induction hypothesis

$$f(z) \leqslant \sum_{i=0}^{n} \beta_i f(x_i) + \frac{n^2 + 3n}{2n + 2}\varepsilon. \qquad (17.4.5)$$

On the other hand

$$\lambda z + (1 - \lambda)x_{n+1} = \alpha_0 x_0 + \cdots + \alpha_n x_n + \alpha_{n+1} x_{n+1} = x,$$

whence by (17.4.1)

$$f(x) \leqslant \lambda f(z) + (1 - \lambda)f(x_{n+1}) + \varepsilon. \tag{17.4.6}$$

Since $\alpha_{n+1} \geqslant \alpha_i$ for $i = 0, \ldots, n$, we have by (17.4.3)

$$1 = \sum_{i=0}^{n+1} \alpha_i \leqslant (n+2)\alpha_{n+1},$$

whence $\alpha_{n+1} \geqslant 1/(n+2)$ and $\lambda = 1 - \alpha_{n+1} \leqslant (n+1)(n+2)$. Relations (17.4.5) and (17.4.6) yield

$$f(x) \leqslant \sum_{i=0}^{n} \alpha_i f(x_i) + \alpha_{n+1} f(x_{n+1}) + \lambda \frac{n^2 + 3n}{2n+2} \varepsilon + \varepsilon,$$

whence

$$f(x) \leqslant \sum_{i=0}^{n+1} \alpha_i f(x_i) + \left(\frac{n+1}{n+2} \frac{n^2 + 3n}{2n+2} + 1 \right) \varepsilon. \tag{17.4.7}$$

Since

$$\frac{n+1}{n+2} \frac{n^2 + 3n}{2n+2} + 1 = \frac{(n+1)^2 + 3(n+1)}{2(n+1) + 2},$$

relation (17.4.7) yields (17.4.2) for $n + 1$. Induction completes the proof. $\square$

Lemma 17.4.2. *Let $D \subset \mathbb{R}^N$ be a convex and open set, and let $f : D \to \mathbb{R}$ be an ε-convex function. Then f is locally bounded at every point of D.*

Proof. Let $\bar{x} \in D$. By Remark 5.1.1 there exists an N-dimensionl simplex $S \subset D$ such that $\bar{x} \in \text{int } S$. Let $x_0, \ldots, x_N$ be the vertices of S so that $S = \text{conv}\{x_0, \ldots, x_N\}$. We have $x_0, \ldots, x_N \in D$ and for every $x \in S$ there exist non-negative numbers $\alpha_0, \ldots, \alpha_N$ such that $\alpha_0 + \cdots + \alpha_N = 1$ and $x = \alpha_0 x_0 + \cdots + \alpha_N x_N$. Let

$$M = \max\{f(x_0), \ldots, f(x_N)\}.$$

We have by Lemma 17.4.1

$$f(x) \leqslant \sum_{i=0}^{N} \alpha_i f(x_i) + \frac{N^2 + 3N}{2N+2} \varepsilon \leqslant M + \frac{N^2 + 3N}{2N+2} \varepsilon,$$

which means that f is bounded above on S, and hence also on $\text{int } S$. But $\text{int } S$ is a neighbourhood of $\bar{x}$, and so f is locally bounded above at $\bar{x}$.

Let $B \subset D$ be an open ball centered at $\bar{x}$ and such that f is bounded above on B:

$$f(u) \leqslant K \quad \text{for } u \in B. \tag{17.4.8}$$

Take an arbitrary $x \in B$. Then also $2\bar{x} - x = \bar{x} - (x - \bar{x}) \in B$, and we have by (17.4.1) and (17.4.8)

$$f(\bar{x}) \leqslant \frac{1}{2} f(x) + \frac{1}{2} f(2\bar{x} - x) + \varepsilon \leqslant \frac{1}{2} f(x) + \frac{1}{2} K + \varepsilon,$$

whence

$$f(x) \geqslant 2f(\bar{x}) - K - 2\varepsilon,$$

which means that f is bounded below on B. Consequently f is locally bounded below at $\bar{x}$.   □

Corollary 17.4.1. *Let $D \subset \mathbb{R}^N$ be a convex and open set, and let $f : D \to \mathbb{R}$ be an ε-convex function. Then f is bounded on every compact $C \subset D$.*

Proof. For every $x \in C$ let B_x be an open ball centered at x and contained in D such that f is bounded on B_x. The compact set C can be covered by a finite number of balls B_x, whence f is bounded on C.   □

The next lemma is known in the theory of convex sets as the Carathéodory theorem (cf., e.g., Eggleston [74]).

Lemma 17.4.3. *Let $A \subset \mathbb{R}^N$ be an arbitrary non-empty set. For every $x \in \operatorname{conv} A$ there exist points $x_0, \ldots, x_r \in A$ such that $x \in \operatorname{conv}\{x_0, \ldots, x_r\}$ and $r \leqslant N$.*

Proof. Let $x \in \operatorname{conv} A$. By Theorem 5.1.3 there exist an $n \in \mathbb{N}$, points $x_0, \ldots, x_n \in A$, and non-negative numbers $\alpha_0, \ldots, \alpha_n$ such that $\alpha_0 + \cdots + \alpha_n = 1$ and $x = \alpha_0 x_0 + \cdots + \alpha_n x_n$. If $n \leqslant N$, there is nothing more to prove, so suppose that $n > N$. The conditions

$$\beta_0 x_0 + \cdots + \beta_n x_n = 0, \tag{17.4.9}$$

$$\beta_0 + \cdots + \beta_n = 1, \tag{17.4.10}$$

represent system of $N + 1$ homogeneous liner equations with $n + 1 > N + 1$ unknowns $\beta_0, \ldots, \beta_n$. Consequently this system has a non-trivial solution. In other words, there exist numbers $\beta_0, \ldots, \beta_n$, not all zeroes and such that (17.4.9) and (17.4.10) hold.

Let T be a set of all those real numbers t which satisfy

$$\alpha_i + t\beta_i \geqslant 0 \quad \text{for } i = 0, \ldots, n. \tag{17.4.11}$$

Clearly $0 \in T$, and $T \neq \mathbb{R}$, since at least one of $\beta_0, \ldots, \beta_n$ is different from zero. Let t_0 be a frontier point of T. T is closed, consequently $t_0 \in T$, which means that $t = t_0$ satisfies (17.4.11). Moreover, we have by (17.4.9) and (17.4.10)

$$\sum_{i=0}^{n} (\alpha_i + t_0\beta_i) = \sum_{i=0}^{n} \alpha_i + t_0 \sum_{i=0}^{n} \beta_i = \sum_{i=0}^{n} \alpha_i = 1$$

and

$$\sum_{i=0}^{n} (\alpha_i + t_0\beta_i)x_i = \sum_{i=0}^{n} \alpha_i x_i + t_0 \sum_{i=0}^{n} \beta_i x_i = \sum_{i=0}^{n} \alpha_i x_i = x.$$

Further, at least for one $i = 0, \ldots, n$, we have $\alpha_i + t_0 \beta_i = 0$. Consequently x can be represented as a linear combintion of $n-1$ points from A with non-negative coefficients summing up to 1.

Proceeding so further, we may reduce the number n to N. $\qquad\square$

Corollary 17.4.2. *Let* $A \subset \mathbb{R}^N$ *be an arbitrary non-empty set. Then*

$$\operatorname{conv} A = \Big\{ x \in \mathbb{R}^N \mid x = \alpha_0 x_0 + \cdots + \alpha_N x_N, \ x_i \in A, \ \alpha_i \geqslant 0$$
$$\text{for } i = 0, \ldots, N, \ \sum_{i=0}^{N} \alpha_i = 1 \Big\}. \tag{17.4.12}$$

Proof. Let B denote the set on the right-hand side of (17.4.12). The inclusion $B \subset \operatorname{conv} A$ follows from Lemma 5.1.3, whereas the inclusion $\operatorname{conv} A \subset B$ results from Lemma 17.4.3. (Note that if x is a combination of a smaller number of points, we may always add terms with arbitrary $x_i \in A$ and $\alpha_i = 0$, so as to obtain a linear combination of $N + 1$ points.) $\qquad\square$

Corollary 17.4.3. *Let* $A \subset \mathbb{R}^N$ *be an arbitrary non-empty compact set. Then the set* $\operatorname{conv} A$ *also is compact.*

Proof. Let

$$\Xi = \Big\{ (\alpha_0, \ldots, \alpha_N) \in \mathbb{R}^{N+1} \mid \alpha_0 \geqslant 0, \ldots, \alpha_N \geqslant 0, \ \sum_{i=0}^{N} \alpha_i = 1 \Big\}.$$

Ξ is a closed subset of the compact set $[0, 1]^{N+1}$, and hence it is compact.

Now define a function $F : \Xi \times A^{N+1} \to \mathbb{R}^N$ by

$$F(\alpha_0, \ldots, \alpha_N, x_0, \ldots, x_N) = \alpha_0 x_0 + \cdots + \alpha_N x_N.$$

The function F is continuous and the set $\Xi \times A^{N+1}$ is compact, since Ξ and A are compact, therefore also the set $F(\Xi \times A^{N+1})$ is compact. But by Corollary 17.4.2

$$F(\Xi \times A^{N+1}) = \operatorname{conv} A,$$

and so $\operatorname{conv} A$ is compact. $\qquad\square$

Lemma 17.4.4. *Let* $S \subset \mathbb{R}^N$ *be an arbitrary non-empty set, and let* $K = \operatorname{conv} S$. *For every frontier point* x *of* K *such that* $x \in K$ *there exist points* $x_0, \ldots, x_{N-1} \in S$ *and non-negative numbers* $\alpha_0, \ldots, \alpha_{N-1}$ *such that* $\alpha_0 + \cdots + \alpha_{N-1} = 1$ *and* $x = \alpha_0 x_0 + \cdots + \alpha_{N-1} x_{N-1}$.

Proof. By Theorem 5.1.7 there exists a support hyperplane H $\big((N-1)$-dimensional$\big)$ of K at x. If we had $H \cap S = \varnothing$, then S would lie entirely in one open half-space π of $\mathbb{R}^N$ determined by H and, since π is a convex set, we would have $K = \operatorname{conv} S \subset \pi$, whence $H \cap K = \varnothing$. But this is impossible, since $x \in H \cap K$.

Consequently $H \cap S \neq \varnothing$. We will prove that

$$\operatorname{conv}(H \cap S) = H \cap K. \tag{17.4.13}$$

Since H and K are convex, the set $H \cap K$ is convex by Theorem 5.1.2, and since evidently $H \cap S \subset H \cap K$, we have $\operatorname{conv}(H \cap S) \subset H \cap K$. To prove the converse inclusion, take an arbitrary $y \in H \cap K$. By Theorem 5.1.3 there exist points $y_1, \ldots, y_n \in S$ and positive numbers $\beta_1, \ldots, \beta_n$ such that $\beta_1 + \cdots + \beta_n = 1$ and $\beta_1 y_1 + \cdots + \beta_n y_n = y$. Let the equation of H be

$$At + B = 0,$$

where $A \in \mathbb{R}^N$, $B \in \mathbb{R}$, and A, B are such that the closed half-space determined by H and containing S is described by the condition

$$At + B \geqslant 0.$$

Hence $Ay_i + B \geqslant 0$ for $i = 1, \ldots, n$, and since

$$Ay + B = \sum_{i=1}^{n} \beta_i (Ay_i + B),$$

the condition $y \in H$, i.e., $Ay + B = 0$, implies tht $Ay_i + B = 0$ for $i = 1, \ldots, n$. Consequently $y_i \in H$, and so $y_i \in H \cap S$, $i = 1, \ldots, n$, whence by Theorem 5.1.3 $y \in \operatorname{conv}(H \cap S)$. This proves the inclusion $H \cap K \subset \operatorname{conv}(H \cap S)$, and concludes the proof of (17.4.13).

Now, we may treat H as $\mathbb{R}^{N-1}$. By Corollary 17.4.2, since $x \in H \cap K$, there exist points $x_0, \ldots, x_{N-1} \in H \cap S$ and non-negative numbers $\alpha_0, \ldots, \alpha_{N-1}$ such that $\alpha_0 + \cdots + \alpha_{N-1} = 1$ and $\alpha_0 x_0 + \cdots + \alpha_{N-1} x_{N-1} = x$, which was to be proved. $\square$

The following theorems are due to D. H. Hyers and S. M. Ulam [153] (cf. also Cholewa [43]).

Theorem 17.4.1. *Let $D \subset \mathbb{R}^N$ be a convex and open set, and let $f : D \to \mathbb{R}$ be an ε-convex function. Then there exists a continuous convex function $g : D \to \mathbb{R}$ such that*

$$g(x) \leqslant f(x) \leqslant g(x) + \left(\frac{N^2 + 3N}{2N + 2} + 1 \right) \varepsilon \tag{17.4.14}$$

for all $x \in D$.

Proof. Let $\{C_n\}_{n \in \mathbb{N}}$ be a sequence of convex and compact subsets of D such that

$$C_n \subset C_{n+1}, \quad n \in \mathbb{N}, \tag{17.4.15}$$

and

$$\bigcup_{n=1}^{\infty} \operatorname{int} C_n = D \tag{17.4.16}$$

(cf. Exercise 17.5). Put $f_n = f \mid C_n$, and $S_n = \operatorname{cl} \operatorname{Gr}(f_n)$, $K_n = \operatorname{conv} S_n$. By Corollary 17.4.1 f is bounded on C_n, so every function f_n is bounded, whence S_n is a compact subset of $\mathbb{R}^{N+1}$. By Corollary 17.4.3, also K_n is a compct subset of $\mathbb{R}^{N+1}$, $n \in \mathbb{N}$. For every $n \in \mathbb{N}$ define the function $g_n : C_n \to \mathbb{R}$ by

$$g_n(x) = \inf\{y \in \mathbb{R} \mid (x, y) \in K_n\}.$$

Since K_n is compact, we have $(x, g_n(x)) \in K_n$ for every $x \in C_n$ and $n \in \mathbb{N}$. Consequently $g_n(x)$ is finite and $g_n : C_n \to \mathbb{R}$ is a function.

Take arbitrary $n \in \mathbb{N}$, $x, y \in C_n$, and $\lambda \in [0, 1]$. We have $(x, g_n(x)) \in K_n$, $(y, g_n(y)) \in K_n$, and, since K_n is convex,

$$\big(\lambda x + (1 - \lambda)y, \lambda g_n(x) + (1 - \lambda)g_n(y)\big) = \lambda\big(x, g_n(x)\big) + (1 - \lambda)\big(y, g_n(y)\big) \in K_n.$$

Consequently, by the definition of g_n,

$$g_n\big(\lambda x + (1 - \lambda)y\big) \leqslant \lambda g_n(x) + (1 - \lambda)g_n(y). \tag{17.4.17}$$

This means that $g_n \mid \operatorname{int} C_n$ $\big($by (17.4.15) and (17.4.16) $\operatorname{int} C_n \neq \varnothing$ for n sufficiently large, and by Theorem 5.1.5 for such n the set $\operatorname{int} C_n$ is convex$\big)$ is a continuous and convex function (Theorem 7.1.1). Fix an $n \in \mathbb{N}$. Three cases may occur for $x \in C_n$:

(i) $\big(x, g_n(x)\big) \in \operatorname{Gr}(f_n)$,

(ii) $\big(x, g_n(x)\big) \in S_n \setminus \operatorname{Gr}(f_n)$,

(iii) $\big(x, g_n(x)\big) \in K_n \setminus S_n$.

In case (i) we have $g_n(x) = f_n(x) = f(x)$, and so, in particular,

$$g_n(x) \leqslant f(x) \leqslant g_n(x) + \varepsilon \leqslant g_n(x) + \left(\frac{N^2 + 3N}{2N + 2} + 1\right)\varepsilon. \tag{17.4.18}$$

In case (ii) there exists a sequence $\{(t_m, z_m)\}_{m \in \mathbb{N}}$ of distinct points such that

$$t_m \in C_n, \quad (t_m, z_m) \in \operatorname{Gr}(f_n), \quad m \in \mathbb{N},$$

and

$$\lim_{m \to \infty} t_m = x, \quad \lim_{m \to \infty} z_m = g_n(x). \tag{17.4.19}$$

Note that if points (t_m, z_m), $(t_{m'}, z_{m'}) \in \operatorname{Gr}(f_n)$ are distinct, then necessarily $t_m \neq t_{m'}$, and consequently all points $t_m, m \in \mathbb{N}$, are distinct. Hence $t_m \neq x$ for all $m \in \mathbb{N}$, except possibly one exceptional value of m. Removing this exceptional element from the sequence $\{(t_m, z_m)\}_{m \in \mathbb{N}}$, we do not spoil the remaining properties of this sequence, and we may assume that $t_m \neq x$ for $m \in \mathbb{N}$.

Let $B \subset D$ be a closed ball centered at x and with a radius $r > 0$, and put

$$y_m = x - r\frac{t_m - x}{|t_m - x|}, \quad m \in \mathbb{N}.$$

Thus $y_m \in B$ for $m \in \mathbb{N}$, and

$$x = \lambda_m t_m + (1 - \lambda_m)y_m, \quad m \in \mathbb{N}, \tag{17.4.20}$$

where

$$\lambda_m = \frac{|x - y_m|}{|t_m - y_m|}, \quad m \in \mathbb{N}.$$

Since B is compact, we may assume, replacing possibly the sequences $\{t_m\}_{m \in \mathbb{N}}$ and $\{y_m\}_{m \in \mathbb{N}}$ by suitable subsequences, that the sequence $\{y_m\}_{m \in \mathbb{N}}$ converges

$$\lim_{m \to \infty} y_m = y_0.$$

Hence by (17.4.19)

$$\lim_{m \to \infty} \lambda_m = \frac{|x - y_0|}{|x - y_0|} = 1. \tag{17.4.21}$$

By (17.4.1) and (17.4.20)

$$f(x) \leqslant \lambda_m f(t_m) + (1 - \lambda_m) f(y_m) + \varepsilon = \lambda_m z_m + (1 - \lambda_m) f(y_m) + \varepsilon, \quad m \in \mathbb{N}. \tag{17.4.22}$$

By Corollary 17.4.1 f is bounded on B so that the sequence $\{f(y_m)\}_{m \in \mathbb{N}}$ is bounded. Letting in (17.4.22) $m \to \infty$, we obtain in view of (17.4.19) and (17.4.21)

$$f(x) \leqslant g_n(x) + \varepsilon, \tag{17.4.23}$$

and since $(x, f_n(x)) \in \mathrm{Gr}(f_n) \subset S_n \subset K_n$, we have

$$g_n(x) \leqslant f_n(x) = f(x). \tag{17.4.24}$$

Relations (17.4.23) and (17.4.24) yield (17.4.18), which means that (17.4.18) holds in case (ii), too.

In case (iii), according to Lemma 17.4.4, there exist points $(x_0, u_0), \ldots,$ $(x_N, u_N) \in S_n$ and non-negative numbers $\alpha_0, \ldots, \alpha_N$ such that

$$\sum_{i=0}^{N} \alpha_i = 1, \quad \sum_{i=0}^{N} \alpha_i x_i = x, \quad \sum_{i=0}^{N} \alpha_i u_i = g_n(x). \tag{17.4.25}$$

Suppose that for a certain j we have $u_j \neq g_n(x_j)$. Since $(x_i, u_i) \in S_n \subset K_n$ for $i = 0, \ldots, N$, we must have $u_i \geqslant g_n(x_i)$ for $i = 0, \ldots, N$, and $u_j > g_n(x_j)$. Hence

$$\sum_{i=0}^{N} \alpha_i u_i > \sum_{i=0}^{N} \alpha_i g_n(x_i)$$

except when $\alpha_j = 0$, when we may replace u_j by $g_n(x_j)$ without violating (17.4.25). On the other hand, since g_n satisfies (17.4.17), we have by (17.4.25) and Lemma 5.3.2

$$\sum_{i=0}^{N} \alpha_i u_i = g_n(x) \leqslant \sum_{i=0}^{N} \alpha_i g_n(x_i).$$

This contradiction shows that $u_i = g_n(x_i)$ for $i = 0, \ldots, N$.

Now, since $\big(x_i, g_n(x_i)\big) \in S_n$ for $i = 0, \ldots, N$, we have by (17.4.18) (case (i) and (ii))

$$f(x_i) \leqslant g_n(x_i) + \varepsilon, \quad i = 0, \ldots, N,$$

whence by (17.4.25) and Lemma 17.4.1

$$f(x) \leqslant \sum_{i=0}^{N} \alpha_i f(x_i) + \frac{N^2 + 3N}{2N + 2}\varepsilon \leqslant \sum_{i=0}^{N} \alpha_i \big(g_n(x_i) + \varepsilon\big) + \frac{N^2 + 3N}{2N + 2}\varepsilon$$

$$= \sum_{i=0}^{N} \alpha_i u_i + \left(\frac{N^2 + 3N}{2N + 2} + 1\right)\varepsilon = g_n(x) + \left(\frac{N^2 + 3N}{2N + 2} + 1\right)\varepsilon.$$

The inequality $g_n(x) \leqslant f(x)$ results from the definition of g_n. Consequently for every $x \in C_n$ we have in all cases (i), (ii), (iii)

$$g_n(x) \leqslant f(x) \leqslant g_n(x) + \left(\frac{N^2 + 3N}{2N + 2} + 1\right)\varepsilon. \tag{17.4.26}$$

From (17.4.15) it follows that $\mathrm{Gr}(f_n) \subset \mathrm{Gr}(f_{n+1})$, whence $S_n \subset S_{n+1}$ and $K_n \subset K_{n+1}$ for every $n \in \mathbb{N}$. Hence, for every $x \in C_n$,

$$g_{n+1}(x) = \inf\{y \in \mathbb{R} \mid (x, y) \in K_{n+1}\} \leqslant \inf\{y \in \mathbb{R} \mid (x, y) \in K_n\} = g_n(x).$$

For a fixed $n \in \mathbb{N}$ the sequence $\{g_m \mid \mathrm{int}\, C_n\}_{m \geqslant n}$ is decreasing and bounded below, since f is bounded below on C_n and by (17.4.26) all g_m are uniformly bounded below on C_n. Consequently the sequence $\{g_m \mid \mathrm{int}\, C_n\}$ converges, and by Theorem 7.10.1 its limit is a continuous and convex function. In virtue of (17.4.16) we may define the function $g : D \to \mathbb{R}$ by

$$g(x) = \lim_{m \to \infty} g_m(x),$$

and for every $n \in \mathbb{N}$ the function $g \mid \mathrm{int}\, C_n$ is continuous and convex. Take arbitrary $x, y \in D$. By (17.4.15) and (17.4.16) there exists an $n \in \mathbb{N}$ such that $x, y \in \mathrm{int}\, C_n$. By Theorem 7.1.1 we have for arbitrary $\lambda \in [0, 1]$

$$g\big(\lambda x + (1 - \lambda)y\big) \leqslant \lambda g(x) + (1 - \lambda)g(y). \tag{17.4.27}$$

Thus g satisfies (17.4.27) for all $x, y \in D$ and $\lambda \in [0, 1]$. By Theorem 7.1.1 g is a continuous and convex function.

Letting in (17.4.26) $n \to \infty$, we obtain (17.4.14). $\qquad\square$

Theorem 17.4.2. *Let $D \subset \mathbb{R}^N$ be a convex and open set, and let $f : D \to \mathbb{R}$ be an ε-convex function. Then there exists a continuous and convex function $g_0 : D \to \mathbb{R}$ such that*

$$|f(x) - g_0(x)| \leqslant \frac{1}{2}\left(\frac{N^2 + 3N}{2N + 2} + 1\right)\varepsilon \tag{17.4.28}$$

for all $x \in D$.

Proof. By Theorem 17.4.1 there exists a continuous and convex function $g : D \to \mathbb{R}$ fulfilling (17.4.14) for all $x \in D$. The function

$$g_0(x) = g(x) + \frac{1}{2}\left(\frac{N^2 + 3N}{2N + 2} + 1\right)\varepsilon$$

satisfies the conditions of the theorem.   □

Estimations (17.4.14) and (17.4.28) are not sharp. It can be shown (Hyers-Ulam [153]) that the function g constructed in the proof of Theorem 17.4.1 actully satisfies

$$g(x) \leqslant f(x) \leqslant g(x) + \frac{N^2 + 3N}{2N + 2}\varepsilon,$$

and then the proof of Theorem 17.4.2 shows that a continuous and convex function $g_0 : D \to \mathbb{R}$ exists fulfilling the inequality

$$|f(x) - g_0(x)| \leqslant \frac{N^2 + 3N}{4N + 4}\varepsilon$$

for $x \in D$.

The problem of extending the above results to functions $f : D \to \mathbb{R}$ satisfying in D the inequlity

$$f\left(\frac{x + y}{2}\right) \leqslant \frac{f(x) + f(y)}{2} + \varepsilon$$

remains still open (cf. also Cholewa [43])[2].

17.5 Set ideals

Let X be an arbitrary set (space). A non-empty family $\mathcal{I} \subset \mathcal{P}(X)$ of subsets of X is called an *ideal* iff it satisfies the two conditions:

(i) If $A \in \mathcal{I}$ and $B \subset A$, then $B \in \mathcal{I}$;

(ii) If $A, B \in \mathcal{I}$, then $A \cup B \in \mathcal{I}$.

If condition (ii) is replaced by the stronger condition

(ii') If $A_n \in \mathcal{I}$, $n \in \mathbb{N}$, then $\bigcup_{n=1}^{\infty} A_n \in \mathcal{I}$,

then $\mathcal{I}$ is called a *σ-ideal*. The family $\mathcal{I} = \{\varnothing\}$ consisting only the empty set and the family $\mathcal{I} = \mathcal{P}(X)$ of all the subsets of X are trivial examples of σ-ideals. Due to condition (i) we must have $\mathcal{I} = \mathcal{P}(X)$ whenever $X \in \mathcal{I}$. We wish to exclude such a trivial case. If an ideal [σ-ideal] $\mathcal{I}$ besides (i), (ii) [(i), (ii')] satisfies also the condition

(iii) $X \notin \mathcal{I}$,

---

[2] Added in the 2nd edition by K. Baron: This has recently been solved by Z. Kominek and J. Mrowiec in *Nonstability results in the theory of convex functions*, C. R. Math. Acad. Sci., Soc. R. Can. **28** (2006), 17–23.

it is called *proper*. If we are given a proper ideal $\mathcal{I} \subset \mathcal{P}(X)$ in X, then we say that a condition is satisfied $\mathcal{I}$-*almost everywhere* in X (written $\mathcal{I}$-(a.e.)) iff there exists a set $A \in \mathcal{I}$ such that the condition in question is satisfied for every $x \in X \setminus A$.

Now suppose that X is endowed with an inner operation $+$ such that $(X, +)$ is a (not necessarily commutative) group (cf. 4.5). We say that an ideal [σ-ideal] $\mathcal{I} \subset \mathcal{P}(X)$ is *linearly invariant* iff besides conditions (i), (ii), [(i), (ii')] it satisfies also the condition

(iv) For every $x \in X$ and $A \in \mathcal{I}$ the set $x - A$ belongs to $\mathcal{I}$.

A proper linearly invariant ideal $\bigl($i.e., a family $\mathcal{I} \subset \mathcal{P}(X)$ fulfilling (i)–$(iv)\bigr)$ will in the sequel be referred to as a p.l.i. ideal.

Lemma 17.5.1. *If $(X, +)$ is a group, and $\mathcal{I} \subset \mathcal{P}(X)$ is a linearly invariant ideal, then, for every $x \in X$ and $A \in \mathcal{I}$, we have*

$$-A \in \mathcal{I}, \quad x + A \in \mathcal{I}, \quad A + x \in \mathcal{I}.$$

Proof. $-A = 0 - A \in \mathcal{I}$ by (iv). Hence also $x + A = x - (-A) \in \mathcal{I}$. Finally $A + x = -(-x - A) \in \mathcal{I}$ by what has already been shown. $\qquad\square$

Below we give some most important examples of ideals.

I. Let X be an arbitrary set, and let $\mathcal{I}_{\aleph_0}$ be the family of all at most countable subsets of X. Then $\mathcal{I}_{\aleph_0}$ is a σ-ideal. If $(X, +)$ is a group, then $\mathcal{I}_{\aleph_0}$ is linearly invariant. $\mathcal{I}_{\aleph_0}$ is proper if and only if $\operatorname{card} X > \aleph_0$. If $X = \mathbb{R}^N$, we write $\mathcal{I}_{\aleph_0}^N$ instead of $\mathcal{I}_{\aleph_0}$. $\mathcal{I}_{\aleph_0}^N$ is a p.l.i. σ-ideal.

II. Let X be a metric space, and let $\mathcal{I}_b$ be the family of all bounded subsets of X. Then $\mathcal{I}_b$ is an ideal, but in general not a σ-ideal. If $(X, +)$ is a metric group, and the metric is invariant under translations, then $\mathcal{I}_b$ is linearly invariant. If X is unbounded, then $\mathcal{I}_b$ is proper. If $X = \mathbb{R}^N$, we write $\mathcal{I}_b^N$ instead of $\mathcal{I}_b$. $\mathcal{I}_b^N$ is a p.l.i. ideal.

III. Let X be a topological space, and let $\mathcal{I}_f$ be the family of all subsets of X which are of the first category. Then $\mathcal{I}_f$ is a σ-ideal. If $(X, +)$ is a topological group, then $\mathcal{I}_f$ is linearly invariant. If X is of the second category, then $\mathcal{I}_f$ is proper. If $X = \mathbb{R}^N$, we write $\mathcal{I}_f^N$ instead of $\mathcal{I}_f$. $\mathcal{I}_f^N$ is a p.l.i. ideal.

IV. Let $X = \mathbb{R}^N$, and let $\mathcal{I}_0^N$ be the family of all subsets of $\mathbb{R}^N$ of N-dimensional measure zero. $\mathcal{I}_0^N$ is a p.l.i. σ-ideal.

V. Let $X = \mathbb{R}^N$, and let $\mathcal{I}_m^N$ be the family of all subsets of $\mathbb{R}^N$ of finite N-dimensional outer measure:

$$A \in \mathcal{I}_m^N \Leftrightarrow A \subset \mathbb{R}^N \text{ and } m_e(A) < \infty.$$

$\mathcal{I}_m^N$ is a p.l.i. ideal, but not a σ-ideal.

VI. Let $(X, +)$ be a group (not necessarily commutative), let $B \subset X$ be an arbitrary set, and let

$$\mathcal{J}(B) = \left\{ A \subset X \mid A = \bigcup_{i=1}^{n} [x_i + (B \cup (-B)) + y_i], \ x_i, y_i \in X, \ i = 1, \ldots, n; \ n \in \mathbb{N} \right\}.$$

Put

$$\mathcal{I}(B) = \bigcup_{A \in \mathcal{J}(B)} \mathcal{P}(A).$$

Then $\mathcal{I}(B)$ is the smallest linerly invariant ideal in X containing B. In fact, if $\mathcal{I}$ is the smallest linearly invariant ideal in X containing B, then $\mathcal{I}(B) \subset \mathcal{I}$ by (i), (ii), and Lemma 17.5.1. On the other hand, if $U, V \in \mathcal{I}(B)$, then there exist $n, m \in \mathbb{N}$ and points x_i, y_i, $i = 1, \dots, n + m$, such that

$$U \subset \bigcup_{i=1}^{n} \left[x_i + (B \cup (-B)) + y_i \right], \quad V \subset \bigcup_{i=n+1}^{n+m} \left[x_i + (B \cup (-B)) + y_i \right].$$

Hence

$$U \cup V \subset \bigcup_{i=1}^{n+m} \left[x_i + (B \cup (-B)) + y_i \right] \in \mathcal{J}(B),$$

whence $U \cup V \in \mathcal{I}(B)$. Similarly, if $Z \subset U \subset \bigcup_{i=1}^{n} \left[x_i + (B \cup (-B)) + y_i \right] \in \mathcal{J}(B)$, then $Z \in \mathcal{I}(B)$. Consequently $\mathcal{I}(B)$ fulfils (i) and (ii), and thus it is an ideal. If $x \in X$ and $U \in \mathcal{I}(B)$, then

$$x - U \subset x - \bigcup_{i=1}^{n} \left[x_i + (B \cup (-B)) + y_i \right] = \bigcup_{i=1}^{n} \left[(x - y_i) + (B \cup (-B)) - x_i \right] \in \mathcal{J}(B),$$

and consequently $x - U \in \mathcal{I}(B)$. Thus $\mathcal{I}(B)$ fulfils (iv), which means that $\mathcal{I}(B)$ is linearly invariant. We have also $B \subset B \cup (-B) = 0 + (B \cup (-B)) + 0 \in \mathcal{J}(B)$, whence $B \in \mathcal{I}(B)$. Consequently $\mathcal{I}(B)$ is a linearly invariant ideal containing B, and since $\mathcal{I}$ is the smallest such ideal, we have $\mathcal{I} \subset \mathcal{I}(B)$. Consequently $\mathcal{I} = \mathcal{I}(B)$.

$\mathcal{I}(B)$ is called the linearly invariant ideal *generated by* B. In general $\mathcal{I}(B)$ need not be proper, nor a σ-ideal. Some conditions on a set $B \subset \mathbb{R}^N$ which guarantee that $\mathcal{I}(B)$ is proper were given by M. Sablik [272]. Let us note (Sablik [272], Ger-Kuczma [116]) that a p.l.i. ideal in $\mathbb{R}^N$ may contain quite large sets, e.g., sets of infinite measure and of the second category[3].

Let X be a set (space). For any set $A \subset X \times X$ and $x \in X$ we write

$$A[x] = \{y \in X \mid (x, y) \in A\}$$

(cf. 2.1). Suppose that we are given an ideal $\mathcal{I}_1$ in X and an ideal $\mathcal{I}_2$ in $X \times X$. We say that ideals $\mathcal{I}_1$ and $\mathcal{I}_2$ are *conjugate* iff for every set $A \in \mathcal{I}_2$ we have

$$A[x] \in \mathcal{I}_1 \quad \mathcal{I}_1\text{-(a.e.) in } X, \tag{17.5.1}$$

i.e., iff there exists a set $U \in \mathcal{I}_1$ such that

$$A[x] \in \mathcal{I}_1 \text{ for } x \in X \setminus U.$$

This is an abstract version of the Fubini theorem.

---

[3] Such is, e.g., the ideal $\mathcal{I}(B)$, where $B \subset \mathbb{R}^2$ is the first quadrant.

We may consider the space $\mathbb{R}^{2N}$ as the product $\mathbb{R}^N \times \mathbb{R}^N$. It is immediately seen from the definition that the ideals $\mathcal{I}_{\aleph_0}^N$ and $\mathcal{I}_{\aleph_0}^{2N}$ as well as the ideals $\mathcal{I}_b^N$ and $\mathcal{I}_b^{2N}$ are conjugate. The ideals $\mathcal{I}_0^N$ and $\mathcal{I}_0^{2N}$ are conjugate in virtue of the Fubini theorem, whereas $\mathcal{I}_f^N$ and $\mathcal{I}_f^{2N}$ are conjugate in virtue of Theorem 2.1.7. Concerning $\mathcal{I}_m^N$ and $\mathcal{I}_m^{2N}$ we have the following[4] (de Bruijn [64])

Lemma 17.5.2. *Let α, β be positive numbers, and let $A \subset \mathbb{R}^{2N}$ be a set such that $m_e^{2N}(A) \leqslant \beta$. Then*

$$m_e^{2N}\left(\{x \in \mathbb{R}^N \mid m_e^{2N}(A[x]) \geqslant \beta/\alpha\}\right) \leqslant \alpha. \tag{17.5.2}$$

Proof. By Lemma 3.1.2 there exists a measurable set $B \subset \mathbb{R}^{2N}$ such that $A \subset B$ and $m^{2N}(B) = m_e^{2N}(A)$. By the Fubini theorem almost all[5] sections $B[x]$ are measurable (in the sense of the N-dimensional measure), the function $\varphi(x) = m^N(B[x])$ is defined almost everywhere in $\mathbb{R}^N$ and is measurable, and

$$m^{2N}(B) = \int\limits_{\mathbb{R}^N} m^N(B[x])\,dx.$$

Thus we have, since $m^N(B[x]) \geqslant 0$,

$$\beta \geqslant m_e^{2N}(A) = m^{2N}(B) = \int\limits_{\mathbb{R}^N} m^N(B[x])\,dx$$

$$\geqslant \int\limits_{\{x \in \mathbb{R}^N \mid m^N(B[x]) \geqslant \beta/\alpha\}} m^N(B[x])\,dx$$

$$\geqslant \frac{\beta}{\alpha}\, m^N\left(\{x \in \mathbb{R}^N \mid m^N(B[x]) \geqslant \beta/\alpha\}\right),$$

whence

$$m^N\left(\{x \in \mathbb{R}^N \mid m^N(B[x]) \geqslant \beta/\alpha\}\right) \leqslant \alpha. \tag{17.5.3}$$

Since $A \subset B$, we have $A[x] \subset B[x]$ for every $x \in \mathbb{R}^N$, and

$$m_e^N(A[x]) \leqslant m_e^N(B[x]) = m^N(B[x])$$

for almost all $x \in \mathbb{R}^N$. Hence

$$\{x \in \mathbb{R}^N \mid m_e^N(A[x]) \geqslant \beta/\alpha\} \subset \{x \in \mathbb{R}^N \mid m^N(B[x]) \geqslant \beta/\alpha\}$$

up to a set of N-dimensional measure zero, whence

$$m_e^N\left(\{x \in \mathbb{R}^N \mid m_e^N(A[x]) \geqslant \beta/\alpha\}\right) \leqslant m^N\left(\{x \in \mathbb{R}^N \mid m^N(B[x]) \geqslant \beta/\alpha\}\right),$$

and from (17.5.3) we obtain (17.5.2). $\qquad\square$

---

[4] We add superscripts to the symbols of the Lebesgue measure in order to indicate the dimension.
[5] In the present proof the expressions "almost all", "almost everywhere" are to be understood in the sense of the ideal $\mathcal{I}_0^N$.

Corollary 17.5.1. *The ideals $\mathcal{I}_m^N$ and $\mathcal{I}_m^{2N}$ are conjugate.*

Proof. Take an arbitrary set $A \subset \mathbb{R}^{2N}$, $A \in \mathcal{I}_m^{2N}$. This means that $\beta = m_e^{2N}(A) < \infty$. We have

$$\{x \in \mathbb{R}^N \mid A[x] \notin \mathcal{I}_m^N\} = \{x \in \mathbb{R}^N \mid m_e^N(A[x]) = \infty\} \subset \{x \in \mathbb{R}^N \mid m_e^N(A[x]) \geqslant \beta\},$$

whence by Lemma 17.5.2 ($\alpha = 1$)

$$m_e^N(\{x \in \mathbb{R}^N \mid A[x] \notin \mathcal{I}_m^N\}) \leqslant m_e^N(\{x \in \mathbb{R}^N \mid m_e^N(A[x]) \geqslant \beta\}) \leqslant 1,$$

i.e.,

$$\{x \in \mathbb{R}^N \mid A[x] \notin \mathcal{I}_m^N\} \in \mathcal{I}_m^N, \qquad \text{or} \qquad A[x] \in \mathcal{I}_m^N \qquad \mathcal{I}_m^N\text{-(a.e.) in } \mathbb{R}^N.$$

This means (cf. (17.5.1)) that $\mathcal{I}_m^N$ and $\mathcal{I}_m^{2N}$ are conjugate. $\square$

Let $(X, +)$ be a group (not necessarily commutative), and suppose that we are given a p.l.i. ideal $\mathcal{I}$ in X. With the aid of $\mathcal{I}$ we may define two families of subsets of $X \times X$ (Ger [105]):

$$\Pi(\mathcal{I}) = \{A \subset X \times X \mid A \subset (U \times X) \cup (X \times U), \ U \in \mathcal{I}\},$$
$$\Omega(\mathcal{I}) = \{A \subset X \times X \mid A[x] \in \mathcal{I} \ \ \mathcal{I}\text{-(a.e.) in } X\}.$$

The proof of the following simple lemma is left to the reader.

Lemma 17.5.3. *Let $(X, +)$ be a group, and let $\mathcal{I} \subset \mathcal{P}(X)$ be a p.l.i. ideal [σ-ideal] in X. Then $\Pi(\mathcal{I})$ and $\Omega(\mathcal{I})$ are p.l.i. ideals [σ-ideals] in $X \times X$, and they are both conjugate with $\mathcal{I}$. If $\mathcal{I}_2 \subset \mathcal{P}(X \times X)$ is a p.l.i. ideal in $X \times X$ such that $\mathcal{I}$ and $\mathcal{I}_2$ are conjugate, then $\mathcal{I}_2 \subset \Omega(\mathcal{I})$.*

We have also the following

Lemma 17.5.4. *Let $(X, +)$ be a group, and let $\mathcal{I} \subset \mathcal{P}(X)$ be a p.l.i. ideal in X. If $U \in \mathcal{I}$, then*

$$M = \{(x, y) \in X \times X \mid x + y \in U\} \in \Omega(\mathcal{I}). \tag{17.5.4}$$

Proof. For every $x \in X$ we have

$$M[x] = \{y \in X \mid x + y \in U\} = \{y \in X \mid y \in -x + U\} = -x + U \in \mathcal{I}$$

by Lemma 17.5.1. Consequently $M \in \Omega(\mathcal{I})$. $\square$

Let $(X, +)$ be a group, and let $\mathcal{I} \subset \mathcal{P}(X)$ be a p.l.i. ideal in X, and let $(S, +)$ be a subsemigroup of $(X, +)$ such that

$$S - S = X \tag{17.5.5}$$

and[6]

$$S \notin \mathcal{I}. \tag{17.5.6}$$

---

[6] Condition (17.5.5) does not imply (17.5.6). E.g., if $X = \mathbb{R}^2$, $S = (0, \infty) \times (0, \infty)$ is the first quadrant, and $\mathcal{I} = \mathcal{I}(S)$, then S fulfils (17.5.5) but not (17.5.6). (This example is due to M. Sablik; cf. Ger [112].)

The following three lemmas are due to R. Ger [112].

Lemma 17.5.5. *Let $(X, +)$ be a group, and let $\mathcal{I} \subset \mathcal{P}(X)$ be a p.l.i. ideal in X, and let $(S, +)$ be a subsemigroup of $(X, +)$ fulfilling (17.5.5) and (17.5.6). Then, for every $s, t \in X$, we have*

$$(s + S) \cap (t + S) \notin \mathcal{I}. \tag{17.5.7}$$

Proof. Suppose that (17.5.7) does not hold for some $s, t \in X$. For arbitrary $u, v \in S$ we have (since $(S, +)$ is a subsemigroup) $u + S \subset S$ and $v + S \subset S$, whence

$$(s + u + S) \cap (t + v + S) \subset (s + S) \cap (t + S) \in \mathcal{I}$$

and

$$(s + u + S) \cap (t + v + S) \in \mathcal{I} \tag{17.5.8}$$

by (i). By (17.5.5) we can find $u, v \in S$ such that $-s + t = u - v$, whence $s + u = t + v$, and by (17.5.8)

$$t + v + S \in \mathcal{I}.$$

By Lemma 17.5.1

$$S = -(t + v) + (t + v + S) \in \mathcal{I},$$

which contradicts (17.5.6). $\square$

Corollary 17.5.2. *Under conditions of Lemma* 17.5.5

$$(s + S) \cap (t + S) \neq \varnothing \tag{17.5.9}$$

for arbitrary $s, t \in S$.

Corollary 17.5.2 may also be formulated as follows (cf. 4.5).

Corollary 17.5.3. *Under conditions of Lemma* 17.5.5 *the semigroup $(S, +)$ is left reversible.*

Lemma 17.5.6. *Under conditions of Lemma* 17.5.5 *we have*

$$(S \setminus U) - (S \setminus V) = X \tag{17.5.10}$$

for every $U, V \in \mathcal{I}$.

Proof. Take an $x \in X$ and sets $U, V \in \mathcal{I}$. By (17.5.5) $x = s - t$, where $s, t \in S$. By Lemma 17.5.5

$$[-s + (S \setminus U)] \cap [-t + (S \setminus V)] = [(-s + S) \cap (-t + S)] \setminus [(-s + U) \cup (-t + V)] \notin \mathcal{I},$$

since $(-s + U) \cup (-t + V) \in \mathcal{I}$ by (ii) and Lemma 17.5.1. Thus, in particular,

$$[-s + (S \setminus U)] \cap [-t + (S \setminus V)] \neq \varnothing.$$

Consequently there exists a $y \in [-s + (S \setminus U)] \cap [-t + (S \setminus V)]$, which means that $s + y \in S \setminus U$, $t + y \in S \setminus V$, and

$$x = s - t = (s + y) - (t + y) \in (S \setminus U) - (S \setminus V).$$

This means that $X \subset (S \setminus U) - (S \setminus V)$. Since the converse inclusion is trivial, we obtain hence (17.5.10). $\square$

Lemma 17.5.7. *Under conditions of Lemma 17.5.5, for every $U \in \mathcal{I}$ and every $u, s', t' \in S \setminus U$, there exist $s, t \in S \setminus U$ such that $t \in u + S$, and $s - t = s' - t'$.*

Proof. Taking in Lemma 17.5.6 $V = -u + U$ we obtain

$$(S \setminus U) - \left[S \setminus (-u + U) \right] = X.$$

This means that for every $y \in X$ there exist $z \in S \setminus U$ and $x \in S \setminus (-u + U)$ such that $y = z - x$, i.e., $y + x = z \in S \setminus U$. Take $y = s' - t' + u$ and choose a suitable x. Put $s = s' - t' + u + x$ and $t = u + x$. Then $s = s' - t' + t$, whence $s - t = s' - t'$, and

$$t = u + x \in u + \left[S \setminus (-u + U) \right] = (u + S) \setminus U \subset S \setminus U, \qquad (17.5.11)$$

since $u + S \subset S$, and, on the other hand, (17.5.11) yields $t \in u + S$. Further,

$$s = s' - t' + u + x = y + x \in S \setminus U.$$

Thus s and t have the required properties. $\qquad \square$

Now suppose that X is an arbitrary set (space), and let $\mathcal{I} \subset \mathcal{P}(X)$ be an arbitrary proper σ-ideal in X. For every function $f : X \to [-\infty, \infty]$ and an arbitrary set $B \subset X$, $B \notin \mathcal{I}$, we put (Smajdor [285])

$$\mathcal{I}\text{-}\operatorname*{infess}_{B} f(x) = \sup_{A \in \mathcal{I}} \inf_{B \setminus A} f(x). \qquad (17.5.12)$$

For $X \subset \mathbb{R}^N$ and $\mathcal{I} = \mathcal{I}_0^N \cap \mathcal{P}(X)$ this notion becomes the usual essential infimum of a function.

Lemma 17.5.8. *Let X be an arbitrary set, let $\mathcal{I} \subset \mathcal{P}(X)$ be a proper σ-ideal in X, and let $f : X \to [-\infty, \infty]$ be an arbitrary function. Further, let $B \subset X$, $B \notin \mathcal{I}$, be an arbitrary set. Then there exists a set $A \in \mathcal{I}$ such that*

$$\mathcal{I}\text{-}\operatorname*{infess}_{B} f(x) = \inf_{B \setminus A} f(x). \qquad (17.5.13)$$

Proof. If $\mathcal{I}\text{-}\operatorname*{infess}_{B} f(x) = -\infty$, then by (17.5.12) we have $\inf_{B \setminus A} f(x) = -\infty$ for every $A \in \mathcal{I}$ and (17.5.13) holds with every set $A \in \mathcal{I}$. So let $a = \mathcal{I}\text{-}\operatorname*{infess}_{B} f(x) > -\infty$, and take a sequence $\{a_n\}_{n \in \mathbb{N}}$ of real numbers a_n such that $a_n < a_{n+1} < a$ for $n \in \mathbb{N}$, and $\lim_{n \to \infty} a_n = a$. By (17.5.12) for every $n \in \mathbb{N}$ there exists a set $A_n \in \mathcal{I}$ such that

$$\inf_{B \setminus A_n} f(x) > a_n. \qquad (17.5.14)$$

Put $A = \bigcup_{n=1}^{\infty} A_n$. By (ii') $A \in \mathcal{I}$, and by (17.5.12)

$$a \geqslant \inf_{B \setminus A} f(x). \qquad (17.5.15)$$

On the other hand, by (17.5.14), since $A_n \subset A$, whence $B \setminus A \subset B \setminus A_n$ for $n \in \mathbb{N}$,

$$\inf_{B \setminus A} f(x) \geqslant \inf_{B \setminus A_n} f(x) > a_n,$$

whence, on letting $n \to \infty$,

$$\inf_{B \setminus A} f(x) \geqslant a. \qquad (17.5.16)$$

(17.5.13) results from (17.5.15) and (17.5.16) in view of the definition of a. $\qquad \square$

17.6 Almost additive functions

In the year 1960 P. Erdős [75] raised the following problem: Suppose that a function $f : \mathbb{R} \to \mathbb{R}$ satisfies the relation

$$f(x + y) = f(x) + f(y) \qquad (17.6.1)$$

for almost all $(x, y) \in \mathbb{R}^2$ (in the sense of the planar Lebesgue measure). Does there exist an additive function $g : \mathbb{R} \to \mathbb{R}$ such that

$$f(x) = g(x) \qquad (17.6.2)$$

almost everywhere in $\mathbb{R}$ (in the sense of the linear Lebesgue measure)? A positive answer to this question was given by N. G. de Bruijn [64], and, independently, by W. B. Jurkat [159]. J. L. Denny [66] also proved an analogous result (cf. also Denny [67]). N. G. de Bruijn [64] has put this problem into a more general setting.

Let $(X, +)$ and $(Y, +)$ be groups, and suppose that we are given two p.l.i. ideals $\mathcal{I}_1$ and $\mathcal{I}_2$ in X and in $X \times X$, respectively. A function $f : X \to Y$ is called $\mathcal{I}_2$-*almost additive* iff relation (17.6.1) holds $\mathcal{I}_2$-(a.e.) in $X \times X$. De Bruijn's result was that if the groups $(X, +)$ and $(Y, +)$ are commutative, and if the ideals $\mathcal{I}_1$ and $\mathcal{I}_2$ are conjugate, then every $\mathcal{I}_2$- almost additive function is equal $\mathcal{I}_1$-(a.e.) in X to a function[7] $g : X \to Y$ satisfying (17.6.1) everywhere in $X \times X$. The Erdős' problem is the particular case $X = Y = \mathbb{R}$, $\mathcal{I}_1 = \mathcal{I}_0^1$, $\mathcal{I}_2 = \mathcal{I}_0^2$.

We start with the following

Lemma 17.6.1. *Let $(X, +)$ and $(Y, +)$ be groups (not necessarily commutative), and let $\mathcal{I} \subset \mathcal{P}(X)$ be a p.l.i. ideal in X. Let $g_1, g_2 : X \to Y$ be homomorphisms such that $g_1 = g_2$ $\mathcal{I}$-(a.e.) in X. Then $g_1 = g_2$ in X.*

Proof. Let $S \in \mathcal{I}$ be such tht

$$g_1(x) = g_2(x) \text{ for } x \in X \setminus S. \qquad (17.6.3)$$

Take an arbitrary $x \in X$. By (17.5.ii) and by Lemma 17.5.1 we have $S \cup (-S + x) \in \mathcal{I}$, and consequently we can find a $y \in X \setminus [S \cup (-S + x)]$. Hence by (17.6.3) $g_1(y) = g_2(y)$ and $g_1(x - y) = g_2(x - y)$, whence, since g_1 and g_2 are homomorphisms,

$$g_1(x) = g_1(x - y) + g_1(y) = g_2(x - y) + g_2(y) = g_2(x),$$

which means that $g_1 = g_2$ in X. $\qquad \square$

---

[7] I.e., g is a homomorphism from $(X, +)$ into $(Y, +)$.

R. Ger [111] (cf. also Ger [112]) generalized de Bruijn's result to the case of non-commutative groups. His theorem runs as follows.

Theorem 17.6.1. *Let $(X, +)$ and $(Y, +)$ be groups (not necessarily commutative), and suppose that we are given two conjugate p.l.i. ideals $\mathcal{I}_1$ and $\mathcal{I}_2$ in X and in $X \times X$, respectively. If $f : X \to Y$ is an $\mathcal{I}_2$-almost additive function, then there exists a unique homomorphism $g : X \to Y$ such that (17.6.2) holds $\mathcal{I}_1$-(a.e.) in X.*

Proof. Suppose that (17.6.1) holds for $(x, y) \in (X \times X) \setminus M$, where $M \in \mathcal{I}_2$. Since $\mathcal{I}_1$ and $\mathcal{I}_2$ are conjugate, there exists a set $U \subset X$, $U \in \mathcal{I}_1$, such that

$$M[x] \in \mathcal{I}_1 \text{ for } x \in X \setminus U. \tag{17.6.4}$$

For every $x \in X$ the set $U \cup (-U + x)$ belongs to $\mathcal{I}_1$ (Lemma 17.5.1), and $X \setminus \big[U \cup (-U + x)\big] \neq \varnothing$. So there exists a $w(x)$,

$$w(x) \in X \setminus \big[U \cup (-U + x)\big]. \tag{17.6.5}$$

Relation (17.6.5) means that $w(x) \notin U$ and $x - w(x) \notin U$, whence by (17.6.4) $M\big[w(x)\big] \in \mathcal{I}_1$ and $M\big[x - w(x)\big] \in \mathcal{I}_1$, and consequently also

$$A_x = M\big[w(x)\big] \cup \big(-w(x) + M\big[x - w(x)\big]\big) \in \mathcal{I}_1. \tag{17.6.6}$$

Now we prove that for every $x \in X$ the expression

$$f(x + y) - f(x) \tag{17.6.7}$$

(considered as a function of y) is constant $\mathcal{I}_1$-(a.e.) in X; more exactly, we will prove that for $y \in X \setminus A_x$ expression (17.6.7) does not depend on y. Take a $y \in X \setminus A_x$. Then $y \notin M\big[w(x)\big]$ and $w(x) + y \notin M\big[x - w(x)\big]$, i.e.,

$$\big(w(x), y\big) \notin M \text{ and } \big(x - w(x), w(x) + y\big) \notin M.$$

Hence

$$f(x + y) = f\big(x - w(x)\big) + f\big(w(x) + y\big) = f\big(x - w(x)\big) + f\big(w(x)\big) + f(y),$$

or

$$f(x + y) - f(y) = f\big(x - w(x)\big) + f\big(w(x)\big), \quad y \notin A_x, \tag{17.6.8}$$

and the right-hand side of (17.6.8) does not depend on y.

Now we define the function $g : X \to Y$ by

$$g(x) = f\big(x - w(x)\big) + f\big(w(x)\big), \quad x \in X. \tag{17.6.9}$$

We will show that g is homomorphism. Fix arbitrary $u, v \in X$. We may find $x, s, t \in X$ such that

$$x \notin A_{u+v} \cup U \cup \big[-(u + v) + U\big],$$

$$s \notin (-x + A_v) \cup M[x] \cup U \cup \big[-(v + x) + U\big],$$

$$t \notin \big[-(v + x + s) + A_u\big] \cup M[v + x + s] \cup M[s] \cup \big(-s + M[u + v + x]\big),$$

because by (17.5.ii), by Lemma 17.5.1, and by (17.6.4) and (17.6.6) the right-hand sides of the above relations belong to $\mathcal{I}_1$. Put $y = x + s$, $z = v + x + s + t$. Then $y \notin A_v$ and $z \notin A_u$, whereas x has been chosen so that $x \notin A_{u+v}$. Thus by (17.6.8) and (17.6.9)

$$g(u) = f(u+z) - f(z), \ g(v) = f(v+y) - f(y), \ g(u+v) = f(u+v+x) - f(x). \quad (17.6.10)$$

Further, $s \notin M[x]$, i.e., $(x, s) \notin M$, and $t \notin M[v+x+s]$, i.e., $(v+x+s, t) \notin M$. This implies that

$$f(y) = f(x + s) = f(x) + f(s)$$

and

$$f(z) = f(v + x + s + t) = f(v + x + s) + f(t) = f(v + y) + f(t),$$

whence

$$-f(x) + f(y) = f(s), \quad -f(v + y) + f(z) = f(t). \quad (17.6.11)$$

Finally, $t \notin M[s]$, i.e., $(s, t) \notin M$, and $s + t \notin M[u + v + x]$, i.e., $(u + v + x, s + t) \notin M$. Hence

$$f(s) + f(t) = f(s+t), \quad f(u+v+x) + f(s+t) = f(u+v+x+s+t) = f(u+z). \quad (17.6.12)$$

Now, by (17.6.10), (17.6.11) and (17.6.12) we have

$$
\begin{aligned}
g(u &+ v) - g(v) - g(u) \\
&= f(u + v + x) - f(x) - [f(v + y) - f(y)] - [f(u + z) - f(z)] \\
&= f(u + v + x) - f(x) + f(y) - f(v + y) + f(z) - f(u + z) \\
&= f(u + v + x) + f(s) + f(t) - f(u + z) \\
&= f(u + v + x) + f(s + t) - f(u + z) = 0,
\end{aligned}
$$

i.e.,

$$g(u + v) = g(u) + g(v).$$

This holds for arbitrary $u, v \in X$, which means that g is a homomorphism.

Now take an arbitrary $x \in X \setminus U$. By (17.6.4) and (17.6.6) $A_x \cup M[x] \in \mathcal{I}_1$, so we may find a $y \in X \setminus (A_x \cup M[x])$. Since $y \notin A_x$, we have by (17.6.8) and (17.6.9)

$$g(x) = f(x + y) - f(y), \quad (17.6.13)$$

whereas, since $y \notin M[x]$, i.e., $(x, y) \notin M$, relation (17.6.1) holds. (17.6.13) and (17.6.1) yield (17.6.2). Thus (17.6.2) certainly holds for all $x \in X \setminus U$, i.o., $\mathcal{I}_1$ (a.c.) in X. The uniqueness results from Lemma 17.6.1. $\square$

Conversely, we have the following

Theorem 17.6.2. *Let $(X, +)$ and $(Y, +)$ be groups (not necessarily commutative), and suppose that we are given a p.l.i. ideal $\mathcal{I}$ in X. If $f : X \to Y$ is a function, and $g : X \to Y$ is a homomorphism, and if (17.6.2) holds $\mathcal{I}$-(a.e.) in X, then f is $\Omega(\mathcal{I})$-almost additive.*

Proof. Let (17.6.2) hold for $x \in X \setminus U$, where $U \in \mathcal{I}$. Let

$$A = \{(x, y) \in X \times X \mid x \in U \text{ or } y \in U \text{ or } x + y \in U\} = (U \times X) \cup (X \times U) \cup M,$$

where M is defined by (17.5.4). Clearly $(U \times X) \cup (X \times U) \in \Omega(\mathcal{I})$, whence $A \in \Omega(\mathcal{I})$ in virtue of Lemmas 17.5.3 and 17.5.4. For $(x, y) \in (X \times X) \setminus A$ we have $f(x) = g(x)$, $f(y) = g(y)$ and $f(x + y) = g(x + y)$, whence

$$f(x + y) = g(x + y) = g(x) + g(y) = f(x) + f(y),$$

so (17.6.1) holds for $(x, y) \in (X \times X) \setminus A$, i.e., $\Omega(\mathcal{I})$-(a.e.) in $X \times X$. $\qquad \square$

The result of Theorem 17.6.1 can further be strengthened in the case of some particular ideals $\mathcal{I}_2$. Suppose that we are given a p.l.i. ideal $\mathcal{I}$ in X. If a function $f : X \to Y$ is $\Pi(\mathcal{I})$-almost additive, then there exists a set $U \in \mathcal{I}$ such that (17.6.1) holds for all $(x, y) \in (X \times X) \setminus [(U \times X) \cup (X \times U)]$, or, in other words, for all $x, y \in X \setminus U$. Conversely, if there exists a set $U \in \mathcal{I}$ such that (17.6.1) holds for all $x, y \in X \setminus U$, then f is $\Pi(\mathcal{I})$-almost additive. Actually we can say more, as is seen from the following theorem of S. Hartman [137] (cf. also de Bruijn [64]):

Theorem 17.6.3. *Let $(X, +)$ and $(Y, +)$ be groups (not necessarily commutative), and suppose that we are given a p.l.i. ideal $\mathcal{I}$ in X. If a function $f : X \to Y$ fulfils (17.6.1) for all $x, y \in X \setminus U$, where $U \in \mathcal{I}$, then f is a homomorphism.*

Proof. As pointed out above, f is $\Pi(\mathcal{I})$-almost additive. $\Pi(\mathcal{I})$ is a p.l.i. ideal in $X \times X$ and $\mathcal{I}$ and $\Pi(\mathcal{I})$ are conjugate (Lemma 17.5.3). By Theorem 17.6.1 there exists a homomorphism $g : X \to Y$ such that (17.6.2) holds $\mathcal{I}$-(a.e.) in X, i.e., for $x \in X \setminus S$, where $S \in \mathcal{I}$. Take an arbitrary $x \in X$. By Lemma 17.5.1 $-(S \cup U) \in \mathcal{I}$ and $-x + (S \cup U) \in \mathcal{I}$, whence also $\left[-(S \cup U) \cup \left(-x + (S \cup U) \right) \right] \in \mathcal{I}$. Thus there exists a $y \in X \setminus \left[-(S \cup U) \cup \left(-x + (S \cup U) \right) \right]$. This means that $-y \in S$ and $x + y \notin S$, whence

$$f(-y) = g(-y), \quad f(x + y) = g(x + y),$$

and $-y \notin U$, $x + y \notin U$, whence

$$f(x) = f(x + y - y) = f(x + y) + f(-y).$$

Hence

$$f(x) = g(x + y) + g(-y) = g(x + y - y) = g(x),$$

since g is a homomorphism. Thus $f = g$ in X, which means that f is a homomorphism. $\qquad \square$

Theorem 17.6.3 can be formulated as follows.

Corollary 17.6.1. *Let $(X, +)$ and $(Y, +)$ be groups (not necessarily commutative), and suppose that we are given a p.l.i. ideal $\mathcal{I}$ in X. If a function $f : X \to Y$ is $\Pi(\mathcal{I})$-almost additive, then f is a homomorphism.*

Corollary 17.6.1 shows that in Theorem 17.6.2 $\Omega(\mathcal{I})$ cannot be replaced by an arbitrary p.l.i. ideal $\mathcal{I}_2$ in $X \times X$.

A similar conclusion can also be obtained from the following (de Bruijn [64])

Theorem 17.6.4. *Let* $f : \mathbb{R}^N \to \mathbb{R}$ *be an* $\mathcal{I}_m^{2N}$*-almost additive function. Then* f *is* $\mathcal{I}_0^{2N}$*-almost additive.*

Proof. Suppose that (17.6.1) holds for all $(x, y) \in \mathbb{R}^{2N} \setminus M$, where $M \in \mathcal{I}_m^{2N}$, and choose a $\beta < \infty$ such that $m_e^{2N}(M) < \beta$. Take an arbitrary $\alpha > 0$. By Lemma 17.5.2

$$m_e^N\left(\{x \in \mathbb{R}^N \mid m_e^N(M[x]) \geqslant \beta/\alpha\}\right) \leqslant \alpha.$$

Let $U = \{x \in \mathbb{R}^N \mid m_e^N(M[x]) \geqslant \beta/\alpha\}$. Then

$$m_e^N(U) \leqslant \alpha, \tag{17.6.14}$$

and

$$m_e^N(M[x]) \leqslant \beta/\alpha \quad \text{for } x \in \mathbb{R}^N \setminus U. \tag{17.6.15}$$

Take an arbitrary $x \in \mathbb{R}^N$. Since

$$m_e^N\left(U \cup (x - U)\right) \leqslant m_e^N(U) + m_e^N(x - U) = 2m_e^N(U) \leqslant 2\alpha < \infty,$$

there exists a $w(x)$ such that $w(x) \notin U$ and $x - w(x) \notin U$. Hence for the set A_x defined by (17.6.6) we have

$$m_e^N(A_x) \leqslant 2\beta/\alpha. \tag{17.6.16}$$

The argument in the proof of Theorem 17.6.1 shows that (17.6.8) holds. Define the function $g : \mathbb{R}^N \to \mathbb{R}$ by (17.6.9) $\left(X = \mathbb{R}^N\right)$. By (17.6.14) and (17.6.16) $U \in \mathcal{I}_m^N$ and $A_x \in \mathcal{I}_m^N$ (for arbitrary $x \in \mathbb{R}^N$), and so we may repeat the argument from the proof of Theorem 17.6.1 to show that g is a homomorphism. For every $x \subset \mathbb{R}^N \setminus U$ we have by (17.6.15) $M[x] \in \mathcal{I}_m^N$, whence $A_x \cup M[x] \in \mathcal{I}_m^N$, and there exists a $y \in \mathbb{R}^N \setminus \left(A_x \cup M[x]\right)$ so that we have (17.6.13) and (17.6.1). Hence $g(x) = f(x)$ for $x \in \mathbb{R}^N \setminus U$, i.e., $\mathcal{I}_m^N$-(a.e.) in $\mathbb{R}^N$. By Lemma 17.6.1 the function g is independent of U, and hence of α. Letting in (17.6.14) $\alpha \to 0$, we obtain (17.6.2) $\mathcal{I}_0^N$-(a.e.) in $\mathbb{R}^N$.

Let $U \subset \mathbb{R}^N$ be a set of measure zero such that

$$f(x) = g(x) \text{ for } x \in \mathbb{R}^N \setminus U.$$

Let $T : \mathbb{R}^{2N} \to \mathbb{R}^{2N}$ be the linear transform

$$T(x, y) = (x, y - x), \quad x, y \in \mathbb{R}^N.$$

Put

$$Z = \{(x, y) \in \mathbb{R}^{2N} \mid x + y \in U\}, \quad Z_0 = \{(x, y) \in \mathbb{R}^{2N} \mid y \in U\} = \mathbb{R}^N \times U.$$

Thus Z_0 is Lebesgue measurable and $m^{2N}(Z_0) = m^N(\mathbb{R}^N)m^N(U) = 0$. By Corollary 3.2.1 the set $Z = T(Z_0)$ is Lebesgue measurable and $m^{2N}(Z) = m^{2N}(Z_0) = 0$. Since evidently $m^{2N}\left[(U \times \mathbb{R}^N) \cup (\mathbb{R}^N \times U)\right] = 0$, we have $(U \times \mathbb{R}^N) \cup (\mathbb{R}^N \times U) \cup Z \in \mathcal{I}_0^{2N}$, and the argument in the proof of Theorem 17.6.2 shows that f is $\mathcal{I}_0^{2N}$-almost additive. $\square$

In the case where $X = \mathbb{R}^N$, $Y = \mathbb{R}$, a homomorphism from X to Y is an additive function. In such a case we can infer from a regularity of f to the continuity of g fulfilling (17.6.2) $\mathcal{I}$-(a.e.), at least for some p.l.i. ideals $\mathcal{I}$. The reader will easily formulate the corresponding results (cf. also Exercises 17.12 and 17.13).

Finally let us observe that the questions considered in the present section belong to the area of conditional Cauchy equations, or Cauchy equations on restricted domains (cf. the end of 13.6).

17.7   Almost polynomial functions

Let $\mathcal{I}$ be a proper ideal in $\mathbb{R}^{2N}$. We say that a function $f : \mathbb{R}^N \to \mathbb{R}$ is $\mathcal{I}$-almost polynomial of order p iff

$$\Delta_y^{p+1} f(x) = 0 \quad \mathcal{I}\text{-(a.e.) in } \mathbb{R}^{2N}. \tag{17.7.1}$$

In virtue of Corollary 15.1.2 relation (17.7.1) can also be written as

$$\sum_{k=0}^{p+1} (-1)^{p+1-k} \binom{p+1}{k} f(x + ky) = 0 \quad \mathcal{I}\text{-(a.e.) in } \mathbb{R}^{2N}, \tag{17.7.2}$$

or, equivalently, in the form

$$(-1)^{p+1} f(x) = \sum_{k=1}^{p+1} (-1)^{p-k} \binom{p+1}{k} f(x + ky) \quad \mathcal{I}\text{-(a.e.) in } \mathbb{R}^{2N},$$

whence, after dividing both the sides by $(-1)^{p+1}$ and observing that $(-1)^{-k-1} = (-1)^{k-1}$,

$$f(x) = \sum_{k=1}^{p+1} (-1)^{k-1} \binom{p+1}{k} f(x + ky) \quad \mathcal{I}\text{-(a.e.) in } \mathbb{R}^{2N}. \tag{17.7.3}$$

In the present section we are going to prove analogues of Theorems 17.6.1–17.6.4 for almost polynomial functions (Ger [99]).

Theorem 17.7.1. *Let $\mathcal{I}$ be a p.l.i. ideal in $\mathbb{R}^N$ fulfilling the condition*

$$\text{If } A \in \mathcal{I} \text{ and } a \in \mathbb{R}, \text{ then } aA \in \mathcal{I}. \tag{i}$$

If $g : \mathbb{R}^N \to \mathbb{R}$ is a polynomial function of order p, and $f : \mathbb{R}^N \to \mathbb{R}$ is a function such that

$$f(x) = g(x) \quad \mathcal{I}\text{-(a.e.) in } \mathbb{R}^N, \tag{17.7.4}$$

then f is an $\Omega(\mathcal{I})$-almost polynomial function of order p.

Proof. Suppose that

$$f(x) = g(x) \text{ for } x \in \mathbb{R}^N \setminus S, \tag{17.7.5}$$

where $S \in \mathcal{I}$. For every fixed $k \in \mathbb{N} \cup \{0\}$ the set

$$A_k = \{(x, y) \in \mathbb{R}^{2N} \mid x + ky \in S\}$$

belongs to $\Omega(\mathcal{I})$. In fact, if $k = 0$, then

$$A_0 = \{(x, y) \in \mathbb{R}^{2N} \mid x \in S\} = S \times \mathbb{R}^N,$$

which is in $\Omega(\mathcal{I})$. If $k \neq 0$, then

$$A_k = \left\{(x, y) \in \mathbb{R}^{2N} \mid ky \in S - x\right\} = \left\{(x, y) \in \mathbb{R}^{2N} \mid y \in \frac{1}{k}(S - x)\right\},$$

and so $A_k[x] = \frac{1}{k}(S - x) \in \mathcal{I}$ by (i) and by Lemma 17.5.1, for every $x \in \mathbb{R}^N$. This means that $A_k \in \Omega(\mathcal{I})$. Consequently also $\bigcup\limits_{k=0}^{p+1} A_k \in \Omega(\mathcal{I})$.

If $(x, y) \in \mathbb{R}^{2N} \setminus \bigcup\limits_{k=0}^{p+1} A_k$, then $x + ky \notin S$ for $k = 0, \ldots, p + 1$, and by (17.7.5)

$$f(x + ky) = g(x + ky) \text{ for } k = 0, \ldots, p + 1.$$

Hence, since g is a polynomial function,

$$\sum_{k=0}^{p+1} (-1)^{p+1-k} \binom{p+1}{k} f(x + ky) = \sum_{k=0}^{p+1} (-1)^{p+1-k} \binom{p+1}{k} g(x + ky) = 0.$$

Thus $\Delta_y^{p+1} f(x) = 0$ for $(x, y) \in \mathbb{R}^{2N} \setminus \bigcup\limits_{k=0}^{p+1} A_k$, i.e., $\Omega(\mathcal{I})$-(a.e.) in $\mathbb{R}^{2N}$. $\qquad\square$

Lemma 17.7.1. *Let $\mathcal{I}$ be a p.l.i. ideal in $\mathbb{R}^N$ fulfilling condition (i). If $g_1, g_2 : \mathbb{R}^N \to \mathbb{R}$ are polynomial functions of order p such that*

$$g_1(x) = g_2(x) \quad \mathcal{I}\text{-(a.e.) in } \mathbb{R}^N,$$

then $g_1 = g_2$ in $\mathbb{R}^N$.

Proof. Let

$$g_1(x) = g_2(x) \text{ for } x \in \mathbb{R}^N \setminus S, \tag{17.7.6}$$

where $S \in \mathcal{I}$. Take an arbitrary $x \in \mathbb{R}^N$ and a $y \in \mathbb{R}^N \setminus \bigcup\limits_{k=1}^{p+1} \frac{1}{k}(S - x)$. Then $x + ky \notin S$ for $k = 1, \ldots, p + 1$, and by (17.7.6) $g_1(x + ky) = g_2(x + ky)$, $k = 1, \ldots, p + 1$. Hence (comp. formula (17.7.3))

$$g_1(x) = \sum_{k=1}^{p+1} (-1)^{k-1} \binom{p+1}{k} g_1(x + ky) = \sum_{k=1}^{p+1} (-1)^{k-1} \binom{p+1}{k} g_2(x + ky) = g_2(x).$$

This shows that $g_1 = g_2$ in $\mathbb{R}^N$. $\qquad\square$

Theorem 17.7.2. *Let $\mathcal{I}_1$ and $\mathcal{I}_2$ be conjugte p.l.i. ideals in $\mathbb{R}^N$ and $\mathbb{R}^{2N}$, respectively, such that $\mathcal{I}_1$ fulfils condition (i). If $f : \mathbb{R}^N \to \mathbb{R}$ is an $\mathcal{I}_2$-almost polynomial function of order p, then there exists a unique polynomial function $g : \mathbb{R}^N \to \mathbb{R}$ of order p such that (17.7.4) holds with $\mathcal{I} = \mathcal{I}_1$.*

Proof. Suppose that

$$\Delta_y^{p+1} f(x) = 0 \text{ for } (x, y) \in \mathbb{R}^{2N} \setminus M, \tag{17.7.7}$$

where $M \in \mathcal{I}_2$. Since $\mathcal{I}_1$ and $\mathcal{I}_2$ are conjugate, there exists a set $U \subset \mathbb{R}^N$, $U \in \mathcal{I}_1$, such that

$$M[x] \in \mathcal{I}_1 \text{ for } x \in \mathbb{R}^N \setminus U. \tag{17.7.8}$$

By (i) and Lemma 17.5.1 we have for every $x \in \mathbb{R}^N$

$$A_x = \bigcup_{k=1}^{p+1} \frac{1}{k} (U - x) \in \mathcal{I}_1. \tag{17.7.9}$$

Thus there exists a function $\varphi : \mathbb{R}^N \to \mathbb{R}^N$ such that

$$\varphi(x) = 0 \quad \text{for } x \notin U, \tag{17.7.10}$$

$$\varphi(x) \notin A_x \quad \text{for } x \in U. \tag{17.7.11}$$

We put

$$g(x) = \sum_{k=1}^{p+1} (-1)^{k-1} \binom{p+1}{k} f\big(x + k\varphi(x)\big). \tag{17.7.12}$$

Since $0 = (1-1)^{p+1} = \sum_{k=0}^{p+1} \binom{p+1}{k}(-1)^k$, we have $\sum_{k=1}^{p+1}(-1)^{k-1}\binom{p+1}{k} = 1$, whence by (17.7.10), for $x \in \mathbb{R}^N \setminus U$

$$g(x) = \sum_{k=1}^{p+1} (-1)^{k-1} \binom{p+1}{k} f(x) = f(x).$$

Thus $f(x) = g(x)$ for $x \in \mathbb{R}^N \setminus U$, i.e., $\mathcal{I}_1$-(a.e.) in $\mathbb{R}^N$. It remains to show that g is a polynomial function of order p. To this aim we show first that, for every $x \in \mathbb{R}^N$,

$$g(x) = \sum_{k=1}^{p+1} (-1)^{k-1} \binom{p+1}{k} f(x + ky) \text{ for } y \notin A_x. \tag{17.7.13}$$

Take an arbitrary $x \in \mathbb{R}^N$. If $x \in U$, then $\varphi(x) \notin A_x$ by (17.7.11). If $x \notin U$, then by (17.7.10) $\varphi(x) = 0$. If we had $0 \in A_x$, then there would exist a $k \in \{1, \ldots, p+1\}$ such that $0 \in \frac{1}{k}(U - x)$ (cf. (17.7.9)), whence $0 \in U - x$, and $x \in U$, contrary to the supposition. Thus $\varphi(x) \notin A_x$ for every $x \in \mathbb{R}^N$. Take an arbitrary $y \notin A_x$. By (17.7.9) we have

$$x + k\varphi(x) \notin U, \; x + ky \notin U \text{ for } k = 1, \ldots, p+1, \qquad (17.7.14)$$

whence by (17.7.8) $M[x + k\varphi(x)] \in \mathcal{I}_1$ and $M[x + ky] \in \mathcal{I}_1$ for $k = 1, \ldots, p+1$. Thus there exists a $z \in \mathbb{R}^N$ such that

$$z \notin \bigcup_{k=1}^{p+1} \frac{1}{k} \left[(M[x + ky] - \varphi(x)) \cup (M[x + k\varphi(x)] - y) \right]. \qquad (17.7.15)$$

Relation (17.7.15) implies that $\varphi(x) + kz \notin M[x + ky]$ and $y + kz \notin M[x + k\varphi(x)]$, $k = 1, \ldots, p+1$, i.e., $(x + ky, \varphi(x) + kz) \notin M$ and $(x + k\varphi(x)y + kz) \notin M$ for $k = 1, \ldots, p+1$. Hence, by (17.7.7) and (17.7.3),

$$f(x + i\varphi(x)) = \sum_{k=1}^{p+1} (-1)^{k-1} \binom{p+1}{k} f(x + i\varphi(x) + k(y + iz)), \quad i = 1, \ldots, p+1,$$

or

$$f(x + i\varphi(x)) = \sum_{k=1}^{p+1} (-1)^{k-1} \binom{p+1}{k} f(x + ky + i(\varphi(x) + kz)), \quad i = 1, \ldots, p+1.$$
$$(17.7.16)$$

Similarly, we have by (17.7.7) and (17.7.3)

$$f(x + ky) = \sum_{i=1}^{p+1} (-1)^{i-1} \binom{p+1}{i} f(x + ky + i(\varphi(x) + kz)), \quad k = 1, \ldots, p+1. \quad (17.7.17)$$

We have by (17.7.12) and (17.7.16)

$$g(x) = \sum_{i=1}^{p+1} (-1)^{i-1} \binom{p+1}{i} f(x + i\varphi(x))$$

$$= \sum_{i=1}^{p+1} (-1)^{i-1} \binom{p+1}{i} \sum_{k=1}^{p+1} (-1)^{k-1} \binom{p+1}{k} f(x + ky + i(\varphi(x) + kz))$$

$$= \sum_{k=1}^{p+1} (-1)^{k-1} \binom{p+1}{k} \sum_{i=1}^{p+1} (-1)^{i-1} \binom{p+1}{i} f(x + ky + i(\varphi(x) + kz)),$$

whence (17.7.13) results in view of (17.7.17).

Now we take arbitrary $u, v \in \mathbb{R}^N$. Choose an $x \in \mathbb{R}^N \setminus A_u$ (cf. (17.7.9)), and

$$y \in \mathbb{R}^N \setminus \left\{ A_u \cup \bigcup_{k=1}^{p+1} \frac{1}{k} \left[(A_{u+kv} - x) \cup (M[u + kx] - v) \right] \right\}. \qquad (17.7.18)$$

Since $x \notin A_u$, we have by (17.7.9) $u + kx \notin U$ for $k = 1, \ldots, p+1$, and by (17.7.8) $M[u + kx] \in \mathcal{I}_1$ for $k = 1, \ldots, p+1$ so that the set in brackets on the right-hand side of (17.7.18) belongs to $\mathcal{I}_1$ and such a choice of y is possible. By (17.7.18) we have $x + ky \notin A_{u+kv}$ for $k = 0, \ldots, p+1$ (for $k = 0$ this follows from the choice of x), whence by (17.7.13)

$$g(u + kv) = \sum_{i=1}^{p+1} (-1)^{i-1} \binom{p+1}{i} f\big(u + kv + i(x + ky)\big) \qquad (17.7.19)$$

$$= \sum_{i=1}^{p+1} (-1)^{i-1} \binom{p+1}{i} f\big(u + ix + k(v + iy)\big), \quad k = 0, \ldots, p+1.$$

Similarly (17.7.18) implies that $v + iy \notin M[u + ix]$, i.e., $(u + ix, v + iy) \notin M$ for $i = 1, \ldots, p+1$. Hence by (17.7.7) and (17.7.2)

$$\sum_{k=0}^{p+1} (-1)^{p+1-k} \binom{p+1}{k} f\big(u + ix + k(v + iy)\big) = 0, \quad i = 1, \ldots, p+1. \qquad (17.7.20)$$

Now we have by (17.7.19), (17.7.20), and Corollary 15.1.2,

$$\Delta_v^{p+1} g(u) = \sum_{k=0}^{p+1} (-1)^{p+1-k} \binom{p+1}{k} g(u + kv)$$

$$= \sum_{k=0}^{p+1} (-1)^{p+1-k} \binom{p+1}{k} \sum_{i=1}^{p+1} (-1)^{i-1} \binom{p+1}{i} f\big(u + ix + k(v + iy)\big)$$

$$= \sum_{i=1}^{p+1} (-1)^{i-1} \binom{p+1}{i} \sum_{k=0}^{p+1} (-1)^{p+1-k} \binom{p+1}{k} f\big(u + ix + k(v + iy)\big) = 0.$$

This means that g is a polynomial function of order p. The uniqueness of g results from Lemma 17.7.1. $\square$

Theorem 17.7.3. *Let $\mathcal{I}$ be a p.l.i. ideal in $\mathbb{R}^N$ fulfilling condition (i). If a function $f : \mathbb{R}^N \to \mathbb{R}$ fulfils the condition*

$$\Delta_y^{p+1} f(x) = 0 \qquad (17.7.21)$$

for all $x, y \in \mathbb{R}^N \setminus U$ ($p \in \mathbb{N}$ fixed), where $U \in \mathcal{I}$, then f is a polynomial function of order p.

Proof. The function f is $\Pi(\mathcal{I})$-almost polynomial. It follows by Theorem 17.7.2 that there exists a polynomial function $g : \mathbb{R}^N \to \mathbb{R}$, of order p, such that (17.7.5) holds with an $S \in \mathcal{I}$. Relation (17.7.21) can equivalently be written as (cf. (17.7.3))

$$f(x + y) = \frac{1}{p+1} \left\{ f(x) + \sum_{k=2}^{p+1} (-1)^k \binom{p+1}{k} f(x + ky) \right\}, \qquad (17.7.22)$$

and (17.7.22) holds for all $x, y \in \mathbb{R}^N \setminus U$.

Take an arbitrary $x \in \mathbb{R}^N$, and choose an $\bar{x}$ such that

$$\bar{x} \in \mathbb{R}^N \setminus \left\{ (S \cup U) \cup (x - U) \cup \bigcup_{k=2}^{p+1} \frac{1}{1-k}(S - kx) \right\}. \tag{17.7.23}$$

Put $y = x - \bar{x}$. By (17.7.23) $\bar{x} \in S$, $\bar{x} + ky = \bar{x} + k(x - \bar{x}) = kx + (1 - k)\bar{x} \notin S$, $k = 2, \dots, p+1$, whence by (17.7.5)

$$f(\bar{x}) = g(\bar{x}), \quad f(\bar{x} + ky) = g(\bar{x} + ky), \quad k = 2, \dots, p+1. \tag{17.7.24}$$

Further, again by (17.7.23), $\bar{x} \notin U$ and $y = x - \bar{x} \notin U$, whence by (17.7.22) we obtain in view of (17.7.24)

$$\begin{aligned}
f(x) = f(\bar{x} + y) &= \frac{1}{p+1} \left\{ f(\bar{x}) + \sum_{k=2}^{p+1} (-1)^k \binom{p+1}{k} f(\bar{x} + ky) \right\} \\
&= \frac{1}{p+1} \left\{ g(\bar{x}) + \sum_{k=2}^{p+1} (-1)^k \binom{p+1}{k} g(\bar{x} + ky) \right\} = g(\bar{x} + y) = g(x),
\end{aligned}$$

since g is a polynomial function of order p. Thus $f = g$ in $\mathbb{R}^N$, i.e., f is a polynomial function of order p. $\qquad\square$

Theorem 17.7.4. *Let $f : \mathbb{R}^N \to \mathbb{R}$ be an $\mathcal{I}_m^{2N}$-almost polynomial function of order p. Then f is an $\mathcal{I}_0^{2N}$-almost polynomial function of order p.*

Proof. Suppose that (17.7.21) holds for $(x, y) \in \mathbb{R}^{2N} \setminus M$, where $m_e^{2N}(M) < \infty$. Choose a β such that $m_e^{2N}(M) < \beta < \infty$, and take an arbitrary $\alpha > 0$. Let

$$U = \{ x \in \mathbb{R}^N \mid m_e^N (M[x]) \geqslant \beta/\alpha \}.$$

By Lemma 17.5.2

$$m_e^N(U) \leqslant \alpha. \tag{17.7.25}$$

By exactly the same argument as in the proof of Theorem 17.7.2 we show that there exists a unique polynomial function $g : \mathbb{R}^N \to \mathbb{R}$, of order p, such that $f(x) = g(x)$ for $x \in \mathbb{R}^N \setminus U$, i.e., $\mathcal{I}_m^N$-(a.e.) in $\mathbb{R}^N$. By Lemma 17.7.1 the function g is independent of α. Letting in (17.7.25) $\alpha \to 0$ we see that $f(x) = g(x)$ $\mathcal{I}_0^N$-(a.e.) in $\mathbb{R}^N$. An argument as at the end of the proof of Theorem 17.6.4 shows that f is an $\mathcal{I}_0^{2N}$-almost polynomial function of order p. $\qquad\square$

17.8 Almost convex functions

In order to extend the most important results of 17.6 to the case of convex functions we must make stronger assumptions about the occurring ideals. Suppose that $\mathcal{I}_1$ is an ideal in $\mathbb{R}^N$, $\mathcal{I}_2$ is an ideal in $\mathbb{R}^{2N}$, and let $T : \mathbb{R}^{2N} \to \mathbb{R}^{2N}$ be a linear transform

$$T(x, y) = \frac{1}{2}(x + y, x - y), \quad x, y \in \mathbb{R}^N. \tag{17.8.1}$$

We assume that $\mathcal{I}_1$ and $\mathcal{I}_2$ fulfil the conditions:

(i) If $A \in \mathcal{I}_1$ and $a \in \mathbb{R}$, then $aA \in \mathcal{I}_1$;

(ii) If $A \in \mathcal{I}_2$, then $T(A) \in \mathcal{I}_2$, where T is given by (17.8.1).

Observe that the ideals $\mathcal{I}_1 = \mathcal{I}_0^N$, $\mathcal{I}_2 = \mathcal{I}_0^{2N}$ fulfil conditions (i), (ii) in virtue of Corollaries 3.2.1 and 3.2.2. Similarly, the ideals $\mathcal{I}_1 = \mathcal{I}_f^N$, $\mathcal{I}_2 = \mathcal{I}_f^{2N}$ fulfil conditions (i), (ii), since the multiplication by a number $a \neq 0$, and transform T, are homeomorphisms. (For $a = 0$ (i) is trivial).

Lemma 17.8.1. *Let $\mathcal{I}_1$ be a p.l.i. σ-ideal in $\mathbb{R}^N$ fulfilling condition (i). If $D \subset \mathbb{R}^N$ is a non-empty open set, then $D \notin \mathcal{I}_1$.*

Proof. Suppose that there exists a non-empty open set $D \subset \mathbb{R}^N$ such that $D \in \mathcal{I}_1$. Let $u \in D$. By Lemma 17.5.1 $D - u \in \mathcal{I}_1$, and by (i) $n(D - u) \in \mathcal{I}_1$ for all $n \in \mathbb{N}$, whence, since $\mathcal{I}_1$ is a σ-ideal,

$$\bigcup_{n=1}^{\infty} n(D - u) \in \mathcal{I}_1. \qquad (17.8.2)$$

Take an arbitrary $x \in \mathbb{R}^N$. Since $D - u$ is a neighborhood of zero in $\mathbb{R}^N$, there exists an $n \in N$ such that $x/n \in D - u$, whence $x \in n(D - u)$. This shows that

$$\bigcup_{n=1}^{\infty} n(D - u) = \mathbb{R}^N,$$

and so (17.8.2) implies that $\mathbb{R}^N \in \mathcal{I}_1$, which contradicts the fact that $\mathcal{I}_1$ is proper. $\square$

Lemma 17.8.2. *Let $\mathcal{I}_1$ and $\mathcal{I}_2$ be conjugate p.l.i. ideals in $\mathbb{R}^N$ and $\mathbb{R}^{2N}$, respectively, and let $A \in \mathcal{I}_2$. Then*

$$U_x = \big\{ h \in \mathbb{R}^N \mid (x, x + h) \in A \big\} \in \mathcal{I}_1 \qquad \mathcal{I}_1\text{-}(a.e.) \ in \ \mathbb{R}^N.$$

Proof. $h \in U_x$ means $(x, x + h) \in A$, or $x + h \in A[x]$, i.e., $h \in A[x] - x$. Thus

$$U_x = A[x] - x, \qquad x \in \mathbb{R}^N. \qquad (17.8.3)$$

Since $A \in \mathcal{I}_2$ and $\mathcal{I}_1$ and $\mathcal{I}_2$ are conjugate, we have $A[x] \in \mathcal{I}_1$ $\mathcal{I}_1$-(a.e.) in $\mathbb{R}^N$ whence, by Lemma 17.5.1, also $A[x] - x \in \mathcal{I}_1$ $\mathcal{I}_1$-(a.e.) in $\mathbb{R}^N$. Hence by (17.8.3) $U_x \in \mathcal{I}_1$ $\mathcal{I}_1$-(a.e.) in $\mathbb{R}^N$. $\square$

Lemma 17.8.3. *Let $\mathcal{I}_1$ and $\mathcal{I}_2$ be conjugate p.l.i. ideals in $\mathbb{R}^N$ and $\mathbb{R}^{2N}$, respectively, and assume that $\mathcal{I}_2$ fulfils condition (ii). Let $A \in \mathcal{I}_2$. Then*

$$V_x = \big\{ h \in \mathbb{R}^N \mid (x + h, x - h) \in A \big\} \in \mathcal{I}_1 \qquad \mathcal{I}_1\text{-}(a.e.) \ in \ \mathbb{R}^N. \qquad (17.8.4)$$

Proof. Since transform (17.8.1) is invertible, we have $(u, v) \in A$ if and only if $T(u, v) \in T(A)$, for arbitrary $u, v \in \mathbb{R}^N$. Thus $h \in V_x$ means $(x + h, x - h) \in A$, which is equivalent to $T(x+h, x-h) \in T(A)$, or, by (17.8.1), to $(x, h) \in T(A)$, i.e., $h \in T(A)[x]$. Consequently

$$V_x = T(A)[x], \qquad x \in \mathbb{R}^N. \qquad (17.8.5)$$

By (ii) $T(A) \in \mathcal{I}_2$, whence $T(A)[x] \in \mathcal{I}_1$ $\mathcal{I}_1$-(a.e.) in $\mathbb{R}^N$, since $\mathcal{I}_1$ and $\mathcal{I}_2$ are conjugate. (17.8.4) now results from (17.8.5). □

Lemma 17.8.4. *Let $\mathcal{I}_1$ be a proper σ-ideal in $\mathbb{R}^N$, let $D \subset \mathbb{R}^N$, $D \notin \mathcal{I}_1$, and let $f : D \to [-\infty, \infty)$ and $g : D \to [-\infty, \infty)$ be functions such that*

$$f(x) = g(x) \qquad \mathcal{I}_1\text{-(a.e.) in } D. \tag{17.8.6}$$

Then

$$\mathcal{I}_1\text{-}\inf_{D}\text{ess } f(x) = \mathcal{I}_1\text{-}\inf_{D}\text{ess } g(x). \tag{17.8.7}$$

Proof. Let $U \in \mathcal{I}_1$ be such that

$$f(x) = g(x) \qquad \text{for } x \in D \setminus U. \tag{17.8.8}$$

By Lemma 17.5.8 there exists a set $A \in \mathcal{I}_1$ such that

$$\mathcal{I}_1\text{-}\inf_{D}\text{ess } f(x) = \inf_{D \setminus A} f(x). \tag{17.8.9}$$

We have $A \cup U \in \mathcal{I}_1$, whence by (17.8.9)

$$\mathcal{I}_1\text{-}\inf_{D}\text{ess } f(x) = \inf_{D \setminus A} f(x) \leqslant \inf_{D \setminus (A \cup U)} f(x) \leqslant \mathcal{I}_1\text{-}\inf_{D}\text{ess } f(x),$$

whence

$$\inf_{D \setminus (A \cup U)} f(x) = \mathcal{I}_1\text{-}\inf_{D}\text{ess } f(x). \tag{17.8.10}$$

By (17.8.8) $f(x) = g(x)$ for $x \in D \setminus (A \cup U) \subset D \setminus U$, whence by (17.8.10)

$$\mathcal{I}_1\text{-}\inf_{D}\text{ess } f(x) = \inf_{D \setminus (A \cup U)} f(x) = \inf_{D \setminus (A \cup U)} g(x) \leqslant \mathcal{I}_1\text{-}\inf_{D}\text{ess } g(x).$$

Since the roles of f and g are symmetric, we may interchange f and g in the above inequality, whence (17.8.7) results. □

Lemma 17.8.5. *Let $D \subset \mathbb{R}^N$ be a convex and open set, and for every $x \in D$ put*[8]

$$D_x = \{h \in \mathbb{R}^N \mid x - h \in D \text{ and } x + h \in D\} = (x - D) \cap (D - x). \tag{17.8.11}$$

Let $\mathcal{I}_1$ be a p.l.i. σ-ideal in $\mathbb{R}^N$, and let $f : D \to [-\infty, \infty)$ and $g : D \to [-\infty, \infty)$ be functions fulfilling (17.8.6). Then, for every $x \in D$,

$$\mathcal{I}_1\text{-}\inf_{h \in D_x}\text{ess } \frac{1}{2}\left[f(x+h) + f(x-h)\right] = \mathcal{I}_1\text{-}\inf_{h \in D_x}\text{ess } \frac{1}{2}\left[g(x+h) + g(x-h)\right].$$

---

[8] Thus D_x is a non-empty (since $0 \in D_x$) open set, whence by Lemma 17.8.1 $D_x \notin \mathcal{I}_1$.

Proof. Let $U \in \mathcal{I}_1$ be such that (17.8.8) holds. We have $f(x + h) = g(x + h)$ for $x + h \in D \backslash U$, i.e., for $h \in (D \backslash U) - x = (D - x) \backslash (U - x)$. Similarly, $f(x - h) = g(x - h)$ for $h \in (x - D) \backslash (x - U)$. If $h \in D_x \backslash [(U - x) \cup (x - U)]$, then $h \in D_x = (D - x) \cap (x - D)$ and $h \notin (U - x) \cap (x - U)$, whence $h \in [(D - x) \backslash (U - x)] \cap [(x - D) \backslash (x - U)]$, and $f(x + h) = g(x + h)$, $f(x - h) = g(x - h)$. Consequently

$$\frac{1}{2} \left[f(x + h) + f(x - h) \right] = \frac{1}{2} \left[g(x + h) + g(x - h) \right] \qquad \text{for } h \in D_x \backslash [(U - x) \cup (x - U)].$$

By Lemma 17.5.1 $U - x \in \mathcal{I}_1$ and $x - U \in \mathcal{I}_1$, whence also $(U - x) \cup (x - U) \in \mathcal{I}_1$. Thus the lemma follows from Lemma 17.8.4. $\square$

Lemma 17.8.6. *Let $D \subset \mathbb{R}^N$ be a convex and open set, let $\mathcal{I}_1$ be a p.l.i. σ-ideal in $\mathbb{R}^N$ fulfilling condition (i), and let $f : D \to \mathbb{R}$ be a convex function. Then*

$$f(x) = \mathcal{I}_1\text{-infess}_{h \in D_x} \frac{1}{2} \left[f(x + h) + f(x - h) \right]. \tag{17.8.12}$$

Proof. Fix an $x \in D$, and let $A \in \mathcal{I}_1$ be arbitrary. By (i)

$$B = \bigcup_{n=0}^{\infty} 2^n \left[A \cup (-A) \right] \in \mathcal{I}_1.$$

For arbitrary $h \in D_x \backslash B$ we have[9] $\pm 2^{-n} h \in D_x \backslash A$, $n = 0, 1, 2 \ldots$. By Theorem 6.1.1 (cf. also Exercise 6.2)

$$\lim_{n \to \infty} f(x \pm 2^{-n} h) = f(x),$$

whence

$$\lim_{n \to \infty} \frac{1}{2} \left[f(x + 2^{-n} h) + f(x - 2^{-n} h) \right] = f(x),$$

and

$$\inf_{h \in D_x \backslash A} \frac{1}{2} \left[f(x + h) + f(x - h) \right] \leqslant f(x). \tag{17.8.13}$$

On the other hand, we have by the convexity of f

$$f(x) \leqslant \frac{1}{2} \left[f(x + h) + f(x - h) \right]$$

for every $h \in D_x$, whence

$$f(x) \leqslant \inf_{h \in D_x \backslash A} \frac{1}{2} \left[f(x + h) + f(x - h) \right]. \tag{17.8.14}$$

---

[9] By Theorems 5.1.1 and 5.1.2 D_x is a convex set ($D_x \neq \varnothing$, since $0 \in D_x$), and clearly D_x is symmetric so that if $h \in D_x$, then also $-h \in D_x$. Hence, for every $h \in D_x$ and $n \in \mathbb{N}$,

$$\pm 2^{-n} h = 2^{-n} (\pm h) + (1 - 2^{-n}) 0 \in D_x.$$

By (17.8.13) and (17.8.14)

$$f(x) = \inf_{h \in D_x \setminus A} \frac{1}{2} \left[f(x+h) + f(x-h) \right].$$

This is valid for arbitrary $A \in \mathcal{I}_1$. Taking the supremum over all $A \in \mathcal{I}_1$, we obtain (17.8.12). $\qquad \square$

Corollary 17.8.1. *Let* $D \subset \mathbb{R}^N$ *be a convex and open set, let* $\mathcal{I}_1$ *be a p.l.i. σ-ideal in* $\mathbb{R}^N$ *fulfilling condition (i), and let* $f : D \to \mathbb{R}$ *and* $g : D \to \mathbb{R}$ *be convex functions such that (17.8.6) holds. Then* $f = g$ *in* D.

Proof. By Lemmas 17.8.6 and 17.8.5

$$f(x) = \mathcal{I}_1\text{-}\underset{h \in D_x}{\text{infess}} \frac{1}{2} \left[f(x+h) + f(x-h) \right]$$

$$= \mathcal{I}_1\text{-}\underset{h \in D_x}{\text{infess}} \frac{1}{2} \left[g(x+h) + g(x-h) \right] = g(x)$$

for every $x \in D$. $\qquad \square$

Now let $D \subset \mathbb{R}^N$ be a convex and open set, and let $f : D \to [-\infty, \infty)$ be a function. Let $\mathcal{I}_2$ be a proper ideal in $\mathbb{R}^{2N}$. If

$$f \left(\frac{x+y}{2} \right) \leqslant \frac{f(x) + f(y)}{2} \qquad \mathcal{I}_2\text{-(a.e.) in } D \times D,$$

then f is called $\mathcal{I}_2$-*almost convex* in D.

Lemma 17.8.7. *Let* $D \subset \mathbb{R}^N$ *be a convex and open set, and suppose that we are given conjugate p.l.i. σ-ideals* $\mathcal{I}_1$ *and* $\mathcal{I}_2$ *in* $\mathbb{R}^N$ *and* $\mathbb{R}^{2N}$, *respectively, such that* $\mathcal{I}_1$ *fulfils condition (i). If* $f : D \to [-\infty, \infty)$ *is an* $\mathcal{I}_2$-*almost convex function fulfilling (17.8.12), then*

$$f \left(\frac{x+y}{2} \right) \leqslant \frac{f(x) + f(y)}{2} \tag{17.8.15}$$

for every $x, y \in D$. *In particular, if* f *is finite, then it is convex.*

Proof. By Lemma 17.5.8 for every $x \in D$ there exists a set $A_x \in \mathcal{I}_1$ such that

$$\mathcal{I}_1\text{-}\underset{h \in D_x}{\text{infess}} \frac{1}{2} \left[f(x+h) + f(x-h) \right] = \inf_{h \in D_x \setminus A_x} \frac{1}{2} \left[f(x+h) + f(x-h) \right],$$

whence by (17.8.12)

$$f(x) \leqslant \frac{1}{2} \left[f(x+h) + f(x-h) \right] \quad \text{for } h \in D_x \setminus A_x. \tag{17.8.16}$$

Put

$$Z = \left\{ (x,y) \in D \times D \mid f \left(\frac{x+y}{2} \right) > \frac{f(x) + f(y)}{2} \right\}. \tag{17.8.17}$$

Since f is $\mathcal{I}_2$-almost convex, $Z \in \mathcal{I}_2$, and consequently, since $\mathcal{I}_1$ and $\mathcal{I}_2$ are conjugate, $Z[x] \in \mathcal{I}_1$ $\mathcal{I}_1$-(i.e.) in D. Thus there exists a set $U \in \mathcal{I}_1$ such that

$$Z[x] \in \mathcal{I}_1 \qquad \text{for } x \in D \setminus U. \tag{17.8.18}$$

For every $x \in D$ we choose an arbitrary $a_x > f(x)$ and we put

$$B_x = \left\{ h \in D_x \ \Big| \ \frac{1}{2} \left[f(x+h) + f(x-h) \right] < a_x \right\}.$$

Suppose that for an $x \in D$ we have $B_x \in \mathcal{I}_1$. For $h \in D_x \setminus B_x$ we have

$$\frac{1}{2} \left[f(x+h) + f(x-h) \right] \geqslant a_x,$$

whence by (17.8.12)

$$\mathcal{I}_1\text{-}\operatorname*{infess}_{h \in D_x} \frac{1}{2} \left[f(x+h) + f(x-h) \right] \geqslant \inf_{h \in D_x \setminus B_x} \frac{1}{2} \left[f(x+h) + f(x-h) \right]$$

$$\geqslant a_x > f(x) = \mathcal{I}_1\text{-}\operatorname*{infess}_{h \in D_x} \frac{1}{2} \left[f(x+h) + f(x-h) \right].$$

The contradiction obtained shows that $B_x \notin \mathcal{I}_1$ for $x \in D$, and thus, for every $x \in D$ and every set $V \in \mathcal{I}_1$, we have

$$B_x \setminus V \neq \varnothing, \quad x \in D, \quad V \in \mathcal{I}_1. \tag{17.8.19}$$

Now take arbitrary $x, y \in D$, and put $z = \frac{1}{2}(x+y) \in D$. By Lemma 17.5.1 and by (17.8.19) there exists an

$$h' \in B_x \setminus \left[(x - U) \cup (U - x) \right]. \tag{17.8.20}$$

In particular, $x - h' \notin U$ and $x + h' \notin U$, whence by (17.8.18) $Z[x - h'] \in \mathcal{I}_1$ and $Z[x + h'] \in \mathcal{I}_1$. Again by Lemma 17.5.1 and by (17.8.19) there exists an

$$h'' \in B_y \setminus \left[(y - Z[x - h']) \cup (Z[x + h'] - y) \cup (2A_z - h') \right]. \tag{17.8.21}$$

Hence $y - h'' \notin Z[x - h']$ and $y + h'' \notin Z[x + h']$, i.e., $(x - h', y - h'') \notin Z$ and $(x + h', y + h'') \notin Z$. Hence by (17.8.17)

$$f\left(z - \frac{h' + h''}{2}\right) \leqslant \frac{f(x - h') + f(y - h'')}{2}, \tag{17.8.22}$$

$$f\left(z + \frac{h' + h''}{2}\right) \leqslant \frac{f(x + h') + f(y + h'')}{2}, \tag{17.8.23}$$

Further, it follows from (17.8.21) that $\frac{1}{2}(h' + h'') \notin A_z$, whereas by (17.8.20) and (17.8.21) $h' \in B_x \subset D_x$, $h'' \in B_y \subset D_y$. This means that $x - h', x + h' \in D$ and

$y - h'', y + h'' \in D$, whence, since D is convex, $z - \frac{1}{2}(h' + h'') = \frac{1}{2}[(x - h') + (y - h'')] \in D$, and $z + \frac{1}{2}(h' + h'') = \frac{1}{2}[(x + h') + (y + h'')] \in D$. Consequently $\frac{1}{2}(h' + h'') \in D_z$, and thus $\frac{1}{2}(h' + h'') \in D_z \setminus A_z$. Hence by (17.8.16)

$$f(z) \leqslant \frac{1}{2}\left[f\left(z + \frac{h' + h''}{2} \right) + f\left(z - \frac{h' + h''}{2} \right) \right]. \tag{17.8.24}$$

Since $h' \in B_x$ and $h'' \in B_y$, we have

$$\frac{1}{2}[f(x + h') + f(x - h')] < a_x, \tag{17.8.25}$$

$$\frac{1}{2}[f(y + h'') + f(y - h'')] < a_y. \tag{17.8.26}$$

Relations (17.8.24), (17.8.22), (17.8.23), (17.8.25) and (17.8.26) yield

$$f(z) \leqslant \frac{1}{4}[f(x - h') + f(y - h'') + f(x + h') + f(y + h'')] < \frac{1}{2}(a_x + a_y).$$

Letting $a_x \to f(x)$, $a_y \to f(y)$, we obtain hence (17.8.15) in view of the definition of z. $\qquad\square$

With the notation from Lemma 17.8.7, we put

$$g(x) = \mathcal{I}_1\text{-}\operatorname*{infess}_{h \in D_x} \frac{1}{2}[f(x + h) + f(x - h)], \qquad x \in D. \tag{17.8.27}$$

Then

$$g(x) = \inf_{h \in D_x \setminus A_x} \frac{1}{2}[f(x + h) + f(x - h)] < \infty.$$

Thus g is a function $g : D \to [-\infty, \infty)$.

Lemma 17.8.8. *Let $D \subset \mathbb{R}^N$ be a convex and open set, and suppose that we are given conjugate p.l.i. σ-ideals $\mathcal{I}_1$ and $\mathcal{I}_2$ in $\mathbb{R}^N$ and $\mathbb{R}^{2N}$, respectively, fulfilling conditions (i) and (ii). If $f : D \to \mathbb{R}$ is an $\mathcal{I}_2$-almost convex function, and $g : D \to [-\infty, \infty)$ is defined by (17.8.27), then (17.8.6) holds.*

Proof. Let Z denote set (17.8.17) so that $Z \in \mathcal{I}_2$, and let

$$U_x = \{h \in D_x \mid (x, x + h) \in Z\}, \qquad V_x = \{h \in D_x \mid (x + h, x - h) \in Z\}.$$

By Lemmas 17.8.2 and 17.8.3 $U_x \in \mathcal{I}_1$ and $V_x \in \mathcal{I}_1$ $\mathcal{I}_1$-(a.e) in D, i.e., there exists a set $S \in \mathcal{I}_1$ such that

$$U_x \in \mathcal{I}_1 \quad \text{and} \quad V_x \in \mathcal{I}_1 \quad \text{for } x \in D \setminus S. \tag{17.8.28}$$

Fix an arbitrary set $A \in \mathcal{I}_1$ and put for every $x \in D \setminus S$

$$B_x = \bigcup_{n=0}^{\infty} 2^n \left[A \cup (-A) \cup U_x \cup (-U_x) \cup V_x \right].$$

By (i) and (17.8.28) we have $B_x \in \mathcal{I}_1$ $(x \in D \setminus S)$, since $\mathcal{I}_1$ is a σ-ideal. For every $h \in D_x \setminus B_x$ we have

$$\pm 2^{-n} h \notin A, \qquad \pm 2^{-n} h \notin U_x, \qquad 2^{-n} h \notin V_x, \qquad n = 0, 1, 2, \ldots \qquad (17.8.29)$$

In particular, $h \notin V_x$, whence $(x + h, x - h) \notin Z$, and by (17.8.17)

$$f(x) \leqslant \frac{1}{2} \left[f(x + h) + f(x - h) \right].$$

Consequently, in view of (17.8.27),

$$f(x) \leqslant \inf_{h \in D_x \setminus B_x} \frac{1}{2} \left[f(x + h) + f(x - h) \right]$$

$$\leqslant \mathcal{I}_1\text{-}\operatorname*{infess}_{h \in D_x} \frac{1}{2} \left[f(x + h) + f(x - h) \right] = g(x). \qquad (17.8.30)$$

On the other hand, we have by (17.8.29) for $h \in D_x \setminus B_x$

$$(x, x \pm 2^{-n} h) \notin Z, \qquad n = 0, 1, 2, \ldots,$$

whence by (17.8.17)

$$f(x \pm 2^{-(n+1)} h) = f\left(\frac{x + x \pm 2^{-n} h}{2} \right) \leqslant \frac{f(x) + f(x \pm 2^{-n} h)}{2}, \qquad n = 0, 1, 2, \ldots$$

Hence

$$f(x \pm 2^{-(n+1)} h) - f(x) \leqslant \frac{1}{2} \left[f(x \pm 2^{-n} h) - f(x) \right], \qquad n = 0, 1, 2, \ldots,$$

and

$$\limsup_{n \to \infty} \left[f(x \pm 2^{-n} h) - f(x) \right] = \limsup_{n \to \infty} \left[f(x \pm 2^{-(n+1)} h) - f(x) \right]$$

$$\leqslant \frac{1}{2} \limsup_{n \to \infty} \left[f(x \pm 2^{-n} h) - f(x) \right].$$

Hence

$$\left[\limsup_{n \to \infty} f(x \pm 2^{-n} h) \right] - f(x) = \limsup_{n \to \infty} \left[f(x \pm 2^{-n} h) - f(x) \right] \leqslant 0,$$

or

$$\limsup_{n \to \infty} f(x \pm 2^{-n} h) \leqslant f(x).$$

Hence also

$$\limsup_{n\to\infty} \frac{1}{2}\left[f(x+2^{-n}h) + f(x-2^{-n}h)\right] \leqslant f(x). \tag{17.8.31}$$

By (17.8.29) $\pm 2^{-n}h \notin A$, $n = 0, 1, 2, \ldots$, and so we obtain from (17.8.31)

$$\inf_{h\in D_x\setminus A} \frac{1}{2}\left[f(x+h) + f(x-h)\right] \leqslant f(x),$$

and, since A has been arbitrary set from $\mathcal{I}_1$, (cf. (17.8.27)),

$$g(x) = \mathcal{I}_1\text{-}\underset{h\in D_x}{\operatorname{infess}}\, \frac{1}{2}\left[f(x+h) + f(x-h)\right] =$$

$$= \sup_{A\in\mathcal{I}_1} \inf_{h\in D_x\setminus A} \frac{1}{2}\left[f(x+h) + f(x-h)\right] \leqslant f(x). \tag{17.8.32}$$

Inequalities (17.8.30) and (17.8.32) yield the equality $f(x) = g(x)$, valid for $x \in D\setminus S$, i.e., $\mathcal{I}_1$-(a.e.) in D. $\qquad\square$

Lemma 17.8.9. *Let $D \subset \mathbb{R}^N$ be a convex and open set, and suppose that we are given a p.l.i. ideal $\mathcal{I}_1$ in $\mathbb{R}^N$ fulfilling condition (i). If $f : [-\infty, \infty)$ is an $\Omega(\mathcal{I}_1)$-almost convex function, and $g : D \to [-\infty, \infty)$ is a function fulfilling (17.8.6), then g is $\Omega(\mathcal{I}_1)$-almost convex.*

Proof. Let

$$f(x) = g(x) \qquad \text{for } x \in D\setminus S, \tag{17.8.33}$$

where $S \in \mathcal{I}_1$, and put

$$A = \{(x,y) \in D\times D \mid x \in S, \text{ or } y \in S, \text{ or } x+y \in 2S\} = (S\times D)\cup(D\times S)\cup M,$$

where M is defined by (17.5.4) with $U = 2S \in \mathcal{I}_1$ (cf. (i)). Clearly $(S\times D)\cup(D\times S) \in \Omega(\mathcal{I}_1)$, whereas $M \in \Omega(\mathcal{I}_1)$ in virtue of Lemma 17.5.4. Hence $A \in \Omega(\mathcal{I}_1)$. Let Z be the set defined by (17.8.17). Since f is $\Omega(\mathcal{I}_1)$-almost convex, $Z \in \Omega(\mathcal{I}_1)$, whence also $A\cup Z \in \Omega(\mathcal{I}_1)$. For $(x,y) \in (D\times D)\setminus(A\cup Z)$ we have

$$f(x) = g(x), \qquad f(y) = g(y), \qquad f\left(\frac{x+y}{2}\right) = g\left(\frac{x+y}{2}\right),$$

and

$$f\left(\frac{x+y}{2}\right) \leqslant \frac{f(x) + f(y)}{2}.$$

Hence

$$g\left(\frac{x+y}{2}\right) = f\left(\frac{x+y}{2}\right) \leqslant \frac{f(x) + f(y)}{2} = \frac{g(x) + g(y)}{2}$$

holds for $(x,y) \in (D\times D)\setminus(A\cup Z)$, i.e., $\Omega(\mathcal{I}_1)$-(a.e.) in $D\times D$. $\qquad\square$

The following two theorems (Kuczma [178])are analogues of Theorems 17.6.2 and 17.6.1.

Theorem 17.8.1. *Let $D \subset \mathbb{R}^N$ be a convex and open set, and let $\mathcal{I}_1$ be a p.l.i. ideal in $\mathbb{R}^N$ fulfilling condition (i). If $g : D \to \mathbb{R}$ is a convex function, and $f : D \to [-\infty, \infty)$ is a function such that (17.8.6) holds, then f is $\Omega(\mathcal{I}_1)$-almost convex.*

Proof. We always have $\varnothing \in \Omega(\mathcal{I}_1)$, so g is, in particular, $\Omega(\mathcal{I}_1)$-almost convex. The theorem results now from Lemma 17.8.9 (the roles of f and g are interchanged). $\square$

Theorem 17.8.2. *Let $D \subset \mathbb{R}^N$ be a convex and open set, and suppose that we are given conjugate p.l.i. σ-ideals $\mathcal{I}_1$ and $\mathcal{I}_2$ in $\mathbb{R}^N$ and $\mathbb{R}^{2N}$, respectively, fulfilling conditions (i) and (ii). If $f : D \to \mathbb{R}$ is an $\mathcal{I}_2$-almost convex function, then there exists a unique convex function $g : D \to \mathbb{R}$ such that $f(x) = g(x) \mathcal{I}_1$-(a.e.) in D.*

Proof. Define $g : D \to [-\infty, \infty)$ by (17.8.27). By Lemma 17.8.8 relation (17.8.6) holds. By Lemma 17.8.5 and (17.8.27)

$$g(x) = \mathcal{I}_1\text{-}\operatorname*{infess}_{h \in D_x} \frac{1}{2} \big[f(x+h) + f(x-h)\big]$$

$$= \mathcal{I}_1\text{-}\operatorname*{infess}_{h \in D_x} \frac{1}{2} \big[g(x+h) + g(x-h)\big]. \qquad (17.8.34)$$

By Lemma 17.5.3 we have $\mathcal{I}_2 \subset \Omega(\mathcal{I}_1)$, whence it follows that f is $\Omega(\mathcal{I}_1)$-almost convex. By (17.8.6) and Lemma 17.8.9 the function g is $\Omega(\mathcal{I}_1)$-almost convex. Using Lemmas 17.8.7 and 17.5.3 we infer from (17.8.34) that

$$g\left(\frac{x+y}{2}\right) \leqslant \frac{g(x) + g(y)}{2} \qquad (17.8.35)$$

for all $x, y \in D$.

In virtue of (17.8.6) there exists a set $S \in \mathcal{I}_1$ such that (17.8.33) holds. Take an arbitrary $x \in D$. We may find a $y \in D \setminus [S \cup (2S - x)]$. Then $y \in D \setminus S$ and, since D is convex, $\dfrac{x+y}{2} \in D \setminus S$. Hence $g(y) = f(y)$, $g\left(\dfrac{x+y}{2}\right) = f\left(\dfrac{x+y}{2}\right)$, and by (17.8.35)

$$g(x) \geqslant 2g\left(\frac{x+y}{2}\right) - g(y) = 2f\left(\frac{x+y}{2}\right) - f(y) > -\infty.$$

Thus g is finite, and hence $\big($cf. (17.8.35)$\big)$ it is a convex function $g : D \to \mathbb{R}$. The uniqueness of g results from Corollary 17.8.1. $\square$

17.9　Almost subadditive functions

The results concerning almost subadditive functions (Ger [110]) are at least satisfactory among those in this chapter. Let $\mathcal{I}_2$ be a p.l.i. ideal in $\mathbb{R}^{2N}$ fulfilling the following condition. Let $T : \mathbb{R}^{2N} \to \mathbb{R}^{2N}$ be the linear transform:

$$T(x, y) = (x + y, y), \qquad x, y \in \mathbb{R}^N. \qquad (17.9.1)$$

We assume that:

(i) For every $A \in \mathcal{I}_2$ also $T(A) \in \mathcal{I}_2$, where T is given by (17.9.1).

However, let us observe that most important ideals in $\mathbb{R}^{2N}$ (listed in 17.5) fulfil condition (i).

Let $C \subset \mathbb{R}^N$ be a set fulfilling condition (16.1.3). (This means that $(C, +)$ is a subsemigroup of $(\mathbb{R}^N, +)$). A function $f : C \to \mathbb{R}$ is called $\mathcal{I}_2$-*almost subadditive* in C iff the inequality

$$f(x + y) \leqslant f(x) + f(y) \tag{17.9.2}$$

holds $\mathcal{I}_2$-(a.e.) in $C \times C$. For every $x \in C$ write

$$C_x = C \cap (x - C) = \{\, h \in C \mid x - h \in C \,\}. \tag{17.9.3}$$

We will need the following

Lemma 17.9.1. *Let $C \in \mathbb{R}^N$ be a set fulfilling (16.1.3). For every $x, y \in C$ and $h \in C_x$, we have*

$$C_y \subset C_{x+y} - h. \tag{17.9.4}$$

Proof. Take arbitrary $x, y \in C$, $h \in C_x$, and $t \in C_y$. By (17.9.3) $t, y - t, h, x - h \in C$, whence also

$$t + h \in C \quad \text{and} \quad y - t + x - h \in C,$$

since C fulfils (16.1.3). Hence $t \in C - h$ and $t \in (x + y - C) - h$, i.e.,

$$t \in \big[C \cap (x + y - C) \big] - h = C_{x+y} - h.$$

Hence (17.9.4) follows.   □

In the sequel we assume that $\mathcal{I}_1$ is a p.l.i. σ-ideal in $\mathbb{R}^N$ fulfilling the condition

$$C_x \notin \mathcal{I}_1 \quad \text{for every } x \in C, \tag{17.9.5}$$

where C_x is defined by (17.9.3). Condition (17.9.5) implies, since $C_x \subset C$, that also

$$C \notin \mathcal{I}_1. \tag{17.9.6}$$

Lemma 17.9.2. *Let $C \subset \mathbb{R}^N$ be a set fulfilling (16.1.3), and suppose that we are given conjugate p.l.i. σ-ideal $\mathcal{I}_1$ in $\mathbb{R}^N$ fulfilling condition (17.9.5) and p.l.i. ideal $\mathcal{I}_2$ in $\mathbb{R}^{2N}$. Let $f : C \to \mathbb{R}$ be an $\mathcal{I}_2$-almost subadditive function. Then the function $\varphi : C \to \mathbb{R}$ defined by*[10]

$$\varphi(x) = -\mathcal{I}_1\text{-}\operatorname*{intess}_{h \in C} \big[f(h) - f(x + h) \big], \qquad x \in C, \tag{17.9.7}$$

is subadditive, and fulfils the condition

$$\varphi(x) \leqslant f(x) \qquad \mathcal{I}_1\text{-(a.e.) in } C. \tag{17.9.8}$$

---

[10] The expression $-\mathcal{I}_1\text{-}\operatorname*{infess}_{B} \big[- F(x) \big]$ is also denoted by $\mathcal{I}_1\text{-}\operatorname*{supess}_{B} F(x) = \inf_{A \in \mathcal{I}_1} \sup_{B \setminus A} F(x)$, and is referred to as the *essential supremum* (with respect to $\mathcal{I}_1$) of F on the set $B \notin \mathcal{I}_1$.

Proof. By Lemma 17.5.8 for every $x \in C$ there exists a set $A_x \in \mathcal{I}_1$ such that

$$\varphi(x) = \sup_{h \in C \setminus A_x} \left[f(x+h) - f(h) \right]. \tag{17.9.9}$$

Take arbitrary $x, y \in C$. Then we have by (17.9.9)

$$\varphi(x) \geqslant f(x+h) - f(h) \qquad \text{for } h \in C \setminus A_x, \tag{17.9.10}$$
$$\varphi(y) \geqslant f(y+k) - f(k) \qquad \text{for } k \in C \setminus A_y. \tag{17.9.11}$$

Let $B = A_x \cup (A_y - x)$ so that $B \in \mathcal{I}_1$. By (17.9.6) $C \setminus B \neq \varnothing$, and thus there exists a $t \in C \setminus B$. For such a t we have $t \notin A_x$ and $x + t \notin A_y$, whence, taking in (17.9.10) and (17.9.11) $h = 1$ and $k = x + t$, we get

$$\varphi(x) \geqslant f(x+t) - f(t), \qquad \varphi(y) \geqslant f(y+x+t) - f(x+t),$$

and

$$\varphi(x) + \varphi(y) \geqslant f(x+y+t) - f(t).$$

Hence

$$\varphi(x) + \varphi(y) \geqslant \sup_{t \in C \setminus B} \left[f(x+y+t) - f(t) \right]$$
$$= - \inf_{t \in C \setminus B} \left[f(t) - f(x+y+t) \right]$$
$$\geqslant \inf_{B \in \mathcal{I}_1} \left(- \inf_{t \in C \setminus B} \left[f(t) - f(x+y+t) \right] \right)$$
$$= - \sup_{B \in \mathcal{I}_1} \inf_{t \in C \setminus B} \left[f(t) - f(x+y+t) \right]$$
$$= - \mathcal{I}_1\text{-}\operatorname*{infess}_{t \in C} \left[f(t) - f(x+y+t) \right] = \varphi(x+y).$$

Thus φ is subadditive.

Since f is $\mathcal{I}_2$-almost subadditive, there exists a set $M \in \mathcal{I}_2$ such that (17.9.2) holds for all $(x, y) \in (C \times C) \setminus M$. Since $\mathcal{I}_1$ and $\mathcal{I}_2$ are conjugate, there exists a set $U \subset \mathbb{R}^N$, $U \in \mathcal{I}_1$, such that

$$M[x] = \{ y \in \mathbb{R}^N \mid (x, y) \in M \} \in \mathcal{I}_1 \qquad \text{for } x \in \mathbb{R}^N \setminus U.$$

Take an $x \in C \setminus U \subset \mathbb{R}^N \setminus U$. Then for $h \in C \setminus M[x]$ we have $f(x+h) \leqslant f(x) + f(h)$, and

$$\varphi(x) = \mathcal{I}_1\text{-}\operatorname*{infess}_{h \in C} \left[f(h) - f(x+h) \right]$$
$$= - \inf_{A \in \mathcal{I}_1} \sup_{h \in C \setminus A} \left[f(x+h) - f(h) \right]$$
$$\leqslant \sup_{h \in C \setminus M[x]} \left[f(x+h) - f(h) \right]$$
$$\leqslant \sup_{h \in C \setminus M[x]} \left[f(x) + f(h) - f(h) \right] = f(x).$$

Thus φ satisfies (17.9.8).

It remains to show that φ is finite. Take an arbitrary $x \in C$ and an $h \in C_x \setminus [U \cup (x - U)]$, where C_x is given by (17.9.3). Such an h exists in virtue of (17.9.5). Then $h \in C$, $x - h \in C$, $h \notin U$, $x - h \notin U$. As we have just proved, this implies that $\varphi(x) \leqslant f(x)$ and $\varphi(x - h) \leqslant f(x - h)$. Since φ is subadditive, we have

$$\varphi(x) \leqslant \varphi(x - h) + \varphi(h) \leqslant f(x - h) + f(h) < \infty.$$

On the other hand, by (17.9.9) we have for arbitrary $h \in C \setminus A_x$

$$\varphi(x) \geqslant f(x + h) - f(h) > -\infty.$$

Thus φ is finite.   □

Lemma 17.9.3. *Let $C \subset \mathbb{R}^N$ be a set fulfilling (16.1.3), and suppose that we are given conjugate p.l.i. σ-ideal $\mathcal{I}_1$ in $\mathbb{R}^N$ fulfilling condition (17.9.5) and p.l.i. ideal $\mathcal{I}_2$ in $\mathbb{R}^{2N}$ fulfilling condition (i). Let $f : C \to \mathbb{R}$ be an $\mathcal{I}_2$-almost subadditive function. Then the function $\Phi : C \to \mathbb{R}$ defined by*

$$\Phi(x) = \mathcal{I}_1\text{-}\operatorname*{infess}_{h \in C_x} \left[f(x - h) - f(h) \right], \qquad x \in C, \tag{17.9.12}$$

is subadditive and fulfils the condition

$$f(x) \leqslant \Phi(x) \qquad \mathcal{I}_1\text{-}(a.e.) \ in \ C. \tag{17.9.13}$$

Proof. By Lemma 17.5.8 for every $x \in C$ there exists a set $A_x \in \mathcal{I}_1$ such that

$$\Phi(x) = \inf_{h \in C_x \setminus A_x} \left[f(x - h) + f(h) \right]. \tag{17.9.14}$$

Relation (17.9.14) implies that

$$\Phi(x) < \infty \qquad \text{for } x \in C. \tag{17.9.15}$$

Let $M \in \mathcal{I}_2$ and $U \in \mathcal{I}_1$ have the same meaning as in the proof of Lemma 17.9.2.

Take arbitrary $x, y \in C$ and $a > \Phi(x)$, $b > \Phi(y)$ (cf. (17.9.15)). Since $U \cup (x - U) \in \mathcal{I}_1$ and $C_x \notin \mathcal{I}_1$ (cf. (17.9.5)), we have by (17.9.12)

$$a > \Phi(x) \geqslant \inf_{h \in C_x \setminus [U \cup (x - U)]} \left[f(x - h) + f(h) \right],$$

and so there exists an $h \in C_x \setminus [U \cup (x - U)]$ such that

$$f(x - h) + f(h) < a. \tag{17.9.16}$$

By Lemma 17.9.1 we have (17.9.4), whence also

$$C_y \setminus \left[M[h] \cup (y - M[x - h]) \cup (A_{x+y} - h) \right] \subset$$
$$\subset (C_{x+y} - h) \setminus \left[M[h] \cup (y - M[x - h]) \cup (A_{x+y} - h) \right]$$
$$= \left[(C_{x+y} \setminus A_{x+y}) - h \right] \setminus \left[M[h] \cup (y - M[x - h]) \right].$$

Hence, since $h \notin U$ and $x - h \notin U$,

$$b > \Phi(y) \geqslant \inf_{k \in C_y \setminus [M[h] \cup (y - M[x-h]) \cup (A_{x+y} - h)]} \left[f(y - k) + f(k) \right]$$

$$\geqslant \inf_{k \in [(C_{x+y} \setminus A_{x+y}) - h] \setminus [M[h] \cup (y - M[x-h])]} \left[f(y - k) + f(k) \right],$$

and so there exists a $k \in \left[(C_{x+y} \setminus A_{x+y}) - h \right] \setminus \left[M[h] \cup (y - M[x - h]) \right]$ such that

$$f(y - k) + f(k) < b. \tag{17.9.17}$$

We have by (17.9.14)

$$\Phi(x + y) = \inf_{t \in C_{x+y} \setminus A_{x+y}} \left[f(x + y - t) + f(t) \right],$$

whence, since $h + k \in C_{x+y} \setminus A_{x+y}$, we get

$$\Phi(x + y) \leqslant f\big((x + y) - (h + k)\big) + f(h + k). \tag{17.9.18}$$

Since $k \notin M[h]$, i.e., $(h, k) \notin M$, we have

$$f(h + k) \leqslant f(h) + f(k), \tag{17.9.19}$$

and since $k \notin y - M[x - h]$, i.e., $(x - h, y - k) \notin M$, we have

$$f\big((x + y) - (h + k)\big) = f(x - h + y - k) \leqslant f(x - h) + f(y - k). \tag{17.9.20}$$

Relations (17.9.18), (17.9.20), (17.9.19), (17.9.16) and (17.9.17) imply

$$\Phi(x + y) \leqslant f(x - h) + f(y - k) + f(h) + f(k) < a + b,$$

whence, on letting $a \to \Phi(x)$ and $b \to \Phi(y)$, we obtain

$$\Phi(x + y) \leqslant \Phi(x) + \Phi(y).$$

Consequently Φ is subadditive.

By (i) we have $T(M) \in \mathcal{I}_2$, where T is given by (17.9.1), whence $T(M)[x] \in \mathcal{I}_1$ $\mathcal{I}_1$-(a.e.) in $\mathbb{R}^N$. Consequently there exists a set $V \in \mathcal{I}_1$ such that

$$T(M)[x] \in \mathcal{I}_1 \qquad \text{for } x \notin V. \tag{17.9.21}$$

But $h \in T(M)[x]$ means $(x, h) \in T(M)$, or, equivalently, $T^{-1}(x, h) \in M$. According to (17.9.1) $T^{-1}(x, h) = (x - h, h)$. Therefore $\big($cf. (17.9.21)$\big)$

$$\{ h \in \mathbb{R}^N \mid (x - h, h) \in M \} \in \mathcal{I}_1 \qquad \text{for } x \notin V. \tag{17.9.22}$$

Now take arbitrary $x \in C \setminus V$ and $a > \Phi(x)$. Then by (17.9.22)

$$W = \{ h \in C_x \mid (x - h, h) \in M \} \subset \{ h \in \mathbb{R}^N \mid (x - h, h) \in M \} \in \mathcal{I}_1,$$

whence by (17.9.12)

$$a > \Phi(x) \geqslant \inf_{h \in C_x \setminus W} \left[f(x - h) + f(h) \right].$$

So we can find an $h \in C_x \setminus W$ such that

$$f(x - h) + f(h) < a. \tag{17.9.23}$$

But since $h \notin W$, we have $(x - h, h) \notin M$, i.e.,

$$f(x - h) + f(h) \geqslant f(x - h + h) = f(x). \tag{17.9.24}$$

Relations (17.9.23) and (17.9.24) yield

$$f(x) < a,$$

whence on letting $a \to \Phi(x)$ we obtain $f(x) \leqslant \Phi(x)$. Hence (17.9.13) results.

It remains to show that Φ is finite. Take an arbitrary $x \in C$. Since $V \in \mathcal{I}_1$, we have $V - x \in \mathcal{I}_1$, and by (17.9.6) $C \setminus (V - x) \notin \mathcal{I}_1$. Therefore there exists a $y \in C \setminus (V - x)$. This means that $x + y \notin V$, whereas by (16.1.3) $x + y \in C$. Thus $x + y \in C \setminus V$, which implies, as we have just shown, that

$$\Phi(x + y) \geqslant f(x + y) > -\infty.$$

Since Φ is subadditive, we have

$$-\infty < \Phi(x + y) \leqslant \Phi(x) + \Phi(y),$$

which shows that $\Phi(x) > -\infty$. This together with (17.9.15) implies that Φ is finite. $\square$

In order to prove a weak analogue of Theorem 17.6.1 we must make still stronger assumptions. For any function $f : \mathbb{R}^N \to \mathbb{R}$ and any $\varepsilon > 0$ put

$$A_f(\varepsilon) = \{ x \in \mathbb{R}^N \mid f(x) < \varepsilon \text{ and } f(-x) < \varepsilon \}. \tag{17.9.25}$$

Theorem 17.9.1. *Let $\mathcal{I}_1$ and $\mathcal{I}_2$ be conjugate p.l.i. σ-ideal in $\mathbb{R}^N$ and p.l.i. in $\mathbb{R}^{2N}$ fulfilling condition (i), respectively. Let $f : \mathbb{R}^N \to \mathbb{R}$ be an $\mathcal{I}_2$-almost subadditive function fulfilling the condition*

$$A_f(\varepsilon) \notin \mathcal{I}_1 \qquad \text{for every } \varepsilon > 0, \tag{17.9.26}$$

where $A_f(\varepsilon)$ is given by (17.9.25). Then there exists a subadditive function $g : \mathbb{R}^N \to \mathbb{R}$ such that

$$f(x) = g(x) \qquad \mathcal{I}_1\text{-(a.e) in } \mathbb{R}^N. \tag{17.9.27}$$

Proof. In this case $C = \mathbb{R}^N$, whence by (17.9.3)

$$C_x = C \cap (x - C) = \mathbb{R}^N \cap (x - \mathbb{R}^N) = \mathbb{R}^N \cap \mathbb{R}^N = \mathbb{R}^N \notin \mathcal{I}_1,$$

since $\mathcal{I}_1$ is proper. Hence condition (17.9.5) is satisfied, and we can use Lemmas 17.9.2 and 17.9.3. Let φ and Φ be defined by (17.9.7) and (17.9.12), respectively, with $C = \mathbb{R}^N$. Our theorem will be proved when we show that

$$\varphi(x) = \Phi(x) \qquad \mathcal{I}_1\text{-(a.e.) in } \mathbb{R}^N.$$

Then we may take as g any of φ and Φ. Then g is subadditive and satisfies (17.9.27) in virtue of Lemmas 17.9.2 and 17.9.3. Clearly we have by (17.9.8) and (17.9.13)

$$\varphi(x) \leqslant \Phi(x) \qquad \mathcal{I}_1\text{-(a.e.) in } \mathbb{R}^N.$$

Thus

$$S = \{x \in \mathbb{R}^N \mid \varphi(x) > \Phi(x)\} \in \mathcal{I}_1.$$

We will show that $\varphi(x) = \Phi(x)$ for $x \notin S$. For an indirect proof suppose that there exists an $x \notin S$ such that $\varphi(x) \neq \Phi(x)$. By the definition of S we must have $\varphi(x) < \Phi(x)$, and we can find an $\varepsilon > 0$ such that

$$\varphi(x) + 2\varepsilon < \Phi(x). \tag{17.9.28}$$

Put

$$Z_0 = \{h \in \mathbb{R}^N \mid f(x + h) - f(h) + 2\varepsilon > f(x - h) + f(h)\},$$

and suppose that $Z_0 \notin \mathcal{I}_1$. Let A_1 and A_2 be sets from $\mathcal{I}_1$ such that

$$\mathcal{I}_1\text{-}\operatorname*{infess}_{h \in \mathbb{R}^N} \left[f(h) - f(x + h) \right] = \inf_{h \in \mathbb{R}^N \setminus A_1} \left[f(h) - f(x + h) \right],$$

$$\mathcal{I}_1\text{-}\operatorname*{infess}_{h \in \mathbb{R}^N} \left[f(x - h) + f(h) \right] = \inf_{h \in \mathbb{R}^N \setminus A_2} \left[f(x - h) + f(h) \right],$$

(cf. Lemma 17.5.8). Then we have $A_1 \cup A_2 \in \mathcal{I}_1$, whence $Z_0 \setminus (A_1 \cup A_2) \notin \mathcal{I}_1$. For $h \in Z_0 \setminus (A_1 \cup A_2)$ we have

$$f(x + h) - f(h) \leqslant \sup_{h \in Z_0 \setminus A_1} \left[f(x + h) - f(h) \right]$$

$$\leqslant \sup_{h \in \mathbb{R}^N \setminus A_1} \left[f(x + h) - f(h) \right]$$

$$= - \inf_{h \in \mathbb{R}^N \setminus A_1} \left[f(h) - f(x + h) \right]$$

$$= -\mathcal{I}_1\text{-}\operatorname*{infess}_{h \in \mathbb{R}^N} \left[f(h) - f(x + h) \right] = \varphi(x),$$

whence by (17.9.28)

$$f(x + h) - f(h) + 2\varepsilon \leqslant \varphi(x) + 2\varepsilon < \Phi(x).$$

Further, for such h,

$$f(x - h) + f(h) \geqslant \inf_{h \in Z_0 \backslash A_2} \left[f(x - h) + f(h) \right]$$

$$\geqslant \inf_{h \in \mathbb{R}^N \backslash A_2} \left[f(x - h) + f(h) \right]$$

$$= \mathcal{I}_1\text{-}\operatorname*{infess}_{h \in \mathbb{R}^N} \left[f(x - h) + f(h) \right] = \Phi(x).$$

Thus, for $h \in Z_0 \backslash (A_1 \cup A_2)$,

$$f(x + h) - f(h) + 2\varepsilon < f(x - h) + f(h),$$

which contradicts the definition on Z_0. Consequently Z_0 belongs to $\mathcal{I}_1$, and so also the set

$$Z = Z_0 \cup (-Z_0).$$

By (17.9.26) $A_f(\varepsilon) \backslash Z \notin \mathcal{I}_1$, and hence there exists an $h \in A_f(\varepsilon) \backslash Z$. For such an h we have $\varepsilon - f(h) > 0$, $\varepsilon - f(-h) > 0$, $f(x + h) - f(h) + 2\varepsilon \leqslant f(x - h) + f(h)$, and $f(x - h) - f(-h) + 2\varepsilon \leqslant f(x + h) + f(-h)$, whence

$$f(x + h) < f(x + h) - f(h) + \varepsilon \leqslant f(x - h) + f(h) - \varepsilon < f(x - h)$$
$$< f(x - h) - f(-h) + \varepsilon \leqslant f(x + h) + f(-h) - \varepsilon < f(x + h),$$

which is impossible. Consequently we must have $\varphi(x) = \Phi(x)$. $\qquad\square$

Note that there is no uniqueness statement in Theorem 17.9.1.

Condition (17.9.26) is certainly fulfilled if the limit $\lim_{x \to 0} f(x)$ exists and is non-positive (cf. Exercise 17.7), but it can also be fulfilled for very irregular f (cf. Ger [110]).

If the conditions of Theorem 17.9.1 are not fulfilled, then the estimation

$$\varphi(x) \leqslant f(x) \leqslant \Phi(x) \qquad \mathcal{I}_1\text{-(a.e.) in } C$$

resulting from Lemmas 17.9.2 and 17.9.3, can be very rough (cf. Exercise 17.15).

Exercises

1. Let $M \in \mathbb{N}$ be fixed, and let $E = \{t \in \mathbb{R} \mid t = n/M, n = 0, \ldots, M\}$. Let $f : \mathbb{R}^N \to \mathbb{R}$ be a function such that

$$f(x + y) - f(x) - f(y) \in E \qquad \text{for every } x, y \in \mathbb{R}^N. \tag{$*$}$$

Show that f can be represented as $f = g + f_0$, where $g : \mathbb{R}^N \to \mathbb{R}$ is additive, and $f_0 : \mathbb{R}^N \to \mathbb{R}$ fulfils $(*)$ and the condition $|f_0(x)| \leqslant 1$ for $x \in \mathbb{R}^N$. (Forti [89].)

2. Let $f, g : \mathbb{R}^N \to \mathbb{R}$ be functions such that for a certain $p \in \mathbb{N}$ and $\varepsilon > 0$

$$|\Delta_h^p f(x) - g(h)| \leqslant \varepsilon$$

holds for all $x, h \in \mathbb{R}^N$. Show that there exist a polynomial function $\gamma : \mathbb{R}^N \to \mathbb{R}$ of order p and a symmetric p-additive function $F : \mathbb{R}^{pN} \to \mathbb{R}$ such that

$$|f(x) - \gamma(x)| \leqslant 2^{p+2}\varepsilon, \quad |g(x) - \varphi(x)| \leqslant \left(1 + 2^{p+2}\right)\varepsilon \text{ for all } x \in \mathbb{R}^N,$$

where $\varphi : \mathbb{R}^N \to \mathbb{R}$ is the diagonalization of F (Albert-Baker [11]).

3. Let $f : \mathbb{R}^N \to \mathbb{R}$ be a function such that for a certain $p \in \mathbb{N}$ and $\varepsilon > 0$

$$|\Delta_h^p f(x) - p! f(h)| \leqslant \varepsilon$$

holds for all $x, h \in \mathbb{R}^N$. Show that there exists a unique symmetric p-additive function $F : \mathbb{R}^{pN} \to \mathbb{R}$ such that

$$|f(x) - \varphi(x)| \leqslant \left(1 + 2^{p+2}\right)\varepsilon \qquad \text{for all } x \in \mathbb{R}^N,$$

where $\varphi : \mathbb{R}^N \to \mathbb{R}$ is the diagonalization of F (Albert-Baker [11]).

4. Let $D \subset \mathbb{R}^N$ be a convex and open set, and put, for every $\varepsilon > 0$,

$$D_\varepsilon = \left\{x \in \mathbb{R}^N \mid d\left(x, \mathbb{R}^N \setminus D\right) \geqslant \varepsilon\right\},$$

where d is the Euclidean distance in $\mathbb{R}^N$. Show that if $D_\varepsilon \neq \varnothing$, then D_ε is convex.

5. Let $D \subset \mathbb{R}^N$ be an open and convex set. Show that there exists a sequence $\{C_n\}_{n \in \mathbb{N}}$ of convex and compact sets such that $C_n \subset C_{n+1}$ for $n \in \mathbb{N}$, and $\bigcup_{n=1}^{\infty} \text{int } C_n = D$.

[Hint: Let B_n, $n \in \mathbb{N}$, be the closed ball centered at the origin and with the radius n. Consider the sets $B_n \cap D_{1/n}$.]

6. Show that if $A \subset \mathbb{R}^N$ is a closed, but not compact set, then the set $\text{conv } A$ need not be closed.

[Hint: Take $A = \text{Gr}(f)$, where $f : \mathbb{R}^N \to \mathbb{R}$ is given by $f(x) = e^{-|x|}$, $x \in \mathbb{R}^{N-1}$.]

7. Let $\mathcal{I}$ be a p.l.i. σ-ideal in $\mathbb{R}^N$. Show that $\mathcal{I}$ contains no non-empty open sets.

8. Let $(X, +)$ be a group, and let $\mathcal{I}$ be a p.l.i. ideal in X. Show that if the sets $P, Q \subset X$ are such that $Q \notin \mathcal{I}$ and $X \setminus P \in \mathcal{I}$, then $P + Q = X$. (Ger [106]).

9. Let $(X, +)$ be a group, and let $(K, +)$ be a subgroup of $(X, +)$. Let $\mathcal{I}$ be a p.l.i. ideal in X. Show that if $X \setminus K \in \mathcal{I}$, then $K = X$. (Ger [106].)

10. Let $(X, +)$ be a group, let $(K, +)$ be a subgroup of $(X, +)$, and let $K' = X \setminus K$. Let $\mathcal{I}$ be a p.l.i. ideal in X. Show that if $S, T \in \mathcal{I}$, then $(K \setminus S) + (K' \setminus T) = K'$. (Ger [106].)

11. Let $(X, +)$ and $(Y, +)$ be abelian groups, and for every set $A \subset X \times X$ write

$$A_T = \{(x, y) \in X \times X \mid (y, x) \in A\}.$$

Let $\mathcal{I}_1$ and $\mathcal{I}_2$ be conjugate p.l.i. ideals in X and in $X \times X$, respectively, such that $A_T \in \mathcal{I}_2$ for every set $A \in \mathcal{I}_2$. Let $f, g, h : X \to Y$ be functions such that

$$f(x + y) = g(x) + h(y) \qquad \mathcal{I}_2\text{-(a.e.) in } X \times X.$$

Show that there exists a unique homomorphism $F : X \to Y$ and constants $a, b \in Y$ such that

$$
\begin{aligned}
f(x) &= F(x) + a + b & & \mathcal{I}_1\text{-(a.e.) in } X, \\
g(x) &= F(x) + a & & \mathcal{I}_1\text{-(a.e.) in } X, \\
h(x) &= F(x) + b & & \mathcal{I}_1\text{-(a.e.) in } X.
\end{aligned}
$$

(Ger [105]).

[Hint: Let $M_0 = \{(x, y) \in X \times X \mid f(x + y) \neq g(x) + h(y)\}$, and let $M = M_0 \cup (M_0)_T$. There exists a set $U \in \mathcal{I}_1$ such that $M[x] \in \mathcal{I}_1$ for $x \in X \setminus U$. Take an $x_0 \in X \setminus U$. Then $g(x_0) + h(x) = g(x) + h(x_0) = f(x_0 + x)$ for $x \notin M[x_0]$. With $a' = -h(x_0)$, $b' = -g(x_0)$, this yields $g(x) = f(x_0 + x) + a'$, $h(x) = f(x_0 + x) + b'$ $\mathcal{I}_1$-(a.e.) in X. Hence $f(x + y) = f(x_0 + x) + f(x_0 + y) + a' + b'$ $\Omega(\mathcal{I}_1)$-(a.e.) in $X \times X$. The function $\Phi(x) = f(2x_0 + x) + a' + b'$ is $\Omega(\mathcal{I}_1)$-almost additive.]

12. Let $D \subset \mathbb{R}^N$ be a convex and open set, and let $f : D \to \mathbb{R}$ be a measurable $\mathcal{I}_0^{2N}$-almost convex function. Show that there exists a continuous convex function $g : D \to \mathbb{R}$ such that $f = g$ $\mathcal{I}_0^N$-(a.e.) in D.

13. Let $D \subset \mathbb{R}^N$ be a convex and open set, and let $f : D \to \mathbb{R}$ be an $\mathcal{I}_f^{2N}$-almost convex function which is bounded above on a set $T \subset D$, where T is of the second category and has the Baire property. Show that there exists a continuous convex function $g : D \to \mathbb{R}$ such that $f = g$ $\mathcal{I}_0^N$-(a.e.) in D.

14. Let $C \subset \mathbb{R}^N$ be a set fulfilling (16.1.3), and let $f : C \to \mathbb{R}$ be a subadditive function. Let $\mathcal{I}$ be a p.l.i. ideal in $\mathbb{R}^N$, and let $f : C \to \mathbb{R}$ be a function such that $f = g$ $\mathcal{I}$-(a.e.) in C. Show that f is $\Omega(\mathcal{I})$-almost subadditive.

15. The function $f : \mathbb{R} \to \mathbb{R}$ given by $f(x) = 3 + \sin x$ is subadditive, and hence, in particular, I_0^{2N}-almost subadditive. Show that for this particular f the functions φ and Φ given by (17.9.7) and (17.9.12), respectively, ($\mathcal{I}_1 = \mathcal{I}_0^N$), are $\varphi(x) = 2|\sin \frac{x}{2}|$ and $\Phi(x) = 6 - 2|\sin \frac{x}{2}|$. (Ger [110].)

16. With the notation and assumptions of Lemmas 17.9.2 and 17.9.3, show that the function $\tilde{\varphi} : C \to \mathbb{R}$ given by $\tilde{\varphi}(x) = \varphi(x) + \mathcal{I}_1\text{-}\underset{h \in C}{\operatorname{infess}} \left[f(h) - \varphi(h) \right]$ is subadditive and satisfies $\varphi \leqslant \tilde{\varphi} \leqslant f$ $\mathcal{I}_1$-(a.e.) in C. Similarly, if $c = \min \left\{ \mathcal{I}_1\text{-}\underset{h \in C}{\operatorname{infess}} \left[\Phi(h) - f(h) \right], \underset{(s,t) \in C \times C}{\inf} \left[\Phi(s) + \Phi(t) - \Phi(s + t) \right] \right\}$, then the function $\tilde{\Phi} : C \to \mathbb{R}$ given by $\tilde{\Phi} = \Phi - c$ is subadditive and satisfies $f \leqslant \tilde{\Phi} < \Phi$ $\mathcal{I}_1$-(a.e.) in C. (Ger [110].)

Chapter 18

Extensions of Homomorphisms

18.1 Commutative divisible groups

Let $(X, +)$ be a group, and $(S, +)$ a subsemigroup (i.e., a semigroup such that $S \subset G$ and the operation $+$ is the same as in X; cf. 4.5). Let $(Y, +)$ ba another group, and let $g : S \to Y$ be a homomorphism[1]. The problem with which we shall deal in 18.1–18.4 is the following. Does there exist a homomorphism $f : X \to Y$ such that $f \mid S = g$? The main result in this section (cf. Dhombres-Ger [70], Balcerzyk [18]) reads as follows.

Theorem 18.1.1. *Let $(X, +)$ and $(Y, +)$ be commutative groups such that Y is divisible, let $(S, +)$ be a subsemigroup of $(X, +)$, and let $g : S \to Y$ be a homomorphism. Then there exists a homomorphism $f : X \to Y$ such that $f \mid S = g$.*

Proof. Let $\mathcal{R}$ be the family of all couples (R, h) such that $(R, +)$ is a subsemigroup of $(X, +)$, and $h : R \to Y$ is a homomorphism:

$$h(x + y) = h(x) + h(y) \qquad (18.1.1)$$

for all $x, y \in R$. An order can be defined in $\mathcal{R}$ in the usual way: $(R_1, h_1) \prec (R_2, h_2)$ iff $(R_1, +)$ is a subsemigroup of $(R_2, +)$ and $h_2 \mid R_1 = h_1$. It is easy to see that every chain $\mathcal{L} \subset \mathcal{R}$ has an upper bound in $\mathcal{R}$: this upper bound is the couple (R_0, h_0) such that $R_0 = \bigcup_{(R, h) \in \mathcal{L}} R$ and $h_0 \mid R = h$ on every R such that $(R, h) \in \mathcal{L}$. By the Lemma of Kuratowski-Zorn (Theorem 1.8.1) there exists in $\mathcal{R}$ a maximal element (T, f) such that $(S, g) \prec (T, f)$. The proof will be complete if we show that $T = X$.

So suppose that $T \neq X$ (of course, necessarily $T \subset X$), and let $x \in X \setminus T$. Consider the set

$$R = \{a \in X \mid a = b + nx,\ b \in T,\ n \in \mathbb{N}_0 = \mathbb{N} \cup \{0\}\}.$$

Let $a, a' \in R$ so that $a = b + nx$, $a' = b' + n'x$, $b, b' \in T$, $n, n' \in \mathbb{N}_0$. Then $a + a' = b + b' + (n + n')x$, where $b + b' \in T$ (since $(T, +)$ is a semigroup) and $n + n' \in \mathbb{N}_0$. Consequently $a + a' \in R$, which shows that $(R, +)$ is a semigroup, and obviously a

---
[1] This means that g satisfies $g(x + y) = g(x) + g(y)$ for all $x, y \in S$.

subsemigroup of $(X, +)$. Now we define an element $y \in Y$. If $(\mathbb{N}x) \cap (T - T) = \varnothing$, then y may be an arbitrary element of Y. If there is an $n \in \mathbb{N}$ such that $nx \in T - T$, i.e., $nx = t_1 - t_2$ with $t_1, t_2 \in T$, then

$$y = \frac{f(t_1) - f(t_2)}{n}, \tag{18.1.2}$$

which is meaningful, since Y is divisible. To show that definition (18.1.2) is correct, we must show that whenever $nx = t_1 - t_2$, $mx = s_1 - s_2$, with $n, m \in \mathbb{N}$, $t_1, t_2, s_1, s_2 \in T$, then

$$\frac{f(t_1) - f(t_2)}{n} = \frac{f(s_1) - f(s_2)}{m}. \tag{18.1.3}$$

But then $mt_1 - mt_2 = mnx = ns_1 - ns_2$, whence

$$mt_1 + ns_2 = ns_1 + mt_2. \tag{18.1.4}$$

It follows by induction from (4.5.15) that

$$f(pu) = pf(u)$$

for every $p \in \mathbb{N}$ and $u \in T$. Hence and by (4.5.15) we get from (18.1.4)

$$mf(t_1) + nf(s_2) = nf(s_1) + mf(t_2),$$

and (18.1.3) follows. Consequently definition (18.1.2) is correct.

Now we define a map $h : R \to Y$. If $a \in R$, $a = b + nx$ with $b \in T$, $n \in \mathbb{N}_0$, we put

$$h(a) = f(b) + ny. \tag{18.1.5}$$

Again we must show that definition (18.1.5) is correct. This amounts to showing that whenever

$$b + nx = b' + n'x \tag{18.1.6}$$

with $b, b' \in T$, $n, n' \in \mathbb{N}_0$, then

$$f(b) + ny = f(b') + n'y. \tag{18.1.7}$$

If $b = b'$, $n = n'$, this is obvious. If one of b, n is not equal to the corresponding element b', n', then the other must be unequal, too, as results from (18.1.6). So now we may assume that $n \neq n'$, say $n > n'$. (If $n' > n$, we interchange the roles of n and n'). From (18.1.6) we get

$$(n - n')x = b' - b \in T - T. \tag{18.1.8}$$

$n - n' \in \mathbb{N}$, so (18.1.8) means that $(\mathbb{N}x) \cap (T - T) \neq \varnothing$. But then we have in virtue of (18.1.2)

$$\frac{f(b') - f(b)}{n - n'} = y, \tag{18.1.9}$$

and (18.1.7) results. Thus definition (18.1.5) is correct.

Let $a, a' \in R$, say, $a = b + nx$, $a' = b' + n'x$, $b, b' \in T$, $n, n' \in \mathbb{N}_0$. Then $a + a' = b + b' + (n + n')x$ (note that X is commutative), whence by (18.1.5)

$$h(a + a') = f(b + b') + (n + n')x = f(b) + f(b') + nx + n'x$$
$$= f(b) + nx + f(b') + n'x = h(a) + h(a'),$$

since Y is commutative and f is a homomorphism. Consequently h is a homomorphism from R into Y.

Finally note that $T \subset R$, for every $b \in T$ may be written as $b = b + 0x \in R$. Consequently $(T, +)$ is a subsemigroup of $(R, +)$. For $b \in T$ ($b = b + 0x$) we have by (18.1.5)

$$h(b) = f(b) + 0y = f(b).$$

This means that $h \mid T = f$.

Consequently $(R, h) \in \mathcal{R}$ and $(T, f) \prec (R, h)$, whereas clearly $(T, f) \neq (R, h)$. But this contradicts the maximality of (T, f). Thus we must have $T = X$, i.e., f is a homomorphism from X into Y. Since $(S, g) \prec (T, f)$, we have $f \mid S = g$. This completes the proof. $\qquad\qquad\square$

Note that the theorem fails to hold if $(Y, +)$ is not divisible. E.g., take $X = Y = \mathbb{Z}$, $S = 2\mathbb{Z}$ (the set of even integers). Those sets endowed with the operation $+$ of the usual addition are commutative groups, but $(Y, +) = (\mathbb{Z}, +)$ is not divisible. Let $g : 2\mathbb{Z} \to \mathbb{Z}$ be the homomorphism defined by $g(2k) = k$ for $k \in \mathbb{Z}$. There does not exist a homomorphism $f : \mathbb{Z} \to \mathbb{Z}$ such that $f \mid 2\mathbb{Z} = g$, for otherwise we would have $f(1) + f(1) = f(2) = g(2) = 1$, i.e., $2f(1) = 1$, which has no solution in $\mathbb{Z}$.

It can be shown (Dhombres-Ger [70]) that if S does not generate X, then the extension f of g furnished by Theorem 18.1.1 is not unique. On the other hand, if S generates X, then f is unique, as follows easily from Corollary 18.2.1 below.

18.2 The simplest case of S generating X

In this section we consider the case where S generates X in such a way that

$$X = S - S. \tag{18.2.1}$$

(It follows from (18.2.1) that S generates X, for then any group containing S must contain also $S - S = X$.) The results are due to Aczél-Baker-Djoković-Kannappan-Radó [8].

We start with a lemma.

Lemma 18.2.1. *Let $(X, +)$ and $(Y, +)$ be groups, and let $(S, +)$ be a subsemigroup of $(X, +)$ such that (18.2.1) holds. Let $g : S \to Y$ be a homomorphism:*

$$g(x + y) = g(x) + g(y) \tag{18.2.2}$$

for all $x, y \in S$. Then, if $x, y, z, x', y', z' \in S$ are such that

$$x - y + z = x' - y' + z', \tag{18.2.3}$$

then

$$g(x) - g(y) + g(z) = g(x') - g(y') + g(z'). \tag{18.2.4}$$

Proof. By (18.2.1) an arbitrary $u \in X$ may be represented as $u = t - s$, with $t, s \in S$. Hence $u + s = t \in S$. So (18.2.1) implies that

(i) for every $u \in X$ there exists an $s \in S$ such that $u + s \in S$.

So, in particular, taking $u = -y + z$, we may find an $s' \in S$ such that

$$-y + z + s' \in S. \tag{18.2.5}$$

Similarly, we can find an $s'' \in S$ such that

$$-y' + z' + s' + s'' \in S. \tag{18.2.6}$$

With $s = s' + s'' \in S$ relations (18.2.5) and (18.2.6) yield

$$-y + z + s \in S, \qquad -y' + z' + s \in S. \tag{18.2.7}$$

On the other hand, relation (18.2.3) implies

$$x - y + z + s = x' - y' + z' + s. \tag{18.2.8}$$

Now, if $u \in S$ and $-u + v \in S$ $\bigl($which implies that $v = u + (-u + v) \in S\bigr)$, then

$$g(-u + v) = -g(u) + g(v), \tag{18.2.9}$$

since taking in (18.2.2) $x = u$, $y = -u + v$, we obtain $g(v) = g(u) + g(-u + v)$, which implies (18.2.9).

It follows by (18.2.8) and (18.2.2) that

$$g(x) + g(-y + z + s) = g(x') + g(-y' + z' + s).$$

Applying now (18.2.9) with $u = y \in S$ and $v = z + s$ $\bigl($by (18.2.7) $-u + v \in S\bigr)$, and also with $u = y' \in S$, $v = z' + s$ $\bigl($again $-u + v \in S$ in virtue of (18.2.7)$\bigr)$, we obtain hece

$$g(x) - g(y) + g(z + s) = g(x') - g(y') + g(z' + s).$$

Hence, again by (18.2.2)

$$g(x) - g(y) + g(z) + g(s) = g(x') - g(y') + g(z') + g(s). \tag{18.2.10}$$

Cancelling in (18.2.10) $g(s)$ yields (18.2.4).   □

Theorem 18.2.1. *Let $(X, +)$ and $(Y, +)$ be groups, and let $(S, +)$ be a subsemigroup of $(X, +)$ such that (18.2.1) holds. Let $g : S \to Y$ be a homomorphism. Then there exists a unique homomorphism $f : X \to Y$ such that $f \mid S = g$.*

Proof. Every $x \in X$ can be represented as $x = t - s$ with $t, s \in S$. For such an x define $f(x)$ as

$$f(x) = g(t) - g(s). \tag{18.2.11}$$

To prove that this definition is correct we must show that whenever

$$t - s = t' - s', \qquad t, s, t', s' \in S,$$

then

$$g(t) - g(s) = g(t') - g(s').$$

But this follows from Lemma 18.2.1 on choosing $x = t$, $y = s + s$, $z = s$, $x' = t'$, $y' = s' + s'$, $z' = s'$. Thus formula (18.2.11) unambiguously defines a map $f : X \to Y$. We will show that f is a homomorphism.

Take arbitrary $x, y \in X$, $x = t - s$, $y = t' - s'$, $t, s, t', s' \in S$. Then $x + y = t - s + t' - s'$ is an element of X, and so $x + y = t'' - s''$ with $t'', s'' \in S$. Hence we get

$$t - s + t' = t'' - s'' + s',$$

and, by Lemma 18.2.1,

$$g(t) - g(s) + g(t') = g(t'') - g(s'') + g(s'),$$

or

$$g(t'') - g(s'') = g(t) - g(s) + g(t') - g(s'). \tag{18.2.12}$$

According to definition (18.2.11)

$$f(x) = g(t) - g(s), \qquad f(y) = g(t') - g(s'), \tag{18.2.13}$$

and

$$f(x + y) = g(t'') - g(s''). \tag{18.2.14}$$

Relations (18.2.13), (18.2.14) and (18.2.12) imply that

$$f(x + y) = f(x) + f(y), \tag{18.2.15}$$

which means that f is a homomorphism.

Let $x \in S$. Then $x = (x + x) - x$, whence by (18.2.11) and (18.2.2)

$$f(x) = g(x + x) - g(x) = g(x) + g(x) - g(x) = g(x).$$

Consequently $f \mid S = g$.

In order to prove the uniqueness, let $h : X \to Y$ be an arbitrary homomorphism such that $h \mid S = g$. Take an arbitrary $x \in X$. Then $x = t - s$ with $t, s \in S$, whence $x + s = t$. Relation (18.1.1) implies that

$$h(t) = h(x) + h(s),$$

whence

$$h(x) = h(t) - h(s) = g(t) - g(s) = f(x),$$

since $h \mid S = g$ and f satisfies (18.2.11). This shows that h coincides with f on the whole of X, i.e., f is unique. $\qquad \square$

Theorems 18.2.1 and 4.5.1 yield the following

Corollary 18.2.1. *Let $(X, +)$ be a commutative group, let $(Y, +)$ be a group, and let $(S, +)$ be a subsemigroup of $(X, +)$ generating X. Let $g : S \to Y$ be a homomorphism. Then there exists a unique homomorphism $f : X \to Y$ such that $f \mid S = g$.*

Similarly, we obtain by Lemma 4.5.6

Corollary 18.2.2. *Let $(X, +)$ and $(Y, +)$ be groups, and let $(S, +)$ be a subsemigroup of $(X, +)$ such that for every $x \in X$, $x \neq 0$, either $x \in S$, or $-x \in S$ (or both). Let $g : S \to Y$ be a homomorphism. Then there exists a unique homomorphism $f : X \to Y$ such that $f \mid S = g$.*

Corollary 18.2.2 generalizes a result of Aczél-Erdős [9] (cf. also 13.5). Theorem 18.2.1 is also related to Theorem 4.4.1.

18.3   A generalization

Now we are going to consider the case, where

$$X = \sum_{i=0}^{n} (-1)^i S \tag{18.3.1}$$

(Grząślewicz-Sikorski [125], Martin [221]). Here n is a fixed positive integer, $n \geqslant 2$. It is easy to see that also in case (18.3.1) S generates X. For $n = 1$ (18.3.1) reduces to (18.2.1). But now ($n \geqslant 2$) it is no longer true that every homomorphism $g : S \to Y$ can be extended to a homomorphism $f : X \to Y$. Some additional assumptions are necessary.

We will include also case (18.2.1) into our considerations, thus allowing n to be 1. In this way Theorem 18.2.1 will become a particular case of Theorem 18.3.1 below.

But again first we must prove a lemma.

Lemma 18.3.1. *Let $(X, +)$ and $(Y, +)$ be groups, and let $(S, +)$ be a subsemigroup of $(X, +)$ such that relation (18.3.1) holds for a certain fixed $n \in \mathbb{N}$. Let $g : S \to Y$ be a homomorphism. Assume that for every $x_0, \ldots, x_{n-1}, y_0, \ldots, y_{n-1} \in S$, whenever*

$$\sum_{i=0}^{n-1} (-1)^i x_i = \sum_{i=0}^{n-1} (-1)^i y_i, \tag{18.3.2}$$

then also[2]

$$\sum_{i=0}^{n-1} (-1)^i g(x_i) = \sum_{i=0}^{n-1} (-1)^i g(y_i). \tag{18.3.3}$$

Then, for arbitrary $k \in \mathbb{N}_0 = \mathbb{N} \cup \{0\}$ and for arbitrary $x_0, \ldots, x_n, y_0, \ldots, y_{n+k} \in S$, whenever

$$\sum_{i=0}^{n} (-1)^i x_i = \sum_{i=0}^{n+k} (-1)^i y_i, \tag{18.3.4}$$

---

[2] For $n = 1$ this condition becomes: $x = y$ implies $g(x) = g(y)$, so it is trivially fulfilled.

then also

$$\sum_{i=0}^{n}(-1)^i g(x_i) = \sum_{i=0}^{n+k}(-1)^i g(y_i).\tag{18.3.5}$$

Proof. We proceed by induction on k. Let $k = 0$. Take arbitrary points $x_0,\ldots,x_n$, $y_0,\ldots,y_n \in S$ such that

$$\sum_{i=0}^{n}(-1)^i x_i = \sum_{i=0}^{n}(-1)^i y_i.\tag{18.3.6}$$

It follows that

$$x_0 + \sum_{i=1}^{n}(-1)^i x_i - (-1)^n y_n = \sum_{i=0}^{n-1}(-1)^i y_i.\tag{18.3.7}$$

The expression following x_0 on the left-hand side of (18.3.7) is an element of X, so by (18.3.1) there exist $z_0,\ldots,z_n \in S$ such that

$$\sum_{i=1}^{n}(-1)^i x_i - (-1)^n y_n = \sum_{i=0}^{n}(-1)^i z_i.\tag{18.3.8}$$

Again, (18.3.8) may be written as

$$\sum_{i=2}^{n}(-1)^i x_i - (-1)^n y_n - (-1)^n z_n = x_1 + z_0 + \sum_{i=1}^{n-1}(-1)^i z_i.\tag{18.3.9}$$

From (18.3.7) and (18.3.8) we obtain

$$x_0 + \sum_{i=0}^{n}(-1)^i z_i = \sum_{i=0}^{n-1}(-1)^i y_i,$$

whence

$$x_0 + z_0 + \sum_{i=1}^{n-1}(-1)^i z_i = \sum_{i=0}^{n-2}(-1)^i y_i + (-1)^{n-1} y_{n-1} - (-1)^n z_n.\tag{18.3.10}$$

Now we must consider separately the cases where n is even or odd. First assume that n is even. Then $-(-1)^n y_n - (-1)^n z_n = -y_n - z_n = -(z_n + y_n)$, whereas $(-1)^{n-1} y_{n-1} - (-1)^n z_n = -y_{n-1} - z_n = -(z_n + y_{n-1})$. Thus we obtain from (18.3.9) and (18.3.10)

$$\sum_{i=2}^{n}(-1)^i x_i - (z_n + y_n) = (x_1 + z_0) + \sum_{i=1}^{n-1}(-1)^i z_i,$$

$$(x_0 + z_0) + \sum_{i=1}^{n-1}(-1)^i z_i = \sum_{i=0}^{n-2}(-1)^i y_i - (z_n + y_{n-1}).$$

By assumption (18.3.3) we get hence

$$\sum_{i=2}^{n}(-1)^i g(x_i) - g(z_n + y_n) = g(x_1 + z_0) + \sum_{i=1}^{n-1}(-1)^i g(z_i),$$

$$g(x_0 + z_0) + \sum_{i=1}^{n-1}(-1)^i g(z_i) = \sum_{i=0}^{n-2}(-1)^i g(y_i) - g(z_n + y_{n-1}),$$

or, since g is a homomorphism,

$$\sum_{i=1}^{n}(-1)^i g(x_i) - g(y_n) = \sum_{i=0}^{n}(-1)^i g(z_i), \quad g(x_0) + \sum_{i=0}^{n}(-1)^i g(z_i) = \sum_{i=0}^{n-1}(-1)^i g(y_i).$$

Hence we obtain, eliminating $\sum_{i=0}^{n}(-1)^i g(z_i)$,

$$\sum_{i=0}^{n}(-1)^i g(x_i) = \sum_{i=0}^{n}(-1)^i g(y_i), \tag{18.3.11}$$

i.e., (18.3.5) for $k = 0$.

Now let n be odd. Then $-(-1)^n y_n - (-1)^n z_n = y_n + z_n$, whereas $(-1)^{n-1} y_{n-1} - (-1)^n z_n = y_{n-1} + z_n$ and we obtain from (18.3.9) and (18.3.10)

$$\sum_{i=2}^{n}(-1)^i x_i + (y_n + z_n) = (x_1 + z_0) + \sum_{i=1}^{n-1}(-1)^i z_i,$$

$$(x_0 + z_0) + \sum_{i=1}^{n-1}(-1)^i z_i = \sum_{i=0}^{n-2}(-1)^i y_i + (y_{n-1} + z_n),$$

and by assumption (18.3.3)

$$\sum_{i=2}^{n}(-1)^i g(x_i) + g(y_n + z_n) = g(x_1 + z_0) + \sum_{i=1}^{n-1}(-1)^i g(z_i),$$

$$g(x_0 + z_0) + \sum_{i=1}^{n-1}(-1)^i g(z_i) = \sum_{i=0}^{n-2}(-1)^i g(y_i) + g(y_{n-1} + z_n),$$

or, since g is a homomorphism,

$$\sum_{i=1}^{n}(-1)^i g(x_i) + g(y_n) = \sum_{i=0}^{n}(-1)^i g(z_i), \quad g(x_0) + \sum_{i=0}^{n}(-1)^i g(z_i) = \sum_{i=0}^{n-1}(-1)^i g(y_i),$$

whence, eliminating $\sum_{i=0}^{n}(-1)^i g(z_i)$, we obtain (18.3.11), i.e., (18.3.5) for $k = 0$.

Now assume that the lemma is true for a $k \in \mathbb{N}_0$, and let $x_0, \ldots, x_n, y_0, \ldots,$ y_{n+k+1} be arbitrary points from S such that

$$\sum_{i=0}^{n}(-1)^i x_i = \sum_{i=0}^{n+k+1}(-1)^i y_i. \tag{18.3.12}$$

$\sum_{i=1}^{n+k+1}(-1)^i y_i$ is an element of X, so, according to (18.3.1), there exist points $z_0, \ldots,$ $z_n \in S$ such that

$$\sum_{i=1}^{n+k+1}(-1)^i y_i = \sum_{i=0}^{n}(-1)^i z_i. \tag{18.3.13}$$

Now take an arbitrary $u \in S$. Again we must distinguish two cases, according as $n + k$ is even or odd.

Suppose that $n + k$ is even. We have $u + y_{n+k+1} \in S$ (since S is a semigroup) and $u \in S$, and so (18.3.13) may be written as

$$\sum_{i=2}^{n+k}(-1)^i y_i - (u + y_{n+k+1}) + u = (y_1 + z_0) + \sum_{i=1}^{n}(-1)^i z_i.$$

By the induction hypothesis it follows

$$\sum_{i=2}^{n+k}(-1)^i g(y_i) - g(u + y_{n+k+1}) + g(u) = g(y_1 + z_0) + \sum_{i=1}^{n}(-1)^i g(z_i),$$

or, since g is a homomorphism,

$$\sum_{i=1}^{n+k+1}(-1)^i g(y_i) = \sum_{i=0}^{n}(-1)^i g(z_i). \tag{18.3.14}$$

If $n + k$ is odd, we have $y_{n+k+1} + u \in S$ and $u \in S$, and so (18.3.13) may be written as

$$\sum_{i=2}^{n+k}(-1)^i y_i + (y_{n+k+1} + u) - u = (y_1 + z_0) + \sum_{i=1}^{n}(-1)^i z_i,$$

and by the induction hypothesis

$$\sum_{i=2}^{n+k}(-1)^i g(y_i) + g(y_{n+k+1} + u) - g(u) = g(y_1 + z_0) + \sum_{i=1}^{n}(-1)^i g(z_i),$$

and, since g is a homomorphism, we arrive again at formula (18.3.14).

Thus (18.3.14) is valid in both cases. We get by (18.3.12) and (18.3.13)

$$\sum_{i=0}^{n}(-1)^i x_i = (y_0 + z_0) + \sum_{i=1}^{n}(-1)^i z_i.$$

Applying to this our lemma for $k = 0$ (which has already been proved), we infer hence that

$$\sum_{i=0}^{n}(-1)^{i}g(x_i) = g(y_0 + z_0) + \sum_{i=1}^{n}(-1)^{i}g(z_i),$$

or, since g is a homomorphism,

$$\sum_{i=0}^{n}(-1)^{i}g(x_i) = g(y_0) + \sum_{i=0}^{n}(-1)^{i}g(z_i). \tag{18.3.15}$$

Eliminating $\sum_{i=0}^{n}(-1)^{i}g(z_i)$ from (18.3.14) and (18.3.15), we obtain

$$\sum_{i=0}^{n}(-1)^{i}g(x_i) = \sum_{i=0}^{n+k+1}(-1)^{i}g(y_i),$$

i.e., (18.3.5) for $k + 1$. Induction ends the proof.   □

Theorem 18.3.1. *Let $(X, +)$ and $(Y, +)$ be groups, and let $(S, +)$ be a subsemigroup of $(X, +)$ such that relation (18.3.1) holds for a certain $n \in \mathbb{N}$. Let $g : S \to Y$ be a homomorphism. There exists a homomorphism $f : X \to Y$ such that $f \mid S = g$ if and only if for every $x_0, \ldots, x_{n-1}, y_0, \ldots, y_{n-1} \in S$ relation (18.3.2) implies (18.3.3). When this is the case, the homomorphism f is unique.*

Proof. First assume that condition (18.3.2) implies (18.3.3). It follows by (18.3.1) that for every $x \in X$ there exist $x_0, \ldots, x_n \in S$ such that

$$x = \sum_{i=0}^{n}(-1)^{i}x_i. \tag{18.3.16}$$

For such an x define $f(x)$ as

$$f(x) = \sum_{i=0}^{n}(-1)^{i}g(x_i). \tag{18.3.17}$$

It follows by Lemma 18.3.1 ($k = 0$) that this definition is correct. So f is a mapping $f : X \to Y$. In order to show that f is a homomorphism take arbitrary $x, y \in X$. There exist $x_0, \ldots, x_n, y_0, \ldots, y_n, z_0, \ldots, z_n \in S$ such that

$$x = \sum_{i=0}^{n}(-1)^{i}x_i, \qquad y = \sum_{i=0}^{n}(-1)^{i}y_i, \qquad z = \sum_{i=0}^{n}(-1)^{i}z_i. \tag{18.3.18}$$

According to definition (18.3.17)

$$f(x) = \sum_{i=0}^{n}(-1)^{i}g(x_i), \qquad f(y) = \sum_{i=0}^{n}(-1)^{i}g(y_i), \qquad f(x+y) = \sum_{i=0}^{n}(-1)^{i}g(z_i).$$

$$\tag{18.3.19}$$

Clearly we have by (18.3.18)

$$\sum_{i=0}^{n}(-1)^i z_i = \sum_{i=0}^{n}(-1)^i x_i + \sum_{i=0}^{n}(-1)^i y_i. \qquad (18.3.20)$$

If n is odd, this implies in virtue of Lemma 18.3.1 ($k = n$)

$$\sum_{i=0}^{n}(-1)^i g(z_i) = \sum_{i=0}^{n}(-1)^i g(x_i) + \sum_{i=0}^{n}(-1)^i g(y_i),$$

i.e., by (18.3.19), we obtain (18.2.15). If n is even, we write (18.3.20) in the form

$$\sum_{i=0}^{n}(-1)^i x_i = \sum_{i=0}^{n}(-1)^i z_i + \sum_{i=0}^{n}(-1)^{i+1} y_{n-i},$$

and, using again Lemma 18.3.1 and relation (18.3.19), we arrive at $f(x) = f(x+y) - f(y)$, which yields (18.2.15). Thus (18.2.15) holds in both cases, which shows that f is a homomorphism.

Now take an arbitrary $x \in S$. If n is even, we have

$$x = \sum_{i=0}^{n}(-1)^i x_i,$$

whence by (18.3.17)

$$f(x) = \sum_{i=0}^{n}(-1)^i g(x) = g(x).$$

If n is odd, we have

$$x = (x + x) + \sum_{i=1}^{n}(-1)^i x,$$

and by (18.3.17)

$$f(x) = g(x + x) + \sum_{i=1}^{n}(-1)^i g(x) = g(x) + g(x) - g(x) = g(x).$$

Thus in either case $f(x) = g(x)$, which means that $f \mid S = g$.

Now assume that a homomorphism $f : X \to Y$ such that $f \mid S = g$ exists. It follows from (18.2.15) by induction that for every $m \in \mathbb{N}$ and every $u_1, \ldots, u_m \in X$ we have

$$f(u_1 + \cdots + u_m) = f(u_1) + \cdots + f(u_m).$$

Also, (18.2.15) implies easily that $f(-x) = -f(x)$ so that $f((-1)^i x) = (-1)^i f(x)$ for every $i \in \mathbb{N}_0$ and $x \in X$. Now, relation (18.3.2) implies

$$\sum_{i=0}^{n-1}(-1)^i f(x_i) = f\left(\sum_{i=0}^{n-1}(-1)^i x_i\right) = f\left(\sum_{i=0}^{n-1}(-1)^i y_i\right) = \sum_{i=0}^{n-1}(-1)^i f(y_i),$$

whence, taking into account the fact that $f \mid S = g$ and that $x_0, \ldots, x_{n-1}, y_0, \ldots, y_{n-1} \in S$, we obtain (18.3.3).

Let $h : X \to Y$ be an arbitrary homomorphism such that $h \mid S = g$. Take an arbitrary $x \in X$. Then x has a representation (18.3.16) with $x_0, \ldots, x_n \in S$. By the same argument as above we show that then

$$h(x) = \sum_{i=0}^{n} (-1)^i h(x_i) = \sum_{i=0}^{n} (-1)^i g(x_i),$$

since $h \mid S = g$. So in view of (18.3.17) $h = f$, which proofs the uniqueness of f and completes the proof of the theorem. $\qquad\square$

18.4   Further extension theorems

The results in this section (Grząślewicz [121]) are closely related to the problems treated in 18.1–18.3. Here we will assume that $(S, +)$ is a subgroup of $(X, +)$.

Theorem 18.4.1. *Let $(X, +)$ and $(Y, +)$ be groups, and let $(S, +)$ be a subgroup of $(X, +)$. Let $g : S \to Y$ be a homomorphism. There exists a homomorphism $f : X \to Y$ such that $f \mid S = g$ and $f(X) = g(S)$ if and only if there exists a normal subgroup $(X_0, +)$ of $(X, +)$ such that*

$$\mathrm{Ker}\, g = S \cap X_0, \tag{18.4.1}$$

$$X = S + X_0. \tag{18.4.2}$$

Here $\mathrm{Ker}\, g = g^{-1}(0) = \{x \in S \mid g(x) = 0\}$ denotes the kernel of g.

Proof. Suppose that a normal subgroup $(X_0, +)$ of $(X, +)$ fulfilling conditions (18.4.1) and (18.4.2) exists. Let elements of X/X_0 be denoted by $\langle x \rangle$, $x \in X$. By Lemma 4.5.7 $(\mathrm{Ker}\, g, +)$ is a normal subgroup of $(S, +)$. Let elements of $S/\mathrm{Ker}\, g$ be denoted by $[x]$, $x \in S$. In virtue of Theorem 4.5.3 $(X/X_0, +)$ and $(S/\mathrm{Ker}\, g, +)$ are groups, where the operations $+$ are defined according to (4.5.14).

By (18.4.1), (18.4.2) and Theorem 4.6.3 $S/\mathrm{Ker}\, g$ is a selective subpartition of X/X_0. Therefore for every $\langle x \rangle \in X/X_0$ there exists a unique class $[u] \in S/\mathrm{Ker}\, g$ such that $[u] \subset \langle x \rangle$. With these notations we define a function $f : X \to Y$ as

$$f(x) = g(u). \tag{18.4.3}$$

From what has just been told and from Corollary 4.5.4 it follows that this definition is unambiguous. It is also immediately seen from (18.4.3) that $f(X) \subset g(S)$. Conversely, for every $t \in g(S)$ there exists a $u \in S$ such that $t = g(u)$. Since $S/\mathrm{Ker}\, g$ is a subpartition of X/X_0, there exists an $x \in X$ such that $[u] \subset \langle x \rangle$. But according to (18.4.3) we have then $f(x) = g(u) = t$. So $t \in f(X)$, and $g(S) \subset f(X)$. Consequently $f(X) = g(S)$.

Let $x \in S$. Then, of course $x \in [x] \cap \langle x \rangle$, and consequently $[x] \cap \langle x \rangle \neq \varnothing$. By Lemma 4.6.1 we then have $[x] \subset \langle x \rangle$. But according to definition (18.4.3) this implies that $f(x) = g(x)$. Thus $f \mid S = g$.

Take arbitrary $x, y \in X$, and let $u, v \in S$ be such that $[u] \subset \langle x \rangle$, $[v] \subset \langle y \rangle$. This means, in particular, that $u \in \langle x \rangle$, $v \in \langle y \rangle$, and so $\langle u \rangle = \langle x \rangle$, $\langle v \rangle = \langle y \rangle$, whence it follows that $\langle x + y \rangle = \langle x \rangle + \langle y \rangle = \langle u \rangle + \langle v \rangle = \langle u + v \rangle$. Thus $u + v \in \langle x + y \rangle$, and since evidently $u + v \in [u+v]$, we get $[u+v] \cap \langle x+y \rangle \neq \varnothing$. By Lemma 4.6.1 $[u+v] \subset \langle x+y \rangle$. Hence

$$f(x + y) = g(u + v) = g(u) + g(v) = f(x) + f(y).$$

This shows that f is a homomorphism.

Conversely, suppose that there exists a homomorphism $f : X \to Y$ with the required properties. Put $X_0 = \mathrm{Ker}\, f = \{x \in X \mid f(x) = 0\}$. By Lemma 4.5.7 $(X_0, +)$ is a normal subgroup of $(X, +)$. $x \in \mathrm{Ker}\, g$ means $x \in S$ and $g(x) = 0$, whence, since $f \mid S = g$, also $f(x) = 0$, and $x \in \mathrm{Ker}\, f = X_0$. Thus $\mathrm{Ker}\, g \subset X_0$. By definition $\mathrm{Ker}\, g \subset S$. So

$$\mathrm{Ker}\, g \subset S \cap X_0. \tag{18.4.4}$$

Now take an arbitrary $x \in S \cap X_0$. Then $x \in X_0 = \mathrm{Ker}\, f$ and so $f(x) = 0$. But since also $x \in S$, we have $f(x) = g(x)$, whence $g(x) = 0$, and $x \in \mathrm{Ker}\, g$. Thus $S \cap X_0 \subset \mathrm{Ker}\, g$, which together with (18.4.4) proves relation (18.4.1).

Since both, S and X_0, are contained in X, we have $S + X_0 \subset X + X \subset X$. On the other hand, take an arbitrary $x \in X$. Since $f(X) = g(S)$, there exists a $u \in S$ such that $g(u) = f(x)$, whence, since $f \mid S = g$, we obtain $f(u) = f(x)$. Since f is a homomorphism, we have

$$f(x) = f\big(u + (-u + x)\big) = f(u) + f(-u + x)$$

so that $f(-u + x) = 0$ and $-u + x \in \mathrm{Ker}\, f = X_0$. Hence $x = u + (-u + x) \in S + X_0$ and $X \subset S + X_0$, and ultimately we get relation (18.4.2). $\qquad \square$

Theorem 18.4.2. *Let $(X, +)$ and $(Y_0, +)$ be groups, and let $(S, +)$ be a subgroup of $(X, +)$. Let $g : S \to Y_0$ be a homomorphism from S onto Y_0. Then there exists a group $(Y, +)$ and a homomorphism $f : X \to Y$ such that $(Y_0, +)$ is a subgroup of $(Y, +)$ and $f \mid S = g$ if and only if there exists a normal subgroup $(X_0, +)$ of $(X, +)$ such that relation (18.4.1) holds.*

Proof. Suppose that there exists a normal subgroup of $(X_0, +)$ of $(X, +)$ such that relation (18.4.1) holds. We will employ the notation used in the proof of Theorem 18.4.1.

It follows from (18.4.1) that $\mathrm{Ker}\, g \subset X_0$, whence, by Theorem 4.6.1, $S/\mathrm{Ker}\, g$ is a subpartition of X/X_0. Hence for every $[u] \in S/\mathrm{Ker}\, g$ there exists a (unique) $\langle x \rangle \in X/X_0$ such that $[u] \subset \langle x \rangle$. Therefore

$$\mathrm{card}\, S/\mathrm{Ker}\, g \leqslant \mathrm{card}\, X/X_0 \tag{18.4.5}$$

On the other hand, by Lemma 4.5.8, to every $[u] \in S/\mathrm{Ker}\, g$ there exists exactly one $t \in Y_0$ such that $g([u]) = \{t\}$. Conversely, since the map $g : S \to Y_0$ is onto, for every $t \in Y_0$ there exists a unique $[u] \in S/\mathrm{Ker}\, g$ such that $g([u]) = \{t\}$. Thus g induces a one-to-one mapping between the elements of $S/\mathrm{Ker}\, g$ and these of Y_0. Let this mapping be $\alpha : S/\mathrm{Ker}\, g \to Y_0$, which is thus defined for $[u] \in S/\mathrm{Ker}\, g$:

$$\alpha([u]) = t \in Y_0 \text{ iff } g([u]) = \{t\}. \tag{18.4.6}$$

As a consequence we obtain that

$$\operatorname{card}(S/\operatorname{Ker} g) = \operatorname{card} Y_0. \tag{18.4.7}$$

By (18.4.5) and (18.4.7) we obtain

$$\operatorname{card} Y_0 \leqslant \operatorname{card}(X/X_0).$$

Let Y be an arbitrary set such that $Y_0 \subset Y$ and

$$\operatorname{card} Y = \operatorname{card}(X/X_0).$$

So we may define a new mapping $\varphi : X/X_0 \to Y$ (one-to-one and onto) in such a way that, whenever for some $u \in S$ and $x \in X$ we have $[u] \subset \langle x \rangle$, then

$$\varphi(\langle x \rangle) = \alpha([u]). \tag{18.4.8}$$

(Note that by Theorem 4.6.2 $S/\operatorname{Ker} g$ is a semiselective subpartition of X/X_0, and so for a given $\langle x \rangle \in X/X_0$ there may exist at most one $[u] \in S/\operatorname{Ker} g$ such that $[u] \subset \langle x \rangle$. Thus condition (18.4.8) is meaningful.)

Now we define an operation $\oplus$ in Y. Take arbitrary $t, s \in Y$. There exist unique classes $\langle x \rangle, \langle y \rangle \in X/X_0$ such that

$$t = \varphi(\langle x \rangle), \quad s = \varphi(\langle y \rangle). \tag{18.4.9}$$

We put

$$t \oplus s = \varphi(\langle x + y \rangle). \tag{18.4.10}$$

It is easy to check that the operation $\oplus$ thus defined $(Y, \oplus)$ is a group[3]. Now suppose that $t, s \in Y_0$. Then there exist classes $[u], [v] \in S/\operatorname{Ker} g$ such that $t = \alpha([u])$, $s = \alpha([v])$. Since $S/\operatorname{Ker} g$ is a subpartition of X/X_0, there exist classes $\langle x \rangle, \langle y \rangle \in X/X_0$ such that

$$[u] \subset \langle x \rangle, \quad [v] \subset \langle y \rangle, \tag{18.4.11}$$

whence, by (18.4.8), $\varphi(\langle x \rangle) = \alpha([u]) = t$, $\varphi(\langle y \rangle) = \alpha([v]) = s$. According to (18.4.6) this means that $g([u]) = \{t\}$, $g([v]) = \{s\}$. In particular,

$$g(u) = t, \quad g(v) = s.$$

Hence $g(u + v) = g(u) + g(v) = t + s$ (here the operation $+$ on the right-hand side denotes the operation originally existing in Y_0), whence, again by (18.4.6),

$$\alpha([u + v]) = t + s. \tag{18.4.12}$$

Further, (18.4.11) implies that $u \in \langle x \rangle$ and $v \in \langle y \rangle$, whence $\langle u \rangle = \langle x \rangle$ and $\langle v \rangle = \langle y \rangle$. Thus $u + v \in \langle u + v \rangle = \langle u \rangle + \langle v \rangle = \langle x \rangle + \langle y \rangle = \langle x + y \rangle$. Since evidently $u + v \in [u + v]$, we

---

[3] The neutral element of $(Y, \oplus)$ is $\varphi(\langle 0 \rangle)$, and, to a given $t = \varphi(\langle x \rangle)$ the inverse element is given by $\varphi(\langle -x \rangle)$.

get hence $[u+v] \cap \langle x+y \rangle \neq \varnothing$, whence it follows by Lemma 4.6.1 that $[u+v] \subset \langle x+y \rangle$. By (18.4.8) $\varphi(\langle x+y \rangle) = \alpha([u+v])$, and by (18.4.10) $\alpha([u+v]) = t \oplus s$. Now we obtain by (18.4.12)

$$t \oplus s = t + s,$$

which shows that on Y_0 both the operations, $\oplus$ and $+$, coincide. Thus in the sequel, without a fear of ambiguity, we may use the same symbol $+$ to denote both these operations.

The argument above shows that $(Y_0, +)$ is a subgroup of $(Y, +)$.

Now we define a mapping $f : X \to Y$ as follows:

$$f(x) = \varphi(\langle x \rangle). \tag{18.4.13}$$

For arbitrary $x, y \in X$ we have according to (18.4.13), (18.4.10) and (18.4.9)

$$f(x+y) = \varphi(\langle x+y \rangle) = \varphi(\langle x \rangle) + \varphi(\langle y \rangle) = f(x) + f(y).$$

So f is a homomorphism.

If $x \in S$, then $x \in [x] \cap \langle x \rangle$, whence by Lemma 4.6.1 $[x] \subset \langle x \rangle$, and by (18.4.8) and (18.4.13)

$$f(x) = \alpha([x]). \tag{18.4.14}$$

But in virtue of (18.4.6) $\alpha([x]) = g(x)$, whence by (18.4.14) $f(x) = g(x)$. Thus $f \mid S = g$.

The proof of necessity of condition (18.4.1) is identical as the corresponding part of the proof of Theorem 18.4.1. $\qquad \square$

Theorem 18.4.3. *Let $(X, +)$ and $(Y, +)$ be groups, and let $(S, +)$ be a subgroup of $(X, +)$ such that with $S' = X \setminus S$ the following conditions are fulfilled:*

(i) For every $x, y \in S'$ we have $x + y \in S$.

(ii) There exists a $u \in S'$ such that[4] $2u = 0$ and for every $x \in X$ we have $x + u = u + x$.

Let $g : S \to Y$ be a homomorphism. Then there exists a homomorphism $f : X \to Y$ such that $f \mid S = g$.

Proof. Put for $x \in X$

$$f(x) = \begin{cases} g(x), & \text{if } x \in S, \\ g(x+u) & \text{if } x \in S', \end{cases} \tag{18.4.15}$$

It is obvious from definition (18.4.15) that $f \mid S = g$. In order to prove that f is a homomorphism we must distinguish four cases. Take arbitrary $x, y \in S$.

1. $x \in S, y \in S$. Then also $x + y \in S$, and by (18.4.15)

$$f(x) = g(x), \quad f(y) = g(y), \quad f(x+y) = g(x+y),$$

---

[4] So u is an element of order 2. Similarly, it follows from condition (i) that $(S, +)$ is a subgroup of $(X, +)$ of index 2 (cf. Lemma 4.5.4; cf. also Exercise 4.7).

and
$$f(x + y) = g(x + y) = g(x) + g(y) = f(x) + f(y),$$

since g is a homomorphism.

2. $x \in S$, $y \in S'$. Then $x + y \in S'$ in virtue of Lemma 4.5.3, whence

$$f(x) = g(x), \quad f(y) = g(y + u), \quad f(x + y) = g(x + y + u),$$

and

$$f(x + y) = g(x + y + u) = g(x) + g(y + u) = f(x) + f(y).$$

3. $x \in S'$, $y \in S$. Then again $x + y \in S'$ (Lemma 4.5.3), whence

$$f(x) = g(x + u), \quad f(y) = g(y), \quad f(x + y) = g(x + y + u).$$

and by (ii)

$$f(x + y) = g(x + y + u) = g(x + u + y) = g(x + u) + g(y) = f(x) + f(y).$$

4. $x \in S'$, $y \in S'$. Then by (i) $x + y \in S$ and

$$f(x) = g(x + u), \quad f(y) = g(y + u), \quad f(x + y) = g(x + y).$$

By (ii)

$$f(x+y) = g(x+y) = g(x+y+2u) = g(x+u+y+u) = g(x+u)+g(y+u) = f(x)+f(y).$$

Consequently (18.2.15) holds in all cases, which means that f is a homomorphism. $\quad\square$

K. Dankiewicz and Z. Moszner [52] showed that Theorem 18.4.3 can be derived directly from Theorem 18.4.1. They also proved a number of related results.

We terminate this section with the following simple theorem (cf., e.g., Kuczma [183]).

Theorem 18.4.4. *Let $(X, +)$ be a commutative group, and let $(S, +)$ be a subgroup of $(X, +)$ such that, for a certain $n \in \mathbb{N}$ and every $x \in X$, we have $nx \in S$. Let $(Y, +)$ be a commutative group in which the division by n is performable, and let $g : S \to Y$ be a homomorphism. Then there exists a unique homomorphism $f : X \to Y$ such that $f \mid S = g$. This homomorphism is given by*

$$f(x) = \frac{g(nx)}{n}, \quad x \in X. \tag{18.4.16}$$

Proof. Let $g : S \to Y$ be a homomorphism, and let f be given by (18.4.16). The definition is meaningful, since $nx \in S$ for every $x \in X$ and the division by n is performable in Y. Take arbitrary $x, y \in X$. Since X is commutative, we have $n(x + y) = nx + ny$, and since Y is commutative,

$$n\left(\frac{g(nx)}{n} + \frac{g(ny)}{y}\right) = n\frac{g(nx)}{n} + n\frac{g(ny)}{n} = g(nx) + g(ny),$$

whence

$$\frac{g(nx) + g(ny)}{n} = \frac{g(nx)}{n} + \frac{g(ny)}{n}.$$

Consequently

$$f(x+y) = \frac{g(n(x+y))}{n} = \frac{g(nx+ny)}{n} = \frac{g(nx)+g(ny)}{n} = \frac{g(nx)}{n} + \frac{g(ny)}{n} = f(x) + f(y).$$

Thus f is a homomorphism.

Now, we have by induction from (18.2.2)

$$g(mx) = mg(x), \quad m \in \mathbb{N}, \ x \in S,$$

whence, in particular, $g(nx) = ng(x)$ for $x \in S$. Hence, if $x \in S$, we have by (18.4.16)

$$f(x) = \frac{g(nx)}{n} = \frac{ng(x)}{n} = g(x).$$

This shows that $f \mid S = g$.

Finally, if $h : X \to Y$ is an arbitrary homomorphism such that $h \mid S = g$, then, similarly as above, we have $h(nx) = nh(x)$ for $x \in X$. As $nx \in S$ and $h \mid S = g$, this yields

$$g(nx) = nh(x), \quad x \in X. \tag{18.4.17}$$

Relations (18.4.17) and (18.4.16) yield

$$h(x) = \frac{g(nx)}{n} = f(x), \quad x \in X.$$

This shows that h coincides with f on the whole of X, which proves the uniqueness of f, and also formula (18.4.16). $\qquad\square$

18.5 Cauchy equation on a cylinder

Let $(X,+)$ and $(Y,+)$ be arbitrary groups, and let $Z \subset Y$ be an arbitrary non-empty set. Suppose that a function $f : X \to Y$ satisfies the equation

$$f(x + y) = f(x) + f(y) \tag{18.5.1}$$

for all $x \in X$, $y \in Z$. Then (18.5.1) is a restricted Cauchy equation, the condition restricting the validity of the equation being $(x, y) \in X \times Z$. A set of the form $X \times Z$, $\varnothing \neq Z \subset X$, is called a *cylinder*, therefore the present problem is referred to as the *Cauchy equation on a cylinder*.

Let $f : X \to Y$ be an arbitrary mapping. The set

$$N_f = \{ y \in X \mid f(x + y) = f(x) + f(y) \text{ for all } x \in X \} \tag{18.5.2}$$

is called the *Cauchy nucleus* of f (cf. Grząślewicz-Powązka-Tabor[124]). The Cauchy equation (18.5.1) on a cylinder $X \times Z$ simply means that $Z \subset N_f$.

The following lemma (Kuczma [183], Dhombres-Ger [70]) is very useful.

Lemma 18.5.1. *Let $(X, +)$ and $(Y, +)$ be groups, and let $f : X \to Y$ be an arbitrary mapping. Then either $N_f = \varnothing$, or $(N_f, +)$ is a subgroup of $(X, +)$.*

Proof. Suppose that $N_f \neq \varnothing$. Then there exists a $y \in N_f$. With this y (18.5.1) holds for all $x \in X$. Taking $x = 0$, we get $f(y) = f(0) + f(y)$, whence $f(0) = 0$. Now again, with y an arbitrary element of N_f and $x = -y$, we get from (18.5.1)

$$0 = f(0) = f(-y) + f(y),$$

whence

$$f(-y) = -f(y) \text{ for every } y \in N_f. \tag{18.5.3}$$

We have for arbitrary $x \in X$ and $y \in N_f$

$$f(x) = f\big((x - y) + y\big) = f(x - y) + f(y),$$

whence

$$f(x - y) = f(x) - f(y), \tag{18.5.4}$$

or by (18.5.3)

$$f\big(x + (-y)\big) = f(x) + f(-y). \tag{18.5.5}$$

Relation (18.5.5) holds for every $x \in X$, which means that $-y \in N_f$ whenever $y \in N_f$. Hence we have, for arbitrary $x \in X$ and $u, v \in N_f$,

$$f\big(x + (u - v)\big) = f\big((x + u) + (-v)\big) = f(x + u) + f(-v),$$

and by (18.5.3)

$$f\big(x + (u - v)\big) = f(x + u) - f(v) = f(x) + f(u) - f(v),$$

since $u \in N_f$. Relation (18.5.4) implies that $f(u - v) = f(u) - f(v)$, whence

$$f\big(x + (u - v)\big) = f(x) + f(u - v).$$

This holds for all $x \in X$ so that $u - v \in N_f$. Lemma 4.5.1 completes the proof.   □

If $N_f \neq \varnothing$, then relation (18.5.1) holds, in particular, for all $x, y \in N_f$. This means that $f \mid N_f$ is a homomorphism. Therefore, instead of (18.5.1), we will investigate the equation

$$f(x + y) = f(x) + g(y) \tag{18.5.6}$$

valid for $x \in X$, $y \in S$, where $(S, +)$ is a subgroup of $(X, +)$, and $g : S \to Y$ is a homomorphism (Tabor [312], Dhombres-Ger [70]).

Theorem 18.5.1. *Let $(X, +)$ and $(Y, +)$ be groups, let $(S, +)$ be a subgroup of $(X, +)$, and let $g : S \to Y$ be a homomorphism. If $f : X \to Y$ is a function such that (18.5.6) holds for every $x \in X$, $y \in S$, then there exist a lifting $\xi : X/S \to X$ and a mapping $h : X/S \to Y$ such that*

$$f(x) = h\big([x]\big) + g\big(-\xi([x]) + x\big), \quad x \in X. \tag{18.5.7}$$

Conversely, every function f of form (18.5.7), where $h : X/S \to Y$ is an arbitrary function and $\xi : X/S \to X$ is an arbitrary lifting, satisfies equation (18.5.6) for all $x \in X$, $y \in S$.

Proof. Let $\xi : X/S \to X$ be a quite arbitrary lifting [5]. Define a function $h : X/S \to Y$ by

$$h([x]) = f(\xi([x])), \quad [x] \in X/S. \tag{18.5.8}$$

According to definition, a lifting is characterized by the property that $[\xi([x])] = [x]$. In other words, the elements $\xi([x])$ and x generate the same equivalence class (left coset), i.e., they are equivalent with respect to S. This means that $-\xi([x]) + x \in S$. Hence, writing x as $x = \xi([x]) + (-\xi([x]) + x)$, we get by (18.5.6) and (18.5.8)

$$f(x) = f(\xi([x])) + g(-\xi([x]) + x) = h([x]) + g(-\xi([x]) + x),$$

which proves relation (18.5.7).

Conversely, suppose that the function f has form (18.5.7), where $h : X/S \to Y$ is an arbitrary function, and $\xi : X/S \to X$ is an arbitrary lifting. Take arbitrary $x \in X$, $y \in S$. Then we have $-x + (x + y) = y \in S$, which means that x and $x + y$ are equivalent with respect to S. Thus $[x] = [x + y]$. Hence we have by (18.5.7), since $g : S \to Y$ is a homomorphism,

$$\begin{aligned}
f(x + y) &= h([x + y]) + g(-\xi([x + y]) + x + y) \\
&= h([x]) + g(-\xi([x]) + x + y) = h([x]) + g(-\xi([x]) + x) + g(y) \\
&= f(x) + g(y),
\end{aligned}$$

i.e., f satisfies equation (18.5.6). $\qquad\square$

Hence we derive the following result (Tabor [312], Dhombres-Ger [70]):

Theorem 18.5.2. *Let $(X, +)$ and $(Y, +)$ be groups, and let $(S, +)$ be a subgroup of $(X, +)$. If $f : X \to Y$ is a function such that (18.5.1) holds for every $x \in X$, $y \in S$, then there exist a homomorphism $g : S \to Y$, a lifting $\xi : X/S \to X$, and a mapping $h : X/S \to Y$ such that (18.5.7) holds and*

$$h(S) = g(\xi[S]). \tag{18.5.9}$$

Conversely, every function f of form (18.5.7), where $g : S \to Y$ is a homomorphism, $\xi : X/S \to X$ is a lifting, and $h : X/S \to Y$ is a mapping fulfilling condition (18.5.9), satisfies (18.5.1) for every $x \in X$, $y \in S$.

Proof. Suppose that a function $f : X \to Y$ satisfies equation (18.5.1) for $x \in X$, $y \in S$. As we have pointed out after the proof of Lemma 18.5.1, the function $g = f \mid S$ is a homomorphism, and it is readily seen that f satisfies equation (18.5.6) for $x \in X$, $y \in S$. In virtue of Theorem 18.5.1 there exist a lifting $\xi : X/S \to X$ and a mapping

---

[5] As is clear from the final remarks in 4.5, the existence of such a lifting results from the Axiom of Choice.

$h : X/S \to Y$ such that relation (18.5.7) holds. It remains to show that also condition (18.5.9) is fulfilled.

Since $g = f \mid S$ is a homomorphism, we have (by (18.2.2)) $f(0) = g(0) = 0$. Hence, setting in (18.5.7) $x = 0$, and taking into account the fact that $[0] = S$, we get

$$0 = h(S) + g\big(-\xi(S)\big).$$

Of course, $\xi(S) \in S$, and, g being homomorphism, $g\big(-\xi(S)\big) = -g\big(\xi(S)\big)$. Hence (18.5.9) results.

Conversely, suppose that f is given by (18.5.7), where $g : S \to Y$ is a homomorphism, $\xi : X/S \to X$ is a lifting, and $h : X/S \to Y$ is a mapping fulfilling condition (18.5.9). By Theorem 18.5.1 f satisfies equation (18.5.6) with g occurring in (18.5.7). If $x \in S$, then $[x] = S$, and also $\xi([x]) \in S$. Hence

$$g\big(-\xi([x]) + x\big) = -g\big(\xi([x])\big) + g(x) = -g\big(\xi(S)\big) + g(x),$$

and we get by (18.5.7) and (18.5.9)

$$f(x) = g(x) \quad \text{for } x \in S.$$

Thus (18.5.1) results from (18.5.6). $\square$

Theorem 18.5.2 yields the general solution of equation (18.5.1) on the cylinder $X \times Z$. Let $(S, +)$ be the subgroup of $(X, +)$ generated by Z. Then clearly $Z \subset S \subset N_f$ (Lemma 18.5.1) so that every solution of (18.5.1) on the cylinder $X \times Z$ must also be a solution of (18.5.1) on the cylinder $X \times S$, and hence must be of form (18.5.7). Conversely, every solution of (18.5.1) on $X \times S$ is also a solution of (18.5.1) on $X \times Z$.

From Theorems 18.5.1 and 18.5.2 we deduce two corollaries (Tabor [312], Kuczma [183]).

Corollary 18.5.1. *Let $(X, +)$ and $(Y, +)$ be groups, and let $(S, +)$ be a subgroup of $(X, +)$. Let $g_0 : X \to Y$ be a homomorphism, and let $g = g_0 \mid S$. If $f : X \to Y$ is a function such that (18.5.6) holds for every $x \in X$, $y \in S$, then there exists a mapping $h : X/S \to Y$ such that*

$$f(x) = h\big([x]\big) + g_0(x). \tag{18.5.10}$$

Conversely, every function of form (18.5.10), where $h : X/S \to Y$ is an arbitrary function, satisfies equation (18.5.6) for all $x \in X$, $y \in S$.

Proof. Suppose that $f : X \to Y$ satisfies equation (18.5.6) for all $x \in X$, $y \in S$. By Theorem 18.5.1 there exist a mapping $h_0 : X/S \to Y$ and a lifting $\xi : X/S \to X$ such that

$$f(x) = h_0\big([x]\big) + g\big(-\xi([x]) + x\big) = h_0\big([x]\big) + g_0\big(-\xi([x]) + x\big)$$
$$= h_0\big([x]\big) + g_0\big(-\xi([x])\big) + g_0(x).$$

With $h(u) = h_0(u) + g_0\big(-\xi(u)\big)$, $u \in X/S$, we obtain hence (18.5.10).

Conversely, suppose that f is given by (18.5.10), where $h : X/S \to Y$ is an arbitrary function. Then, for $x \in X$, $y \in S$, we have $-(x+y) + x = -y - x + x = -y \in S$, which means that $x + y$ and x are equivalent with respect to S, whence $[x+y] = [x]$, and

$$f(x+y) = h([x+y]) + g_0(x+y) = h([x]) + g_0(x) + g_0(y)$$
$$= h([x]) + g_0(x) + g(y) = f(x) + g(y),$$

and so (18.5.6) holds. $\square$

Corollary 18.5.2. *Let $(X, +)$ and $(Y, +)$ be groups, and let $(S, +)$ be a subgroup of $(X, +)$. Let $f : X \to Y$ be a function such that (18.5.1) holds for all $x \in X$, $y \in S$. Suppose that there exists[6] a homomorphism $g_0 : X \to Y$ such that $g_0 \mid S = f \mid S$. Then there exists a mapping $h : X/S \to Y$ such that (18.5.10) holds. Conversely, every function of form (18.5.10), where $h : X/S \to Y$ is an arbitrary function and $g_0 : X \to Y$ is a homomorphism such that $g_0 \mid S = f \mid S$, satisfies equation (18.5.1) for all $x \in X$, $y \in S$.*

Proof. Let $f : X \to Y$ satisfy (18.5.1) for all $x \in X$, $y \in S$. Put $g = f \mid S$. Then g_0 is a homomorphism from X to into Y, $g_0 \mid S = g$, and f satisfies (18.5.6) for $x \in X$, $y \in S$. By Corollary 18.5.1 there exists a mapping $h : X/S \to Y$ such that (18.5.10) holds. Conversely, if f has form (18.5.10), then, by Corollary 18.5.1, f satisfies (18.5.6) for $x \in X$, $y \in S$. But for $y \in S$ we have $g(y) = f(y)$, so (18.5.6) reduces to (18.5.1). $\square$

One can also consider the problems symmetric to those dealt with the present section: the Cauchy equation (18.5.1) on a cylinder $Z \times X$, and the equation

$$f(x+y) = g(x) + f(y) \tag{18.5.11}$$

for $x \in S$, $y \in X$. Such problems can be reduced to those treated in the present section as follows. Introduce in X and Y new operations $\oplus$ defined in the following way:

$$x \oplus y = y + x \quad \text{for } x, y \in X,$$

$$t \oplus s = s + t \quad \text{for } t, s \in Y.$$

With these operations $(X, \oplus)$ and $(Y, \oplus)$ are groups, and, whenever $(S, +)$ is a subgroup of $(X, +)$, $(S, \oplus)$ is a subgroup of $(X, \oplus)$. Then equation (18.5.1) on the cylinder $Z \times X$ is equivalent to the equation

$$f(x \oplus y) = f(x) \oplus f(y)$$

for $x \in X$, $y \in Z$, whereas equation (18.5.11) ($x \in S$, $y \in X$) is equivalent to

$$f(x \oplus y) = f(x) \oplus g(y)$$

for $x \in X$, $y \in S$.

---

[6] Some conditions for the existence of such a homomorphism are found in 18.1 and 18.4. Note that $f \mid S$ in a homomorphism from S into Y.

18.6 Cauchy nucleus

Let $(X, +)$ and $(Y, +)$ be arbitrary groups, and let $f : X \to Y$ be an arbitrary function. According to Lemma 18.5.1 the Cauchy nucleus N_f of f, if non-empty, is a subgroup of X. A natural question arises as to whether every subgroup of X is the Cauchy nucleus of a certain mapping $f : X \to Y$? An answer to this question (Grząślewicz-Powązka-Tabor [124]) is the subject matter of this section.

Theorem 18.6.1. *Let $(X, +)$ and $(Y, +)$ be arbitrary groups, and let $(S, +)$ be a subgroup of $(X, +)$. If the group Y contains an element $t \neq 0$ which is not of order two, then there exists a function $f : X \to Y$ such that $N_f = S$.*

Proof. The assumption on Y means that $2t \neq 0$. Put

$$f(x) = \begin{cases} 0, & \text{for } x \in S \\ t, & \text{for } x \in X \setminus S. \end{cases} \tag{18.6.1}$$

Take arbitrary $x \in X$, $y \in S$. Then either $x \in S$, or $x \in X \setminus S$. In the former case we have $x + y \in S$, and $f(x) = f(y) = f(x + y) = 0$ so that (18.5.1) holds. If $x \in X \setminus S$, then also $x + y \in X \setminus S$ (Lemma 4.5.3), and $f(x) = f(x + y) = t$, whereas $f(y) = 0$, and again (18.5.1) holds. Thus (18.5.1) holds for every $x \in X$, $y \in S$, which shows that

$$S \subset N_f. \tag{18.6.2}$$

Now take an arbitrary $x \in X \setminus S$. Then, by Lemma 4.5.3, also $-x \in X \setminus S$, and, since obviously $0 \in S$,

$$f(-x + x) = f(0) = 0 \neq 2t = f(-x) + f(x)$$

so that $x \notin N_f$. In other words,

$$N_f \subset S, \tag{18.6.3}$$

which, together with (18.6.2), yields the desired equality $N_f = S$. $\square$

Theorem 18.6.2. *Let $(X, +)$ and $(Y, +)$ be arbitrary groups such that $\operatorname{card} Y > 1$, and let $(S, +)$ be a subgroup of $(X, +)$, which is not of index 2. Then there exists a function $f : X \to Y$ such that $N_f = S$.*

Proof. Since $\operatorname{card} Y > 1$, then there exists a $t \in Y$ such that

$$t \neq 0. \tag{18.6.4}$$

We define the function $f : X \to Y$ by (18.6.1). The same argument as in the proof of Theorem 18.6.1 shows that relation (18.5.1) holds for all $x \in X$, $y \in S$, whence we obtain (18.6.2).

If $S = X$, then (18.6.2) implies $N_f = S$. If $S \neq X$, then take an arbitrary $y \in X \setminus S$. Since S is not of index 2, there exists an $x \in X \setminus S$ such that $x + y \in X \setminus S$ (cf. Exercise 4.7). Then by (18.6.1) and (18.6.4)

$$f(x + y) = t \neq t + t = f(x) + f(y)$$

whence $y \notin N_f$. Hence we get (18.6.3), which together with (18.6.2) yields $N_f = S$. $\square$

If $\operatorname{card} Y = 1$, then the only element of Y is 0, whence $f = 0$ is the only function $f : X \to Y$. This function fulfills (18.5.1) for all $x, y \in X$ so that $N_f = X$, and no proper subgroup of X can be the Cauchy nucleus of any $f : X \to Y$. If $\operatorname{card} Y > 1$, then the only case not covered by Theorems 18.6.1 and 18.6.2 is when every element of Y has order 2, and S is a subgroup of X of index 2. Then not every subgroup of X is the Cauchy nucleus of an $f : X \to Y$ (cf. Theorem 18.6.3 and Example 18.6.1 below).

Lemma 18.6.1. *Let $(X, +)$ be a group. If every element $x \neq 0$ of X has order 2, then $(X, +)$ is commutative.*

Proof. The assumption about X amounts to the fact $2x = 0$, or

$$x = -x \text{ for every}^7 x \in X . \tag{18.6.5}$$

Take arbitrary $x, y \in X$. By (18.6.5) $x + y = -(x + y) = -y - x = y + x$, which means that $(X, +)$ is commutative. $\qquad \square$

Let $(X, +)$ and $(Y, +)$ be groups. A homomorphism $f : X \to Y$ is called an *isomorphism* iff it is one-to-one and onto. If an isomorphism $f : X \to Y$ exists, then $(X, +)$ and $(Y, +)$ are said to be *isomorphic*.

Theorem 18.6.3. *Let $(X, +)$ and $(Y, +)$ be groups such that every element $t \neq 0$ of Y has order 2, and let $(S, +)$ be a subgroup of $(X, +)$ whose index is 2. Then there exists a function $f : X \to Y$ such that $N_f = S$ if and only if there exists a normal subgroup $(T, +)$ of $(S, +)$ such that the group $(S/T, +)$ is isomorphic to a subgroup of $(Y, +)$, and either there exist $x \in S$, $u \in X \setminus S$ such that*

$$x - u - x + u \notin T, \tag{18.6.6}$$

or there exist $x \in X \setminus S$, and $u \in X \setminus S$ such that

$$-u + x - u - x \notin T. \tag{18.6.7}$$

Proof. Take arbitrary $x, y \in X \setminus S$. Then also $-x, -y \in X \setminus S$ (cf. Lemma 4.5.3), whence $-x + y = (-x) - (-y) \in S$, since the index of s is two. Thus every $x, y \in X \setminus S$ are equivalent with respect to S, which means that the set X/S consists of two classes only: S and $X \setminus S$.

Suppose that there exists a function $f : X \to Y$ such that $N_f = S$. By Theorem 18.5.2 there exist a lifting $\xi : X/S \to X$, a homomorphism $g : S \to Y$, and a mapping $h : X/S \to Y$ fulfilling condition (18.5.9) and such that (18.5.7) holds. For $x \in S$ we have $[x] = S$ and

$$f(x) = h([x]) + g(-\xi([x]) + x) = h(S) + g(-\xi(S) + x) = h(S) + g(-\xi(S)) + g(x),$$

whence, since g is a homomorphism, and so $g(-\xi(S)) = -g(\xi(S))$, we obtain by (18.5.9) $f(x) = g(x)$. So in view of the preceding remarks, f can be written as

$$f(x) = \begin{cases} g(x) & \text{for } x \in S, \\ h(X \setminus S) + g(-u + x) & \text{for } x \in X \setminus S, \end{cases} \tag{18.6.8}$$

---

^{7} For $x = 0$ formula (18.6.5) is evident.

where $u = \xi(X \setminus S) \in X \setminus S$. Take a $y \in X \setminus S$. In virtue of the condition $N_f = S$ there exists an $x \in X$ such that

$$f(x+y) \neq f(x) + f(y). \tag{18.6.9}$$

If $x \in S$, then $x + y \in X \setminus S$, since otherwise we would have $-x \in S$ and $y = (-x) + (x+y) \in S$, and so we get by (18.6.8) and (18.6.9)

$$h(X \setminus S) + g(-u+x+y) \neq g(x) + h(X \setminus S) + g(-u+y) \tag{18.6.10}$$
$$= h(X \setminus S) + g(x) + g(-u+y),$$

since according to Lemma 18.6.1 $(Y, +)$ is commutative. Thus we get by (18.6.10)

$$g(-u+x+y) \neq g(x) + g(-u+y) = g(x-u+y). \tag{18.6.11}$$

If $x \in X \setminus S$, then $x + y \in S$ (Corollary 4.5.1) and we get by (18.6.8) and (18.6.9) and Lemma 18.6.1

$$g(x+y) \neq h(X \setminus S) + g(-u+x) + h(X \setminus S) + g(-u+y) \tag{18.6.12}$$
$$= 2h(X \setminus S) + g(-u+x) + g(-u+y).$$

But $h(X \setminus S) \in Y$ is an element of order 2. So we get from (18.6.12)

$$g(x+y) \neq g(-u+x) + g(-u+y) = g(-u+x-u+y), \tag{18.6.13}$$

g being a homomorphism.

Put $T = \mathrm{Ker}\, g$. By Lemma 4.5.7 $(T, +)$ is a normal subgroup of $(S, +)$. By Lemma 4.5.8 the function g is constant on every class from S/T (in the sequel, these classes will be denoted by $\langle x \rangle$, $x \in S$), so the formula

$$\gamma(\langle x \rangle) = g(x)$$

unambiguously defines a mapping $\gamma : S/T \to Y$. If we put $Y_0 = \gamma(S/T) = g(S)$, this mapping clearly is one-to-one (cf. Lemma 4.5.8) and onto Y_0. Take arbitrary $u, v \in Y_0$. Then there exist $x, y \in S$ such that $u = \gamma(\langle x \rangle) = g(x)$, $v = \gamma(\langle y \rangle) = g(y)$, and

$$u - v = g(x) - g(y) = g(x-y) = \gamma(\langle x - y \rangle).$$

So $u - v \in Y_0$, and we infer from Lemma 4.5.1 that $(Y_0, +)$ is a subgroup of $(Y, +)$. Finally, we have for arbitrary $\langle x \rangle, \langle y \rangle \in S/T$

$$\gamma(\langle x \rangle + \langle y \rangle) = \gamma(\langle x + y \rangle) = g(x+y) = g(x) + g(y) = \gamma(\langle x \rangle) + \gamma(\langle y \rangle),$$

which means that $\gamma : S/T \to Y_0$ is a homomorphism, and being one-to-one and onto, is an isomorphism. Thus the groups $(S/T, +)$ and $(Y_0, +)$ are isomorphic.

Now, condition (18.6.11) means that

$$0 \neq g(x-u+y) - g(-u+x+y) = g\big((x-u+y) - (-u+x+y)\big)$$
$$= g(x-u+y-y-x+u) = g(x-u-x+u),$$

i.e., $x - u - x + u \notin \mathrm{Ker}\, g = T$. So we get condition (18.6.6).

Condition (18.6.13) means that

$$0 \neq g(-u+x-u+y) - g(x+y) = g\big((-u+x-u+y) - (x+y)\big)$$
$$= g(-u+x-u+y-y-x) = g(-u+x-u-x),$$

i.e., $-u+x-u-x \notin \operatorname{Ker} g = T$. So we get condition (18.6.7).

Conversely, suppose that there exists a normal subgroup $(T, +)$ of $(S, +)$ such that $(S/T, +)$ is isomorphic to a subgroup $(Y_0, +)$ of $(Y, +)$ and that there exist either $x \in S$, $u \in X \backslash S$ fulfilling (18.6.6), or, $x, u \in X \backslash S$ fulfilling (18.6.7). Let $\varphi : S/T \to Y_0$ be the isomorphism, and let $\pi : S \to S/T$ be the canonical homomorphism (cf. Lemma 4.5.9). Put $g = \varphi \circ \pi$. It is easily seen that $g : S \to Y_0 \subset Y$ is a homomorphism. We have $\operatorname{Ker} g = g^{-1}(0) = \pi^{-1}\big(\varphi^{-1}(0)\big)$. But since φ is one-to-one, we have $\varphi^{-1}(0) = \{\langle 0 \rangle\}$, whence $\operatorname{Ker} g = \pi^{-1}(\langle 0 \rangle) = \pi^{-1}(T) = T$.

Let condition (18.6.6) hold for certain $x \in S$, $u \in X \setminus S$. Take arbitrary $y_0 \in Y$ and $v \in S$. The formulas

$$\begin{cases} \xi(S) = v, & h(S) = g(v) \\ \xi(X \setminus S) = u, & h(X \setminus S) = y_0, \end{cases} \tag{18.6.14}$$

define a lifting $\xi : X/S \to X$ and a mapping $h : X/S \to Y$ fulfilling condition (18.5.9), and formula (18.5.7) defines a function $f : X \to Y$ satisfying, according to Theorem 18.5.2, equation (18.5.1) on the cylinder $X \times S$. Thus, for this particular f, inclusion (18.6.2) holds.

By the same argument, as in the first part of the proof, the function f can be written in form (18.6.8).

Take an arbitrary $y \in X \setminus S$. Let x and u denote points occurring in condition (18.6.6). Then $x + y \in X \setminus S$, and $(x - u + y), (-u + x + y) \in S$. Moreover, by (18.6.6),

$$(x - u + y) - (-u + x + y) = x - u - x + u \notin T = \operatorname{Ker} g$$

so that

$$g(x - u + y) - g(-u + x + y) = g\big((x - u + y) - (-u + x + y)\big) \neq 0,$$

and consequently we get, in turn (18.6.11), (18.6.10) and (18.6.9). This means that $y \notin N_f$, whence relation (18.6.3) follows, which together with (18.6.2) implies that $N_f = S$.

Now suppose that condition (18.6.7) holds for certain $x, u \in X \backslash S$. Take arbitrary $y_0 \subset Y$, $v \subset S$, and define the lifting $\xi : X/S \to X$ and the mapping $h : X/S \to Y$ fulfilling (18.5.9) by formulas (18.6.14). Then formula (18.5.7), which again can be written in form (18.6.8), defines a function $f : X \to Y$ satisfying the Cauchy equation (18.5.1) on the cylinder $X \times S$, whence inclusion (18.6.2) follows.

Take an arbitrary $y \in X \setminus S$. Let x and u denote points occurring in condition (18.6.7). Then $x + y \in S$, $-u + x - u + y \in S$, and, by (18.6.7),

$$(-u + x - u + y) - (x + y) = -u + x - u - x \notin T = \operatorname{Ker} g.$$

Hence

$$g(-u+x-u+y) - g(x+y) = g\big((-u+x-u+y)-(x+y)\big) \neq 0,$$

and consequently we get, in turn, (18.6.13) , (18.6.12) and (18.6.9). This means that $y \notin N_f$, whence relation (18.6.3) follows, which together with (18.6.2) implies that $N_f = S$. $\square$

Example 18.6.1. Let $(X,+)$ and $(Y,+)$ be arbitrary groups in which every element has[8] order 2, and let $(S,+)$ be a subgroup of $(X,+)$, whose index is two. By Lemma 18.6.1 all these groups are commutative. So we have $x - u - x + u = 0 \in T$ and $-u + x - u - x = 0 \in T$ for every subgroup $(T,+)$ of $(S,+)$ and for every x, $u \in X$. By Theorem 18.6.3 there does not exist a function $f : X \to Y$ such that $N_f = S$.

Example 18.6.2. Let $X = \mathbb{Z}$, and let $S = 2\mathbb{Z}$ be the set of all even integers. With the usual addition of numbers $(X,+)$ is a commutative group, and $(S,+)$ is its subgroup, whose index is two. Let $T = 4\mathbb{Z}$ be the set of all integers divisible by 4. $(T,+)$ is a subgroup of $(S,+)$, and all these groups being commutative, $(T,+)$ is a normal subgroup of $(S,+)$. Put $Y = S/T$. Then $(Y,+)$ is group isomorphic to $(S/T,+)$; the identity function yields an isomorphism of $(S/T,+)$ onto $(Y,+)$. Take $u = 1 \in X \setminus S$. Then $-u + x - u - x = -2 \notin T$ for every $x \in X$. It follows by Theorem 18.6.3 that there exists a function $f : X \to Y$ such that $N_f = S$.

18.7   Theorem of Ger

Let $(X,+)$ and $(Y,+)$ be arbitrary groups, and let $(S,+)$ be a subsemigroup of $(X,+)$ such that

$$X = S - S. \tag{18.7.1}$$

Further, suppose that we are given two conjugate p.l.i. ideals $\mathcal{I}_1$ and $\mathcal{I}_2$ in X and in $X \times X$, respectively (cf. 17.5), and

$$S \notin \mathcal{I}_1. \tag{18.7.2}$$

Finally, suppose that we are given a function $f : S \to Y$ such that

$$f(x+y) = f(x) + f(y) \quad \mathcal{I}_2\text{-(a.e.) in } S \times S. \tag{18.7.3}$$

Let

$$M = \{(x,y) \in S \times S \mid f(x+y) \neq f(x) + f(x)\}. \tag{18.7.4}$$

By (18.7.3) we have $M \in \mathcal{I}_2$. Since $\mathcal{I}_1$ and $\mathcal{I}_2$ are conjugate, there exists a set $U \in \mathcal{I}_1$ such that

$$M[x] = \{y \in X \mid (x,y) \in M\} \in \mathcal{I}_1 \quad \text{for } x \notin U. \tag{18.7.5}$$

---

[8] There exist quite large groups in which every element has order 2. Let $n \in \mathbb{N}$, let X be the set of all $n \times n$ diagonal matrices with only $+1$ and -1 on the main diagonal. If $\cdot$ denotes the multiplication of matrices, then $(X, \cdot)$ is a group, every element of which has order 2, and card $X = 2^n$. Similarly, there exist groups $(X,+)$ every element of which has order 2 and such that the set X is infinite.

Now, under the conditions specified above, and with the above notation, we have the following

Lemma 18.7.1. *For every* $x, y, u, v \in S \setminus U$ *the equality* $x - y = u - v$ *implies* $f(x) - f(y) = f(u) - f(v)$.

Proof. Take $x, y, u, v \in S \setminus U$ such that $x - y = u - v$. By Lemma 17.5.5 $(y + S) \cap (v + S) \notin \mathcal{I}_1$, whence also $(-v + y + S) \cap S = -v + [(y + S) \cap (v + S)] \notin \mathcal{I}_1$. On the other hand, since $x, u \notin U$, we have $M[x], M[u] \in \mathcal{I}_1$, whence also $(-v + y + M[x]) \cup M[u] \in \mathcal{I}_1$, and

$$[-v + y + (S \setminus M[x])] \cap (S \setminus M[u])$$
$$= [(-v + y + S) \cap S] \setminus ((-v + y + M[x]) \cup M[u]) \notin \mathcal{I}_1.$$

Similarly, $M[v] \cup (-v + y + M[y]) \in \mathcal{I}_1$, whence

$$A = (([-v + y + (S \setminus M[x])] \cap (S \setminus M[u]))) \setminus [M[v] \cup (-v + y + M[y])] \notin \mathcal{I}_1,$$

and, in particular, $A \neq \varnothing$. Thus there exists an $s \in A$. For such an s, with the notation

$$z = -y + v + s, \tag{18.7.6}$$

we have

$$s \in S, \ s \notin M[u], \ s \notin M[v], \ z \in S, \ z \notin M[x], \ z \notin M[y],$$

whence

$$(u, s) \notin M, \ (v, s) \notin M, \ (x, z) \notin M, \ (y, z) \notin M. \tag{18.7.7}$$

On the other hand, we have

$$x - y + v + s = u - v + v + s = u + s,$$

i.e.,

$$x + z = u + s.$$

Hence by (18.7.4) and (18.7.7)

$$f(x) + f(z) = f(x + z) = f(u + s) = f(u) + f(s). \tag{18.7.8}$$

Relation (18.7.6) yields $y + z = v + s$, whence by (18.7.4) and (18.7.7)

$$f(y) + f(z) = f(y + z) = f(v + s) = f(v) + f(s). \tag{18.7.9}$$

Now, (18.7.8) and (18.7.9) yield that $f(x) - f(y) = f(u) - f(v)$. $\qquad \square$

The following result is due to R. Ger [112]:

Theorem 18.7.1. *Let* $(X, +)$ *and* $(Y, +)$ *be arbitrary groups, and let* $(S, +)$ *be a sub-semigroup of* $(X, +)$. *Further, let* $\mathcal{I}_1$ *and* $\mathcal{I}_2$ *be conjugate p.l.i. ideals in* X *and in* $X \times X$, *respectively, and suppose that conditions* (18.7.1) *and* (18.7.2) *hold. If* $f : S \to Y$ *is a function fulfilling* (18.7.3), *then there exists a unique homomorphism* $g : X \to Y$ *such that*

$$g \mid S = f \quad \mathcal{I}_1\text{-(a.e.) in } S. \tag{18.7.10}$$

Proof. Let the set $M \in \mathcal{I}_2$ be defined by (18.7.4) and $U \in \mathcal{I}_1$ by (18.7.5). Take a $z \in X$. By Lemma 17.5.6 there exist $x, y \in S \setminus U$ such that $z = x - y$. Define $g(z)$ as

$$g(z) = f(x) - f(y). \tag{18.7.11}$$

In virtue of Lemma 18.7.1, definition (18.7.11) is independent of the choice of $x, y \in S \setminus U$ fulfilling $z = x - y$. Consequently formula (18.7.11) unambiguously defines a function $g : X \to Y$.

Now we show that g is a homomorphism. Take arbitrary $x, y \in X$. There exist $s', t', u, v, p, q \in S \setminus U$ such that

$$x = s' - t', \quad y = u - v, \quad x + y = p - q. \tag{18.7.12}$$

By Lemma 17.5.7 there exist $s, t \in S \setminus U$ such that

$$t \in u + S \tag{18.7.13}$$

and

$$s - t = s' - t' = x. \tag{18.7.14}$$

By (18.7.12) and (18.7.14) $p - q = x + y = s - t + u - v$, whence

$$s - t + u = p - q + v. \tag{18.7.15}$$

By Lemma 17.5.5 $(t + S) \cap (y + q + S) \notin \mathcal{I}_1$, whence

$$(-y + t + S) \cap (q + S) = -y + \left[(t + S) \cap (y + q + S)\right] \notin \mathcal{I}_1,$$

and, since by (18.7.12) $-y + t = v - u + t$, we have $(v - u + t + S) \cap (q + S) \notin \mathcal{I}_1$, whence also

$$A = (-u + t + S) \cap (-v + q + S) = -v + \left[(v - u + t + S) \cap (q + S)\right] \notin \mathcal{I}_1. \tag{18.7.16}$$

On the other hand, since $p, q, s, t, u, v \notin U$, we have according to (18.7.5) $M[p], M[q], M[s], M[t], M[u], M[v] \in \mathcal{I}_1$, and

$$B = \left(-v + q + M[p]\right) \cup \left(-v + q + M[q]\right) \cup \left(-u + t + M[s]\right) \tag{18.7.17}$$
$$\cup \left(-u + t + M[t]\right) \cup M[u] \cup M[v] \in \mathcal{I}_1,$$

and (cf. (18.7.16) and (18.7.17)) $A \setminus B \notin \mathcal{I}_1$. In particular, $A \setminus B \neq \varnothing$, which means that there exists a $w \in A \setminus B$. Thus, since by (18.7.13) $-u + t \in S$,

$$w \in -u + t + S \subset S,$$

and with $z_1 = -t + u + w$, $z_2 = -q + v + w$, we have

$$z_1 \in S \setminus \left(M[s] \cup M[t]\right), \quad z_2 \in S \setminus \left(M[p] \cup M[q]\right). \tag{18.7.18}$$

Moreover, $w \notin M[u] \cup M[v]$, whence

$$(u, w) \notin M \quad \text{and} \quad (v, w) \notin M, \tag{18.7.19}$$

whereas by (18.7.18)

$$(s, z_1) \notin M, \ (t, z_1) \notin M, \ (p, z_2) \notin M, \ (q, z_2) \notin M. \tag{18.7.20}$$

Relation (18.7.15) yields $s + z_1 = p + z_2$, whence by (18.7.20) and (18.7.4)

$$f(s) + f(z_1) = f(s + z_1) = f(p + z_2) = f(p) + f(z_2). \tag{18.7.21}$$

On the other hand, by the definition of z_1 and z_2 we have $t + z_1 = u + w$ and $q + z_2 = v + w$, whence by (18.7.19), (18.7.20) and (18.7.4)

$$f(t) + f(z_1) = f(t + z_1) = f(u + w) = f(u) + f(w),$$

and

$$f(q) + f(z_2) = f(q + z_2) = f(v + w) = f(v) + f(w),$$

whence we get by (18.7.21)

$$f(s) - f(t) + f(u) + f(w) = f(p) - f(q) + f(v) + f(w)$$

and

$$f(p) - f(q) = f(s) - f(t) + f(u) + f(v). \tag{18.7.22}$$

According to (18.7.12), (18.7.14) and definition (18.7.11), relation (18.7.22) means

$$g(x + y) = g(x) + g(y),$$

i.e., g is a homomorphism.

Now we prove relation (18.7.10). Take an arbitrary $x \in S \setminus U$. In particular, $x \in X$, whence by (18.7.1) there exist $s, t \in S$ such that $x = s - t$. We have $U \in \mathcal{I}_1$, and, by (18.7.5), $M[x] \in \mathcal{I}_1$, whence also $U \cup M[x] \in \mathcal{I}_1$, and

$$(-s + U) \cup \big(-t + (U \cup M[x])\big) \in \mathcal{I}_1,$$

and in virtue of Lemma 17.5.5

$$\big(-s + [S \setminus U]\big) \cap \big(-t + [S \setminus (U \cup M[x])]\big)$$
$$= (-s + S) \cap (-t + S) \setminus \big[(-s + U) \cup [-t + (U \cup M[x])]\big] \notin \mathcal{I}_1. \tag{18.7.23}$$

In particular, set (18.7.23) is non-empty, and thus it contains an element y. Such a y fulfils

$$s + y \in S \setminus U, \quad t + y \in S \setminus (U \cup M[x]), \tag{18.7.24}$$

and clearly

$$(s + y) - (t + y) = s + y - y - t = s - t = x. \tag{18.7.25}$$

By definition (18.7.11)
$$g(x) = f(s + y) - f(t + y).$$ (18.7.26)

On the other hand, we get by (18.7.25) $x + (t + y) = s + y$, whereas by (18.7.24) we have $(x, t + y) \notin M$, whence by (18.7.4)

$$f(x) + f(t + y) = f(s + y),$$

i.e.,

$$f(x) = f(s + y) - f(t + y).$$ (18.7.27)

Relations (18.7.26) and (18.7.27) yield

$$g(x) = f(x),$$ (18.7.28)

and (18.7.28) holds for every $x \in X \setminus U$, i.e., $\mathcal{I}_1$-(a.e.) in S. This proves relation (18.7.10).

It remains to show the uniqueness of g. Suppose that homomorphisms $g_1, g_2 : X \to Y$ both fulfil (18.7.10). Thus there exist sets $V_1, V_2 \in \mathcal{I}_1$ such that $g_1(x) = f(x)$ for $x \in S \setminus V_1$ and $g_2(x) = f(x)$ for $x \in S \setminus V_2$. We have $V = V_1 \cup V_2 \in \mathcal{I}_1$, and

$$g_1(x) = g_2(x) = f(x) \quad \text{for } x \in S \setminus V.$$ (18.7.29)

Take arbitrary $x \in X$. By Lemma 17.5.6 there exist $s, t \in S \setminus V$ such that $x = s - t$. Since g_1 and g_2 are homomorphisms, we have according to (18.7.29)

$$g_1(x) = g_1(s - t) = g_1(s) - g_1(t) = g_2(s) - g_2(t) = g_2(s - t) = g_2(x).$$

Consequently $g_1 = g_1$ in X, which proves the uniqueness of g. $\square$

Let us observe that if in Theorem 18.7.1 we take $\mathcal{I}_1 = \{\varnothing\}$ and $\mathcal{I}_2 = \{\varnothing\}$, then we obtain Theorem 18.2.1, and if in Theorem 18.7.1 we take $S = X$, then we obtain Theorem 17.6.1. Consequently both, Theorem 18.2.1 and Theorem 17.6.1, are particular cases of Theorem 18.7.1. This indicates at the great generality (and hence also importance) of Ger's Theorem (Theorem 18.7.1 above).

18.8   Inverse additive functions

Let $(S, +)$ be a semigroup, and let $(F; +, \cdot)$ be a field. A function $f : S \to F$ is said to be *inverse additive* iff

$$\frac{1}{f(x + y)} = \frac{1}{f(x)} + \frac{1}{f(y)}$$ (18.8.1)

holds for all $x, y \in S$ such that

$$f(x + y) \neq 0, \ f(x) \neq 0, \ f(y) \neq 0.$$ (18.8.2)

In order to bring equation (18.8.1) closer to Cauchy's functional equation, we introduce the operation $\hat{\ }$ defined for functions $f : S \to F$ as follows

$$\hat{f}(x) = \begin{cases} 1/f(x) & \text{if } f(x) \neq 0, \\ 0 & \text{if } f(x) = 0, \end{cases} \quad x \in S.$$

Thus $\hat{f}$ again is a function $\hat{f} : S \to F$, and the following lemma is obvious.

Lemma 18.8.1. *Let $(S, +)$ be a semigroup, let $(F; +, \cdot)$ be a field. The function $f_0 : S \to F$ is inverse additive if and only if the function $\hat{f}_0$ satisfies the Cauchy equation*

$$f(x + y) = f(x) + f(y) \tag{18.8.3}$$

for all $x, y \in S$ such that f_0 fulfils (18.8.2).

The passage back from $\hat{f}_0$ to f_0 is well possible, because the operation $\,\hat{}\,$ is involutory: $\hat{\hat{f}} = f$. In other words, if $g = \hat{f}$, then $f = \hat{g}$.

The problem of solving equation (18.8.1) under condition (18.8.2) was raised by J. Aczél [6], and was dealt with by R. Ger-M. Kuczma [116], and R. Ger [112]. Equation (18.8.1) (or, equivalently, (18.8.3)) with condition (18.8.2) cannot be solved unless we make some assumptions about the set

$$Z = f^{-1}(0) = \{x \in S \mid f(x) = 0\}. \tag{18.8.4}$$

In fact, if the set (18.8.4) is very large, then condition (18.8.2) is very seldom (possibly never) fulfilled, and equation (18.8.1) furnishes very few, or even no, information about the behaviour of the function f. This is well seen from the following example (Ger-Kuczma [116]):

Example 18.8.1. Let $S = F = \mathbb{R}$ (with the ordinary operations of addition and multiplication). We define by induction a sequence of intervals $J_n = (a_n, b_n) \subset \mathbb{R}$. We put $J_1 = (a_1, b_1) = (1, 2)$, and we define the sequences $\{a_n\}$ and $\{b_n\}$ by recurrence as follows

$$a_{n+1} = 2b_n, \quad b_{n+1} = 2b_n + 1, \quad n \in \mathbb{N}.$$

We clearly have

$$1 = a_1 < b_1 < a_2 < b_2 < \cdots < a_n < b_n < a_{n+1} < b_{n+1}, \ldots, n \in \mathbb{N},$$

and

$$b_n = a_n + 1, \quad n \in \mathbb{N}.$$

Put

$$U = \bigcup_{n=1}^{\infty} J_n,$$

and take arbitrary $x, y \in U$. There exist $k, m \in \mathbb{N}$ such that $x \in J_k$, $y \in J_m$, and let, e.g., $k \leqslant m$. We have

$$a_k < x < b_k, \quad a_m < y < b_m,$$

whence

$$b_m = a_m + 1 \leqslant a_m + a_k < x + y < b_m + b_k \leqslant 2b_m = a_{m+1}.$$

Consequently, $b_m < x + y < a_{m+1}$, which means that $x + y$ does not belong to any interval J_n, and hence does not belong to U. Consequently the set U has the property

$$x, y \in U \text{ implies } x + y \notin U. \tag{18.8.5}$$

Define a function $f :\to \mathbb{R}$ as

$$f(x) = \begin{cases} 0 & \text{for } x \in \mathbb{R} \setminus U, \\ \text{arbitrary} & \text{for } x \in U. \end{cases} \tag{18.8.6}$$

According to (18.8.5), condition (18.8.2) cannot be fulfilled for any $x, y \in \mathbb{R}$, and thus function (18.8.6) is inverse additive. Other examples are found in Ger-Kuczma [116].

Instead of equation (18.8.1), we consider first equation (18.8.3) under condition (18.8.2). Of course, (18.8.3) with (18.8.2) is a conditional Cauchy equation (cf. the end of 13.6). Actually, now we do not need the operation of multiplication in the range of f, and so we may consider equation (18.8.3) with (18.8.2) for functions $f : S \to Y$, where $(Y, +)$ is an arbitrary group.

Lemma 18.8.2. *Let $(S, +)$ be a semigroup, let $(Y, +)$ be a group, and let $Z \subset S$ be an arbitrary set. Further, let $h : S \to Y$ be a homomorphism. Then the function*

$$f(x) = \begin{cases} 0 & \text{for } x \in Z, \\ h(x) & \text{for } x \in S \setminus Z, \end{cases} \tag{18.8.7}$$

satisfies equation (18.8.3) with condition (18.8.2).

Proof. Take arbitrary $x, y \in S$ fulfilling (18.8.2). In view of (18.8.7) this means that $x, y, x + y \in S \setminus Z$, and consequently $f(x) = h(x)$, $f(y) = h(y)$, $f(x + y) = h(x + y)$. Now (18.8.3) results from the fact that h is a homomorphism.   □

In order to prove a result in the converse direction we will need the following (Ger [112]).

Lemma 18.8.3. *Let $(X, +)$ be a group, and let $(S, +)$ be a subsemigroup of $(X, +)$ such that (18.7.1) holds. Suppose that a set $Z \subset S$ satisfies the condition*

for every $k \in \mathbb{N}$ and for every $s, s_1, \ldots, s_k, t_1, \ldots, t_k \in S$ there   (18.8.8)
exists a $t \in S + s$ such that $t_i + t \notin Z + s_i$, $s_i \notin Z + t_i + t$, $i = 1, \ldots, k$.

Then the linearly invariant ideal $\mathcal{I}(Z)$ generated[9] by Z is proper. More exactly,

$$S \notin \mathcal{I}(Z). \tag{18.8.9}$$

Proof. For an indirect proof suppose that $S \in \mathcal{I}(Z)$, i.e., there exist a $k \in \mathbb{N}$, and $x_1, \ldots, x_k, y_1, \ldots, y_k \in X$ such that

$$S \subset \bigcup_{i=1}^{k} [x_i + (Z \cup (-Z)) + y_i].$$

According to (18.7.1) to every $x \in Z$ there exists an $s' \in S$ such that $z + s' \in S$ (cf. the proof of Lemma 18.2.1, condition (i)). We define $s'_i \in S$, $i = 1, \ldots, k$, as

---
[9] Cf. 17.5

follows: $s_1' \in S$ is such that $y_1 + s_1' \in S$, and if we know $s_1', \ldots, s_m'$, $1 \leqslant m < k$, then $s_{m+1}' \in S$ is such that $(y_{m+1} + s_1' + \cdots + s_m') + s_{m+1}' \in S$. Thus

$$y_i + \sum_{j=1}^{i} s_j' \in S, \quad i = 1, \ldots, k,$$

and with $s = s_1' + \cdots + s_k' \in S$, also

$$s_i = y_i + s = \left(y_i + \sum_{j=1}^{i} s_j' \right) + \sum_{j=i+1}^{k} s_i' \in S, \quad i = 1, \ldots, k.$$

We have

$$S + s \subset \bigcup_{i=1}^{k} [x_i + (Z \cup (-Z)) + s_i],$$

whence

$$-s - S \subset \bigcup_{i=1}^{k} [-s_i + (Z \cup (-Z)) - x_i]. \qquad (18.8.10)$$

Repeating the above construction, we may find a $\tilde{t} \in S$ such that

$$t_i = -x_i + \tilde{t} \in S, \quad i = 1, \ldots, k,$$

whence by (18.8.10)

$$-s - S + \tilde{t} \subset \bigcup_{i=1}^{k} [-s_i + (Z \cup (-Z)) + t_i],$$

or

$$-\tilde{t} + S + s \subset \bigcup_{i=1}^{k} [-t_i + (Z \cup (-Z)) + s_i]. \qquad (18.8.11)$$

Now take an arbitrary $u \in S$. Then $v = \tilde{t} + u \in S$, whence $u = -\tilde{t} + v \in -\tilde{t} + S$. Consequently $S \subset -\tilde{t} + S$, and $S + s \subset -\tilde{t} + S + s$, and by (18.8.11)

$$S + s \subset \bigcup_{i=1}^{k} [-t_i + (Z \cup (-Z)) + s_i].$$

Therefore for every $t \in S + s$ there exists an $i \in \mathbb{N}$, $1 \leqslant i \leqslant k$, such that

$$t \in -t_i + Z + s_i \quad \text{or} \quad t \in -t_i - Z + s_i,$$

i.e.,

$$t_i + t \in Z + s_i \quad \text{or} \quad s_i \in Z + t_i + t.$$

But it contradicts (18.8.8). □

Corollary 18.8.1. *Let $(X, +)$ be a group, and let $(S, +)$ be a subsemigroup of $(X, +)$ such that (18.7.1) holds. Let $Z \subset S$ be an arbitrary set. Then conditions (18.8.8) and (18.8.9) are equivalent.*

Proof. That (18.8.8) implies (18.8.9) is the contents of Lemma 18.8.3. Conversely, suppose that (18.8.9) holds, and take arbitrary $s, s_1, \ldots, s_k, t_1, \ldots, t_k \in S$. The set $\bigcup\limits_{i=1}^{k} \left[-t_i + (Z \cup (-Z)) + s_i \right]$ belong to $\mathcal{I}(Z)$, and since by (18.8.9) $S + s \notin \mathcal{I}(Z)$, there exists a t,

$$t \in (S + s) \setminus \bigcup_{i=1}^{k} [-t_i + (Z \cup (-Z)) + s_i].$$

This means that $t \in S+s$, and $t_i+t \notin Z+s_i$, $s_i \notin Z+t_i+t$, $i = 1, \ldots, k$. Consequently condition (18.8.8) is fulfilled. $\square$

The importance of condition (18.8.8) lies in that it is expressed in terms of S only, and it is meaningful even if we do not know the group $(X, +)$, and hence also the ideal $\mathcal{I}(Z)$. An example of verifying condition (18.8.8) may be found in Ger [112].

Theorem 18.8.1. *Let $(S, +)$ be a left reversible cancellative semigroup, and let $(Y, +)$ be a group. Suppose that a function $f : S \to Y$ satisfies (18.8.3) for $x, y \in S$ fulfilling (18.8.2). If the set (18.8.4) fulfils condition (18.8.8), then there exists a homomorphism $h : S \to Y$ such that (18.8.7) holds.*

Proof. By Theorem 4.5.2 there exists a group $(X, +)$ such that $(S, +)$ is a subsemigroup of $(X, +)$, and condition (18.7.1) holds. By Lemma 18.8.3 condition (18.8.9) holds. Thus $\mathcal{I}_1 = \mathcal{I}(Z)$ is a p.l.i. ideal in X fulfilling (18.7.2). Put

$$M = \{(x, y) \in S \times S \mid x \in Z \text{ or } y \in Z \text{ or } x + y \in Z\}$$
$$= (Z \times S) \cup (S \times Z) \cup \{(x, y) \in S \times S \mid x + y \in Z\}.$$

Clearly $Z \times S \in \Omega(\mathcal{I}_1)$ and $S \times Z \in \Omega(\mathcal{I}_1)$, because we have

$$(Z \times S)[x] = \begin{cases} S & \text{for } x \in Z \in \mathcal{I}_1, \\ \varnothing \in \mathcal{I}_1 & \text{otherwise,} \end{cases}$$

and

$$(S \times Z)[x] = \begin{cases} Z \in \mathcal{I}_1 & \text{for } x \in S, \\ \varnothing \in \mathcal{I}_1 & \text{otherwise,} \end{cases}$$

so that $(Z \times S)[x] \in \mathcal{I}_1$ $\mathcal{I}_1$-(a.e.) in X, and $(S \times Z)[x] \in \mathcal{I}_1$ for all $x \in X$. By Lemma 17.5.4 $\{(x, y) \in S \times S \mid x + y \in Z\} \subset \{(x, y) \in X \times X \mid x + y \in Z\} \in \Omega(\mathcal{I}_1)$, whence $\{(x, y) \in S \times S \mid x + y \in Z\} \in \Omega(\mathcal{I}_1)$ and $M \in \Omega(\mathcal{I}_1)$. Write $\mathcal{I}_2 = \Omega(\mathcal{I}_1)$. By Lemma 17.5.3 $\mathcal{I}_1$ and $\mathcal{I}_2$ are conjugate p.l.i. ideals.

Now, if $(x, y) \in (S \times S) \setminus M$, then $x \notin Z, y \notin Z$, and $x+y \notin Z$, which means that condition (18.8.2) is fulfilled, and (18.8.3) holds. Consequently the function f satisfies equation (18.8.3) for all $(x, y) \in (S \times S) \setminus M$, i.e., $\mathcal{I}_2$-(a.e.) in $S \times S$. By Theorem 18.7.1 there exists a unique homomorphism $g : X \to Y$ such that (18.7.10) holds. Write

$$E = \{x \in S \mid f(x) \neq g(x)\}.$$

By (18.7.10) $E \in \mathcal{I}_1$, whence also $T = E \cup Z \in \mathcal{I}_1$. By (18.7.1) and Lemma 17.5.6 we have

$$X = (S \setminus T) - (S \setminus T). \tag{18.8.12}$$

Take an arbitrary $x \in S \setminus Z \subset X$. By (18.8.12) there exist $u, v \in S \setminus T$ such that $x = u - v$. Thus $u = x + v$, and since $u, v, x \notin Z$, we have $f(u) \neq 0$, $f(v) \neq 0$, $f(x) \neq 0$, and by (18.8.3)

$$f(u) = f(x + v) = f(x) + f(v),$$

whence

$$f(x) = f(u) - f(v). \tag{18.8.13}$$

Since also $u, v \notin E$, we have $f(u) = g(u)$, $f(v) = g(v)$, and by (18.8.13)

$$f(x) = g(u) - g(v) = g(u - v) = g(x),$$

since g is a homomorphism. Consequently

$$f(x) = g(x) \quad \text{for } x \in S \setminus Z. \tag{18.8.14}$$

Now, if we put $h = g \mid S$, then $h : S \to Y$ is a homomorphism, and (18.8.7) is a consequence of (18.8.4) and (18.8.14). $\qquad\square$

Theorem 18.8.2. *Let $(S, +)$ be a left reversible and cancellative semigroup, and let $(F; +, \cdot)$ be a field. Let $f : S \to F$ be an inverse additive function. If the set (18.8.4) fulfils condition (18.8.8), then there exists a homomorphism[10] $h : S \to F$ such that*

$$f(x) = \begin{cases} 0 & \text{for } x \in Z, \\ \dfrac{1}{h(x)} & \text{for } x \in S \setminus Z. \end{cases}$$

Proof. This results immediately from Theorem 18.8.1 and Lemma 18.8.1. $\qquad\square$

18.9 Concluding remarks

Even in so large a book it was impossible even to touch upon, leave alone to discuss more thoroughly, many topics connected with the Cauchy equation, or convex functions. So we did not considered extensions to more general spaces like, e.g., topological vector spaces, since otherwise we would have to include the whole functional analysis. Cauchy's equation on restricted domains (conditional Cauchy equations) have only perfunctorily been mentioned. The interested reader will find further informations and references in Dhombres [68], Dhombres-Ger [69], [70], Kuczma [187], L. Paganoni-S. Paganoni Marzegalli [252].

Also of various generalizations of convex functions only p-convex functions have been discussed in the present book. Concerning may other generalizations consult the references quoted in 5.3. In connection with convex functions cf. also the monographs

---

[10] When we speak about a homomorphism $h : S \to F$, we have in mind only additive structure of F, i.e., F is considered as a group $(F, +)$.

Rockafellar [268] and Roberts-Varberg [267] as well as the article Beckenbach [22], and the numerous references given therein.

Unfortunately, the lack of place detained us from discussing here extremely interesting in generalizations of additive functions and convex functions to stochastic processes. Cf. Nagy [235] and Nikodem [238].

Also set-valued additive functions and convex functions (Fifer [79], Godini [118], Henney [140], Nikodem [239], [240], [242], [241], Rådström [261]) have not found place in this book.

A generalization of additive functions to the case of relations is due to Száz-Száz [308]. Another generalization of additive functions was dealt with by Aleksandrov [12] and Forti [86], [87].

There may exist further results pertinent to the material dealt with in this book, which has been consciously or unconsciously omitted. We suspect that if we wanted to include here absolutely everything related to additive functions and convex functions, this book would have never been finished, because new results would arise faster than we could proceed with writing. So we must have stopped at some moment.

This moment has just come, and we trust that, imperfect as it is, this book will prove useful to many readers. If such is the case, our task will be accomplished.

Exercises

1. Let X be the set of all differentiable functions $\varphi : \mathbb{R} \to \mathbb{R}$ such that $\varphi(\mathbb{R}) = \mathbb{R}$, $\varphi(0) = 0$ and $\varphi'(x) > 0$ for all $x \in \mathbb{R}$. Let $S = \{\varphi \in X \mid \varphi'(0) > 1\}$. With $\circ$ denoting the composition of functions, show that $(X, \circ)$ is a group and $(S, \circ)$ is its subsemigroup. Let $g : S \to \mathbb{R} \setminus \{0\}$ (where $\mathbb{R} \setminus \{0\}$ is considered as a multiplicative group) be given by $g(\varphi) = \varphi'(0)$. Show that $f(\varphi) = \varphi'(0)$, $\varphi \in X$, defines the unique homomorphism $f : X \to \mathbb{R} \setminus \{0\}$ such that $f \mid S = g$.

2. Let X be the set of all non-singular $n \times n$ matrices, n odd, and let $\cdot$ denote the usual multiplication of matrices. Then $(X, \cdot)$ is a group. Put $S = \{A \in X \mid \det A > 0\}$. Then $(S, \cdot)$ is a subgroup of $(X, \cdot)$. Let $g : S \to \mathbb{R} \setminus \{0\}$ (where $\mathbb{R} \setminus \{0\}$ is considered as a multiplicative group) be given by $g(A) = \gamma(\det A)$, where $\gamma : (0, \infty) \to (0, \infty)$ is a function such that $\gamma(uv) = \gamma(u)\gamma(v)$ for all $u, v \in (0, \infty)$. Using Theorem 18.4.3 show that there exists a homomorphism $f : X \to \mathbb{R} \setminus \{0\}$ such that $f \mid S = g$. Show that there exist exactly two such homomorphisms.

3. Let $X = \{(a, b, c, d) \in \mathbb{Q}^4 \mid ac \neq 0\}$, and define a binary operation $\cdot$ in X by $(a, b, c, d) \cdot (a', b', c', d') = (aa', ab' + b, cc', dc' + d')$. Then $(X, \cdot)$ is a group. Put $S = \{(a, b, c, d) \in X \mid a, b, c, d \in \mathbb{Z}\}$. Then $(S, \cdot)$ is a subsemigroup of $(X, \cdot)$. Show that $\left(1, \dfrac{1}{2}, 1, 0\right) \notin S \cdot S^{-1}$ (and so $X \neq S \cdot S^{-1}$), $\left(1, 0, 1, \dfrac{1}{2}\right) \notin S^{-1} \cdot S$ (and so $X \neq S^{-1} \cdot S$), but $(1, 0, u, v) \cdot (z, 0, w, 0)^{-1} \cdot (x, y, 1, 0) = \left(\dfrac{x}{z}, \dfrac{y}{z}, \dfrac{u}{w}, \dfrac{v}{w}\right)$ for all $x, y, z, u, v, w \in \mathbb{Z}$ such that $xzuw \neq 0$ (and so $X = S \cdot S^{-1} \cdot S$) (Benz [27]).

Bibliography

[1] A. Abian, *An example of a nonmeasurable set*, Boll. Un. Mat. Ital. (4) **1** (1968), 366–368.

[2] J. Aczél, *Über eine Klasse von Funktionalgleichungen*, Comment. Math. Helv. **21** (1948), 247–252.

[3] J. Aczél, *Miszellen über Funktionalgleichungen. I*, Math. Nachr. **19** (1958), 87–99.

[4] J. Aczél, *The general solution of two functional equations by reduction to functions additive in two variables and with the aid of Hamel bases*, Glasnik Mat.-Fiz. Astronom. Ser. II Društvo Mat. Fiz. Hrvatske **20** (1965), 65–73.

[5] J. Aczél, *Lectures on Functional Equations and Their Applications*, Mathematics in Science and Engineering, vol. 19, Academic Press, New York–London, 1966.

[6] J. Aczél, *Problem (P141)*, Aequationes Math. **12** (1975), 303.

[7] J. Aczél, *Remark (P178R3)*, Aequationes Math. **19** (1979), 286.

[8] J. Aczél, J. A. Baker, D. Ž. Djoković, Pl. Kannappan, and F. Radó, *Extensions of certain homomorphisms of subsemigroups to homomorphisms of groups*, Aequationes Math. **6** (1971), 263–271.

[9] J. Aczél and P. Erdős, *The non-existence of a Hamel-basis and the general solution of Cauchy's functional equation for non-negative numbers*, Publ. Math. Debrecen **12** (1965), 253–263.

[10] J. Aczél and P. Erdős, *Problem (P37)*, Aequationes Math. **2** (1969), 378.

[11] M. Albert and J. A. Baker, *Functions with bounded nth differences*, Ann. Polon. Math. **43** (1983), no. 1, 93–103.

[12] A. D. Aleksandrov, *A certain generalization of the functional equation $f(x + y) = f(x) + f(y)$.*, Sibirsk. Mat. Ž. **11** (1970), 264–278.

[13] A. Alexiewicz and W. Orlicz, *Remarque sur l'équation fonctionelle $f(x + y) = f(x) + f(y)$*, Fund. Math. **33** (1945), 314–315.

[14] J. Anastassiadis, *Fonctions semi-monotones et semi-convexes et solutions d'une équation fonctionnelle*, Bull. Sci. Math. (2) **76** (1952), 148–160.

[15] J. A. Baker, *The stability of the cosine equation*, Proc. Amer. Math. Soc. **80** (1980), no. 3, 411–416.

[16] J. A. Baker, J. Lawrence, and F. Zorzitto, *The stability of the equation $f(x + y) = f(x)f(y)$*, Proc. Amer. Math. Soc. **74** (1979), no. 2, 242–246.

[17] J. A. Baker and S. L. Segal, *On a problem of Kemperman concerning Hamel functions*, Aequationes Math. **2** (1968), 114–115.

[18] S. Balcerzyk, *Wstęp do algebry homologicznej*, Biblioteka Matematyczna, Tom 34. [Mathematical Library, Vol. 34]. Państwowe Wydawnictwo Naukowe, Warsaw, 1970.

[19] S. Banach, *Sur l'équation fonctionnelle* $f(x+y) = f(x) + f(y)$, Fund. Math. **1** (1920), 123–124.

[20] K. Baron and R. Ger, *On Mikusiński-Pexider functional equation*, Colloq. Math. **28** (1973), 307–312.

[21] E. F. Beckenbach, *Generalized convex functions*, Bull. Amer. Math. Soc. **43** (1937), no. 6, 363–371.

[22] E. F. Beckenbach, *Convex functions*, Bull. Amer. Math. Soc. **54** (1948), 439–460.

[23] E. F. Beckenbach, *Convexity* (unpublished).

[24] E. F. Beckenbach and R. Bellman, *Inequalities*, 2nd rev. ed., Springer Verlag, Ergebnisse der Mathematik und ihrer Grenzgebiete. Neue Folge, Band 30, New York, 1965.

[25] E. F. Beckenbach and R. H. Bing, *On generalized convex functions*, Trans. Amer. Math. Soc. **58** (1945), 220–230.

[26] W. Benz, *Remark (P178R1)*, Aequationes Math. **20** (1980), 304.

[27] W. Benz, *Remark (P190S1)*, Aequationes Math. **20** (1980), 305.

[28] B. Bereanu, *Partial monotonicity and (J)-convexity*, Rev. Roumaine Math. Pures Appl. **14** (1969), 1085–1087.

[29] G. M. Bergman, *P178S1*, Aequationes Math. **23** (1981), 312–313.

[30] F. Bernstein and G. Doetsch, *Zur Theorie der konvexen Funktionen*, Math. Ann. **76** (1915), no. 4, 514–526.

[31] E. Berz, *Sublinear functions on R*, Aequationes Math. **12** (1975), no. 2/3, 200–206.

[32] D. Blanuša, *The functional equation* $f(x+y-xy) + f(xy) = f(x) + f(y)$, Aequationes Math. **5** (1970), 63–67.

[33] H. Blumberg, *On convex functions*, Trans. Amer. Math. Soc. **20** (1919), no. 1, 40–44.

[34] Ralph P. Boas, Jr., *The Jensen-Steffensen inequality*, Univ. Beograd. Publ. Elektrotehn. Fak. Ser. Mat. Fiz. (1970), no. 302-319, 1–8.

[35] T. Bonnesen and W. Fenchel, *Theorie der konvexen Körper*, Berlin, 1934.

[36] É. Borel, *Éléments de la Théorie des Ensembles*, Éditions Albin Michel, Paris, 1949.

[37] N. Bourbaki, *Élements de mathématique. XII. Premierère partie: Les structures fondamentales de l'analyse. Livre IV: Fonctions d'une variable réelle. (Théorie élémentaire). Chapitre IV: Équations différentielles. Chapitre V: Étude locale des fonctions. Chapitre VI: Développements tayloriens généralisés; formule sommatoire d'Euler-Maclaurin. Chapitre VII: La fonction gamma*, Actualités Sci. Ind., no. 1132, Hermann et Cie., Paris, 1951.

[38] A. M. Bruckner, J. G. Ceder, and M. Weiss, *Uniform limits of Darboux functions*, Colloq. Math. **15** (1966), 65–77.

[39] C. Burstin, *Die Spaltung des Kontinuums in c im L. Sinne ninchtmessbare Mengen*, Sitzungsber. Akad. Wiss. Wien, Math. nat. Klasse, Abt. IIa **125** (1916), 209–217.

[40] C. Carathéodory, *Vorlesungen über reelle Funktionen*, Leipzig-Berlin, 1918.

[41] A.-L. Cauchy, *Cours d'analyse de l'École Royale Polytechnique.*, Première Partie, Analyse algébrique, Paris, 1821 [Oeuvres (2) **3**, Paris, 1897].

[42] P. W. Cholewa, *The stability of the sine equation*, Proc. Amer. Math. Soc. **88** (1983), no. 4, 631–634.

[43] P. W. Cholewa, *Remarks on the stability of functional equations*, Aequationes Math. **27** (1984), no. 1-2, 76–86.

[44] P. W. Cholewa, *The stability problem for a generalized Cauchy type functional equation*, Rev. Roumaine Math. Pures Appl. **29** (1984), no. 6, 457–460.

[45] Z. Ciesielski, *A note on some inequalities of Jensen's type*, Ann. Polon. Math. **4** (1958), 269–274.

[46] Z. Ciesielski, *Some properties of convex functions of higher orders*, Ann. Polon. Math. **7** (1959), 1–7.

[47] A. H. Clifford and G. B. Preston, *The algebraic theory of semigroups. Vol. I*, Mathematical Surveys, No. 7, American Mathematical Society, Providence, R.I., 1961.

[48] P. J. Cohen, *The independence of the continuum hypothesis. II*, Proc. Nat. Acad. Sci. U.S.A. **51** (1964), 105–110.

[49] R. Cooper, *The converse of the Cauchy–Hölder inequality and the solutions of the inequality $g(x + y) \leq g(x) + g(y)$*, Proc. London Math. Soc. (2) **26** (1927), 415–432.

[50] Á. Császár, *Sur les ensembles et les fonctions convexes* (in Hungarian), Mat. Lapok **9** (1958), 273–282.

[51] S. Czerwik, *C-convex solutions of a linear functional equation*, Uniw. Śląski w Katowicach Prace Naukowe, No. 87 Prace Mat. No. 6, (1975), 49–53.

[52] K. Dankiewicz and Z. Moszner, *Prolongements des homomorphismes et des solutions de l'équation de translation*, Rocznik Naukowo-Dydaktyczny WSP w Karkowie, Prace Mat. **10** (1982), 27–44.

[53] G. Darboux, *Sur la composition des forces en statique*, Bull. Sci. Math. (1) **9** (1875), 281 288.

[54] Z. Daróczy, *Notwendige und hinreichende Bedingungen für die Existenz von nichtkonstanten Lösungen linearer Funktionalgleichungen*, Acta Sci. Math. Szeged **22** (1961), 31–41.

[55] Z. Daróczy, *On a class of bilinear functional equations* (in Hungarian), Mat. Lapok **15** (1964), 52–86.

[56] Z. Daróczy, *Über die Funktionalgleichung $f(xy) + f(x + y - xy) = f(x) + f(y)$*, Publ. Math. Debrecen **16** (1969), 129–132.

[57] Z. Daróczy, *On the general solution of the functional equation $f(x + y - xy) + f(xy) = f(x) + f(y)$*, Aequationes Math. **6** (1971), 130–132.

[58] Z. Daróczy and K. Győry, *Die Cauchysche Funktionalgleichung über diskrete Mengen*, Publ. Math. Debrecen **13** (1900), 249–255.

[59] Z. Daróczy and L. Losonczi, *Über die Erweiterung der auf einer Punktmenge additiven Funktionen*, Publ. Math. Debrecen **14** (1967), 239–245.

[60] T. M. K. Davison, *The complete solution of Hosszú's functional equation over a field*, Aequationes Math. **11** (1974), 273–276.

[61] T. M. K. Davison, *On the functional equation $f(m + n - mn) + f(mn) = f(m) + f(n)$*, Aequationes Math. **11** (1974), 206–211.

[62] T. M. K. Davison, *On Hosszú's functional equation*, Mathematics Institute, University of Warwick, Coventry, England, April 1976.

[63] T. M. K. Davison and L. Redlin, *Hosszú's functional equation over rings generated by their units*, Aequationes Math. **21** (1980), no. 2-3, 121–128.

[64] N. G. de Bruijn, *On almost additive functions*, Colloq. Math. **15** (1966), 59–63.

[65] E. Deák, *Über konvexe und interne Funktionen, sowie eine gemeinsame Verallgemeinerung von beiden*, Ann. Univ. Sci. Budapest. Eötvös Sect. Math. **5** (1962), 109–154.

[66] J. L. Denny, *Sufficient conditions for a family of probabilities to be exponential*, Proc. Nat. Acad. Sci. U.S.A. **57** (1967), 1184–1187.

[67] J. L. Denny, *Cauchy's equation and sufficient statistics on arcwise connected spaces*, Ann. Math. Statist. **41** (1970), 401–411.

[68] J. Dhombres, *Some aspects of functional equations*, With a Thai preface by Wirun Bu nsambatí, Chulalongkorn University, Department of Mathematics, Bangkok, 1979, Lecture Notes.

[69] J. Dhombres and R. Ger, *Équations de Cauchy conditionnelles*, C. R. Acad. Sci. Paris Sér. A-B **280** (1975), A513–A515.

[70] J. G. Dhombres and R. Ger, *Conditional Cauchy equations*, Glas. Mat. Ser. III **13(33)** (1978), no. 1, 39–62.

[71] D. Ž. Djoković, *A representation theorem for $(X_1 - 1)(X_2 - 1) \cdots (X_n - 1)$ and its applications*, Ann. Polon. Math. **22** (1969/1970), 189–198.

[72] L. Dubikajtis, C. Ferens, R. Ger, and M. Kuczma, *On Mikusiński's functional equation*, Ann. Polon. Math. **28** (1973), 39–47.

[73] P. Dubreil, *Sur les problèmes d'immersion et la théorie des modules*, C. R. Acad. Sci. Paris **216** (1943), 625–627.

[74] H. G. Eggleston, *Convexity*, Cambridge Tracts in Mathematics and Mathematical Physics 47, Cambridge University Press, New York, 1958.

[75] P. Erdős, *P 310*, Colloq. Math. **7** (1960), 311.

[76] P. Erdős, *On some properties of Hamel bases*, Colloq. Math. **10** (1963), 267–269.

[77] P. Erdős and S. Marcus, *Sur la décomposition de l'espace euclidien en ensembles homogènes*, Acta Math. Acad. Sci. Hungar **8** (1957), 443–452.

[78] I. Fenyő, *On the general solution of a functional equation in the domain of distributions*, Aequationes Math. **3** (1969), 236–246.

[79] Z. Fifer, *Set-valued Jensen functional equation*, Rev. Roumaine Math. Pures Appl. **31** (1986), no. 4, 297–302.

[80] T. Figiel, *The functional equation $f(x + y) = f(x) + f(y)$* (in Polish), Wiadom. Mat. (2) **11** (1969), 15–18 (1969).

[81] P. Fischer, *Sur l'équivalence des équations fonctionnelles $f(x + y) = f(x) + f(y)$ et $f^2(x+y) = [f(x) + f(y)]^2$*, Ann. Fac. Sci. Univ. Toulouse (4) **30** (1966), 71–74 (1968).

[82] P. Fischer, *Problème*, Aequationes Math. **1** (1968), 300.

[83] P. Fischer and Gy. Muszély, *Generalizations of a certain kind for Cauchy's functional equation* (in Hungarian), Mat. Lapok **16** (1965), 67–75.

[84] P. Fischer and Gy. Muszély, *On some new generalizations of the functional equation of Cauchy*, Canad. Math. Bull. **10** (1967), 197–205.

[85] P. Fischer and Z. Słodkowski, *Christensen zero sets and measurable convex functions*, Proc. Amer. Math. Soc. **79** (1980), no. 3, 449–453.

[86] G. L. Forti, *On the functional equation $f(L+x) = f(L)+f(x)$*, Istit. Lombardo Accad. Sci. Lett. Rend. A **111** (1977), no. 2, 296–302 (1978).

[87] G. L. Forti, *Bounded solutions with zeros of the functional equation $f(L+x) = f(L)+ f(x)$*, Boll. Un. Mat. Ital. A (5) **15** (1978), no. 1, 248–256.

[88] G. L. Forti, *An existence and stability theorem for a class of functional equations*, Stochastica **4** (1980), no. 1, 23–30.

[89] G. L. Forti, *On an alternative functional equation related to the Cauchy equation*, Aequationes Math. **24** (1982), no. 2-3, 195–206.

[90] M. Fréchet, *Pri la fukncia equacio $f(x + y) = f(x) + f(y)$*, Enseignement Math. **15** (1913), 390–393.

[91] M. Fréchet, *A propos d'un article sur l'équation fonctionnelle $f(x + y) = f(x) + f(y)$*, Enseignement Math. **16** (1914), 136.

[92] M. Fréchet, *Les polynômes abstraits*, J. Math. Pures Appl. (9) **8** (1929), 71–92.

[93] L. Fuchs, *A new proof of an inequality of Hardy-Littlewood-Pólya*, Mat. Tidsskr. B. **1947** (1947), 53–54.

[94] Z. Gajda, *On some properties of Hamel bases connected with the continuity of polynomial functions*, Aequationes Math. **27** (1984), no. 1-2, 57–75.

[95] Z. Gajda, *Christensen measurability of polynomial functions and convex functions of higher orders*, Ann. Polon. Math. **47** (1986), no. 1, 25–40.

[96] C. F. Gauss, *Theoria motus corporum coelestium*, Hamburg, 1809 [Werke VII, Leipzig, 1906].

[97] R. Ger, *Some remarks on convex functions*, Fund. Math. **66** (1969/1970), 255–262.

[98] R. Ger, *Some new conditions of continuity of convex functions*, Mathematica (Cluj) **12(35)** (1970), 271–277.

[99] R. Ger, *On almost polynomial functions*, Colloq. Math. **24** (1971/72), 95–101.

[100] R. Ger, *On some properties of polynomial functions*, Ann. Polon. Math. **25** (1971/72), 195–203.

[101] R. Ger, *Convex functions of higher orders in Euclidean spaces*, Ann. Polon. Math. **25** (1971/72), 293–302.

[102] R. Ger, *Note on convex functions bounded on regular hypersurfaces*, Demonstratio Math. **6** (1973), 97–103, Collection of articles dedicated to Stanisław Gołąb on his 70th birthday, I.

[103] R. Ger, *Thin sets and convex functions*, Bull. Acad. Polon. Sci. Sér. Sci. Math. Astronom Phys. **21** (1973), 413–416.

[104] R. Ger, *n-convex functions in linear spaces*, Aequationes Math. **10** (1974), 172–176.

[105] R. Ger, *On some functional equations with a restricted domain, I.; II.*, Fund. Math. **89**, (1975), 131–149.; **98**, (1978), no. 3, 250–272.

[106] R. Ger, *Certain functional equations with a restricted domain* (in Polish), Uniw. Śląski w Katowicach Prace Naukowe, no. 132, (1976), 36.

[107] R. Ger, *On a method of solving of conditional Cauchy equations*, Univ. Beograd. Publ. Elektrotehn. Fak. Ser. Mat. Fiz., no. 544-576, (1976), 159–165.

[108] R. Ger, *Functional equations with a restricted domain*, Rend. Sem. Mat. Fis. Milano **47** (1977), 175–184.

[109] R. Ger, *On an alternative functional equation*, Aequationes Math. **15** (1977), no. 2-3, 145–162.

[110] R. Ger, *Almost subadditive functions*, General Inequalities, 1 (Oberwolfach, 1976) (E. F. Beckenbach, ed.), International Series of Numerical Mathematics, vol. 41, Birkhäuser, Basel, 1978, pp. 159–167.

[111] R. Ger, *Note on almost additive functions*, Aequationes Math. **17** (1978), no. 1, 73–76.

[112] R. Ger, *Almost additive functions on semigroups and a functional equation*, Publ. Math. Debrecen **26** (1979), no. 3-4, 219–228.

[113] R. Ger, *Homogeneity sets for Jensen-convex functions*, General Inequalities, 2 (Oberwolfach, 1978) (E. F. Beckenbach, ed.), International Series of Numerical Mathematics, vol. 47, Birkhäuser, Basel, 1980, pp. 193–201.

[114] R. Ger, *Almost approximately additive mappings*, General Inequalities, 3 (Oberwolfach, 1981) (E. F. Beckenbach and W. Walter, eds.), International Series of Numerical Mathematics, vol. 64, Birkhäuser, Basel, 1983, pp. 263–276.

[115] R. Ger and M. Kuczma, *On the boundedness and continuity of convex functions and additive functions*, Aequationes Math. **4** (1970), 157–162.

[116] R. Ger and M. Kuczma, *On inverse additive functions*, Boll. Un. Mat. Ital. (4) **11** (1975), no. 3, 490–495.

[117] E. Głowacki and M. Kuczma, *Some remarks on Hosszú's functional equation on integers*, Uniw. Śląski w Katowicach Prace Nauk.-Prace Mat. no. 9, (1979), 53–63.

[118] G. Godini, *Set-valued Cauchy functional equation*, Rev. Roumaine Math. Pures Appl. **20** (1975), no. 10, 1113–1121.

[119] S. Gołąb and L. Losonczi, *Über die Funktionalgleichung der Funktion Arccosinus. I. Die lokalen Lösungen*, Publ. Math. Debrecen **12** (1965), 159–174.

[120] J. W. Green and W. Gustin, *Quasiconvex sets*, Canadian J. Math. **2** (1950), 489–507.

[121] A. Grząślewicz, *On extensions of homomorphisms*, Aequationes Math. **17** (1978), no. 2-3, 199–207.

[122] A. Grząślewicz, *Some remarks to additive functions*, Math. Japon. **23** (1978/79), no. 5, 573–578.

[123] A. Grząślewicz, *On the solution of the system of functional equations related to quadratic functionals*, Glas. Mat. Ser. III **14(34)** (1979), no. 1, 77–82.

[124] A. Grząślewicz, Z. Powązka, and J. Tabor, *On Cauchy's nucleus*, Publ. Math. Debrecen **25** (1978), no. 1-2, 47–51.

[125] A. Grząślewicz and P. Sikorski, *On some homomorphisms in Ehresmann groupoids*, Wyż. Szkoła Ped. Krakow. Rocznik Nauk.-Dydakt. Prace Mat. No. 9, (1979), 55–66.

[126] A. Guerraggio and L. Paganoni, *On a class of convex functions* (in Italian), Riv. Mat. Univ. Parma (4) **4** (1978), 239–245 (1979).

[127] J. Hadamard, *Étude sur les propriétés des fonctions entières et en particulier d'une fonction considérée par Riemann*, J. Math. Pures Appl. **58** (1893), 171–215.

[128] H. Hahn and A. Rosenthal, *Set Functions*, The University of New Mexico Press, Albuquerque, N. M., 1948.

[129] P. R. Halmos, *Measure Theory*, Springer-Verlag, New York-Heidelberg-Berlin, 1974.

[130] P. R. Halmos, *Naive set theory*, Reprint of the 1960 edition, Undergraduate Texts in Mathematics, Springer-Verlag, New York-Heidelberg, 1974.

[131] I. Halperin, *Non-finite solutions of the equation* $f(x + y) = f(x) + f(y)$, Bull. Amer. Math. Soc. **54** (1948), 1063.

[132] I. Halperin, *Non-measurable sets and the equation* $f(x+y) = f(x)+f(y)$, Proc. Amer. Math. Soc. **2** (1951), 221–224.

[133] J. D. Halpern, *Bases in vector spaces and the axiom of choice*, Proc. Amer. Math. Soc. **17** (1966), 670–673.

[134] G. Hamel, *Eine Basis aller Zahlen und die unstetigen Lösungen der Funktionalgleichung:* $f(x + y) = f(x) + f(y)$, Math. Ann. **60** (1905), no. 3, 459–462.

[135] G. H. Hardy, J. E. Littlewood, and G. Pólya, *Some simple inequalities satisfied by convex functions*, Messenger Math. **58** (1928/29), 145–152.

[136] G. H. Hardy, J. E. Littlewood, and G. Pólya, *Inequalities*, 2nd ed., Cambridge University Press, 1952.

[137] S. Hartman, *A remark on Cauchy's equation*, Colloq. Math. **8** (1961), 77–79.

[138] S. Hartman and J. Mikusiński, *Teoria miary i całki Lebesgue'a*, Państwowe Wydawnictwo Naukowe, Warsaw, 1957.

[139] O. Haupt and G. Aumann, *Differenzial- und Integralrechnung*, Berlin, 1938.

[140] D. Henney, *Set-valued additive functions*, Riv. Mat. Univ. Parma (2) **9** (1968), 43–46.

[141] I. E. Highberg, *A note on abstract polynomials in complex spaces*, J. Math. Pures Appl. (9) **16** (1937), 307–314.

[142] E. Hille and R. S. Phillips, *Functional analysis and semi-groups*, American Mathematical Society Colloquium Publications, vol. 31, American Mathematical Society, Providence, R. I., 1957, rev. ed.

[143] O. Hölder, *Über einen Mittelwerthssatz*, Nachr. Ges. Wiss. Göttingen (1889), 38–47.

[144] S. Horinouchi and Pl. Kannappan, *On the system of functional equations* $f(x + y) = f(x) + f(y)$ *and* $f(xy) = p(x)f(y) + q(y)f(x)$, Aequationes Math. **6** (1971), 195–201.

[145] M. Hosszú, *On an alternative functional equation* (in Hungarian), Mat. Lapok **14** (1963), 98–102.

[146] M. Hosszú, *On the Fréchet's functional equation*, Bul. Inst. Politehn. Iaşi (N.S.) **10** (**14**) (1064), no. 1-2, 27–28.

[147] M. Hosszú, *A remark on the dependence of functions*, Zeszyty Naukowe Uniwersytetu Jagiellońskiego, Prace Matematyczne **14** (1970), 127–129.

[148] M. Hukuhara, *Sur la fonction convexe*, Proc. Japan Acad. **30** (1954), 683–685.

[149] W. Hurewicz, *Zur Theorie der analytischen Mengen*, Fund. Math. **15** (1930), 4–17.

[150] D. H. Hyers, *On the stability of the linear functional equation*, Proc. Natl. Acad. Sci. U.S.A. **27** (1941), 222–224.

[151] D. H. Hyers, *Transformations with bounded mth differences*, Pacific J. Math. **11** (1961), 591–602.

[152] D. H. Hyers, *A note on Fréchet's definition of "polynômes abstraits"*, Houston J. Math. **4** (1978), no. 3, 359–362.

[153] D. H. Hyers and S. M. Ulam, *Approximately convex functions*, Proc. Amer. Math. Soc. **3** (1952), 821–828.

[154] T. J. Jech, *The axiom of choice*, Studies in Logic and the Foundations of Mathematics, Vol. 75., North-Holland Publishing Co., Amsterdam-London; American Elsevier Publishin Co., Inc., New York, 1973.

[155] J. L. W. V. Jensen, *Om konvekse funktioner og uligheder imellem middelvaerdier*, Nyt. Tidsskrift for Mathematik **16B** (1905), 49–69.

[156] J. L. W. V. Jensen, *Sur les fonctions convexes et les inégalités entre les valeurs moyennes*, Acta Math. **30** (1906), no. 1, 175–193.

[157] F. B. Jones, *Connected and disconnected plane sets and the functional equation $f(x) + f(y) = f(x + y)$*, Bull. Amer. Math. Soc. **48** (1942), 115–120.

[158] F. B. Jones, *Measure and other properties of a Hamel basis*, Bull. Amer. Math. Soc. **48** (1942), 472–481.

[159] W. B. Jurkat, *On Cauchy's functional equation*, Proc. Amer. Math. Soc. **16** (1965), 683–686.

[160] M. Kac, *Une remarque sur les équations fonctionnelles*, Comment. Math. Helv. **9** (1936), no. 1, 170–171.

[161] Pl. Kannappan, *Remark (P178R1*, Aequationes Math. **19** (1979), 283.

[162] Pl. Kannappan and M. Kuczma, *On a functional equation related to the Cauchy equation*, Ann. Polon. Math. **30** (1974), 49–55.

[163] Pl. Kannappan and S. Kurepa, *Some relations between additive functions, I.; II.*, Aequationes Math. **4** (1970), 163–176; **6**, (1971), 46–58.

[164] J. Karamata, *Sur une inégalité relative aux fonctions convexes*, Publ. Math. Univ. Belgrade **1** (1932), 145–148.

[165] J. H. B. Kemperman, *A general functional equation*, Trans. Amer. Math. Soc. **86** (1957), 28–56.

[166] J. H. B. Kemperman, *Problem*, Aequationes Math. **1** (1968), 303.

[167] J. H. B. Kemperman, *On the regularity of generalized convex functions*, Trans. Amer. Math. Soc. **135** (1969), 69–93.

[168] J. H. B. Kemperman, *The dual of the cone of all convex functions on a vector space*, Aequationes Math. **13** (1975), no. 1/2, 103–119.

[169] H. Kestelman, *On the functional equation $f(x + y) = f(x) + f(y)$*, Fund. Math. **34** (1947), 144–147.

[170] H. Kestelman, *Automorphisms of the field of complex numbers*, Proc. London Math. Soc. (2) **53** (1951), 1–12.

[171] B. Kominek and Z. Kominek, *On some classes connected with the continuity of additive and Q-convex functions*, Uniw. Śląski w Katowicach Prace Nauk.-Prace Mat. No. 8, (1978), 60–63.

[172] Z. Kominek, *On the sum and difference of two sets in topological vector spaces*, Fund. Math. **71** (1971), no. 2, 165–169.

[173] Z. Kominek, *On the continuity of Q-convex functions and additive functions*, Aequationes Math. **23** (1981), no. 2-3, 146–150.

[174] Z. Kominek, *Some remarks on the set classes A_C^N and B_C^N*, Comment. Math. Prace Mat. **23** (1983), no. 1, 49–52.

[175] M. A. Krasnosel′skiĭ and J. B. Rutickiĭ, *Convex functions and Orlicz spaces*, Translated from the first Russian edition by Leo F. Boron, P. Noordhoff Ltd., Groningen, 1961.

[176] M. Kuczma, *Note on convex functions*, Ann. Univ. Sci. Budapest. Eötvös. Sect. Math. **2** (1959), 25–26.

[177] M. Kuczma, *Functional equations in a single variable*, Monografie Matematyczne, Tom 46, Państwowe Wydawnictwo Naukowe, Warsaw, 1968.

[178] M. Kuczma, *Almost convex functions*, Colloq. Math. **21** (1970), 279–284.

[179] M. Kuczma, *Some remarks about additive functions on cones*, Aequationes Math. **4** (1970), 303–306.

[180] M. Kuczma, *Some remarks on convexity and monotonicity*, Rev. Roumaine Math. Pures Appl. **15** (1970), 1463–1469.

[181] M. Kuczma, *Convex functions*, Functional Equations and Inequalities, Edizioni Cremonese, Roma, 1971, [Corso tenuto a La Mendola (Trento) dal 20 al 28 agosto 1970], pp. 195–213.

[182] M. Kuczma, *Note on additive functions of several variables*, Uniw. Śląski w Katowicach – Prace Mat. **2** (1972), 49–51.

[183] M. Kuczma, *Cauchy's functional equation on a restricted domain*, Colloq. Math. **28** (1973), 313–315.

[184] M. Kuczma, *On some set classes occurring in the theory of convex functions*, Comment. Math. Prace Mat. **17** (1973), 127–135.

[185] M. Kuczma *Additive functions and the Egorov theorem*, General Inequalities, 1 (Oberwolfach, 1976), (E. F. Beckenbach, ed.), International Series of Numerical Mathematics, vol. 41, Birkhäuser, Basel, 1978, pp. 169–173.

[186] M. Kuczma, *On some alternative functional equations*, Aequationes Math. **17** (1978), no. 2-3, 182–198.

[187] M. Kuczma, *Functional equations on restricted domain*, Aequationes Math. **18** (1978), 1–35.

[188] M. Kuczma, *On some properties of Erdős sets*, Colloq. Math. **48** (1984), no. 1, 127–134.

[189] M. Kuczma, *On some analogies between measure and category and their applications in the theory of additive functions*, Ann. Math. Sil. (1985), no. 13, 155–162.

[190] M. Kuczma and J. Smítal, *On measures connected with the Cauchy equation*, Aequationes Math. **14** (1976), no. 3, 421–428.

[191] M. E. Kuczma, *On discontinuous additive functions*, Fund. Math. **66** (1969/1970), 383–392.

[192] M. E. Kuczma, *A generalization of Steinhaus' theorem to coordinatewise measure preserving binary transformations*, Colloq. Math. **36** (1976), no. 2, 241–248.

[193] M. E. Kuczma, *Differentiation of implicit functions and Steinhaus' theorem in topological measure spaces*, Colloq. Math. **39** (1978), no. 1, 95–107, 189.

[194] M. E. Kuczma and M. Kuczma, *An elementary proof and an extension of a theorem of Steinhaus*, Glasnik Mat. Ser. III **6(26)** (1971), 11–18.

[195] W. Kulpa, *On the existence of maps having graphs connected and dense*, Fund. Math. **76** (1972), no. 3, 207–211.

[196] C. Kuratowski, *Topologie. Vol. I*, Monografie Matematyczne, Tom 20, Państwowe Wydawnictwo Naukowe, Warsaw, 1958, 4ème éd.

[197] K. Kuratowski, *Une méthode d'élimination des nombres transfinis des raisonnements mathématiques*, Fund. Math. **3** (1922), 76–108.

[198] K. Kuratowski and A. Mostowski, *Set theory*, With an introduction to descriptive set theory, Translated from the 1966 Polish original, revised ed., North-Holland Publishing Co., Amsterdam-New York-Oxford; PWN-Polish Scientific Publishers, Warsaw, 1976, Studies in Logic and the Foundations of Mathematics, Vol. 86.

[199] S. Kurepa, *Convex functions*, Glasnik Mat.-Fiz. Astr. Ser. II. **11** (1956), 89–94.

[200] S. Kurepa, *Note on the difference set of two measurable sets in E^n*, Glasnik Mat.-Fiz. Astronom. Društvo Mat. Fiz. Hrvatske Ser. II **15** (1960), 99–105.

[201] S. Kurepa, *The Cauchy functional equation and scalar product in vector spaces*, Glasnik Mat.-Fiz. Astronom. Ser. II Društvo Mat. Fiz. Hrvatske **19** (1964), 23–36.

[202] S. Kurepa, *Remarks on the Cauchy functional equation*, Publ. Inst. Math. (Beograd) (N.S.) **5(19)** (1965), 85–88.

[203] I. B. Lacković and M. R. Stanković, *On Hadamard's integral inequality for convex functions*, Univ. Beograd. Publ. Elektrotehn. Fak. Ser. Mat. Fiz. (1973), no. 412–460, 89–92.

[204] K. Lajkó, *Applications of extensions of additive functions*, Aequationes Math. **11** (1974), 68–76.

[205] H. Läuchli, *Auswahlaxiom in der Algebra*, Comment. Math. Helv. **37** (1962/1963), 1–18.

[206] A. M. Legendre, *Eléments de géometrie*, Paris, 1791. Note II.

[207] V. K. Lim, *A note on an inequality*, Nanta Math. **5** (1971), no. 1, 38–40.

[208] S. Łojasiewicz, *Wstęp do teorii funkcji rzeczywistych*, With a collection of problems compiled by W. Mlak and Z. Opial, Biblioteka Matematyczna, Tom 46. [Mathematics Library, Vol. 46], Państwowe Wydawnictwo Naukowe, Warsaw, 1973. English edition: John Wiley and Sons Ltd., Chichester, 1988.

[209] L. Losonczi, *Bestimmung aller nichtkonstanten Lösungen von linearen Funktionalgleichungen*, Acta Sci. Math. (Szeged) **25** (1964), 250–254.

[210] N. Lusin, *Sur un probléme de M. Baire*, C. R. Acad. Sci. Paris **158** (1914), 1258–1261.

[211] N. Lusin, *Sur les ensembles analytiques*, Fund. Math. **10** (1927), 1–95.

[212] N. Lusin and W. Sierpiński, *Sur une décomposition d'un intervalle en une infinité non dénombrable d'ensembles non mesurables*, C. R. Acad. Sci. Paris **165** (1917), 422–424.

[213] N. Lusin and M. Souslin, *Sur une définition des ensembles mesurables B sans nombres transfinis*, C. R. Acad. Sci. Paris **164** (1917), 88.

[214] I. Makai, *Über invertierbare Lösungen der additiven Cauchy-Funktionalgleichung*, Publ. Math. Debrecen **16** (1969), 239–243.

[215] S. Marcus, *Sur les fonctions de Hamel* (in Romanian), Acad. R. P. Romîne. Bul. Şti. Secţ. Şti. Mat. Fiz. **8** (1956), 517–528.

[216] S. Marcus, *Sur un problème de la théorie de la mesure de H. Steinhaus et S. Ruziewicz*, Bull. Acad. Polon. Sci. Cl. III. **4** (1956), 197–199.

[217] S. Marcus, *Sur une classe de fonctions définies par des inégalités, introduite par M. Á. Császár*, Acta Sci. Math. Szeged **19** (1958), 192–218.

[218] S. Marcus, *Généralisation, aux fonctions de plusieurs variables, des théorèmes de Alexander Ostrowski et de Masuo Hukuhara concernant les fonctions convexes (J)*, J. Math. Soc. Japan **11** (1959), 171–176.

[219] E. Marczewski, *On measurability and Baire property* (in Polish), C. R. I Congrès Math. des Pays Slaves, Warszawa, 1929, Książnica Atlas, Warszawa, 1930, pp. 297–309.

[220] E. Marczewski, *Remarques sur les fonctions de Hamel*, Colloq. Math. **1** (1948), 249–250.

[221] S. C. Martin, *Extensions and decompositions of homomorphisms of semigroups*, manuscript.

[222] S. Mazur and W. Orlicz, *Grundlegende Eigenschaften der polynomischen Operationen I., II.*, Studia Math. **5** (1934), 50–68, 179–189.

[223] M. A. McKiernan, *On vanishing nth ordered differences and Hamel bases*, Ann. Polon. Math. **19** (1967), 331–336.

[224] M. R. Mehdi, *On convex functions*, J. London Math. Soc. **39** (1964), 321–326.

[225] D. S. Mitrinović, *The Steffensen inequality*, Univ. Beograd. Publ. Elektrotehn. Fak. Ser. Mat. Fiz. No. **247–273** (1969), 1–14.

[226] D. S. Mitrinović, *Analytic inequalities*, In cooperation with P. M. Vasić, Springer-Verlag, New York–Berlin, 1970.

[227] E. Mohr, *Beitrag zur Theorie der konvexen Funktionen*, Math. Nachr. **8** (1952), 133–148.

[228] E. Moldovan, *Sur une généralisation de la notion de convexité* (in Romanian), Acad. R. P. Romîne. Fil. Cluj. Stud. Cerc. Şti. Ser. I **6** (1955), no. 3-4, 65–73.

[229] E. Moldovan, *Propriétés des fonctions convexes généralisées* (in Romanian), Acad. R. P. Romîne. Fil. Cluj. Stud. Cerc. Mat. **8** (1957), 21–35.

[230] E. Moldovan, *O pojmu konveksnih funkcija*, Matematička biblioteka **42** (1969), 25–40.

[231] J. Mościcki, *Note on characterizations of some set-classes connected with the continuity of additive functions*, Uniw. Śląski w Katowicach Prace Nauk.-Prace Mat. (1982), no. 12, 47–52.

[232] Z. Moszner, *Sur une hypothèse au sujet des fonctions subadditives*, Aequationes Math. **2** (1969), 380–386.

[233] Z. Moszner, *Sur la stabilité de l'équation d'homomorphisme*, Aequationes Math. **29** (1985), no. 2-3, 290–306.

[234] H. P. Mulholland, *On generalizations of Minkowski's inequality in the form of a triangle inequality*, Proc. London Math. Soc. (2) **51** (1950), 294–307.

[235] B. Nagy, *On a generalization of the Cauchy equation*, Aequationes Math. **11** (1974), 165–171.

[236] M. A. Naĭmark, *Normed rings* (in Russian), Izdat. "Nauka", Moscow, 1968, Second edition, revised.

[237] I. P. Natanson, *Theory of functions of a real variable* (in Russian), Gosudarstv. Izdat. Tehn-Teor. Lit., Moscow-Leningrad, 1950.

[238] K. Nikodem, *On convex stochastic processes*, Aequationes Math. **20** (1980), no. 2-3, 184–197.

[239] K. Nikodem, *On additive set-valued functions*, Rev. Roumaine Math. Pures Appl. **26** (1981), no. 7, 1005–1013.

[240] K. Nikodem, *Additive set valued functions in Hilbert spaces*, Rev. Roumaine Math. Pures Appl. **28** (1983), no. 3, 239–242.

[241] K. Nikodem, *On Jensen's functional equation for set-valued functions*, Rad. Mat. **3** (1987), no. 1, 23–33.

[242] K. Nikodem, *On midpoint convex set-valued functions*, Aequationes Math. **33** (1987), no. 1, 46–56.

[243] A. Nishiyama and S. Horinouchi, *On a system of functional equations*, Aequationes Math. **1** (1968), 1–5.

[244] N. E. Nörlund, *Vorlesungen über Differenzialrechnung*, Berlin, 1924.

[245] F. Obreanu, *La puissance de certaines classes de fonctions*, Duke Math. J. **14** (1947), 377–380.

[246] O. Øre, *Linear equations in non-commutative fields*, Ann. of Math. (2) **32** (1931), no. 3, 463–477.

[247] W. Orlicz and Z. Ciesielski, *Some remarks on the convergence of functionals on bases*, Studia Math. **16** (1958), 335–352.

[248] A. Ostrowski, *Über die Funktionalgleichung der Exponentialfunktion und verwandte Funktionalgleichungen*, Jber. Deutsch. Math.-Verein **38** (1929), 54–62.

[249] A. Ostrowski, *Zur Theorie der konvexen Funktionen*, Comment. Math. Helv. **1** (1929), no. 1, 157–159.

[250] J. C. Oxtoby, *Measure and category. A survey of the analogies between topological and measure spaces*, Graduate Texts in Mathematics, Vol. 2., Springer-Verlag, New York-Berlin, 1971.

[251] L. Paganoni, *Misure de Cauchy, Ostrowski e Steinhaus sull'asse reale*, Atti. Accad. Sci. Torino Cl. Sci. Fis. Mat. Natur. **109** (1975), no. 1-2, 145–155.

[252] L. Paganoni and St. Paganoni Marzegalli, *Cauchy's functional equation on semigroups*, Fund. Math. **110** (1980), no. 1, 63–74.

[253] M. Petrović, *Sur une équation fonctionnelle*, Publ. Math. Univ. Belgrade **1** (1932), 149–156.

[254] H. W. Pexider, *Notiz über Funktionaltheoreme*, Monatsh. Math. Phys. **14** (1903), no. 1, 293–301.

[255] R. S. Phillips, *An inversion formula for Laplace transforms and semi-groups of linear operators*, Ann. of Math. (2) **59** (1954), 325–356.

[256] S. Piccard, *Sur des ensembles parfaits*, Mém. Univ. Neuchâtel, vol. 16, Secrétariat de l'Université, Neuchâtel, 1942.

[257] Gy. Pólya and G. Szegő, *Aufgaben und Lehrsätze aus der Analysis, Vol. I*, Die Grundlehren der mathematischen Wissenschaften, Band 19, Springer, Berlin, 1925.

[258] J. Ponstein, *Seven kinds of convexity*, SIAM Rev. **9** (1967), 115–119.

[259] T. Popoviciu, *Sur quelques propriétés des fonctions d'une ou de deux variables réelles*, Mathematica (Cluj) **8** (1934), 1–85.

[260] T. Popoviciu, *Les fonctions convexes*, Hermann et Cie, Paris, 1944.

[261] H. Rådström, *One-parameter semigroups of subsets of a real linear space*, Ark. Mat. **4** (1960), 87–97 (1960).

[262] H. Rasiowa, *Wstęp do matematyki współczesnej*, Biblioteka Matematyczna, Tom 30, Państowe Wydawnictwo Naukowe, Warsaw, 1971. English edition: North-Holland Publishin Co., Amsterdam-London; American Elsevier Publishing Co., Inc., New York, 1973.

[263] J. Rätz, *On the homogeneity of additive mappings*, Aequationes Math. **14** (1976), no. 1/2, 67–71.

[264] J. Rätz *On approximately additive mappings*, General Inequalities, 2 (Oberwolfach, 1978) (E. F. Beckenbach, ed.), International Series in Numerical Mathematics, vol. 47, Birkhäuser, Basel, 1980, pp. 233–251.

[265] D. Rees, *On the group of a set of partial transformations*, J. London Math. Soc. **22** (1947), 281–284 (1948).

[266] M. Riesz, *Court exposé des propriétés principales de la mesure de Lebesgue*, Ann. Soc. Polon. Math. **25** (1952), 298–308 (1953).

[267] A. W. Roberts and D. E. Varberg, *Convex Functions*, Pure and Applied Mathematics, Vol. 57, Academic Press, New York–London, 1973.

[268] R. T. Rockafellar, *Convex analysis*, Princeton Mathematical Series, No. 28, Princeton University Press, Princeton, N.J., 1970.

[269] R. A. Rosenbaum, *Sub-additive functions*, Duke Math. J. **17** (1950), 227–247.

[270] F. Rothberger, *Problem (P178)*, Aequationes Math. **19** (1979), 300.

[271] S. Ruziewicz, *Une application de l'équation fonctionnelle $f(x + y) = f(x) + f(y)$ à la décomposition de la droite en ensembles superposables, non-mesurables*, Fund. Math. **5** (1924), 92–95.

[272] M. Sablik, *On proper linearly invariant ideals of sets*, Glas. Mat. Ser. III **14(34)** (1979), no. 1, 41–50.

[273] W. Sander, *Verallgemeinerungen eines Satzes von S. Piccard*, Manuscripta Math. **16** (1975), no. 1, 11–25.

[274] W. Sander, *Verallgemeinerungen eines Satzes von H. Steinhaus* Fund. Math. **1** (1920), 93–104, Manuscripta Math. **18** (1976), no. 1, 25–42. Erratum: Manuscripta Math., **20** (1977), no. 1, 101–103.

[275] W. Sander, *A remark on convex functions*, manuscript.

[276] W. Sierpiński, *Sur la question de la mesurabilité de la base de Hamel*, Fund. Math. **1** (1920), 105–111.

[277] W. Sierpiński, *Sur un problème concernant les ensembles mesurables superficiellement*, Fund. Math. **1** (1920), 112–115.

[278] W. Sierpiński, *Sur l'équation fonctionnelle $f(x + y) = f(x) + f(y)$*, Fund. Math. **1** (1920), 116–122.

[279] W. Sierpiński, *Sur les fonctions convexes mesurables*, Fund. Math. **1** (1920), 125–128.

[280] W. Sierpiński, *Sur une propriété des fonctions de M. Hamel*, Fund. Math. **5** (1924), 334–336.

[281] W. Sierpiński, *Sur un ensemble non dénombrable, dont toute image continue est de mesure nulle*, Fund. Math. **11** (1928), 302–304.

[282] W. Sierpiński, *La base de M. Hamel et la propriété de Baire*, Publ. Math. Univ. Belgrade **4** (1935), 221–225.

[283] W. Sierpiński, *Un théorème de la théorie générale de ensembles et ses conséquences*, Fund. Math. **24** (1935), 8–11.

[284] W. Sierpiński, *Les ensembles projectifs et analytiques*, Mémor. Sci. Math., no. 112, Gauthier-Villars, Paris, 1950.

[285] A. Smajdor, *On monotonic solutions of some functional equations*, Dissertationes Math. Rozprawy Mat. **82** (1971), 58pp.

[286] J. Smítal, *On the functional equation $f(x + y) = f(x) + f(y)$*, Rev. Roumaine Math. Pures Appl. **13** (1968), 555–561.

[287] J. Smítal, *On boundedness and discontinuity of additive functions*, Fund. Math. **76** (1972), no. 3, 245–253.

[288] J. Smítal, *A necessary and sufficient condition for continuity of additive functions*, Czechoslovak Math. J. **26(101)** (1976), no. 2, 171–173.

[289] J. Smítal, *On convex functions bounded below*, Aequationes Math. **14** (1976), no. 3, 345–350.

[290] J. Smítal, *On a problem of Aczél and Erdős concerning Hamel bases*, Aequationes Math. **28** (1985), no. 1-2, 135–137.

[291] J. Smítal and L. Snoha, *Generalization of a theorem of S. Piccard*, Acta Math. Univ. Comenian. **37** (1980), 173–181.

[292] R. M. Solovay, *A model of set-theory in which every set of reals is Lebesgue measurable*, Ann. of Math. (2) **92** (1970), 1–56.

[293] M. Souslin, *Sur une définition des ensembles mesurables B sans nombres transfinis*, C. R. Acad. Sci. Paris **164** (1917), 89.

[294] L. R. Stanković and I. B. Lacković, *Some remarks on the paper: "A note on an inequality"*, Nanta Math. **5** (1971), no. 1, 38–40 of V. K. Lim, Univ. Beograd. Publ. Elektrotehn. Fak. Ser. Mat. Fiz. (1974), no. 461-497, 51–54.

[295] J. F. Steffensen, *On certain inequalities between mean values, and their application to actuarial problems*, Skand. Aktuarietidskr. (1918), 82–97.

[296] J. F. Steffensen, *On certain inequalities and methods of approximation*, J. Institue Actuaries **51** (1919), 274–297.

[297] H. Steinhaus, *Sur les distances dans les points des ensembles de mesure positive*, Fund. Math. **1** (1920), 93–104.

[298] O. Stolz, *Grundzüge der Differenzial- und Integralrechnung, Vol. I*, Teubner, Leipzig, 1893.

[299] H. Światak, *On the functional equations*
$$f_1(x_1 + \cdots + x_n)^2 = [\sum_{(i_1, \cdots, i_n)} f_1(x_{i_1}) \cdots f_n(x_{i_n})]^2,$$
Ann. Univ. Sci. Budapest. Eötvös Sect. Math. **10** (1967), 49–52.

[300] H. Światak, *On the functional equation $f(x + y - xy) + f(xy) = f(x) + f(y)$*, Mat. Vesnik **5(20)** (1968), 177–182.

[301] H. Światak, *Remarks on the functional equation $f(x + y - xy) + f(xy) = f(x) + f(y)$*, Aequationes Math. **1** (1968), 239–241.

[302] H. Światak, *On the equivalence of some functional equations*, Publ. Techn. Univ. Miskolc **30** (1970), 275–279.

[303] H. Światak, *On the functional equation $f(x + y)^2 = [f(x) + f(y)]^2$*, Publ. Techn. Univ. Miskolc **30** (1970), 307–308.

[304] H. Światak, *A proof of the equivalence of the equation $f(x+y-xy)+f(xy) = f(x)+f(y)$ and Jensen's functional equation*, Aequationes Math. **6** (1971), 24–29.

[305] H. Światak, *On alternative functional equations*, Aequationes Math. **15** (1977), no. 1, 35–47.

[306] H. Światak and M. Hosszú, *Notes on functional equations of polynomial form*, Publ. Math. Debrecen **17** (1970), 61–66 (1971).

[307] H. Światak and M. Hosszú, *Remarks on the functional equation $e(x, y)f(xy) = f(x) + f(y)$*, Publ. Techn. Univ. Miskolc **30** (1970), 323–325.

[308] Á. Száz and G. Száz, *Additive relations*, Publ. Math. Debrecen **20** (1973), 259–272 (1974).

[309] L. Székelyhidi, *The general representation of an additive function on an open point set*, (in Hungarian), Magyar Tud. Akad. Mat. Fiz. Oszt. Közl. **21** (1972), 503–509.

[310] L. Székelyhidi, *Remark on a paper of M. A. McKiernan: "On vanishing nth-ordered differences and Hamel bases"*, Ann. Polon. Math. **36** (1979), no. 3, 245–247.

[311] K. Szymiczek, *Note on semigroup homomorphisms*, Uniw. Śląski w Katowicach – Prace Mat. **3** (1973), 75–78.

[312] K. Szymiczek, *Solution of Cauchy's functional equation on a restricted domain*, Colloq. Math. **33** (1975), no. 2, 203–208.

[313] J. Tabor, *On mappings preserving the stability of the Cauchy functional equation*, Wyż. Szkota Ped. Krakow. Rocznik Nauk-Dydakt. Prace Mat. No. 12 (1987), 139–147.

[314] A. Tarski, *Axiomatic and algebraic aspects of two theorems on sums of cardinals*, Fund. Math. **35** (1948), 79–104.

[315] M. Tomić, *Théorème de Gauss relatif au centre de gravité et son application* (in Serbian), Bull. Soc. Math. Phys. Serbie **1** (1949), 31–40.

[316] P. M. Vasić and I. B. Lacković, *On an inequality for convex functions*, Univ. Beograd. Publ. Elektrotehn. Fak. Ser. Mat. Fiz. (1974), no. 461-497, 63–66.

[317] B. A. Vertgeĭm and G. Š. Rubinšteĭn, *The definition of quasi-convex functions* (in Russian), Mathematical Programming (Russian), Izdat. "Nauka", Moscow, 1966, pp. 121–134.

[318] E. Vincze, *Beitrag zur Theorie der Cauchyschen Funktionalgleichungen*, Arch. Math. **15** (1964), 132–135.

[319] E. Vincze, *Solutions of alternative functional equations* (in Hungarian), Mat. Lapok **15** (1964), 179–195.

[320] E. Vincze, *Über eine Verallgemeinerung der Cauchyschen Funktionalgleichung*, Funkcial. Ekvac **6** (1964), 55–62.

[321] E. Vincze, *Über eine Klasse der alternativen Funktionalgleichungen*, Aequationes Math. **2** (1969), 364–365.

[322] J. E. Wetzel, *On the functional inequality $f(x+y) \geq f(x) f(y)$*, Amer. Math. Monthly **74** (1967), 1065–1068.

[323] H. Whitney, *On functions with bounded nth differences*, J. Math. Pures Appl. (9) **36** (1957), 67–95.

[324] H. Whitney, *On bounded functions with bounded nth differences*, Proc. Amer. Math. Soc. **10** (1959), 480–481.

[325] O. Zariski and P. Samuel, *Commutative algebra, Volume I*, With the cooperation of I. S. Cohen, The University Series in Higher Mathematics, D. Van Nostrand Company, Inc., Princeton, New Jersey, 1958.

[326] E. Zermelo, *Beweis, daß jede Menge wohlgeordnet werden kann*, Math. Ann. **59** (1904), no. 4, 514–516.

[327] M. Zorn, *A remark on method in transfinite algebra*, Bull. Amer. Math. Soc. **41** (1935), 667–670.

[328] A. Zygmund, *Trigonometric series. Vol. I, II*, third ed., With a foreword by Robert A. Fefferman, Cambridge Mathematical Library, Cambridge University Press, Cambridge, 2002.

Index of Symbols

Subject Index

Index of Names

Abian, A., 282, 285
Aczél, J., xiii, 128, 138, 282, 304, 338, 343, 351, 355, 364, 367, 369, 382, 537, 540, 565
Albert, M., 449, 483, 485–487, 532
Aleksandrov, A. D., 570
Alexiewicz, A., 241
Anastassiadis, J., 335
Aumann, G., 131

Baker, J. A., 413, 449, 483, 485–487, 532, 537
Balcerzyk, S., 535
Banach, S., 241
Baron, K., xiii, xiv, 376
Beckenbach, E. F., 131, 197, 211, 226, 570
Bellman, R., 197, 211
Benz, W., 338, 570
Bereanu, B., 335
Bergman, G. M., 338
Bernstein, F., 131, 155
Berz, E., 259, 456, 471, 473, 475
Bing, R. N., 131
Blanuša, D., 374
Blumberg, H., 131, 241
Boas, R. P. Jr., 201, 205
Bonnesen, T., 117, 241
Borel, E.[11], 63
Bourbaki, N., 131
Bruckner, A. M., 322
Burstin, C., 285, 331

Carathéodory, C., 47
Cauchy, A. L., 128, 343

---

[11] The name of Borel is not quoted if it occurs in the context *Borel sets*. Similarly, the name of Baire is not quoted here as it occurs only in the context *Baire property*.

Ceder, J. G., 322
Cholewa, P. W., 483, 490, 494, 498
Ciesielski, Z., 42, 205, 438
Clifford, A. H., 92
Cohen, P. J., 12, 16
Cooper, R., 456
Császár, A., 250
Czerwik, S., 131

Dankiewicz, K., 550
Darboux, G., 402
Daróczy, Z., 369, 372, 374, 382
Davison, T. M. K., 374, 376
de Bruijn, N. G., 501, 505, 506, 508, 509
Deák, E., 131
Denny, J. L., 505
Dhombres, J., xiii, 374, 535, 537, 552, 553, 569
Djoković, D. Ž., 446, 537
Doetsch, G., 131, 155
Dubikajtis, L., 42, 376
Dubreil, P., 92

Eggleston, H. G., 117, 127, 492
Erdős, P., 63, 282, 288, 295, 304, 328, 329, 331, 369, 505, 540

Fenchel, W., 117, 241
Fenyő, I., 374
Ferens, C., 42, 376
Fifer, Z., 570
Figiel, T., 241
Fischer, P., 241, 380, 473
Forti, G. L., 483, 531, 570
Fréchet, M., 241, 453
Fuchs, L., 211

Gajda, Z., 439, 451
Gauss, C. F., 128